REDOX REGULATION OF DIFFERENTIATION AND DE-DIFFERENTIATION

OXIDATIVE STRESS AND DISEASE

Series Editors
Enrique Cadenas, MD, PhD
University of Southern California School of Pharmacy
Los Angeles, California

Helmut Sies, MD
Heinrich-Heine-Universität Düsseldorf
Düsseldorf, Germany

Nutrition and Epigenetics
edited by Emily Ho and Frederick Domann

Lipid Oxidation in Health and Disease
edited by Corinne M. Spickett and Henry Jay Forman

Diversity of Selenium Functions in Health and Disease
edited by Regina Brigelius-Flohé and Helmut Sies

Mitochondria in Liver Disease
edited by Derick Han and Neil Kaplowitz

Fetal and Early Postnatal Programming and Its Influence on Adult Health
edited by Mulchand S. Patel and Jens H. Nielsen

Biomedical Application of Nanoparticles
edited by Bertrand Rihn

The Biology of the First 1,000 Days
edited by Crystal D. Karakochuk, Kyly C. Whitfield, Tim J. Green, and Klaus Kraemer

Hydrogen Peroxide Metabolism in Health and Disease
edited by Margreet C.M. Vissers, Mark Hampton, and Anthony J. Kettle

Glutathione
edited by Leopold Flohé

Vitamin C: Biochemistry and Function
edited by Margreet C.M. Vissers and Qi Chen

Cancer and Vitamin C
edited by Margreet C.M. Vissers and Qi Chen

Mammalian Heme Peroxidases: Diverse Roles in Health and Disease
edited by Clare Hawkins and William M. Nauseef

Redox Regulation of Differentiation and De-Differentiation
edited by Carsten Berndt and Christopher Horst Lillig

For more information about this series, please visit:
www.crcpress.com/Oxidative-Stress-and-Disease/book-series/CRCOXISTRDIS

REDOX REGULATION OF DIFFERENTIATION AND DE-DIFFERENTIATION

Edited by
Carsten Berndt
Christopher Horst Lillig

CRC Press
Taylor & Francis Group
Boca Raton London New York

CRC Press is an imprint of the
Taylor & Francis Group, an **informa** business

First edition published 2022
by CRC Press
6000 Broken Sound Parkway NW, Suite 300, Boca Raton, FL 33487-2742

and by CRC Press
2 Park Square, Milton Park, Abingdon, Oxon, OX14 4RN

ISBN: 978-0-367-89566-2 (hbk)
ISBN: 978-1-032-06842-8 (pbk)
ISBN: 978-1-003-20409-1 (ebk)

DOI: 10.4324/9781003204091

Typeset in Joanna
by Apex CoVantage, LLC

CONTENTS

v

Supersulfide-Mediated
Signaling during Differentiation
and De-Differentiation

Redox Medicine

PREFACE

Cells constantly receive and process signals from their environment. Both the processes of differentiation and de-differentiation rely on the adequate processing of such signals. In general, signals must be sensed by a receptor, passed on via transducer proteins and second messenger molecules, and act on effector proteins that realize biological responses that must include the termination of the original signal and/or its transduction. Today, the redox modifications of proteins, such as cysteinyl-disulfides, persulfides, and sulfenic acids, or methionyl-sulfoxides are recognized as key mechanisms in signal transduction and the regulation of cell functions. The aim of this book, which is part of the series "Oxidative Stress and Disease," edited by Enrique Cadenas and Helmut Sies, is to provide a state-of-the-art point of reference on the role of redox regulation mechanisms in various differentiation and de-differentiation processes. The different sections address the following: (i) The development of trypanosomatids, zebrafish, and plant model organisms. (ii) The differentiation of vertebrate tissues (i.e., the role of hydrogen peroxide and S-nitrosylation in the development and regeneration of different tissues and organs). (iii) Signaling pathways and mechanisms implied in differentiation and longevity. Proteomic techniques, the role of glutathione, NRF2, iron, and hydrogen peroxide are discussed in various models, including cytoskeletal dynamics, neurogenesis, and aging. (iv) Signaling pathways in differentiation and de-differentiation by means of sulfide-signaling, selenocysteine-containing proteins, and hypoxia during cancer development. (v) Potential applications in redox medicine (e.g., in the form of cold atmospheric plasma treatment and nutrition intervention). Taken together, this collection of reviews provides insights into the various roles and mechanisms of redox modifications in all phases of cell signaling and signal transduction that are relevant to the differentiation and transformation of cells. The growing mechanistic understanding of this topic will certainly contribute to new potential strategies addressing urgent medical conditions and biological problems.

With this book, we hope to increase the recognition of redox signaling as an essential part of cellular processes in the scientific community, and welcome we experienced PhDs and MDs and also young scientists to the circle of redox aficionados. As an introduction, the book starts with a historical overview on thiol redox regulation.

CARSTEN BERNDT
Düsseldorf, Germany
CHRISTOPHER HORST LILLIG
Greifswald, Germany

CONTRIBUTORS

TAKAAKI AKAIKE
Department of Environmental Medicine and
    Molecular Toxicology
Tohoku University Graduate School of Medicine
Sendai, Japan

ORHAN AKTAS
Department of Neurology
Heinrich-Heine-Universität Düsseldorf
Düsseldorf, Germany

IRÈNE AMBLARD
Center for Interdisciplinary Research in Biology
    (CIRB), Collège de France
Sorbonne Université
Paris, France

SANDER BEKESCHUS
Leibniz Institute for Plasma Science and
    Technology
Universität Greifswald
Greifswald, Germany

CHRISTOPHE BELIN
Laboratoire Génome et Développement des Plantes
Université Perpignan Via Domitia
Perpignan, France

CARSTEN BERNDT
Department of Neurology
Heinrich-Heine-Universität Düsseldorf
Düsseldorf, Germany

LARS BRÄUTIGAM
Karolinska Institute
Stockholm, Sweden

BOB B. BUCHANAN
Department of Plant & Microbial Biology
University of California, Berkeley
Berkeley, California

MARCELO A. COMINI
Laboratory Redox Biology of Trypanosomes
Institut Pasteur de Montevideo
Montevideo, Uruguay

AVILIEN DARD
Laboratoire Génome et Développement des Plantes
Université Perpignan Via Domitia
Perpignan, France

BRANDON M. DAVIES
Department of Physiology and Developmental
    Biology
Brigham Young University
Provo, Utah

MANUELA GELLERT
Institute for Biochemistry and Molecular Biology
University of Greifswald
Greifswald, Germany

ILORA GHOSH
School of Environmental Sciences
Jawaharlal Nehru University
New Delhi, India

CHRISTIAN GONZÁLEZ-BILLAULT
Department of Biology
Universidad de Chile
Santiago, Chile

KRISTIN HAMRE
Institute of Marine Research
Bergen, Norway

JASON M. HANSEN
Department of Physiology and Developmental
    Biology
Brigham Young University
Provo, Utah

ALAIN JOLIOT
Center for Interdisciplinary Research in Biology
    (CIRB), Collège de France
CNRS, INSERM, PSL Research University
Paris, France

ANNA P. KIPP
Department of Molecular Nutritional Physiology
Institute of Nutritional Sciences
Friedrich Schiller University Jena
Jena, Germany

CHRISTIAN KROLL
Department of Neurology
Heinrich-Heine-Universität Düsseldorf
Düsseldorf, Germany

DIKSHA KULSHRESHTHA
Special Centre for Molecular Medicine
Jawaharlal Nehru University
New Delhi, India

CHRISTOPHER HORST LILLIG
Institute for Biochemistry and Molecular Biology
University of Greifswald
Greifswald, Germany

QUN LIN
Department of Therapeutic Radiology
Yale School of Medicine
New Haven, Connecticut

STUART A. LIPTON
Department of Neurosciences
University of California, San Diego
School of Medicine
La Jolla, California

TETSURO MATSUNAGA
Department of Environmental Medicine and
    Molecular Toxicology
Tohoku University Graduate School of Medicine
Sendai, Japan

MIKAEL MOLIN
Department of Biology and Biological
    Engineering
Chalmers University of Technology
Gothenburg, Sweden

JOSHUA E. MONSIVAIS
Department of Physiology and Developmental
    Biology
Brigham Young University
Provo, Utah

MASANOBU MORITA
Department of Environmental Medicine and
    Molecular Toxicology
Tohoku University Graduate School of Medicine
Sendai, Japan

HOZUMI MOTOHASHI
Department of Gene Expression Regulation
Tohoku University
Sendai, Japan

CHINMAY K. MUKHOPADHYAY
Special Centre for Molecular Medicine
Jawaharlal Nehru University
New Delhi, India

ERNESTO MUÑOZ-PALMA
Department of Biology
Universidad de Chile
Santiago, Chile

SHOHEI MURAKAMI
Department of Gene Expression Regulation
Tohoku University
Sendai, Japan

TOMOHIRO NAKAMURA
Departments of Molecular Medicine and
    Neuroscience and Neuroscience
    Translational Center
The Scripps Research Institute
La Jolla, California

CHANG-KI OH
Departments of Molecular Medicine and
    Neuroscience and Neuroscience
    Translational Center
The Scripps Research Institute
La Jolla, California

LUCÍA PIACENZA
Departamento de Bioquímica
Universidad de la República
Montevideo, Uruguay

GEREON POSCHMANN
Institute of Molecular Medicine
Heinrich-Heine-Universität Düsseldorf
Düsseldorf, Germany

TIM PROZOROVSKI
Department of Neurology
Heinrich-Heine-Universität Düsseldorf
Düsseldorf, Germany

RAFAEL RADI
Departamento de Bioquímica
Universidad de la República
Montevideo, Uruguay

CHRISTINE RAMPON
Center for Interdisciplinary Research in Biology
    (CIRB)
Collège de France
CNRS, INSERM, PSL Research University
Paris, France

JEAN-PHILIPPE REICHHELD
Laboratoire Génome et Développement des Plantes
Université Perpignan Via Domitia
Perpignan, France

SOFIE REMØ
Institute of Marine Research
Bergen, Norway

CLARA ORTEGÓN SALAS
Institute for Biochemistry and Molecular Biology
University of Greifswald
Greifswald, Germany

KATRIN SCHRÖDER
Institute for Cardiovascular Physiology
Goethe-University
Frankfurt, Germany

HELMUT SIES
Institute of Biochemistry and Molecular Biology I
Heinrich-Heine-Universität Düsseldorf
Düsseldorf, Germany

GABRIELA SPECKER
Departamento de Bioquímica
Universidad de la República
Montevideo, Uruguay

TSUYOSHI TAKATA
Department of Environmental Medicine and
    Molecular Toxicology
Tohoku University Graduate School of Medicine
Sendai, Japan

KENNETH D. TEW
Department of Cell and Molecular Pharmacology
    and Experimental Therapeutics
Medical University of South Carolina
Charleston, South Carolina

MARION THAUVIN
Center for Interdisciplinary Research in Biology
    (CIRB)
Collège de France
CNRS, INSERM, PSL Research University
Paris, France

MICHEL VOLOVITCH
Center for Interdisciplinary Research in Biology
    (CIRB)
Collège de France
CNRS, INSERM, PSL Research University
Paris, France

SOPHIE VRIZ
Center for Interdisciplinary Research in Biology
    (CIRB)
Collège de France
CNRS, INSERM, PSL Research University
Paris, France

RUNE WAAGBØ
Institute of Marine Research
Bergen, Norway

SAMEEKSHA YADAV
Special Centre for Molecular Medicine
Jawaharlal Nehru University
New Delhi, India

ZHONG YUN
Department of Therapeutic Radiology
Yale School of Medicine
New Haven, Connecticut

LEILEI ZHANG
Department of Cell and Molecular Pharmacology
    and Experimental Therapeutics
Medical University of South Carolina
Charleston, South Carolina

XU ZHANG
Departments of Molecular Medicine and
    Neuroscience and Neuroscience
    Translational Center
The Scripps Research Institute
La Jolla, California

Contributors

Introduction

CHAPTER ONE

Thiol Redox Regulation

A BRIEF HISTORICAL OVERVIEW

Carsten Berndt, Bob B. Buchanan, Christopher Horst Lillig, and Helmut Sies

CONTENTS

1.1 OXIDATIVE STRESS AND REDOX SIGNALING

Uncontrolled or uncontrollable oxidation of proteins, lipids, and nucleotides via reactive oxygen/nitrogen/sulfur species leads to damage and death of cells. This paradigm was summarized in the concept of oxidative stress formulated by one of us (H.S.) in 1985[1]. Although oxidative stress and the accompanying cell damage via increased formation of reactive species are linked to almost all pathological situations, the presence of low levels of reactive species, such as hydrogen peroxide or nitric oxide under non-pathological conditions, is also important for physiological signaling[2]. An updated definition of oxidative stress acknowledges these two sides of reactive species by distinguishing between oxidative distress (pathological damaging) and oxidative eustress (physiological signaling)[3]. The identification of the role of nitric oxide as a signaling molecule/second messenger formed the basis for the 1998 Nobel Prize awarded to Robert F. Furchgott, Louis J. Ignarro, and Ferid Murad for work in the 1970s and 1980s [4–7]. Hydrogen peroxide was described as a second messenger as early as 1974[8], but its role in cellular signaling has gained greater visibility only recently. This is surprising, because Otto Warburg described at the beginning of the 20th century the importance of oxidation for physiological processes, namely development and differentiation. Warburg followed the development of sea urchin eggs after fertilization and demonstrated that an oxidative burst is necessary for proper development and that abolishing the burst inhibited development [9,10].

These signaling events are specific and local and do not affect the total cellular redox state because cellular compartments differ in their individual redox states[11]. The concept of redox compartmentation goes back to 1958, when Bücher and Klingenberg described the connections between metabolic pathways and several redox systems in different compartments of the cell[12]. Moreover, redox signaling does not depend on chemical equilibria between reduced and oxidized molecules or oxidants and antioxidants. This might be

DOI: 10.4324/9781003204091-2

surprising for some, but this has been known for many years. In 1928, Leonor Michaelis wrote:

> The concept of a definite, finite redox potential and of a definite, finite redox system is limited to reversible systems and moreover to reversible systems at equilibrium. An attempt to apply the theory of redox potentials to physiological materials is met with the difficulty that oxidation in tissues is, on the whole, an irreversible process, impossible of ending in a state of equilibrium: such equilibrium would be contradictory to life[13].

Two years before Michaelis wrote this statement, Kendall and Nord concluded from their experiments that thiol-disulfide systems are reversible[14]. Today we know that redox signaling depends on reversible posttranslational oxidative thiol modifications regulated by enzymes. Reactivity of small molecules such as hydrogen peroxide or glutathione is limited without enzymes such as peroxiredoxins or glutaredoxins (see below) [15,16]. In fact, biological functions have not been found to depend solely on the amount of glutathione or the ratio between reduced and oxidized glutathione[17]. Thus, redox signaling is driven and controlled by non-equilibrium thermodynamics and enzymatic activities but not by equilibrium thermodynamics.

1.2 REDOX REGULATION OF ENZYMES

Redox regulatory enzymes contain cysteines and/ or selenocysteines in their active or regulatory sites.

Glutathione peroxidase (Flohé et al. and Rotruck et al.) and protein A of the glycine reductase system (Turner and Stadtman) were identified as the first selenoproteins in three publications in 1973 [18–20]. Selenocysteine was identified in protein A in 1976[21]. The human genome encodes 25 selenoproteins[22] and 214,000 cysteines in proteins[23]. Roughly 15% of these cysteines are redox active, suggesting that many or most cellular signaling pathways contain a thiol as a redox-regulated element.

1.2.1 Thiols

In terms of inorganic chemistry, H_2S is the first identified thiol. It was first mentioned in 1700 by Bernardino Ramazzini[24]; Carl Wilhelm Scheele analyzed the gas that he described as "stinky sulfur air" in 1777[25]; and the chemical composition was discovered by Claude Louis Berthollet in 1798[26]. At the end of the 19th century, Joseph de Rey-Pailhade described the reduction of elementary sulfur to H_2S by tissue extracts and named the responsible molecule "philothione" (see below). A paper published in 1942 mentioned the production of H_2S during the transsulfuration pathway in the liver of mammals[27], and in 1996 Hideo Kimura's group indicated that endogenous H_2S affects physiological functions[28].

The predominant biological thiol, cysteine, was discovered in 1884 by Eugen Baumann after the reduction of cystine[29], the disulfide that was isolated

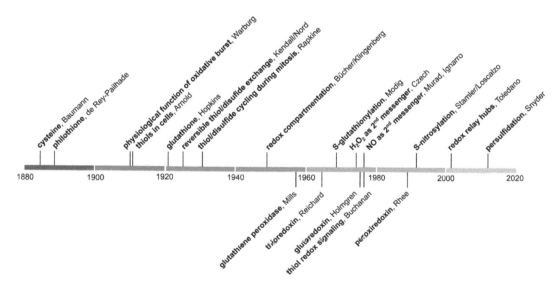

Figure 1.1 Milestones in research of thiol redox regulation.

in 1810[30]. Roughly 100 years later, Vinzenz Arnold demonstrated that thiols occur in tissue extracts[31]. In 1888, Joseph de Rey-Pailhade isolated a sulfur-containing substance in various biological samples, which he named "philothione"[32]. Frederick Gowland Hopkins rediscovered this compound containing cysteine in 1921 and named it "glutathione"[33]. The structure as a tripeptide, γ-Glu-Cys-Gly, was established between 1929 and 1935 by several groups using different methods [34–37] and by synthesis of the molecule[37]. The second sulfur-containing amino acid, methionine, was discovered and described as another source of SH groups in 1922[38]. Several years later, the importance of oxidation and reduction of SH groups for enzymatic activity was shown, with urease as an early example[39]. In 1931, Louis Rapkine uncovered the existence of a thiol-disulfide cycle during mitosis, thereby indicating that the reduction and oxidation of thiol groups might affect cellular processes[40]. In the 1940s and 1950s, Eleazar Guzman Barron described the oxidation of protein thiols following ionizing radiation and the accompanying inhibition of enzymes such as yeast alcohol dehydrogenase[41].

Protein thiols can undergo a variety of oxidative modifications. As mentioned above, thiol-disulfide exchange was postulated as early as 1926[14]. Later, thiols were found to be reversibly S-nitrosylated, persulfidated (S-sulfurated, S-sulfhydrated), oxidized to sulfenic acids, and able to form mixed disulfides (e.g., by S-glutathionylation). At the beginning of the 1950s, GSH was found to serve as a regulator of enzymatic activity for enzymes such as glyceraldehyde-3-phosphate dehydrogenase[42]. In 1968, Hans Modig demonstrated that glutathione forms mixed disulfides with protein cysteines within cells[43]. Remarkably, knowledge regarding persulfidation developed during the same period. In 1966, Kato et al. reported that the activity of serine dehydratase changed upon persulfidation[44]. However, intracellular persulfidation under physiological conditions within cells was not described until 2009 in research carried out by the group led by Solomon Snyder[45]. Somewhat earlier, in 1992, S-nitrosylation was described by Jonathan Stamler and Joseph Loscalzo working on endothelium-derived relaxing factor[46]. These different posttranslational modifications contribute to redox signaling. Today we know that pathways throughout biology are controlled by so-called thiol switches acting on specific enzymes. The first example of these was provided in 1977 in photosynthesis experiments conducted by Ricardo Wolosiuk and one of us (B.B.B.) with the oxidoreductase thioredoxin and fructose 1,6-bisphosphatase, an enzyme of the Calvin-Benson cycle[47].

1.2.2 Reduction of Enzymes

Key enzymes in the reduction of reversible post-translational oxidative thiol modifications include oxidoreductases of the thioredoxin family, namely thioredoxins (Trxs) and glutaredoxins (Grxs). Both names were introduced by the laboratory of Peter Reichard and Arne Holmgren in Sweden.

1.2.2.1 Thioredoxin

Trx activity in yeast was published in 1960 and 1961 as enzyme II in the reduction of methionine sulfoxide to methionine[48] and fraction C of the sulfate-reducing system, respectively [48,49]. The name "thioredoxin" was introduced in 1964 by Peter Reichard's group for an enzyme providing electrons for ribonucleotide reductase in *Escherichia coli*[50]. In 1971, it was shown that Trx is present in mammals[51], and three years later it was found to function in the regulation of photosynthetic enzymes in vitro [47,52]. The sequence and structure of Trx was elucidated in Stockholm, thereby providing the characteristics of the Trx-fold, the CPYC active site motif, and the opposite located cis-proline [53,54].

Today, we know that Trx functions not only in the regulation of photosynthesis and the formation of deoxyribonucleotides, but that it also controls a spectrum of cellular processes in all kingdoms of life[55]. Trx is able to reduce disulfides as well as S-nitrosylated and persulfidated thiols[56]. Oxidized Trx was originally shown to be reduced by the flavoenzyme thioredoxin reductase, an NADP-linked enzyme discovered in 1964[39]. We now know that Trx can be reduced by several other enzymes, including ferredoxin-thioredoxin reductase in oxygenic photosynthesis[47] and ferredoxin-dependent flavin thioredoxin reductase (FFTR) in fermentation[57].

Recently, both *Arabidopsis thaliana* Trx and mammalian Trx1 were characterized as FeS cluster–coordinating proteins, demonstrating that Grxs are not the only FeS cluster–containing oxidoreductases of the Trx family [58,59].

1.2.2.2 Glutaredoxin

Originally, Grx was shown by Arne Holmgren to serve as an alternate electron donor for ribonucleotide reductase in an *E. coli* mutant lacking Trx[60]. Two years earlier, in 1974, Grx from rabbit liver was described as a disulfide-reducing enzyme named thioltransferase by the group led by Bengt Mannervik[61]. The activity of Grxs in the reduction of disulfides and S-glutathionylated thiols depends on GSH and the two cysteinyl residues in the active site. In these cases, the enzyme follows a monothiol or a dithiol mechanism, respectively [62,63]. It was not until 1994 that Grxs were identified in plants[64]. Now we know that plant genomes encode around 20 Grxs[65]. Some of them, namely, the CC-type Grxs identified in 2004[66], function in flower development[67] and are specific for land plants. In 1999, Enrique Herrero and his group identified a highly conserved subfamily characterized by only one cysteine in the active site[68]. The same group described these so-called monothiol Grxs as important regulators of iron homeostasis and FeS cluster biosynthesis[69].

Grxs that lack a prolyl residue in their CxxC active site motif were characterized as FeS-proteins in 2005[70]. This protein family coordinates the FeS cluster using non-covalently bound GSH as a non-protein ligand, making Grxs unique FeS-proteins [71–73]. Another remarkable unique feature is that some of these FeS clusters may serve as redox sensors controlling the thiol redox regulating activity of these enzymes.

1.2.3 Oxidation of Enzymes

Under conditions of a low ratio between GSH and its disulfide (GSSG), Grxs are able to glutathionylate thiols. However, in contrast to reducing enzymes, the number of identified thiol oxidases is very limited. In the frame of this short overview, we will focus on H_2O_2-mediated thiol oxidation. Hydrogen peroxide was characterized in 1818 by Louis Jacques Thénard as *eau oxygenée*[74], but it took until 1970 to prove its production within cells[75]. The different theoretical possibilities and experimental findings on how H_2O_2 could oxidize thiols are summarized in[76]. These possibilities can be narrowed down to (i) direct oxidation or (ii) indirect oxidation via redox relay hubs. Such hubs could be peroxiredoxins (Prxs) because of their very high second-order rate constant for H_2O_2 and by their high abundance

in cells. Prxs were named peroxidoxin[77], thioredoxin peroxidase[78], and thiol specific antioxidant (TSA)[79]. In fact, Prx has been known since 1968 under the name torin[80], but its function in protection of cells against oxidative damage was first described in *Saccharomyces cerevisiae* in 1988[79]. The finding in 2011 that yeast cells lacking Prxs fail to show a transcriptional or a translational response following H_2O_2 treatment suggests a central role of the proteins in H_2O_2-mediated signaling[81]. Similar results were obtained with the cytosol of mammalian cells lacking Prxs 1 and 2, where H_2O_2-induced protein thiol oxidation is abolished[82]. In human cells, Prx1 is essential for H_2O_2-mediated oxidation and activation of apoptosis signaling kinase 1[83], whereas Prx2 transmits oxidation of the transcription factor STAT3[84].

Prxs are not the only transducers of H_2O_2-induced oxidation. Of note, the first identified relay is GPx3/Orp1, which bridges the oxidation and activation of the transcription factor Yab1 via H_2O_2 in *S. cerevisiae*[85]. Trx reduces both GPx3 as well as Yab1 and turns off the pathway[85]. Following this result obtained by the group of Michel Toledano, several other GPxs, a protein family described for the first time in 1957[86], were shown to oxidize proteins within different organisms [87–89].

The second sulfur-containing amino acid, methionine, was reported to be oxidized in some proteins, such as actin, specifically by monooxygenases—namely members of the MICAL protein family[90]. Furthermore, the resulting methionine sulfoxides are specifically reduced by methionine sulfoxide reductases[91]. Whether methionine oxidation also depends on a relay system or whether cysteine residues can be oxidized by monooxygenases is an open question[92].

1.3 CONCLUSION

More and more researchers and clinicians are becoming aware of the importance and function of reactive oxygen species as second messengers and of redox regulation in general. While the field is being more appreciated, much remains to be learned. The identification of oxidases regulating the oxidation of specific protein thiols emerges as a particularly important area to be investigated in the future. However, before starting new studies, an examination of the history of the field can be helpful, because several apparently "novel" observations may have already been described years ago.

REFERENCES

[1] Sies, H., Ed. Oxidative stress: introductory remarks. In *Oxidative Stress*, ed. H Sies, pp. 1–8; London, Academic Press, **1985**.

[2] Sies, H.; Jones, D. P. Reactive oxygen species (ROS) as pleiotropic physiological signalling agents. *Nature Reviews Molecular Cell Biology* **2020**, 21, 363–383. doi:10.1038/s41580-020-0230-3.

[3] Sies, H.; Berndt, C.; Jones, D. P. Oxidative stress. *Annual Review of Biochemistry* **2017**, 86, 715–748. doi:10.1146/annurev-biochem-061516-045037.

[4] Arnold, W. P.; Mittal, C. K.; Katsuki, S.; Murad, F. Nitric oxide activates guanylate cyclase and increases guanosine 3':5'-cyclic monophosphate levels in various tissue preparations. *Proceedings of the National Academy of Sciences of the United States of America* **1977**, 74, 3203–3207. doi:10.1073/pnas.74.8.3203.

[5] Furchgott, R. F.; Zawadzki, J. V. The obligatory role of endothelial cells in the relaxation of arterial smooth muscle by acetylcholine. *Nature* **1980**, 288, 373–376. doi:10.1038/288373a0.

[6] Gruetter, C. A.; Barry, B. K.; McNamara, D. B.; Gruetter, D. Y.; Kadowitz, P. J.; Ignarro, L. Relaxation of bovine coronary artery and activation of coronary arterial guanylate cyclase by nitric oxide, nitroprusside and a carcinogenic nitrosoamine. *Journal of Cyclic Nucleotide Research* **1979**, 5, 211–224.

[7] Ignarro, L. J.; Buga, G. M.; Wood, K. S.; Byrns, R. E.; Chaudhuri, G. Endothelium-derived relaxing factor produced and released from artery and vein is nitric oxide. *Proceedings of the National Academy of Sciences of the United States of America* **1987**, 84, 9265–9269. doi:10.1073/pnas.84.24.9265.

[8] Czech, M. P.; Lawrence, J. C.; Lynn, W. S. Evidence for the involvement of sulfhydryl oxidation in the regulation of fat cell hexose transport by insulin. *Proceedings of the National Academy of Sciences of the United States of America* **1974**, 71, 4173–4177. doi:10.1073/pnas.71.10.4173.

[9] Warburg, O. Beobachtungen über die Oxydationsprozesse im Seeigelei. *Z. Physiol. Chem.* **1908**, 57, 1–16.

[10] Warburg, O. Über die Oxydation in lebenden Zellen nach Versuchen am Seeigelei. *Z. Physiol. Chem.* **1910**, 66, 305–340.

[11] Go, Y.-M.; Jones, D. P. Redox compartmentalization in eukaryotic cells. *Biochimica et biophysica acta* **2008**, 1780, 1273–1290. doi:10.1016/j.bbagen.2008.01.011.

[12] Bücher, T.; Klingenberg, M. Wege des Wasserstoffs in der lebendigen Organisation. *Angew. Chem.* **1958**, 70, 552–570.

[13] Michaelis, L.; Flexner, L. B. oxidation-reduction systems of biological significance: I. the reduction potential of cysteine: its measurement and significance. *Journal of Biological Chemistry* **1928**, 79, 689–722.

[14] Kendall, E. C.; Nord, F. F. Reversible oxidation-reduction systems of cysteine-cystine and reduced and oxidized glutathione. *Journal of Biological Chemistry* **1926**, 69, 295–337.

[15] Winterbourn, C. C. The biological chemistry of hydrogen peroxide. *Methods Enzymol* **2013**, 528, 3–25.

[16] Berndt, C.; Lillig, C. H.; Flohé, L. Redox regulation by glutathione needs enzymes. *Frontiers in Pharmacology* **2014**, 5, 168. doi:10.3389/fphar.2014.00168.

[17] Flohé, L. The fairytale of the GSSG/GSH redox potential. *Biochimica et biophysica acta* **2013**, 1830, 3139–3142. doi:10.1016/j.bbagen.2012.10.020.

[18] Flohé, L.; Günzler, W. A.; Schock, H. H. Glutathione peroxidase: a selenoenzyme. *FEBS Letters* **1973**, 32, 132–134. doi:10.1016/0014-5793(73)80755-0.

[19] Rotruck, J. T.; Pope, A. L.; Ganther, H. E.; Swanson, A. B.; Hafeman, D. G.; Hoekstra, W. G. Selenium: biochemical role as a component of glutathione peroxidase. *Science (New York, N.Y.)* **1973**, 179, 588–590. doi:10.1126/science.179.4073.588.

[20] Turner, D. C.; Stadtman, T. C. Purification of protein components of the clostridial glycine reductase system and characterization of protein A as a selenoprotein. *Archives of Biochemistry and Biophysics* **1973**, 154, 366–381. doi:10.1016/0003-9861(73)90069-6.

[21] Cone, J. E.; Del Río, R. M.; Davis, J. N.; Stadtman, T. C. Chemical characterization of the selenoprotein component of clostridial glycine reductase: identification of selenocysteine as the organoselenium moiety. *Proceedings of the National Academy of Sciences of the United States of America* **1976**, 73, 2659–2663. doi:10.1073/pnas.73.8.2659.

[22] Lobanov, A. V.; Hatfield, D. L.; Gladyshev, V. N. Eukaryotic selenoproteins and selenoproteomes. *Biochimica et biophysica acta* **2009**, 1790, 1424–1428. doi:10.1016/j.bbagen.2009.05.014.

[23] Go, Y.-M.; Jones, D. P. The redox proteome. *Journal of Biological Chemistry* **2013**, 288, 26512–26520. doi:10.1074/jbc.R113.464131.

[24] Ramazzini, B. *De Morbis Artificum Diatriba*; Modena, **1700**.

[25] Scheele, C. W. Die stinckende Schwefel Luft. In *Chemische Abhandlung von der Luft und dem Feuer*, § 97, pp. 149–155, **1777**.

[26] Berthollet, C. L. Observations sur l'Hydrogène Sulfuré. *Ann. Chim.* **1798**, 25, 233–272.

[27] Binkley, F.; du Vigneaud, V. The formation of cysteine from homocysteine and serine by liver tissue of rats. *Journal of Biological Chemistry* **1942**, 144, 507–511.

[28] Abe, K.; Kimura, H. The possible role of hydrogen sulfide as an endogenous neuromodulator. *Journal of Neuroscience* **1996**, 16, 1066–1071. doi:10.1523/JNEUROSCI.16-03-01066.1996.

[29] Baumann, E. Über Cystin und Cystein. *Zeitschrift für Physiologische Chemie* **1884**, 8, 299–305.

[30] Wollaston, W. H. On cystic oxide, a new species of urinary calculus. *Philosophical Transactions of the Royal Society* **1810**, 100, 223–230.

[31] Arnold, V. Eine Farbenreaktion von Eiweißkörpern mit Nitroprussidnatrium. *Zeitschrift für Physiologische Chemie* **1911**, 70, 300–325.

[32] Rey-Pailhade, J. de. Sur un corps d'origine organique hydrogénant le soufre 1 à froid. *Comptes rendus de l'Académie des Sciences* **1888**, 106, 1683–1684.

[33] Hopkins, F. G. On an autoxidisable constituent of the cell. *Biochemical Journal* **1921**, 15, 286–305.

[34] Hopkins, F. G. On glutathione: a reinvestigation. *Journal of Biological Chemistry* **1929**, 84, 269–320. doi:10.1016/S0021-9258(18)77062-2.

[35] Kendall, E. C.; McKenzie, B. F.; Mason, H. L. A study of glutathione. *Journal of Biological Chemistry* **1929**, 84, 657–674. doi:10.1016/S0021-9258(18)77022-1.

[36] Pirie, N. W.; Pinhey, K. G. The titration curve of glutathione. *Journal of Biological Chemistry* **1929**, 84, 321–333. doi:10.1016/S0021-9258(18)77063-4.

[37] Harington, C. R.; Mead, T. H. Synthesis of glutathione. *Biochemical Journal* **1935**, 29, 1602–1611. doi:10.1042/bj0291602.

[38] Mueller, J. H. A new sulphur-containing amino acid isolated from casein. *Experimental Biology and Medicine* **1922**, 19, 161–163. doi:10.3181/00379727-19-75.

[39] Hellerman, L.; Perkins, M. E., Clark, W. M. Urease activity as influenced by oxidation and reduction. *Proceedings of the National Academy of Sciences of the United States of America* **1933**, 19, 855–860. doi:10.1073/pnas.19.9.855.

[40] Rapkine, L. Su les processus chimiques au cours de la division cellulaire. *Annales de Physiologie et de Physicochimie Biologique* **1931**, 7, 382–418.

[41] Barron, E. S. G. Oxidation of some oxidation-reduction systems by oxygen at high pressures. *Archives of Biochemistry and Biophysics* **1955**, 59, 502–510. doi:10.1016/0003-9861(55)90516-6.

[42] Krimsky, I.; Racker, E. Glutathione, a prosthetic group of glyceraldehyde-3-phosphate dehydrogenase. *Journal of Biological Chemistry* **1952**, 198, 721–729.

[43] Modig, H. Cellular mixed disulphides between thiols and proteins, and their possible implication for radiation protection. *Biochemical Pharmacology* **1968**, 17, 177–186. doi:10.1016/0006-2952(68)90321-3.

[44] Kato, A.; Ogura, M.; Suda, M. Control mechanism in the rat liver enzyme system converting L-methionine to L-cystine. 3. Noncompetitive inhibition of cystathionine synthetase-serine dehydratase by elemental sulfur and competitive inhibition of cystathionase-homoserine dehydratase by L-cysteine and L-cystine. *Journal of Biochemistry* **1966**, 59, 40–48. doi:10.1093/oxfordjournals.jbchem.a128256.

[45] Mustafa, A. K.; Gadalla, M. M.; Sen, N.; Kim, S.; Mu, W.; Gazi, S. K.; Barrow, R. K.; Yang, G.; Wang, R.; Snyder, S. H. H2S signals through protein S-sulfhydration. *Science Signaling* **2009**, 2, ra72. doi:10.1126/scisignal.2000464.

[46] Stamler, J. S.; Simon, D. I.; Osborne, J. A.; Mullins, M. E.; Jaraki, O.; Michel, T.; Singel, D. J.; Loscalzo, J. S-nitrosylation of proteins with nitric oxide: synthesis and characterization of biologically active compounds. *Proceedings of the National Academy of Sciences of the United States of America* **1992**, 89, 444–448. doi:10.1073/pnas.89.1.444.

[47] Wolosiuk, R. A.; Buchanan, B. B. Thioredoxin and glutathione regulate photosynthesis in chloroplasts. *Nature* **1977**, 266, 565–567.

[48] Black, S.; Harte, E. M.; Hudson, B.; Wartofsky, L. A specific enzymatic reduction of L (−) methionine sulfoxide and a related non-specific reduction of disulfides. *Journal of Biological Chemistry* **1960**, 235.

[49] Asahi, T.; Bandurski, R. S.; Wilson, L. G. Yeast sulfate-reducing system. II. Enzymatic reduction of protein disulfide. *Journal of Biological Chemistry* **1961**, 236, 1830–1835.

[50] Laurent, T. C.; Moore, E. C.; Reichard, P. Enzymatic synthesis of deoxyribonucleotides.

iv. isolation and characterization of thioredoxin, the hydrogen donor from Escherichia coli B. Journal of Biological Chemistry **1964**, 239, 3436–3444.

[51] Engström, N. E.; Holmgren, A.; Larsson, A.; Söderhäll, S. Isolation and characterization of calf liver thioredoxin. Journal of Biological Chemistry **1974**, 249, 205–210.

[52] Holmgren, A. Photosynthetic regulatory protein from rabbit liver is identical with thioredoxin. FEBS Letters **1977**, 82, 351–354. doi:10.1016/0014-5793(77)80619-4.

[53] Holmgren, A. Thioredoxin. 6. The amino acid sequence of the protein from Escherichia coli B. European Journal of Biochemistry **1968**, 6, 475–484. doi:10.1111/j.1432-1033.1968.tb00470.x.

[54] Holmgren, A.; Söderberg, B. O.; Eklund, H.; Brändén, C. I. Three-dimensional structure of Escherichia coli thioredoxin-S2 to 2.8 A resolution. Proceedings of the National Academy of Sciences of the United States of America **1975**, 72, 2305–2309. doi:10.1073/pnas.72.6.2305.

[55] Hanschmann, E.-M.; Godoy, J. R.; Berndt, C.; Hudemann, C.; Lillig, C. H. Thioredoxins, glutaredoxins, and peroxiredoxins: molecular mechanisms and health significance: from cofactors to antioxidants to redox signaling. Antioxidants & Redox Signaling **2013**, 19, 1539–1605. doi:10.1089/ars.2012.4599.

[56] Dóka, É.; Pader, I.; Bíró, A.; Johansson, K.; Cheng, Q.; Ballagó, K.; Prigge, J. R.; Pastor-Flores, D.; Dick, T. P.; Schmidt, E. E.; et al. A novel persulfide detection method reveals protein persulfide- and polysulfide-reducing functions of thioredoxin and glutathione systems. Science Advances **2016**, 2, e1500968. doi:10.1126/sciadv.1500968.

[57] Buey, R. M.; Fernández-Justel, D.; Pereda, J. M. de; Revuelta, J. L.; Schürmann, P.; Buchanan, B. B.; Balsera, M. Ferredoxin-linked flavoenzyme defines a family of pyridine nucleotide-independent thioredoxin reductases. Proceedings of the National Academy of Sciences of the United States of America **2018**, 115, 12967–12972. doi:10.1073/pnas.1812781115.

[58] Zannini, F.; Roret, T.; Przybyla-Toscano, J.; Dhalleine, T.; Rouhier, N.; Couturier, J. Mitochondrial Arabidopsis thaliana TRXo isoforms bind an iron-sulfur cluster and reduce NFU proteins in vitro. Antioxidants (Basel, Switzerland) **2018**, 7. doi:10.3390/antiox7100142.

[59] Berndt, C.; Hanschmann, E.-M.; Urbainsky, C.; Jordt, L. M.; Müller, C. S.; Bodnar, Y.; Schipper, S.; Handorf, O.; Nowack, R.; Moulis, J.-M.; et al. FeS-cluster coordination of vertebrate thioredoxins regulates suppression of hypoxia-induced factor 2α through iron regulatory protein 1, BioRxiv **2020**.

[60] Holmgren, A. Hydrogen donor system for Escherichia coli ribonucleoside-diphosphate reductase dependent upon glutathione. Proceedings of the National Academy of Sciences of the United States of America **1976**, 73, 2275–2279. doi:10.1073/pnas.73.7.2275.

[61] Eriksson, S.; Askelöf, P.; Axelsson, K.; Carlberg, I.; Guthenberg, C.; Mannervik, B. Resolution of glutathione-linked enzymes in rat liver and evaluation of their contribution to disulfide reduction via thiol-disulfide interchange. Acta Chemica Scandinavica. Series B: Organic Chemistry and Biochemistry **1974**, 28, 922–930. doi:10.3891/acta.chem.scand.28b-0922.

[62] Bushweller, J. H.; Aslund, F.; Wüthrich, K.; Holmgren, A. Structural and functional characterization of the mutant Escherichia coli glutaredoxin (C14→S) and its mixed disulfide with glutathione. Biochemistry **1992**, 31, 9288–9293. doi:10.1021/bi00153a023.

[63] Holmgren, A. Glutathione-dependent synthesis of deoxyribonucleotides. Characterization of the enzymatic mechanism of Escherichia coli glutaredoxin. Journal of Biological Chemistry **1979**, 254, 3672–3678.

[64] Minakuchi, K.; Yabushita, T.; Masumura, T.; Ichihara, K.'i.; Tanaka, K. Cloning and sequence analysis of a cDNA encoding rice glutaredoxin. FEBS Letters **1994**, 337, 157–160. doi:10.1016/0014-5793(94)80264-5.

[65] Meyer, Y.; Belin, C.; Delorme-Hinoux, V.; Reichheld, J.-P.; Riondet, C. Thioredoxin and glutaredoxin systems in plants: molecular mechanisms, crosstalks, and functional significance. Antioxidants & Redox Signaling **2012**, 17, 1124–1160. doi:10.1089/ars.2011.4327.

[66] Lemaire, S. D. The glutaredoxin family in oxygenic photosynthetic organisms. Photosynthesis Research **2004**, 79, 305–318. doi:10.1023/B:PRES.0000017174.60951.74.

[67] Xing, S.; Rosso, M. G.; Zachgo, S. ROXY1, a member of the plant glutaredoxin family, is required for petal development in Arabidopsis thaliana. Development (Cambridge, England) **2005**, 132, 1555–1565. doi:10.1242/dev.01725.

[68] Rodríguez-Manzaneque, M. T.; Ros, J.; Cabiscol, E.; Sorribas, A.; Herrero, E. Grx5 glutaredoxin plays a central role in protection against protein oxidative damage in Saccharomyces cerevisiae.

Molecular and Cellular Biology **1999**, 19, 8180–8190. doi:10.1128/mcb.19.12.8180.

[69] Rodríguez-Manzaneque, M. T.; Tamarit, J.; Bellí, G.; Ros, J.; Herrero, E. Grx5 is a mitochondrial glutaredoxin required for the activity of iron/sulfur enzymes. *Molecular Biology of the Cell* **2002**, 13, 1109–1121. doi:10.1091/mbc.01-10-0517.

[70] Lillig, C. H.; Berndt, C.; Vergnolle, O.; Lönn, M. E.; Hudemann, C.; Bill, E.; Holmgren, A. Characterization of human glutaredoxin 2 as iron-sulfur protein: a possible role as redox sensor. *Proceedings of the National Academy of Sciences of the United States of America* **2005**, 102, 8168–8173. doi:10.1073/pnas.0500735102.

[71] Berndt, C.; Hudemann, C.; Hanschmann, E.-M.; Axelsson, R.; Holmgren, A.; Lillig, C. H. How does iron-sulfur cluster coordination regulate the activity of human glutaredoxin 2? *Antioxidants & Redox Signaling* **2007**, 9, 151–157. doi:10.1089/ars.2007.9.151.

[72] Johansson, C.; Kavanagh, K. L.; Gileadi, O.; Oppermann, U. Reversible sequestration of active site cysteines in a 2Fe-2S-bridged dimer provides a mechanism for glutaredoxin 2 regulation in human mitochondria. *Journal of Biological Chemistry* **2007**, 282, 3077–3082. doi:10.1074/jbc.M608179200.

[73] Rouhier, N.; Unno, H.; Bandyopadhyay, S.; Masip, L.; Kim, S.-K.; Hirasawa, M.; Gualberto, J. M.; Lattard, V.; Kusunoki, M.; Knaff, D. B.; et al. Functional, structural, and spectroscopic characterization of a glutathione-ligated 2Fe-2S cluster in poplar glutaredoxin C1. *Proceedings of the National Academy of Sciences of the United States of America* **2007**, 104, 7379–7384. doi:10.1073/pnas.0702268104.

[74] Thénard, L. J. Observations sur des nouvelles combinaisons entre l'oxigéne et divers acides. *Annales de Chimie et de Physique* **1818**, 8, 306–312.

[75] Sies, H.; Chance, B. The steady state level of catalase compound I in isolated hemoglobin-free perfused rat liver. *FEBS Letters* **1970**, 11, 172–176. doi:10.1016/0014-5793(70)80521-x.

[76] Stöcker, S.; van Laer, K.; Mijuskovic, A.; Dick, T. P. The conundrum of hydrogen peroxide signaling and the emerging role of peroxiredoxins as redox relay hubs. *Antioxidants & Redox Signaling* **2018**, 28, 558–573. doi:10.1089/ars.2017.7162.

[77] Chae, H. Z.; Robison, K.; Poole, L. B.; Church, G.; Storz, G.; Rhee, S. G. Cloning and sequencing of thiol-specific antioxidant from mammalian brain: alkyl hydroperoxide reductase and thiol-specific antioxidant define a large family

of antioxidant enzymes. *Proceedings of the National Academy of Sciences of the United States of America* **1994**, 91, 7017–7021. doi:10.1073/pnas.91.15.7017.

[78] Chae, H. Z.; Chung, S. J.; Rhee, S. G. Thioredoxin-dependent peroxide reductase from yeast. *Journal of Biological Chemistry* **1994**, 269, 27670–27678.

[79] Kim, K.; Kim, I. H.; Lee, K. Y.; Rhee, S. G.; Stadtman, E. R. The isolation and purification of a specific "protector" protein which inhibits enzyme inactivation by a thiol/Fe(III)/O2 mixed-function oxidation system. *Journal of Biological Chemistry* **1988**, 263, 4704–4711.

[80] Harris, J. R. Release of a macromolecular protein component from human erythrocyte ghosts. *Biochimica et Biophysica Acta (BBA)—Biomembranes* **1968**, 150, 534–537. doi:10.1016/0005-2736(68)90157-0.

[81] Fomenko, D. E.; Koc, A.; Agisheva, N.; Jacobsen, M.; Kaya, A.; Malinouski, M.; Rutherford, J. C.; Siu, K.-L.; Jin, D.-Y.; Winge, D. R.; et al. Thiol peroxidases mediate specific genome-wide regulation of gene expression in response to hydrogen peroxide. *Proceedings of the National Academy of Sciences of the United States of America* **2011**, 108, 2729–2734. doi:10.1073/pnas.1010721108.

[82] Stöcker, S.; Maurer, M.; Ruppert, T.; Dick, T. P. A role for 2-Cys peroxiredoxins in facilitating cytosolic protein thiol oxidation. *Nature Chemical Biology* **2018**, 14, 148–155. doi:10.1038/nchembio.2536.

[83] Jarvis, R. M.; Hughes, S. M.; Ledgerwood, E. C. Peroxiredoxin 1 functions as a signal peroxidase to receive, transduce, and transmit peroxide signals in mammalian cells. *Free Radical Biology & Medicine* **2012**, 53, 1522–1530. doi:10.1016/j.freeradbiomed.2012.08.001.

[84] Sobotta, M. C.; Liou, W.; Stöcker, S.; Talwar, D.; Oehler, M.; Ruppert, T.; Scharf, A. N. D.; Dick, T. P. Peroxiredoxin-2 and STAT3 form a redox relay for H_2O_2 signaling. *Nature Chemical Biology* **2015**, 11, 64–70. doi:10.1038/nchembio.1695.

[85] Delaunay, A.; Pflieger, D.; Barrault, M.-B.; Vinh, J.; Toledano, M. B. A thiol peroxidase is an H_2O_2 receptor and redox-transducer in gene activation. *Cell* **2002**, 111, 471–481. doi:10.1016/s0092-8674(02)01048-6.

[86] MILLS, G. C. Hemoglobin catabolism. I. Glutathione peroxidase, an erythrocyte enzyme which protects hemoglobin from oxidative breakdown. *Journal of Biological Chemistry* **1957**, 229, 189–197.

[87] Conrad, M.; Moreno, S. G.; Sinowatz, F.; Ursini, F.; Kölle, S.; Roveri, A.; Brielmeier, M.; Wurst, W.; Maiorino, M.; Bornkamm, G. W. The nuclear form of phospholipid hydroperoxide glutathione peroxidase is a protein thiol peroxidase contributing to sperm chromatin stability. *Molecular and Cellular Biology* **2005**, 25, 7637–7644. doi:10.1128/MCB.25.17.7637-7644.2005.

[88] van Nguyen, D.; Saaranen, M. J.; Karala, A.-R.; Lappi, A.-K.; Wang, L.; Raykhel, I. B.; Alanen, H. I.; Salo, K. E. H.; Wang, C.-C.; Ruddock, L. W. Two endoplasmic reticulum PDI peroxidases increase the efficiency of the use of peroxide during disulfide bond formation. *Journal of Molecular Biology* **2011**, 406, 503–515. doi:10.1016/j.jmb.2010.12.039.

[89] Gutscher, M.; Sobotta, M. C.; Wabnitz, G. H.; Ballikaya, S.; Meyer, A. J.; Samstag, Y.; Dick, T. P. Proximity-based protein thiol oxidation by H_2O_2-scavenging peroxidases. *Journal of Biological Chemistry* **2009**, 284, 31532–31540. doi:10.1074/jbc.M109.059246.

[90] Lee, B. C.; Péterfi, Z.; Hoffmann, F. W.; Moore, R. E.; Kaya, A.; Avanesov, A.; Tarrago, L.; Zhou, Y.; Weerapana, E.; Fomenko, D. E.; *et al.* MsrB1 and MICALs regulate actin assembly and macrophage function via reversible stereoselective methionine oxidation. *Molecular Cell* **2013**, 51, 397–404. doi:10.1016/j.molcel.2013.06.019.

[91] Brot, N.; Weissbach, L.; Werth, J.; Weissbach, H. Enzymatic reduction of protein-bound methionine sulfoxide. *Proceedings of the National Academy of Sciences of the United States of America* **1981**, 78, 2155–2158. doi:10.1073/pnas.78.4.2155.

[92] Ortegón Salas, C.; Schneider, K.; Lillig, C. H.; Gellert, M. Signal-regulated oxidation of proteins via MICAL. *Biochemical Society Transactions* **2020**, 48, 613–620. doi:10.1042/BST20190866.

Development of (In)Vertebrate Model Organisms

CHAPTER TWO

Redox Regulation of Plant Development

Jean-Philippe Reichheld, Avilien Dard, and Christophe Belin

CONTENTS

2.1 INTRODUCTION

The development of plant organs is dependent on the maintenance and the sequential activation of stem cells embedded in meristematic tissues. Cell division and differentiation cycles determine the shape and the functions of the different plant organs. While endogenous factors like plant hormones are major determinants of plant development programs, they are also largely influenced by their environment. Understanding the mechanisms by which a plant perceives its environment to adapt its development is a major question for plant biologists. One of these mechanisms relies on regulation by oxidoreduction (redox). Redox regulation is mediated by reactive oxygen species (ROS), which are by-products of oxygen metabolism. ROS include free radical species like superoxides ($O_2^{\cdot-}$), hydroxyl radicals ($OH^{\cdot}$), or nitric oxide ($NO^{\cdot}$), and nonradical species like hydrogen peroxide (H_2O_2) and peroxynitrite ($ONOO^-$) (Sies et al., 2017). In plants, major sources of ROS are photosynthetic and respiratory chains in chloroplasts and mitochondria (Figure 2.1).

ROS are also generated by plasma membrane NADPH oxidases and peroxisomal xanthine oxidases and exert fine-tune regulation of developmental process depending on the location of their generation. ROS generation in challenged tissues can trigger oxidative distress, damaging cells or inducing cell death (Foyer and Noctor, 2016; Choudhury et al., 2017). Oxidative eustress also performs important signaling functions by inducing post-translational modifications (PTM) and by regulating protein redox state. Plant cells display a large panel of ROS-scavenging enzymes like catalases, peroxidases, and superoxide dismutases. They also generate compounds that reverse ROS-induced oxidations. Among these compounds are antioxidant molecules like glutathione and ascorbate, which both play important roles as cofactors for thiol reduction enzymes like peroxidases and reductases (Noctor et al., 2018; Rahantaniana et al., 2017). Glutathione and ascorbate are themselves reduced by glutathione reductases (GRs) and dehydroascorbate reductases (DHARs). Thioredoxins (TRXs) and glutaredoxins (GRXs) are key thiol reduction enzymes. They act as reducing power of metabolic enzymes and ROS scavenging systems but they also regulate thiol-based post-transcriptional redox modifications

DOI: 10.4324/9781003204091-4

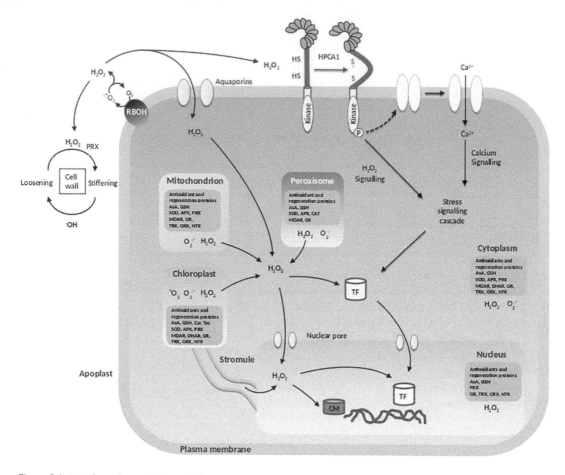

Figure 2.1 ROS-dependent cellular metabolism.

ROS generation by reduction of molecular oxygen occurs at several subcellular and extracellular sites. Extracellular ROS generated by RBOH and peroxidases plays a major role in cell wall flexibility and control cell elongation. The most stable ROS, H_2O_2, can move from apoplast to the intracellular space by diffusion through aquaporins. It can also be sensed by the HPCA1-type receptor kinase to trigger Ca^{2+} translocation and signaling cascades. The chloroplast is the main site of singlet oxygen formation whereas superoxides are generated in all cell compartments, where they are rapidly dismuted to H_2O_2 by SODs. H_2O_2 can move from the different compartments in which it is mainly produced to alter cytosolic and nuclear redox states, which can be perceived by receptor proteins like transcription factors (TF) or chromatin modifiers (CM) in the nucleus. Gene expression will be modified by altered activity or translocation of these factors, which may or may not themselves be redox modified. Extensive panels of plant antioxidative systems and thiols regeneration are found within the cell. The figure summarizes available information for *Arabidopsis* on the subcellular localization of antioxidative enzymes and related proteins. The information is not exhaustive, and other proteins may be involved.

APX, ascorbate peroxidase. CAT, catalase. CM, chromatin modifier. DHAR, dehydroascorbate reductase. GR, glutathione reductase. GRX, glutaredoxin. HPCA1, hydrogen peroxide-induced Ca^{2+} increases 1. MDAR, monodehydroascorbate reductase. NTR, NADPH-thioredoxin reductase. PRX, peroxiredoxin. RBOH, respiratory burst oxidase homologue. SOD, superoxide dismutase. TF, transcription factor. TRX, thioredoxin. O_2, Oxygen.$^1O_2$, singlet oxygen. OH^-, hydroxide ion. H_2O_2, hydrogen peroxide. SH, reduced thiol. S-S, disulfide bond. AsA, ascorbic acid. Ca^{2+}, calcium cation. Car, carotenoid. GSH, glutathione. Toc, tocopherol.

(Figure adapted from Noctor et al. 2018.)

in proteins (Meyer et al., 2012). Oxidized TRXs are generally reduced by NADPH-dependent thioredoxin reductases (NTRs), whereas the reduction of GRXs is dependent on glutathione. Due to their multifunctional thiol reduction capacities, TRXs and GRXs have been involved in many metabolic functions, controlling plant developmental programs and acting as key signaling molecules in response to abiotic and biotic stresses (Meyer et al., 2009; Rouhier et al., 2015).

In this chapter, we summarize the current knowledge on how redox regulation influences plant development programs and how it mediates plant perception of environmental constraints (Table 2.1).

2.2 CELL DIVISION AND DIFFERENTIATION/MERISTEMS

Plant growth and development is intimately dependent on founder cell proliferation in meristematic zones. The shoot apical meristem (SAM) is established during embryogenesis and gives rise to all aboveground tissues of a plant. The SAM is composed of different cell layers with contrasted cell division capacities.

The central zone (CZ) contains three layers (L1, L2, and L3) of slowly dividing stem cells, while the organizing center (OC) found underneath maintains no dividing stem cells. Surrounding the CZ, the peripheral zone (PZ) contains rapidly dividing cells and gives rise to new boundary organs, such as leaf primordia (Figure 2.2A). By percepting the environmental constraints, ROS and the redox status are involved in both triggering cell proliferation and regulating cell cycle progression. Oxidative stress conditions are known to block cell cycle progression at specific checkpoints (Reichheld et al., 1999; Vernoux et al., 2000) and different forms of ROS have antagonistic roles in plant stem cell regulation. $O_2^{\cdot-}$ enriched in stem cells activate WUSCHEL to inhibit division of stem cells, whereas H_2O_2 is more abundant in the peripheral zone and promotes stem cell differentiation (Zeng et al., 2017). A balance between $O_2^{\cdot-}$ and H_2O_2 is also observed in the root apical meristem (RAM). $O_2^{\cdot-}$ are enriched in the cell proliferation zone, whereas H_2O_2 accumulate in the cell differentiation zone (Dunand et al., 2007; Tsukagoshi et al., 2010; Figure 2.2B). In the root meristem transition zone, the transcription factor UPBEAT1 (UPB1) directly regulates the expression

of a set of peroxidases that modulate the $O_2^{\cdot-}/$ H_2O_2 balance (Tsukagoshi et al., 2010). Further emphasizing the redox regulation of meristematic cell proliferation, the DNA binding activity TEOSINTHE BRANCHED/CYCLOIDEA/PCF (TCP) transcription is affected by a Cys oxidation in its binding and dimerization domain. TCP transcription factors regulate meristematic cell proliferation by inducing expression of cell cycle genes like CYCA2, CYCB1, and RETINOBLASTOMA-RELATED genes (Li et al., 2005, 2012; Viola et al., 2013).

Indeed, very little is known about the redox-sensitive proteins that influence cell proliferation and cell cycle progression. Based on Cys-residue accessibility predictions, different bioinformatics and redox proteomics approaches have identified potential redox-sensitive cell cycle regulators (Foyer et al., 2018; Martins et al., 2018). Some candidates are the DNA synthesis enzyme RIBONUCLEOTIDE REDUCTASE subunits, which need disulfide reduction for their catalytic activity in different organisms, such as Escherichia coli, yeast, and mammals (Gon et al., 2006; Zahedi Avval and Holmgren, 2009), or the CYCLIN-DEPENDENT KINASE A (CDKA1), which activity on cell cycle progression is inhibited by Tyr-nitration by RNS in maize (Méndez et al., 2020).

Studies on proteins regulating thiol redox status have also highlighted their role in maintaining meristem activity. Among key actors are thiols reduction systems, for which glutathione and thioredoxin play major functions. For example, mutants deficient in glutathione biosynthesis or reduction mutants show major meristematic defects. The ROOTMERISTEMLESS1 (rml1) mutant accumulating low levels of glutathione shows an aborted RAM unable to maintain stem cell fate, associated with downregulation of cell cycle gene expression and an arrest of cell cycle progression at the G1/S transition (Howden et al., 1995; Vernoux et al., 2000; Schnaubelt et al., 2015) (Table 2.1). Other evidence points to the key roles of the glutathione nuclear/cytosolic distribution and the redox cycle to regulate cell cycle progression (Pellny et al., 2009; Diaz Vivancos et al., 2010; Diaz-Vivancos et al., 2015; García-Giménez et al., 2013a; Schnaubelt et al., 2015).

While mutants inactivated in glutathione biosynthesis affect root meristem development, TRX are able to replace glutathione for SAM development. In Arabidopsis, a ntra ntrb mutants inactivated in both thioredoxin reductases or mutants cad2

TABLE 2.1

Developmental Defect in Redox Mutants

Protein	Gene locus	Plant species	Subcellular localization	Phenotype	References
ROS and NOS metabolism					
RBOHC	At5g51060	*Arabidopsis thaliana*	Plasma membrane	Root hair defective	Foreman et al., 2003
RBOHD	At5g47910	*A. thaliana*	Plasma membrane	Tubulin formation	Yao et al., 2011
RBOHD/RBOHF	At5g47910/ At1g64060	*A. thaliana*	Plasma membrane	Early emergence of LR and enhanced density of LR	Li et al., 2015
RBOHE	At1g19230	*A. thaliana*	Plasma membrane	Aborted pollen and reduced fertility	Xie et al., 2014
RBOHH/RBOHJ	At5g60010/ At3g45810	*A. thaliana*	Plasma membrane	Root hair defect. Reduced fertility and impaired pollen tube growth	Mangano et al., 2017; Kaya et al., 2014
CAT2	At4g23100	*A. thaliana*	Peroxisome	Delayed growth and hyponastic leaves	Queval et al., 2007
MSD1	At3g10920	*A. thaliana*	Mitochondrion	Defect in embryo sac development	Martin et al., 2013
APX1	At1g07890	*A. thaliana*	Cytosol	Reduced growth and embryo defect	Pagnussat et al., 2005
APX6	At4g32320	*A. thaliana*	Cytosol	Reduced germination	Chen et al., 2014
GSNOR	At5g43940	*A. thaliana*	Cytosol	Reduced root length, defects in stem and trichome branching, reduced fertility, loss of heat acclimation loss of apical dominance, reduced hypocotyl elongation, decreased silique size and seed production	Espunya et al., 2006; Xu et al., 2013; Lee et al., 2008; Kwon et al., 2012
NOX1		*A. thaliana*	Chloroplast	Reduced growth, late flowering	He et al., 2004
NIA1/NIA2/NOA1-2		*A. thaliana*	Cytosol	Reduced size, fertility, and seed germination. Increased dormancy and stomatal closure	Lozano-Juste and León, 2010
Ascorbate and glutathione metabolism					
VTC1	At2g39770	*A. thaliana*	Cytosol/nucleus	Early flowering and senescence	Barth et al., 2010
VTC2/VTC3/VTC4	At4g26850/ At3g02870	*A. thaliana*	Cytosol/nucleus	Early flowering and senescence	Kotchoni et al., 2009
VTC1/VTC2	At2g39770/ At4g26850	*A. thaliana*	Cytosol/nucleus	Seedling lethal	Dowdle et al., 2007

Protein	Gene locus	Plant species	Subcellular localization	Phenotype	References
GSH1	At4g23100	A. thaliana	Chloroplast	Knockout mutant embryo lethal, strong allele shows severely impaired post-embryonic growth and weak alleles, defects in pollen and root development and auxins transport/signaling	Cairns et al., 2006; Howden et al., 1995; Vernoux et al., 2000; Ball et al., 2004; Bashandy et al., 2010; Parisy et al., 2007; Shanmugam et al., 2012; Trujillo-Hernandez et al., 2020
GSH2	At5g27380	A. thaliana	Chloroplast/cytosol	Seedling lethal	Pasternak et al., 2008
GR1	At3g24170	A. thaliana	Cytosol/peroxisome/nucleus	No growth phenotype. Slow growth in gr1 cat1. gr1 ntra ntrb is pollen lethal.	Marty et al., 2009; Mhamdi et al., 2010
GR2	At3g54660	A. thaliana	Chloroplasts/mitochondrion	KO is embryo lethal. Weak alleles have defects in root growth and in root apical meristem maintenance. Inactivation of mitochondrial GR2 and NTRA/NTRB is plant lethal.	Tzafrir et al., 2004; Yu et al., 2013; Marty et al., 2019
Thioredoxins, glutaredoxins, and targets					
NTRA/NTRB	At2g17420/At4g35460	A. thaliana	Cytosol/mitochondrion/nucleus	Growth defect and reduced fertility	Reichheld et al., 2007
FTR catalytic subunit	At2g04700	A. thaliana	Chloroplast	Abnormal chloroplast development	Wang et al., 2013
NTRC	At2g41680	A. thaliana	Chloroplast	Retarded growth of shoots and roots, defects in lateral root formation	Pérez-Ruiz et al., 2006; Lepistö et al., 2009; Kirchsteiger et al., 2012
TRXm3	At2g15570	A. thaliana	Chloroplast	Embryo lethal, impaired meristem development	Benitez-Alfonso et al., 2009
TRXm	Os12g08730	Oryza sativa	Chloroplast	Abnormal chloroplast development	Chi et al., 2008
TRXm1/2/4	At1g03680/At4g03520/At3g15360	A. thaliana	Chloroplast	Pale-green leaves and reduced stability of PSII	Wang et al., 2013
TRXz	At3g06730	A. thaliana	Chloroplast	Albino phenotype	Meng et al., 2010; Arsova et al., 2010
	AY500242	Nicotiana benthamiana	Chloroplast	Albino phenotype	Meng et al., 2010
TRXh9	At3g08710	A. thaliana	Plasma membrane	Dwarf plants with short roots and small, yellowish leaves	Meng et al., 2010
TRXo1	At2g35010	A. thaliana	Mitochondrion	Accelerated germination	Ortiz-Espín et al., 2017
NRX1	At1g60420	A. thaliana	Cytosol/nucleus	Reduced pollen fertility and fitness	Qin et al., 2009; Marchal et al., 2014

(Continued)

TABLE 2.1

Developmental Defect in Redox Mutants

Protein	Gene locus	Plant species	Subcellular localization	Phenotype	References
THL1/THL2	AAB53694/ AAB53695	*Brassica napus*	Cytosol	Self-incompatibility rejection response with reduced pollen adhesion, germination, and pollen tube growth	Cabrillac et al., 2001; Haffani et al., 2004
GRXC1/GRXC2	At5g63030/ At5g40370	*A. thaliana*	Cytosol	Embryo lethal	Riondet et al., 2012
GRXS13	At1g03850	*A. thaliana*	Cytosol/nucleus	Reduced plant growth	Laporte et al., 2012
GRXS17	At4g04950	*A. thaliana*	Cytosol/nucleus	Compromised meristems and LR development, growth arrest, and delayed bolting. Hypersensitivity to elevated temperature and altered auxin perception.	Cheng et al., 2C11; Knuesting et al., 2015; Martins et al., 2020
ROXY1	At3g02000	*A. thaliana*	Cytosol/nucleus	Impaired petal primordia initiation and petal development	Xing et al., 2005
ROXY2	At5g14070	*A. thaliana*	Cytosol/nucleus	Impaired anther development	Xing and Zachgo, 2008
ROXY4	At3g62950	*A. thaliana*	Cytosol/nucleus	Impaired anther development	Hou et al., 2008
MIL1	Os07g05630	*O. sativa*	Cytosol/nucleus	Defective meiosis	Hong et al., 2012
MSCA1	CAX52135	*Zea mays*	Cytosol/nucleus	Impaired anther development	Kelliher and Walbot, 2012; Timofejeva et al., 2013
GRXS3/4/5/8 ROXY11/12/13/15	At4g15700/ At4g15690/ At4g15680/ At4g15660	*A. thaliana*	Cytosol/nucleus	Increased PR length and decreased sensitivity to nitrate	Patterson et al., 2016
PDI1	At2g47470	*A. thaliana*	Cytosol	Defect in embryo development	Pagnussat et al., 2005
GPX3	Os02g44500	*O. sativa*	Mitochondrion	Reduced root and shoot development	Passaia et al., 2014
GPX5	At3g63080	*A. thaliana*	Plasma membrane	Arrested embryo development	Pagnussat et al., 2005
GPX1/GPX7	At2g25080/ At4g31870	*A. thaliana*	Chloroplast	Altered leaf cells and chloroplast morphology	Chang et al., 2009
MSRB2	EF144171	*Capsicum annuum*	Chloroplast	Reduced shoot development	Oh et al., 2010
ECB1	At4g28590	*A. thaliana*	Chloroplast	Albino phenotype, impaired chloroplast development	Yua et al., 2014
HCF164	At4g37200	*A. thaliana*	Chloroplast	Impaired accumulation of cytochrome b6f complex subunits	Lennartz et al., 2001

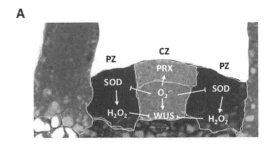

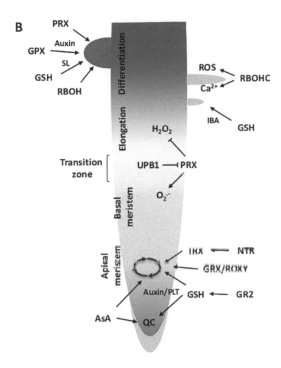

Figure 2.2 Redox regulation of apical meristems.

(A) In the SAM, high levels of $O_2^{\bullet-}$ accumulate in the stem cells of the central zone (CZ, green and orange colors) by repressing superoxide dismutases (SOD) to maintain WUS expression and plant stem cell fate. Stem cell–specific peroxidases (PRX) are expressed to maintain the optimal low level of H_2O_2 in stem cells and fine-tune the $O_2^{\bullet-}$ levels by negative feedback. In the peripheral zone (PZ), H_2O_2 is generated by SOD to promote stem cell differentiation partially by repressing WUS activity. (Green, central zone/stem cells; orange, organizing center; brown, rib zone; red, peripheral zone; blue, leaf primordia.) (B) In the root tip, $O_2^{\bullet-}$ accumulates in the meristem (blue area), whereas H_2O_2 accumulates in the elongation zone (red area). UBP1 represses expression of peroxidases in the transition zone, regulating the ROS balance between the cell proliferation and differentiation zones. The activity of the apical meristem and the quiescent center (QC) are maintained by glutathione (GSH) and ascorbic acid (AsA) as well as different thiol reduction enzymes, thioredoxins (TRX), and glutaredoxins (GRX/ROXY), as well as their respective reductases (NTR and GR2), acting through the auxin pathway (Auxin/PLT). In the differentiation zone, root hair (gray pins) elongation is controlled by RBOHC, which drives ROS and Ca^{2+} gradients in the root hair tip, and GSH in an IBA-dependent way. Lateral root (brown dome) development is controlled by a panel of peroxidases (PRX, GPX), oxidases (RBOH), and GSH, partially by affecting the auxin and strigolactone pathways.

AsA, ascorbic acid. Ca^{2+}, calcium cation. CZ, central zone. GPX, glutathione peroxidase. GR, glutathione reductase. GRX, glutaredoxin. GSH, glutathione. H_2O_2, hydrogen peroxide. IBA, indole butyric acid. MDAR, monodehydroascorbate reductase. NTR, NADPH-thioredoxin reductase. PLT, PLETHORA. PRX, peroxiredoxin. PZ, peripheral zone. QC, quiescent center. RBOH, respiratory burst oxidase homologue. ROS, reactive oxygen species. SL, strigolactone. SOD, superoxide dismutase. TRX, thioredoxin. UPB1, UPBEAT 1. WUS, *WUSCHEL*.

having lower capacities of glutathione have a normal vegetative development. However, the *ntra ntrb cad2* mutant combining both thioredoxin reductase and glutathione biosynthesis mutation abolishes flower meristem development, producing a PIN-like meristem structure similar to auxin transport mutants (Bashandy et al., 2010). Similarly, association of the *ntra ntrb* mutant with the strong glutathione biosynthesis mutant ROOTMERISTEMLESS1 (*rml1*) completely abolishes SAM development (Reichheld et al., 2007).

Accumulation of oxidized glutathione in distinct GLUTATHIONE REDUCTASE (GR) mutants affects different aspects of plant development. Knockout mutation of the chloroplastic GR2 isoform leads to embryo lethality in *Arabidopsis*, whereas the mitochondrial and the cytosolic glutathione reductions can be backed up by thioredoxin reductases and only triple mutants *ntra ntrb gr* exhibit developmental defects (Marty et al., 2009; Marty et al., 2019). Interestingly, the knocked-down EMS-mutant *miao* with the GR activity being reduced to ~50% displays strong inhibition of root growth and severe defects in the RAM by affecting the auxin/ PLETHORA (PLT) signaling pathways (Yu et al., 2013). Pointing to the role of ascorbic acid (AsA) in meristem development, the AsA biosynthesis mutant *vtc1*, with a higher H_2O_2 level, exhibits abnormal QC and periclinal cell division patterns in the RAM (Kka et al., 2018; Figure 2.2B).

The contribution of the thiols reductases thioredoxin and glutaredoxins in meristematic activities has been documented. In *Arabidopsis*, the involvement of the chloroplastic TRXm3 in meristem maintenance was attributed to its role in the redox regulation of callose deposition and symplastic permeability. Here, the *trxm3* mutant could not maintain a functional SAM because of callose accumulation in the meristem (Benitez-Alfonso et al., 2009). In *Arabidopsis*, an atypical mitochondrial thioredoxin DCC1 was shown to affect shoot regeneration by interacting with CARBONIC ANHYDRASE2 (CA2), an essential subunit of the respiratory chain NADH dehydrogenase complex I. Inactivation of DCC1 results in mitochondrial ROS accumulation that is supposed to affect shoot regeneration. As the disulfide reduction activity of DCC1 is not established yet, the exact nature of the redox regulation is still to be established (Zhang et al., 2018) (Table 2.1).

Glutaredoxins that use glutathione as a cofactor have also been involved in SAM functions. For example, *Arabidopsis* plants deficient in the iron-sulfur cluster containing GRXS17 are sensitive to high temperatures and long-day photoperiods, resulting in compromised shoot apical meristem and auxin transport (Cheng et al., 2011; Knuesting et al., 2015). GRXS17 also interacts with the NF-YC11/NC2a transcription factor, which participates in plant developmental processes in relation to the photoperiod duration, and *nf-yc11/nc2a* and *grxS17* plants share similar developmental characteristics in long-day conditions (Knuesting et al., 2015). In maize, the glutaredoxin ABBERANT PHYLLOTAXY (ABPHYL2) regulates shoot meristem and phyllotaxy and interacts with the b-zip TGA transcription factor FASCIATED EAR4 (Pautler et al., 2015; Yang et al., 2015).

2.3 REDOX REGULATION OF REPRODUCTIVE DEVELOPMENT

Works on reproductive development have defined redox status as a determinant of germ cell fate (Zafra et al., 2016; Luria et al., 2019; García-Quirós et al., 2020). In maize, hypoxia naturally occurring in growing anther tissues is necessary to maintain germ cell fate. Decreased oxygen (or H_2O_2) activates germ cell formation, whereas oxidizing environments inhibit germ cell specification and cause ectopic differentiation in anther tissues. The plant-specific glutaredoxin MSCA1 protects redox environment in anther cells (Kelliher and Walbot, 2012).

Indeed, different members of the same GRX clade are acting in distinct aspects of flower development. In *Arabidopsis*, ROXY1/GRXC7 regulate petal formation and growth (Xing et al., 2005) and take part in the ABCE flower development model, which specifies the development of the different flower whorls through the activity of specific MADS-box transcription factors (Coen and Meyerowitz, 1991). On the one hand, the ROXY1 gene is a direct target of the floral development master regulators MADS-box APETALA3, PISTILLATA, and AGAMOUS (Wuest et al., 2012; Ó'Maoiléidigh et al., 2013). On the other hand, ROXY1 redox-regulates the activity of the TGA transcription factor PERIANTHIA, possibly through reduction of a specific Cys localized in a transactivation domain (Li et al., 2009; Li et al., 2011).

In *Arabidopsis*, ROXY1/GRXC7 and ROXY2/GRXC8 are redundantly involved in anther development by reducing specific Cys residues in TGA9 and TGA10 transcription factors (Xing and Zachgo, 2008;

Murmu et al., 2010). In the roxy1 roxy2 mutant, both anther lobe and pollen mother cell differentiations are defective. Lack of ROXY1 and ROXY2 function affects a large number of anther genes at the transcriptional level downstream of the early-acting anther gene SPOROCYTELESS/NOZZLE and upstream of DYSFUNCTIONAL TAPETUM1, controlling tapetum development (Xing and Zachgo, 2008).

Thiol reductase systems also act on other aspects of pollen development. In Arabidopsis, the ntra ntrb mutant has a decreased pollen fitness rather than a pollen development deficiency (Reichheld et al., 2007). Further works on the NTR-dependent nuclear thioredoxin NRX1 also revealed decreased pollen fertility in the nrx1 mutant, which was caused by defective pollen tube growth in the pistil (Qin et al., 2009; Marchal et al., 2014). Finally, the ntra ntrb grl triple mutant is fully pollen sterile, suggesting that both TRX and GSH reduction systems act redundantly on pollen fertility (Marty et al., 2009).

2.4 REDOX CONTROL OF SEED ESTABLISHMENT, DORMANCY, AND GERMINATION

In flowering plants, double fertilization occurring in the ovule gives rise to a diploid embryo developing from the zygote, surrounded by a triploid storage tissue, the endosperm. Seed development can be divided into three phases, starting with an active phase of cell divisions that builds up the body plan of the embryo and structures the endosperm. During mid-phase, intense and numerous biochemical activities allow the accumulation of food reserves, dehydration-permissive compounds, and toxic protective molecules. Finally, seeds of most Angiosperm species desiccate and acquire dormancy, characterized by a very low residual metabolic activity. After seed dispersal and dormancy release, imbibition strongly reactivates metabolism and the breakdown of food reserves. This allows germination to take place, and the subsequent embryo develops into a novel autotrophic seedling (Figure 2.3).

Glutathione plays essential functions during seed maturation, because null alleles of GR2 (encoding the plastid localized glutathione reductase), GSH1 (encoding the first limiting step of glutathione biosynthesis), and both GRXC1 and GRXC2 (encoding redundant glutaredoxins) are embryo lethal in Arabidopsis (Tzafrir et al., 2004; Cairns et al., 2006; Riondet et al., 2012; Marty et al., 2019). gsh1 homozygous embryos can initiate their developmental

program and grow normally but start bleaching after the torpedo stage. This suggests a redox control of the metabolic machinery responsible for the accumulation of reserves and protective compounds. Such regulation also occurs in the rice endosperm, where the first committed enzyme in starch synthesis, ADP-glucose pyrophosphorylase, is activated by reduction, probably via a TRXh member (Tuncel et al., 2014). Accordingly, seed tissues display increasing reducing conditions while accumulating reserves (De Gara et al., 2003; Ferreira et al., 2012).

During the ultimate phase of seed development, dehydration and acquisition of dormancy are accompanied by the dramatic drop of metabolic activities, including reducing pathways. Dry dormant seeds are characterized by an overall oxidized status, while residual ROS production is thought to be very limited in both embryo and endosperm (De Gara et al., 2003; Bailly et al., 2008).

Upon dormancy release and imbibition, a brutal and dual redox switch occurs. On the one hand, ROS and RNS concentrations increase rapidly, via the recovery of metabolism and the activation of dedicated producing systems (Bailly et al., 2008). This rise in ROS and RNS is critical for dormancy breakdown and for germination to take place (Lozano-Juste and León, 2010; Mhamdi and Van Breusegem, 2018). ROS allows mRNA oxidation, which in turn regulates translation and selective protein oxidation, thus adapting the proteome for successful germination (El-Maarouf-Bouteau et al., 2013). In contrast, an excess of ROS could be toxic and alter the embryo viability. ROS and RNS homeostasis is therefore tightly regulated inside the borders of an "oxidative window" convenient for the germination process (Bailly et al., 2008; Bailly, 2019). On the other hand, a massive reduction of thiol-containing proteins also occurs during seed rehydration, mainly though the action of the TRX system, both in cytosol and mitochondria (Alkhalfioui et al., 2007; Nietzel et al., 2019). Hence, the subtle redox control of seed proteome, by the concomitant oxidation and reduction of different subsets of proteins, is a central component of seed to seedling transition (Figure 2.3).

ABA is a key phytohormone that promotes seed maturation and dormancy and inhibits germination. ROS/RNS and ABA negatively control each other to fine-tune seed germination. Indeed, RNS-deficient seeds are hypersensitive to ABA (Lozano-Juste and León, 2010), while the master

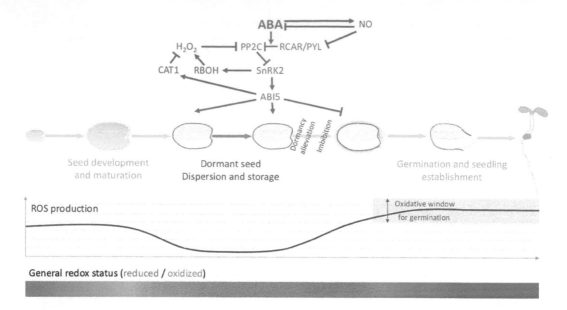

Figure 2.3 Redox regulation of seed development, dormancy, and germination.

ROS generation is proportional to metabolic activities, with high ROS levels accompanying seed development followed by a rapid drop during the latest stage of seed maturation. These low levels are characteristic of dry dormant seeds, and the recovery of higher ROS production is necessary for dormancy breakdown and germination, although it should be kept inside a non-toxic, permissive window. In parallel, reciprocal changes appear in the global reduction status of thiol-containing proteins, with an oxidized proteome being characteristic of dry dormant seeds. Finally, abscisic acid (ABA) is a critical plant hormone that promotes seed dormancy and negatively regulates seed germination. NO and H_2O_2 modulate ABA biosynthesis or early signaling, while they are themselves regulated by ABA-dependent pathways. This allows the integration of different exogenous and endogenous signals to properly adapt seed germination and the establishment of a novel autotrophic plantlet.

ABA, abscisic acid. ABI5, abscisic acid insensitive 5. CAT1, Catalase 1. H_2O_2, hydrogen peroxide. NO, nitric oxide. PP2C, protein phosphatase 2C. RBOH, respiratory burst oxidase homologue. RCAR/PYL, REGULATORY COMPONENTS OF ABA RECEPTORS (RCAR)/PYRABACTIN RESISTANCE1 (PYR1)/PYR1-LIKE (PYL). SnRK2, SNF1-related protein kinase 2.

(Adapted from Bentsink and Kornneef, *The Arabidopsis Book*, 2008.)

ABA-activated transcription factor ABI5 directly binds to the promoter of *CATALASE1* to regulate H_2O_2 homeostasis (Bi et al., 2017). In contrast, H_2O_2 has also been reported as an important second messenger in ABA signaling, involving its production by NADPH oxidases (Pei et al., 2000). The negative regulators of ABA signaling, ABI1 and ABI2 phosphatases, are inactivated by the oxidation of some of their cysteines. They could be direct targets of H_2O_2 release, thus allowing ABA responses to take place (Meinhard and Grill, 2001; Meinhard et al., 2002).

2.5 REDOX CONTROL OF ROOT DEVELOPMENT

The development and adaptation of an optimal root system is critical for plant anchoring and nutrition.

Canonical root system architecture depends on root growth and root branching. Cell divisions in the root apical meristem and cellular elongation both support root growth and precede cellular differentiation that allows the root to become mature and fully functional. Root branching first relies on the specification and initiation of lateral root primordia (LRP), mainly in the pericycle layer that surrounds the central stele. The development of the primordium and its protrusion through overlying tissues leads to the emergence of the new lateral root from the primary root. In addition to this canonical root system, adventitious roots develop from aerial tissues, most often from stems, and help the plants to adapt to their environment. Finally, while it is not the case for *Arabidopsis*, most plant species develop root-derived symbioses with microbial partners, such as mycorrhiza or nodulation.

Many studies support a critical role for ROS in regulating most of the steps of root system development. In the root apical meristem, the balance between the two major ROS, superoxide and hydrogen peroxide, is associated with the transition from the distal meristematic zone, where $O_2^{\cdot-}$ is predominant, to the basal maturing zone, in which H_2O_2 predominates (Dunand et al., 2007). This gradient regulates the proper transition from cellular proliferation to differentiation and is tightly controlled by the bHLH transcription factor UPBEAT1, which downregulates the expression of peroxidases in the transition zone (Tsukagoshi et al., 2010; Figure 2.2B). The role of ROS in regulating cell differentiation is also demonstrated in trichoblast development, because RBOHC localization at the tip of the root hair drives ROS and calcium gradients, thus allowing exocytosis and cellular growth (Foreman et al., 2003). In addition, ROS and peroxidases are involved in the regulation of LRP development and emergence, but not their initial specification (Manzano et al., 2014; Fernández-Marcos et al., 2017). Moreover, LRP protrusion requires cell wall remodeling of surrounding tissues, which is controlled by NADPH oxidase derived ROS production (Orman-Ligeza et al., 2016). Adventitious root development also depends on ROS homeostasis in rice. In response to the pressure of the growing root, mechano-sensing of surrounding tissues leads to ethylene signaling via ROS production, which initiates cell death to let the root protrude (Steffens et al., 2012). Finally, RBOH-derived ROS, downstream of Nod factor perception, regulate nodulation in legumes via the control of infection thread progression, nodule organogenesis, function, and senescence (reviewed in Montiel et al., 2016).

Glutathione is also critical for root system development. This is supported by the very severe root development phenotypes of the rml1 (root meristemless) allele of GSH1 (Vernoux et al., 2000) and of the mutant in the plastidial GR2 (Yu et al., 2013). The glutathione-deficient cad2 mutant (a weak allele of GSH1) is impaired in lateral root and root hair growth in response to the auxin indole butyric acid (Trujillo-Hernandez et al., 2020). Some mutants in glutathione-dependent pathways also confirm its role during root development, such as the glutathione peroxidases GPX1 and GPX7, which mediate auxin control of LR development, or the glutaredoxin GRXS17, which regulates root apical meristems (Passaia et al., 2014; Martins et al., 2020).

LR analyses also establish a link between strigolactones (SLs) and the GSH pool that occurs in a MAX2-dependent manner (Marquez-Garcia et al., 2014). The ntra ntrb cad2 triple mutant also displays an important reduction in root growth, affecting both primary and secondary roots more severely than cad2 or ntra ntrb mutants (Bashandy et al., 2010). This indicates that thiol reducing systems redundantly control root developmental pathways.

Many works have reported redox regulation systems as pivotal regulators of root development in response to endogenous or exogenous signals. First, the plastidial NTRC controls root growth and LR formation, thus adapting the root system development to the primary metabolism of plants (Kirchsteiger et al., 2012).

Glutathione and glutaredoxins appear as critical modulators of root system development in response to nutrient availability. Indeed, glutathione controls the regulation of root hair elongation in response to phosphorus starvation (Trujillo-Hernandez et al., 2020). Moreover, a cluster of glutaredoxins from the class III subfamily (ROXY class) redundantly mediate nitrate inhibition of primary root growth downstream of cytokinins perception (Patterson et al., 2016). Recent works based on ROXY15 overexpression lines tend to suggest that this control of nitrate-dependent transcriptional and developmental responses occurs via binding to the transcription factors TGA1 and TGA4 (Ehrary et al., 2020). More generally, the ROXY genes could be major adaptors of root development to nutrient availability because nitrate starvation— and also phosphorus, potassium, sulfur, and iron deprivations—transcriptionally regulate some of these genes (Jung et al., 2013). Very recently, ROXY7 and ROXY8 have been proposed as mobile signals regulated by the nitrogen status of shoots, which enhance nitrate transport by regulating the expression of NRT2.1 in roots (Ota et al., 2020). Altogether, these data suggest that ROXY glutaredoxins are very promising redox components that adapt root development and functions to nutrient availability in soils and the nutrition status of plants. Finally, ROXY1 was recently found to act also in the RAM, where it interacts with RNA polymerase II subunits (Maß et al., 2020).

Redox systems are also involved in mediating root responses to other stress conditions. In Arabidopsis, the NADPH oxidases RBOHD and RBOHF mediate the ABA-dependent reduction of root sensitivity to auxin, while the rice OsTRXh1 is required for root system adaptation to salt stress and ABA by regulating the apoplastic redox homeostasis (Jiao et al., 2013; Zhang et al., 2011). The GRXS17 control of root development is exacerbated and extended to lateral root growth in case of heat stress by protecting root meristems from cell death through its holdase activity (Martins et al., 2020). Finally, oxygen-deficient conditions activate RBOHH-dependent production of ROS in rice via the ethylene-dependent CDPK5 and CDPK13 kinases. These ROS trigger programmed cell death in cortical cells to create the adapted aerenchyma (Yamauchi et al., 2017).

2.6 DEVELOPMENTAL ADAPTATION TO STRESS

Adaptation of plant development to environmental constraints involves redox regulation as well. Light and temperature have a major impact on physiological and metabolic changes in plants. Similar to the immune system, they trigger local and systemic signals and induce defense responses. ROS and redox regulation are involved in both signaling and defense priming. Upon various stress conditions, generation of ROS in the apoplast is challenging local responses that are likely initiated by extracellular sensors of hydrogen peroxide, like the recently characterized hydrogen peroxide-induced Ca^{2+} increases 1 (HPCA1) receptor kinase (Wu et al., 2020) and by diffusion of H_2O_2 through aquaporins (Rodriguez et al., 2017) during stomatal closure (Figure 2.1). Stress-induced oxidative burst also initiate systemic responses called systemic acquired resistance (SAR) in the case of biotic stress or systemic acquired acclimation (SAA) for abiotic stress. A number of different signaling mechanisms were implicated in this response, including a cell-to-cell transmission of ROS, calcium (Ca^{2+}), and electrical signals (Gilroy et al., 2016).

The RESPIRATORY BURST OXIDASE HOMOLOG (RBOH) type of NADPH oxidases is a key player in both local and systemic responses. The well-characterized members RBOHD and RBOHF are implicated in a variety of stress-related and developmental responses in plants, including stomatal closure induced by abscisic acid, pathogenesis-related responses, and systemic signaling to stresses such as wounding, drought, and salt (Chaouch et al., 2012; Kwak et al., 2003; Miller et al., 2009). They are also both required for local and systemic ROS

signaling at the vascular bundles during light stress (Zandalinas et al., 2020).

The role of ROS scavenging systems in relaying environmental signals is highlighted by inactivating the peroxisomal H_2O_2-scavenging enzyme CATALASE2 (CAT2). The cat2 mutant exhibits a conditional light-dependent phenotype (Queval et al., 2007). In the cat2 mutant, light intensity affects leaf bending, revealing a crosstalk between H_2O_2 and auxin signaling that is mediated by changes in glutathione redox status (Gao et al., 2014; Kerchev et al., 2015). While the glutathione redox pool is perturbed under both under long-day (LD) and short-day (SD) conditions, only LD triggers a salicylic acid (SA)–dependent programmed cell death (PCD; Queval et al., 2007, 2009, 2012). Interestingly, a mutation in SHORT-ROOT (SHR) rescued the cell death phenotype of cat2 plants, further integrating developmental and stress response pathways (Waszczak et al., 2016). Plants deficient in the chloroplastic thioredoxin reductase NTRC exhibit stunted growth, particularly under SD, due to the fact that metabolic and antioxidant reactions performed by NTRC are particularly important during the night (Michalska et al., 2009; Lepistö et al., 2009, 2012). Accordingly, NTRC also maintains redox homeostasis in non-photosynthetic chloroplasts in the root, and ntrc mutant exhibits a lack in root auxin biosynthesis and lateral root growth inhibition (Kirchsteiger, 2012).

Light and temperature changes are major factors signaling daily and seasonal changes and leading to ROS generation (Mittler et al., 2012; Choudhury et al., 2017). Indeed, plants are capable of integrating signals simultaneously generated during light and temperature stress combination, and systemic ROS signaling is involved in transmitting those stresses (Zandalinas et al., 2020). Interestingly, thiol reductases like the glutaredoxin GRXS17 regulate meristem development and flowering time in response to temperature and photoperiod-dependent redox regulation, potentially through interaction with the transcription factor NF-YC11 (Knuesting et al., 2015). GRXS17 was also involved in heat stress tolerance through its redox-dependent holdase activity (Martins et al., 2020), a function shared with other thiol peroxidases and reductases (Muthuramalingam et al., 2009; Lee et al., 2009; Park et al., 2009; Chae et al., 2013). The paradigm for a single protein playing a dual role in regulating plant development and in stress defense responses was also revealed for the auxin biosynthesis monooxygenase YUCCA6 (YUC6), which displays thiol-reductase activity and improves plant drought stress tolerance (Cha et al., 2015).

Oxygen sensing by the N-end rule pathway plays an important role in signaling various environmental constraints, such as hypoxia occurring in waterlogged plants or during photomorphogenesis processes following seedling establishment after germination or in shoot apical meristem activity (Licausi et al., 2011; Gibbs et al., 2011; Abbas et al., 2015; Weits et al., 2019). In this pathway, a penultimate N-terminal Cys residue (Cys2) is exposed by removal of the initial methionine by a Met aminopeptidase. In the presence of nitric oxide or oxygen, the Cys can be oxidized by plant cysteine oxidases and further arginylated by Arg-aminotransferases (Weits et al., 2014; Graciet et al., 2009). Finally, proteins labeled with an N-terminal Arg can then be ubiquitinated by the single-subunit E3 ligase PROTEOLYSIS 6 (PRT6) and thereby targeted to proteasomal degradation. Hypoxia inhibits the N-rule pathway and the proteolysis of substrates of transcription factors regulating water stress and photomorphogenesis shoot meristems activities.

2.7 EPIGENETIC AND REDOX REGULATION

The crucial role of epigenetics in plant development is exemplified by the embryo lethality of null mutants and the dramatic and pleiotropic developmental defects of many mutants affected in epigenetic processes. However, redox regulation of epigenetic processes has mostly been addressed in mammals, but this field is still poorly explored in plants (Delorme-Hinoux et al., 2016; Shen et al., 2016; García-Giménez et al., 2017). The conservation of these epigenetic mechanisms in living organisms makes it likely that such regulation also occurs in plants. In mammals, oxidative stress alters global histone modification, histone acetylation, and DNA methylation, affecting chromatin conformation and transcription (Ito et al., 2004; Chen et al., 2006; Ago et al., 2008; Nott et al., 2008; Zhou et al., 2008; Doyle and Fitzpatrick, 2010; Niu et al., 2015). During brain development, neurotrophic factors induce S-nitrosylation at conserved Cys of HDAC2 in neurons, resulting in changes of histone modification and gene expression (Nott et al., 2008). Upon cardiac hypertrophy, a ROS/TRX-dependent redox switch of key Cys residues

affects nuclear trafficking of a class II HDAC and subsequent gene expression (Ago et al., 2008). In plants, among the large family of HDAC (Pandey et al., 2002), members of the class I RPD-3 like HDAC were identified as redox-sensitive proteins during early salicylate and flg22 response (Liu et al., 2015). NO-induced HDAC inhibition is proposed to operate in plant stress response by facilitating the stress-induced transcription of genes (Mengel et al., 2017; Ageeva-Kieferle et al., 2021).

Some key metabolic enzymes involved in DNA methylation are candidates for redox regulation. S-adenosyl methionine (SAM), which provide precursors for DNA and histone methylation, is among these enzymes (Shen et al., 2016). The DNA demethylases Repressor of Silencing1 (ROS1) and Demeter-like (DME, DML2, DML3) enzymes, which remove methylated bases from the DNA backbone, are other nuclear candidates (Zhu, 2009). These enzymes contain an iron-sulfur (Fe-S) cluster, which might be susceptible to oxidation by ROS. Moreover, different members of the cytosolic Fe-S cluster assembly machinery (i.e., MET18, AE7) are involved in DNA methylation, likely because they affect the nuclear DNA demethylase's Fe-S cluster metabolism (Liu et al., 2015; Duan et al., 2015). Therefore, all these examples show an emerging link between redox regulation and epigenetic regulation.

In mammals, histone H3, the core constituent of nucleosome, is glutathionylated on a conserved Cys residue (García-Giménez et al., 2013b). Histone H3 glutathionylation fluctuates with cell proliferation and aging. This produces structural changes that affect nucleosome stability and lead to more open chromatin structures and modifications of gene expression (Xu et al., 2014; García-Giménez et al., 2017; García-Giménez et al., 2013b). Recently, Attar et al. (2020) revealed another role of eukaryotic histone H3-H4 tetramer in addition to chromatin compaction or epigenetic regulation. In *Xenopus laevis*, the H3-H4 tetramer binds Cu^{2+} through its Cys residue and acts as a Cu^{2+}/Cu^{1+} oxidoreductase dependent on NADPH and usable for mitochondrial respiration and ROS detoxification enzymes (Attar et al., 2020).

Small RNAs (siRNA and miRNA) are other important regulators of gene expression and are involved in various developmental and stress response processes in eukaryotic cells (D'Ario et al., 2017; Leisegang et al., 2017). The biogenesis of small RNAs is coordinated by DICER-LIKE

(DCL) and RNASE THREE-LIKE (RTL) endonucleases, which process almost all classes of double-stranded RNA precursors. In *Arabidopsis*, members of the DCL and RTL families were shown to be redox regulated through disulfide bonds or S-glutathionylation of conserved Cys in their RNA binding domains, therefore affecting their RNase III activity. Indeed, RTL1 activity can be restored by GRXs, indicating that redox switch controls small RNA biogenesis (Comella et al., 2008; Charbonnel et al., 2017).

2.8 CONCLUSION

Research performed in the last two decades has evidenced the panel ROS effects on plant growth and development. This has mainly been highlighted by the characterization of knockout mutants in the large-panel ROS metabolism, scavenging, or oxidoreduction systems. The functions of almost all plant organs, tissues, and cell types as well as principal plant development programs are affected by oxidoreduction processes. This gives high credence to the notion of redox control as a major hub regulating plant growth and development in response to environmental constraints. However, much is still to be learned about the targets of oxidoreduction regulation. This is a challenging task, mostly due to the high dynamic of redox modifications. Techniques dedicated to analyze redox modifications on proteins like redox proteomics are still difficult to manage in *planta* due to high fluctuations of the modifications in different tissues or under various stress conditions and due to the limitation in detection of low abundance target proteins like receptors, transcription factors, and other signaling molecules. Interestingly, recent breakthrough research has been generated by forward genetic screens, like the recent identification of the HCPA receptor kinase/sensor of H_2O_2. By the way, genetically based fluorescent imaging is a promising field to design new genetic screens and to decipher redox modifications at the cellular and subcellular levels. Much is still to discover on the crosstalks between different redox pathways but also on the redundancy between different members of multigenic families in the regulation of plant development and stress response pathways. Good examples are within the multigenic families of thioredoxin, glutaredoxins, or peroxidases, which exhibit several dozen genes, most of them yet uncharacterized.

2.9 ACKNOWLEDGMENTS

This work was supported in part by the Centre National de la Recherche Scientifique, the University of Perpignan Via Domitia, and the Agence Nationale de la Recherche (ANR-REPHARE 19-CE12-0027 and ANR-RoxRNase 20-CE12-0025). This project was funded through Labex AGRO (under I-Site Muse framework) and coordinated by the Agropolis Fondation (Flagship project CalClim, 1802-002). This study is set within the framework of the Laboratoires d'Excellence (LabEx) TULIP (ANR-10-LABX-41). Avilien Dard is supported by a PhD grant from the Université de Perpignan Via Domitia (Ecole Doctorale Energie et Environnement ED305).

REFERENCES

Abbas M, Berckhan S, Rooney DJ, Gibbs DJ, Vicente Conde J, Sousa Correia C, Bassel GW, Marín-de la Rosa N, León J, Alabadí D, et al. (2015) Oxygen sensing coordinates photomorphogenesis to facilitate seedling survival. Curr Biol **25**: 1483–1488

Ageeva-Kieferle A, Georgii E, Winkler B, Ghirardo A, Albert A, Hüther P, Mengel A, Becker C, Schnitzler JP, Durner J, et al. (2021) Nitric oxide coordinates growth, development, and stress response via histone modification and gene expression. Plant Physiol May 18: kiab222

Ago T, Liu T, Zhai P, Chen W, Li H, Molkentin JD, Vatner SF, Sadoshima J (2008) A redox-dependent pathway for regulating class II HDACs and cardiac hypertrophy. Cell **133**: 978–993

Alkhalfioui F, Renard M, Vensel WH, Wong J, Tanaka CK, Hurkman WJ, Buchanan BB, Montrichard F (2007) Thioredoxin-linked proteins are reduced during germination of *Medicago truncatula* seeds. Plant Physiol **144**: 1559–1579

Arsova B, Hoja U, Wimmelbacher M, Greiner E, Üstün Ş, Melzer M, Petersen K, Lein W, Börnke F (2010) Plastidial thioredoxin z interacts with two fructokinase-like proteins in a thiol-dependent manner: evidence for an essential role in chloroplast development in *Arabidopsis* and *Nicotiana benthamiana*. The Plant Cell **22**: 1498–1515

Attar N, Campos OA, Vogelauer M, Cheng C, Xue Y, Schmollinger S, Salwinski L, Mallipeddi NV, Boone BA, Yen L, et al. (2020) The histone H3-H4 tetramer is a copper reductase enzyme. Science **369**: 59–64

Bailly C (2019) The signalling role of ROS in the regulation of seed germination and dormancy. Biochem J **476**: 3019–3032

Bailly C, El-Maarouf-Bouteau H, Corbineau F (2008) From intracellular signaling networks to cell death: the dual role of reactive oxygen species in seed physiology. C R Biol **331**: 806–814

Ball L, Accotto G-P, Bechtold U, Creissen G, Funck D, Jimenez A, Kular B, Leyland N, Mejia-Carranza J, Reynolds H, et al. (2004) Evidence for a direct link between glutathione biosynthesis and stress defense gene expression in *Arabidopsis*. Plant Cell **16**: 2448–2462

Barth C, Gouzd ZA, Steele HP, Imperio RM (2010) A mutation in GDP-mannose pyrophosphorylase causes conditional hypersensitivity to ammonium, resulting in *Arabidopsis* root growth inhibition, altered ammonium metabolism, and hormone homeostasis. J Exp Bot **61**: 379–394

Bashandy T, Guilleminot J, Vernoux T, Caparros-Ruiz D, Ljung K, Meyer Y, Reichheld J-P (2010) Interplay between the NADP-linked thioredoxin and glutathione systems in *Arabidopsis* auxin signaling. Plant Cell **22**: 376–391

Benitez-Alfonso Y, Cilia M, San Roman A, Thomas C, Maule A, Hearn S, Jackson D (2009) Control of *Arabidopsis* meristem development by thioredoxin-dependent regulation of intercellular transport. Proc Natl Acad Sci USA **106**: 3615–3620

Bentsink L, Koornneef M (2008) Seed dormancy and germination. Arabidopsis Book **6**: e0119

Bi C, Ma Y, Wu Z, Yu Y-T, Liang S, Lu K, Wang X-F (2017) *Arabidopsis* ABI5 plays a role in regulating ROS homeostasis by activating CATALASE 1 transcription in seed germination. Plant Mol Biol **94**: 197–213

Cabrillac D, Cock JM, Dumas C, Gaude T (2001) The S-locus receptor kinase is inhibited by thioredoxins and activated by pollen coat proteins. Nature **410**: 220–223

Cairns NG, Pasternak M, Wachter A, Cobbett CS, Meyer AJ (2006) Maturation of *Arabidopsis* seeds is dependent on glutathione biosynthesis within the embryo. Plant Physiol **141**: 446–455

Cha JY, Kim WY, Kang SB, Kim JI, Baek D, Jung IJ, Kim MR, Li N, Kim HJ, Nakajima M, et al. (2015) A novel thiol-reductase activity of Arabidopsis YUC6 confers drought tolerance independently of auxin biosynthesis. Nat Commun **6**: 8041

Chae HB, Moon JC, Shin MR, Chi YH, Jung YJ, Lee SY, Nawkar GM, Jung HS, Hyun JK, Kim WY, et al. (2013) Thioredoxin reductase type C (NTRC) orchestrates enhanced thermotolerance to Arabidopsis by its redox-dependent holdase chaperone function. Mol Plant **6**: 323–336

Chang CCC, Slesak I, Jordá L, Sotnikov A, Melzer M, Miszalski Z, Mullineaux PM, Parker JE, Karpinska B, Karpinski S (2009) *Arabidopsis* chloroplastic glutathione peroxidases play a role in cross talk between photooxidative stress and immune responses. Plant Physiol **150**: 670–683

Chaouch S, Queval G, Noctor G (2012) AtRbohF is a crucial modulator of defence-associated metabolism and a key actor in the interplay between intracellular oxidative stress and pathogenesis responses in *Arabidopsis*. Plant J **69**: 613–627

Charbonnel C, Niazi AK, Elvira-Matelot E, Nowak E, Zytnicki M, de Bures A, Jobet E, Opsomer A, Shamandi N, Nowotny M, et al. (2017) The siRNA suppressor RTL1 is redox-regulated through glutathionylation of a conserved cysteine in the double-stranded-RNA-binding domain. Nucleic Acids Res **45**: 11891–11907. doi:10.1093/nar/gkx820

Chen C, Letnik I, Hacham Y, Dobrev P, Ben-Daniel B-H, Vanková R, Amir R, Miller G (2014) ASCORBATE PEROXIDASE6 protects *Arabidopsis* desiccating and germinating seeds from stress and mediates cross talk between reactive oxygen species, abscisic acid, and auxin. Plant Physiol **166**: 370–383

Chen H, Ke Q, Kluz T, Yan Y, Costa M (2006) Nickel ions increase histone H3 lysine 9 dimethylation and induce transgene silencing. Mol Cell Biol **26**: 3728–3737

Cheng N-H, Liu J-Z, Liu X, Wu Q, Thompson SM, Lin J, Chang J, Whitham SA, Park S, Cohen JD, et al. (2011) *Arabidopsis* monothiol glutaredoxin, AtGRXS17, is critical for temperature-dependent postembryonic growth and development via modulating auxin response. J Biol Chem **286**: 20398–20406

Chi YH, Moon JC, Park JH, Kim H-S, Zulfugarov IS, Fanata WI, Jang HH, Lee JR, Lee YM, Kim ST, et al. (2008) Abnormal chloroplast development and growth inhibition in rice thioredoxin m knock-down plants. Plant Physiol **148**: 808–817

Choudhury FK, Rivero RM, Blumwald E, Mittler R (2017) Reactive oxygen species, abiotic stress and stress combination. Plant J **90**: 856–867

Coen ES, Meyerowitz EM (1991) The war of the whorls: genetic interactions controlling flower development. Nature **353**: 31–37

Comella P, Pontvianne F, Lahmy S, Vignols F, Barbezier N, Debures A, Jobet E, Brugidou E, Echeverria M, Sáez-Vásquez J (2008) Characterization of a ribonuclease III-like protein required for cleavage of the pre-rRNA in the 3'ETS in *Arabidopsis*. Nucleic Acids Res **36**: 1163–1175

D'Ario M, Griffiths-Jones S, Kim M (2017) Small RNAs: big impact on plant development. Trends Plant Sci 22: 1056–1068

De Gara L, de Pinto MC, Moliterni VMC, D'Egidio MG (2003) Redox regulation and storage processes during maturation in kernels of *Triticum durum*. J Exp Bot 54: 249–258

Delorme-Hinoux V, Bangash SAK, Meyer AJ, Reichheld J-P (2016) Nuclear thiol redox systems in plants. Plant Sci 243: 84–95

Diaz-Vivancos P, de Simone A, Kiddle G, Foyer CH (2015) Glutathione-linking cell proliferation to oxidative stress. Free Radic Biol Med 89: 1154–1164

Diaz-Vivancos P, Wolff T, Markovic J, Pallardó FV, Foyer CH (2010) A nuclear glutathione cycle within the cell cycle. Biochem J 431: 169–178

Dowdle J, Ishikawa T, Gatzek S, Rolinski S, Smirnoff N (2007) Two genes in *Arabidopsis thaliana* encoding GDP-L-galactose phosphorylase are required for ascorbate biosynthesis and seedling viability. Plant J 52: 673–689

Doyle K, Fitzpatrick FA (2010) Redox signaling, alkylation (carbonylation) of conserved cysteines inactivates class I histone deacetylases 1, 2, and 3 and antagonizes their transcriptional repressor function. J Biol Chem 285: 17417–17424

Duan C-G, Wang X, Tang K, Zhang H, Mangrauthia SK, Lei M, Hsu C-C, Hou Y-J, Wang C, Li Y, et al. (2015) MET18 connects the cytosolic iron-sulfur cluster assembly pathway to active DNA demethylation in *Arabidopsis*. PLoS Genet 11: e1005559

Dunand C, Crèvecoeur M, Penel C (2007) Distribution of superoxide and hydrogen peroxide in *Arabidopsis* root and their influence on root development: possible interaction with peroxidases. New Phytol 174: 332–341

Ehrary A, Rosas M, Carpinelli S, Davalos O, Cowling C, Fernandez F, Escobar M (2020) Glutaredoxin AtGRXS8 represses transcriptional and developmental responses to nitrate in *Arabidopsis thaliana* roots. Plant Direct 4: e00227

El-Maarouf-Bouteau H, Meimoun P, Job C, Job D, Bailly C (2013) Role of protein and mRNA oxidation in seed dormancy and germination. Front Plant Sci 4: 77

Espunya MC, Díaz M, Moreno-Romero J, Martínez MC (2006) Modification of intracellular levels of glutathione-dependent formaldehyde dehydrogenase alters glutathione homeostasis and root development. Plant Cell Environ 29: 1002–1011

Fernández-Marcos M, Desvoyes B, Manzano C, Liberman LM, Benfey PN, Del Pozo JC, Gutierrez C (2017) Control of *Arabidopsis* lateral root primordium boundaries by MYB36. New Phytol 213: 105–112

Ferreira MSL, Samson M-F, Bonicel J, Morel M-H (2012) Relationship between endosperm cells redox homeostasis and glutenin polymers assembly in developing durum wheat grain. Plant Physiol Biochem 61: 36–45

Foreman J, Demidchik V, Bothwell JHF, Mylona P, Miedema H, Torres MA, Linstead P, Costa S, Brownlee C, Jones JDG, et al. (2003) Reactive oxygen species produced by NADPH oxidase regulate plant cell growth. Nature 422: 442–446

Foyer CH, Noctor G (2016) Stress-triggered redox signalling: what's in pROSpect? Plant Cell Environ 39: 951–964

Foyer CH, Wilson MH, Wright MH (2018) Redox regulation of cell proliferation: bioinformatics and redox proteomics approaches to identify redox-sensitive cell cycle regulators. Free Radic Biol Med 122: 137–149

Gao X, Yuan HM, Hu YQ, Li J, Lu YT (2014) Mutation of Arabidopsis CATALASE2 results in hyponastic leaves by changes of auxin levels. Plant Cell Environ 37: 175–188

García-Giménez JL, Markovic J, Dasí F, Queval G, Schnaubelt D, Foyer CH, Pallardó FV (2013a) Nuclear glutathione. Biochim Biophys Acta 1830: 3304–3316

García-Giménez JL, Òlaso G, Hake SB, Bönisch C, Wiedemann SM, Markovic J, Dasí F, Gimeno A, Pérez-Quilis C, Palacios O, et al. (2013b) Histone h3 glutathionylation in proliferating mammalian cells destabilizes nucleosomal structure. Antioxid Redox Signal 19: 1305–1320

García-Giménez JL, Romá-Mateo C, Pérez-Machado G, Peiró-Chova L, Pallardó FV (2017) Role of glutathione in the regulation of epigenetic mechanisms in disease. Free Radic Biol Med 112: 36–48

García-Quirós E, Alché J de D, Karpinska B, Foyer CH (2020) Glutathione redox state plays a key role in flower development and pollen vigour. J Exp Bot 71: 730–741

Gibbs DJ, Lee SC, Isa NM, Gramuglia S, Fukao T, Bassel GW, Correia CS, Corbineau F, Theodoulou FL, Bailey-Serres J, et al. (2011) Homeostatic response to hypoxia is regulated by the N-end rule pathway in plants. Nature 479: 415–418

Gilroy S, Białasek M, Suzuki N, Górecka M, Devireddy AR, Karpiński S, Mittler R (2016) ROS, Calcium, and Electric Signals: Key Mediators of Rapid Systemic Signaling in Plants. Plant Physiol 171: 1606–1615

Gon S, Faulkner MJ, Beckwith J (2006) In vivo requirement for glutaredoxins and thioredoxins in the reduction of the ribonucleotide reductases of *Escherichia coli*. Antioxid Redox Signal **8**: 735–742

Graciet E, Walter F, Maoiléidigh DÓ, Pollmann S, Meyerowitz EM, Varshavsky A, Wellmer F (2009) The N-end rule pathway controls multiple functions during *Arabidopsis* shoot and leaf development. PNAS **106**: 13618–13623

Haffani YZ, Gaude T, Cock JM, Goring DR (2004) Antisense suppression of thioredoxin h mRNA in *Brassica napus cv.Westar* pistils causes a low level constitutive pollen rejection response. Plant Mol Biol **55**: 619–630

He Y, Tang R-H, Hao Y, Stevens RD, Cook CW, Ahn SM, Jing L, Yang Z, Chen L, Guo F, et al. (2004) Nitric oxide represses the *Arabidopsis* floral transition. Science **305**: 1968–1971

Hong L, Tang D, Zhu K, Wang K, Li M, Cheng Z (2012) Somatic and reproductive cell development in rice anther is regulated by a putative glutaredoxin. Plant Cell **24**: 577–588

Hou X, Hu W-W, Shen L, Lee LYC, Tao Z, Han J-H, Yu H (2008) Global identification of DELLA target genes during *Arabidopsis* flower development. Plant Physiol **147**: 1126–1142

Howden R, Andersen CR, Goldsbrough PB, Cobbett CS (1995) A cadmium-sensitive, glutathione-deficient mutant of *Arabidopsis thaliana*. Plant Physiol **107**: 1067–1073

Ito K, Hanazawa T, Tomita K, Barnes PJ, Adcock IM (2004) Oxidative stress reduces histone deacetylase 2 activity and enhances IL-8 gene expression: role of tyrosine nitration. Biochem Biophys Res Commun **315**: 240–245

Jiao Y, Sun L, Song Y, Wang L, Liu L, Zhang L, Liu B, Li N, Miao C, Hao F (2013) AtrbohD and AtrbohF positively regulate abscisic acid-inhibited primary root growth by affecting Ca^{2+} signalling and auxin response of roots in *Arabidopsis*. J Exp Bot **64**: 4183–4192

Jung YJ, Chi YH, Chae HB, Shin MR, Lee ES, Cha J-Y, Paeng SK, Lee Y, Park JH, Kim WY, et al. (2013) Analysis of *Arabidopsis* thioredoxin-h isotypes identifies discrete domains that confer specific structural and functional properties. Biochem J **456**: 13–24

Kaya H, Nakajima R, Iwano M, Kanaoka MM, Kimura S, Takeda S, Kawarazaki T, Senzaki E, Hamamura Y, Higashiyama T, et al. (2014) Ca^{2+}-activated reactive oxygen species production by *Arabidopsis* RbohH and RbohJ is essential for proper pollen tube tip growth. Plant Cell **26**: 1069–1080

Kelliher T, Walbot V (2012) Hypoxia triggers meiotic fate acquisition in maize. Science **337**: 345–348

Kerchev P, Mühlenbock P, Denecker J, Morreel K, Hoeberichts FA, Van Der Kelen K, Vandorpe M, Nguyen L, Audenaert D, Van Breusegem F (2015) Activation of auxin signalling counteracts photorespiratory H_2O_2-dependent cell death. Plant Cell Environ **38**: 253–265

Kirchsteiger K, Ferrández J, Pascual MB, González M, Cejudo FJ (2012) NADPH thioredoxin reductase C is localized in plastids of photosynthetic and non-photosynthetic tissues and is involved in lateral root formation in *Arabidopsis*. Plant Cell **24**: 1534–1548

Kka N, Rookes J, Cahill D (2018) The influence of ascorbic acid on root growth and the root apical meristem in *Arabidopsis thaliana*. Plant Physiol Biochem **129**: 323–330

Knuesting J, Riondet C, Maria C, Kruse I, Bécuwe N, König N, Berndt C, Tourrette S, Guilleminot-Montoya J, Herrero E, et al. (2015) *Arabidopsis* glutaredoxin S17 and its partner, the nuclear factor Y subunit C11/negative cofactor 2α, contribute to maintenance of the shoot apical meristem under long-day photoperiod. Plant Physiol **167**: 1643–1658

Kotchoni SO, Larrimore KE, Mukherjee M, Kempinski CF, Barth C (2009) Alterations in the endogenous ascorbic acid content affect flowering time in *Arabidopsis*. Plant Physiol **149**: 803–815

Kwak JM, Mori IC, Pei Z-M, Leonhardt N, Torres MA, Dangl JL, Bloom RE, Bodde S, Jones JDG, Schroeder JI (2003) NADPH oxidase AtrbohD and AtrbohF genes function in ROS-dependent ABA signaling in *Arabidopsis*. EMBO J **22**: 2623–2633

Kwon E, Feechan A, Yun B-W, Hwang B-H, Pallas JA, Kang J-G, Loake GJ (2012) AtGSNOR1 function is required for multiple developmental programs in *Arabidopsis*. Planta **236**: 887–900

Laporte D, Olate E, Salinas P, Salazar M, Jordana X, Holuigue L (2012) Glutaredoxin GRXS13 plays a key role in protection against photooxidative stress in *Arabidopsis*. J Exp Bot **63**: 503–515

Lee JR, Lee SS, Jang HH, Lee YM, Park JH, Park SC, Moon JC, Park SK, Kim SY, Lee SY, et al. (2009) Heat-shock dependent oligomeric status alters the function of a plant-specific thioredoxin-like protein, AtTDX. Proc Natl Acad Sci U S A **106**: 5978–5983

Lee U, Wie C, Fernandez BO, Feelisch M, Vierling E (2008) Modulation of nitrosative stress by S-nitrosoglutathione reductase is critical for thermotolerance and plant growth in *Arabidopsis*. Plant Cell **20**: 786–802

Leisegang MS, Schröder K, Brandes RP (2017) Redox regulation and noncoding RNAs. Antioxid Redox Signal. doi:10.1089/ars.2017.7276

Lennartz K, Plücken H, Seidler A, Westhoff P, Bechtold N, Meierhoff K (2001) HCF164 encodes a thioredoxin-like protein involved in the biogenesis of the cytochrome b(6)f complex in *Arabidopsis*. Plant Cell **13**: 2539–2551

Lepistö A, Kangasjärvi S, Luomala E-M, Brader G, Sipari N, Keränen M, Keinänen M, Rintamäki E (2009) Chloroplast NADPH-thioredoxin reductase interacts with photoperiodic development in *Arabidopsis*. Plant Physiol **149**: 1261–1276

Lepistö A, Rintamäki E (2012) Coordination of plastid and light signaling pathways upon development of Arabidopsis leaves under various photoperiods. Mol Plant **5**: 799–816

Li C, Potuschak T, Colón-Carmona A, Gutiérrez RA, Doerner P (2005) *Arabidopsis* TCP20 links regulation of growth and cell division control pathways. Proc Natl Acad Sci USA **102**: 12978–12983

Li N, Sun L, Zhang L, Song Y, Hu P, Li C, Hao FS (2015) AtrbohD and AtrbohF negatively regulate lateral root development by changing the localized accumulation of superoxide in primary roots of *Arabidopsis*. Planta **241**: 591–602

Li S, Gutsche N, Zachgo S (2011) The ROXY1 C-terminal L**LL motif is essential for the interaction with TGA transcription factors. Plant Physiol **157**: 2056–2068

Li S, Lauri A, Ziemann M, Busch A, Bhave M, Zachgo S (2009) Nuclear activity of ROXY1, a glutaredoxin interacting with TGA factors, is required for petal development in *Arabidopsis thaliana*. Plant Cell **21**: 429–441

Li ZY, Li B, Dong AW (2012) The Arabidopsis transcription factor AtTCP15 regulate endoreduplication by modulating expression of key cell-cycle genes. Mol Plant **5**: 270–280

Licausi F, Kosmacz M, Weits DA, Giuntoli B, Giorgi FM, Voesenek LACJ, Perata P, van Dongen JT (2011) Oxygen sensing in plants is mediated by an N-end rule pathway for protein destabilization. Nature **479**: 419–422

Liu P, Zhang H, Yu B, Xiong L, Xia Y (2015) Proteomic identification of early salicylate- and flg22-responsive redox-sensitive proteins in *Arabidopsis*. Sci Rep **5**: 8625

Lozano-Juste J, León J (2010) Enhanced abscisic acid mediated responses in nia1nia2noa1-2 triple mutant impaired in NIA/NR- and AtNOA1-dependent nitric oxide biosynthesis in *Arabidopsis*. Plant Physiol **152**: 891–903

Luria G, Rutley N, Lazar I, Harper JF, Miller G (2019) Direct analysis of pollen fitness by flow cytometry: implications for pollen response to stress. Plant J **98**: 942–952

Mangano S, Denita-Juarez SP, Choi H-S, Marzol E, Hwang Y, Ranocha P, Velasquez SM, Borassi C, Barberini ML, Aptekmann AA, et al. (2017) Molecular link between auxin and ROS-mediated polar growth. PNAS **114**: 5289–5294

Manzano C, Pallero-Baena M, Casimiro I, Rybel BD, Orman-Ligeza B, Isterdael GV, Beeckman T, Draye X, Casero P, Pozo JC del (2014) The emerging role of reactive oxygen species signaling during lateral root development. Plant Physiol **165**: 1105–1119

Marchal C, Delorme-Hinoux V, Bariat L, Siala W, Belin C, Saez-Vasquez J, Riondet C, Reichheld J-P (2014) NTR/NRX define a new thioredoxin system in the nucleus of *Arabidopsis thaliana* cells. Mol Plant **7**: 30–44

Marquez-Garcia B, Njo M, Beeckman T, Goormachtig S, Foyer CH (2014) A new role for glutathione in the regulation of root architecture linked to strigolactones. Plant Cell Environ **37**: 488–498

Martin MV, Fiol DF, Sundaresan V, Zabaleta EJ, Pagnussat GC (2013) Oiwa, a female gametophytic mutant impaired in a mitochondrial manganese-superoxide dismutase, reveals crucial roles for reactive oxygen species during embryo sac development and fertilization in *Arabidopsis*. Plant Cell **25**: 1573–1591

Martins L, Knuesting J, Bariat L, Dard A, Freibert SA, Marchand C, Young D, Dung NHT, Voth W, de Bures A, et al. (2020) Redox modification of the Fe-S glutaredoxin GRXS17 activates holdase activity and protects plants from heat stress. Plant Physiol. doi:10.1104/pp.20.00906

Martins L, Trujillo-Hernandez JA, Reichheld J-P (2018) Thiol based redox signaling in plant nucleus. Front Plant Sci **9**: 705

Marty L, Bausewein D, Müller C, Bangash SAK, Moseler A, Schwarzländer M, Müller-Schüssele SJ, Zechmann B, Riondet C, Balk J, et al. (2019) *Arabidopsis* glutathione reductase 2 is indispensable in plastids, while mitochondrial glutathione is safeguarded by additional reduction and transport systems. New Phytol **224**: 1569–1584

Marty L, Siala W, Schwarzländer M, Fricker MD, Wirtz M, Sweetlove LJ, Meyer Y, Meyer AJ, Reichheld J-P, Hell R (2009) The NADPH-dependent thioredoxin system constitutes a functional backup for cytosolic glutathione reductase in *Arabidopsis*. Proc Natl Acad Sci USA **106**: 9109–9114

Maß L, Holtmannspötter M, Zachgo S (2020) Dual-color 3D-dSTORM colocalization and quantification of ROXY1 and RNAPII variants throughout the transcription cycle in root meristem nuclei. Plant J. doi:10.1111/tpj.14986

Meinhard M, Grill E (2001) Hydrogen peroxide is a regulator of ABI1, a protein phosphatase 2C from *Arabidopsis*. FEBS Lett **508**: 443–446

Meinhard M, Rodriguez PL, Grill E (2002) The sensitivity of ABI2 to hydrogen peroxide links the abscisic acid-response regulator to redox signalling. Planta **214**: 775–782

Méndez AAE, Mangialavori IC, Cabrera AV, Benavides MP, Vázquez-Ramos JM, Gallego SM (2020) Tyrnitration in maize CDKA;1 results in lower affinity for ATP binding. Biochim Biophys Acta Proteins Proteom **1868**: 140479

Meng L, Wong JH, Feldman LJ, Lemaux PG, Buchanan BB (2010) A membrane-associated thioredoxin required for plant growth moves from cell to cell, suggestive of a role in intercellular communication. Proc Natl Acad Sci USA **107**: 3900–3905

Mengel A, Ageeva A, Georgii E, Bernhardt J, Wu K, Durner J, Lindermayr C (2017) Nitric oxide modulates histone acetylation at stress genes by inhibition of histone deacetylases. Plant Physiol **173**: 1434–1452

Meyer Y, Belin C, Delorme-Hinoux V, Reichheld J-P, Riondet C (2012) Thioredoxin and glutaredoxin systems in plants: molecular mechanisms, crosstalks, and functional significance. Antioxid Redox Signal **17**: 1124–1160

Meyer Y, Buchanan BB, Vignols F, Reichheld J-P (2009) Thioredoxins and glutaredoxins: unifying elements in redox biology. Annu Rev Genet **43**: 335–367

Mhamdi A, Hager J, Chaouch S, Queval G, Han Y, Taconnat L, Saindrenan P, Gouia H, Issakidis-Bourguet E, Renou J-P, et al. (2010) *Arabidopsis* GLUTATHIONE REDUCTASE1 plays a crucial role in leaf responses to intracellular hydrogen peroxide and in ensuring appropriate gene expression through both salicylic acid and jasmonic acid signaling pathways. Plant Physiol **153**: 1144–1160

Mhamdi A, Van Breusegem F (2018) Reactive oxygen species in plant development. Development. doi:10.1242/dev.164376

Michalska J, Zauber H, Buchanan BB, Cejudo FJ, Geigenberger P (2009) NTRC links built-in thioredoxin to light and sucrose in regulating starch synthesis in chloroplasts and amyloplasts. Proc Natl Acad Sci U S A **106**: 9908–9913

Miller G, Schlauch K, Tam R, Cortes D, Torres MA, Shulaev V, Dangl JL, Mittler R (2009) The plant NADPH oxidase RBOHD mediates rapid systemic signaling in response to diverse stimuli. Sci Signal **2**: ra45–ra45

Mittler R, Finka A, Goloubinoff P (2012) How do plants feel the heat? Trends Biochem Sci **37**: 118–125

Montiel J, Arthikala M-K, Cárdenas L, Quinto C (2016) Legume NADPH oxidases have crucial roles at different stages of nodulation. Int J Mol Sci. doi:10.3390/ijms17050680

Murmu J, Bush MJ, DeLong C, Li S, Xu M, Khan M, Malcolmson C, Fobert PR, Zachgo S, Hepworth SR (2010) *Arabidopsis* basic leucine-zipper transcription factors TGA9 and TGA10 interact with floral glutaredoxins ROXY1 and ROXY2 and are redundantly required for anther development. Plant Physiol **154**: 1492–1504

Muthuramalingam M, Seidel T, Laxa M, Nunes de Miranda SM, Gärtner F, Ströher E, Kandlbinder A, Dietz KJ (2009) Multiple redox and non-redox interactions define 2-Cys peroxiredoxin as a regulatory hub in the chloroplast. Mol Plant **2**: 1273–1288

Nietzel T, Mostertz J, Ruberti C, Née G, Fuchs P, Wagner S, Moseler A, Müller-Schüssele SJ, Benamar A, Poschet G, et al. (2019) Redox-mediated kickstart of mitochondrial energy metabolism drives resource-efficient seed germination. Proc Natl Acad Sci USA. doi:10.1073/pnas.1910501117

Niu Y, DesMarais TL, Tong Z, Yao Y, Costa M (2015) Oxidative stress alters global histone modification and DNA methylation. Free Radical Biol Med **82**: 22–28

Noctor G, Reichheld JP, Foyer C (2018) ROS-related redox regulation and signaling in plants. Semin Cell Dev Biol **80**: 3–12. doi:10.1016/j.semcdb.2017.07.013

Nott A, Watson PM, Robinson JD, Crepaldi L, Riccio A (2008) S-Nitrosylation of histone deacetylase 2 induces chromatin remodelling in neurons. Nature **455**: 411–415

Oh S-K, Baek K-H, Seong ES, Joung YH, Choi G-J, Park JM, Cho HS, Kim EA, Lee S, Choi D (2010) CaMsrB2, pepper methionine sulfoxide reductase B2, is a novel defense regulator against oxidative stress and pathogen attack. Plant Physiol **154**: 245–261

Ó'Maoiléidigh DS, Wuest SE, Rae L, Raganelli A, Ryan PT, Kwasniewska K, Das P, Lohan AJ, Loftus B, Graciet E, et al. (2013) Control of reproductive floral organ identity specification in *Arabidopsis* by the C function regulator AGAMOUS. Plant Cell **25**: 2482–2503

Orman-Ligeza B, Parizot B, de Rycke R, Fernandez A, Himschoot E, Van Breusegem F, Bennett MJ, Périlleux C, Beeckman T, Draye X (2016) RBOH-mediated ROS production facilitates lateral root emergence in *Arabidopsis*. Development **143**: 3328–3339

Ortiz-Espín A, Iglesias-Fernández R, Calderón A, Carbonero P, Sevilla F, Jiménez A (2017) Mitochondrial AtTrxo1 is transcriptionally regulated by AtbZIP9 and AtAZF2 and affects seed germination under saline conditions. J Exp Bot **68**: 1025–1038

Ota R, Ohkubo Y, Yamashita Y, Ogawa-Ohnishi M, Matsubayashi Y (2020) Shoot-to-root mobile CEPD-like 2 integrates shoot nitrogen status to systemically regulate nitrate uptake in *Arabidopsis*. Nat Commun **11**: 641

Pagnussat GC, Yu H-J, Ngo QA, Rajani S, Mayalagu S, Johnson CS, Capron A, Xie L-F, Ye D, Sundaresan V (2005) Genetic and molecular identification of genes required for female gametophyte development and function in *Arabidopsis*. Development **132**: 603–614

Pandey R, Müller A, Napoli CA, Selinger DA, Pikaard CS, Richards EJ, Bender J, Mount DW, Jorgensen RA (2002) Analysis of histone acetyltransferase and histone deacetylase families of *Arabidopsis thaliana* suggests functional diversification of chromatin modification among multicellular eukaryotes. Nucleic Acids Res **30**: 5036–5055

Parisy V, Poinssot B, Owsianowski L, Buchala A, Glazebrook J, Mauch F (2007) Identification of PAD2 as a gamma-glutamylcysteine synthetase highlights the importance of glutathione in disease resistance of *Arabidopsis*. Plant J **49**: 159–172

Park SK, Jung YJ, Lee JR, Lee YM, Jang HH, Lee SS, Park JH, Kim SY, Moon JC, Lee SY, et al. (2009) Heat-shock and redox-dependent functional switching of an h-type Arabidopsis thioredoxin from a disulfide reductase to a molecular chaperone. Plant Physiol **150**: 552–561

Passaia G, Queval G, Bai J, Margis-Pinheiro M, Foyer CH (2014) The effects of redox controls mediated by glutathione peroxidases on root architecture in *Arabidopsis thaliana*. J Exp Bot **65**: 1403–1413

Pasternak M, Lim B, Wirtz M, Hell R, Cobbett CS, Meyer AJ (2008) Restricting glutathione biosynthesis to the cytosol is sufficient for normal plant development. Plant J **53**: 999–1012

Patterson K, Walters LA, Cooper AM, Olvera JG, Rosas MA, Rasmusson AG, Escobar MA (2016) Nitrate-regulated glutaredoxins control *Arabidopsis* primary root growth. Plant Physiol **170**: 989–999

Pautler M, Eveland AL, LaRue T, Yang F, Weeks R, Lunde C, Je BI, Meeley R, Komatsu M, Vollbrecht E, et al. (2015) FASCIATED EAR4 encodes a bZIP transcription factor that regulates shoot meristem size in maize. Plant Cell **27**: 104–120

Pei ZM, Murata Y, Benning G, Thomine S, Klüsener B, Allen GJ, Grill E, Schroeder JI (2000) Calcium channels activated by hydrogen peroxide mediate abscisic acid signalling in guard cells. Nature **406**: 731–734

Pellny TK, Locato V, Vivancos PD, Markovic J, De Gara L, Pallardó FV, Foyer CH (2009) Pyridine nucleotide cycling and control of intracellular redox state in relation to poly (ADP-ribose) polymerase activity and nuclear localization of glutathione during exponential growth of *Arabidopsis* cells in culture. Mol Plant **2**: 442–456

Pérez-Ruiz JM, Spínola MC, Kirchsteiger K, Moreno J, Sahrawy M, Cejudo FJ (2006) Rice NTRC is a high-efficiency redox system for chloroplast protection against oxidative damage. Plant Cell **18**: 2356–2368

Qin Y, Leydon AR, Manziello A, Pandey R, Mount D, Denic S, Vasic B, Johnson MA, Palanivelu R (2009) Penetration of the stigma and style elicits a novel transcriptome in pollen tubes, pointing to genes critical for growth in a pistil. PLoS Genet. doi:10.1371/journal.pgen.1000621

Queval G, Issakidis-Bourguet E, Hoeberichts FA, Vandorpe M, Gakière B, Vanacker H, Miginiac-Maslow M, Van Breusegem F, Noctor G (2007) Conditional oxidative stress responses in the *Arabidopsis* photorespiratory mutant cat2 demonstrate that redox state is a key modulator of daylength-dependent gene expression, and define photoperiod as a crucial factor in the regulation of H_2O_2-induced cell death. Plant J **52**: 640–657

Queval G, Neukermans J, Vanderauwera S, Van Breusegem F, Noctor G (2012) Day length is a key regulator of transcriptomic responses to both CO(2) and H(2)O(2) in Arabidopsis. Plant Cell Environ **35**: 374–387

Queval G, Thominet D, Vanacker H, Miginiac-Maslow M, Gakière B, Noctor G (2009) H_2O_2-activated up-regulation of glutathione in Arabidopsis involves induction of genes encoding enzymes involved in cysteine synthesis in the chloroplast. Mol Plant **2**: 344–356

Rahantaniana MS, Li S, Chatel-Innocenti G, Tuzet A, Issakidis-Bourguet E, Mhamdi A, Noctor G (2017) Cytosolic and chloroplastic DHARs cooperate in the induction of the salicylic acid pathway by oxidative stress. Plant Physiol **174**: 956–971

Reichheld J-P, Khafif M, Riondet C, Droux M, Bonnard G, Meyer Y (2007) Inactivation of thioredoxin reductases reveals a complex interplay between thioredoxin and glutathione pathways in *Arabidopsis* development. Plant Cell **19**: 1851–1865

Reichheld JP, Vernoux T, Lardon F, Van Montagu M, Inze D (1999) Specific checkpoints regulate plant cell cycle progression in response to oxidative stress. Plant J **17**: 647–656

Riondet C, Desouris JP, Montoya JG, Chartier Y, Meyer Y, Reichheld J-P (2012) A dicotyledon-specific glutaredoxin GRXC1 family with dimer-dependent redox regulation is functionally redundant with GRXC2. Plant Cell Environ **35**: 360–373

Rodriguez O, Reshetnyak G, Grondin A, Saijo Y, Leonhardt N, Maurel C, Verdoucq L (2017) Aquaporins facilitate hydrogen peroxide entry into guard cells to mediate ABA- and pathogen-triggered stomatal closure. Proc Natl Acad Sci U S A **114**: 9200–9205

Rouhier N, Cerveau D, Couturier J, Reichheld J-P, Rey P (2015) Involvement of thiol-based mechanisms in plant development. Biochim Biophys Acta **1850**: 1479–1496

Schnaubelt D, Queval G, Dong Y, Diaz-Vivancos P, Makgopa ME, Howell G, De Simone A, Bai J, Hannah MA, Foyer CH (2015) Low glutathione regulates gene expression and the redox potentials of the nucleus and cytosol in *Arabidopsis thaliana*. Plant Cell Environ **38**: 266–279

Shanmugam V, Tsednee M, Yeh K-C (2012) ZINC TOLERANCE INDUCED BY IRON 1 reveals the importance of glutathione in the cross-homeostasis between zinc and iron in *Arabidopsis thaliana*. Plant J **69**: 1006–1017

Shen Y, Issakidis-Bourguet E, Zhou D-X (2016) Perspectives on the interactions between metabolism, redox, and epigenetics in plants. J Exp Bot **67**: 5291–5300

Sies H, Berndt C, Jones DP (2017) Oxidative stress. Annu Rev Biochem **86**: 715–748

Steffens B, Kovalev A, Gorb SN, Sauter M (2012) Emerging roots alter epidermal cell fate through mechanical and reactive oxygen species signaling. The Plant Cell **24**: 3296–3306

Timofejeva L, Skibbe DS, Lee S, Golubovskaya I, Wang R, Harper L, Walbot V, Cande WZ (2013) Cytological characterization and allelism testing of anther developmental mutants identified in a screen of maize male sterile lines. G3 (Bethesda) **3**: 231–249

Trujillo Hernandez JA, Bariat L, Enders TA, Strader LC, Reichheld J-P, Belin C (2020) A Glutathione-dependent control of IBA pathway supports *Arabidopsis* root system adaptation to phosphate deprivation. J Exp Bot. doi:10.1093/jxb/eraa195

Tsukagoshi H, Busch W, Benfey PN (2010) Transcriptional regulation of ROS controls transition from proliferation to differentiation in the root. Cell **143**: 606–616

Tuncel A, Cakir B, Hwang S-K, Okita TW (2014) The role of the large subunit in redox regulation of the rice endosperm ADP-glucose pyrophosphorylase. FEBS J **281**: 4951–4963

Tzafrir I, Pena-Muralla R, Dickerman A, Berg M, Rogers R, Hutchens S, Sweeney TC, McElver J, Aux G, Patton D, et al. (2004) Identification of genes required for embryo development in *Arabidopsis*. Plant Physiol **135**: 1206–1220

Vernoux T, Wilson RC, Seeley KA, Reichheld JP, Muroy S, Brown S, Maughan SC, Cobbett CS, Van Montagu M, Inzé D, et al. (2000) The ROOT MERISTEMLESS1/CADMIUM SENSITIVE2 gene defines a glutathione-dependent pathway involved in initiation and maintenance of cell division during postembryonic root development. Plant Cell **12**: 97–110

Viola IL, Güttlein LN, Gonzalez DH (2013) Redox modulation of plant developmental regulators from the class I TCP transcription factor family. Plant Physiol **162**: 1434–1447

Wang P, Liu J, Liu B, Feng D, Da Q, Wang P, Shu S, Su J, Zhang Y, Wang J, et al. (2013) Evidence for a role of chloroplastic m-type thioredoxins in the biogenesis of photosystem II in *Arabidopsis*. Plant Physiol **163**: 1710–1728

Waszczak C, Kerchev PI, Mühlenbock P, Hoeberichts FA, Kelen KVD, Mhamdi A, Willems P, Denecker J, Kumpf RP, Noctor G, et al. (2016) SHORT-ROOT deficiency alleviates the cell death phenotype of the *Arabidopsis* catalase2 mutant under photorespiration-promoting conditions. Plant Cell **28**: 1844–1859

Weits DA, Giuntoli B, Kosmacz M, Parlanti S, Hubberten H-M, Riegler H, Hoefgen R, Perata P, van Dongen JT, Licausi F (2014) Plant cysteine oxidases control the oxygen-dependent branch of the N-end-rule pathway. Nat Commun **5**: 3425

Weits DA, Kunkowska AB, Kamps NCW, Portz KMS, Packbier NK, Nemec Venza Z, Gaillochet C, Lohmann JU, Pedersen O, van Dongen JT, et al. (2019) An apical hypoxic niche sets the pace of shoot meristem activity. Nature **569**: 714–717

Wu F, Chi Y, Jiang Z, Xu Y, Xie L, Huang F, Wan D, Ni J, Yuan F, Wu X, et al. (2020) Hydrogen peroxide sensor HPCA1 is an LRR receptor kinase in Arabidopsis. Nature. **578**: 577–581

Wuest SE, O'Maoileidigh DS, Rae L, Kwasniewska K, Raganelli A, Hanczaryk K, Lohan AJ, Loftus B, Graciet E, Wellmer F (2012) Molecular basis for the specification of floral organs by APETALA3 and PISTILLATA. Proc Natl Acad Sci U S A **109**: 13452–13457

Xie H-T, Wan Z-Y, Li S, Zhang Y (2014) Spatiotemporal production of reactive oxygen species by NADPH oxidase is critical for tapetal programmed cell death and pollen development in *Arabidopsis*. Plant Cell **26**: 2007–2023

Xing S, Rosso MG, Zachgo S (2005) ROXY1, a member of the plant glutaredoxin family, is required for petal development in *Arabidopsis thaliana*. Development **132**: 1555–1565

Xing S, Zachgo S (2008) ROXY1 and ROXY2, two *Arabidopsis* glutaredoxin genes, are required for anther development. Plant J **53**: 790–801

Xu S, Guerra D, Lee U, Vierling E (2013) S-nitrosoglutathione reductases are low-copy number, cysteine-rich proteins in plants that control multiple developmental and defense responses in *Arabidopsis*. Front Plant Sci **4**: 430

Xu Y-M, Du J-Y, Lau ATY (2014) Posttranslational modifications of human histone H3: an update. Proteomics **14**: 2047–2060

Yamauchi T, Yoshioka M, Fukazawa A, Mori H, Nishizawa NK, Tsutsumi N, Yoshioka H, Nakazono M (2017) An NADPH oxidase RBOH functions in rice roots during lysigenous aerenchyma formation under oxygen-deficient conditions. The Plant Cell **29**: 775–790

Yang F, Bui HT, Pautler M, Llaca V, Johnston R, Lee B, Kolbe A, Sakai H, Jackson D (2015) A maize glutaredoxin gene, abphyl2, regulates shoot meristem size and phyllotaxy. Plant Cell **27**: 121–131

Yao L-L, Zhou Q, Pei B-L, Li Y-Z (2011) Hydrogen peroxide modulates the dynamic microtubule cytoskeleton during the defence responses to *Verticillium dahliae* toxins in *Arabidopsis*. Plant Cell Environ **34**: 1586–1598

Yu X, Pasternak T, Eiblmeier M, Ditengou F, Kochersperger P, Sun J, Wang H, Rennenberg H, Teale W, Paponov I, et al. (2013) Plastid-localized glutathione reductase2–regulated glutathione redox status is essential for *Arabidopsis* root apical meristem maintenance. The Plant Cell **25**: 4451–4468

Yua Q-B, Ma Q, Kong M-M, Zhao T-T, Zhang X-L, Zhou Q, Huang C, Chong K, Yang Z-N (2014) AtECB1/MRL7, a thioredoxin-like fold protein with disulfide reductase activity, regulates chloroplast gene expression and chloroplast biogenesis in *Arabidopsis thaliana*. Mol Plant **7**: 206–217

Zafra A, Rejón JD, Hiscock SJ, Alché J de D (2016) Patterns of ROS accumulation in the stigmas of angiosperms and visions into their multi-functionality in plant reproduction. Front Plant Sci **7**: 1112

Zahedi Avval F, Holmgren A (2009) Molecular mechanisms of thioredoxin and glutaredoxin as hydrogen donors for Mammalian s phase ribonucleotide reductase. J Biol Chem **284**: 8233–8240

Zandalinas SI, Fichman Y, Devireddy AR, Sengupta S, Azad RK, Mittler R (2020) Systemic signaling during abiotic stress combination in plants. Proc Natl Acad Sci U S A **117**: 13810–13820

Zeng J, Dong Z, Wu H, Tian Z, Zhao Z (2017) Redox regulation of plant stem cell fate. EMBO J **36**: 2844–2855

Zhang C-J, Zhao B-C, Ge W-N, Zhang Y-F, Song Y, Sun D-Y, Guo Y (2011) An apoplastic h-type thioredoxin is involved in the stress response through regulation of the apoplastic reactive oxygen species in rice. Plant Physiol **157**: 1884–1899

Zhang H, Zhang TT, Liu H, Shi DY, Wang M, Bie XM, Li XG, Zhang XS (2018) Thioredoxin-mediated ROS homeostasis explains natural variation in plant regeneration. Plant Physiol **176**: 2231–2250

Zhou X, Sun H, Ellen TP, Chen H, Costa M (2008) Arsenite alters global histone H3 methylation. Carcinogenesis **29**: 1831–1836

Zhu J-K (2009) Active DNA demethylation mediated by DNA glycosylases. Annu Rev Genet **43**: 143–166

Thiol-Disulphide Redox Signalling/Control during the Life Cycle of Pathogenic Trypanosomatids

Gabriela Specker, Lucía Piacenza, Rafael Radi, and Marcelo A. Comini

CONTENTS

3.1 TRYPANOSOMATIDS: GENERAL CONSIDERATIONS AND DIFFERENTIATION PROCESS DURING THE LIFE CYCLE

Trypanosomatids are unicellular protists and obligatory parasites with the ability to infect a single (monoxenic: invertebrate or vertebrate) or two different (dixenic: invertebrate and vertebrate) hosts (Lukeš et al. 2018). Several of the digenetic trypanosomatids are responsible for severe and mortal diseases in humans and livestock. Among them, African sleeping sickness, Chagas disease, and leishmaniasis affect millions of people inhabiting mainly tropical regions of the world (Stuart et al. 2008), whereas animal trypanosomiasis has an important economic impact in rural areas (Giordani et al. 2016). For the pathogenic species, the invertebrate host is a hematophagous insect from the Order Hemiptera or Diptera that acts as a transmission vector. Depending on the trypanosomatid species, parasites display tropism for different host tissues or systems (Pereira et al. 2019).

In order to successfully accomplish this complex life cycle, the parasite must undergo important structural and metabolic changes for adapting to physical (e.g., temperature, pH, osmolality) and biochemical/biological (e.g., nutritional starvation, digestive enzymes, host defences) conditions that are markedly different in each host (Figure 3.1).

For instance, the parasite is exposed to significant temperature, pH, and osmolality fluctuations, and the supply of nutrients is variable in the insect stage (Santos et al. 2011; Kollien et al. 2001; Weiss et al. 2019; Roditi and Lehane 2008). Importantly, upon blood meal trypanosomatids must withstand the action of the insect's antimicrobial peptides, lectins, and digestive enzymes. In addition, the insect gut holds high levels of reactive oxygen and nitrogen species as part of the defence mechanism and to control microbiota (Diaz-Albiter et al. 2012; Hao et al. 2003; MacLeod, Maudlin et al. 2007; Macleod, Darby et al. 2007). Only a few parasites survive this challenging environment and are able to migrate to organs (e.g., salivary glands or distal gut), where they differentiate into highly infective forms for mammals.

Worth noting, in trypanosomatids, genes are expressed as long polycistronic units (Clayton 2019),

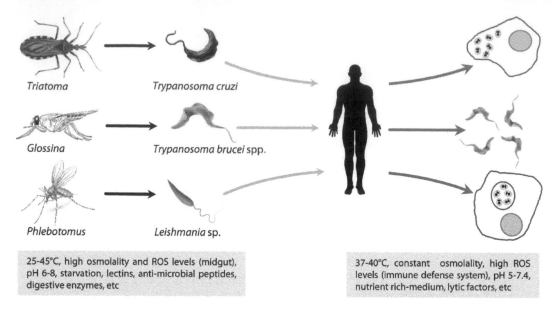

| 25-45°C, high osmolality and ROS levels (midgut), pH 6-8, starvation, lectins, anti-microbial peptides, digestive enzymes, etc | 37-40°C, constant osmolality, high ROS levels (immune defense system), pH 5-7.4, nutrient rich-medium, lytic factors, etc |

Figure 3.1 Trypanosomatids' life cycle.

Each trypanosomatid genus/species is able to survive in a specific hematophagous insect, which acts as a transmission vector of highly infective stages (black and grey arrows) to mammalian hosts (i.e., humans). Upon successful infection of mammals, the pathogens differentiate to less virulent but persistent stages that show a species-specific tropism for host cells and tissues (dark grey arrows). *Trypanosoma cruzi* and *Leishmania* sp. are mainly intracellular pathogens that colonize several cell types or only macrophages, respectively. Inside the host cells, *T. cruzi* resides in cytosol whereas *Leishmania* remains in the phagolysosome. In the course of the infection, parasites are released to the extracellular medium. *Trypanosoma brucei* is an extracellular pathogen that colonizes the hemolymphatic system and cerebrospinal and interstitial fluids of the host. All extracellular stages of these pathogens are capable of infecting new cells/tissues or, upon differentiation, the insect vectors (during blood meal). The major environmental conditions and factors that trypanosomatids face during their developmental stages in invertebrate and vertebrate hosts are highlighted in boxes.

and chromatin remodelling and other epigenetic factors occurring during differentiation facilitate the activation of transcription units that are stage specific (Jensen et al. 2009; Queiroz et al. 2009; Berná et al. 2017). This feature allows the differentiated parasites to rapidly respond to the environmental conditions and renders posttranscriptional and posttranslational modifications as major mechanisms to control protein expression and activity.

In contrast to the hostile environment in the insect vectors, the mammalian host offers a nutrient-rich and physicochemical stable medium for parasite survival. Nonetheless, the parasites must face the innate and adaptive immune response, which involves humoral and cellular mechanisms, among which the activity of phagocytic cells stands out for its importance for infection control.

In order to overcome these challenges, trypanosomatids evolved very complex adaptive and evading mechanisms that are life stage specific. For instance,

during transition from the insect stage to the mammalian infective form, trypanosomatids populate their surface with glycoproteins that favour host cell attachment and invasion or evasion of immune response (i.e., complement, antibodies) and reconfigure their energetic metabolism (Hannaert et al. 2003; Cazzulo 1984; Cazzulo 1992; Berná et al. 2017; Marchese et al. 2018; Shiratsubaki et al. 2020; Silber et al. 2009; Saunders et al. 2014; Singh et al. 2016). Finally, the transcriptomic and translational machinery is downregulated (Christiano et al. 2017), and parasites stop proliferating until finding their niche in the new host.

Interestingly, the content of proteins from the thiol-redox system has been reported to increase during the differentiation to infective stages of different trypanosomatids (Comini et al. 2007; Piacenza et al. 2013; Sardar et al. 2013; Zago et al. 2016). Although there is overwhelming evidence supporting a protective role of this metabolism against the oxidative defence mounted by the host during infection, redox-regulatory functions have recently emerged for different components of this system.

3.2 THIOL-REDOX-DEPENDENT METABOLISM OF TRYPANOSOMATIDS

In all aerobic organisms, thiol-redox homeostasis and signalling are governed by an electron transport chain that involves coupling the reducing power of, generally, NADPH to low molecular weight thiols (e.g., glutathione, GSH) and different oxidoreductases (e.g., glutaredoxin, Grx; thioredoxin, Trx) via specialized reductases (e.g., thioredoxin reductase, TrxR; glutathione reductase, GR). Grx, Trx, and low molecular weight thiols act as redox transducing elements that modify the redox state and, thereby, the activity of different effector proteins (e.g., peroxidases, kinases, transcription factors). Among the biochemical specializations evolved by trypanosomatids, the thiol-redox metabolism stands out for diverging from those ubiquitous to most eukaryotes. Compared to mammals, trypanosomatids are devoid of gene coding for GR and TrxR (Berriman et al. 2005; Ivens et al. 2005; El-Sayed et al. 2005), they lack catalase and GSH-dependent peroxidase (Boveris and Stoppani 1977; Boveris et al. 1980), and they have a lower GSH content and set of peroxidases (Krauth-Siegel and Comini 2008). Instead, their redox system depends on the utilization of a unique low molecular weight dithiol: N^1,N^8 bisglutathionylspermidine, also known as trypanothione (T(SH)$_2$; Ariyanayagam and Fairlamb 2001; Fairlamb and Cerami 1985; Figure 3.2A). In order to warrant an efficient use of T(SH)$_2$ as a substrate, several redox enzymes of the parasites, such as trypanothione reductase (TR) and Grx, underwent specific structural adaptations (Manta et al. 2018). Moreover, these parasites are endowed with a multipurpose oxidoreductase, tryparedoxin (TXN), which shares features with Grx and Trx but uses T(SH)$_2$ as a reducing substrate (Nogoceke et al. 1997; Lüdemann et al. 1998). Although equipped with Trx (Currier et al. 2019; H. Schmidt and Krauth-Siegel 2003; Piattoni et al. 2006; Krauth-Siegel and Schmidt 2002), TXN plays the housekeeping role in thiol-redox homeostasis (Comini et al. 2007; Romao et al. 2009). Furthermore, all developmental stages of trypanosomatids express two classes of peroxidases (Castro and Tomás 2008): 2-Cys-peroxiredoxins (Prx; Castro et al. 2002;

Piacenza et al. 2008; Trujillo et al. 2004; Wilkinson et al. 2000) and glutathione peroxidase-type TXN-dependent peroxidases (Px; Diechtierow and Krauth-Siegel 2011; Wilkinson, Taylor, Touitha et al. 2002). *T. cruzi* and *Leishmania*, but not *T. brucei*, also express a third class of peroxidase: a hybrid type A heme peroxidase (Adak and Datta 2005; Hugo et al. 2017; Wilkinson, Obado, et al. 2002). The peroxidases fulfil an essential protective function against a wide spectrum of reactive species derived from the partial reduction of oxygen or from nitric oxide (Trujillo et al. 2004; Bogacz et al. 2020; Bogacz and Krauth-Siegel 2018; Castro et al. 2002; Hiller et al. 2014).

3.3 THIOL-DISULPHIDE-MEDIATED REDOX CONTROL AND SIGNAL TRANSDUCTION

Thiol-disulfide redox signalling, redox control, and redox homeostasis are intricately related processes that, despite sharing molecular components, can be distinguished by the extent of their action and physiological consequences (Jones and Sies 2015). Redox homeostasis refers to the mechanisms engaged in maintaining the overall redox status of a cell/organism, which, among other factors, depends on energy metabolism and the surrounding environment.

Redox control implies the dynamic regulation of the activity of different macromolecules and is mediated by oxidoreductases or low molecular weight thiols. In general, this regulatory activity is target and compartment specific. Redox signalling involves a primary stimulus (endogenous or exogenous) that triggers a signal in the cell that will be conveyed through a discrete and specific subset of redox elements to direct a specific cellular response (Jones 2010; Figure 3.2B). Similar to the redox control mechanisms, redox-signalling pathways are compartmentalized but, at variance with it, crosstalk between compartments may occur via either the signalling molecules or the translocation of macromolecules to the organelles.

In a thiol-disulphide signalling pathway, the *second messenger* is usually a low molecular weight oxidant with the capacity to diffuse in aqueous media or permeate membranes and modify reversibly the redox state of cysteine residues in the *sensor* molecule. Taking this into account, the candidate molecule that would lead to this signalling mechanism could be nitric oxide ($^\bullet$NO), hydrogen peroxide (H$_2$O$_2$),

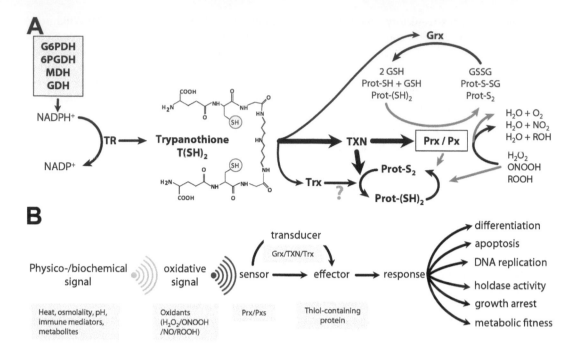

Figure 3.2 (A) Thiol-redox system of trypanosomatids. Several metabolic pathways (G6PDH, glucose 6-phosphate dehydrogenase; 6PGDH, 6-phosphogluconate dehydrogenase; MDH, malate dehydrogenase; GDH, glutamate dehydrogenase) contribute to NADPH biosynthesis in a stage-specific manner. The reducing power of NADPH is employed by trypanothione reductase (TR) to maintain trypanothione (bis-glutathionylspermidine) in its reduced form (T(SH)$_2$). Trypanothione very efficiently reduces glutaredoxins (Grx), thioredoxins (Trx), and tryparedoxin (TXN). Grx can reduce different types of disulfides: glutathione disulphide (GSSG), glutathione-protein (Prot-S-SG), and protein disulphides (Prot-S2). Trx can reduce protein disulphides but their potential targets are unknown. TXN has a dominant role in thiol-redox metabolism by reducing the oxidized forms of different types of peroxidases (Prx, peroxiredoxins; Px, glutathione-type tryparedoxin peroxidases) and several proteins (Prot). The peroxidases have specificity for decomposing different types of peroxides (H$_2$O$_2$, hydrogen peroxide; ONOOH, peroxynitrite; ROOH, alkyl hydroperoxides). The blue lines denote potential contributions of oxidants (peroxides or oxidized peroxidases) to signalling pathways by mediating the oxidation of target-proteins. (B) Components of a canonical redox signalling pathway. Candidate molecules and effects linked to oxidative signalling in trypanosomatids.

peroxynitrite,[1] or organic peroxides (ROOH). Trypanosomatids encounter these species both in the insect vector and the vertebrate host and can also be generated at the parasite cytosol and mitochondrion in different contexts (de Almeida Nogueira et al. 2011; Green et al. 1991; Wilkinson et al. 2006). The reversible modification of the sensor's redox active thiols induced by these oxidants includes S-nitrosylation, oxidation to sulfenic and sulfinic acid, or to disulphide (homo or mixed; Barford 2004). Protein S-thiolation can occur either by the reaction between the partially oxidized protein sulfhydryl or by thiol-disulfide exchange reactions. The mixed disulphide of GSH with protein thiols, also called glutathionylation, is a dynamic posttranslational modification that modulates cellular processes. Although thermodynamically not favoured due to its dithiol character, T(SH)$_2$ may potentially form short-lived mixed disulphides with specific molecular targets (Ulrich et al. 2017).

The sensor molecules are enzymes with high selectivity and kinetics to react and outcompete non-specific reactions of the signal with endogenous molecules. Peroxidases qualify as top sensor molecules for abundance, wide subcellular distribution, specificity, and high reaction rates versus oxidants (Paulsen and Carroll 2010; Brigelius-Flohé and Flohé 2011). The sensor molecules may act directly on the effector elements or on intermediate transducer molecules that will further transmit the signal to the effector.

Trypanosomatids Prx and GPx are distributed in different subcellular compartments and have capacity to detect minor fluctuations in the levels of H$_2$O$_2$ and ONOOH, or ROOH, respectively

(Diechtierow and Krauth-Siegel 2011; Piacenza et al. 2008; Schlecker et al. 2005; Tetaud et al. 2001; Wilkinson et al. 2000; Wilkinson et al. 2003). Studies addressing the involvement of these peroxidases in redox signalling pathways are lagging behind, but recent reports provided some clues towards their stage-specific roles beyond oxidant protection and are discussed in the next sections.

Grx are implicated in promoting and reversing the glutathionylation of proteins and may play roles in both redox signalling or control pathways (Allen and Mieyal 2012). The development of fluorescent protein-based redox biosensors coupled to this oxidoreductase revealed that in cells, Grx is able to detect nM concentrations of GSSG and to transfer very efficiently (in nanoseconds) the oxidizing equivalents to target proteins (Gutscher et al. 2008; Østergaard et al. 2004). Given that trypanosomatids have significant concentrations of GSH, possess Grx with deglutathionylating activity (Ceylan et al. 2010; Márquez et al. 2014), and lack GR (which may contribute to accumulation of GSSG) as well as an efficient transcriptional regulation of gene expression, it was hypothesized that thiolation may play a posttranslational regulatory role in protein function (Comini et al. 2013; Krauth-Siegel and Comini 2008). Efforts have been made to determine whether this modification occurs in parasites. In vitro studies with the recombinant form of different T. brucei proteins, whose activity depends on cysteine residues, showed that monothiol Grx-1 (1CGrx1), Prx, and a Trx were susceptible to glutathionylation (Melchers et al. 2007). For Prx, either any, but not both, catalytic cysteines (Schlecker et al. 2007) is glutathionylated, which is suggestive of a protective rather than a regulatory role for this modification. In contrast, 1CGrx1 and Trx were modified at non-active-site cysteines, whose function on protein activity remains unknown. Further proteomic studies performed on parasites subjected to oxidative stress originated exogenously (diamide, H_2O_2, or hypochlorite) or endogenously (inhibition of T(SH)$_2$ biosynthesis) showed that several T. brucei proteins undergo reversible thiolation. Only three proteins were S-thiolated under all stress conditions, which points to an oxidant-specific response. The most relevant targets included surface glycoproteins, structural proteins from the flagellum, microtubule-associated proteins, cathepsin-like proteases, kinases, several proteins from the redox metabolism (TXN, Prx, and protein disulfide isomerase) and chaperones,

as well as a regulator of mRNA metabolism (TSR1). Interestingly, TSR1 is the major point of regulation of gene expression in trypanosomatids. Strikingly, trypanothiolation of Prx was also detected, and it was hypothesized that this modification may serve as a general protective mechanism because the hydrolysis of protein-bound T(SH)$_2$ could generate GSH and monoglutathionylspermidine, which may react with other protein thiols, preventing their overoxidation (Ulrich et al. 2017). Nonetheless, the relevance of trypanothione in regulating protein activity by thiolation needs to be further investigated.

In T. cruzi, the possibility that tryparedoxins (TXN I and TXN II) mediate signal transduction or exert redox control by thiol-disulphide exchanges was addressed by an interactomic approach. Using single-cysteine mutants of these oxidoreductases as bait, several covalent protein partners belonging to the polyamine metabolism, oxidative metabolism, cytoskeleton, and protein translation machinery, in addition to components of the parasite antioxidant systems, were identified (Arias et al. 2015; Piñeyro, Parodi-Talice et al. 2011). Interestingly, many of these interactors are homologues to those reported to interact with Trx in other organisms, which supports the idea that TXNs can have redox regulatory roles.

3.4 REDOX SIGNALLING/CONTROL DURING CELL PROLIFERATION AND DIFFERENTIATION

Different bioreductive processes such as DNA and protein biosynthesis demand reducing power. For example, the conversion of ribonucleotides into their deoxy form involves a reductive step catalysed by ribonucleotide reductase (RnR) and assisted by Trx, TXN or, alternatively, Grx as electron donors. Thus, slight changes in the flux of reducing power towards this pathway may have consequences in cell cycle progression, virulence, and even differentiation.

In vitro, RnR activity was compromised at increasing ratios of oxidized versus reduced T(SH)$_2$ due to inhibition of TXN (Dormeyer et al. 2001). In line with this, downregulation of TXN as well as inhibition of T(SH)$_2$ biosynthesis but not silencing of Grx1 led to growth arrest in infective T. brucei (Ceylan et al. 2010; Comini et al. 2007; Comini et al. 2004; Mesías et al. 2019; Sousa et al. 2014). This supports that the T(SH)$_2$/TXN couple plays a major regulatory role in parasite proliferation. Interestingly, glucose 6-phosphate dehydrogenase, one of the major suppliers of cellular NADPH (Figure 3.2A), from *T. cruzi* was upregulated and showed increased product formation in infective stages and upon oxidative challenge (in vitro and in vivo), respectively (Igoillo-Esteve and Cazzulo 2006; Ortíz et al. 2019). Altogether, this suggests that under scenarios of metabolic restriction and oxidative stress, trypanosomes will halt cell division and other bioreductive reactions to spare or divert reducing power to support more life-critical functions (Figure 3.3).

In trypanosomatids, the replication of the mitochondrial DNA (called kinetoplast) precedes that of the nucleus and is perfectly synchronized with the cell cycle (Woodward and Gull 1990). The universal minicircle binding protein (UMSBP) is a CCHC-type zinc-finger protein that participates in the initiation of kinetoplast DNA replication and the segregation of kinetoplast DNA networks (Milman et al. 2007). A body of evidences suggest that the cysteine residues forming the two zinc-finger domain of UMSBP function as a redox switch modulating kinetoplast DNA replication. The first observations performed with the trypanosomatid-related organism *Crithidia fasciculata* showed that (i) oxidation or alkylation of UMSBP abolishes binding to DNA, (ii) treatment with $T(SH)_2$/TXN recovered the reduced and active form of UMSBP, and (iii) binding of UMSBP to DNA correlated to its redox state in vivo and in a cell cycle–dependent manner (Onn et al. 2004; Sela et al. 2008). Further biological investigations carried in the insect stage of *T. brucei* demonstrated that overexpression of a mitochondrial Prx or cytochrome b5 reductase-like (Cb5RL) caused oxidation of UMSBP and loss of mitochondrial DNA, and that this phenotype could be rescued by simultaneous overexpression of TXN (Motyka et al. 2006). A later study showed that the binding but not the dissociation of UMSBP from the DNA is the redox-sensitive step (Sela and Shlomai 2009). Almost identical results were obtained in L. *donovani* (Singh et al. 2016). None of these studies investigated whether the sensor element Prx may oxidize directly UMSBP, but the authors speculated that the aforementioned phenotypes were caused by competition of oxidized Prx and Cb5RL for the reducing equivalents of TXN (Figure 3.3).

Although the signalling pathways are not fully understood, it is clear that reactive oxygen and nitrogen species play major roles in parasite differentiation and adaptation to different hosts. In this respect, neutralization of insect-derived oxidants by addition of different low molecular antioxidant molecules to the blood meal has been shown to inhibit dramatically the cell death of *T. brucei*, promoting the establishment of infections in tsetse flies (Macleod, Maudlin et al. 2007). However, a certain threshold of endogenously produced oxidants appears to be required for the development of both the insect and mammalian stages of trypanosomatids. For instance, iron-mediated regulation of H_2O_2 levels has been shown to control the differentiation of non-infective promastigotes into the virulent amastigotes in *Leishmania* (Mittra et al. 2013). In this signalling pathway, iron superoxide dismutase (FeSOD) has been shown to fulfil a pivotal role by generating H_2O_2 that signals the differentiation process (Figure 3.3). Also, the transformation of the *Leishmania* insect stage to a highly infectious form is accompanied by an increased ability of the parasite to resist H_2O_2 and survive intracellularly in macrophages. The increased refractoriness of the parasite to oxidative damage has been shown to be triggered by different stimuli, such as metabolic starvation and heat shock (Zarley et al. 1991), and to involve upregulation of enzymes from the hydroperoxide metabolizing system (Bhattacharya et al. 2009).

The relevance of reactive oxygen and nitrogen species for trypanosomatid biology was further demonstrated for African trypanosomes. For instance, 'NO has been shown necessary for the differentiation of trypanosomes in the tsetse midgut (Macleod, Darby et al. 2007). Furthermore, expression of a *C. fasciculata* or human catalase compromised *T. brucei* infection of the tsetse fly midgut and attenuated parasitemia in the mammalian host, respectively (Horáková et al. 2020). Thus, although catalase activity was beneficial in vitro by conferring resistance to exogenous H_2O_2, it turned detrimental under in vivo conditions. This interesting finding somehow agrees with the lineage-specific absence of catalase genes in digenetic trypanosomatids and raises the hypothesis of an evolutionary specialization to facilitate optimal adaptation to both hosts.

Mitochondrial metabolic remodelling during the life cycle of trypanosomatids, and in particular for African trypanosomes, is of upmost importance. A recent study disclosed the relevance of redox signalling in the differentiation process of *T. brucei* (Doleželová et al. 2020). Parasite differentiation throughout the different stages colonizing the insect was induced by overexpression of a

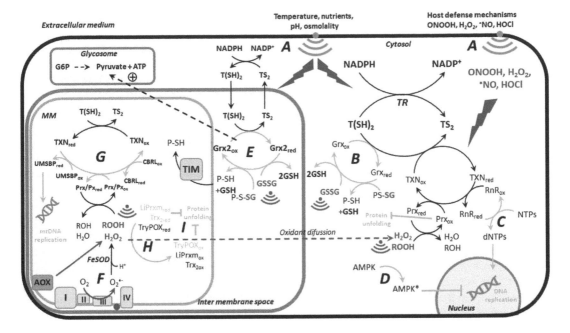

Figure 3.3 Overview of major thiol-disulphide redox signalling pathways in trypanosomatids.

The scheme integrates the redox-regulated or redox-signalled pathways discussed in the text for different trypanosomatid species and life stages. Different stimuli/signals from the extracellular medium trigger intracellular signals in the parasite that actionize different cellular responses. (A) Biochemical and biophysical stimuli from the surrounding environment induce an intracellular oxidizing environment with increased level of glutathionylated proteins and low molecular weight thiols (glutathione and trypanothione disulphide: GSSG and TS_2) that can further act as transducer/signalling molecules. (B) In the cytosol, glutaredoxin (Grx1) (de)glutathionylates target proteins and maintains the glutathione homeostasis. Peroxiredoxins (Prx) neutralize different oxidants (ONOOH and H_2O_2) and, mainly in their oxidized forms, act as holdases protecting proteins from stress-induced unfolding. (C) The synthesis of deoxynucleotides and DNA replication is regulated by the cellular reducing power via the couple trypanothione/tryparedoxin (TXN). (D) Changes in the AMP/ATP ratio activate mitochondrial metabolism with formation of H_2O_2 that leads to phosphorylation (*) of AMP kinase (AMPK) that will promote differentiation of bloodstream *T. brucei* to a quiescent stage. (E) In the mitochondrial intermembrane space (IMS), a glutaredoxin (Grx2) contributes to thiol-redox homeostasis, by likely acting on a component of the respiratory complex because its downregulation repressed mitochondrial energy metabolism and increased metabolic utilization of glucose in the insect stage of *T. brucei*. (F) In the mitochondrial respiratory chain, electron leakage to oxygen leads to formation of superoxide anion ($O_2^{\cdot-}$), which is rapidly converted into H_2O_2 by iron-superoxide dismutase (FeSOD). FeSOD and H_2O_2 have been shown to promote differentiation of *Leishmania* parasites to highly infective forms. During differentiation to mammal infective stages occurring in the insect, an alternative oxidase (AOX) is activated and leads to production of H_2O_2, which acts as a signalling molecule. (G) In the mitochondrial matrix (MM), cytochrome b5 reductase (CBRL) and different peroxidases (Prx/PX) may compete with the UMSBP for the reducing power of TXN. Upon thiol oxidation, UMSBP is unable to bind and initiate the replication of the mitochondrial DNA in *Leishmania* and *T. brucei*. (H) Mitochondrial peroxiredoxins (Prx) remove and are signalled by H_2O_2, and probably regulate the activity of the translocase import membrane (TIM) complex affecting the mitochondrial energy metabolism of the insect stage of *T. brucei*. (I) Different mitochondrial proteins (*Leishmania infantum* peroxiredoxin, LiPrxm; *T. brucei* thioredoxin 2, Trx2; and the bacteria homologue of heat shock protein 33, TrypOX) present redox-regulated chaperone activity.

single RNA binding protein 6 (RBP6; Kolev et al. 2012) and was characterized by an immediate redirection of electron flow from the cytochrome-mediated pathway to an alternative oxidase (AOX), and by upregulation of proline consumption, complex II activity, and certain TCA cycle enzymes. In consequence, mitochondrial membrane was hyperpolarized, and superoxide and H_2O_2 levels

increased (Figure 3.3). Interestingly, H_2O_2 appeared to act as signalling molecule dictating developmental progression because ectopic expression of catalase halted the in vitro–induced differentiation (Doleželová et al. 2020).

A redox signalling event appears to be behind the activation of an AMP-activated kinase (AMPK) that triggers the differentiation from a proliferative to a quiescent stage in bloodstream *T. brucei* (Barquilla et al. 2012; Saldivia et al. 2016). In the proliferative stage of bloodstream parasites, the production of energy by the mitochondrion is repressed, whereas it is fully functional in the short-stumpy form (quiescent form) allowing the pathogen to metabolize the nutrients available in the insect host (Timms et al. 2002). Concomitant with an increase in the AMP/ATP ratio, mitochondrial-derived oxidants have been shown to activate AMPK phosphorylation (Saldivia et al. 2016; Figure 3.3). Interestingly, time-course treatment with H_2O_2 recapitulated the effect observed under physiological conditions and complementary experiments suggested that AMPK is the sensor element reacting directly with H_2O_2. However, the fact that TXN and different TXN-dependent peroxidases co-immunoprecipitated with the AMPK core complex (Saldivia et al. 2016) indicates that these oxidoreductases may contribute to the signalling as sensor/transducer elements.

3.5 REDOX SIGNALLING LINKED TO THERMOTOLERANCE AND APOPTOSIS

Temperature changes are major physical signals that commit parasites to differentiation or apoptosis. For instance, a temperature drop of almost 15°C occurring upon ingestion of a blood meal by the tsetse fly primes non-dividing parasites to cis-aconitate/citrate-mediated differentiation (Engstler and Boshart 2004). In trypanosomatids, heat shock inhibits transcription and translation (Comini et al. 2012; Názer et al. 2012) and, in some species, it also induces apoptosis within the mammalian host, which is necessary to control cell population and minimize immune reaction (Moreira et al. 1996; De Souza et al. 2003).

The mechanisms involved in sensing and transducing the changes in temperature into biochemical signals are unknown. However, a potential link between heat shock and a thiol-disulphide perturbation that may trigger redox signalling emerged upon the finding that a slight temperature shift (37°C to 39°C) induced oxidization of $T(SH)_2$ and increased the level of S-glutathionylated proteins in bloodstream *T. brucei* (Ebersoll et al. 2018). Taking into consideration that biosynthesis of low molecular thiols is not impaired under heat shock, a shortfall or inactivation of any component from the peroxide-metabolizing pathway or, eventually, an increase in the production of oxidants may account for the observed thiol oxidation.

As discussed previously, Grx mediate protein (de)glutathionylation and have been shown to play distinct roles in the sensitivity of various cells towards oxidative stressors (Allen and Mieyal 2012; Fernandes and Holmgren 2004; Mailloux and Treberg 2016). Interestingly, deletion or downregulation of both Grx encoded by *T. brucei* conferred thermotolerance in a stage-specific fashion (Musunda et al. 2015; Ebersoll et al. 2018). In the case of the cytosolic Grx1, the protein was dispensable for the insect stage of the parasite, even under several stressful growth conditions. For bloodstream parasites, Grx1 deletion was accompanied with an impaired deglutathionylation capacity and an increased resistance to long-term growth at 2°C above the physiological culture temperature (37°C). Thus, Grx1 appears to be part of a signalling mechanism that modulates cell response to heat and involves protein (de)glutathionylation as one or more regulatory steps (Ebersoll et al. 2018; Figure 3.3). It is worth recalling that trypanosomatids are endowed with a manifold of heat-shock proteins (Requena et al. 2015), and proteomic studies performed in *T. brucei* exposed to different oxidants revealed the S-thiolation of several heat-shock proteins (Ulrich et al. 2017). Notably, the HSP70 from diverse organisms (Fratelli et al. 2002; Fratelli et al. 2003; Michelet et al. 2008) has been shown to be susceptible to specific glutathionylation. In fact, glutathionylation of HSP70 from rat hepatocytes activates its chaperon activity, which was proposed to be an adaptation to stress conditions (e.g., oxidizing milieu and low ATP concentration; Hoppe et al. 2004). Thus, it is tempting to speculate that some chaperones from trypanosomatids may be effectors of Grx in this signalling pathway.

Grx2 is expressed and localizes in the mitochondrial intermembrane space of both life stages of *T. brucei* and, under normal culture conditions, proved essential only for the insect stage (Ceylan et al. 2010; Ebersoll et al. 2018). As observed for Grx1, Grx2 did not play a crucial role in protecting infective trypanosomes against exogenous oxidative stressors. A striking molecular difference is that the indispensable function of Grx2 in the insect stage requires only the N-terminal cysteine of the active site, whereas in the bloodstream form both redox active cysteines are necessary to attenuate cell growth at 39°C. Because deglutathionylation by Grx relies only on a monothiol catalytic mechanism (Deponte 2013), it is then very likely that

Grx2 function in the insect form of the parasite is to deglutathionylate target proteins. Eventually Grx2 may contribute to GSH/GSSG homeostasis in the intermembrane space as proposed for the cytosolic Grx1 (Manta et al. 2019; Figure 3.3). Grx2 appears to participate in a life-cycle checkpoint during the transition from the insect to the mammalian stage because depletion of the protein led to growth arrest and a moderate rewiring of their energetic metabolism towards a more efficient use of glucose. Thus, physiological conditions affecting Grx2 redox activity may trigger this signalling event. The molecular target of Grx2 remains unknown, but the authors speculated that it might be a component of the respiratory chain (Ebersoll et al. 2018). The fact that bloodstream parasites lacking the single or both Grx displayed an identical phenotype suggests that these two proteins work concertedly within a redox signalling process that represses parasite growth at high temperature.

T. cruzi possesses a single Grx that is closely related to T. brucei Grx2 and displays deglutathionylation activity (Márquez et al. 2014). Overexpression of the protein in the insect form of the parasite yielded cells resistant to exogenous H_2O_2 yet induced apoptotic phenotypes as well. The molecular events triggering apoptosis were not investigated; however, under normal culture conditions of the insect form of T. cruzi, some apoptosis-related proteins were identified as partners of Prx (Peloso et al. 2016). The single mitochondrion of trypanosomatids is recognized to be pivotal in orchestrating cellular apoptosis upon specific biophysical signals and via a process that involves the formation of superoxide (de Castro et al. 2017; Figarella et al. 2005; Figarella et al. 2006; Mehta and Shaha 2006; Piacenza et al. 2007). In eukaryotes, the activity of several components of apoptosis-signalling pathways is susceptible to glutathionylation and regulation by Grxs (Allen and Mieyal 2012). Probably, Grx plays a similar regulatory role in T. cruzi, where differentiation to the infective stages has been associated with massive insect-stage parasite death via apoptosis (Ameisen et al. 1995).

In the bloodstream form of T. brucei, oxidants were implicated as intermediates in the signalling pathway driving apoptosis induced by prostaglandins (D2 and J series; Figarella et al. 2005; Figarella et al. 2006). Prostaglandins are produced mainly by the non-replicative form of the parasite that infects mammals and were proposed to play a key role as regulators of parasite population in the host.

Again, the components of this signalling pathway and the contribution of the parasite thiol-redox metabolism are yet unknown.

3.6 REDOX REGULATION OF HOLDASE/ CHAPERONE ACTIVITY

The stress response involves the expression of molecular chaperones (Shonhai et al. 2011) that assist folding or assembly without being part of its final structure (Kim et al. 2013; Jakob et al. 1999). Chaperones can be divided into ATP-dependent and ATP-independent, the so-called holdases (Folgueira and Requena 2007; Kim et al. 2013). Oxidative stress produces protein damage and unfolding, which in many organisms is counteracted by the holdase activity of typical 2-Cys Prx (Perkins et al. 2015; Jang et al. 2004). This dual (peroxidase and holdase) activity is interdependent (Wood et al. 2002) because oxidized forms of Prx, which have mainly dimeric conformations, do not exhibit holdase activity, whereas their reduced or overoxidized forms, which form decamers and/or higher molecular weight oligomers, display holdase activity (Rhee and Woo 2011).

The more extensively characterized holdase from trypanosomatids is the mitochondrial Prx from L. infantum (LiPrx1m). LiPrx1m is constitutively expressed throughout the life cycle of L. infantum (Teixeira et al. 2015), and gene-deletion proved detrimental for the survival of the parasite in vitro and in vivo. In the case of the insect promastigotes, the mutant parasites were highly sensitive to a heat shock that mimics the shift occurring during the invasion of the vertebrate host (25°C to 37°C). The phenotype was associated with a higher level of protein aggregation at 37°C. The LiPrx1m null mutants were also unable to thrive in mice competent or defective in oxidative phagocyte response, except if an ectopic copy of LiPrx1m lacking the peroxidatic cysteine (Cp) is co-expressed. This clearly indicated that the peroxidase activity of LiPrx1m is dispensable for parasite survival (Castro et al. 2011). Nonetheless, only reduced LiPRXm displays chaperone activity, whereas its oxidized form loses the ability to prevent aggregation of client proteins (Figure 3.3). Interestingly, the chaperone activity was stimulated by an increase in temperature. The authors concluded that LiPrx1m functions as a mitochondrial chaperone reservoir that allows the parasite to deal successfully with protein unfolding conditions during the transition from insect to the mammalian hosts and to generate viable parasites capable of perpetuating infection (Teixeira et al. 2015).

In *T. cruzi*, the peroxidase activity and the role in infection of the cytosolic (TcCPX) and mitochondrial (TcMPX) Prx have been characterized previously (Alvarez et al. 2011; Piacenza et al. 2008; Piñeyro, Arcari et al. 2011). Similar to most Prx that act as holdases, TcCPX has a toroid-shaped decameric structure in its active and reduced form (Piñeyro et al. 2005). In vitro, TcCPX slowed down the aggregation of malate dehydrogenase. In contrast to LiPrx1m, all redox species of TcCPX have holdase activity, with the oxidized or overoxidized forms displaying higher efficiency than the reduced form in preventing protein aggregation (Figure 3.3). This difference may relate to the quaternary structure adopted by (over)oxidized TcCPX, which aggregates into high molecular weight complexes (Piñeyro et al. 2019). It is interesting to note that unlike many Prx, the oxidation of TcCXP does not cause the complete transition from decamer to dimer. It is proposed that the overoxidation of Cp is a major regulation of the transition from peroxidase activity to holdase activity (Conway and Lee 2015), given that overoxidation prevents the formation of dimers and inhibits peroxidase activity. In the case of trypanosomatids Prx, overoxidation of Cp is not necessary for the acquisition of holdase activity because Cp mutants of both LiPrx1m and TcCPX retained full holdase activity (Piñeyro et al. 2019; Teixeira et al. 2015). Future studies should investigate the putative chaperone activity of TcCPX in vivo because protein overoxidation was observed upon treatment with peroxynitrite (Piacenza et al. 2008) and Prx expression was upregulated in the insect stage subjected to heat shock (Pérez-Morales et al. 2012).

In *T. brucei*, the cytosolic (TbcPrx) and mitochondrial Prx (TbmPrx) also appear to play different functions. TbcPrx is essential for both life stages of the parasites by contributing to maintain redox homeostasis (Wilkinson et al. 2003). In contrast, TbmPrx has a redox-independent activity that is necessary for maintaining proliferation, normal morphology, and mitochondrial function in the insect stage (Bogacz et al. 2020) In fact, TbmPrx deficiency causes a growth defect that was not associated with an impaired redox state or oxidant-detoxifying capacity of non-infective parasites. For bloodstream trypanosomes, TbmPrx was dispensable unless the parasites were incubated at 39°C,

mimicking fever or infection of animals having higher body temperature (i.e., cows). At variance with the phenomena observed in *Leishmania*, this heat shock does not result in protein aggregation in *T. brucei* defective in TbmPrx but has been shown to induce a more oxidizing redox milieu in bloodstream cells (Ebersoll et al. 2018). The authors speculated that TbmPrx is assisting the import of proteins to the mitochondria by interacting or regulating the activity of the mitochondrial translocase of the inner membrane (TIM) complex (Smith et al. 2018; Figure 3.3). This function is essential for maintaining a robust mitochondrial metabolism in the insect but not in the bloodstream stage. However, the differentiation to the quiescent short-stumpy form entails the activation of the mitochondrial metabolism with production of superoxide and H_2O_2 (Saldivia et al. 2016; Doležalová et al. 2020), and the more oxidizing environment may act as a differentiation signal to activate mitochondrial metabolism. Such a pathway would not be operative in bloodstream parasites lacking TbmPrx, which may explain the incapacity of the mutant cell line to withstand heat-shock. These hypotheses as well as the potential activity of TbmPrx as a molecular chaperone deserve further investigation.

More recently, a mitochondrial thioredoxin-like protein (Trx2) with holdase activity has been characterized in *T. brucei* (Currier et al. 2019). In contrast to the dispensable cytosolic Trx (Schmidt et al. 2002), Trx2 is essential for the proliferation, resistance to prolonged heat stress, and infectivity of *T. brucei*. In vitro, Trx2 prevented the aggregation of client proteins by maintaining and transferring them in a folding-competent state to more specialized refolding systems. Resembling the behaviour of LiPrx1m (Currier et al. 2019), the chaperone/holdase activity of Trx2 was redox regulated, where the reduced form but not the oxidized one was active (Figure 3.3). In addition, an increase in temperature induced a conformational change in Trx2 that activates its holdase function. The deleterious phenotype caused by Trx2 downregulation was exacerbated in the insect stage, suggesting a predominant role of the protein in this developmental stage. Trx2 is not involved in the oxidative stress response, and in agreement with the in vitro behaviour, a wild-type or cysteine-less Trx2 mutant was able to rescue the growth of Trx2-KO parasites. The major biological role of the protein is likely to protect mitochondrial proteins against aggregation. The redox signal and/or transducer elements that inactivate (oxidation) or restore the active (reduced) Trx2 are unknown.

T. brucei also expresses an homologue of the bacteria redox chaperone HSP33 (Graf et al. 2004; Ilbert et al. 2007), which was named TrypOX (Aramin et al. 2020). Similar to HSP33, TrypOX is redox regulated and shows capacity to bind and deliver client proteins to folding systems. Although the in vivo signals that trigger TrypOX activation were not characterized, it was shown that oxidation of cysteines in the redox domain activates the chaperone activity through conformational changes that expose protein-hydrophobic regions that bind client proteins (Figure 3.3). Silencing TrypOX in T. brucei resulted in growth inhibition upon mild oxidative stress induced by H_2O_2 in a temperature-independent manner. This phenotype was ascribed to a TrypOX-dependent decrease in the abundance of enzymes from the antioxidant systems that may impair peroxide detoxification (Aramin et al. 2020).

3.7 CONCLUDING REMARKS

Overwhelming evidence highlights the importance of thiol-disulphide systems for the ability of trypanosomatids to cope with the different environments they encounter throughout their life cycle; in particular for maintaining redox homeostasis. Recently, redox-regulatory functions have emerged for different components of this system, which are proposed to be involved in controlling the activity of certain pathways and in transducing signals triggered by endogenous or exogenous oxidant species. In this respect, efforts have been made to establish whether modifications by low molecular weight thiols play a posttranslational regulatory role on protein function, which led to the identification of proteins that undergo thiolation in trypanosomatids. Further work is required to fully explore the role of this modification in parasites' life cycle. Moreover, several oxidoreductases and peroxidases of trypanosomatids have been reported to function in signal transduction or thiol-disulphide exchanges that provide kinetic redox control to the functions of several proteins. Although host-derived physiological oxidants are known to be involved in controlling infection by these pathogens, it is only recently these species have been found to play an essential role in the development and differentiation of the parasites, by triggering the (de)activation of signalling pathways.

Interestingly, novel functions as holdase were disclosed for proteins related to the thiol-redox metabolism. The holdase activity of thioredoxin-like peroxiredoxins and small heat-shock proteins appears to be redox regulated and contributes to protect parasite proteins against aggregation under stress conditions (heat shock, oxidizing environment). Yet the partners and precise functions for the holdase activity during differentiation remain unknown and should be investigated.

Undoubtedly, several members of the thiol-disulphide metabolism have specific regulatory functions beyond maintaining thiol-redox homeostasis and its involvement in signalling pathways occurring during differentiation and in the early phase of the adaptation to the hosts may be even more important than initially thought.

NOTE

1 Peroxynitrite refers to the peroxynitrite anion $(ONOO^-)$ and its conjugated acid peroxynitrous acid (pKa = 6.8). The species that reacts with thiols is ONOOH.

REFERENCES

Adak, Subrata, and Alok K. Datta. 2005. "*Leishmania major* Encodes an Unusual Peroxidase That Is a Close Homologue of Plant Ascorbate Peroxidase: A Novel Role of the Transmembrane Domain." *Biochemical Journal* 390 (2): 465–474. https://doi.org/10.1042/BJ20050311.

Allen, Erin M. G., and John J. Mieyal. 2012. "Protein-Thiol Oxidation and Cell Death: Regulatory Role of Glutaredoxins." *Antioxidants and Redox Signaling*. https://doi.org/10.1089/ars.2012.4644.

Alvarez, María Noel, Gonzalo Peluffo, Lucía Piacenza, and Rafael Radi. 2011. "Intraphagosomal Peroxynitrite as a Macrophage-Derived Cytotoxin against Internalized *Trypanosoma cruzi*: Consequences for Oxidative Killing and Role of Microbial Peroxiredoxins in Infectivity." *Journal of Biological Chemistry*. https://doi.org/10.1074/jbc.M110.167247.

Ameisen, Jean Claude, Thierry Idziorek, Odile Billaut-Mulot, Marc Loyens, Jean Pierre Tissier, Arnaud Potentier, and Ali Ouaissi. 1995. "Apoptosis in a Unicellular Eukaryote (*Trypanosoma cruzi*): Implications for the Evolutionary Origin and Role of Programmed Cell Death in the Control of Cell Proliferation, Differentiation and Survival." *Cell Death and Differentiation*. https://doi.org/10.1016/0169-4758(96)80652-1.

Aramin, Samar, Rosi Fassler, Vaibhav Chikne, Mor Goldenberg, Tal Arian, Liat Kolet Eliaz, Oded Rimon, Oren Ram, Shulamit Michaeli, and Dana Reichmann. 2020. "TrypOx, a Novel Eukaryotic Homolog of the Redox-Regulated Chaperone Hsp33 in *Trypanosoma brucei*." *Frontiers in Microbiology*. https://doi.org/10.3389/fmicb.2020.01844.

Arias, Diego G., María Dolores Piñeyro, Alberto A. Iglesias, Sergio A. Guerrero, and Carlos Robello. 2015. "Molecular Characterization and Interactome Analysis of *Trypanosoma cruzi* Tryparedoxin II." *Journal of Proteomics* 120: 95–104. https://doi.org/10.1016/j.jprot.2015.03.001.

Ariyanayagam, Mark R., and Alan H. Fairlamb. 2001. "Ovothiol and Trypanothione as Antioxidants in Trypanosomatids." *Molecular and Biochemical Parasitology*. https://doi.org/10.1016/S0166-6851(01)00285-7.

Barford, David. 2004. "The Role of Cysteine Residues as Redox-Sensitive Regulatory Switches." *Current Opinion in Structural Biology*. https://doi.org/10.1016/j.sbi.2004.09.012.

Barquilla, Antonio, Manuel Saldivia, Rosario Diaz, Jean Mathieu Bart, Isabel Vidal, Enrique Calvo, Michael N. Hall, and Miguel Navarro. 2012. "Third Target of Rapamycin Complex Negatively Regulates Development of Quiescence in *Trypanosoma brucei*." *Proceedings of the National Academy of Sciences of the United States of America*. https://doi.org/10.1073/pnas.1210465109.

Berná, Luisa, Maria Laura Chiribao, Gonzalo Greif, Matias Rodriguez, Fernando Alvarez-Valin, and Carlos Robello. 2017. "Transcriptomic Analysis Reveals Metabolic Switches and Surface Remodeling as Key Processes for Stage Transition in *Trypanosoma cruzi*." *PeerJ*. https://doi.org/10.7717/peerj.3017.

Berriman, Matthew, Elodie Ghedin, Christiane Hertz-Fowler, Gaëlle Blandin, Hubert Renauld, Daniella C. Bartholomeu, Nicola J. Lennard, et al.. 2005. "The Genome of the African Trypanosome *Trypanosoma brucei*." *Science*. https://doi.org/10.1126/science.1112642.

Bhattacharya, Arijit, Arunima Biswas, and Pijush K. Das. 2009. "Role of a Differentially Expressed CAMP Phosphodiesterase in Regulating the Induction of Resistance against Oxidative Damage in *Leishmania donovani*." *Free Radical Biology and Medicine*. https://doi.org/10.1016/j.freeradbiomed.2009.08.025.

Bogacz, Marta, Natalie Dirdjaja, Benedikt Wimmer, Carina Habich, and R. Luise Krauth-Siegel. 2020. "The Mitochondrial Peroxiredoxin Displays Distinct Roles in Different Developmental Stages of African Trypanosomes." *Redox Biology* 34 (April): 101547. https://doi.org/10.1016/j.redox.2020.101547.

Bogacz, Marta, and R. Luise Krauth-Siegel. 2018. "Tryparedoxin Peroxidase-Deficiency Commits Trypanosomes to Ferroptosis-Type Cell Death." *ELife*. https://doi.org/10.7554/eLife.37503.

Boveris, A., H. Sies, E. E. Martino, R. Docampo, J. F. Turrens, and A. O. Stoppani. 1980. "Deficient Metabolic Utilization of Hydrogen Peroxide in *Trypanosoma cruzi*." *The Biochemical Journal*. https://doi.org/10.1042/bj1880643.

Boveris, A., and A. O. M. Stoppani. 1977. "Hydrogen Peroxide Generation in *Trypanosoma cruzi*." *Experientia*. https://doi.org/10.1007/BF01920148.

Brigelius-Flohé, Regina, and Leopold Flohé. 2011. "Basic Principles and Emerging Concepts in the Redox Control of Transcription Factors." *Antioxidants and Redox Signaling*. https://doi.org/10.1089/ars.2010.3534.

Castro, Helena, Heike Budde, Leopold Flohé, Birgit Hofmann, Heinrich Lünsdorf, Joseph Wissing, and Ana M. Toms. 2002. "Specificity and Kinetics of a Mitochondrial Peroxiredoxin of *Leishmania infantum*." *Free Radical Biology and Medicine*. https://doi.org/10.1016/S0891-5849(02)01088-2.

Castro, Helena, Filipa Teixeira, Susana Romao, Mariana Santos, Tânia Cruz, Manuela Flórido, Rui Appelberg, Pedro Oliveira, Frederico Ferreira-da-Silva, and Ana M. Tomás. 2011. "*Leishmania* Mitochondrial Peroxiredoxin Plays a Crucial Peroxidase-Unrelated Role during Infection: Insight into Its Novel Chaperone Activity." *PLoS Pathogens*. https://doi.org/10.1371/journal.ppat.1002325.

Castro, Helena, and Ana M. Tomás. 2008. "Peroxidases of Trypanosomatids." *Antioxidants and Redox Signaling*. https://doi.org/10.1089/ars.2008.2050.

Cazzulo, Juan José. 1984. "Protein and Amino Acid Catabolism in *Trypanosoma cruzi*." *Comparative Biochemistry and Physiology*. https://doi.org/10.1016/0305-0491(84)90381-X.

Cazzulo, Juan José. 1992. "Aerobic Fermentation of Glucose by Trypanosomatids." *The FASEB Journal*. https://doi.org/10.1096/fasebj.6.13.1397837.

Ceylan, Sevgi, Vera Seidel, Nicole Ziebart, Carsten Berndt, Natalie Dirdjaja, and R. Luise Krauth-Siegel. 2010. "The Dithiol Glutaredoxins of African Trypanosomes Have Distinct Roles and Are Closely Linked to the Unique Trypanothione Metabolism." *Journal of Biological Chemistry*. https://doi.org/10.1074/jbc.M110.165860.

Christiano, Romain, Nikolay G. Kolev, Huafang Shi, Elisabetta Ullu, Tobias C. Walther, and Christian Tschudi. 2017. "The Proteome and Transcriptome of the Infectious Metacyclic Form of *Trypanosoma* brucei Define Quiescent Cells Primed for Mammalian Invasion." *Molecular Microbiology*. https://doi.org/10.1111/mmi.13754.

Clayton, Christine. 2019. "Regulation of Gene Expression in Trypanosomatids: Living with Polycistronic Transcription." *Open Biology*. https://doi.org/10.1098/rsob.190072.

Comini, Marcelo A., Sergio A. Guerrero, Simon Haile, Ulrich Menge, Heinrich Lünsdorf, and Leopold Flohé. 2004. "Valdiation of *Trypanosoma brucei* Trypanothione Synthetase as Drug Target." *Free Radical Biology and Medicine*. https://doi.org/10.1016/j.freeradbiomed.2004.02.008.

Comini, Marcelo A., R. Luise Krauth-Siegel, and Massimo Bellanda. 2013. "Mono- and Dithiol Glutaredoxins in the Trypanothione-Based Redox Metabolism of Pathogenic Trypanosomes." *Antioxidants and Redox Signaling*. https://doi.org/10.1089/ars.2012.4932.

Comini, Marcelo A., R. Luise Krauth-Siegel, and Leopold Flohé. 2007. "Depletion of the Thioredoxin Homologue Tryparedoxin Impairs Antioxidative Defence in African Trypanosomes." *Biochemical Journal*. https://doi.org/10.1042/BJ20061341.

Comini, Marcelo A., Andrea Medeiros, and Bruno Manta. 2012. "Stress Response in the Infective Stage of *Trypanosoma brucei*." Edited by Jose M. Requena. *Stress Response in Microbiology*, Caister Academic Press, UK.

Conway, Myra E., and Christopher Lee. 2015. "The Redox Switch That Regulates Molecular Chaperones." *Biomolecular Concepts*. https://doi.org/10.1515/bmc-2015-0015.

Currier, Rachel B., Kathrin Ulrich, Alejandro E. Leroux, Natalie Dirdjaja, Matías Deambrosi, Mariana Bonilla, Yasar Luqman Ahmed, et al.. 2019. "An Essential Thioredoxin-Type Protein of *Trypanosoma brucei* Acts as Redox-Regulated Mitochondrial Chaperone." *PLoS Pathogens*. https://doi.org/10.1371/journal.ppat.1008065.

Deponte, Marcel. 2013. "Glutathione Catalysis and the Reaction Mechanisms of Glutathione-Dependent Enzymes." *Biochimica et Biophysica Acta—General Subjects*. https://doi.org/10.1016/j.bbagen.2012.09.018.

de Almeida Nogueira, Natália Pereira de, Cintia Fernandes de Souza, Francis Monique de Souza Saraiva, Pedro Elias Sultano, Sergio Ranto Dalmau, Roberta Eitler Bruno, Renata de Lima Sales Gonçalves, et al.. 2011. "Heme-Induced ROS in *Trypanosoma cruzi* Activates Camkii-Like That Triggers Epimastigote Proliferation. One Helpful Effect of ROS." *PLoS ONE*. https://doi.org/10.1371/journal.pone.0025935.

de Castro, Emanuella, Thamile Luciane Reus, Alessandra Melo de Aguiar, Andrea Rodrigues Ávila, and Tatiana de Arruda Campos Brasil de Souza. 2017. "Procaspase-Activating Compound-1 Induces Apoptosis in *Trypanosoma cruzi*." *Apoptosis*. https://doi.org/10.1007/s10495-017-1428-5.

Diaz-Albiter, Hector, Mauricio R. V. Sant'Anna, Fernando A. Genta, and Rod J. Dillon. 2012. "Reactive Oxygen Species-Mediated Immunity against *Leishmania mexicana* and *Serratia marcescens* in the Phlebotomine Sand Fly *Lutzomyia longipalpis*." *Journal of Biological Chemistry*. https://doi.org/10.1074/jbc.M112.376095.

Diechtierow, Michael, and R. Luise Krauth-Siegel. 2011. "A Tryparedoxin-Dependent Peroxidase Protects African Trypanosomes from Membrane Damage." *Free Radical Biology and Medicine*. https://doi.org/10.1016/j.freeradbiomed.2011.05.014.

Doleželová, Eva, Michaela Kunzová, Mario Dejung, Michal Levin, Brian Panicucci, Clément Regnault, Christian J Janzen, Michael P. Barrett, Falk Butter, and Alena Zíková. 2020. "Cell-Based and Multi-Omics Profiling Reveals Dynamic Metabolic Repurposing of Mitochondria to Drive Developmental Progression of *Trypanosoma brucei*." *PLoS Biology*. https://doi.org/10.1371/journal.pbio.3000741.

Dormeyer, Matthias, Nina Reckenfelderbäumer, Heike Lüdemann, and R. Luise Krauth-Siegel. 2001. "Trypanothione-Dependent Synthesis of Deoxyribonucleotides by *Trypanosoma brucei* Ribonucleotide Reductase." *Journal of Biological Chemistry*. https://doi.org/10.1074/jbc.M010352200.

Ebersoll, Samantha, Blessing Musunda, Torsten Schmenger, Natalie Dirdjaja, Mariana Bonilla, Bruno Manta, Kathrin Ulrich, Marcelo A. Comini, and R. Luise Krauth-Siegel. 2018. "A Glutaredoxin in the Mitochondrial Intermembrane Space Has Stage-Specific Functions in the Thermo-Tolerance and Proliferation of African Trypanosomes." *Redox Biology*. https://doi.org/10.1016/j.redox.2018.01.011.

El-Sayed, Najib M., Peter J. Myler, Daniella C. Bartholomeu, Daniel Nilsson, Gautam Aggarwal, Anh Nhi Tran, Elodie Ghedin, et al.. 2005. "The Genome Sequence of *Trypanosoma cruzi*, Etiologic Agent of Chagas Disease." *Science*. https://doi.org/10.1126/science.1112631.

Engstler, Markus, and Michael Boshart. 2004. "Cold Shock and Regulation of Surface Protein Trafficking Convey Sensitization to Inducers of

Stage Differentiation in *Trypanosoma brucei*." *Genes and Development*. https://doi.org/10.1101/gad.323404.

Fairlamb, Alan H., and Anthony Cerami. 1985. "Identification of a Novel, Thiol-Containing Co-Factor Essential for Glutathione Reductase Enzyme Activity in Trypanosomatids." *Molecular and Biochemical Parasitology*. https://doi.org/10.1016/0166-6851(85)90037-4.

Fernandes, Aristi Potamitou, and Arne Holmgren. 2004. "Glutaredoxins: Glutathione-Dependent Redox Enzymes with Functions Far Beyond a Simple Thioredoxin Backup System." *Antioxidants and Redox Signaling*. https://doi.org/10.1089/152308604771978354.

Figarella, K., M. Rawer, N. L. Uzcategui, B. K. Kubata, K. Lauber, F. Madeo, S. Wesselborg, and Michael Duszenko. 2005. "Prostaglandin D2 Induces Programmed Cell Death in *Trypanosoma brucei* Bloodstream Form." *Cell Death and Differentiation*. https://doi.org/10.1038/sj.cdd.4401564.

Figarella, K., N. L. Uzcategui, A. Beck, C. Schoenfeld, B. K. Kubata, F. Lang, and Michael Duszenko. 2006. "Prostaglandin-Induced Programmed Cell Death in *Trypanosoma brucei* Involves Oxidative Stress." *Cell Death and Differentiation*. https://doi.org/10.1038/sj.cdd.4401862.

Folgueira, Cristina, and Jose M. Requena. 2007. "A Postgenomic View of the Heat Shock Proteins in Kinetoplastids." *FEMS Microbiology Reviews*. https://doi.org/10.1111/j.1574-6976.2007.00069.x.

Fratelli, Maddalena, Hans Demol, Magda Puype, Simona Casagrande, Ivano Eberini, Mario Salmona, Valentina Bonetto, et al.. 2002. "Identification by Redox Proteomics of Glutathionylated Proteins in Oxidatively Stressed Human T Lymphocytes." *Proceedings of the National Academy of Sciences of the United States of America*. https://doi.org/10.1073/pnas.052592699.

Fratelli, Maddalena, Hans Demol, Magda Puype, Simona Casagrande, Pia Villa, Ivano Eberini, Joel Vandekerckhove, Elisabetta Gianazza, and Pietro Ghezzi. 2003. "Identification of Proteins Undergoing Glutathionylation in Oxidatively Stressed Hepatocytes and Hepatoma Cells." *Proteomics*. https://doi.org/10.1002/pmic.200300436.

Giordani, Federica, Liam J. Morrison, Tim G. Rowan, Harry P. De Koning, and Michael P. Barrett. 2016. "The Animal Trypanosomiases and Their Chemotherapy: A Review." *Parasitology*. https://doi.org/10.1017/S0031182016001268.

Graf, Paul C. F., Maria Martinez-Yamout, Stephen VanHaerents, Hauke Lilie, H. Jane Dyson, and Ursula Jakob. 2004. "Activation of the Redox-Regulated Chaperone Hsp33 by Domain Unfolding." *Journal of Biological Chemistry*. https://doi.org/10.1074/jbc.M401764200.

Green, S. J., C. A. Nacy, and M. S. Meltzer. 1991. "Cytokine-Induced Synthesis of Nitrogen Oxides in Macrophages: A Protective Host Response to Leishmania and Other Intracellular Pathogens." *Journal of Leukocyte Biology*. https://doi.org/10.1002/jlb.50.1.93.

Gutscher, Marcus, Anne Laure Pauleau, Laurent Marty, Thorsten Brach, Guido H. Wabnitz, Yvonne Samstag, Andreas J. Meyer, and Tobias P. Dick. 2008. "Real-Time Imaging of the Intracellular Glutathione Redox Potential." *Nature Methods*. https://doi.org/10.1038/nmeth.1212.

Hannaert, Véronique, Frédéric Bringaud, Fred R. Opperdoes, and Paul A. M. Michels. 2003. "Evolution of Energy Metabolism and Its Compartmentation in Kinetoplastida." *Kinetoplastid Biology and Disease*. https://doi.org/10.1186/1475-9292-2-11.

Hao, Zhengrong, Irene Kasumba, and Serap Aksoy. 2003. "Proventriculus (Cardia) Plays a Crucial Role in Immunity in Tsetse Fly (Diptera: Glossinidiae)." *Insect Biochemistry and Molecular Biology*. https://doi.org/10.1016/j.ibmb.2003.07.001.

Hiller, Corinna, Amrei Nissen, Diego Benítez, Marcelo A. Comini, and R. Luise Krauth-Siegel. 2014. "Cytosolic Peroxidases Protect the Lysosome of Bloodstream African Trypanosomes from Iron-Mediated Membrane Damage." *PLoS Pathogens*. https://doi.org/10.1371/journal.ppat.1004075.

Hoppe, George, Yuh Cherng Chai, John W. Crabb, and Jonathan Sears. 2004. "Protein S-Glutathionylation in Retinal Pigment Epithelium Converts Heat Shock Protein 70 to an Active Chaperone." *Experimental Eye Research*. https://doi.org/10.1016/j.exer.2004.02.001.

Horáková, Eva, Drahomíra Faktorová, Natalia Kraeva, Binnypreet Kaur, Jan Van Den Abbeele, Vyacheslav Yurchenko, and Julius Lukeš. 2020. "Catalase Compromises the Development of the Insect and Mammalian Stages of *Trypanosoma brucei*." *FEBS Journal*. https://doi.org/10.1111/febs.15083.

Hugo, Martín, Alejandra Martínez, Madia Trujillo, Damián Estrada, Mauricio Mastrogiovanni, Edlaine Linares, Ohara Augusto, et al.. 2017. "Kinetics, Subcellular Localization, and Contribution to Parasite Virulence of a *Trypanosoma cruzi* Hybrid Type A Heme Peroxidase (TcAPx-CcP)." *Proceedings of the National Academy of Sciences of the United States of America*. https://doi.org/10.1073/pnas.1618611114.

Igoillo-Esteve, Mariana, and Juan José Cazzulo. 2006. "The Glucose-6-Phosphate Dehydrogenase from *Trypanosoma cruzi*: Its Role in the Defense of the Parasite against Oxidative Stress." *Molecular and Biochemical Parasitology.* https://doi.org/10.1016/j.molbiopara.2006.05.009.

Ilbert, Marianne, Janina Horst, Sebastian Ahrens, Jeannette Winter, Paul C. F. Graf, Hauke Lilie, and Ursula Jakob. 2007. "The Redox-Switch Domain of Hsp33 Functions as Dual Stress Sensor." *Nature Structural and Molecular Biology.* https://doi.org/10.1038/nsmb1244.

Ivens, Alasdair C., Christopher S. Peacock, Elizabeth A. Worthey, Lee Murphy, Gautam Aggarwal, Matthew Berriman, Ellen Sisk, et al.. 2005. "The Genome of the Kinetoplastid Parasite, *Leishmania major.*" *Science.* https://doi.org/10.1126/science.1112680.

Jakob, Ursula, Wilson Muse, Markus Eser, and James C. A. Bardwell. 1999. "Chaperone Activity with a Redox Switch." *Cell.* https://doi.org/10.1016/S0092-8674(00)80547-4.

Jang, Ho Hee, Kyun Oh Lee, Yong Hun Chi, Bae Gyo Jung, Soo Kwon Park, Jin Ho Park, Jung Ro Lee, et al.. 2004. "Two Enzymes in One: Two Yeast Peroxiredoxins Display Oxidative Stress-Dependent Switching from a Peroxidase to a Molecular Chaperone Function." *Cell.* https://doi.org/10.1016/j.cell.2004.05.002.

Jensen, Bryan C., Dhileep Sivam, Charles T. Kifer, Peter J. Myler, and Marilyn Parsons. 2009. "Widespread Variation in Transcript Abundance within and across Developmental Stages of *Trypanosoma brucei.*"*BMC Genomics.*https://doi.org/10.1186/1471-2164-10-482.

Jones, D. P. 2010. "Redox Sensing: Orthogonal Control in Cell Cycle and Apoptosis Signalling." In *Journal of Internal Medicine.* https://doi.org/10.1111/j.1365-2796.2010.02268.x.

Jones, Dean P., and Helmut Sies. 2015. "The Redox Code." *Antioxidants and Redox Signaling.* https://doi.org/10.1089/ars.2015.6247.

Kim, Yujin E., Mark S. Hipp, Andreas Bracher, Manajit Hayer-Hartl, and F. Ulrich Hartl. 2013. "Molecular Chaperone Functions in Protein Folding and Proteostasis." *Annual Review of Biochemistry.* https://doi.org/10.1146/annurev-biochem-060208-092442.

Kolev, Nikolay G., Kiantra Ramey-Butler, George A. M. Cross, Elisabetta Ullu, and Christian Tschudi. 2012. "Developmental Progression to Infectivity in *Trypanosoma brucei* Triggered by an RNA-Binding Protein." *Science.* https://doi.org/10.1126/science.1229641.

Kollien, Astrid H., Thorsten Grospietsch, Torsten Kleffmann, Irene Zerbst-Boroffka, and Günter A.

Schaub. 2001. "Ionic Composition of the Rectal Contents and Excreta of the Reduviid Bug *Triatoma infestans.*" *Journal of Insect Physiology.* https://doi.org/10.1016/S0022-1910(00)00170-0.

Krauth-Siegel, R. Luise, and Marcelo A. Comini. 2008. "Redox Control in Trypanosomatids, Parasitic Protozoa with Trypanothione-Based Thiol Metabolism." *Biochimica et Biophysica Acta—General Subjects.* https://doi.org/10.1016/j.bbagen.2008.03.006.

Krauth-Siegel, R. Luise, and Heide Schmidt. 2002. "Trypanothione and Tryparedoxin in Ribonucleotide Reduction." *Methods in Enzymology.* https://doi.org/10.1016/S0076-6879(02)47025-5.

Lüdemann, Heike, Matthias Dormeyer, Christian Sticherling, Dirk Stallmann, Hartmut Follmann, and R. Luise Krauth-Siegel. 1998. "*Trypanosoma brucei* Tryparedoxin, a Thioredoxin-like Protein in African Trypanosomes." *FEBS Letters.* https://doi.org/10.1016/S0014-5793(98)00793-5.

Lukeš, Julius, Anzhelika Butenko, Hassan Hashimi, Dmitri A. Maslov, Jan Votýpka, and Vyacheslav Yurchenko. 2018. "Trypanosomatids Are Much More than Just Trypanosomes: Clues from the Expanded Family Tree." *Trends in Parasitology.* https://doi.org/10.1016/j.pt.2018.03.002.

MacLeod, Ewan Thomas, Ian Maudlin, Alistair C. Darby, and Sue C. Welburn. 2007. "Antioxidants Promote Establishment of Trypanosome Infections in Tsetse." *Parasitology.* https://doi.org/10.1017/S0031182007002247.

Macleod, Ewan Thomas, Alistair C. Darby, Ian Maudlin, and Sue C. Welburn. 2007. "Factors Affecting Trypanosome Maturation in Tsetse Flies." *PLoS ONE.* https://doi.org/10.1371/journal.pone.0000239.

Mailloux, Ryan J., and Jason R. Treberg. 2016. "Protein S-Glutathionlyation Links Energy Metabolism to Redox Signaling in Mitochondria." *Redox Biology.* https://doi.org/10.1016/j.redox.2015.12.010.

Manta, Bruno, Mariana Bonilla, Lucía Fiestas, Mattia Sturlese, Gustavo Salinas, Massimo Bellanda, and Marcelo A. Comini. 2018. "Polyamine-Based Thiols in Trypanosomatids: Evolution, Protein Structural Adaptations, and Biological Functions." *Antioxidants and Redox Signaling.* https://doi.org/10.1089/ars.2017.7133.

Manta, Bruno, Matías N. Möller, Mariana Bonilla, Matías Deambrosi, Karin Grunberg, Massimo Bellanda, Marcelo A. Comini, and Gerardo Ferrer-Sueta. 2019. "Kinetic Studies Reveal a Key Role of a Redox-Active Glutaredoxin in the Evolution of the Thiol-Redox Metabolism of Trypanosomatid Parasites." *Journal of*

Biological Chemistry. https://doi.org/10.1074/jbc.RA118.006366.

Marchese, Letícia, Janaina De Freitas Nascimento, Flávia Silva Damasceno, Frédéric Bringaud, Paul A. M. Michels, and Ariel Mariano Silber. 2018. "The Uptake and Metabolism of Amino Acids, and Their Unique Role in the Biology of Pathogenic Trypanosomatids." Pathogens. https://doi.org/10.3390/pathogens7020036.

Márquez, Vanina E., Diego G. Arias, Maria L. Chiribao, Paula Faral-Tello, Carlos Robello, Alberto A. Iglesias, and Sergio A. Guerrero. 2014. "Redox Metabolism in Trypanosoma cruzi. Biochemical Characterization of Dithiol Glutaredoxin Dependent Cellular Pathways." Biochimie. https://doi.org/10.1016/j.biochi.2014.07.027.

Mehta, Ashish, and Chandrima Shaha. 2006. "Mechanism of Metalloid-Induced Death in Leishmania spp.: Role of Iron, Reactive Oxygen Species, Ca^{2+}, and Glutathione." Free Radical Biology and Medicine. https://doi.org/10.1016/j.freeradbiomed.2006.01.024.

Melchers, Johannes, Natalie Dirdjaja, Thomas Ruppert, and R. Luise Krauth-Siegel. 2007. "Glutathionylation of Trypanosomal Thiol Redox Proteins." Journal of Biological Chemistry. https://doi.org/10.1074/jbc.M608140200.

Mesías, Andrea C., Nisha J. Garg, and M. Paola Zago. 2019. "Redox Balance Keepers and Possible Cell Functions Managed by Redox Homeostasis in Trypanosoma cruzi." Frontiers in Cellular and Infection Microbiology. https://doi.org/10.3389/fcimb.2019.00435.

Michelet, Laure, Mirko Zaffagnini, Hélène Vanacker, Pierre Le Maréchal, Christophe Marchand, Michael Schroda, Stéphane D. Lemaire, and Paulette Decottignies. 2008. "In Vivo Targets of S-Thiolation in Chlamydomonas reinhardtii." Journal of Biological Chemistry. https://doi.org/10.1074/jbc.M802331200.

Milman, Neta, Shawn A. Motyka, Paul T. Englund, Derrick Robinson, and Joseph Shlomai. 2007. "Mitochondrial Origin-Binding Protein UMSBP Mediates DNA Replication and Segregation in Trypanosomes." Proceedings of the National Academy of Sciences of the United States of America. https://doi.org/10.1073/pnas.0706858104.

Mittra, Bidyottam, Mauro Cortez, Andrew Haydock, Gowthaman Ramasamy, Peter J. Myler, and Norma W. Andrews. 2013. "Iron Uptake Controls the Generation of Leishmania Infective Forms through Regulation of ROS Levels." Journal of Experimental Medicine. https://doi.org/10.1084/jem.20121368.

Moreira, Maria Elisabete C., Hernando A. Del Portillo, Regina V. Milder, Jose Mario F. Balanco, and Marcello A. Barcinski. 1996. "Heat Shock Induction of Apoptosis in Promastigotes of the Unicellular Organism Leishmania (Leishmania) amazonensis." Journal of Cellular Physiology. https://doi.org/10.1002/(SICI)1097-4652(199605)167:2<305::AID-JCP15>3.0.CO;2-6.

Motyka, Shawn A., Mark E. Drew, Gokben Yildirir, and Paul T. Englund. 2006. "Overexpression of a Cytochrome B5 Reductase-like Protein Causes Kinetoplast DNA Loss in Trypanosoma brucei." Journal of Biological Chemistry. https://doi.org/10.1074/jbc.M602880200.

Musunda, Blessing, Diego Benítez, Natalie Dirdjaja, Marcelo A. Comini, and R. Luise Krauth-Siegel. 2015. "Glutaredoxin-Deficiency Confers Bloodstream Trypanosoma brucei with Improved Thermotolerance." Molecular and Biochemical Parasitology. https://doi.org/10.1016/j.molbiopara.2016.02.001.

Názer, Ezequiel, Ramiro E. Verdún, and Daniel O. Sánchez. 2012. "Severe Heat Shock Induces Nucleolar Accumulation of MRNAs in Trypanosoma cruzi." PLoS ONE. https://doi.org/10.1371/journal.pone.0043715.

Nogoceke, Everson, Daniel U. Gommel, Michael Kieß, Henryk M. Kalisz, and Leopold Flohé. 1997. "A Unique Cascade of Oxidoreductases Catalyses Trypanothione-Mediated Peroxide Metabolism in Crithidia fasciculata." Biological Chemistry. https://doi.org/10.1515/bchm.1997.378.8.827.

Onn, Itay, Neta Milman-Shtepel, and Joseph Shlomai. 2004. "Redox Potential Regulates Binding of Universal Minicircle Sequence Binding Protein at the Kinetoplast DNA Replication Origin." Eukaryotic Cell. https://doi.org/10.1128/EC.3.2.277-287.2004.

Ortíz, Cecilia, Horacio Botti, Alejandro Buschiazzo, and Marcelo A. Comini. 2019. "Glucose-6-Phosphate Dehydrogenase from the Human Pathogen Trypanosoma cruzi Evolved Unique Structural Features to Support Efficient Product Formation." Journal of Molecular Biology. https://doi.org/10.1016/j.jmb.2019.03.023.

Østergaard, Henrik, Christine Tachibana, and Jakob R. Winther. 2004. "Monitoring Disulfide Bond Formation in the Eukaryotic Cytosol." Journal of Cell Biology. https://doi.org/10.1083/jcb.200402120.

Paulsen, Candice E., and Kate S. Carroll. 2010. "Orchestrating Redox Signaling Networks through Regulatory Cysteine Switches." ACS Chemical Biology. https://doi.org/10.1021/cb900258z.

Peloso, E. F., L. Dias, R. M. L. Queiroz, A. F. P. Paes Leme, C. N. Pereira, C. M. Carnielli, C. C. Werneck, M. V. Sousa, C. A. O. Ricart, and F. R. Gadelha. 2016. "Trypanosoma cruzi Mitochondrial Tryparedoxin Peroxidase is Located throughout the Cell and Its Pull down Provides One Step towards the

Understanding of Its Mechanism of Action." *Biochimica et Biophysica Acta—Proteins and Proteomics.* https://doi.org/10.1016/j.bbapap.2015.10.005.

Pereira, Sara Silva, Sandra Trindade, Mariana De Niz, and Luisa M. Figueiredo. 2019. "Tissue Tropism in Parasitic Diseases." *Open Biology.* https://doi.org/10.1098/rsob.190036.

Pérez-Morales, Deyanira, Humberto Lanz-Mendoza, Gerardo Hurtado, Rodrigo Martínez-Espinosa, and Bertha Espinoza. 2012. "Proteomic Analysis of Trypanosoma cruzi Epimastigotes Subjected to Heat Shock." *Journal of Biomedicine and Biotechnology.* https://doi.org/10.1155/2012/902803.

Perkins, Arden, Kimberly J. Nelson, Derek Parsonage, Leslie B. Poole, and P. Andrew Karplus. 2015. "Peroxiredoxins: Guardians against Oxidative Stress and Modulators of Peroxide Signaling." *Trends in Biochemical Sciences.* https://doi.org/10.1016/j.tibs.2015.05.001.

Piacenza, Lucía, Florencia Irigoín, María Noel Alvarez, Gonzalo Peluffo, Martin C Taylor, John M Kelly, Shane R Wilkinson, and Rafael Radi. 2007. "Mitochondrial Superoxide Radicals Mediate Programmed Cell Death in Trypanosoma cruzi: Cytoprotective Action of Mitochondrial Iron Superoxide Dismutase Overexpression." *Biochemical Journal* 403: 323–334. https://doi.org/10.1042/BJ20061281.

Piacenza, Lucía, Gonzalo Peluffo, María Noel Alvarez, John M. Kelly, Shane R. Wilkinson, and Rafael Radi. 2008. "Peroxiredoxins Play a Major Role in Protecting Trypanosoma cruzi against Macrophage- and Endogenously-Derived Peroxynitrite." *Biochemical Journal.* https://doi.org/10.1042/BJ20071138.

Piacenza, Lucía, Gonzalo Peluffo, María Noel Alvarez, Alejandra Martínez, and Rafael Radi. 2013. "Trypanosoma cruzi Antioxidant Enzymes as Virulence Factors in Chagas Disease." *Antioxidants and Redox Signaling.* https://doi.org/10.1089/ars.2012.4618.

Piattoni, Claudia V., Víctor S. Blancato, Hilario Miglietta, Alberto A. Iglesias, and Sergio A. Guerrero. 2006. "On the Occurrence of Thioredoxin in Trypanosoma cruzi." *Acta Tropica.* https://doi.org/10.1016/j.actatropica.2005.10.005.

Piñeyro, María Dolores, Talia Arcari, Carlos Robello, Rafael Radi, and Madia Trujillo. 2011. "Tryparedoxin Peroxidases from Trypanosoma cruzi: High Efficiency in the Catalytic Elimination of Hydrogen Peroxide and Peroxynitrite." *Archives of Biochemistry and Biophysics.* https://doi.org/10.1016/j.abb.2010.12.014.

Piñeyro, María Dolores, Diego Arias, Alejandro Ricciardi, Carlos Robello, and Adriana Parodi-Talice. 2019. "Oligomerization Dynamics and Functionality of Trypanosoma cruzi Cytosolic Tryparedoxin Peroxidase as Peroxidase and Molecular Chaperone." *Biochimica et Biophysica Acta—General Subjects.* https://doi.org/10.1016/j.bbagen.2019.06.013.

Piñeyro, Maria Dolores, Adriana Parodi-Talice, Magdalena Portela, Diego G. Arias, Sergio A. Guerrero, and Carlos Robello. 2011. "Molecular Characterization and Interactome Analysis of Trypanosoma cruzi Tryparedoxin 1." *Journal of Proteomics.* https://doi.org/10.1016/j.jprot.2011.04.006.

Piñeyro, María Dolores, Juan Carlos Pizarro, Fernando Lema, Otto Pritsch, Alfonso Cayota, Graham A. Bentley, and Carlos Robello. 2005. "Crystal Structure of the Tryparedoxin Peroxidase from the Human Parasite Trypanosoma cruzi." *Journal of Structural Biology.* https://doi.org/10.1016/j.jsb.2004.12.005.

Queiroz, Rafael, Corinna Benz, Kurt Fellenberg, Jörg D. Hoheisel, and Christine Clayton. 2009. "Transcriptome Analysis of Differentiating Trypanosomes Reveals the Existence of Multiple Post-Trans_criptional Regulons." *BMC Genomics.* https://doi.org/10.1186/1471-2164-10-495.

Requena, Jose M., Ana M. Montalvo, and Jorge Fraga. 2015. "Molecular Chaperones of Leishmania: Central Players in Many Stress-Related and -Unrelated Physiological Processes." *BioMed Research International.* https://doi.org/10.1155/2015/301326.

Rhee, Sue Goo, and Hyun Ae Woo. 2011. "Multiple Functions of Peroxiredoxins: Peroxidases, Sensors and Regulators of the Intracellular Messenger H_2O_2, and Protein Chaperones." *Antioxidants and Redox Signaling.* https://doi.org/10.1089/ars.2010.3393.

Roditi, Isabel, and Michael J. Lehane. 2008. "Interactions between Trypanosomes and Tsetse Flies." *Current Opinion in Microbiology.* https://doi.org/10.1016/j.mib.2008.06.006.

Romao, Susana, Helena Castro, Carla Sousa, Sandra Carvalho, and Ana M. Tomás. 2009. "The Cytosolic Tryparedoxin of Leishmania infantum Is Essential for Parasite Survival." *International Journal for Parasitology.* https://doi.org/10.1016/j.ijpara.2008.11.009.

Saldivia, Manuel, Gloria Ceballos-Pérez, Jean Mathieu Bart, and Miguel Navarro. 2016. "The AMPKα1 Pathway Positively Regulates the Developmental Transition from Proliferation to Quiescence in Trypanosoma brucei." *Cell Reports.* https://doi.org/10.1016/j.celrep.2016.09.041.

Santos, Vânia C., Cássio A. Nunes, Marcos H. Pereira, and Nelder F. Gontijo. 2011. "Mechanisms of PH Control in the Midgut of Lutzomyia longipalpis: Roles for Ingested Molecules and Hormones." *Journal of Experimental Biology.* https://doi.org/10.1242/jeb.051490.

Sardar, Abul Hasan, Sudeep Kumar, Ashish Kumar, Bidyut Purkait, Sushmita Das, Abhik Sen, Manish Kumar, et al.. 2013. "Proteome Changes Associated with *Leishmania donovani* Promastigote Adaptation to Oxidative and Nitrosative Stresses." *Journal of Proteomics.* https://doi.org/10.1016/j.jprot.2013.01.011.

Saunders, Eleanor C., William W. Ng, Joachim Kloehn, Jennifer M. Chambers, Milica Ng, and Malcolm J. McConville. 2014. "Induction of a Stringent Metabolic Response in Intracellular Stages of *Leishmania mexicana* Leads to Increased Dependence on Mitochondrial Metabolism." *PLoS Pathogens.* https://doi.org/10.1371/journal.ppat.1003888.

Schlecker, Tanja, Marcelo A. Comini, Johannes Melchers, Thomas Ruppert, and R. Luise Krauth-Siegel. 2007. "Catalytic Mechanism of the Glutathione Peroxidase-Type Tryparedoxin Peroxidase of *Trypanosoma brucei.*" *Biochemical Journal.* https://doi.org/10.1042/BJ20070259.

Schlecker, Tanja, Armin Schmidt, Natalie Dirdjaja, Frank Voncken, Christine Clayton, and R. Luise Krauth-Siegel. 2005. "Substrate Specificity, Localization, and Essential Role of the Glutathione Peroxidase-Type Tryparedoxin Peroxidases in *Trypanosoma brucei.*" *Journal of Biological Chemistry.* https://doi.org/10.1074/jbc.M413338200.

Schmidt, Armin, Christine E. Clayton, and R. Luise Krauth-Siegel. 2002. "Silencing of the Thioredoxin Gene in *Trypanosoma brucei brucei.*" *Molecular and Biochemical Parasitology.* https://doi.org/10.1016/S0166-6851(02)00215-3.

Schmidt, Heide, and R. Luise Krauth-Siegel. 2003. "Functional and Physicochemical Characterization of the Thioredoxin System in *Trypanosoma brucei.*" *Journal of Biological Chemistry.* https://doi.org/10.1074/jbc.M305338200.

Sela, Dotan, and Joseph Shlomai. 2009. "Regulation of UMSBP Activities through Redox-Sensitive Protein Domains." *Nucleic Acids Research.* https://doi.org/10.1093/nar/gkn927.

Sela, Dotan, Nurit Yaffe, and Joseph Shlomai. 2008. "Enzymatic Mechanism Controls Redox-Mediated Protein-DNA Interactions at the Replication Origin of Kinetoplast DNA Minicircles." *Journal of Biological Chemistry.* https://doi.org/10.1074/jbc.M804417200.

Shiratsubaki, Isabel S., Xin Fang, Rodolpho O. O. Souza, Bernhard O. Palsson, Ariel M. Silber, and Jair L. Siqueira-Neto. 2020. "Genome-Scale Metabolic Models Highlight Stage-Specific Differences in Essential Metabolic Pathways in *Trypanosoma cruzi.*" *PLoS Neglected Tropical Diseases.* https://doi.org/10.1371/journal.pntd.0008728.

Shonhai, Addmore, Alexander G. Maier, Jude M. Przyborski, and Gregory L. Blatch. 2011. "Intracellular Protozoan Parasites of Humans: The Role of Molecular Chaperones in Development and Pathogenesis." *Protein & Peptide Letters* 18 (2): 143–157. https://doi.org/10.2174/092986611794475002.

Silber, Ariel M., Renata R. Tonelli, Camila G. Lopes, Narcisa Cunha-e-Silva, Ana Cláudia T. Torrecilhas, Robert I. Schumacher, Walter Colli, and Maria Júlia M. Alves. 2009. "Glucose Uptake in the Mammalian Stages of *Trypanosoma cruzi.*" *Molecular and Biochemical Parasitology.* https://doi.org/10.1016/j.molbiopara.2009.07.006.

Singh, Ruby, Bidyut Purkait, Kumar Abhishek, Savita Saini, Sushmita Das, Sudha Verma, Abhishek Mandal, et al.. 2016. "Universal Minicircle Sequence Binding Protein of *Leishmania donovani* Regulates Pathogenicity by Controlling Expression of Cytochrome-B." *Cell and Bioscience.* https://doi.org/10.1186/s13578-016-0072-z.

Smith, Joseph T., Ujjal K. Singha, Smita Misra, and Minu Chaudhuri. 2018. "Divergent Small Tim Homologues Are Associated with TbTim17 and Critical for the Biogenesis of TbTim17 Protein Complexes in *Trypanosoma brucei.*" *MSphere.* https://doi.org/10.1128/msphere.00204-18.

Sousa, André F., Ana G. Gomes-Alves, Diego Benítez, Marcelo A. Comini, Leopold Flohé, Timo Jaeger, Joana Passos, Friedrich Stuhlmann, Ana M. Tomás, and Helena Castro. 2014. "Genetic and Chemical Analyses Reveal That Trypanothione Synthetase but Not Glutathionylspermidine Synthetase Is Essential for *Leishmania infantum.*" *Free Radical Biology and Medicine.* https://doi.org/10.1016/j.freeradbiomed.2014.05.007.

De Souza, E. M., T. C. Araújo-Jorge, C. Bailly, A. Lansiaux, M. M. Batista, G. M. Oliveira, and M. N. C. Soeiro. 2003. "Host and Parasite Apoptosis Following *Trypanosoma cruzi* Infection in In Vitro and in Vivo Models." *Cell and Tissue Research.* https://doi.org/10.1007/s00441-003-0782-5.

Stuart, Ken, Reto Brun, Simon Croft, Alan Fairlamb, Ricardo E. Gürtler, Jim McKerrow, Steve Reed, and Rick Tarleton. 2008. "Kinetoplastids: Related Protozoan Pathogens, Different Diseases." *Journal of Clinical Investigation.* https://doi.org/10.1172/JCI33945.

Teixeira, Filipa, Helena Castro, Tânia Cruz, Eric Tse, Philipp Koldewey, Daniel R. Southworth, Ana M. Tomás, and Ursula Jakob. 2015. "Mitochondrial Peroxiredoxin Functions as Crucial Chaperone Reservoir in *Leishmania infantum.*" *Proceedings of the National*

Academy of Sciences of the United States of America. https://doi.org/10.1073/pnas.1419682112.

Tetaud, Emmanuel, Christiane Giroud, Alan R. Prescott, David W. Parkin, Dominique Baltz, Nicolas Biteau, Théo Baltz, and Alan H. Fairlamb. 2001. "Molecular Characterization of Mitochondrial and Cytosolic Trypanothione-Dependent Tryparedoxin Peroxidases in Trypanosoma brucei." Molecular and Biochemical Parasitology. https://doi.org/10.1016/S0166-6851(01)00320-6.

Timms, Mark W., Frederick J. Van Deursen, Edward F. Hendriks, and Keith R. Matthews. 2002. "Mitochondrial Development during Life Cycle Differentiation of African Trypanosomes: Evidence for a Kinetoplast-Dependent Differentiation Control Point." Molecular Biology of the Cell. https://doi.org/10.1091/mbc.E02-05-0266.

Trujillo, Madia, Heike Budde, María Dolores Piñeyro, Matthias Stehr, Carlos Robello, Leopold Flohé, and Rafael Radi. 2004. "Trypanosoma brucei and Trypanosoma cruzi Tryparedoxin Peroxidases Catalytically Detoxify Peroxynitrite via Oxidation of Fast Reacting Thiols." Journal of Biological Chemistry. https://doi.org/10.1074/jbc.M404317200.

Ulrich, Kathrin, Caroline Finkenzeller, Sabine Merker, Federico Rojas, Keith Matthews, Thomas Ruppert, and R. Luise Krauth-Siegel. 2017. "Stress-Induced Protein S-Glutathionylation and S-Trypanothionylation in African Trypanosomes—A Quantitative Redox Proteome and Thiol Analysis." Antioxidants and Redox Signaling. https://doi.org/10.1089/ars.2016.6947.

Weiss, Brian L., Michele A. Maltz, Aurélien Vigneron, Yineng Wu, Katharine S. Walter, Michelle B. O'Neill, Jingwen Wang, and Serap Aksoy. 2019. "Colonization of the Tsetse Fly Midgut with Commensal Kosakonia cowanii Zambiae Inhibits Trypanosome Infection Establishment." PLoS Pathogens. https://doi.org/10.1371/journal.ppat.1007470.

Wilkinson, Shane R., David Horn, S. Radhika Prathalingam, and John M. Kelly. 2003. "RNA Interference Identifies Two Hydroperoxide Metabolizing Enzymes That Are Essential to the Bloodstream Form of the African Trypanosome." Journal of Biological Chemistry. https://doi.org/10.1074/jbc.M303035200.

Wilkinson, Shane R., Samson O. Obado, Isabel L. Mauricio, and John M. Kelly. 2002. "Trypanosoma cruzi Expresses a Plant-like Ascorbate-Dependent Hemoperoxidase Localized to the Endoplasmic Reticulum." Proceedings of the National Academy of Sciences of the United States of America. https://doi.org/10.1073/pnas.202422899.

Wilkinson, Shane R., S. Radhika Prathalingam, Martin C. Taylor, Aiyaz Ahmed, David Horn, and John M. Kelly. 2006. "Functional Characterisation of the Iron Superoxide Dismutase Gene Repertoire in Trypanosoma brucei." Free Radical Biology and Medicine. https://doi.org/10.1016/j.freeradbiomed.2005.06.022.

Wilkinson, Shane R., Martin C. Taylor, Said Touitha, Isabel L. Mauricio, David J. Meyer, and John M. Kelly. 2002. "TcGPXII, a Glutathione-Dependent Trypanosoma cruzi Peroxidase with Substrate Specificity Restricted to Fatty Acid and Phospholipid Hydroperoxides, Is Localized to the Endoplasmic Reticulum." Biochemical Journal. https://doi.org/10.1042/BJ20020038.

Wilkinson, Shane R., Nigel J. Temperton, Angeles Mondragon, and John M. Kelly. 2000. "Distinct Mitochondrial and Cytosolic Enzymes Mediate Trypanothione-Dependent Peroxide Metabolism in Trypanosoma cruzi." Journal of Biological Chemistry. https://doi.org/10.1074/jbc.275.11.8220.

Wood, Zachary A., Leslie B. Poole, Roy R. Hantgan, and P. Andrew Karplus. 2002. "Dimers to Doughnuts: Redox Sensitive Oligomerization of 2-Cysteine Peroxiredoxins." Biochemistry. https://doi.org/10.1021/bi012173m.

Woodward, R., and K. Gull. 1990. "Timing of Nuclear and Kinetoplast DNA Replication and Early Morphological Events in the Cell Cycle of Trypanosoma brucei." Journal of Cell Science.

Zago, María Paola, Yashoda M. Hosakote, Sue jie Koo, Monisha Dhiman, María Dolores Piñeyro, Adriana Parodi-Talice, Miguel A. Basombrio, Carlos Robello, and Nisha J. Gargc. 2016. "TcI Isolates of Trypanosoma cruzi Exploit the Antioxidant Network for Enhanced Intracellular Survival in Macrophages and Virulence in Mice." Infection and Immunity. https://doi.org/10.1128/IAI.00193-16.

Zarley, J. H., B. E. Britigan, and M. E. Wilson. 1991. "Hydrogen Peroxide-Mediated Toxicity for Leishmania donovani chagasi Promastigotes: Role of Hydroxyl Radical and Protection by Heat Shock." Journal of Clinical Investigation. https://doi.org/10.1172/JCI115461.

CHAPTER FOUR

Redox Regulation during Zebrafish Development

Lars Bräutigam and Carsten Berndt

CONTENTS

4.1 INTRODUCTION

In the early 19th century, Scottish physician Francis Hamilton traveled to India to study the fauna along the river Ganges. In the river Kosi near the Ganges, he found a "beautiful fish" that he described as "a Cyprinus of the Danio kind with several blue and silver stripes on each side."[1] Hamilton would be truly surprised to see that dozens of zebrafish conferences, from large international meetings with thousands of participants to specialized workshops on advanced techniques, are arranged every year and that the beautiful Cyprinus of the Danio kind" is now home to almost 1,400 laboratories worldwide and an essential part of thousands of scientific research articles per year.

But how did this small tropical fish, home to the floodplains of the Indian subcontinent and well adapted to the constantly changing environmental conditions there, find its way into research laboratories throughout the world?

4.2 THE BEGINNINGS OF THE WORLD CAREER OF THE ZEBRAFISH

After the initial report by Hamilton in 1822, the zebrafish was largely forgotten and was barely used as an ornamental fish, if at all.

In the late 1960s, George Streisinger—a phage geneticist working at the University of Oregon—was looking for a model system that allowed him to dissect the genetics of neural development in vertebrates. Although other fish species like *Fundulus* or the medaka had been used in research before, his choice fell on the zebrafish, most likely for practical reasons such as easy breeding, fast development, and embryo transparency. Over the years, Streisinger developed basic tools to dissect development and genetics in zebrafish. His work culminated in a groundbreaking 1981 publication in *Nature*, in which he described the cloning of zebrafish.[2] At the same time, a close colleague of Streisinger, Charles Kimmel, took advantage of the optical clarity and rapid development of zebrafish

DOI: 10.4324/9781003204091-6

embryos and established unprecedented lineage maps of almost all major cell types of the zebrafish embryo.[3] After Streisinger's death in 1984, Kimmel and other zebrafish researchers at the University of Oregon carried on studying neural and lineage development in zebrafish embryos, and in 1990 the first scientific meeting on zebrafish as a research model was held. Soon after, the first book on zebrafish husbandry was published and the Zebrafish International Resource Center (ZIRC) was founded. ZIRC is a genetic repository that today holds more than 30,000 individual zebrafish lines. With the pace accelerating, Kimmel published the zebrafish staging series in 1995, a crucial step that allowed for standardizing conditions between experiments and laboratories.[4] With these advances, Christiane Nüsslein-Volhard and her former graduate student, Wolfgang Driever, took note of the new emerging model in their study of zebrafish. For this, they repeated the massive genetic screen Nüsslein-Volhard had earlier carried out on *Drosophila*, for which she was awarded the Nobel Prize in Physiology or Medicine in 1995. This "big screen" resulted in the isolation of over 4,000 mutants and it was published in a single collection of 37 screening papers in a complete volume of the journal *Development*.[5] This seminal screen was the starting point for new, groundbreaking techniques such as antisense morpholinos for knocking down the function of single genes, new gene-editing tools, and the tol2 system that allows the establishment of almost any transgenic line. In parallel, the zebrafish genome was mapped and then fully sequenced. Today, the 11th assembly is available, and following unprecedented revelations of vertebrate development, the zebrafish has established itself as a crucial platform for applied and translational research as well as large drug discovery screens.

4.3 THE ZEBRAFISH AS A RESEARCH MODEL FOR HUMAN PHYSIOLOGY AND PATHOLOGY

The zebrafish is a small, bony fish (2–4 cm) found mainly in monsoon areas characterized by high seasonal variation. Nevertheless, zebrafish can easily be housed and bred under constant laboratory conditions. Besides the easy and cheap husbandry, natural traits of the zebrafish make it ideal for scientific research: (i) Zebrafish are not seasonal breeders, and eggs for experiments can be obtained

year-round. (ii) Females can lay hundreds of eggs in a single mating. (iii) Eggs and early embryos are completely transparent and develop outside the mother, which allows scientists to observe the most delicate cellular and subcellular structures in living individuals. (iv) The development of early zebrafish embryos is very rapid; the heart starts beating at less than 30 hours post-fertilization (hpf) and the larvae is freely swimming when barely 2 days old (Figure 4.1).

Those characteristics, together with the genetic accessibility and unique imaging properties, propelled the creation of thousands of transgenic lines in which a specific cell lineage can be followed over time in living animals. The zebrafish became the model of choice for the dissection of complex developmental and differentiational processes in vertebrates. One of the most prominent transgenic strains, the fli1a:EGFP line, provided unprecedented insights into physiological blood vessel development in vertebrates, an area that had been technically challenging to explore.[6] For the first time in a living vertebrate, scientists could follow the differentiation and migration of vascular endothelial cells from their origin in the ventral mesoderm to the lateral plate mesoderm, from where they spread along the trunk and form the first major vessels in the zebrafish. Another astonishing example of how complex biological processes can be followed in real time in the zebrafish is wound healing and regeneration. Timothy Mitchison and his group established zebrafish with a genetically encoded probe for hydrogen peroxide; they provided unique real-time and in vivo evidence that rapid generation of H_2O_2 after wounding is responsible for the quick recruitment of lymphocytes, which in turn starts the wound-healing process.[7] Besides wound healing, the zebrafish also possesses a remarkable potential to regenerate tissues and organs. Again, the availability of transgenic lines and the translucence of zebrafish embryos are unique traits of that animal model that provided a fertile ground for studying how fins can completely regenerate after amputation, or how the de-differentiation of cardiomyocytes allows for the astonishing capacity to regenerate cardiac structures. After deciphering the underlying genetics for those differentiation and de-differentiation processes, it turned out that the mechanisms are highly conserved between zebrafish and human.

The outstanding experimental possibilities the zebrafish provides and the conservation of

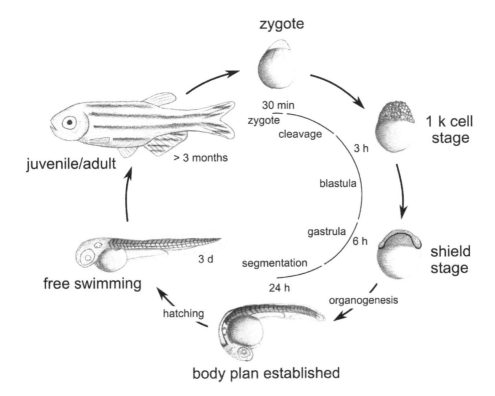

zygote

1 k cell stage

shield stage

body plan established

free swimming

juvenile/adult

30 min
zygote

cleavage

3 h

blastula

gastrula

6 h

segmentation

24 h

organogenesis

> 3 months

3 d

hatching

Figure 4.1 Generation cycle of the zebrafish.

Thirty minutes after fertilization, the first cell appears from which the whole fish develops. After 3 hours, at the 1k-cell stage, embryonic transcription starts and soon thereafter, gastrulation begins and the three germ layers form. At 24 hours after fertilization, most major organs have been formed and started to function. The heart starts beating 30 hours after fertilization. The zebrafish embryo hatches at 2–3 days of age and becomes a free-swimming larvae that is, from 5 days on, dependent on external food sources. The zebrafish is sexually mature within approximately 3 months and can live up to 3 years in captivity.

developmental and regenerative mechanisms are key factors that promoted the establishment of clinically relevant models for human diseases and their treatments. Zebrafish disease models proved to be so influential that the Zebrafish Disease Models Society was founded in 2014; the society promotes knowledge sharing, collaborations, and international conferences on this topic. Today, there are genetic or inducible zebrafish models available for almost any major health issue including psychological disorders, neurodegenerative and cardiovascular diseases, and cancer. Zebrafish tumor models have been highly influential in understanding genetic and environmental factors for tumorigenesis and tumor-host interactions but also for finding new cures.

Cancer cells can easily be transplanted into zebrafish embryos, either into the yolk sac or orthotopic locations. The optical clarity of the embryos and the readily available reporter lines provide a unique possibility to follow tumor growth and migration as well as real-time interaction with components of the primary immune system and tumor angiogenesis. Deciphering the details of tumor growth and finding new antitumorigenic molecules were highly facilitated because zebrafish embryos can be kept immobilized in 96-well plates for long periods of time. Tumor-bearing embryos can be exposed to new candidate drugs, and automated imaging solutions are not only able to follow tumor growth and metastasis but also to determine key physiological factors such as the heartbeat.[8] The zebrafish is the only vertebrate that can be used in high-throughput drug discovery pipelines in vivo.

4.4 REDOX RESEARCH IN ZEBRAFISH

Redox changes are essential for differentiation of cells and development of whole organisms. Already at the beginning of the 20th century, Otto Warburg described the importance of oxidation during development of sea urchins.[9,10] That was one of the first findings suggesting physiological roles of reactive oxygen species, which are still mainly seen as purely damaging molecules during different pathological situations. However, the importance of these reactive species in the modulation of signaling pathways controlling several cellular processes such as differentiation/development is increasingly being recognized, and it is part of the latest definition of oxidative stress.[11] This definition distinguishes between oxidative distress (the pathophysiological damaging part) and oxidative eustress (the physiological part). Oxidative eustress/redox signaling functions via reversible oxidative posttranslational thiol modifications such as thiol-disulfide exchanges, S-glutathionylation, nitrosylation, or persulfidation induced by reactive oxygen, nitrogen, and sulfur species. As described above, the zebrafish is the prominent model for vertebrate development. Surprisingly, the number of studies regarding redox regulation in this important model organism is quite limited. Only a few publications describe the role of proteins controlling enzyme activities via regulation of reversible thiol modifications, the prerequisite for redox signaling (see Section 4.3). There is a greater number of publications describing the impact of general changes of the redox state, such as hypoxic conditions on developmental processes (e.g., brain[12,13] or vascular development[14,15]). Taking advantage of the transparency of zebrafish embryos, several studies characterizing probes for these general redox changes were performed in this model organism (see Section 4.2). Fast development is another advantage of the zebrafish model, allowing the characterization of molecules manipulating redox signaling and/or the redox state in the context of embryonic development (see Section 4.1).

4.4.1 Drug-Based Modulation of Redox Signaling

Redox systems can be altered in vitro (or within cells) by the addition of chemicals or small-molecule inhibitors. The zebrafish embryo, with its small size, optical clarity, and fast embryonic development, provides a unique platform to explore those compounds in a living vertebrate. Moreover, the availability of transgenic reporter lines offers the possibility to dissect redox signaling pathways in the context of tissue and organ development as well as regeneration.

Once the chorion (eggshell) has been removed enzymatically or manually, zebrafish embryos can be exposed to chemicals by simple immersion. Most chemicals are readily taken up; however, uptake speed, route, and concentration accumulated inside the embryo may differ between compound classes and their chemical properties. It needs also to be considered that cellular redox signaling pathways are intricate networks of oxidation and/or reduction processes that are taking place in *distinct* compartments of a cell, organ, or tissue. Importantly, redox signaling events in most cases modulate highly specific redox switches without altering the global redox status of a cell. However, most externally administered small molecules alter the *global* cellular redox status. Nevertheless, the zebrafish model has provided a pivotal platform to dissect the role of redox signaling pathways in tissue differentiation[16] and regeneration.[17] Roehl and colleagues modulated hydrogen peroxide production from NADPH oxidases with DPI (diphenylene iodonium) and MCI196 and thereby provided in vivo evidence that redox signaling cascades can repurpose developmental signaling networks to regeneration of damaged tissue.

Besides modulating H_2O_2 levels indirectly or directly with tert-butylhydrogen peroxide (tBOOH), glutathione (GSH) levels can be altered in living zebrafish embryos using an array of different compounds including N-acetyl cysteine (NAC) and L-buthionine-(S,R)-sulfoximine (BSO). A selection of several other molecules tested in zebrafish is provided in Table 4.1.

Due to the genetic accessibility of the zebrafish model, redox systems can easily be modulated on the gene or transcript level. Antisense morpholinos are a powerful tool to transiently knock down the expression of target proteins. Among others, the expression of iNOS,[18] Grx 5,[19] Grx2,[20,21] Grx3,[22] and Trx1[23] has been modulated, which has shed light on the role of redox signaling in development and differentiation. Besides, several components of redox systems have been knocked out in zebrafish using CrispR-Cas9.[24,25]

4.4.2 Visualization of Redox Signaling during Development and Regeneration

Zebrafish were used to establish and characterize several redox probes. Here, we can only present a small selection of the available publications. More than 100 probes detecting different redox active

TABLE 4.1
Selection of Small Molecules Used to Alter Redox Systems in Zebrafish

Compound	Observation	Publication
Diphenylene iodonium	Inhibitor of flavoenzymes	[17]
Hydrogen peroxide	General oxidant	[56]
N-acetyl cysteine	Modulation of GSH levels	[56]
Tert-butylhydroperoxide	General oxidant	[28]
BSO	Inhibition of GSH synthesis	[57]
L-hydroxyglutaric acid	Oxidative damage of neurons	[57]
2-oxothiazolidine-4-carboxylat	Modulation of GSH levels	[57]
Ascorbic acid	Water-soluble antioxidant	[57]
Alpha-tocopherol	Fat-soluble antioxidant	[57]

molecules in zebrafish have been published in recent years.

Many different probes were used to measure GSH. GSH is highly regulated during development, especially during organogenesis in vertebrates including zebrafish;[26] for more information, see Chapter 10. In zebrafish, GSH was visualized, for example, with the excited-state intramolecular proton transfer–based fluorescent probe BTFMD[27] and monochlorobimane.[28]

Thiols in proteins and small molecules such as GSH can be stained in living fish with MIPY-DNBS, a cyan fluorescent probe consisting of an imidazole pyridine derivative and 2,4-dinitrobenzensufonate[29] or an iridium (III) complex probe.[30]

Probes to detect reactive species were also measured in zebrafish. In principle, probes for almost all reactive species (oxygen, nitrogen, sulfur) were tested for their functionality in zebrafish. Using the zebrafish model, H_2O_2 was introduced as an important factor during wound healing and sensory axon regeneration.[7,31,32] Different H_2O_2-specific probes were established; for example, a combination of selenamorpholine and a BODIPY fluorophore to detect specifically lysosomal H_2O_2 in zebrafish.[33] The function of a variety of different probes reacting with superoxide anion or hypochlorous acid—important reactive species for the function of the immune system in zebrafish[34]—was investigated in zebrafish, such as fluorescent probes based on phosphinate[35] or consisting of a naphthalimide and a silicone small molecule,[36] respectively. Formation of the reactive nitrogen species nitric oxide and peroxynitrite were visualized, for example, via fluorescent metal complex–based or boronate-based probes.[37,38] Moreover, probes detecting reactive

sulfur species (e.g., hydrogen sulfide) were established in zebrafish.[39]

In addition, already established and widely used genetically encoded redox sensors are in use in zebrafish. Reduction-oxidation sensitive green fluorescent protein (roGFP) coupled to Grx1 was introduced into the genome of zebrafish, forming the ZebROS line to detect the ratio between reduced and oxidized GSH (GSSG).[40] A similar line expressing a sensor in which roGFP was exchanged to the red fluorescent roCherry was also established.[41] Furthermore, zebrafish lines encoding the H_2O_2 sensor HyPer also have been generated.[42]

4.4.3 Redox Signaling via Redoxins

Oxidoreductases of the thioredoxin family, namely, thioredoxins (Trxs), glutaredoxins (Grxs), and peroxiredoxins (Prxs), are key players in redox signaling via reduction or oxidation of thiols and/or hydroperoxide.[43] All these proteins, summarized by the term redoxins, share the Trx fold—consisting of a central four- or five-stranded β-sheet surrounded by three to four α-helices[44]—and are present in mitochondria and cytosol in vertebrates.[43] Here, we will summarize a few publications dedicated to specific thiol redox signaling.

Zebrafish express both the cytosolic Trx1 and the mitochondrial Trx2. Trx1 mRNA is mainly present in the brain of developing zebrafish, especially olfactory epithelia cells, Mauthner, and serotonergic raphe cells.[23] The knockdown of Trx1 leads to a smaller medulla oblongata and an expanded ventricle.[23] However, specific Trx1-dependent signaling cascades have not been described to date.

Embryos lacking the mitochondrial Trx2 show dysregulation of Bcl, Bax, and puma, leading to mitochondrial dysfunction, hepatic cell death, and subsequently defects in liver development.[45] Thus Trx activity is needed for proper development of the brain and liver.

The next described members of the redoxins are the GSH-dependent Grxs. As described earlier, GSH is an important redox circuit and regulates during development. However, not a single biological function is linked to GSH itself; the function of GSH depends on enzymes,[46] such as microsomal glutathione transferase 1 (MGST1)—an enzyme important for hematopoiesis during zebrafish development[47]—or Grxs. Most likely, changes in the ratio of reduced GSH and oxidized glutathione (GSSG) lead to glutathionylation of proteins—a Grx-dependent reversible oxidative posttranslational modification—and their activation/inactivation.[26] Grxs are divided into two subfamilies: monothiol and dithiol.[48] Vertebrates express two monothiol Grxs (Grxs 3 and 5) and two dithiol Grxs (Grxs 1 and 2).[48] Monothiol Grxs are important regulators of iron homeostasis and FeS-cluster biosynthesis,[49] and knockdown of the two monothiol Grxs induces heme deficiency in zebrafish embryos.[19,22] Small structural differences distinguish between activities of monothiol and dithiol Grxs.[50,51] Exchange of five amino acids forming a loop structure close to the active site changed the function of Grxs, whereas the respective Grx5 mutant lost its function during heme formation, the Grx2 mutant gained this function in zebrafish.[51] Here we will focus on the oxidoreductase function of dithiol Grxs during zebrafish development. Zebrafish embryos lacking Grx2 display impaired brain development.[52] This was the first described redoxin-dependent phenotype (2011), although Grx5 was the first investigated Grx in zebrafish (2005). Knockdown of Grx2 inhibits formation of the axonal scaffold via dysregulation of the formation and reduction of a disulfide in collapsin response mediator protein 2, an effector protein of semaphorin 3a signaling.[52,53] The lack of the axonal scaffold led to cell death of all tested neuronal subtypes 24 hpf. In addition, Grx2 knockdown affects development of the cardiovascular system. Blood circulation within the body of zebrafish embryos is absent in fish lacking Grx2.[20] This phenotype is based on two described mechanisms: disturbed formation of both the heart and the vasculature.

Heart formation, especially formation of the right angle between ventricle and atrium of the two-chamber zebrafish heart that is important for the pumping efficiency, is disturbed by delayed arrival of cardiac neural crest cells.[20] Migration of these neural crest cells depends on deglutathionylation of actin by Grx2. (De)glutathionylation is also essential for the development of a functioning vasculature.[21] The redox state of a specific cysteine residue in Sirtuin 1, a protein deacetylase, regulates its enzymatic activity, most likely because the bound GSH blocks the binding of the cofactor NAD to the Sirtuin 1 active site. Sirtuin 1 without the (de)glutathionylated cysteine loses redox regulation of its catalytic activity. Because knockdown of Grx2 induced no general oxidative damage, both brain and cardiovascular development depend on specific redox regulation via the enzymatic activity of Grx2.[21,52]

Prxs are able to detoxify H_2O_2 but seem to be also necessary for the specific oxidative modification induced by H_2O_2.[54] Six Prxs have been identified in the zebrafish genome. Prx1 is a Trx-dependent cytosolic protein that is highly conserved among vertebrates and expressed in developing vessels in zebrafish embryos.[55] Zebrafish embryos lacking Prx1 displayed formation of pericardial edema and impaired sprouting of intersegmental vessels, leading to impaired circulation.[55] These Prx1 defects are most likely connected to Notch and BMP signaling.

In summary, oxidoreductase activity of Grxs, Trxs, and Prxs is essential for the development of functioning neuronal and vascular systems as well as proper brain and liver structures (Figure 4.2).

4.5 CONCLUSIONS

The similarities between developmental, pathological, and regenerative processes, as well as the respective underlying mechanisms of zebrafish and humans, make zebrafish a highly valuable and recognized model for differentiation and de-differentiation. Although various compounds and probes manipulating and visualizing redox active molecules were described in zebrafish, a surprisingly low number of studies investigate the impact of specific redox signaling in this important model organism. Further studies will improve our knowledge of vertebrate and thereby human developmental processes, providing new therapeutic targets regarding developmental diseases and also regeneration and cancer progression.

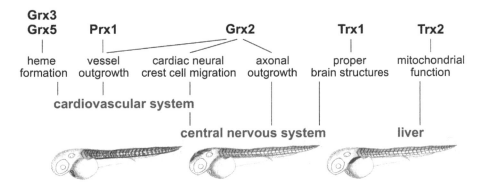

Figure 4.2 Summary of developmental processes regulated by redoxins. Published functions of thioredoxins (Trx), glutaredoxins (Grx), and peroxiredoxins (Prx) during zebrafish development (see text for details).

REFERENCES

1 Hamilton, F. *An account of the fishes found in the River Ganges and its branches.* (Printed for A. Constable and Company; etc., 1822).

2 Streisinger, G., Walker, C., Dower, N., Knauber, D. & Singer, F. Production of clones of homozygous diploid zebra fish (*Brachydanio rerio*). *Nature* **291**, 293–296, doi:10.1038/291293a0 (1981).

3 Kimmel, C. B. & Law, R. D. Cell lineage of zebrafish blastomeres. I. Cleavage pattern and cytoplasmic bridges between cells. *Developmental Biology* **108**, 78–85, doi:10.1016/0012-1606(85)90010-7 (1985).

4 Kimmel, C. B., Ballard, W. W., Kimmel, S. R., Ullmann, B. & Schilling, T. F. Stages of embryonic development of the zebrafish. *Developmental Dynamics* **203**, 253–310, doi:10.1002/aja.1002030302 (1995).

5 Nusslein-Volhard, C. The zebrafish issue of Development. *Development* **139**, 4099–4103, doi:10.1242/dev.085217 (2012).

6 Lawson, N. D. & Weinstein, B. M. In vivo imaging of embryonic vascular development using transgenic zebrafish. *Developmental Biology* **248**, 307–318, doi:10.1006/dbio.2002.0711 (2002).

7 Niethammer, P., Grabher, C., Look, A. T. & Mitchison, T. J. A tissue-scale gradient of hydrogen peroxide mediates rapid wound detection in zebrafish. *Nature* **459**, 996–999, doi:10.1038/nature08119 (2009).

8 MacRae, C. A. & Peterson, R. T. Zebrafish as tools for drug discovery. *Nature Reviews Drug Discovery* **14**, 721–731, doi:10.1038/nrd4627 (2015).

9 Warburg, O. Beobachtungen über die Oxydationsprozesse im Seeigelei. *Journal of Biological Chemistry* **57**, 1, doi:10.1515/bchm2.1908.57.1-2.1 (1908).

10 Warburg, O. Über die Oxydationen in lebenden Zellen nach Versuchen am Seeigelei. *Journal of Biological Chemistry* **66**, 305, doi:10.1515/bchm2.1910.66.4-6.305 (1910).

11 Sies, H., Berndt, C. & Jones, D. P. Oxidative stress. *Annual Review of Biochemistry* **86**, 715–748, doi:10.1146/annurev-biochem-061516-045037 (2017).

12 Yang, L. Q., Chen, M., Zhang, J. T., Ren, D. T. & Hu, B. Hypoxia delays oligodendrocyte progenitor cell migration and myelin formation by suppressing Bmp2b signaling in larval zebrafish. *Frontiers in Cell Neuroscience* **12**, 348, doi:10.3389/fncel.2018.00348 (2018).

13 Jensen, L. D. et al. VEGF-B-Neuropilin-1 signaling is spatiotemporally indispensable for vascular and neuronal development in zebrafish. *Proceedings of the National Academy of Sciences of the United States of America* **112**, E5944–5953, doi:10.1073/pnas.1510245112 (2015).

14 Gerri, C. et al. Hif-1alpha regulates macrophage-endothelial interactions during blood vessel development in zebrafish. *Nature Communications* **8**, 15492, doi:10.1038/ncomms15492 (2017).

15 Watson, O. et al. Blood flow suppresses vascular Notch signalling via dll4 and is required for angiogenesis in response to hypoxic signalling. *Cardiovascular Research* **100**, 252–261, doi:10.1093/cvr/cvt170 (2013).

16 Timme-Laragy, A. R., Hahn, M. E., Hansen, J. M., Rastogi, A. & Roy, M. A. Redox stress and signaling during vertebrate embryonic development: Regulation and responses. *Seminars in Cell*

and *Developmental Biology* **80**, 17–28, doi:10.1016/j.semcdb.2017.09.019 (2018).

17 Romero, M. M. G., McCathie, G., Jankun, P. & Roehl, H. H. Damage-induced reactive oxygen species enable zebrafish tail regeneration by repositioning of Hedgehog expressing cells. *Nature Communications* **9**, 4010, doi:10.1038/s41467-018-06460-2 (2018).

18 Peng, G. *et al.* Nitric oxide-dependent biodegradation of graphene oxide reduces inflammation in the gastrointestinal tract. *Nanoscale* **12**, 16730–16737, doi:10.1039/d0nr03675g (2020).

19 Wingert, R. A. *et al.* Deficiency of glutaredoxin 5 reveals Fe-S clusters are required for vertebrate haem synthesis. *Nature* **436**, 1035–1039, doi:10.1038/nature03887 (2005).

20 Berndt, C., Poschmann, G., Stuhler, K., Holmgren, A. & Brautigam, L. Zebrafish heart development is regulated via glutaredoxin 2 dependent migration and survival of neural crest cells. *Redox Biology* **2**, 673–678, doi:10.1016/j.redox.2014.04.012 (2014).

21 Brautigam, L. *et al.* Glutaredoxin regulates vascular development by reversible glutathionylation of sirtuin 1. *Proceedings of the National Academy of Sciences of the United States of America* **110**, 20057–20062, doi:10.1073/pnas.1313753110 (2013).

22 Haunhorst, P. *et al.* Crucial function of vertebrate glutaredoxin 3 (PICOT) in iron homeostasis and hemoglobin maturation. *Molecular Biology of the Cell* **24**, 1895–1903, doi:10.1091/mbc.E12-09-0648 (2013).

23 Yang, L. *et al.* Functions of thioredoxin1 in brain development and in response to environmental chemicals in zebrafish embryos. *Toxicology Letters* **314**, 43–52, doi:10.1016/j.toxlet.2019.07.009 (2019).

24 Katsouda, A., Peleli, M., Asimakopoulou, A., Papapetropoulos, A. & Beis, D. Generation and characterization of a CRISPR/Cas9-induced 3-mst deficient zebrafish. *Biomolecules* **10**, doi:10.3390/biom10020317 (2020).

25 Yamashita, A. *et al.* Increased susceptibility to oxidative stress-induced toxicological evaluation by genetically modified nrf2a-deficient zebrafish. *Journal of Pharmacological and Toxicology Methods* **96**, 34–45, doi:10.1016/j.vascn.2018.12.006 (2019).

26 Hansen, J. M. & Harris, C. Glutathione during embryonic development. *Biochimica et Biophysica Acta* **1850**, 1527–1542, doi:10.1016/j.bbagen.2014.12.001 (2015).

27 Zheng, Y. L. *et al.* Rational design of an ESIPT-based fluorescent probe for selectively monitoring glutathione in live cells and zebrafish. *Spectrochimica Acta Part A: Molecular and Biomolecular Spectroscopy* **238**, 118429, doi:10.1016/j.saa.2020.118429 (2020).

28 Rastogi, A., Clark, C. W., Conlin, S. M., Brown, S. E. & Timme-Laragy, A. R. Mapping glutathione utilization in the developing zebrafish (*Danio rerio*) embryo. *Redox Biology* **26**, 101235, doi:10.1016/j.redox.2019.101235 (2019).

29 Chen, S., Hou, P., Sun, J., Wang, H. & Liu, L. Recognition of thiols in living cells and zebrafish using an Imidazo[1,5-alpha]pyridine-derivative indicator. *Molecules* **24**, doi:10.3390/molecules24183328 (2019).

30 Du, Z. *et al.* Iridium(III) Complex-based activatable probe for phosphorescent/time-gated luminescent sensing and imaging of cysteine in mitochondria of live cells and animals. *Chemistry* **25**, 1498–1506, doi:10.1002/chem.201805079 (2019).

31 Rieger, S. & Sagasti, A. Hydrogen peroxide promotes injury-induced peripheral sensory axon regeneration in the zebrafish skin. *PLoS Biology* **9**, e1000621, doi:10.1371/journal.pbio.1000621 (2011).

32 Yoo, S. K., Starnes, T. W., Deng, Q. & Huttenlocher, A. Lyn is a redox sensor that mediates leukocyte wound attraction in vivo. *Nature* **480**, 109–112, doi:10.1038/nature10632 (2011).

33 Xu, C. & Qian, Y. A selenamorpholine-based redox-responsive fluorescent probe for targeting lysosome and visualizing exogenous/endogenous hydrogen peroxide in living cells and zebrafish. *Journal of Materials Chemistry B* **7**, 2714–2721, doi:10.1039/c8tb03010c (2019).

34 Phan, Q. T. *et al.* Neutrophils use superoxide to control bacterial infection at a distance. *PLoS Pathogens* **14**, e1007157, doi:10.1371/journal.ppat.1007157 (2018).

35 Zhang, J. *et al.* A phosphinate-based near-infrared fluorescence probe for imaging the superoxide radical anion in vitro and in vivo. *Chemical Communications (Cambridge)* **52**, 2679–2682, doi:10.1039/c5cc09976e (2016).

36 Zhang, Y. *et al.* Novel fluorescent probe with a bridged Si-O-Si bond for the reversible detection of hypochlorous acid and biothiol amino acids in live cells and zebrafish. *Analyst* **144**, 5075–5080, doi:10.1039/c9an00844f (2019).

37 Wang, L., Zhang, J., An, X. & Duan, H. Recent progress on the organic and metal complex-based

fluorescent probes for monitoring nitric oxide in living biological systems. *Organic and Biomolecular Chemistry* **18**, 1522–1549, doi:10.1039/c9ob02561h (2020).

38 Palanisamy, S. et al. In vitro and in vivo imaging of peroxynitrite by a ratiometric boronate-based fluorescent probe. *Biosensors Bioelectronics* **91**, 849–856, doi:10.1016/j.bios.2017.01.027 (2017).

39 Naha, S., Thirumalaivasan, N., Garai, S., Wu, S. P. & Velmathi, S. Nanomolar detection of H2S in an aqueous medium: Application in endogenous and exogenous imaging of HeLa cells and zebrafish. *ACS Omega* **5**, 19896–19904, doi:10.1021/acsomega.0c02963 (2020).

40 Brautigam, L. et al. Hypoxic signaling and the cellular redox tumor environment determine sensitivity to MTH1 inhibition. *Cancer Research* **76**, 2366–2375, doi:10.1158/0008-5472.CAN-15-2380 (2016).

41 Shokhina, A. G. et al. Red fluorescent redox-sensitive biosensor Grx1-roCherry. *Redox Biology* **21**, 101071, doi:10.1016/j.redox.2018.101071 (2019).

42 Bilan, D. S. et al. HyPer-3: A genetically encoded H(2)O(2) probe with improved performance for ratiometric and fluorescence lifetime imaging *ACS Chemical Biology* **8**, 535–542, doi:10.1021/cb300625g (2013).

43 Hanschmann, E. M., Godoy, J. R., Berndt, C., Hudemann, C. & Lillig, C. H. Thioredoxins, glutaredoxins, and peroxiredoxins—molecular mechanisms and health significance: From cofactors to antioxidants to redox signaling. *Antioxidants & Redox Signaling* **19**, 1539–1605, doi:10.1089/ars.2012.4599 (2013).

44 Martin, J. L. Thioredoxin: A fold for all reasons. *Structure* **3**, 245–250, doi:10.1016/s0969-2126(01)00154-x (1995).

45 Zhang, J. et al. The mitochondrial thioredoxin is required for liver development in zebrafish. *Current Molecular Medicine* **14**, 772–782, doi:10.2174/1566524014666140724103927 (2014).

46 Berndt, C., Lillig, C. H. & Flohé, L. Redox regulation by glutathione needs enzymes. *Frontiers in Pharmacology* **5**, 168, doi:10.3389/fphar.2014.00168 (2014).

47 Brautigam, L. et al. MGST1, a GSH transferase/peroxidase essential for development and hematopoietic

stem cell differentiation. *Redox Biology* **17**, 171–179, doi:10.1016/j.redox.2018.04.013 (2018).

48 Lillig, C. H., Berndt, C. & Holmgren, A. Glutaredoxin systems. *Biochimica et Biophysica Acta* **1780**, 1304–1317, doi:10.1016/j.bbagen.2008.06.003 (2008).

49 Berndt, C. & Lillig, C. H. Glutathione, Glutaredoxins, and Iron. *Antioxidants & Redox Signaling* **27**, 1235–1251, doi:10.1089/ars.2017.7132 (2017).

50 Liedgens, L. et al. Quantitative assessment of the determinant structural differences between redox-active and inactive glutaredoxins. *Nature Communications* **11**, 1725, doi:10.1038/s41467-020-15441-3 (2020).

51 Trnka, D. et al. Molecular basis for the distinct functions of redox-active and FeS-transfering glutaredoxins. *Nature Communications* **11**, 3445, doi:10.1038/s41467-020-17323-0 (2020).

52 Brautigam, L. et al. Vertebrate-specific glutaredoxin is essential for brain development. *Proceedings of the National Academy of Sciences of the United States of America* **108**, 20532–20537, doi:10.1073/pnas.1110085108 (2011).

53 Gellert, M. et al. Identification of a dithiol-disulfide switch in collapsin response mediator protein 2 (CRMP2) that is toggled in a model of neuronal differentiation. *Journal of Biological Chemistry* **288**, 35117–35125, doi:10.1074/jbc.M113.521443 (2013).

54 Stocker, S., Maurer, M., Ruppert, T. & Dick, T. P. A role for 2-Cys peroxiredoxins in facilitating cytosolic protein thiol oxidation. *Nature Chemical Biology* **14**, 148–155, doi:10.1038/nchembio.2536 (2018).

55 Huang, P. C. et al. Prdx1-encoded peroxiredoxin is important for vascular development in zebrafish. *FEBS Letters* **591**, 889–902, doi:10.1002/1873-3468.12604 (2017).

56 Formella, I. et al. Real-time visualization of oxidative stress-mediated neurodegeneration of individual spinal motor neurons in vivo. *Redox Biology* **19**, 226–234, doi:10.1016/j.redox.2018.08.011 (2018).

57 Parng, C., Ton, C., Lin, Y. X., Roy, N. M. & McGrath, P. A zebrafish assay for identifying neuroprotectants in vivo. *Neurotoxicology and Teratology* **28**, 509–516, doi:10.1016/j.ntt.2006.04.003 (2006).

Development/
Differentiation of
Vertebrate Tissues

The Differential Roles of NOX-Derived Reactive Oxygen Species in Development, Function, and Dysfunction of the Nervous System

Ernesto Muñoz-Palma and Christian González-Billault

CONTENTS

5.1 INTRODUCTION

The oxidative stress theory established that a buildup of reactive oxygen species (ROS) is responsible for accumulated redox modifications that affect the structure and function of the building blocks in a cell: proteins, lipids, and nucleic acids. Such abnormal accumulation of damaged macromolecules is induced by an imbalance between the production of oxidant species and enzymatic and molecular cell defense mechanisms (Sies, 1997; Sies and Jones, 2020). However, it has become apparent that in addition to these accumulated and deleterious features, ROS play a fundamental role as signaling molecules and are implicated in several cell functions that include cell proliferation, differentiation, migration, and stress adaptation. The previous paradigms concerning oxidative stress concepts have been recently challenged, and two emerging concepts currently define the physiological roles of ROS, denoted as oxidative eustress, and the abnormal accumulation of molecular damage owing to non-physiological levels of ROS, termed oxidative

distress (Sies, 2017; Sies and Jones, 2020). In this chapter, we describe evidence supporting a physiological role of ROS (particularly H_2O_2) derived from the activity of NADPH oxidase (NOX) and discuss its different functions in the life of a neuron. We provide sections addressing the role of H_2O_2 and redox signaling in several aspects related to neurogenesis and differentiation, neuronal development, and the role of H_2O_2 in neurotransmission and synaptic plasticity. We also highlight how dysfunctional physiological ROS production can lead to known difficulties/disease in the nervous system and point to potential therapeutic strategies.

ROS correspond to a set of reactive molecules derived from the sequential reduction of molecular oxygen. In living systems, the three main ROS are superoxide anions (O_2^-), hydrogen peroxide (H_2O_2) and hydroxyl radicals (OH-) (Bigarella et al., 2014). These molecules display different properties in terms of their molecular stability, diffusion, specificity, and reactivity. However, independent of the ROS source, most superoxide anions dismute into

DOI: 10.4324/9781003204091-8

H2O2, either enzymatically by superoxide dismutase (SOD) or spontaneously (Winterbourn, 2008). In contrast to the high reactivity of superoxide anions and hydroxyl radicals, H_2O_2 is less reactive and more stable, which allows it to have enhanced diffusivity that could modify target proteins that are distal from the source of ROS production (Forman et al., 2010). The different properties of ROS suggest that they can be engaged in different signal transduction cascades owing to the fact that they can react with proteins, lipids, and nucleic acids. However, for the purposes of this chapter, we will primarily focus on the role of redox signaling linked to the oxidization of specific cysteine residues of proteins that affect their structure and function.

In the nervous system, ROS can be generated by different sources, especially mitochondria, which is the main cellular organelle involved in ATP production using oxygen, and therefore produces ROS as a by-product from metabolism (Kirkinezos and Moraes, 2001). Other sources also produce ROS, such as neuronal nitric oxidase synthase (nNOS; Forstermann and Sessa, 2012) and monoamine oxidase (MAO; Riederer et al., 1987). However, the main source of regulated ROS production is mediated by NOX, which is expressed in several cell and tissue types, including those of the nervous system (Bedard and Krause, 2007).

5.2 NADPH OXIDASES IN THE NERVOUS SYSTEM

NOX is a family of membrane-bound enzymes whose primary catalytic function, depending on the member of the family, is the generation of superoxide or H_2O_2 (Sies and Jones, 2020). These enzymes catalyze the transfer of electrons from intracellular reduced NADPH as an electron donor towards extracellular molecular oxygen as the electron acceptor, forming primarily the superoxide anion (Bedard and Krause, 2007). However, this primary enzymatic product can be rapidly dismuted into H_2O_2, the main ROS signaling molecule, whose transport into the cell is facilitated through the water channels, the aquaporins (Figure 5.1). In most eukaryotic cells, the NOX family is composed of seven members, named NOX1–5 and dual oxidase 1 and 2 (DUOX1–2).

All catalytic subunits of the NOX family possess an NADPH oxidase domain characterized by six (seven for DUOX1-2) membrane-spanning alpha-helical domains containing two heme groups, a

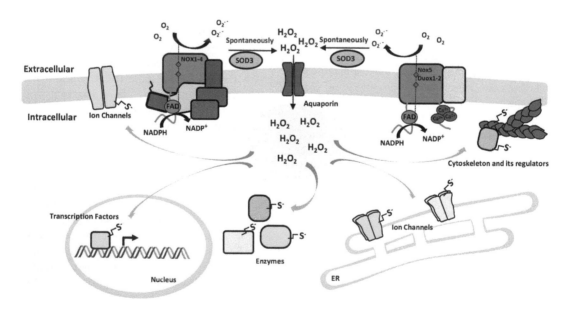

Figure 5.1 Controlled hydrogen peroxide production by NADPH oxidase complexes modulates several cellular processes. Upon activation, NOX promotes the production of extracellular superoxide anions that spontaneously or enzymatically dismute into H_2O_2. NOX-derived H_2O_2 translocates into the cytoplasm through aquaporins. Once inside the cell, H_2O_2 can reach and oxidize specific thiols of cysteine residues of a wide range of proteins, such as transcription factors, enzymes, ion channels (at the plasma and endoplasmic membranes), the cytoskeleton, and its regulators, among others, thus regulating a wide range of cellular processes.

region for flavin adenine dinucleotide (FAD), and a NADPH binding site in the cytosolic C-terminus (Kawahara and Lambeth, 2007). Despite the similarity in their structural core, the NOX family members differ in their subunit requirements, activation mechanisms, and characteristic tissue distribution. Whereas NOX1, NOX2, and NOX3 require cytosolic subunits for full activation, NOX4 generates H_2O_2 constitutively, and NOX5 and DUOX require raised intracellular Ca^{2+} concentrations (Buvelot et al., 2019; Nayernia et al., 2014). NOX isoforms have been found in several cells and tissue, with specific expression and distribution patterns, including in the nervous system (Bedard and Krause, 2007; Nayernia et al., 2014; Wilson et al., 2018). The NOX2 isoform, originally termed gp91[phox], is the best characterized member of this family. At the plasma membrane, NOX2 interacts and binds to the p22[phox] subunit, a protein that brings together and stabilizes the NOX2 complex. There are other regulatory cytosolic proteins that are needed for NOX2 activation: p47[phox] corresponds to the organizer subunit, p40[phox] is a regulator subunit, and p67[phox] and the small GTPase Rac1 act as activator subunits (Buvelot et al., 2019). Thus, after intracellular or extracellular stimuli, the assembly of the complex leads to superoxide anion production, which dismutes spontaneously

or enzymatically into H_2O_2, the main physiological ROS with molecular signaling features.

5.3 PROTEIN REDOX MODIFICATION BY H_2O_2

ROS function as signaling molecules acting on specific targets. The main biological targets of H_2O_2 are cysteine residues in proteins. At physiological pH, most cysteine residues in proteins are protonated (Cys-SH), showing a weak reaction with H_2O_2. However, oxidative modification only takes place on the thiolate anion form (Cys-S-), which is the ionized and reactive cysteine residue (Figure 5.2). These cysteines are characterized by a low ionization constant (pKa) that depends on its protein environment (Bilan and Belousov, 2018; Paulsen and Carroll, 2013).

Thus, H_2O_2 can efficiently oxidize a thiolate anion generating sulfenic acid (Cys-SOH) on specific cysteine residues in a variety of proteins (Figure 5.1). This oxidation not only depends on its pKa value but also on the pH inside the cellular compartments, thiol accessibility, the steric orientation of H_2O_2 and thiol in the reaction site, and its stabilization of the reaction transition state (Nagy, 2013; Salsbury et al., 2008). Upon cysteine oxidation, these modifications lead to local or global conformational rearrangements that can alter structure, catalysis, and protein function (Wani et al., 2014),

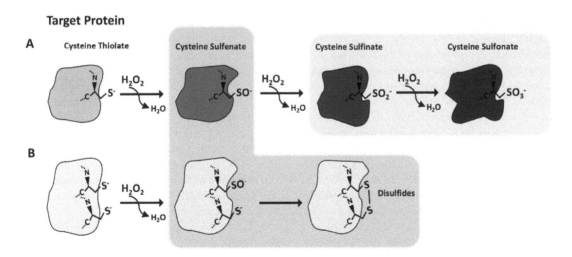

Figure 5.2 Oxidative modification of cysteine residues in proteins by hydrogen peroxide.

(A) Target proteins containing specific cysteine residues (thiolate) can be oxidized by H_2O_2 forming cysteine sulfenate. This reversible modification can alter the structure and function of proteins, thus controlling intracellular signaling (shaded in dark grey). In the presence of excess H_2O_2, such as under conditions of oxidative stress, the cysteine sulfenate can be hyperoxidized, resulting in cysteine sulfinate and sulfonate, which irreversibly and negatively alters protein function (shaded in light grey). (B) Oxidized cysteine can react with a neighboring cysteine, resulting in a disulfide bridge, which also alters the structure and function of proteins (shaded in dark grey).

leading to the modulation of protein signaling and cellular function. Importantly, these modifications are reversible processes due to the action of anti-oxidant enzymes such as SOD, peroxiredoxins, glutathione peroxidases, and catalase as well as the molecule, glutathione. Altogether, these antioxidant systems maintain steady-state intracellular H_2O_2 levels in the nanomolar to low micromolar range, suitable for physiological cell function (Sies, 2017). However, in the presence of excess or accumulation of H_2O_2 (oxidative stress), the oxidized cysteine can be irreversibly hyperoxidized, leading to cysteine sulfinate and sulfonate forms, which is detrimental for protein function (Figure 5.2).

The reversibility of redox modification of cysteine thiols represents an important property in their function as a binary redox switch, regulating protein function, interactions, and localization (Paulsen and Carroll, 2013), in addition to other well-known posttranslational modifications, such as phosphorylation and acetylation. Therefore, the covalent redox modification of proteins, being both site-specific and reversible, is an essential molecular mechanism to convert an oxidant signal into a biological response.

5.4 ROLE OF ROS IN THE NERVOUS SYSTEM

The brain, the most complex organ in mammals, consumes over 20% of the total oxygen used by an organism to support ATP production and thus neuronal activity (Cobley et al., 2018). Therefore, it is highly exposed to increasingly oxidant environments, and so the generation of ROS that can produce redox modifications must be tightly regulated in order to maintain normal neuronal functions. Moreover, neurons are postmitotic cells that need to cope with cumulative effects of pro-oxidant molecules including H_2O_2. There is substantial evidence showing that the expression and participation of the NOX family is important in practically all cellular activities affecting the development and function of the nervous system (Wilson et al., 2018). In the following sections, we will focus on several processes in which ROS have an important role, such as neural proliferation and differentiation, neuritogenesis, neuronal development, and neurotransmission.

5.5 ROLE OF HYDROGEN PEROXIDE IN NEUROGENESIS AND DIFFERENTIATION

During brain development, the neural stem cells (NSCs) and progenitors undergo proliferation and/

or differentiation to produce functional brain tissue (Joseph and Hermanson, 2010). A growing body of literature highlights the role of ROS production by NOX complexes during neurogenesis (Forsberg et al., 2013; Ostrakhovitch and Semenikhin, 2013; Vieira et al., 2011), demonstrating an important role in the differentiation of several cell lineages from their respective precursors including embryonic, adult, and induced pluripotent stem cells (Chaudhari et al., 2014; Borquez et al., 2016).

Early evidence of ROS participation in neurogenesis derived from the finding that neuroblastoma cells differentiate into neurons in high oxygen concentrations in vitro (Nissen et al., 1973). In addition, hyperoxia (50% O_2) resulted in differentiated neuronal phenotypes of PC12 cells by ROS production. Furthermore, these neural features were also observed in the presence of pro-oxidizing agents such as xanthine/xanthine oxidase, which generate H_2O_2 and O_2^-, whereas treatment with antioxidants abolished hyperoxia-induced neurite extension and differentiation (Katoh et al., 1997). In addition, the differentiation of nerve cells via nerve growth factor, retinoic acid, and neuregulin, among others, requires the production of ROS mediated by NOX complexes because the presence of ROS scavengers, NOX inhibitors, or the overexpression of a dominant negative form of the Rac protein abolished neural differentiation (Goldsmit et al., 2001; Suzukawa et al., 2000; Nitti et al., 2010). Altogether, these results highlight that controlled ROS production is necessary to support nerve cell differentiation.

On the other hand, during brain development, changes in the content of ROS can modify the ratio between proliferation and differentiation of neuronal stem and progenitor cells at the ventricular and subventricular zones (Doe, 2008). The reduction of endogenous ROS levels both by the presence of free radical–scavenging agents, as well as the inhibition of the NOX complex using apocynin, significantly inhibits proliferation of neural stem/progenitor cells, whereas treatment with SIN-1, a peroxynitrite generator, promotes the proliferation activity of embryonic hippocampal-derived neural progenitor cells (Yoneyama et al., 2010), suggesting that endogenous ROS are essential for the proliferation of embryonic neural stem/progenitor cells. Moreover, both NOX2 knockdown in cultured adult hippocampal progenitor cells as well as NOX2 knock-out mice display a decrease in the number of proliferating progenitors in the adult hippocampus, suggesting that ROS production via NOX2 is essential for neurogenesis of neuronal precursor cells (NPCs) in vitro and in vivo (Dickinson et al., 2011).

Another relevant participant in ROS-dependent neurogenesis is the tumor suppressor p53 (Forsberg et al., 2013). It has been described that the loss of function of p53 leads to elevated ROS levels that induce early neurogenesis, whereas restoration of p53 and antioxidant treatment partially reverse the phenotype associated with premature neurogenesis, suggesting that p53 fine-tunes endogenous ROS levels to ensure the appropriate timing of neurogenesis in NPCs (Forsberg et al., 2013). Moreover, in vitro experiments show that the stimulation of cultured embryonic rat cortical cells with fibroblast growth factor 2 (FGF2) promotes the acquisition of the conspicuous cell morphology of neurons that is accompanied by higher ROS levels compared to other neural progenitor or glial cells. This molecular feature can be further utilized to isolate different cell populations that are highly enriched in either cortical neurons or neural progenitor cells (Tsatmali et al., 2005). In addition, ROS were also described to modulate the differentiation of neurons in clonal cortical cultures (Tsatmali et al., 2006). Therefore, a proper balance of ROS production, signaling cascades after oxidative modification of proteins, and tight ROS regulation are critical for the regulation of neural self-renewal and differentiation. Putting all this information together, it is suggested that H_2O_2 production mediated by NOX complexes promotes neural differentiation instead of proliferation.

5.6 ROLE OF HYDROGEN PEROXIDE IN NEURONAL DEVELOPMENT

Brain functions are highly dependent on the acquisition of a polarized morphology by neurons. Such acquisition depends on the presence of two seemingly different cellular domains that differ in their molecular composition, structural features, and functions: the somatodendritic and axonal compartments. Dendrites are the sites in neurons where neural transmission is received. In contrast, the axon is involved in the generation of the action potential and propagation of information in the nervous system. These two domains are formed during early neuronal development at embryonic stages where axonal specification and outgrowth (i.e., neuronal polarization) takes place (Dotti et al., 1988; Takano et al., 2015). Several signaling pathways both in vitro and in vivo have been

described to control such processes. Intrinsic and extrinsic factors have been reported to regulate neuronal polarization including TGFB, neurotrophins such as BDNF and NT-3, and PI3 kinases (Namba et al., 2015). Interestingly, several of these factors have also been implicated in ROS production in models of neurite outgrowth during cell line differentiation. The first indications suggesting a connection between ROS production and neuronal differentiation were obtained two decades ago, when it was demonstrated that several extracellular cues that control neurite protrusion and elongation in immortalized cells rely on ROS produced by the NOX complexes (Goldsmit et al., 2001; Nitti et al., 2010; Suzukawa et al., 2000). However, only recently have some of the mechanisms responsible for such cellular modifications started to be unveiled. Both in embryonic hippocampal neurons and cerebellar granule neurons, it has been shown that NOX2-mediated ROS participated in the establishment of neuronal polarity, neurite growth, and axonal specification (Olguin-Albuerne and Moran, 2015; Wilson et al., 2016; Wilson et al., 2015). First, the NOX complex subunits are expressed and distributed in somatodendritic and axonal compartments during neuronal polarization. Second, both pharmacological treatments with NOX2 inhibitors and genetic expression of a dominant negative version of p22[phox] delayed neuronal polarity acquisition and reduced axonal outgrowth (Wilson et al., 2015). By contrast, overactivation of NOX2 by overexpression of the p47[phox] regulatory subunit strongly enhanced axonal growth (Wilson et al., 2016). Finally, neurons derived from both NOX2- and p47[phox] (ncf1-/-) null mice showed reduced neurite and axon length compared to control neurons (Olguin-Albuerne and Moran, 2015; Wilson et al., 2016). Altogether, these studies suggest that NOX2 activity modulates the redox state and supports early neuronal development.

The mechanisms by which physiological levels of ROS derived from NOX2 sustain neuronal polarization and axonal development have been well-documented. Interestingly, physiological NOX2 activity generates ROS, which promotes calcium release from intracellular stores, such as the endoplasmic reticulum (ER), mediated by the ryanodine (RyRs) and inositol-1,4,5-triphosphate (IP3) receptors. We hypothesized that these changes affecting calcium release from the ER depend on transient redox modification, likely altering cysteine residues sensitive to ROS that are present in these receptors

(Aracena-Parks et al., 2006; Donoso et al., 2011). This idea is supported by the fact that both NOX2 inhibition and neurons derived from p47phox null mice failed to release calcium mediated by RyR (Wilson et al., 2016). In addition, experiments performed in HEK293 cells and cerebellar granule neurons suggest that calcium release mediated by RyRs promotes Rac1 activation (Jin et al., 2005), an essential regulatory subunit of the NOX2 complex (Lambeth, 2004). In turn, Rac1 activation stimulates and maintains NOX2 activity, suggesting the existence of a positive feedback loop that controls neuronal development (Wilson et al., 2016). However, how such a mechanism is coupled to external cues, and the identity of the ligands that promote and maintain such positive feedback, are still unknown.

In parallel, cultured neurons derived from the invertebrate *Aplysia* showed that the reduction of ROS content both by general and specific sources of ROS (like NOX and lipoxygenases) reduced the F-actin content in the peripheral domain of growth cones. In addition, ROS depletion reduced actin polymerization and retrograde actin flow in growth cones while promoting the contractility of actin structures in the growth cone transition zone, impairing growth cone formation and reducing neurite outgrowth. Stimulation of the NOX2 complex increased levels of ROS in the growth cone periphery (Munnamalai and Suter, 2009; Munnamalai et al., 2014). Taken together, these studies reinforce the idea that physiological ROS production derived from the NOX2 complex controls neuronal morphology and, therefore, may induce local reversible changes affecting essential components that regulate the neuronal cytoskeleton (Olguin-Albuerne and Moran, 2015; Wilson et al., 2016; Wilson et al., 2015).

5.7 PHYSIOLOGICAL ROLES OF HYDROGEN PEROXIDE IN NEUROTRANSMISSION

Brain processive functions are heavily dependent on the ability of the neuronal network to transiently modify its structure and connectivity patterns. These structural changes are indeed accompanied by molecular changes that regulate the assembly and function of the synaptic contacts between nerve cells. Neuronal activity–dependent modifications of the strength or efficacy of synaptic transmission at preexisting synapses are associated with the concept of synaptic plasticity (Citri and Malenka, 2008) and include long-term potentiation (LTP) and long-term depression (LTD), where synaptic transmission can be either enhanced or depressed by activity involving either long-lasting increases or decreases in synaptic strength and efficacy, respectively (Blundon and Zakharenko, 2008; Lynch, 2004). These two processes are at the core of learning and memory mechanisms. During the induction and consolidation of LTP or LTD (forms of synaptic plasticity), classical cell signaling cascades have been shown to be required. Most of the synaptic plasticity expression is mediated by NMDA receptors. For instance, in excitatory glutamatergic neurons, NMDA receptor activation is dependent on presynaptic glutamate release, the binding to glutamate receptors, and postsynaptic depolarization (Kaneki et al., 2009; Zucker, 1999). After postsynaptic NMDA receptors are activated, the influx and changes of intracellular Ca^{2+} into the postsynaptic terminal can lead to activation of intracellular signaling cascades involving a number of protein kinases, such as calcium/calmodulin kinase II, extracellular signal-regulated kinase (ERK), cyclic adenosine monophosphate (cAMP)–dependent protein kinase A, and protein kinase C, all of which are involved in the induction and expression of LTP, LTD, and memory formation (Sweatt, 2004; Xia and Storm, 2005).

A current and cumulative body of evidence indicates that physiological levels of ROS, functioning as signaling molecules, regulate signal transduction mainly through reversible oxidation of protein cysteine residues (Cross and Templeton, 2006) and are involved in synaptic plasticity in the nervous system (Kishida and Klann, 2007; Massaad and Klann, 2011; Wilson et al., 2015a). Early evidence showed that exogenous superoxide scavengers, mimicking the activity of SOD, prevented induction of LTP in the CA1 area of rat hippocampal slices (Klann, 1998). On the one hand, hippocampal slices from mice overexpressing both extracellular SOD (EC-SOD) or cytosolic SOD (SOD1) showed deficits in LTP, where superoxide production in the hippocampal slice was insufficient to trigger the induction of LTP or increase H_2O_2 accumulation, respectively (Kamsler and Segal, 2004; Thiels et al., 2000). On the other hand, mice overexpressing mitochondrial SOD (SOD2) exhibited normal LTP, suggesting that mitochondrial superoxide anion production was unlikely to play a role in promoting synaptic plasticity (Hu et al., 2007). Overall, the overexpression of the three different SOD isoforms affects LTP in different ways, suggesting that mechanisms underlying these processes are varied and involve several molecular mechanisms. However, it is clear that superoxide anions are necessary to support the induction of LTP.

A growing body of evidence points to the other important source of ROS production—the NOX complex—as playing an increasingly relevant role in neurotransmission (Beckhauser et al., 2016; Nayernia et al., 2014). First of all, NOX subunits are expressed, localized, and function at synaptic sites of mature neurons of murine brain (Massaad and Klann, 2011; Sorce and Krause, 2009; Tejada-Simon et al., 2005; Vallet et al., 2005). In human, pediatric patients affected by chronic granulomatous disease (CGD), a hereditary syndrome caused by mutations in genes encoding NOX subunits, which is characterized by susceptibility to fungal and bacterial infections due to the failure of phagocytes to generate superoxide and thus kill these microorganisms (Schappi et al., 2008), also display cognitive defects and lower intellectual coefficients (Pao et al., 2004). Moreover, pharmacological NOX inhibition or knockout mice lacking subunits gp91[phox] or p47[phox] (mouse models for human CGD) exhibit impaired hippocampal LTP and memory formation along with mild deficits in spatial memory (Kishida et al., 2006), consistent with reports of cognitive dysfunction in patients with CGD. Furthermore, mice lacking gp91[phox] show impaired LTP and LTD in the primary visual cortex (De Pasquale et al., 2014). Altogether, the deficiency in NOX function affects LTP induction, strongly suggesting that H_2O_2 production and signaling mediated by NOX are necessary for synaptic plasticity and cognitive function. Therefore, the superoxide anion and H_2O_2 are necessary for neural transmission by either potentiating or depressing synaptic plasticity within the physiological ROS range (Hidalgo and Arias-Cavieres, 2016; Massaad and Klann, 2011; Wilson et al., 2018).

The production of ROS mediated by NOX has been shown to respond to specific extracellular and intracellular stimuli in multiple cell types (Zhang et al., 2016). In the nervous system, previous reports have shown that NMDA receptor activation promotes superoxide production (Bindokas et al., 1996) and that it is mediated by the NOX complex in hippocampal and cortical mature neurons, suggesting crosstalk between glutamatergic signaling and NOX activity (Brennan et al., 2009; Girouard et al., 2009; Guemez-Gamboa et al., 2011; Reyes et al., 2012). Additionally, the administration of H_2O_2 in hippocampal slices promotes the phosphorylation of ERK. This could be blocked by antioxidants (Kanterewicz et al., 1998), whereas NOX is required for ERK phosphorylation (and activation) via NMDAR-dependent activation (Kishida et al., 2006). Even though NOX-mediated superoxide anion production has been linked to neurotoxicity after NMDAR activation (Brennan et al., 2009; Patel et al., 1996; Reyes et al., 2012; Suh et al., 2008), and each of these participants is essential for cognitive function, the physiological crosstalk between NMDARs and NOX-mediated ROS production still needs to be further explored.

5.8 CONCLUSIONS

ROS are not merely by-products of metabolism. Their physiological roles in the nervous system position them as key molecular players controlling the structure, function, and dysfunction of brain cells. Regulated production of H_2O_2 is now widely accepted as a signaling mechanism that, like that of other second messengers, is finely controlled by extrinsic and intrinsic pathways in cells. It is therefore very important to consider the crosstalk between the mechanisms involved in H_2O_2 production and those associated with the generation of other signaling molecules such as calcium, IP3, and cAMP/cGMP, among others. It is very likely that some of the mechanisms associated with bona fide production of the most canonical signaling molecules (e.g., calcium) will be coupled to the production of H_2O_2 mediated by the NOX family of enzymes. Therefore, reversible redox modifications affecting key proteins in cells emerge as a novel layer of complexity in cell signaling that we expect will contribute to more sophisticated and integrated cell regulatory mechanisms. Migrating away from the traditional vision that ROS can only produce damage in macromolecules, and its association with oxidative stress, opens up many other possibilities for the development of novel therapeutic strategies to reveal the complex nature of cell responses to redox biology management. While in the past, the prevalent approach was to use broad antioxidant molecules to target the excessive production of ROS in biological systems, such a strategy has proven to be much less successful than originally thought. If we acknowledge that regulated production of ROS, particularly H_2O_2, can serve signaling purposes, it is of utmost importance to identify the physiological ranges required to support the many functions it exerts in different tissues and cells. It is also relevant to identify small molecules or compounds that specifically target the sources involved in the regulated production of H_2O_2. These compounds may be well positioned to enter the drug discovery pipeline to address specific pathological conditions affecting the brain, such as CGD and Alzheimer's disease. Nonetheless, the search for these compounds should always consider the delicate balance between oxidative eustress and distress (i.e., redox homeostasis). Lessons learned in recent years suggest that the NOX family of proteins is involved in the formation and development of the nervous system and in controlling physiological and pathological functions, including nerve regeneration. While previously it was proposed that NOX inhibition could be important to counteract pathophysiological aspects of neurodegenerative diseases, it has become clear that excessive inhibition of NOX-mediated H_2O_2 production has deleterious consequences for neurons and glial cells in the brain and in the peripheral nervous system. The fact that NOX functions can be intertwined with fundamental cellular mechanisms such as cytoskeleton dynamics, cell trafficking, and transport anticipates that novel aspects related to H_2O_2 production in cells will emerge as important signaling hubs to understand the fundamental mechanisms controlling brain functions.

The development of improved genetic and small-molecule compounds to analyze and regulate the H_2O_2 content and concentrations in live cells, tissues, and organisms is a very interesting area of research that may reveal as yet unknown or poorly explored features of redox biology. We anticipate that all of these new developments will lead us to a new era where redox biology will develop into redox biomedicine. Such a new research field will necessarily address not only the mechanistic insights into nerve cell physiology and neurodegenerative diseases but also should focus on understanding brain aging at the cellular, molecular, and system levels.

5.9 ACKNOWLEDGMENTS

We thank Michael Handford for English proofreading and editing. CG-B is supported by ANID/FONDAP/15150012 and ANID/FONDECYT/1180419. EM-P is supported by ANID PhD Fellowships Program ANID/21201556 to EM-P.

REFERENCES

Aracena-Parks, P., Goonasekera, S.A., Gilman, C.P., Dirksen, R.T., Hidalgo, C., and Hamilton, S.L. (2006). Identification of cysteines involved in S-nitrosylation, S-glutathionylation, and oxidation to disulfides in ryanodine receptor type 1. J Biol Chem 281, 40354–40368.

Beckhauser, T.F., Francis-Oliveira, J., and De Pasquale, R. (2016). Reactive oxygen species: physiological and physiopathological effects on synaptic plasticity. J Exp Neurosci 10, 23–48.

Bedard, K., and Krause, K.H. (2007). The NOX family of ROS-generating NADPH oxidases: physiology and pathophysiology. Physiol Rev 87, 245–313.

Bigarella, C.L., Liang, R., and Ghaffari, S. (2014). Stem cells and the impact of ROS signaling. Development 141, 4206–4218.

Bilan, D.S., and Belousov, V.V. (2018). In vivo imaging of hydrogen peroxide with HyPer probes. Antioxid Redox Signal 29, 569–584.

Bindokas, V.P., Jordan, J., Lee, C.C., and Miller, R.J. (1996). Superoxide production in rat hippocampal neurons: selective imaging with hydroethidine. J Neurosci 16, 1324–1336.

Blundon, J.A., and Zakharenko, S.S. (2008). Dissecting the components of long-term potentiation. Neuroscientist 14, 598–608.

Borquez, D.A., Urrutia, P.J., Wilson, C., van Zundert, B., Nuñez, M.T., and Gonzalez-Billault, C. (2016). Dissecting the role of redox signaling in neuronal development. J Neurochem 137, 506–517.

Brennan, A.M., Suh, S.W., Won, S.J., Narasimhan, P., Kauppinen, T.M., Lee, H., Edling, Y., Chan, P.H., and Swanson, R.A. (2009). NADPH oxidase is the primary source of superoxide induced by NMDA receptor activation. Nat Neurosci 12, 857–863.

Buvelot, H., Jaquet, V., and Krause, K.H. (2019). Mammalian NADPH oxidases. Methods Mol Biol 1982, 17–36.

Chaudhari, P., Ye, Z., and Jang, Y.Y. (2014). Roles of reactive oxygen species in the fate of stem cells. Antioxid Redox Signal 20, 1881–1890.

Citri, A., and Malenka, R.C. (2008). Synaptic plasticity: multiple forms, functions, and mechanisms. Neuropsychopharmacology 33, 18–41.

Cobley, J.N., Fiorello, M.L., and Bailey, D.M. (2018). 13 reasons why the brain is susceptible to oxidative stress. Redox Biol 15, 490–503.

Cross, J.V., and Templeton, D.J. (2006). Regulation of signal transduction through protein cysteine oxidation. Antioxid Redox Signal 8, 1819–1827.

De Pasquale, R., Beckhauser, T.F., Hernandes, M.S., and Giorgetti Britto, L.R. (2014). LTP and LTD in the visual cortex require the activation of NOX2. J Neurosci 34, 12778–12787.

Dickinson, B.C., Peltier, J., Stone, D., Schaffer, D.V., and Chang, C.J. (2011). Nox2 redox signaling maintains essential cell populations in the brain. Nat Chem Biol 7, 106–112.

Doe, C.Q. (2008). Neural stem cells: balancing self-renewal with differentiation. Development 135, 1575–1587.

Donoso, P., Sanchez, G., Bull, R., and Hidalgo, C. (2011). Modulation of cardiac ryanodine receptor activity by ROS and RNS. Front Biosci (Landmark Ed) 16, 553–567.

Dotti, C.G., Sullivan, C.A., and Banker, G.A. (1988). The establishment of polarity by hippocampal neurons in culture. J Neurosci 8, 1454–1468.

Forman, H.J., Maiorino, M., and Ursini, F. (2010). Signaling functions of reactive oxygen species. Biochemistry 49, 835–842.

Forsberg, K., Wuttke, A., Quadrato, G., Chumakov, P.M., Wizenmann, A., and Di Giovanni, S. (2013). The tumor suppressor p53 fine-tunes reactive oxygen species levels and neurogenesis via PI3 kinase signaling. J Neurosci 33, 14318–14330.

Forstermann, U., and Sessa, W.C. (2012). Nitric oxide synthases: regulation and function. Eur Heart J 33, 829–837, 837a–837d.

Girouard, H., Wang, G., Gallo, E.F., Anrather, J., Zhou, P., Pickel, V.M., and Iadecola, C. (2009). NMDA receptor activation increases free radical production through nitric oxide and NOX2. J Neurosci 29, 2545–2552.

Goldsmit, Y., Erlich, S., and Pinkas-Kramarski, R. (2001). Neuregulin induces sustained reactive oxygen species generation to mediate neuronal differentiation. Cell Mol Neurobiol 21, 753–769.

Guemez-Gamboa, A., Estrada-Sanchez, A.M., Montiel, T., Paramo, B., Massieu, L., and Moran, J. (2011). Activation of NOX2 by the stimulation of ionotropic and metabotropic glutamate receptors contributes to glutamate neurotoxicity in vivo through the production of reactive oxygen species and calpain activation. J Neuropathol Exp Neurol 70, 1020–1035.

Hidalgo, C., and Arias-Cavieres, A. (2016). Calcium, reactive oxygen species, and synaptic plasticity. Physiology (Bethesda) 31, 201–215.

Hu, D., Cao, P., Thiels, E., Chu, C.T., Wu, G.Y., Oury, T.D., and Klann, E. (2007). Hippocampal long-term potentiation, memory, and longevity in mice that overexpress mitochondrial superoxide dismutase. Neurobiol Learn Mem 87, 372–384.

Jin, M., Guan, C.B., Jiang, Y.A., Chen, G., Zhao, C.T., Cui, K., Song, Y.Q., Wu, C.P., Poo, M.M., and Yuan, X.B. (2005). Ca^{2+}-dependent regulation of rho GTPases triggers turning of nerve growth cones. J Neurosci 25, 2338–2347.

Joseph, B., and Hermanson, O. (2010). Molecular control of brain size: regulators of neural stem cell life, death and beyond. Exp Cell Res 316, 1415–1421.

Kamsler, A., and Segal, M. (2004). Hydrogen peroxide as a diffusible signal molecule in synaptic plasticity. Mol Neurobiol 29, 167–178.

Kaneki, K., Araki, O., and Tsukada, M. (2009). Dual synaptic plasticity in the hippocampus: Hebbian and spatiotemporal learning dynamics. Cogn Neurodyn 3, 153–163.

Kanterewicz, B.I., Knapp, L.T., and Klann, E. (1998). Stimulation of p42 and p44 mitogen-activated protein kinases by reactive oxygen species and nitric oxide in hippocampus. J Neurochem 70, 1009–1016.

Katoh, S., Mitsui, Y., Kitani, K., and Suzuki, T. (1997). Hyperoxia induces the differentiated neuronal phenotype of PC12 cells by producing reactive oxygen species. Biochem Biophys Res Commun 241, 347–351.

Kawahara, T., and Lambeth, J.D. (2007). Molecular evolution of Phox-related regulatory subunits for NADPH oxidase enzymes. BMC Evol Biol 7, 178.

Kirkinezos, I.G., and Moraes, C.T. (2001). Reactive oxygen species and mitochondrial diseases. Semin Cell Dev Biol 12, 449–457.

Kishida, K.T., Hoeffer, C.A., Hu, D., Pao, M., Holland, S.M., and Klann, E. (2006). Synaptic plasticity deficits and mild memory impairments in mouse models of chronic granulomatous disease. Mol Cell Biol 26, 5908–5920.

Kishida, K.T., and Klann, E. (2007). Sources and targets of reactive oxygen species in synaptic plasticity and memory. Antioxid Redox Signal 9, 233–244.

Klann, E. (1998). Cell-permeable scavengers of superoxide prevent long-term potentiation in hippocampal area CA1. J Neurophysiol 80, 452–457.

Lambeth, J.D. (2004). NOX enzymes and the biology of reactive oxygen. Nat Rev Immunol 4, 181–189.

Lynch, M.A. (2004). Long-term potentiation and memory. Physiol Rev 84, 87–136.

Massaad, C.A., and Klann, E. (2011). Reactive oxygen species in the regulation of synaptic plasticity and memory. Antioxid Redox Signal 14, 2013–2054.

Munnamalai, V., and Suter, D.M. (2009). Reactive oxygen species regulate F-actin dynamics in neuronal growth cones and neurite outgrowth. J Neurochem 108, 644–661.

Munnamalai, V., Weaver, C.J., Weisheit, C.E., Venkatraman, P., Agim, Z.S., Quinn, M.T., and Suter, D.M. (2014). Bidirectional interactions between NOX2-type NADPH oxidase and the F-actin cytoskeleton in neuronal growth cones. J Neurochem 130, 526–540.

Nagy, P. (2013). Kinetics and mechanisms of thiol-disulfide exchange covering direct substitution and thiol oxidation-mediated pathways. Antioxid Redox Signal 18, 1623–1641.

Namba, T., Funahashi, Y., Nakamuta, S., Xu, C., Takano, T., and Kaibuchi, K. (2015). Extracellular and intracellular signaling for neuronal polarity. Physiol Rev 95, 995–1024.

Nayernia, Z., Jaquet, V., and Krause, K.H. (2014). New insights on NOX enzymes in the central nervous system. Antioxid Redox Signal 20, 2815–2837.

Nissen, C., Ciesielski-Treska, J., Hertz, L., and Mandel, P. (1973). Regulation of oxygen consumption in neuroblastoma cells: effects of differentiation and of potassium. J Neurochem 20, 1029–1035.

Nitti, M., Furfaro, A.L., Cevasco, C., Traverso, N., Marinari, U.M., Pronzato, M.A., and Domenicotti, C. (2010). PKC delta and NADPH oxidase in retinoic acid-induced neuroblastoma cell differentiation. Cell Signal 22, 828–835.

Olguin-Albuerne, M., and Moran, J. (2015). ROS produced by NOX2 control in vitro development of cerebellar granule neurons development. ASN Neuro 7.

Ostrakhovitch, E.A., and Semenikhin, O.A. (2013). The role of redox environment in neurogenic development. Arch Biochem Biophys 534, 44–54.

Pao, M., Wiggs, E.A., Anastacio, M.M., Hyun, J., DeCarlo, E.S., Miller, J.T., Anderson, V.L., Malech, H.L., Gallin, J.I., and Holland, S.M. (2004). Cognitive function in patients with chronic granulomatous disease: a preliminary report. Psychosomatics 45, 230–234.

Patel, M., Day, B.J., Crapo, J.D., Fridovich, I., and McNamara, J.O. (1996). Requirement for superoxide in excitotoxic cell death. Neuron 16, 345–355.

Paulsen, C.E., and Carroll, K.S. (2013). Cysteine-mediated redox signaling: chemistry, biology, and tools for discovery. Chem Rev 113, 4633–4679.

Reyes, R.C., Brennan, A.M., Shen, Y., Baldwin, Y., and Swanson, R.A. (2012). Activation of neuronal NMDA receptors induces superoxide-mediated oxidative stress in neighboring neurons and astrocytes. J Neurosci 32, 12973–12978.

Riederer, P., Konradi, C., Schay, V., Kienzl, E., Birkmayer, G., Danielczyk, W., Sofic, E., and Youdim, M.B. (1987). Localization of MAO-A and MAO-B in human brain: a step in understanding the therapeutic action of L-deprenyl. Adv Neurol 45, 111–118.

Salsbury, F.R., Jr., Knutson, S.T., Poole, L.B., and Fetrow, J.S. (2008). Functional site profiling and electrostatic analysis of cysteines modifiable to cysteine sulfenic acid. Protein Sci 17, 299–312.

Schappi, M.G., Jaquet, V., Belli, D.C., and Krause, K.H. (2008). Hyperinflammation in chronic granulomatous disease and anti-inflammatory role of the phagocyte NADPH oxidase. Semin Immunopathol 30, 255–271.

Sies, H. (1997). Oxidative stress: oxidants and antioxidants. Exp Physiol 82, 291–295.

Sies, H. (2017). Hydrogen peroxide as a central redox signaling molecule in physiological oxidative stress: oxidative eustress. Redox Biol 11, 613–619.

Sies, H., and Jones, D.P. (2020). Reactive oxygen species (ROS) as pleiotropic physiological signalling agents. Nat Rev Mol Cell Biol 21, 363–383.

Sorce, S., and Krause, K.H. (2009). NOX enzymes in the central nervous system: from signaling to disease. Antioxid Redox Signal 11, 2481–2504.

Suh, S.W., Hamby, A.M., Gum, E.T., Shin, B.S., Won, S.J., Sheline, C.T., Chan, P.H., and Swanson, R.A. (2008). Sequential release of nitric oxide, zinc, and superoxide in hypoglycemic neuronal death. J Cereb Blood Flow Metab 28, 1697–1706.

Suzukawa, K., Miura, K., Mitsushita, J., Resau, J., Hirose, K., Crystal, R., and Kamata, T. (2000). Nerve growth factor-induced neuronal differentiation requires generation of Rac1-regulated reactive oxygen species. J Biol Chem 275, 13175–13178.

Sweatt, J.D. (2004). Mitogen-activated protein kinases in synaptic plasticity and memory. Curr Opin Neurobiol 14, 311–317.

Takano, T., Xu, C., Funahashi, Y., Namba, T., and Kaibuchi, K. (2015). Neuronal polarization. Development 142, 2088–2093.

Tejada-Simon, M.V., Serrano, F., Villasana, L.E., Kanterewicz, B.I., Wu, G.Y., Quinn, M.T., and Klann, E. (2005). Synaptic localization of a functional NADPH oxidase in the mouse hippocampus. Mol Cell Neurosci 29, 97–106.

Thiels, E., Urban, N.N., Gonzalez-Burgos, G.R., Kanterewicz, B.I., Barrionuevo, G., Chu, C.T., Oury, T.D., and Klann, E. (2000). Impairment of long-term potentiation and associative memory in mice that overexpress extracellular superoxide dismutase. J Neurosci 20, 7631–7639.

Tsatmali, M., Walcott, E.C., and Crossin, K.L. (2005). Newborn neurons acquire high levels of reactive oxygen species and increased mitochondrial proteins upon differentiation from progenitors. Brain Res 1040, 137–150.

Tsatmali, M., Walcott, E.C., Makarenkova, H., and Crossin, K.L. (2006). Reactive oxygen species modulate the differentiation of neurons in clonal cortical cultures. Mol Cell Neurosci 33, 345–357.

Vallet, P., Charnay, Y., Steger, K., Ogier-Denis, E., Kovari, E., Herrmann, F., Michel, J.P., and Szanto, I. (2005). Neuronal expression of the NADPH oxidase NOX4, and its regulation in mouse experimental brain ischemia. Neuroscience 132, 233–238.

Vieira, H.L., Alves, P.M., and Vercelli, A. (2011). Modulation of neuronal stem cell differentiation by hypoxia and reactive oxygen species. Prog Neurobiol 93, 444–455.

Wani, R., Nagata, A., and Murray, B.W. (2014). Protein redox chemistry: post-translational cysteine modifications that regulate signal transduction and drug pharmacology. Front Pharmacol 5, 224.

Wilson, C., and Gonzalez-Billault, C. (2015a). Regulation of cytoskeletal dynamics by redox signaling and oxidative stress: implications for neuronal development and trafficking. Front Cell Neuroscie 9, 381.

Wilson, C., Munoz-Palma, E., and Gonzalez-Billault, C. (2018). From birth to death: a role for reactive oxygen species in neuronal development. Semin Cell Dev Biol 80, 43–49.

Wilson, C., Munoz-Palma, E., Henriquez, D.R., Palmisano, I., Nunez, M.T., Di Giovanni, S., and Gonzalez-Billault, C. (2016). A feed-forward mechanism involving the NOX complex and RyR-mediated Ca^{2+} release during axonal specification. J Neurosci 36, 11107–11119.

Wilson, C., Nunez, M.T., and Gonzalez-Billault, C. (2015). Contribution of NADPH oxidase to the establishment of hippocampal neuronal polarity in culture. J Cell Sci 128, 2989–2995.

Winterbourn, C.C. (2008). Reconciling the chemistry and biology of reactive oxygen species. Nat Chem Biol 4, 278–286.

Xia, Z., and Storm, D.R. (2005). The role of calmodulin as a signal integrator for synaptic plasticity. Nat Rev Neurosci 6, 267–276.

Yoneyama, M., Kawada, K., Gotoh, Y., Shiba, T., and Ogita, K. (2010). Endogenous reactive oxygen species are essential for proliferation of neural stem/progenitor cells. Neurochem Int 56, 740–746.

Zhang, J., Wang, X., Vikash, V., Ye, Q., Wu, D., Liu, Y., and Dong, W. (2016). ROS and ROS-mediated cellular signaling. Oxid Med Cell Longev 2016, 4350965.

Zucker, R.S. (1999). Calcium- and activity-dependent synaptic plasticity. Curr Opin Neurobiol 9, 305–313.

CHAPTER SIX

H$_2$O$_2$ in Morphogenesis and Regeneration

Marion Thauvin, Irène Amblard, Alain Joliot,
Michel Volovitch, Christine Rampon, and Sophie Vriz

CONTENTS

6.1 INTRODUCTION

Reactive oxygen species (ROS), in particular hydrogen peroxide (H$_2$O$_2$), first considered as deleterious compounds actually also behave as bona fide signaling molecules (Holmstrom and Finkel 2014; Sies 2017; Sies and Jones 2020). Indeed, the physiological role of H$_2$O$_2$ in living systems has gained increased interest and been investigated in several species (e.g., *Caenorhabditis elegans, Xenopus laevis, Danio rerio*; Back et al. 2012; Knoefler et al. 2012; Gauron et al. 2016; Han et al. 2018). These studies revealed the high dynamics of ROS levels from early development to terminal differentiation, and later on in the adult, during regeneration (Rampon et al. 2018). Recent data showed that H$_2$O$_2$ is a key signaling molecule for fundamental cellular processes such as proliferation, migration, and epigenetic regulation, even if the redox machinery involved in these processes remains to be identified (Sies and Jones 2020). In this chapter, we will first describe the state of the art for quantifying H$_2$O$_2$ levels and then focus on two functions of H$_2$O$_2$, the regulation of the transition from progenitor to differentiated cells and the establishment of morphogen gradients, in the context of development and during adult appendage regeneration.

6.2 H$_2$O$_2$ QUANTIFICATION DURING DEVELOPMENT AND REGENERATION

The study of ROS dynamics in living systems has long been a complex issue, mostly because of their short lifetime, high-speed diffusion, and reactivity with a wide range of cellular components. Due to the growing interest in the physiological roles of ROS, their detection and quantification in living systems became essential for our understanding of redox signaling (Schwarzlander et al. 2016; Kostyuk et al. 2018), and dedicated methods to detect them have been developed in past decades. While chromatography, spectrophotometry, or electron resonance were successfully used for the direct identification of ROS *in vitro*, dynamic *in situ* detection of ROS in living systems relied on the development of quantitative ratiometric fluorescent probes. Among these chemical probes, some are specific to one type of ROS (Andina, Leroux, and Luciani 2017).

Besides synthetic probes, genetically encoded sensors have aroused strong interest because they can be expressed in live organisms and allow the precise targeting of tissues or cell types and can be addressed to specific subcellular structures. The ideal sensor should be bright enough to allow

DOI: 10.4324/9781003204091-9

good spatiotemporal resolution, sensitive enough to detect low concentrations and small fluctuations in the range of ROS physiological levels with linear response, and specific to a chemical species.

There are two classes of encoded redox sensors. The first class is based on variant auto-fluorescent proteins (AFPs) that were turned sensitive to the redox state of their environment. Two cysteine residues positioned close to the chromophore are readily engaged in a disulfide bridge, which affects their spectral properties. The most used representative of this class, roGFP2, was fused to various enzymes involved in redox metabolism—glutaredoxins, peroxidases (e.g., Orp1), or peroxiredoxins—and are presented in excellent reviews (Meyer and Dick 2010; Ezerina, Morgan, and Dick 2014; Van Laer and Dick 2016; Schwarzlander et al. 2016). They proved to be extremely valuable to dissect the intracellular mechanisms of redox signaling but still await further development in the context of physiological processes at the scale of tissue or living animals. The second class of encoded redox sensors was developed 14 years ago by Vsevolod Belousov, who designed the first encoded ratiometric sensor for H_2O_2, HyPer1 (Belousov et al. 2006). Since then, this laboratory provided a variety of H_2O_2 sensors based on the HyPer1 prototype (Ermakova et al. 2014; Bilan and Belousov 2018). The HyPer family is based on a circularly permuted fluorescent protein fused to the regulatory domain of a transcription factor of the OxyR family, previously reported for their restricted sensitivity to H_2O_2 (Choi et al. 2001). Interaction of H_2O_2 with the OxyR regulatory domain modifies the conformation of the fused fluorescent protein and subsequently its spectral properties. The prototype of HyPer sensors, HyPer1, contains the *Escherichia coli* OxyR domain and the circularly permutated YFP (cpYFP; Belousov et al. 2006). Most studies reporting the high spatiotemporal dynamics of H_2O_2 levels during development of living animals were achieved thanks to the HyPer sensors.

HyPer1 was used to study the dynamics of H_2O_2 during development and regeneration in several species. It was first used in C. *elegans*, revealing a strong correlation between H_2O_2 levels and environmental cues (Back et al. 2012; Knoefler et al. 2012). In X. *laevis* embryos, it was shown that fertilization induces mitochondrial H_2O_2 production that is maintained in an oscillatory manner along the cell cycle progression (Han et al. 2018). In D. *rerio* (e.g., zebrafish), transgenic fish expressing HyPer1 were used to investigate H_2O_2 level dynamics during embryonic development and in adult stages (Gauron et al. 2016), allowing full developmental study of H_2O_2 dynamics in vivo at the cellular level. More specifically, H_2O_2 is first detected during gastrulation, opposite to the shield. A burst of H_2O_2 levels is then observed during somitogenesis, predominantly in the head and the tail. From 24 hours post-fertilization (hpf) to 48 hpf, H_2O_2 levels strongly decrease, except in the heart and the nervous system. Starting from 72 hpf, when almost all developmental processes are achieved, H_2O_2 levels are rapidly reduced to the basal levels observed in adult zebrafish. During appendage regeneration in larvae or adults, a burst of H_2O_2 is observed after lesion and is necessary for wounding and regeneration to proceed (Niethammer et al. 2009; Love et al. 2013; Gauron et al. 2013; Meda et al. 2016).

Even though Hyper1 was considered a major breakthrough to detect and quantify H_2O_2 in vivo, it had some drawbacks, responding only to a short range of H_2O_2 concentrations and being sensitive to pH. In particular, its response to H_2O_2 was not suitable for the analysis of H_2O_2 gradients and fast-tracking of H_2O_2 dynamics. Refinement of HyPer1 ended up with the last member of the HyPer family, HyPer7, which almost reaches the goal of the perfect H_2O_2 sensor (Pak et al. 2020). HyPer7 is based on the *Neisseria meningitidis* OxyR domain inserted into a mutated cpGFP. Compared to previous HyPer sensors, the HyPer7 sensor is 15- to 17-fold brighter, pH stable, ultrafast, and significantly more sensitive to low H_2O_2 concentrations (nM range). It is capable of detecting at least 30-fold lower concentrations of H_2O_2, and its response to H_2O_2 variations is 80 times faster compared to previous green and red HyPer sensors (Pak et al. 2020). This new sensor was recently used to decipher the dynamics of H_2O_2 at the subcellular level (in organelles such as nucleus, mitochondria, and plasma membrane protrusions) but also at the tissue-scale level during development or regeneration (Pak et al. 2020; see Figure 6.1).

6.3 H_2O_2 REGULATION OF DIFFERENTIATION

Recent studies have shown that the innervation of the optic tectum by retinal ganglion cells (RGCs) is sensitive to H_2O_2 (Gauron et al. 2016; Weaver et al. 2018). Because it was observed that H_2O_2 levels are highly dynamic during retina differentiation, the putative role of redox signaling in this process was

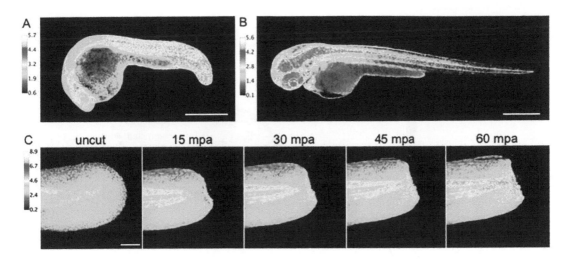

Figure 6.1 H₂O₂-level dynamics during development and regeneration in zebrafish embryos.

Zebrafish embryos were injected with HyPer7 mRNA at the one-cell stage, and H₂O₂ level monitoring was performed during development and after tail fin amputation at 48 hpf. Representative images of H₂O₂ levels in 24 hpf (A) or 48 hpf (B) zebrafish embryos, and during early tail fin regeneration in 48 hpf zebrafish embryos (C). Scale bars: 500 μm (A), 200 μm (B), and 100 μm (C). hpf: hours post-fertilization; mpa: minutes post-amputation.

analyzed. Indeed, the retina is a model of choice to study cell differentiation in vivo, because of its accessibility for in vivo imaging at high resolution, and for the spatial separation of the different steps leading to differentiation at a single time (Wan et al. 2016; see Figure 6.2). Differentiation takes place in the central retina, whereas no differentiation occurs in the ciliary marginal zone (CMZ), which contains both retinal stem cells (RSC) and retinal progenitor cells (RPC). This structure itself can be divided into three compartments characterized by marker gene expression: the most peripheral Rx2 domain corresponds to RSC and RPC, an intermediate CyclinD1 (ccdn1) domain contains proliferating but not yet committed RPC, and the more central Atoh7 domain contains cycling and committed RPC. Taking advantage of these features, Albadri et al. demonstrated that the variation in H₂O₂ levels that likely results from the differential expression patterns of H₂O₂-producing (super oxide dismutase-2 [Sod2]) and H₂O₂-degrading (catalase) enzymes correlates with the different steps of retinal differentiation (Albadri et al. 2019). The central part of the retina, which contains the differentiating cells, presents low H₂O₂ levels whereas the CMZ, which remains in a proliferative state, exhibits higher H₂O₂ levels. In addition, Sod2 was specifically expressed in the Rx2 domain (CMZ marker) whereas catalase expression colocalized

with the Atoh7 domain (central retina marker). To decipher whether the reduction in H₂O₂ levels was the consequence or the trigger of the differentiation process, zebrafish were engineered either for ectopic catalase expression in the stem cell domain or for the elimination of catalase expression in differentiated cells. Early expression of catalase in the proliferative epithelium was sufficient to drive the differentiation of RPCs into photoreceptors, and conversely in a *catalase* null mutant the retina remained proliferative. These results demonstrate that catalase expression is sufficient to induce cell differentiation, providing a striking role for catalase, and consequently H₂O₂ levels, in retina differentiation (Albadri et al. 2019).

It had been previously demonstrated that 9-hydroxystearic acid (9-HSA), a lipid known to be a by-product of lipid peroxidation, inhibits histone deacetylase HDAC1 activity (Parolin et al. 2012). Such inhibition led to Notch and Wnt activation, which is essential for the regulation of stemness/differentiation balance. Albadri et al. demonstrated that lipid peroxidation, known to be induced by the reaction of H₂O₂ with polyunsaturated fatty acids, is increased in the proliferative area corresponding to the CMZ. They further showed that 9-HSA treatment led to a strong reduction in the expression of differentiation markers, leading accordingly to retinal differentiation defects. This work brought

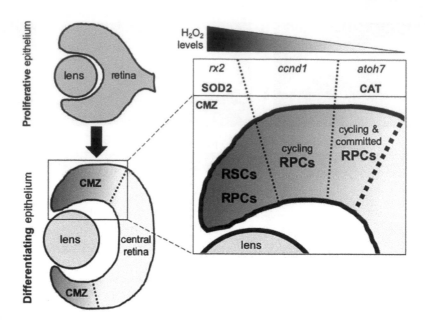

Figure 6.2 H_2O_2 dynamics during retina differentiation.

Left: schematic view of zebrafish eye at 24 hpf (upper left: the whole retinal epithelium is still proliferative) and 36 hpf (lower left: differentiation has started in central retina). Right: enlarged view of the insert in lower left, showing the topographic correlation between H_2O_2 gradient, ciliary marginal zone (CMZ) compartments identified by markers (rx2, ccnd1, atoh7), expression domains for SOD2 and catalase (CAT), and status of retinal cells (retinal stem cells [RSCs], retinal progenitor cells [RPCs]).

new insights into the relationship between H_2O_2 levels and gene expression regulation by chromatin modifiers such as HDAC proteins (Albadri et al. 2019).

6.4 H_2O_2 AND MORPHOGEN SPREADING

Patterning of the developing embryo is an essential feature of development that primarily relies on the graded distribution of morphogens (Sagner and Briscoe 2017). These morphogens activate distinct differentiation programs depending on their concentration. The regulation of morphogen distribution in developmental tissues is thus a key issue to understand how patterns are set up. HyPer sensors revealed a highly dynamic spatiotemporal distribution of H_2O_2 that coincides with the wave of morphogenetic events, suggesting a possible role of H_2O_2 in morphogen distribution. This issue was recently investigated using the homeoprotein Engrailed, a transcription factor shown to also act as a genuine morphogen due to its ability to be secreted and internalized (Spatazza et al. 2013; Di Nardo et al. 2018).

The zebrafish midbrain-hindbrain boundary (MHB) displays high levels of H_2O_2 that cannot be lowered without affecting the topography of tectum innervation by RGCs (Gauron et al. 2016), a process that also requires the graded distribution of the Engrailed proteins (EN in amniotes, Eng in fish; Retaux and Harris 1996; Scholpp and Brand 2001; Nakamura 2001; Omi and Nakamura 2015). The unsuspected cooperation between Engrailed and H_2O_2 in tectum patterning was recently reported in zebrafish (Amblard et al. 2020). At the tissue level, Engrailed is expressed at the MHB and—owing to its ability to transfer between neighboring cells (Rampon et al. 2015)—it diffuses in the midbrain until 26 hours hpf. Engrailed expression precedes the emergence of H_2O_2 gradient in the midbrain at 26 hpf, and perturbation of the redox equilibrium through Nox enzyme inhibition at this stage delays Engrailed spreading and affects its gradient shape.

It was further shown that ex vivo, H_2O_2 has opposite effects on the two steps of Engrailed intercellular transfer, secretion and internalization, which are stimulated by high and low H_2O_2 levels, respectively. Modulation of H_2O_2 concentration

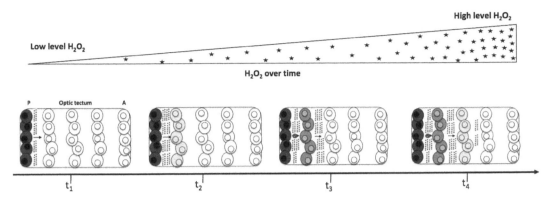

Figure 6.3 H_2O_2 and Engrailed interplay for tectum patterning.

At t1, MHB cells express EN (gray dot), which starts to diffuse in recipient neighboring cells. After being internalized, EN activates the generation of H_2O_2 (increasing over time; black stars [upper panel]; gray shading intensity in cells [lower panel]) (t2). This ROS generation switches cells from EN acceptor to EN donor (t3) and directs the propagation of EN (t4).

can thus switch cells between Engrailed-receiving to Engrailed-providing cells, which can account for the narrower spreading of Engrailed upon H_2O_2 depletion (resulting in restricted secretion) observed in vivo (Amblard et al. 2020).

Surprisingly, Engrailed internalization induces a rise of intracellular H_2O_2 within a few minutes ex vivo (not observed with an internalization-deficient mutant or after pre-incubation with an extracellular neutralizing anti-Engrailed antibody). As Engrailed expression predates H_2O_2 detection, it could thus participate in the reported H_2O_2 rise in the tectum. In line with this hypothesis, neutralizing Engrailed paracrine activity and internalization in vivo strongly alters H_2O_2 distribution in the tectum (Amblard et al. 2020).

Within proteins, cysteine and methionine residues are the main target of H_2O_2 action. The Engrailed protein used in the above studies contains a single cysteine residue that is sensitive to the redox environment and highly conserved across species. The contribution of this cysteine residue was tested using an Engrailed variant lacking the cysteine (EN2C175S). The resulting mutant was strongly impaired both in secretion and internalization and became insensitive to H_2O_2 modulation. No obvious defects were observed in DNA binding and transactivation activities, suggesting that this conserved cysteine residue, located next to the hexapeptide involved in protein-protein interactions, is a specific integrator of the redox regulation exerted by H_2O_2 on Engrailed intercellular transfer (Amblard et al. 2020).

Altogether, these observations led to the proposal of a dynamic co-propagation model for Engrailed and H_2O_2. At 24 hpf, MHB cells express Engrailed, which starts to spread in recipient neighboring cells. Engrailed internalization increases H_2O_2 that in turn converts the recipient cells to Engrailed spreaders for their neighbors. This mechanism recapitulates the diffusion of Engrailed first, followed by the emergence of an H_2O_2 wave (Figure 6.3).

Most, if not all, homeoproteins share the property of intercellular transfer (Spatazza et al. 2013; Lee et al. 2019), and some of them also behave as morphogens (Spatazza et al. 2013). It is tempting to propose that other homeoproteins might also cooperate with H_2O_2 in other developmental contexts. Such cooperation would depend on accessible cysteine residues within the homeoprotein, and their pKa in a given environment, so that the range of redox perturbations as well as the outcomes might differ between developmental processes and between homeoproteins.

6.5 H_2O_2 AND REGENERATIVE PROGRAMS

Reactive oxygen species (ROS) play a key signaling role in plants and animals (Mittler 2017; Rampon et al. 2018). In animals, the regeneration of the missing part requires the succession of different steps. The first one, wound repair, is an immediate response to injury or amputation. The regeneration-induction module begins shortly thereafter, and undifferentiated cells (some of them arising from de-differentiation) are recruited to the wound to form a mass of proliferative cells, the blastema. Ultimately, the missing part is formed by differentiation and morphogenesis. In the last years, the role of ROS has been addressed in the regenerative processes of amphibians, Squamata, Teleost fish, and mammals.

It was demonstrated that the first step of regeneration (wound repair) induces a burst of H_2O_2 in surrounding tissues that is essential for the regenerative program to proceed (Meda et al. 2018) and, in parallel, that H_2O_2 functions as a chemoattractant for recruiting immune cells to the wound (Niethammer et al. 2009; Yoo et al. 2011; Han et al. 2014; Tauzin et al. 2014; Niethammer 2018). In zebrafish larvae, a concentration gradient is produced mainly by dual oxidase (Duox) and is required for the rapid recruitment of leukocytes to the wound (Niethammer et al. 2009). The mechanism by which H_2O_2 attracts leukocytes is partially known. Direct H_2O_2 sensing has been reported to occur in neutrophils through Lyn, an Src family kinase that is activated by transient cysteine oxidation (Yoo et al. 2011). However, this Duox-mediated long-range signaling may require other spatial relay mechanisms besides extracellular H_2O_2 diffusion, which does not spread further than 30 μm (Niethammer 2018). Recently, Katikaneni and collaborators observed extensive lipid peroxidation gradients extending much deeper in the tissue and proposed that they function as a spatial redox relay for inflammatory cues and cell death signals (Katikaneni et al. 2020). How redox sensing by Lyn is transformed into directional migration remains to be elucidated.

Similar to its role in neutrophils, ROS production is also important for early and late recruitment of macrophages to injury (Tauzin et al. 2014). In this study, the authors uncovered the roles of p22phox and the Src family kinase Yrk in macrophage chemotaxis to the wound, and subsequently on neutrophil repulsion by macrophages from this site. During zebrafish adult heart regeneration, H_2O_2 potentially generated from Duox/Nox2 plays a dual role. During the early phase, H_2O_2 production recruits leukocytes to the wound. Later, 14–30 days post-amputation, it also promotes heart regeneration independently of immune cell recruitment (Han et al. 2014). In addition to attracting immune cells to the wound, ROS also induce cell death and proliferative responses. Following zebrafish adult fin amputation, ROS signaling triggers two pathways, sustained apoptosis and JNK activation. Both events are fundamental for the compensatory proliferation

of the epidermal cells and are essential to blastemal formation (Gauron et al. 2013; Meda et al. 2016).

During newt adult brain or axolotl larva tail regeneration, ROS production is also necessary, respectively for neural stem cell proliferation and neurogenesis (Hameed et al. 2015) or spinal cord proliferation (Al Haj Baddar, Chithrala, and Voss 2019). In zebrafish larva and adult tail regeneration, it has been demonstrated that an increase in ROS production induced by tissue lesion stimulates axon growth (Rieger and Sagasti 2011; Meda et al. 2016). In zebrafish larva, this phenomenon is leukocyte independent (Rieger and Sagasti 2011). In adult zebrafish, a feedback loop between nerves and H_2O_2 is observed in which nerves control H_2O_2 levels, which in turn participate in axon regrowth after injury (Meda et al. 2016; Meda, Joliot, and Vriz 2017; Meda et al. 2018).

In gecko adult tail regeneration, Nox2 and its two cytosolic regulatory subunits, p40phox and p47phox, are involved in the production of ROS in the skeleton and promote autophagy through regulation of ULK and MAPK activities (Zhang et al. 2016). H_2O_2 production is also implicated, by an unknown mechanism, in renal progenitor cell proliferation in an acute renal injury model in zebrafish (Chen et al. 2019). During zebrafish heart regeneration, the production of H_2O_2 in the epicardium is essential for elevating the phosphorylation of Erk1/2 MAPK through a de-repression mechanism involving Dusp6. The elevated pErk1/2 in the epicardium might then promote myocyte proliferation (Han et al. 2014).

In mammals, regeneration at adulthood is very limited but here also, H_2O_2 appears to play a key role. After partial hepatectomy in rats, sustained elevated H_2O_2 production initiates a switch from quiescence to proliferation in the hepatocytes via ERK activation and upregulation of cyclin 1 and p-Rb (Bai et al. 2015). Upon acute liver injury in mice, IL-11 is produced by hepatocytes in a ROS-dependent manner and induces the phosphorylation of STAT3 in healthy cells, which results in compensatory proliferation (Nishina et al. 2012). A comparison between regenerative and non-regenerative mammals reveals a correlation between ROS production and regenerative capacities (Simkin et al. 2017; Labit et al. 2018). In the non-regenerative mouse C57BL/6 strain, artificial enhancement of H_2O_2 leads to regeneration (Labit et al. 2018). Recently, Saxena and collaborators isolated primary ear pinna fibroblasts and observed that, contrary to cells from non-regenerative mammals that undergo rapid senescence in response to H_2O_2, cells from regenerative mammals continue to proliferate (Saxena et al. 2019).

The signaling pathways involved in the blastemal growth have been the objects of several studies. Amputation-induced H_2O_2 production is important to activate Wnt/β-catenin, FGF, and Sonic hedgehog signaling during regeneration of xenopus tadpole tail (Love et al. 2013) or zebrafish caudal fin, both in larvae and adults (Meda et al. 2016; Romero et al. 2018). Recently, Ferreira and collaborators demonstrated that an elevated O_2 influx after amputation fuels ROS production and correlates with regeneration efficiency in xenopus. They proposed that O_2 influx and ROS production stabilize HIF1α indirectly via hypoxia that modulates regeneration probably via HSP90 and bioelectric activities (Ferreira et al. 2018). The amputation-induced ROS production is also involved in the acetylation of histone 3 lysine 9 (H3K9ac) in the regenerating notochord at the onset of tail regeneration in xenopus tadpole (Suzuki et al. 2016).

6.6 CONCLUSION

It is now clear that H_2O_2 behaves as a signaling molecule essential for the correct proceedings of development and regeneration in animals (at least in vertebrates) and in plants (Choudhary, Kumar, and Kaur 2020). And we have recently learned how environment and metabolism can act on morphogenesis via this signaling system. We are only starting to decipher how H_2O_2 signal originates and operates within cells, what are the direct molecular targets, and which other signaling pathway(s) cooperate with H_2O_2. Importantly, we just have glimpses on several important issues. How is a short-lived signal originating at the plasma membrane relayed within the receiving cell to other compartments? How is it conveyed between cells, sometimes to a great distance? How is it orchestrated, qualitatively and quantitatively, the crosstalk between H_2O_2 and the other signaling pathways? These questions will certainly give rise to exciting new discoveries in the future.

REFERENCES

Al Haj Baddar, N. W., A. Chithrala, and S. R. Voss. 2019. "Amputation-induced reactive oxygen species signaling is required for axolotl tail regeneration." Dev Dyn 248 (2):189–196. doi:10.1002/dvdy.5.

Albadri, S., F. Naso, M. Thauvin, C. Gauron, C. Parolin, K. Duroure, J. Vougny, J. Fiori, C. Boga, S. Vriz, N. Calonghi, and F. Del Bene. 2019. "Redox signaling via lipid peroxidation regulates retinal progenitor cell differentiation." Dev Cell 50 (1):73–89 e6. doi:10.1016/j.devcel.2019.05.011.

Amblard, I., M. Thauvin, C. Rampon, I. Queguiner, V. V. Pak, V. V. Belousov, A. Prochiantz, M. Volovitch, A. Joliot, and S. Vriz. 2020. "H_2O_2 and Engrailed 2 paracrine activity synergize to shape the Zebrafish optic tectum." Communications Biology in press.

Andina, D., J. C. Leroux, and P. Luciani. 2017. "Ratiometric fluorescent probes for the detection of reactive oxygen species." Chemistry 23 (55):13549–13573. doi:10.1002/chem.201702458.

Back, P., W. H. De Vos, G. G. Depuydt, F. Matthijssens, J. R. Vanfleteren, and B. P. Braeckman. 2012. "Exploring real-time in vivo redox biology of developing and aging Caenorhabditis elegans." Free Radic Biol Med 52 (5):850–859. doi:10.1016/j.freeradbiomed.2011.11.037.

Bai, H., W. Zhang, X. J. Qin, T. Zhang, H. Wu, J. Z. Liu, and C. X. Hai. 2015. "Hydrogen peroxide modulates the proliferation/quiescence switch in the liver during embryonic development and posthepatectomy regeneration." Antioxid Redox Signal 22 (11):921–937. doi:10.1089/ars.2014.5960.

Belousov, V. V., A. F. Fradkov, K. A. Lukyanov, D. B. Staroverov, K. S. Shakhbazov, A. V. Terskikh, and S. Lukyanov. 2006. "Genetically encoded fluorescent indicator for intracellular hydrogen peroxide." Nat Methods 3 (4):281–286. doi:nmeth866 [pii] 10.1038/nmeth866.

Bilan, D. S., and V. V. Belousov. 2018. "In vivo imaging of hydrogen peroxide with HyPer probes." Antioxid Redox Signal 29 (6):569–584. doi:10.1089/ars.2018.7540.

Chen, J., T. Yu, X. He, Y. Fu, L. Dai, B. Wang, Y. Wu, J. He, Y. Li, F. Zhang, J. Zhao, and C. Liu. 2019. "Dual roles of hydrogen peroxide in promoting zebrafish renal repair and regeneration." Biochem Biophys Res Commun 516 (3):680–685. doi:10.1016/j.bbrc.2019.06.052.

Choi, H., S. Kim, P. Mukhopadhyay, S. Cho, J. Woo, G. Storz, and S. E. Ryu. 2001. "Structural basis of the redox switch in the OxyR transcription factor." Cell 105 (1):103–113. doi:10.1016/s0092-8674(01)00300-2.

Choudhary, A., A. Kumar, and N. Kaur. 2020. "ROS and oxidative burst: roots in plant development." Plant Divers 42 (1):33–43. doi:10.1016/j.pld.2019.10.002.

Di Nardo, A. A., J. Fuchs, R. L. Joshi, K. L. Moya, and A. Prochiantz. 2018. "The physiology of homeoprotein transduction." Physiol Rev 98 (4):1943–1982. doi:10.1152/physrev.00018.2017.

Ermakova, Y. G., D. S. Bilan, M. E. Matlashov, N. M. Mishina, K. N. Markvicheva, O. M. Subach, F. V. Subach, I. Bogeski, M. Hoth, G. Enikolopov, and V. V. Belousov. 2014. "Red fluorescent genetically encoded indicator for intracellular hydrogen peroxide." *Nat Commun* 5:5222. doi:10.1038/ncomms6222.

Ezerina, D., B. Morgan, and T. P. Dick. 2014. "Imaging dynamic redox processes with genetically encoded probes." *J Mol Cell Cardiol* 73:43–49. doi:10.1016/j.yjmcc.2013.12.023.

Ferreira, F., V. Raghunathan, G. Luxardi, K. Zhu, and M. Zhao. 2018. "Early redox activities modulate *Xenopus* tail regeneration." *Nat Commun* 9 (1):4296. doi:10.1038/s41467-018-06614-2.

Gauron, C., F. Meda, E. Dupont, S. Albadri, N. Quenech'Du, E. Ipendey, M. Volovitch, F. Del Bene, A. Joliot, C. Rampon, and S. Vriz. 2016. "Hydrogen peroxide (H_2O_2) controls axon pathfinding during zebrafish development." *Dev Biol* 414 (2):133–141.

Gauron, C., C. Rampon, M. Bouzaffour, E. Ipendey, J. Teillon, M. Volovitch, and S. Vriz. 2013. "Sustained production of ROS triggers compensatory proliferation and is required for regeneration to proceed." *Sci Rep* 3:2084. doi:10.1038/srep02084srep02084 [pii].

Hameed, L. S., D. A. Berg, L. Belnoue, L. D. Jensen, Y. Cao, and A. Simon. 2015. "Environmental changes in oxygen tension reveal ROS-dependent neurogenesis and regeneration in the adult newt brain." *Elife* 4. doi:10.7554/eLife.08422.

Han, P., X. H. Zhou, N. Chang, C. L. Xiao, S. Yan, H. Ren, X. Z. Yang, M. L. Zhang, Q. Wu, B. Tang, J. P. Diao, X. Zhu, C. Zhang, C. Y. Li, H. Cheng, and J. W. Xiong. 2014. "Hydrogen peroxide primes heart regeneration with a derepression mechanism." *Cell Res* 29 (4):1091–1107. doi:10.1038/cr.2014.108cr2014108 [pii].

Han, Y., S. Ishibashi, J. Iglesias-Gonzalez, Y. Chen, N. R. Love, and E. Amaya. 2018. "Ca(2+)-induced mitochondrial ROS regulate the early embryonic cell cycle." *Cell Rep* 22 (1):218–231. doi:10.1016/j.celrep.2017.12.042.

Holmstrom, K. M., and T. Finkel. 2014. "Cellular mechanisms and physiological consequences of redox-dependent signalling." *Nat Rev Mol Cell Biol* 15 (6):411–421. doi:10.1038/nrm3801.

Katikaneni, A., M. Jelcic, G. F. Gerlach, Y. Ma, M. Overholtzer, and P. Niethammer. 2020. "Lipid peroxidation regulates long-range wound detection through 5-lipoxygenase in zebrafish." *Nat Cell Biol*. doi:10.1038/s41556-020-0564-2.

Knoefler, D., M. Thamsen, M. Koniczek, N. J. Niemuth, A. K. Diederich, and U. Jakob. 2012. "Quantitative in vivo redox sensors uncover oxidative stress as an early event in life." *Mol Cell* 47 (5):767–776. doi:10.1016/j.molcel.2012.06.016.

Kostyuk, A. I., A. S. Panova, D. S. Bilan, and V. V. Belousov. 2018. "Redox biosensors in a context of multiparameter imaging." *Free Radic Biol Med* 128:23–39. doi:10.1016/j.freeradbiomed.2018.04.004.

Labit, E., L. Rabiller, C. Rampon, C. Guissard, M. Andre, C. Barreau, B. Cousin, A. Carriere, M. A. Eddine, B. Pipy, L. Penicaud, A. Lorsignol, S. Vriz, C. Dromard, and L. Casteilla. 2018. "Opioids prevent regeneration in adult mammals through inhibition of ROS production." *Sci Rep* 8 (1):12170. doi:10.1038/s41598-018-29594-1.

Lee, E. J., N. Kim, J. W. Park, K. H. Kang, W. I. Kim, N. S. Sim, C. S. Jeong, S. Blackshaw, M. Vidal, S. O. Huh, D. Kim, J. H. Lee, and J. W. Kim. 2019. "Global analysis of intercellular homeodomain protein transfer." *Cell Rep* 28 (3):712–722 e3. doi:10.1016/j.celrep.2019.06.056.

Love, N. R., Y. Chen, S. Ishibashi, P. Kritsiligkou, R. Lea, Y. Koh, J. L. Gallop, K. Dorey, and E. Amaya. 2013. "Amputation-induced reactive oxygen species are required for successful *Xenopus* tadpole tail regeneration." *Nat Cell Biol* 15 (2):222–228. doi:10.1038/ncb2659ncb2659 [pii].

Meda, F., C. Gauron, C. Rampon, J. Teillon, M. Volovitch, and S. Vriz. 2016. "Nerves control redox levels in mature tissues through Schwann cells and hedgehog signaling." *Antioxid Redox Signal* 24 (6):299–311. doi:10.1089/ars.2015.6380.

Meda, F., A. Joliot, and S. Vriz. 2017. "Nerves and hydrogen peroxide: how old enemies become new friends." *Neural Regeneration Research* 12 (4):569–569.

Meda, F., C. Rampon, E. Dupont, C. Gauron, A. Mourton, I. Queguiner, M. Thauvin, M. Volovitch, A. Joliot, and S. Vriz. 2018. "Nerves, H_2O_2 and Shh: three players in the game of regeneration." *Semin Cell Dev Biol* 80:65–73. doi:10.1016/j.semcdb.2017.08.015.

Meyer, A. J., and T. P. Dick. 2010. "Fluorescent protein-based redox probes." *Antioxid Redox Signal* 13 (5):621–650. doi:10.1089/ars.2009.2948.

Mittler, R. 2017. "ROS Are Good." *Trends Plant Sci* 22 (1):11–19. doi:10.1016/j.tplants.2016.08.002.

Nakamura, H. 2001. "Regionalization of the optic tectum: combinations of gene expression that define the tectum." *Trends Neurosci* 24 (1):32–39.

Niethammer, P. 2018. "Wound redox gradients revisited." *Semin Cell Dev Biol* 80:13–16. doi:10.1016/j.semcdb.2017.07.038.

Niethammer, P., C. Grabher, A. T. Look, and T. J. Mitchison. 2009. "A tissue-scale gradient of hydrogen peroxide mediates rapid wound detection in zebrafish." *Nature* 459 (7249):996–999. doi:10.1038/nature08119.

Nishina, T., S. Komazawa-Sakon, S. Yanaka, X. Piao, D. M. Zheng, J. H. Piao, Y. Kojima, S. Yamashina, E. Sano, T. Putoczki, T. Doi, T. Ueno, J. Ezaki, H. Ushio, M. Ernst, K. Tsumoto, K. Okumura, and H. Nakano. 2012. "Interleukin-11 links oxidative stress and compensatory proliferation." *Sci Signal* 5 (207):ra5. doi:10.1126/scisignal.20020565/207/ra5 [pii].

Omi, M., and H. Nakamura. 2015. "Engrailed and tectum development." *Dev Growth Differ* 57 (2):135–145. doi:10.1111/dgd.12197.

Pak, V. V., D. Ezeriņa, O. G. Lyublinskaya, B. Pedre, P. A. Tyurin-Kuzmin, N. M. Mishina, M. Thauvin, D. Young, K. Wahni, S. Martinez-Gache, A. D. Demidovich, Y. G. Ermakova, Y. D. Maslova, E. Eroglu, D. S. Bilan, I. Bogeski, T. Michel, S. Vriz, J. Messens, and V. V. Belousov. 2020. "Ultrasensitive genetically encoded indicator for intracellular hydrogen peroxide identifies novel roles for cellular oxidants in cell migration and mitochondrial function." *Cell Metabolism* in press

Parolin, C., N. Calonghi, E. Presta, C. Boga, P. Caruana, M. Naldi, V. Andrisano, L. Masotti, and G. Sartor. 2012. "Mechanism and stereoselectivity of HDAC I inhibition by (R)-9-hydroxystearic acid in colon cancer." *Biochim Biophys Acta* 1821 (10):1334–1340. doi:10.1016/j.bbalip.2012.07.007.

Rampon, C., C. Gauron, T. Lin, F. Meda, E. Dupont, A. Cosson, E. Ipendey, A. Frerot, I. Aujard, T. Le Saux, D. Bensimon, L. Jullien, M. Volovitch, S. Vriz, and A. Joliot. 2015. "Control of brain patterning by Engrailed paracrine transfer: a new function of the Pbx interaction domain." *Development* 142 (10):1840–1849. doi:10.1242/dev.114181.

Rampon, C., M. Volovitch, A. Joliot, and S. Vriz. 2018. "Hydrogen Peroxide and Redox Regulation of Developments." *Antioxidants (Basel)* 7 (11). doi:10.3390/antiox7110159.

Retaux, S., and W. A. Harris. 1996. "Engrailed and retinotectal topography." *Trends Neurosci* 19 (12):542–546.

Rieger, S., and A. Sagasti. 2011. "Hydrogen peroxide promotes injury-induced peripheral sensory axon regeneration in the zebrafish skin." *PLoS Biol* 9 (5):e1000621. doi:10.1371/journal.pbio. 100062110-PLBI-RA-9260R3 [pii].

Romero, M. M. G., G. McCathie, P. Jankun, and H. H. Roehl. 2018. "Damage-induced reactive oxygen species enable zebrafish tail regeneration by repositioning of Hedgehog expressing cells." *Nat Commun* 9 (1):4010. doi:10.1038/s41467-018-06460-2.

Sagner, A., and J. Briscoe. 2017. "Morphogen interpretation: concentration, time, competence, and signaling dynamics." *Wiley Interdiscip Rev Dev Biol* 6 (4). doi:10.1002/wdev.271.

Saxena, S., H. Vekaria, P. G. Sullivan, and A. W. Seifert. 2019. "Connective tissue fibroblasts from highly regenerative mammals are refractory to ROS-induced cellular senescence." *Nat Commun* 10 (1):4400. doi: 10.1038/s41467-019-12398-w.

Scholpp, S., and M. Brand. 2001. "Morpholino-induced knockdown of zebrafish engrailed genes eng2 and eng3 reveals redundant and unique functions in midbrain–hindbrain boundary development." *Genesis* 30 (3):129–133.

Schwarzlander, M., T. P. Dick, A. J. Meyer, and B. Morgan. 2016. "Dissecting Redox Biology Using Fluorescent Protein Sensors." *Antioxid Redox Signal* 24 (13):680–712. doi:10.1089/ars.2015.6266.

Sies, H. 2017. "Hydrogen peroxide as a central redox signaling molecule in physiological oxidative stress: oxidative eustress." *Redox Biol* 11:613–619. doi:10.1016/j.redox.2016.12.035.

Sies, H., and D. P. Jones. 2020. "Reactive oxygen species (ROS) as pleiotropic physiological signalling agents." *Nat Rev Mol Cell Biol*. doi:10.1038/s41580-020-0230-3.

Simkin, J., T. R. Gawriluk, J. C. Gensel, and A. W. Seifert. 2017. "Macrophages are necessary for epimorphic regeneration in African spiny mice." *Elife* 6:e24623. doi:10.7554/eLife.24623.

Spatazza, J., E. Di Lullo, A. Joliot, E. Dupont, K. L. Moya, and A. Prochiantz. 2013. "Homeoprotein signaling in development, health, and disease: a shaking of dogmas offers challenges and promises from bench to bed." *Pharmacol Rev* 65 (1):90–104. doi:10.1124/pr.112.006577pr.112.006577 [pii].

Suzuki, M., C. Takagi, S. Miura, Y. Sakane, M. Suzuki, T. Sakuma, N. Sakamoto, T. Endo, Y. Kamei, Y. Sato, H. Kimura, T. Yamamoto, N. Ueno, and K. T. Suzuki. 2016. "In vivo tracking of histone H3 lysine 9 acetylation in *Xenopus laevis* during tail regeneration." *Genes Cells* 21 (4):358–369. doi:10.1111/gtc.12349.

Tauzin, S., T. W. Starnes, F. B. Becker, P. Y. Lam, and A. Huttenlocher. 2014. "Redox and Src family kinase signaling control leukocyte wound attraction and neutrophil reverse migration." *J Cell Biol* 207 (5):589–598. doi:10.1083/jcb.201408090.

Van Laer, K., and T. P. Dick. 2016. "Utilizing Natural and Engineered Peroxiredoxins As Intracellular Peroxide Reporters." *Mol Cells* 39 (1):46–52. doi:10.14348/molcells.2016.2328.

Wan, Y., A. D. Almeida, S. Rulands, N. Chalour, L. Muresan, Y. Wu, B. D. Simons, J. He, and W. A. Harris.

2016. "The ciliary marginal zone of the zebrafish retina: clonal and time-lapse analysis of a continuously growing tissue." *Development* 143 (7):1099–10107. doi:10.1242/dev.133314.

Weaver, C. J., A. Terzi, H. Roeder, T. Gurol, Q. Deng, Y. F. Leung, and D. M. Suter. 2018. "nox2/cybb deficiency affects zebrafish retinotectal connectivity." *J Neurosci* 38 (26):5854–5871. doi:10.1523/JNEUROSCI.1483-16.2018.

Yoo, S. K., T. W. Starnes, Q. Deng, and A. Huttenlocher. 2011. "Lyn is a redox sensor that mediates leukocyte wound attraction in vivo." *Nature* 480 (7375):109–112. doi:10.1038/nature10632.

Zhang, Q., Y. Wang, L. Man, Z. Zhu, X. Bai, S. Wei, Y. Liu, M. Liu, X. Wang, X. Gu, and Y. Wang. 2016. "Reactive oxygen species generated from skeletal muscles are required for gecko tail regeneration." *Sci Rep* 6:20752. doi:10.1038/srep20752.

CHAPTER SEVEN

Protein S-Nitrosylation in Neuronal Development

Tomohiro Nakamura, Xu Zhang, Chang-ki Oh, and Stuart A. Lipton

CONTENTS

7.1 INTRODUCTION

Endogenously produced reactive nitrogen species (RNS), including nitric oxide (NO)–related species, serve as signal messengers mediating a variety of biological functions throughout the body, including the cardiovascular system, immune system, and central nervous system (CNS). In the brain, basal levels of NO are implicated in physiological processes such as synaptic plasticity, neuronal development, and neuronal survival (1–3). However, hyper- or hypo-production of NO can lead to pathological events including neuroinflammation, decreased neurogenesis, synaptic injury, and neuronal damage, contributing to the etiology of neurodevelopmental and neurodegenerative disorders (4–7). NO synthases (NOS) catalyze the production of NO from L-arginine in the presence of oxygen, NADPH, and essential cofactors such as tetrahydrobiopterin (BH4) and flavin adenine dinucleotide (FAD). Three types of NOS exist in mammalian cells: neuronal NOS (nNOS

or NOS1), inducible NOS (iNOS or NOS2), and endothelial NOS (eNOS or NOS3). Depending on their expression pattern and degree of activation, these NOS enzymes allow NO to execute distinct roles in brain development. For instance, during early brain development, physiological levels of NO facilitate the establishment of neuronal network activity (8–10). In contrast, increased production of NO in adult mouse brains inhibits additional neurogenesis (11, 12). As a downstream effector of NOS, initial studies identified soluble guanylate cyclase, with a consequent increase in production of cGMP (13–15). However, emerging evidence suggests that protein S-nitrosylation, a redox-mediated post-translational modification of a cysteine thiol group (or more properly thiolate anion) on the target protein with NO-related species (most likely in the form of a nitrosonium cation NO^+ intermediate), affects the cellular function and activity of the signaling proteins (16–18). Mechanistically, this nitrosation reaction is thought to occur in the

DOI: 10.4324/9781003204091-10

presence of a catalytic metal ion that accepts the free electron in the outer pi molecular orbital of NO during the reaction process. The resulting formation of an S-nitrosothiol is now known as protein S-nitrosylation and is well recognized as a key mechanism for the signaling actions of NO/RNS. Note that S-nitrosylation of a sulfhydryl group is chemically distinct from tyrosine nitration, resulting for example from peroxynitrite (ONOO$^-$)-mediated redox reactions that form 3-nitrotyrosine adducts on proteins (19) (Figure 7.1).

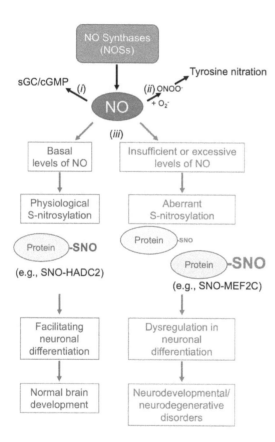

Figure 7.1 NO-mediated signaling modulates neuronal differentiation and development.

Nitric oxide (NO) produced from NO synthases (NOS) can (i) activate soluble guanylate cyclase (sGC) to produce cGMP, (ii) react with superoxide (O$_2^-$) to produce peroxynitrite (ONOO$^-$) and mediate nitration of tyrosine residues, or (iii) trigger protein S-nitrosylation reactions. In general, physiologically produced S-nitrosylated (SNO) proteins (such as SNO-HDAC2) support neuronal differentiation and development, whereas pathologically increased generation of NO-related species leads to aberrant protein S-nitrosylation (e.g., SNO-MEF2C), impairing normal neuronal development.

7.2 PROTEIN S-NITROSYLATION

As stated above, the addition of NO-related species, most likely in the form of a nitrosonium cation (NO$^+$) intermediate, to a critical cysteine thiol (-SH) group (or more properly thiolate anion, -S$^-$) leads to the generation of an S-nitrosothiol (SNO) on a target protein (forming a "SNO-protein") (2, 17, 20–22) (Figure 7.1). Although the vast majority of proteins contain multiple cysteine residues, "SNO formation" is known to take place only at specific sites. The determinants of SNO specificity entail several mechanisms, including (i) the presence of a signature SNO motif comprised of acid-base and hydrophobic amino acids, (ii) proximity of the target protein to the NOS complex, and (iii) protein-protein transnitrosylation involving transfer of the NO group from the thiol of one protein to the thiol of another. The effects of SNO specificity have been extensively discussed in recent review articles (see 16, 17, 23–25). Moreover, generation of SNO-proteins can in many cases be reversed by denitrosylation enzymes including S-nitrosoglutathione (GSNO) reductase (R), SNO-CoA reductase (SCoR), and thioredoxin (Trx) (23, 26, 27). Notably, in some prominent cases, S-nitrosylation-induced conformational changes can facilitate further oxidation reactions of the same cysteine with reactive oxygen species (ROS), resulting in sulfenic acid adducts (-SOH) or more stable oxidative modifications such as sulfinic acid (-SO$_2$H) or irreversible sulfonic acid (-SO$_3$H) (28).

Interestingly, recent studies have revealed that a series of signaling cascades known to affect neuronal differentiation and development, including the histone deacetylase 2 (HDAC2)/cAMP-response element-binding protein (CREB) transcriptional pathway and the myocyte enhancer factor-2C (MEF2C) transcription factor-regulated pathway, are influenced by S-nitrosylation/oxidation (4, 10, 12, 29–31). In this chapter, we focus on these and other S-nitrosylated proteins that affect neurogenesis, neuronal maturation, and neuronal network formation.

7.3 S-NITROSYLATION OF PROTEINS IN THE HDAC/CREB-MEDIATED TRANSCRIPTIONAL PATHWAY AFFECTS NEUROGENESIS

Both genetic and environmental factors are thought to influence the process of brain development (32).

In particular, exposure to environmental toxins, including alcohol, air pollution, infectious agents, and industrial and agricultural chemicals, negatively impacts neuronal differentiation and maturation (33, 34). The effects of these xenobiotics can be mediated at least in part via abnormal production of hormones, growth factors, cytokines, and metabolites (33), which are known to stimulate excessive ROS/RNS production. These reactive molecules modulate a series of signal transduction events, altering gene expression related to neuronal development, differentiation, and survival. Various transcription factors play important roles in this process. Among them, the cAMP and calcium-responsive transcription factor CREB and MEF2 are well-studied transcription factors in this context.

CREB was originally discovered as a transcription factor that binds to the cAMP-response element (CRE) of the promoter region of the gene encoding the hypothalamic peptide somatostatin (35). CREB belongs to the family of leucine zipper transcription factors, featuring the highly conserved bZip (basic region/leucine zipper) domain required for dimerization and DNA binding (36). During neuronal development, CREB promotes dendritic outgrowth in both neuronal activity-dependent and activity-independent fashions. As an example of neuronal activity-dependent regulation of CREB activity, depolarization of neuronal cells evokes calcium influx, leading to the activation of a series of CREB kinases, such as CaMKIV (calcium/calmodulin-dependent protein kinase type IV) (37, 38). CaMKIV phosphorylates CREB at serine-133, resulting in CREB activation. Alternatively, CREB can be activated by neurotrophins, such as BDNF (brain-derived neurotrophic factor), in an NO-dependent manner (39). BDNF increases nNOS activity and also enhances CREB binding to the CRE site located in the promoter region of many genes associated with neuronal differentiation, including c-fos, nNos, vgf (nerve growth factor inducible), and erg1 (early growth response protein 1) (39). Moreover, in mice lacking nNOS expression, CREB-dependent gene expression is impaired, consistent with the notion that NO is a key upstream regulator of CREB activation.

Interestingly, the NO-dependent increase in CREB activity does not involve serine-133 phosphorylation. Rather, immunoblot analysis using an anti-S-nitrosocysteine antibody showed that S-nitrosylated protein(s) are associated with the c-fos promoter (39). Along these lines, a recent study identified Cys300, Cys310 and Cys337 as the potential sites of S-nitrosylation in CREB (30). All of these cysteines reside within the bZip domain and have implications for its DNA binding ability as well as the recruitment of the CREB cofactor CRTC2 (CREB regulated transcription coactivator 2) (40, 41). Thus, these findings point to the possibility that S-nitrosylation of CREB may increase its binding to the CRE site via enhanced homodimerization or interaction with its cofactors. However, additional proteins that regulate CREB function may also be S-nitrosylated, and it remains for future work to sort out the effect of nitrosylation reactions on these CREB partners.

In addition to SNO-CREB, other SNO-proteins also regulate CREB-dependent dendritic outgrowth. For example, S-nitrosylation of HDAC2 and histone-binding protein RBBP7 contributes to chromatin remodeling and dendritogenesis in neurons (29, 30). Histone acetylation inhibits the interaction between histone proteins and DNA, providing a more accessible structure for transcriptional machinery. In contrast, histone deacetylation decreases accessibility of DNA to transcription activators. It has been reported that BDNF, NGF, and possibly other neurotrophins stimulate S-nitrosylation of HDAC2 (Figure 7?). While S-nitrosylation of HDAC2 per se does not affect deacetylation activity, it is required for release from chromatin, thereby facilitating histone acetylation and gene expression. Expression of a non-S-nitrosylatable HDAC2 mutant in cerebrocortical neurons was shown to result in decreased dendritic length compared to neurons transfected with wild-type HDAC2 (29).

Concerning other S-nitrosylated proteins that affect histone acetylation, a recent proteomic study revealed some 614 nuclear targets of S-nitrosylation in rat cerebrocortical neurons, including RBBP7 (30). RBBP7 is a histone-binding protein, which serves as a scaffolding subunit for the assembly of several chromatin-modifying complexes, including the NuRD (nucleosome remodeling deacetylase) complex. The NuRD complex also contains HDAC1, HDAC2, and chromodomain-helicase-DNA-binding protein CHD3 or CHD4, among other proteins (42). S-Nitrosylation of RBBP7 at Cys166 increases its interaction with the CHD4 subunit of NuRD, which may alter the complex assembly on chromatin. Knockdown of RBBP7 and its paralog RBBP4 remarkably decreased dendritic outgrowth in rat cortical neurons,

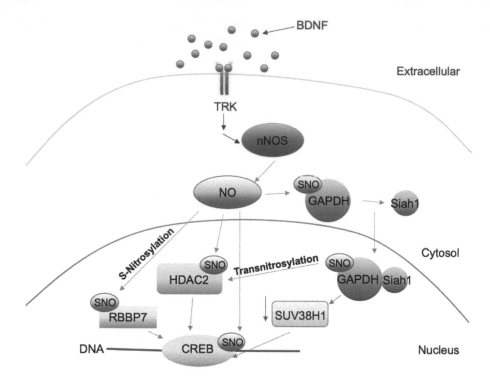

Figure 7.2 BDNF/S-nitrosylation/CREB-dependent dendritic outgrowth machinery.

BDNF promotes nNOS activation, thus increasing NO production, which in turn S-nitrosylates proteins involved in the CREB-dependent dendritic outgrowth pathway, such as GAPDH, HDAC2, RBBP7, and CREB itself. S-Nitrosylation of HDAC2 and histone-binding protein RBBP7 contributes to chromatin remodeling and CREB-dependent dendritogenesis. S-Nitrosylation of CREB increases its binding to the promoter region of the targeted genes. S-Nitrosylation of GAPDH promotes binding to its nuclear translocation partner Siah1, which lowers its turnover rate. Siah1 thus stabilized can facilitate the degradation of SUV39H1 and results in decreased methylation of H3K9. S-Nitrosylated GAPDH can transnitrosylate nuclear proteins, such as HDAC2.

and this effect was rescued by co-transfection with wild-type RBBP7, but not by RBBP7(C166S). Collectively, these studies are consistent with the notion that S-nitrosylation of HDAC2 and RBBP7 facilitate CREB-dependent neurite development through chromatin remodeling.

Another key player that influences the CREB signaling pathway via S-nitrosylation is GAPDH (glyceraldehyde-3-phosphate dehydrogenase). GAPDH was originally discovered as a housekeeping protein in the glycolysis cascade but turns out to be a pleiotropic protein involved in various cellular signaling pathways. For example, GAPDH regulates gene expression via SNO-GAPDH formation and through its transnitrosylation activity, resulting in S-nitrosylation of other proteins (43–47). Although it has no NLS (nuclear localization signal), GAPDH translocates into the nucleus upon S-nitrosylation of its active site Cys150, and

this translocation event is associated with cell death (45). S-Nitrosylation of GAPDH abolishes its enzymatic activity but also promotes recruitment of its nuclear translocation partner, the E3 ubiquitin ligase Siah1 (seven in absentia homolog 1). Siah1 possesses an NLS; thus, binding to GAPDH results in nuclear entry of both of these molecules. Siah1 can ubiquitinate a wide range of substrates, including several nuclear proteins. However, the tissue level of Siah1 under basal conditions is low due to its short half-life, which limits its enzymatic function. When GAPDH binds to Siah1, its breakdown is inhibited, thus stabilizing its activity. Thus, co-translocation of GAPDH and Siah1 into the nucleus promotes Siah1-mediated degradation of nuclear targets (43). The nuclear substrates of Siah1 include SUV39H1 (histone-methylating enzyme suppressor of variegation 3–9 homolog 1), which is a major histone methylating enzyme.

The trimethylation of lysine 9 on histone H3 (H3K9) by SUV39H1 is a hallmark of gene silencing. BDNF treatment, by inducing NO production and S-nitrosylation, increases the nuclear entry of the SNO-GAPDH/Siah1 complex. Subsequent degradation of SUV39H1 by Siah1 decreases methylation of H3K9 and increases its acetylation, leading to activation of CREB target genes, thus promoting dendritic outgrowth (47). In the nucleus, GAPDH also binds and activates P300/CBP (CREB binding protein), enhancing its acetylation activity and facilitating the expression of CREB target genes (48). SNO-GAPDH also serves as an excellent source of protein S-nitrosylation in the nucleus by transnitrosylating several nuclear proteins to alter their function. These nuclear proteins include SIRT1 (deacetylating enzyme sirtuin-1), HDAC2, DNA-PK (DNA-activated protein kinase), and B23/nucleophosmin (44, 49). In fact, GAPDH transnitrosylates SIRT1 more effectively than GSNO, and S-nitrosylation of SIRT1 abrogates its deacetylation activity, leading to elevated acetylation of its substrates, such as transcriptional co-activator PGC1α (proliferator-activated receptor gamma co-activator 1) (44). Additionally, transnitrosylation of HDAC2 by SNO-GAPDH would be expected to increase CREB-dependent dendritic outgrowth, as delineated above. Future mechanistic studies are warranted to determine how SNO-GAPDH differentially regulates two seemingly contradictory effects (i.e., neuronal cell death vs. dendritic outgrowth).

In summary, to promote dendritic outgrowth and neuronal maturation, neurotrophins, excitatory synaptic transmission, and possibly other environmental cues enhance CREB activity through a series of protein S-nitrosylation reactions. Direct S-nitrosylation of CREB has been reported to stimulate its activity, thus enhancing expression of CREB target genes. S-Nitrosylation of chromatin remodeling proteins such as HDAC2, RBBP7 and SIRT1 provides a more accessible platform for the recruitment of CREB transcriptional machinery. Other, as yet unknown, S-nitrosylated partners of CREB are also likely to exist. Additionally, while S-nitrosylated GAPDH plays an important role in neuronal cell death, it can also alter the function of nuclear proteins via transnitrosylation. Further studies in this area are encouraged to reveal a more complete picture of protein S-nitrosylation and its effect on CREB-mediated neuronal outgrowth and development.

7.4 S-NITROSYLATION OF MEF2 AFFECTS NEUROGENESIS AND NEURONAL SURVIVAL

Historically, neurogenesis was thought to occur only when neural stem cells were developing into neurons during embryogenesis. However, over the last few decades, unexpected detection of adult-born neurons in rodents (50), primates (51), and possibly humans (52) suggests that active neurogenesis may continue throughout life in mammals. In these studies, adult neurogenesis in the rodent hippocampus seems most persuasive, but two recent studies reported that hippocampal neurogenesis was undetectable in the adult human brain (53, 54). Thus, additional studies are needed to investigate whether and to what degree neurogenesis exists in the adult human brain (55).

Thus, from the existing evidence, at least in rodents, adult neurogenesis occurs primarily in two regions of the brain: (i) the subgranular zone (SGZ) of the dentate gyrus (DG) in the hippocampus, and (ii) the subventricular zone (SVZ) of the lateral ventricles (56). These newborn neurons integrate into the existing hippocampal circuitry as they differentiate into mature neurons, contributing to learning and spatial memory formation (57). However, adult hippocampal neurogenesis may be impaired in several neurological diseases, including Alzheimer's disease (AD), Parkinson's disease (PD), mood disorders (e.g., anxiety and depression), and epilepsy (58), expediting disease progression.

MEF2 is a member of the MADS (MCM1-agamous-deficiens-serum response factor) family acting as a transcription factor (59). In vertebrates, four MEF2 isotypes are present as MEF2A, B, C, and D, whereas only a single Mef2 gene exists in yeast (Saccharomyces cerevisiae), fruit fly (Drosophila melanogaster), and worm (Caenorhabditis elegans) (Figure 7.3) (59). Our group originally cloned the human Mef2c gene and found it highly expressed in neurons of cerebrocortex, cardiac muscle, macrophages/microglia, and lymphocytes (60–62); MEF2C is also the earliest MEF2 isoform expressed in the developing brain. Differential but partially overlapping expression patterns of the four MEF2 proteins are observed during development and throughout adulthood in the brain (63). MEF2 isoforms are composed of a highly conserved N-terminal MADS domain and a MEF2 domain, which facilitate DNA binding and dimerization, a C-terminal

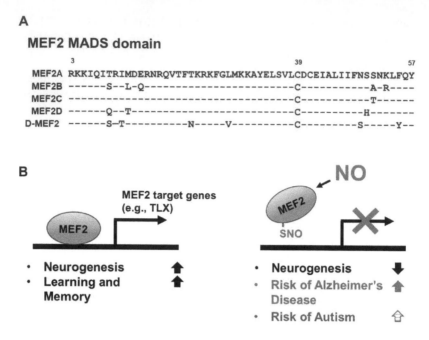

A

MEF2 MADS domain

```
            3                                              39                        57
MEF2A  RKKIQITRIMDERNRQVTFTKRKFGLMKKAYELSVLCDCEIALIIFNSSNKLFQY
MEF2B  ------S--L-Q----------------------C-----------A-R----
MEF2C  ---------------------------------C-----------T------
MEF2D  ------Q--T-----------------------C----------H-------
D-MEF2 ------S-T----------N-----V---------C---------S-----Y--
```

B

Figure 7.3 S-Nitrosylation of MEF2 and its mechanism of action in neurogenesis.

(A) Conserved sequence of MADS domain in human MEF2 isoforms (MEF2A, B, C, and D) and drosophila MEF2 (D-MEF2). S-Nitrosylation site (Cysteine 39 or Cys39) highlighted gray with other conserved amino acids indicated (–). (B) Schematic of SNO-MEF2-mediated inhibition of neurogenesis. S-Nitrosylation of MEF2 decreases its binding to the promoter of target genes such as TLX, impairing adult neurogenesis in the hippocampus. By disrupting neurogenesis, formation of SNO-MEF2 may contribute to the pathogenesis of Alzheimer's disease (AD) and potentially autism spectrum disorder (ASD).

TAD (transactivation domain), which recruits cofactors to regulate transcriptional activity, and a NLS (63). In brain, muscle and heart, homodimers or heterodimers of MEF2 directly bind to an A/T-rich core sequence [5'-CT(A/t)(a/t)AAATAG-3']. Additionally, in the brain, promoters targeted by MEF2 contain an additional extended sequence constraint [5'-TGTTACT(A/t)(a/t)AAATAGA(A/t)-3'] (64). Thus, there is the potential for differential DNA binding of MEF2 in brain vs. other tissues, which may contribute to brain-specific gene regulation during neurogenesis.

Published evidence in rodent models supports the notion that MEF2 is involved in normal development of the nervous system, synaptic plasticity, and neuronal survival. For example, MEF2 enhances neuronal migration (65) and differentiation (66, 67) during embryogenesis. In addition, MEF2 is involved in neuronal activity-dependent cell survival (68, 69). Alternatively, excitotoxic insults or environmental toxins impair the neuroprotective activity of various MEF2 isoforms, in part by causing caspase-mediated degradation of MEF2 isoforms, thus enhancing neuronal cell death (31, 70, 71). MEF2A/C also facilitate the initial formation of dendrites and synapses during neurogenesis, but also the pruning and remodeling of synapses with maturation, strengthening hippocampal-dependent learning and memory (72–76). Moreover, a decrease in MEF2C expression in embryonic mouse brain results in autistic phenotypes (63, 65, 72, 77, 78). More recently, in humans, microdeletions, missense, or nonsense mutations in the human *Mef2c* gene have been found to cause MEF2C haploinsufficiency syndrome (MCHS), an emerging neurodevelopmental disorder characterized by features of autism spectrum disorder (ASD), absence of or little speech output, stereotypical movements, and intellectual disability (ID), and often accompanied by seizures (63, 77).

Additionally, MEF2C drives gene expression that is critical for the development and maturation of A9-type dopaminergic (DA) neurons, the type of neurons initially lost in the substantia nigra pars compacta in PD (65, 66, 79). Along these lines, human embryonic stem cells (hESCs) expressing a constitutively active from of MEF2C (MEF2CA) produce a very high yield of A9 DA neurons in vitro (>90% purity) (31, 66). Moreover, transplantation of MEF2CA-programmed human neural progenitor cells (hNPCs) derived from these hESCs into rat and monkey models of PD improves the survivability of transplanted cells, increases differentiation into A9 DA neurons, and improves behavioral deficits related to parkinsonism without causing hyperproliferation of the implanted cells (66, 79). Preferential differentiation of MEF2CA hNPCs into the A9 DA phenotype rather than other neuronal types (e.g., serotonergic or GABAergic neurons) is anticipated to decrease graft-induced dyskinesia as a cell-based therapy of PD patients. Critically, the anti-apoptotic activity of MEF2CA and the fact that this construct also ameliorates the inhibitory effect of S-nitrosylation on MEF2C activity that is known to occur in the human PD brain bode well for the predicted efficacy of MEF2CA-programmed hNPCs in ameliorating PD signs and symptoms (31, 69). Thus, these aspects of MEF2C neurogenic and anti-apoptotic activity may provide improved neurobehavioral outcomes for PD patients in future clinical transplantation studies.

Besides its effect on neurodevelopment, MEF2 also contributes to adult neurogenesis (4, 80). Although all four MEF2s are extensively expressed in the adult brain (63), differential expression patterns suggest that each MEF2 isoform may play a distinct role at different stages of adult neurogenesis. For instance, MEF2A is highly expressed in adult hippocampal progenitor cells (4), influencing pre- and post-synapse formation of hippocampal neurons (73, 75, 76, 81). Importantly, several post-translational modifications (e.g., sumoylation, acetylation, in addition to S-nitrosylation) modulate the transcriptional activity of MEF2 in developing and adult brains (4, 12, 31, 75, 76). Below, we highlight the effects of S-nitrosylation of MEF2 on adult neurogenesis.

Multiple lines of evidence suggest that RNS/NO is an important regulator of adult hippocampal neurogenesis. Studies using nNOS-deficient mice or NOS inhibitors revealed that NO diminishes adult neurogenesis in mouse brains (11, 82, 83). Mechanistically, NO-related species can S-nitrosylate various MEF2 isoforms (forming SNO-MEF2) at conserved Cys39 in the MADS domain, thus inhibiting DNA binding and resulting in an impairment of its transactivity (4, 12) (Figure 7.3). Formation of SNO-MEF2 (in

particular SNO-MEF2A) disrupts adult hippocampal neurogenesis not only in cultured adult rat hippocampal neural progenitor/stem cells but also in intact mice. In this context, formation of SNO-MEF2 attenuates the expression of nuclear receptor tailless (TLX), a critical mediator of adult neurogenesis (84–86) (Figure 7.3). Interestingly, we observed elevated levels of SNO-MEF2 in animal models of several neurodegenerative disorders, ranging from focal stroke to AD. Moreover, in cell-based and mouse models of AD, expression of non-nitrosylatable mutant MEF2 (whose DNA binding is not affected by NO) restores TLX expression, rescuing the neurogenesis activity of MEF2. Furthermore, we found similar levels of SNO-MEF2 in human AD brains as in mouse models of AD with impaired neurogenesis. These findings are consistent with the notion that pathophysiologically relevant amounts of SNO-MEF2 are present in human AD brain that can impair gene expression related to adult neurogenesis (4, 12).

Additionally, in models of PD, we reported that S-nitrosylation of MEF2C mitigates expression of anti-apoptotic genes such as Bcl-xL and PGC1-α, thus augmenting neuronal cell death (31), but effects of SNO-MEF2 on neurogenesis in PD brains have not yet been reported. Similarly, while nitrosative stress has been linked to neurodevelopmental disorders such as ASD (87, 88), it is as yet unknown if S-nitrosylation of MEF2 and other proteins can effect neurogenesis and other aspects of the pathogenesis of ASD and other neurodevelopmental disorders.

7.5 OTHER SNO-PROTEINS INVOLVED IN NEURONAL DEVELOPMENT AND DIFFERENTIATION

In the developing brain, S-nitrosylation reactions engendered by low levels of NO normally act as physiological mediators of neuronal survival and differentiation. In contrast, excessive generation of NO-related species results in aberrant protein S-nitrosylation, contributing to neurodevelopmental and neurodegenerative disorders. As discussed below, besides the HDAC2, CREB, and MEF2 pathways, S-nitrosylation of additional proteins is believed to affect neuronal differentiation or neurogenesis via other signaling cascades under both physiological and pathological conditions, via formation of SNO-NDEL1, SNO-MAP1B, and SNO-Cdk5/SNO-Drp1.

7.5.1 S-Nitrosylation of NDEL1

Highly expressed in the brain, nuclear distribution element-like 1 or nuclear distribution protein nudE-like 1 (NDEL1) activates a microtubule motor protein, dynein, and thereby regulates microtubule networks during neuronal development (89). NDEL1 also interacts with Disrupted in Schizophrenia 1 (DISC1), a key scaffolding protein that mediates normal neurite outgrowth during cerebrocortical neurodevelopment. However, a schizophrenia-associated DISC1 mutant fails to bind to NDEL1, leading to decreased neurite extension (90). The groups of Snyder, Sawa, and Kamiya recently demonstrated nNOS-dependent S-nitrosylation of NDEL1 at Cys203 during normal dendritic development (91). S-Nitrosylation of NDEL1 maintains dendritic branching in cell-based models of neuronal maturation (Figure 7.4). Moreover, in a mouse model of neurodevelopmental abnormalities due to neonatal ethanol exposure, denitrosylation of NDEL1 leads to dendritic impairment. These findings are consistent with the notion that S-nitrosylation of NDEL1 contributes to the normal development of neuronal processes. While this study provides solid evidence for the physiological role of SNO-NDEL1 in dendritic development, several questions remain unanswered. For instance, it remains unclear if S-nitrosylation affects the physiological function/ activity of NDEL1 (i.e., its interaction with dynein and DISC1). Additionally, the consequences of preventing SNO-NDEL1 formation on animal behavior are still unknown and remain for future studies.

7.5.2 S-Nitrosylation of MAP1B-LC1

Microtubule-associated protein 1B (MAP1B) is known to regulate axonal outgrowth, branching, and retraction via cytoskeletal remodeling during neuronal migration and axon guidance (92, 93). MAP1B is composed of a heavy and light chain (LC1). Prior work has shown that physical interaction of nNOS with MAP1B promotes S-nitrosylation of MAP1B-LC1 at Cys2457, leading to a conformational change of LC1 (94). The S-nitrosylation-induced structural alteration results in increased binding of the MAP1B complex to microtubules. This blocks the action of dynein, suppressing the extension force of microtubules. As a consequence, formation of SNO-MAP1B-LC1 contributes to NO-dependent growth-cone collapse and axon retraction during neuronal development (Figure 7.4).

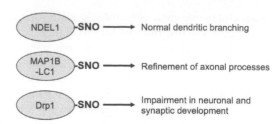

Figure 7.4 Effects of S-nitrosylated NDEL1, MAP1B, and Drp1 on neuronal differentiation and development.

(Top) S-Nitrosylation of NDEL1 facilitates normal dendritic growth, thereby contributing to neuronal maturation during brain development. (Middle) S-Nitrosylation of MAP1B-LC1 increases its binding to microtubules and promotes axonal retraction. This reaction facilitates the refinement of axonal process during brain development. (Bottom) S-Nitrosylation of Drp1 dramatically increases mitochondrial fission, compromising mitochondrial bioenergetics in neuronal processes and synapses. Thus, formation of SNO-Drp1 can impair synaptic structures.

Moreover, the mitochondrial ubiquitin ligase, MITOL (or March5), preferentially binds to the S-nitrosylated form of MAP1B-LC1, ubiquitinating this protein and thus targeting it for proteasomal degradation (95). In this context, S-nitrosylation of MAP1B limits mitochondrial movement in axons via inhibition of dynein, and MITOL acts as a negative regulator of SNO-MAP1B-LC1. Although further work is needed to elucidate the effects of SNO-MAP1B-LC1 in *vivo*, S-nitrosylation of MAP1B-LC1 may thereby contribute to neuronal circuit refinement via axonal retraction in developing brains.

7.5.3 S-Nitrosylation of Drp1 and Cdk5

The balance between fusion and fission of mitochondria (collectively termed mitochondrial dynamics) helps maintain proper morphology and bioenergetics of these organelles within neurons and other cell types. Accordingly, mitochondrial dynamics ensure efficient distribution of mitochondria to different subcellular localizations, including along the length of the axon and in pre- and post-synaptic terminals (96, 97). This process allows mitochondria to provide sufficient energy at the desired locations to facilitate neuronal development and differentiation, synaptic activity, and neuronal survival. Dynamin-related protein 1, Drp1 represents a key GTPase that mediates mitochondrial fission in mammalian cells. Our group and subsequently others demonstrated that β-amyloid (Aβ) oligomers or other forms of pathological stress engender S-nitrosylation of Drp1 at Cys644, resulting in an aberrant increase in GTPase activity. In this manner, SNO-Drp1 leads to excessive mitochondrial fission or fragmentation with cristae damage and consequent bioenergetic failure (98). Because synapse integrity requires substantial energy, formation of SNO-Drp1 ultimately results in synaptic damage, as demonstrated in both cell-based and animal models of AD (5). Notably, expression of non-nitrosylatable mutant Drp1 (C644A) mitigates oligomeric Aβ-induced synaptic damage in cell-based models (5). Moreover, SNO-Drp1 formation is drastically increased in human brains with AD and other neurodegenerative diseases, consistent with the pathophysiological relevance of SNO-Drp1 in these disorders (5, 99). Thus, these findings strongly suggest that SNO-Drp1 plays a causal role in the pathogenesis of AD by contributing to synaptic damage, arguably representing the best neuropathological correlate to cognitive decline (100, 101).

Additionally, during brain development, dysregulation of Drp1 activity causes impairment in neuronal differentiation. For instance, mice with neuron-specific knock out of Drp1 show elongated neuronal mitochondria accompanied by a decrease in the number of dendrites and synapses (102, 103). Conversely, increasing Drp1 activity in cultured NPCs produces severely fragmented mitochondria, compromises mitochondrial bioenergetics, activates the unfolded protein response (UPR), and inhibits neuronal differentiation (104). These results are consistent with the notion that physiological activity of Drp1 mediates normal mitochondrial fission required for proper mitochondria distribution and bioenergetics during neuronal differentiation. Moreover, either enhanced or diminished Drp1 activity results in abnormal neuronal development. Recently, unbiased mass spectrometry-based analysis of the S-nitrosoproteome has revealed the presence of SNO-Drp1 in models of neurodevelopmental disorders, including models of ASD (105). To study the meaning of this result mechanistically, future studies are warranted to examine if and how aberrant S-nitrosylation of Drp1 contributes to disturbances in neuronal differentiation and synapse formation in human neurodevelopmental disorders (Figure 7.4).

Intriguingly, our group demonstrated that S-nitrosylated Cdk5 acts as an S-nitrosylating enzyme towards Drp1 by donating its NO group to Drp1, a reaction termed transnitrosylation from one protein to another (106). The SNO-Drp1 thus formed then leads to mitochondrial fragmentation, bioenergetic compromise, and synaptic damage, as discussed above (106). Additionally, SNO-Cdk5 exhibits pathologically increased kinase activity that also contributes Aβ-induced neuronal damage via decreased dendritic growth and branching (107). Taken together, these results suggest that SNO-Cdk5 operates as a dual-function enzyme (i.e., both a kinase and a nitrosylase) with pathological consequences (106, 107). Future studies are needed to examine whether SNO-Cdk5-dependent transnitrosylation of Drp1 compromises mitochondrial morphology and energy production during neuronal differentiation in addition to the effect of SNO-Cdk5 on synaptic damage in the mature nervous system that is discussed above.

7.6 CONCLUSIONS AND PERSPECTIVE FOR THE FUTURE

In this review, we have discussed the emerging role of protein S-nitrosylation in neuronal differentiation and development in the context of both embryologic and adult neurogenesis. Specifically, we have highlighted S-nitrosylation-mediated pathways that either increase neuronal development via neurite outgrowth (e.g., via SNO-HDAC2, SNO-GAPDH, and SNO-NDEL1) or disrupt dendritic, synaptic, or neurogenic structures (e.g., SNO-MEF2C, SNO-MAP1B-LC1, and SNO-Drp1). These findings are consistent with the notion that signaling pathways affected by protein S-nitrosylation signaling can effect fine-tuning of normal neurodifferentiation and network activity, but, conversely, aberrant protein S-nitrosylation in the face of excessive levels of RNS can disrupt neurogenesis and contribute to synapse destruction.

We anticipate that the list of S-nitrosylated (SNO)-proteins as well as the knowledge of physiological and pathophysiological events associated with S-nitrosylation will continue to grow. Along these lines, mass spectrometry–based approaches have recently been successfully adapted to obtain a more comprehensive view of the S-nitrosoproteome (i.e., the full panoply of SNO-proteins) during normal neurodevelopment and neurodevelopmental disorders (105, 108, 109). Among these studies, Steven Tannenbaum's group has elegantly revealed changes in the "endogenous"

S-nitrosoproteome in a mouse model of autism spectrum disorder (ASD) bearing a Shank3 mutation (105). Nonetheless, the identification of the SNO-proteome affecting neurodifferentiation and development has just begun. It will be important in future studies to uncover endogenous networks of S-nitrosylated proteins in additional models of neurodevelopmental disorders, during normal brain development, and even in human brains. Importantly, these studies may identify molecular mechanisms for S-nitrosylation-modulated neural differentiation and thus potentially facilitate discovery of new therapeutic strategies for neurodevelopmental disorders.

7.7 ACKNOWLEDGMENTS

We are grateful to Dr. Scott R. McKercher (The Scripps Research Institute) for critical reading of the manuscript. This work was supported in part by NIH grants RF1 AG057409, R01 AG056259, R01 NS086890, R01 DA048882, DP1 DA041722 (to S.A.L.), and R01 AG061845 (to T.N.); awards from the California Tobacco-Related Disease Research Program (TDRP 27IR-0010), California Institute for Regenerative Medicine (CIRM) DISC2-1107, and Deerfield Management/Poseidon Innovation to advance disease-curing therapeutics; a Distinguished Investigator Award from the Brain & Behavior Research Foundation (to S.A.L.); and the Michael J. Fox Foundation (to S.A.L. and T.N.).

REFERENCES

1. H. H. Wu, C. V. Williams, S. C. McLoon, Involvement of nitric oxide in the elimination of a transient retinotectal projection in development. *Science* **265**, 1593–1596 (1994).

2. S. A. Lipton, Y. B. Choi, Z. H. Pan, S. Z. Lei, H. S. Chen, N. J. Sucher, J. Loscalzo, D. J. Singel, J. S. Stamler, A redox-based mechanism for the neuroprotective and neurodestructive effects of nitric oxide and related nitroso-compounds. *Nature* **364**, 626–632 (1993).

3. Y. B. Choi, L. Tenneti, D. A. Le, J. Ortiz, G. Bai, H. S. Chen, S. A. Lipton, Molecular basis of NMDA receptor-coupled ion channel modulation by S-nitrosylation. *Nat Neurosci* **3**, 15–21 (2000).

4. S. Okamoto, S. A. Lipton, S-Nitrosylation in neurogenesis and neuronal development. *Biochim Biophys Acta* **1850**, 1588–1593 (2015).

5. D. H. Cho, T. Nakamura, J. Fang, P. Cieplak, A. Godzik, Z. Gu, S. A. Lipton, S-Nitrosylation of Drp1 mediates β-amyloid-related mitochondrial fission and neuronal injury. *Science* **324**, 102–105 (2009).

6. T. Nakamura, L. Wang, C. C. Wong, F. L. Scott, B. P. Eckelman, X. Han, C. Tzitzilonis, F. Meng, Z. Gu, E. A. Holland, A. T. Clemente, S. Okamoto, G. S. Salvesen, R. Riek, J. R. Yates, 3rd, S. A. Lipton, Transnitrosylation of XIAP regulates caspase-dependent neuronal cell death. *Mol. Cell* **39**, 184–195 (2010).

7. T. Uehara, T. Nakamura, D. Yao, Z. Q. Shi, Z. Gu, Y. Ma, E. Masliah, Y. Nomura, S. A. Lipton, S-Nitrosylated protein-disulphide isomerase links protein misfolding to neurodegeneration. *Nature* **441**, 513–517 (2006).

8. C. Cserep, A. Szonyi, J. M. Veres, E. Szabadits, J. de Vente, N. Hajos, T. F. Freund, G. Nyiri, Nitric oxide signaling modulates synaptic transmission during early postnatal development. *Cereb Cortex* **21**, 2065–2074 (2011).

9. A. J. Roskams, D. S. Bredt, T. M. Dawson, G. V. Ronnett, Nitric oxide mediates the formation of synaptic connections in developing and regenerating olfactory receptor neurons. *Neuron* **13**, 289–299 (1994).

10. A. Nott, P. M. Watson, J. D. Robinson, L. Crepaldi, A. Riccio, S-Nitrosylation of histone deacetylase 2 induces chromatin remodelling in neurons. *Nature* **455**, 411–415 (2008).

11. M. A. Packer, Y. Stasiv, A. Benraiss, E. Chmielnicki, A. Grinberg, H. Westphal, S. A. Goldman, G. Enikolopov, Nitric oxide negatively regulates mammalian adult neurogenesis. *Proc Natl Acad Sci USA* **100**, 9566–9571 (2003).

12. S. Okamoto, T. Nakamura, P. Cieplak, S. F. Chan, E. Kalashnikova, L. Liao, S. Saleem, X. Han, A. Clemente, A. Nutter, S. Sances, C. Brechtel, D. Haus, F. Haun, S. Sanz-Blasco, X. Huang, H. Li, J. D. Zaremba, J. Cui, Z. Gu, R. Nikzad, A. Harrop, S. R. McKercher, A. Godzik, J. R. Yates, 3rd, S. A. Lipton, S-Nitrosylation-mediated redox transcriptional switch modulates neurogenesis and neuronal cell death. *Cell Rep* **8**, 217–228 (2014).

13. L. J. Ignarro, G. M. Buga, K. S. Wood, R. E. Byrns, G. Chaudhuri, Endothelium-derived relaxing factor produced and released from artery and vein is nitric oxide. *Proc Natl Acad Sci USA* **84**, 9265–9269 (1987).

14. F. Murad, Cyclic guanosine monophosphate as a mediator of vasodilation. *J Clin Invest* **78**, 1–5 (1986).

15. R. M. Palmer, A. G. Ferrige, S. Moncada, Nitric oxide release accounts for the biological activity of endothelium-derived relaxing factor. *Nature* **327**, 524–526 (1987).

16. D. T. Hess, A. Matsumoto, S. O. Kim, H. E. Marshall, J. S. Stamler, Protein S-nitrosylation: purview and parameters. *Nat Rev Mol Cell Biol* **6**, 150–166 (2005).

17. T. Nakamura, S. Tu, M. W. Akhtar, C. R. Sunico, S. Okamoto, S. A. Lipton, Aberrant protein S-nitrosylation in neurodegenerative diseases. *Neuron* **78**, 596–614 (2013).

18. N. B. Fernhoff, E. R. Derbyshire, M. A. Marletta, A nitric oxide/cysteine interaction mediates the activation of soluble guanylate cyclase. *Proc Natl Acad Sci USA* **106**, 21602–21607 (2009).

19. H. Ischiropoulos, L. Zhu, J. Chen, M. Tsai, J. C. Martin, C. D. Smith, J. S. Beckman, Peroxynitrite-mediated tyrosine nitration catalyzed by superoxide dismutase. *Arch Biochem Biophys* **298**, 431–437 (1992).

20. S. Z. Lei, Z. H. Pan, S. K. Aggarwal, H. S. Chen, J. Hartman, N. J. Sucher, S. A. Lipton, Effect of nitric oxide production on the redox modulatory site of the NMDA receptor-channel complex. *Neuron* **8**, 1087–1099 (1992).

21. J. S. Stamler, S. Lamas, F. C. Fang, Nitrosylation. the prototypic redox-based signaling mechanism. *Cell* **106**, 675–683 (2001).

22. J. S. Stamler, D. I. Simon, J. A. Osborne, M. E. Mullins, O. Jaraki, T. Michel, D. J. Singel, J. Loscalzo, S-Nitrosylation of proteins with nitric oxide: synthesis and characterization of biologically active compounds. *Proc Natl Acad Sci USA* **89**, 444–448 (1992).

23. C. T. Stomberski, D. T. Hess, J. S. Stamler, Protein S-nitrosylation: determinants of specificity and enzymatic regulation of S-nitrosothiol-based signaling. *Antioxid Redox Signal* **30**, 1331–1351 (2019).

24. T. Nakamura, S. A. Lipton, Nitric oxide-dependent protein post-translational modifications impair mitochondrial function and metabolism to contribute to neurodegenerative diseases. *Antioxid Redox Signal* **32**, 817–833 (2020).

25. J. S. Stamler, E. J. Toone, S. A. Lipton, N. J. Sucher, (S)NO signals: translocation, regulation, and a consensus motif. *Neuron* **18**, 691–696 (1997).

26. P. Anand, A. Hausladen, Y. J. Wang, G. F. Zhang, C. Stomberski, H. Brunengraber, D. T. Hess, J. S. Stamler, Identification of S-nitroso-CoA reductases that regulate protein S-nitrosylation. *Proc Natl Acad Sci USA* **111**, 18572–18577 (2014).

27. M. Benhar, M. T. Forrester, J. S. Stamler, Protein denitrosylation: enzymatic mechanisms and cellular functions. *Nat Rev Mol Cell Biol* **10**, 721–732 (2009).

28. Z. Gu, M. Kaul, B. Yan, S. J. Kridel, J. Cui, A. Strongin, J. W. Smith, R. C. Liddington, S. A. Lipton, S-Nitrosylation of matrix metalloproteinases: signaling pathway to neuronal cell death. *Science* **297**, 1186–1190 (2002).

29. A. Nott, J. Nitarska, J. V. Veenvliet, S. Schacke, A. A. Derijck, P. Sirko, C. Muchardt, R. J. Pasterkamp, M. P. Smidt, A. Riccio, S-nitrosylation of HDAC2 regulates the expression of the chromatin-remodeling factor Brm during radial neuron migration. *Proc Natl Acad Sci USA* **110**, 3113–3118 (2013).

30. J. G. Smith, S. G. Aldous, C. Andreassi, G. Cuda, M. Gaspari, A. Riccio, Proteomic analysis of S-nitrosylated nuclear proteins in rat cortical neurons. *Sci Signal* **11** (2018).

31. S. D. Ryan, N. Dolatabadi, S. F. Chan, X. Zhang, M. W. Akhtar, J. Parker, F. Soldner, C. R. Sunico, S. Nagar, M. Talantova, B. Lee, K. Lopez, A. Nutter, B. Shan, E. Molokanova, Y. Zhang, X. Han, T. Nakamura, E. Masliah, J. R. Yates, 3rd, N. Nakanishi, A. Y. Andreyev, S. Okamoto, R. Jaenisch, R. Ambasudhan, S. A. Lipton, Isogenic human iPSC Parkinson's model shows nitrosative stress-induced dysfunction in MEF2-PGC1alpha transcription. *Cell* **155**, 1351–1364 (2013).

32. R. K. Lenroot, J. N. Giedd, The changing impact of genes and environment on brain development during childhood and adolescence: initial findings from a neuroimaging study of pediatric twins. *Dev Psychopathol* **20**, 1161–1175 (2008).

33. U. Meyer, Neurodevelopmental resilience and susceptibility to maternal immune activation. *Trends Neurosci* **42**, 793–806 (2019).

34. P. Mendola, S. G. Selevan, S. Gutter, D. Rice, Environmental factors associated with a spectrum of neurodevelopmental deficits. *Ment Retard Dev Disabil Res Rev* **8**, 188–197 (2002).

35. M. R. Montminy, L. M. Bilezikjian, Binding of a nuclear protein to the cyclic-AMP response element of the somatostatin gene. *Nature* **328**, 175–178 (1987).

36. J. Y. Altarejos, M. Montminy, CREB and the CRTC co-activators: sensors for hormonal and metabolic signals. *Nat Rev Mol Cell Biol* **12**, 141–151 (2011).

37. L. Redmond, A. Ghosh, Regulation of dendritic development by calcium signaling. *Cell Calcium* **37**, 411–416 (2005).

38. M. Sheng, G. McFadden, M. E. Greenberg, Membrane depolarization and calcium induce c-fos transcription via phosphorylation of transcription factor CREB. *Neuron* **4**, 571–582 (1990).

39. A. Riccio, R. S. Alvania, B. E. Lonze, N. Ramanan, T. Kim, Y. Huang, T. M. Dawson, S. H. Snyder, D. D. Ginty, A nitric oxide signaling pathway controls CREB-mediated gene expression in neurons. *Mol Cell* **21**, 283–294 (2006).

40. Q. Luo, K. Viste, J. C. Urday-Zaa, G. Senthil Kumar, W. W. Tsai, A. Talai, K. E. Mayo, M. Montminy, I. Radhakrishnan, Mechanism of CREB recognition and coactivation by the CREB-regulated transcriptional coactivator CRTC2. *Proc Natl Acad Sci USA* **109**, 20865–20870 (2012).

41. I. Goren, E. Tavor, A. Goldblum, A. Honigman, Two cysteine residues in the DNA-binding domain of CREB control binding to CRE and CREB-mediated gene expression. *J Mol Biol* **313**, 695–709 (2001).

42. Y. Xue, J. Wong, G. T. Moreno, M. K. Young, J. Côté, W. Wang, NURD, a Novel Complex with Both ATP-Dependent Chromatin-Remodeling and Histone Deacetylase Activities. *Molecular Cell* **2**, 851–861 (1998).

43. M. R. Hara, N. Agrawal, S. F. Kim, M. B. Cascio, M. Fujimuro, Y. Ozeki, M. Takahashi, J. H. Cheah, S. K. Tankou, L. D. Hester, C. D. Ferris, S. D. Hayward, S. H. Snyder, A. Sawa, S-nitrosylated GAPDH initiates apoptotic cell death by nuclear translocation following Siah1 binding. *Nat Cell Biol* **7**, 665–674 (2005).

44. M. D. Kornberg, N. Sen, M. R. Hara, K. R. Juluri, J. V. Nguyen, A. M. Snowman, L. Law, L. D. Hester, S. H. Snyder, GAPDH mediates nitrosylation of nuclear proteins. *Nat Cell Biol* **12**, 1094–1100 (2010).

45. A. Sawa, A. A. Khan, L. D. Hester, S. H. Snyder, Glyceraldehyde-3-phosphate dehydrogenase: nuclear translocation participates in neuronal and nonneuronal cell death. *Proc Natl Acad Sci USA* **94**, 11669–11674 (1997).

46. N. Sen, M. R. Hara, A. S. Ahmad, M. B. Cascio, A. Kamiya, J. T. Ehmsen, N. Agrawal, L. Hester, S. Dore, S. H. Snyder, A. Sawa, GOSPEL: a neuroprotective protein that binds to GAPDH upon S-nitrosylation. *Neuron* **63**, 81–91 (2009).

47. N. Sen, S. H. Snyder, Neurotrophin-mediated degradation of histone methyltransferase by S-nitrosylation cascade regulates neuronal differentiation. *Proc Natl Acad Sci USA* **108**, 20178–20183 (2011).

48. N. Sen, M. R. Hara, M. D. Kornberg, M. B. Cascio, B. I. Bae, N. Shahani, B. Thomas, T. M. Dawson, V. L. Dawson, S. H. Snyder, A. Sawa, Nitric oxide-induced nuclear GAPDH activates p300/CBP and mediates apoptosis. *Nat Cell Biol* **10**, 866–873 (2008).

49. S. B. Lee, C. K. Kim, K. H. Lee, J. Y. Ahn, S-nitrosylation of B23/nucleophosmin by GAPDH protects cells from the SIAH1-GAPDH death cascade. *J Cell Biol* **199**, 65–76 (2012).

50. J. Altman, Are new neurons formed in the brains of adult mammals? *Science* **135**, 1127–1128 (1962).

51. E. Gould, P. Tanapat, B. S. McEwen, G. Flugge, E. Fuchs, Proliferation of granule cell precursors in the dentate gyrus of adult monkeys is diminished by stress. *Proc Natl Acad Sci USA* **95**, 3168–3171 (1998).

52. P. S. Eriksson, E. Perfilieva, T. Bjork-Eriksson, A. M. Alborn, C. Nordborg, D. A. Peterson, F. H. Gage, Neurogenesis in the adult human hippocampus. *Nat Med* **4**, 1313–1317 (1998).

53. S. F. Sorrells, M. F. Paredes, A. Cebrian-Silla, K. Sandoval, D. Qi, K. W. Kelley, D. James, S. Mayer, J. Chang, K. I. Auguste, E. F. Chang, A. J. Gutierrez, A. R. Kriegstein, G. W. Mathern, M. C. Oldham, E. J. Huang, J. M. Garcia-Verdugo, Z. Yang, A. Alvarez-Buylla, Human hippocampal neurogenesis drops sharply in children to undetectable levels in adults. *Nature* **555**, 377–381 (2018).

54. M. Boldrini, C. A. Fulmore, A. N. Tartt, L. R. Simeon, I. Pavlova, V. Poposka, G. B. Rosoklija, A. Stankov, V. Arango, A. J. Dwork, R. Hen, J. J. Mann, Human hippocampal neurogenesis persists throughout aging. *Cell Stem Cell* **22**, 589–599 e585 (2018).

55. F. H. Gage, Adult neurogenesis in mammals. *Science* **364**, 827–828 (2019).

56. R. Faigle, H. Song, Signaling mechanisms regulating adult neural stem cells and neurogenesis. *Biochim Biophys Acta* **1830**, 2435–2448 (2013).

57. C. Zhao, W. Deng, F. H. Gage, Mechanisms and functional implications of adult neurogenesis. *Cell* **132**, 645–660 (2008).

58. T. Toda, S. L. Parylak, S. B. Linker, F. H. Gage, The role of adult hippocampal neurogenesis in brain health and disease. *Mol Psychiatry* **24**, 67–87 (2019).

59. J. B. Dietrich, The MEF2 family and the brain: from molecules to memory. *Cell Tissue Res* **352**, 179–190 (2013).

60. D. Leifer, D. Krainc, Y. T. Yu, J. McDermott, R. E. Breitbart, J. Heng, R. L. Neve, B. Kosofsky, B. Nadal-Ginard, S. A. Lipton, MEF2C, a MADS/MEF2-family transcription factor expressed in a

laminar distribution in cerebral cortex. *Proc Natl Acad Sci USA* **90**, 1546–1550 (1993).

61. K. Madugula, R. Mulherkar, Z. K. Khan, D. I. Chigbu, D. Patel, E. W. Harhaj, P. Jain, MEF-2 isoforms' (A-D) roles in development and tumorigenesis. *Oncotarget* **10**, 2755–2787 (2019).

62. B. J. Swanson, H. M. Jack, G. E. Lyons, Characterization of myocyte enhancer factor 2 (MEF2) expression in B and T cells: MEF2C is a B cell-restricted transcription factor in lymphocytes. *Mol Immunol* **35**, 445–458 (1998).

63. A. Assali, A. J. Harrington, C. W. Cowan, Emerging roles for MEF2 in brain development and mental disorders. *Curr Opin Neurobiol* **59**, 49–58 (2019).

64. V. Andres, M. Cervera, V. Mahdavi, Determination of the consensus binding site for MEF2 expressed in muscle and brain reveals tissue-specific sequence constraints. *J Biol Chem* **270**, 23246–23249 (1995).

65. H. Li, J. C. Radford, M. J. Ragusa, K. L. Shea, S. R. McKercher, J. D. Zaremba, W. Soussou, Z. Nie, Y. J. Kang, N. Nakanishi, S. Okamoto, A. J. Roberts, J. J. Schwarz, S. A. Lipton, Transcription factor MEF2C influences neural stem/progenitor cell differentiation and maturation in vivo. *Proc Natl Acad Sci USA* **105**, 9397–9402 (2008).

66. E. G. Cho, J. D. Zaremba, S. R. McKercher, M. Talantova, S. Tu, E. Masliah, S. F. Chan, N. Nakanishi, A. Terskikh, S. A. Lipton, MEF2C enhances dopaminergic neuron differentiation of human embryonic stem cells in a parkinsonian rat model. *PLoS One* **6**, e24027 (2011).

67. B. Zhu, R. E. Carmichael, L. Solabre Valois, K. A. Wilkinson, J. M. Henley, The transcription factor MEF2A plays a key role in the differentiation/maturation of rat neural stem cells into neurons. *Biochem Biophys Res Commun* **500**, 645–649 (2018).

68. Z. Mao, A. Bonni, F. Xia, M. Nadal-Vicens, M. E. Greenberg, Neuronal activity-dependent cell survival mediated by transcription factor MEF2. *Science* **286**, 785–790 (1999).

69. S. Okamoto, D. Krainc, K. Sherman, S. A. Lipton, Antiapoptotic role of the p38 mitogen-activated protein kinase-myocyte enhancer factor 2 transcription factor pathway during neuronal differentiation. *Proc Natl Acad Sci USA* **97**, 7561–7566 (2000).

70. S. Okamoto, Z. Li, C. Ju, M. N. Scholzke, E. Mathews, J. Cui, G. S. Salvesen, E. Bossy-Wetzel, S. A. Lipton, Dominant-interfering forms of MEF2 generated by caspase cleavage contribute to NMDA-induced neuronal apoptosis. *Proc Natl Acad Sci USA* **99**, 3974–3979 (2002).

71. Z. Li, S. R. McKercher, J. Cui, Z. Nie, W. Soussou, A. J. Roberts, T. Sallmen, J. H. Lipton, M. Talantova, S. Okamoto, S. A. Lipton, Myocyte enhancer factor 2C as a neurogenic and anti-apoptotic transcription factor in murine embryonic stem cells. *J Neurosci* **28**, 6557–6568 (2008).

72. A. C. Barbosa, M. S. Kim, M. Ertunc, M. Adachi, E. D. Nelson, J. McAnally, J. A. Richardson, E. T. Kavalali, L. M. Monteggia, R. Bassel-Duby, E. N. Olson, MEF2C, a transcription factor that facilitates learning and memory by negative regulation of synapse numbers and function. *Proc Natl Acad Sci USA* **105**, 9391–9396 (2008).

73. S. W. Flavell, C. W. Cowan, T. K. Kim, P. L. Greer, Y. Lin, S. Paradis, E. C. Griffith, L. S. Hu, C. Chen, M. E. Greenberg, Activity-dependent regulation of MEF2 transcription factors suppresses excitatory synapse number. *Science* **311**, 1008–1012 (2006).

74. S. E. Latchney, Y. Jiang, D. P. Petrik, A. J. Eisch, J. Hsieh, Inducible knockout of Mef2a, -c, and -d from nestin-expressing stem/progenitor cells and their progeny unexpectedly uncouples neurogenesis and dendritogenesis in vivo. *FASEB J* **29**, 5059–5071 (2015).

75. A. Shalizi, B. Gaudilliere, Z. Yuan, J. Stegmuller, T. Shirogane, Q. Ge, Y. Tan, B. Schulman, J. W. Harper, A. Bonni, A calcium-regulated MEF2 sumoylation switch controls postsynaptic differentiation. *Science* **311**, 1012–1017 (2006).

76. T. Yamada, Y. Yang, J. Huang, G. Coppola, D. H. Geschwind, A. Bonni, Sumoylated MEF2A coordinately eliminates orphan presynaptic sites and promotes maturation of presynaptic boutons. *J Neurosci* **33**, 4726–4740 (2013).

77. S. Tu, M. W. Akhtar, R. M. Escorihuela, A. Amador-Arjona, V. Swarup, J. Parker, J. D. Zaremba, T. Holland, N. Bansal, D. R. Holohan, K. Lopez, S. D. Ryan, S. F. Chan, L. Yan, X. Zhang, X. Huang, A. Sultan, S. R. McKercher, R. Ambasudhan, H. Xu, Y. Wang, D. H. Geschwind, A. J. Roberts, A. V. Terskikh, R. A. Rissman, E. Masliah, S. A. Lipton, N. Nakanishi, NitroSynapsin therapy for a mouse MEF2C haploinsufficiency model of human autism. *Nat Commun* **8**, 1488 (2017).

78. A. J. Harrington, C. M. Bridges, S. Berto, K. Blankenship, J. Y. Cho, A. Assali, B. M. Siemsen, H. W. Moore, E. Tsvetkov, A. Thielking,

G. Konopka, D. B. Everman, M. D. Scofield, S. A. Skinner, C. W. Cowan, MEF2C hypofunction in neuronal and neuroimmune populations produces MEF2C haploinsufficiency syndrome-like Behaviors in mice. *Biol Psychiatry* (2020).

79. R. Ambasudhan, N. Dolatabadi, A. Nutter, E. Masliah, S. R. McKercher, S. A. Lipton, Potential for cell therapy in Parkinson's disease using genetically programmed human embryonic stem cell-derived neural progenitor cells. *J Comp Neurol* **522**, 2845–2856 (2014).

80. D. Petrik, Y. Jiang, S. G. Birnbaum, C. M. Powell, M. S. Kim, J. Hsieh, A. J. Eisch, Functional and mechanistic exploration of an adult neurogenesis-promoting small molecule. *FASEB J* **26**, 3148–3162 (2012).

81. B. E. Pfeiffer, T. Zang, J. R. Wilkerson, M. Taniguchi, M. A. Maksimova, L. N. Smith, C. W. Cowan, K. M. Huber, Fragile X mental retardation protein is required for synapse elimination by the activity-dependent transcription factor MEF2. *Neuron* **66**, 191–197 (2010).

82. B. Moreno-Lopez, C. Romero-Grimaldi, J. A. Noval, M. Murillo-Carretero, E. R. Matarredona, C. Estrada, Nitric oxide is a physiological inhibitor of neurogenesis in the adult mouse subventricular zone and olfactory bulb. *J Neurosci* **24**, 85–95 (2004).

83. A. Torroglosa, M. Murillo-Carretero, C. Romero-Grimaldi, E. R. Matarredona, A. Campos-Caro, C. Estrada, Nitric oxide decreases subventricular zone stem cell proliferation by inhibition of epidermal growth factor receptor and phosphoinositide-3-kinase/Akt pathway. *Stem Cells* **25**, 88–97 (2007).

84. C. L. Zhang, Y. Zou, W. He, F. H. Gage, R. M. Evans, A role for adult TLX-positive neural stem cells in learning and behaviour. *Nature* **451**, 1004–1007 (2008).

85. H. K. Liu, T. Belz, D. Bock, A. Takacs, H. Wu, P. Lichter, M. Chai, G. Schutz, The nuclear receptor tailless is required for neurogenesis in the adult subventricular zone. *Genes Dev* **22**, 2473–2478 (2008).

86. K. Murai, Q. Qu, G. Sun, P. Ye, W. Li, G. Asuelime, E. Sun, G. E. Tsai, Y. Shi, Nuclear receptor TLX stimulates hippocampal neurogenesis and enhances learning and memory in a transgenic mouse model. *Proc Natl Acad Sci USA* **111**, 9115–9120 (2014).

87. S. M. Colvin, K. Y. Kwan, Dysregulated nitric oxide signaling as a candidate mechanism of fragile X syndrome and other neuropsychiatric disorders. *Front Genet* **5**, 239 (2014).

88. A. Frustaci, M. Neri, A. Cesario, J. B. Adams, E. Domenici, B. Dalla Bernardina, S. Bonassi, Oxidative stress-related biomarkers in autism: systematic review and meta-analyses. *Free Radic Biol Med* **52**, 2128–2141 (2012).

89. T. Shu, R. Ayala, M. D. Nguyen, Z. Xie, J. G. Gleeson, L. H. Tsai, Ndel1 operates in a common pathway with LIS1 and cytoplasmic dynein to regulate cortical neuronal positioning. *Neuron* **44**, 263–277 (2004).

90. Y. Ozeki, T. Tomoda, J. Kleiderlein, A. Kamiya, L. Bord, K. Fujii, M. Okawa, N. Yamada, M. E. Hatten, S. H. Snyder, C. A. Ross, A. Sawa, Disrupted-in-Schizophrenia-1 (DISC-1): mutant truncation prevents binding to NudE-like (NUDEL) and inhibits neurite outgrowth. *Proc Natl Acad Sci USA* **100**, 289–294 (2003).

91. A. Saito, Y. Taniguchi, S. H. Kim, B. Selvakumar, G. Perez, M. D. Ballinger, X. Zhu, J. Sabra, M. Jallow, P. Yan, K. Ito, S. Rajendran, S. Hirotsune, A. Wynshaw-Boris, S. H. Snyder, A. Sawa, A. Kamiya, Developmental Alcohol exposure impairs activity-dependent S-nitrosylation of NDEL1 for neuronal maturation. *Cereb Cortex* **27**, 3918–3929 (2017).

92. E. W. Dent, F. B. Gertler, Cytoskeletal dynamics and transport in growth cone motility and axon guidance. *Neuron* **40**, 209–227 (2003).

93. Y. Takei, J. Teng, A. Harada, N. Hirokawa, Defects in axonal elongation and neuronal migration in mice with disrupted tau and MAP1b genes. *J Cell Biol* **150**, 989–1000 (2000).

94. H. Stroissnigg, A. Trancikova, L. Descovich, J. Fuhrmann, W. Kutschera, J. Kostan, A. Meixner, F. Nothias, F. Propst, S-Nitrosylation of microtubule-associated protein 1B mediates nitric-oxide-induced axon retraction. *Nat Cell Biol* **9**, 1035–1045 (2007).

95. R. Yonashiro, Y. Kimijima, T. Shimura, K. Kawaguchi, T. Fukuda, R. Inatome, S. Yanagi, Mitochondrial ubiquitin ligase MITOL blocks S-nitrosylated MAP1B-light chain 1-mediated mitochondrial dysfunction and neuronal cell death. *Proc Natl Acad Sci USA* **109**, 2382–2387 (2012).

96. D. C. Chan, Mitochondrial fusion and fission in mammals. *Annu Rev Cell Dev Biol* **22**, 79–99 (2006).

97. D. H. Cho, T. Nakamura, S. A. Lipton, Mitochondrial dynamics in cell death and neurodegeneration. *Cell Mol Life Sci* **67**, 3435–3447 (2010).

98. M. J. Barsoum, H. Yuan, A. A. Gerencser, G. Liot, Y. Kushnareva, S. Graber, I. Kovacs, W. D. Lee, J. Waggoner, J. Cui, A. D. White, B. Bossy, J. C. Martinou, R. J. Youle, S. A. Lipton, M. H. Ellisman, G. A. Perkins, E. Bossy-Wetzel, Nitric oxide-induced mitochondrial fission is regulated by dynamin-related GTPases in neurons. *EMBO J* **25**, 3900–3911 (2006).

99. F. Haun, T. Nakamura, A. D. Shiu, D. H. Cho, T. Tsunemi, E. A. Holland, A. R. La Spada, S. A. Lipton, S-nitrosylation of dynamin-related protein 1 mediates mutant huntingtin-induced mitochondrial fragmentation and neuronal injury in Huntington's disease. *Antioxid Redox Signal* **19**, 1173–1184 (2013).

100. R. D. Terry, E. Masliah, D. P. Salmon, N. Butters, R. DeTeresa, R. Hill, L. A. Hansen, R. Katzman, Physical basis of cognitive alterations in Alzheimer's disease: synapse loss is the major correlate of cognitive impairment. *Ann Neurol* **30**, 572–580 (1991).

101. S. T. DeKosky, S. W. Scheff, Synapse loss in frontal cortex biopsies in Alzheimer's disease. correlation with cognitive severity. *Ann Neurol* **27**, 457–464 (1990).

102. N. Ishihara, M. Nomura, A. Jofuku, H. Kato, S. O. Suzuki, K. Masuda, H. Otera, Y. Nakanishi, I. Nonaka, Y. Goto, N. Taguchi, H. Morinaga, M. Maeda, R. Takayanagi, S. Yokota, K. Mihara, Mitochondrial fission factor Drp1 is essential for embryonic development and synapse formation in mice. *Nat Cell Biol* **11**, 958–966 (2009).

103. J. Wakabayashi, Z. Zhang, N. Wakabayashi, Y. Tamura, M. Fukaya, T. W. Kensler, M. Iijima, H. Sesaki, The dynamin-related GTPase Drp1 is required for embryonic and brain development in mice. *J Cell Biol* **186**, 805–816 (2009).

104. C. Vantaggiato, M. Castelli, M. Giovarelli, G. Orso, M. T. Bassi, E. Clementi, C. De Palma, The fine tuning of Drp1-dependent mitochondrial remodeling and autophagy controls neuronal differentiation. *Front Cell Neurosci* **13**, 120 (2019).

105. H. Amal, B. Barak, V. Bhat, G. Gong, B. A. Joughin, X. Wang, J. S. Wishnok, G. Feng, S. R. Tannenbaum, Shank3 mutation in a mouse model of autism leads to changes in the S-nitroso-proteome and affects key proteins involved in vesicle release and synaptic function. *Mol Psychiatry* (2018).

106. J. Qu, T. Nakamura, G. Cao, E. A. Holland, S. R. McKercher, S. A. Lipton, S-Nitrosylation activates Cdk5 and contributes to synaptic spine loss induced by β-amyloid peptide. *Proc Natl Acad Sci USA*, 14330–14335 (2011).

107. P. Zhang, P. C. Yu, A. H. Tsang, Y. Chen, A. K. Fu, W. Y. Fu, K. K. Chung, N. Y. Ip, S-Nitrosylation of cyclin-dependent kinase 5 (cdk5) regulates its kinase activity and dendrite growth during neuronal development. *J Neurosci* **30**, 14366–14370 (2010).

108. A. I. Santos, A. S. Lourenco, S. Simao, D. Marques da Silva, D. F. Santos, A. P. Onofre de Carvalho, A. C. Pereira, A. Izquierdo-Alvarez, E. Ramos, E. Morato, A. Marina, A. Martinez-Ruiz, I. M. Araujo, Identification of new targets of S-nitrosylation in neural stem cells by thiol redox proteomics. *Redox Biol* **32**, 101457 (2020).

109. R. Mnatsakanyan, S. Markoutsa, K. Walbrunn, A. Roos, S. H. L. Verhelst, R. P. Zahedi, Proteome-wide detection of S-nitrosylation targets and motifs using bioorthogonal cleavable-linker-based enrichment and switch technique. *Nat Commun* **10**, 2195 (2019).

CHAPTER EIGHT

NADPH Oxidases in Bone Cells

Katrin Schröder

CONTENTS

8.1 INTRODUCTION

Bones build a skeleton that functions as scaffold for muscles and organs. Although skeletons can persist for millions of years to tell archeologists some details about our ancestors, in a living human being bones are neither uniform nor constantly persistent. During development of a human being, two ways have been identified, of how bone is formed. Facial and cranial bones as well as clavicles are formed via dermal ossification. In this process, islands of mesenchymal stem cells differentiate into osteoblasts and build osteoid (bone matrix). In contrast, long bones, such as the femur and radius, are formed by chondral ossification. In this scenario, mesenchymal stem cells differentiate into chondrocytes and produce cartilage. In a second step, cartilage is mineralized and becomes bone [1]. In adults, every year 10% of the skeleton is renewed. This process is facilitated by osteoclasts that remove old bone matrix and osteoblasts that build new bone. Accordingly, a fine-tuning in formation of osteoclasts and osteoblasts is needed for constant remodeling. The main regulators are

receptors for activation of NFκB (RANK), RANK ligand (RANKL), and osteoprotegerin (OPG). RANKL and its decoy OPG are produced by osteoblasts and osteocytes and regulate the formation of osteoclasts [2].

Osteoclasts derive from macrophage-like precursors. Stimulation with M-CSF increases the expression of RANK, which acts as a receptor for RANKL. RANKL activates transcription factors like PU.1, MITF, c-Fos, and NFATc1, which control the expression of Atp6v0d2, OC-STAMP, and CD9, which eventually facilitate the fusion of precursor cells required for osteoclastogenesis [3]. Resulting from the fusion of three or more precursor cells, mature osteoclasts are large and multinucleated [4]. RANKL further activates the alternative NFκB pathway (p52) [5]. Subsequently, expression of inflammatory cytokines like tumor necrosis factor (TNF) α increases. TNFα then stimulates synthesis of interleukin (IL)-1β and macrophage colony-stimulating factor (M-CSF) that promotes proliferation of macrophages and formation of osteoclasts as pointed out above. Consequently, inflammation forces osteoclast formation and bone degradation.

DOI: 10.4324/9781003204091-11

Like macrophages, mature osteoclasts express a vitronectin receptor, which enables their binding to integrin $\alpha V\beta 3$ at the bone surface. Once attached, they form a ruffled border membrane, where lysosomal secretory vesicles fuse and pump H^+ ions into the extracellular space that demineralize the matrix. Lastly, lysosomal proteases like cathepsin K degrade the remaining protein matrix [3].

Overactive osteoclasts, or an increase in the number of osteoclasts, are the main reasons for osteoporosis. As the ratio of RANKL/OPG determines osteoclastogenesis, it could be used as a predictor for osteoporosis. A high ratio would represent bone formation, whereas a low ratio would indicate bone resorption. Unfortunately, the relationship of RANKL/OPG with bone mineral density (BMD) is controversial: some studies indicate a negative correlation, whereas others show no correlation [6]. Therefore, more studies or just better analysis of existing data is needed in order to clarify the potential of RANKL/OPG as a clinical marker for osteoporosis. Nevertheless, continuing increase of the parathyroid hormone (PTH) in the case of chronic kidney disease is associated with bone loss. PTH not only directly stimulates osteoclasts, but it also upregulates RANKL and downregulates OPG gene expressions, thus increasing the RANKL/OPG ratio [7]. Interestingly, oxidation of the PTH receptor may prevent PTH-promoted internalization and recycling of the PTH receptor [8]. By that mechanism, ROS may protect from PTH-induced bone loss.

Osteoblasts differentiate in vitro from mesenchymal stem cells under the control of bone morphogenic protein (BMP)-2 and Wnt/beta-catenin signaling [9,10]. Osteoblasts build osteoid, the organic matrix of the bone by synthesizing a special dense type I collagen, proteoglycans and proteins such as osteocalcin, osteonectin and osteopontin. For the production of inorganic components, osteoblasts have to occur in organized groups of connected cells, in order to produce calcium-hydroxyapatite and osteocalcin-phosphate, which then is deposited into the osteoid. This process eventually leads to the formation of a strong and dense mineralized bone tissue. Osteoblast activity and thereby bone formation can be regulated and actually visualized. The cells change their shape dependent on their activity state. In the bone forming unit, the osteon, the surface layer of active osteoblasts is cuboidal. In inactive units, the surface osteoblasts

are flattened [11]. Osteoblasts buried in the matrix are called osteocytes.

Different to osteoclasts, osteoblasts are rather injured by TNFα. In rat primary osteoblasts TNFα suppressed cell viability, induced cellular apoptosis, suppressed Runx2 mRNA expression and inhibited alkaline phosphatase activity, which can be interpreted as sign of loss of osteoblast function. Furthermore, TNFα induced the formation of excessive amounts of nitric oxide (NO) by the inducible NO-synthase as well as increased ROS level by NADPH oxidase-activation and mitochondria dysfunction [12]. Antimycin-A increases intracellular ROS level in the MC3T3-E1 osteoblastic cell line by inhibition of the mitochondrial electron transport chain. Antioxidants restore the antimycin-A-induced decrease in PI3K (phosphoinositide 3-kinase), Akt (protein kinase B), and CREB (cAMP-response element-binding protein) activity, thereby maintaining the function of the osteoblast cell line [13–15].

In conclusion, deregulated ROS formation either as consequence of TNFα signaling or disruption of mitochondrial function impairs osteoblast function. This effect, however, is not specific for osteoblasts but rather reflects general signs of oxidative stress and ROS-mediated cell damage.

Osteocytes are the most abundant cells in bone. They are connected to a surface layer of osteoblasts. Feedback from osteocytes, mediated by the Wnt-agonist sclerostin, inhibits osteoblast activity and thereby controls the size of the bone-forming unit as well as bone synthesis. Simultaneously osteoclastogenesis and bone resorption by osteoclasts is reduced by Wnt-signaling [16]. Eventually, a stage of quiescence is reached that can persist for up to 25 years. Accordingly, osteocytes, once formed, represent a relatively static portion of the bone. Osteocytes differentiate out of osteoblasts and although considered terminal differentiated, they are able to de-differentiate [17]. Neither differentiation nor de-differentiation so far has been linked to redox signaling in vivo. However, osteocytes may also stem from mesenchymal stromal cells. Those can differentiate into adipocytes and osteocytes. Stimulation of osteogenesis appears to suppress adipogenesis and vice versa, a process with tight regulation by ROS. Interestingly, adipogenesis is associated with an increased ROS and osteogenesis requires a reduction of ROS formation [18]. In line with that, in vitro differentiation of induced pluripotent stem cells into osteocytes is enhanced in the presence of the antioxidant and sirtuin inhibitor, resveratrol [19].

In young individuals, osteocytes make up to 95% of all bone cells but only 58% in old individuals. This age-associated loss of osteocytes appears to be related to an increase in reactive oxygen species. Interestingly, osteocyte death is dependent on the age of the bone, not on the age of the subject [20]. In fact, most literature on "osteocytes and reactive oxygen species" so far published deals with aging and oxidative stress rather than redox signaling in osteocytes. The source of age-related ROS in osteocytes is still undefined, but a recent study associates the NADPH oxidase Nox4 with osteocyte apoptosis [21]. Further evidence is needed to verify this finding as an in vivo and age-related process. In summary, it appears that osteocyte senescence and death is a consequence of oxidative stress and that ROS may not play a role in osteocyte differentiation or function.

8.2 REACTIVE OXYGEN SPECIES IN BONE HOMEOSTASIS AND OSTEOPOROSIS

The major and best-characterized parameter of bone quality is bone mineral density (BMD). Low BMD is associated with a high total plasma oxidant status [22] and negatively correlates with plasma lipid oxidation, as shown in osteoporotic postmenopausal women when compared to a healthy group [23]. A marker of the plasma oxidant status is homocysteine. In humans, hyperhomocysteinemia is associated with a high level of reactive oxygen species (ROS) and an increased risk of osteoporosis [24]. ROS measured in biological samples usually are superoxide anion ($\cdot O_2^-$), hydrogen peroxide (H_2O_2) [25] and nitric oxide (NO). Cellular ROS sources are small molecules (iron, flavins, or thiols) or organelles like peroxisomes or mitochondria and enzymatic ROS producers. ROS are often considered as by-products of cellular metabolism, which need to be decomposed as fast and rigorously as possible to avoid damage of the cell. Indeed, for that purpose antioxidant systems and ROS-degrading enzymes evolved, including catalase, superoxide dismutase (SOD), and glutathione peroxidases. The chemical-toxic view on ROS is contrasted by the findings that they contribute to cellular signal transduction; for example, ROS can transiently inhibit phosphatases [26]. Such function requires a targeted and tightly controlled formation of ROS, which is facilitated by NADPH oxidases of the Nox family.

8.2.1 NADPH Oxidases

The members of the NADPH oxidase family are seven protein-complexes denominated according to their integral-membrane catalytically active core unit: Nox1 through Nox5, Duox1 and Duox2. They transport electrons, taken from NADPH at one side of the membrane to the other side, where they react with oxygen to form ROS.

Nox2, the best-studied NADPH oxidase, is highly expressed in neutrophils and macrophages, where it is essentially involved in first-line host defense. Additionally, several growth factors like angiotensin II, platelet-derived growth factor (PDGF), or inflammatory cytokines like tumor necrosis factor alpha (TNFα) activate Nox2 and induce its expression [27–29]. The active Nox2 complex requires assembly of p22phox and the cytosolic subunits p47phox, p67phox, and Rac [30]. For more information on the activation of the NADPH oxidase complexes, the interested reader is referred to special reviews [31,32]. Similar to Nox2, Nox1 and Nox3 require p22phox and cytosolic subunits including Rac for their activity. Like Nox2, assembly of the subunits acutely activates Nox1 and 3. In vivo Nox1 and 3 form stable complexes with constitutively active substitutes of p47phox and p67phox, namely NoxO1 and NoxA1 [33,34]. Due to their localization in the plasma membrane [35], Nox1 and 2 generate $\cdot O_2^-$ towards the extracellular space. There, $\cdot O_2^-$ spontaneously, or catalyzed by extracellular superoxide dismutase (ecSOD), converts into H_2O_2. The polar nature of H_2O_2 enables the molecule to freely pass the membrane of the same or a neighbor cell and additionally may use aquaporins as channels [36]. Additionally, direct effects of $\cdot O_2^-$ are possible. A major characteristic of $\cdot O_2^-$ is its relative high reactivity, for example in the reaction with nitric oxide (NO). This reaction will lead to the formation of $ONOO^-$ and limit the NO bioavailability. In a rat model of ovariectomy-induced bone loss, treatment of the animals with simvastatin prevented a large portion of the loss in bone density, which potentially is due to a better NO bioavailability and lower ROS formation [37].

Nox5, Duox1 and Duox2, are independent of Rac and other cytosolic subunits, but contain EF-hands that bind calcium (Ca^{2+}) to activate the enzyme. Duox1 and 2 are required for the thyroid-peroxidase-mediated oxidation of iodine that is involved in the formation of thyroxin

(T4) and triiodothyronine (T3) [38,39]. Nox5 is expressed in a number of different tissues in humans but missing in rodents due to gene deletion. This absence of Nox5 in rodents limits the analysis of its role in vivo. Nevertheless, a number of studies implicated a role for Nox5 in cardiovascular diseases, kidney injury, and cancer [40]. In human sperms Nox5 contributes to the motility of the sperm tail [41]. The importance of these NADPH oxidases for bone homeostasis is unclear. Therefore, this group of NADPH oxidases is only briefly mentioned here for the sake of completeness.

Nox4 is a very special NADPH oxidase that has constitutive activity. It requires p22phox for stability but no other subunits to be active [42]. Nox4 is able to directly produce H_2O_2 through the subsequent transfer of two electrons through the membrane onto oxygen [43]. Due to its constitutive activity, Nox4-dependent ROS formation is controlled by its expression level. Cellular stress and TGFβ have been identified as major inducers of Nox4 expression [44]. Once expressed, Nox4 mainly contributes to adaptive signaling and has been suggested to contribute to differentiation of several cells [45–47]. Especially in cells of mesenchymal origin, a rise in Nox4 expression appears to be a prerequisite for the process of differentiation [48].

Most NADPH oxidases exhibit a cell- and organ-specific expression pattern. Especially, Nox1, Nox2, and Nox4 are expressed in bone cells. An overview of these three NADPH oxidases is provided in the scheme in Figure 8.1.

8.3 NADPH OXIDASE IN BONE REMODELING

8.3.1 Nox2

Nox2 maintains unspecific host defense by producing extremely high amounts of ROS in leukocytes to kill fungi and bacteria, the so-called oxidative burst. This function is also exercised in bone marrow–resident polymorphonuclear neutrophils [49]. However, Nox2 has several other functions in bone specific cells.

Osteoclasts, as pointed out above, derive from bone marrow macrophages [50]. Accordingly, as in macrophages, Nox2 is the major NADPH oxidase in those cells. In osteoclast precursor cells, M-CSF transiently increases the intracellular level of ROS in a Nox2-dependent manner. This

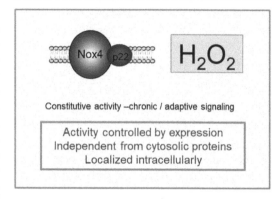

Figure 8.1 General overview of the NADPH oxidases most relevant in bone remodeling.

In bone, the most relevant NADPH oxidases are Nox1, Nox2, and Nox4. Nox1 and Nox2 primarily generate superoxide anions. The activity of Nox1 and Nox2 is acutely inducible and requires an assembly with cytosolic subunits. Therefore, Nox1 and Nox2 mediate acute signal transduction, which is important for proliferation. In case of an overactivation, the large amounts of $\cdot O_2^-$ formed by those NADPH oxidases contribute to inflammation. Nox4, in contrast, produces hydrogen peroxide; constitutively and independent from any cytosolic subunits. The level of H_2O_2 produced by this enzyme is relatively low and the cell adapts to Nox4-derived H_2O_2. Such adaptation leads to cellular quiescence or even differentiation. $\cdot O_2^-$, superoxide anion; H_2O_2, hydrogen peroxide.

contributes to the activation of Erk1/2, which controls the activity of the transcription factors PU.1 and MITF and subsequently expression of RANK [51]. Accordingly, Nox2 knock-out bone marrow–derived macrophages display significantly reduced levels of Erk1/2 phosphorylation upon stimulation with M-CSF [52]. In vitro, native murine macrophages of the RAW264.7 cell line present with highly abundant Nox2 mRNA, whereas stimulation with RANKL attenuates Nox2 mRNA expression over time, which is accompanied by a osteoclast-like differentiation [53]. In conclusion, Nox2 appears to be important for the very first steps in osteoclast formation, while later on a decrease in Nox2 expression appears to be necessary.

Osteoblast formation can be determined by the demands defined by mechanic load of the bone. ROS produced by Nox2 may play a role in several of the pathways related to mechanosensitive bone remodeling, such as in MAPKinase or NFκB activation [54,55]. A major second messenger in mechanotransduction is Ca^{2+}. Fluid shear stress–induced ROS generation by Nox2 targets TRPV4 to elicit a Ca^{2+} influx. The elevation of intracellular Ca^{2+} activates CaMKII, which eventually reduces the abundance of sclerostin (inhibitor of osteoblasts, thereby mechanic load of the bones contributes to osteoblast activity and bone formation [56]. An increase in intracellular Ca^{2+} activates calcium-sensitive kinases such as PKCα and PKCβ, which activate Nox2 mediated $\bullet O_2^-$ formation [57]. Nox2 contributes to Ca^{2+} oscillations in myocytes and mast cells [58,59]. It cannot be excluded that this is an essential function of Nox2 in bone cells as well. In summary, a feedforward regulation from intracellular Ca^{2+} oscillation to Nox2 further to Ca^{2+} appears possible in bones. No stop signal is needed, as mechanical load will have less effect once the bone adapts.

Bone mineralization and osteoblast proliferation and differentiation can be induced by vitamin D in rats and humans, whereas murine osteoblasts have been reported to be inhibited by vitamin D [60]. Accordingly, vitamin D has been used for decades in the treatment of postmenopausal osteoporosis, as it increases bone density and reduces the rate of fractures [61]. In the human osteoblast cell line SaOS2 vitamin D treatment increased the formation of ROS within 1 hour and rapidly increased proliferation as measured by the production of new DNA [62]. Vitamin D–induced maturation of osteoblasts' state is mediated by activation of protein kinase C (PKC) [63], which is one of the most potent activators of the organizing subunit of the NADPH oxidase Nox2. Mice of the C57Bl/6 strain lose bone density from the age of 10 weeks on. It is uncertain but likely that the high content of vitamin D in most mouse food has an impact here. Bone density of p47phox knockout mice is increased at the age of 6 weeks but decreased at the age of 2 years, when compared to wildtype mice. It is possible that deficiency in the Nox2 complex results in fewer osteoclasts in young mice, which may compensate for reduced osteoblast formation

due to vitamin D. The late loss of bone was attributed to an increased senescence-associated secretory phenotype of osteoblasts and an increase in inflammation [64]. Indeed, if vitamin D still prevents osteoblastogenesis and inflammation forces osteoclastogenesis, the result is lower bone density. It is likely that inflammation first destroys cartilage, which is followed by a loss in bone density as a secondary effect. This view is supported by the fact that in arthritis induced by intra-articular antigen or bovine serum albumin (BSA) injection in mice, p47phox-mediated ROS production is not essential for inflammation-mediated bone loss [65].

8.3.2 Nox1

The activity of Nox1, similar to that of Nox2, with the difference that it is activated *in vivo* by the organizer NoxO1 and the activator NoxA1 instead of p67phox and p47phox [31]. The constitutive activity of NoxA1 and NoxO1 facilitates a constant $\bullet O_2^-$ formation by Nox1.

Whether Nox1 contributes to bone turnover is discussed controversially. At least our lab did not identify any bone phenotype in Nox1 knockout mice [45].

Nevertheless, bone marrow–derived macrophages react with an acute increase in Nox1-derived ROS formation within 10 minutes upon RANKL stimulation [66]. This effect is mediated by a TRAF6-dependent recruitment of Rac1 to the membrane. In contrast, others found that bone marrow–derived macrophages from Nox1 or Nox2 knockout mice generated similar level of ROS in response to RANKL and differentiate into osteoclasts to the same extent as wild-type bone marrow–derived macrophages [67]. In RAW264.7 macrophages Rac1 was downregulated in the course of RANKL-induced osteoclastogenesis and Nox2 rapidly disappears, whereas Nox1 was upregulated [68].

An interesting study shows an increased expression of Nox1, NoxA1, and NoxO1 in chondrocyte maturation and the process of endochondral ossification in murine femurs. Nox1 appears to play a role in cartilage formation and bone matrix deposition [69].

8.3.3 Nox4 in Bone Cells

Nox4 is expressed in murine bone marrow–derived *osteoclasts*, and its expression is increased in the course of osteoclastogenesis [70]. In RAW264.7

macrophages Nox1 substitutes for Nox4. Accordingly, studies in those cells fail to find a role of Nox4 in osteoclastogenesis. In bone marrow–derived macrophages primed with M-CSF, there is a distinct pattern of Nox expression: Nox2 expression is 1000 times higher than that of Nox1 whereas Nox4 is undetectable. Upon stimulation with RANKL and thereby induction of osteoclastogenesis, Nox4 is upregulated and becomes detectable [67,71]. This increase in Nox4 expression is needed for osteoclast differentiation in vivo and in vitro [45]. Accordingly, Nox4-deficient mice present with higher bone density when compared to wild-type littermates. In the course of ovariectomy-induced osteoporosis, Nox4 expression was increased in wildtype mice and bone loss was prevented by a Nox4 inhibitor as well as by acute genetic deletion of Nox4. Although the experiments were performed in mice, they apply to humans as well. A small nucleotide polymorphism (SNP) associated with an elevated Nox4 expression correlates with a lower bone density in humans, and human osteoclasts express Nox4. As Nox2 in osteoblasts, Nox4 controls the Ca^{2+} level in osteoclasts. Nox4 deficient cells fail to raise their intracellular Ca^{2+} level. Consequently, NFATc1 and AP-1 activity is not increased over a certain threshold upon RANKL stimulation. Eventually, Nox4 deficient cells are quite unsuccessful in becoming osteoclasts. The connection between endothelial cells and osteoclasts is not obvious; however, both cell types originate from similar or even the same progenitors and share some central characteristics. Others found that carbon monoxide, the product of heme oxygenase-1, inhibits osteoclastogenesis and protects mice from ovariectomy-induced bone loss by [72]. Heme oxygenase-1 expression in turn depends on the activity of the transcription factor Nrf2, and Nrf2 deficiency promotes RANKL-induced osteoclast differentiation [73]. Nox4 maintains Nrf2 expression as well as its downstream target hemeoxygenase-1 [74]. Taken together, Nox4 promotes RANKL-induced osteoclastogenesis by enabling an increase in intracellular Ca^{2+} and by maintenance of carbon monoxide formation via heme-oxygenase 1 and Nrf2.

An interesting fact is that Nox4 deficiency in mice promotes IL4 and IL10-induced M2 polarization of macrophages in vitro [75]. Besides promotion of the M2 phenotype, IL4 and IL10 inhibit osteoclastogenesis. IL4 stimulation of precursor cells leads to a STAT6-dependent inhibition of NFκB, which inhibits RANKL-induced osteoclast differentiation [76]. In murine macrophages, Nox4 deficiency promotes activation of NFκB and inhibits the activity of STAT6.

Osteoblast precursors such as 2T3 cells express Nox4 already under basal conditions. Differentiation of 2T3 osteoblast precursor cells into mature osteoblasts is stimulated by bone morphogenic protein 2 (BMP-2). BMP-2 induces a rapid increase in activation of NADPH oxidases and subsequent ROS formation. NAC and the unspecific flavin-protein and NADPH oxidase inhibitor diphenylene iodonium (DPI) not only inhibit BMP-2-induced ROS formation but also BMP-2-induced alkaline phosphatase expression, needed for maintenance of the osteoblast phenotype. In a positive feedback loop, BMP-2 stimulates its own expression. NAC, DPI, and a dominant negative Nox4 interrupt this [77]. Osteoblast precursors, mesenchymal stem cells, can differentiate into adipocytes or osteoblasts, with adipogenesis inhibiting osteoblastogenesis and vice versa. Nox4 contributes to the differentiation of mesenchymal cells into adipocytes [78]. In fact, insulin-induced differentiation of murine 3T3- and human mesenchymal stem cells into adipocytes is prevented upon downregulation of Nox4 with siRNA. Interestingly, Nox4 is also involved in osteoblastogenesis. Bone marrow cultures from Nox4-deficient mice formed fewer osteoblastic colonies than wildtype cultures [79]. The authors however suggest a role for Nox4 in the maintenance of mesenchymal progenitor cell populations rather than in osteoblastogenesis. This aspect is also emphasized in a study with osteoblastic MC3T3-E1 cells. In those cells, dexamethasone-induced apoptosis was associated with an increase in Nox4 expression leading to the activation of the intrinsic mitochondrial apoptosis pathway [80]. Potentially, Nox4 has no direct effect on osteoblastogenesis but rather regulates their survival.

Similar to Nox1, Nox4 is increased in the process of endochondral ossification, where it is likely involved in chondrocyte maturation and bone matrix formation [69].

8.4 CONCLUDING REMARKS

NADPH oxidases influence bone homeostasis and remodeling. The individual members of the NADPH oxidase family regulate differentiation and survival of the major bone cells. Acute ROS formation is facilitated by Nox2. This NADPH oxidase supports osteoblast formation. In contrast, prolonged ROS formation by Nox4 supports formation of bone resorbing osteoclasts.

8.5 ABBREVIATIONS

Bone morphogenic protein 2 (BMP-2); interferon (IFN); interleukin (IL); N-acetyl-cysteine (NAC); osteoprotegerin (OPG); receptor activator of nuclear factor (NF)-κB receptor (RANK); receptor activator of NFκB ligand (RANKL); six-transmembrane epithelial antigen of prostate (Steap); transforming growth factor (TGF); tumor necrosis factor (TNF).

8.6 ACKNOWLEDGMENT

This research was supported by grants from the Deutsche Forschungsgemeinschaft (DFG) (SCHR1241/1-1 and SFB815 TP1).

REFERENCES

[1] B. Hatami Kia, J.R.G. Mendes, H.-D. Müller, P. Heimel, R. Gruber, Bone-conditioned medium obtained from calvaria, mandible, and tibia cause an equivalent TGF-β1 response in vitro, J. Craniofac. Surg. 29 (2018) 553–557. https://doi.org/10.1097/SCS.0000000000004251.

[2] D. Lane, I. Matte, C. Laplante, P. Garde-Granger, C. Rancourt, A. Piché, Osteoprotegerin (OPG) activates integrin, focal adhesion kinase (FAK), and Akt signaling in ovarian cancer cells to attenuate TRAIL-induced apoptosis, J. Ovarian Res. 6 (2013) 82. https://doi.org/10.1186/1757-2215-6-82.

[3] B.F. Boyce, Advances in the regulation of osteoclasts and osteoclast functions, J. Dent. Res. 92 (2013) 860–867. https://doi.org/10.1177/0022034513500306.

[4] M. Asagiri, H. Takayanagi, The molecular understanding of osteoclast differentiation, Bone 40 (2007) 251–264. https://doi.org/10.1016/j.bone.2006.09.023.

[5] B.F. Boyce, Y. Xiu, J. Li, L. Xing, Z. Yao, NF-κB-mediated regulation of osteoclastogenesis, Endocrinol. Metab. (Seoul) 30 (2015) 35–44. https://doi.org/10.3803/EnM.2015.30.1.35.

[6] B. Parveen, A. Parveen, D. Vohora, Biomarkers of osteoporosis: an update, Endocr. Metab. Immune Disord. Drug Targets 19 (2019) 895–912. https://doi.org/10.2174/1871530319666190204165207.

[7] J.C. Huang, T. Sakata, L.L. Pfleger, M. Bencsik, B.P. Halloran, D.D. Bikle, R.A. Nissenson, PTH differentially regulates expression of RANKL and OPG, J. Bone Miner. Res. 19 (2004) 235–244. https://doi.org/10.1359/JBMR.0301226.

[8] J.A. Ardura, V. Alonso, P. Esbrit, P.A. Friedman, Oxidation inhibits PTH receptor signaling and trafficking, Biochem. Biophys. Res. Commun. 482 (2017) 1019–1024. https://doi.org/10.1016/j.bbrc.2016.11.150.

[9] M. Zaidi, Skeletal remodeling in health and disease, Nat. Med. 13 (2007) 791–801. https://doi.org/10.1038/nm1593.

[10] J.J. Pinzone, B.M. Hall, N.K. Thudi, M. Vonau, Y.-W. Qiang, T.J. Rosol, J.D. Shaughnessy, The role of Dickkopf-1 in bone development, homeostasis, and disease, Blood 113 (2009) 517–525. https://doi.org/10.1182/blood-2008-03-145169.

[11] A. Rutkovskiy, K.-O. Stensløkken, I.J. Vaage, Osteoblast differentiation at a glance, Med. Sci. Monit. Basic Res. 22 (2016) 95–106. https://doi.org/10.12659/msmbr.901142.

[12] W.-W. Cai, M.-H. Zhang, Y.-S. Yu, J.-H. Cai, Treatment with hydrogen molecule alleviates TNFα-induced cell injury in osteoblast, Mol. Cell. Biochem. 373 (2013) 1–9. https://doi.org/10.1007/s11010-012-1450-4.

[13] E.M. Choi, Luteolin protects osteoblastic MC3T3-E1 cells from antimycin A-induced cytotoxicity through the improved mitochondrial function and activation of PI3K/Akt/CREB, Toxicol. In Vitro 25 (2011) 1671–1679. https://doi.org/10.1016/j.tiv.2011.07.004.

[14] E.M. Choi, Y.S. Lee, Protective effect of apocynin on antimycin A-induced cell damage in osteoblastic MC3T3-E1 cells, J. Appl. Toxicol. 32 (2012) 714–721. https://doi.org/10.1002/jat.1689.

[15] E.M. Choi, Y.S. Lee, Involvement of PI3K/Akt/CREB and redox changes in mitochondrial defect of osteoblastic MC3T3-E1 cells, Toxicol. In Vitro 25 (2011) 1085–1088. https://doi.org/10.1016/j.tiv.2011.03.022.

[16] J. Albers, J. Keller, A. Baranowsky, F.T. Beil, P. Catala-Lehnen, J. Schulze, M. Amling, T. Schinke, Canonical Wnt signaling inhibits osteoclastogenesis independent of osteoprotegerin, J. Cell Biol. 200 (2013) 537–549. https://doi.org/10.1083/jcb.201207142.

[17] N. Sawa, H. Fujimoto, Y. Sawa, J. Yamashita, Alternating differentiation and dedifferentiation between mature osteoblasts and osteocytes, Sci. Rep. 9 (2019) 13842. https://doi.org/10.1038/s41598-019-50236-7.

[18] F. Atashi, A. Modarressi, M.S. Pepper, The role of reactive oxygen species in mesenchymal stem cell adipogenic and osteogenic differentiation: a review, Stem Cells Dev. 24 (2015) 1150–1163. https://doi.org/10.1089/scd.2014.0484.

[19] C.-L. Kao, L.-K. Tai, S.-H. Chiou, Y.-J. Chen, K.-H. Lee, S.-J. Chou, Y.-L. Chang, C.-M. Chang, S.-J. Chen, H.-H. Ku, H.-Y. Li, Resveratrol promotes osteogenic differentiation and protects against dexamethasone damage in murine induced pluripotent stem cells, Stem Cells Dev. 19 (2010) 247–258. https://doi.org/10.1089/scd.2009.0186.

[20] S.C. Manolagas, A.M. Parfitt, What old means to bone, Trends Endocrinol. Metabol. 21 (2010) 369–374. https://doi.org/10.1016/j.tem.2010.01.010.

[21] S.L. Werner, R. Sharma, K. Woodruff, D. Horn, S.E. Harris, Y. Gorin, D.-Y. Lee, R. Hua, S. Gu, R.J. Fajardo, S.L. Habib, J.X. Jiang, CSF-1 in Osteocytes Inhibits Nox4-mediated oxidative stress and promotes normal bone homeostasis, JBMR Plus 4 (2020) e10080. https://doi.org/10.1002/jbm4.10080.

[22] O. Altindag, O. Erel, N. Soran, H. Celik, S. Selek, Total oxidative/anti-oxidative status and relation to bone mineral density in osteoporosis, Rheumatol. Int. 28 (2008) 317–321. https://doi.org/10.1007/s00296-007-0452-0.

[23] O.F. Sendur, Y. Turan, E. Tastaban, M. Serter, Antioxidant status in patients with osteoporosis: a controlled study, Joint Bone Spine 76 (2009) 514–518. https://doi.org/10.1016/j.jbspin.2009.02.005.

[24] M.S. LeBoff, R. Narweker, A. LaCroix, L. Wu, R. Jackson, J. Lee, D.C. Bauer, J. Cauley, C. Kooperberg, C. Lewis, A.M. Thomas, S. Cummings, Homocysteine levels and risk of hip fracture in postmenopausal women, J. Clin. Endocrinol. Metabol. 94 (2009) 1207–1213. https://doi.org/10.1210/jc.2008-1777.

[25] K. Schröder, NADPH oxidases in redox regulation of cell adhesion and migration, Antioxid. Redox Signal. 20 (2014) 2043–2058. https://doi.org/10.1089/ars.2013.5633.

[26] K. Schröder, A. Kohnen, A. Aicher, E.A. Liehn, T. Büchse, S. Stein, C. Weber, S. Dimmeler, R.P. Brandes, NADPH oxidase Nox2 is required for hypoxia-induced mobilization of endothelial progenitor cells, Circ. Res. 105 (2009) 537–544. https://doi.org/10.1161/CIRCRESAHA.109.205138.

[27] T. Adachi, H. Togashi, A. Suzuki, S. Kasai, J. Ito, K. Sugahara, S. Kawata, NAD(P)H oxidase plays a crucial role in PDGF-induced proliferation of hepatic stellate cells, Hepatology 41 (2005) 1272–1281. https://doi.org/10.1002/hep.20719.

[28] C.-W. Lee, C.-C. Lin, I.-T. Lee, H.-C. Lee, C.-M. Yang, Activation and induction of cytosolic phospholipase A2 by TNF-α mediated through Nox2, MAPKs, NF-κB, and p300 in human tracheal smooth muscle cells, J. Cell. Physiol. 226 (2011) 2103–2114. https://doi.org/10.1002/jcp.22537.

[29] J.L. Wilkinson-Berka, I. Rana, R. Armani, A. Agrotis, Reactive oxygen species, Nox and angiotensin II in angiogenesis: implications for retinopathy, Clin. Sci. 124 (2013) 597–615. https://doi.org/10.1042/CS20120212.

[30] M. Margaritis, F. Sanna, C. Antoniades, Statins and oxidative stress in the cardiovascular system, Curr. Pharm. Des. (2017). https://doi.org/10.2174/1381612823666170926130338.

[31] K. Schröder, N. Weissmann, R.P. Brandes, Organizers and activators: Cytosolic Nox proteins impacting on vascular function, Free Radic. Biol. Med. 109 (2017) 22–32. https://doi.org/10.1016/j.freeradbiomed.2017.03.017.

[32] K. Miyano, H. Sumimoto, Role of the small GTPase Rac in p22phox-dependent NADPH oxidases, Biochimie 89 (2007) 1133–1144. https://doi.org/10.1016/j.biochi.2007.05.003.

[33] B. Bánfi, R.A. Clark, K. Steger, K.-H. Krause, Two novel proteins activate superoxide generation by the NADPH oxidase NOX1, J. Biol. Chem. 278 (2003) 3510–3513. https://doi.org/10.1074/jbc.C200613200.

[34] N. Ueno, R. Takeya, K. Miyano, H. Kikuchi, H. Sumimoto, The NADPH oxidase Nox3 constitutively produces superoxide in a p22phox-dependent manner: its regulation by oxidase organizers and activators, J. Biol. Chem. 280 (2005) 23328–23339. https://doi.org/10.1074/jbc.M414548200.

[35] I. Helmcke, S. Heumüller, R. Tikkanen, K. Schröder, R.P. Brandes, Identification of structural elements in Nox1 and Nox4 controlling localization and activity, Antioxid. Redox Signal. 11 (2009) 1279–1287. https://doi.org/10.1089/ARS.2008.2383.

[36] I. Al Ghouleh, G. Frazziano, A.I. Rodriguez, G. Csányi, S. Maniar, C.M. St Croix, E.E. Kelley, L.A. Egaña, G.J. Song, A. Bisello, Y.J. Lee, P.J. Pagano, Aquaporin 1, Nox1, and Ask1 mediate oxidant-induced smooth muscle cell hypertrophy, Cardiovasc. Res. 97 (2013) 134–142. https://doi.org/10.1093/cvr/cvs295.

[37] H. Yin, Z.-G. Shi, Y.-S. Yu, J. Hu, R. Wang, Z.-P. Luan, D.-H. Guo, Protection against osteoporosis by statins is linked to a reduction of oxidative stress and restoration of nitric oxide formation in aged and ovariectomized rats, Eur. J. Pharmacol. 674 (2012) 200–206. https://doi.org/10.1016/j.ejphar.2011.11.024.

[38] M. Katsuyama, K. Matsuno, C. Yabe-Nishimura, Physiological roles of NOX/NADPH oxidase, the superoxide-generating enzyme, J. Clin. Biochem. Nutr. 50 (2012) 9–22. https://doi.org/10.3164/jcbn.11-06SR.

[39] K. Bedard, K.-H. Krause, The NOX family of ROS-generating NADPH oxidases: physiology and pathophysiology, Physiol. Rev. 87 (2007) 245–313. https://doi.org/10.1152/physrev.00044.2005.

[40] R.M. Touyz, A. Anagnostopoulou, F. Rios, A.C. Montezano, L.L. Camargo, NOX5: Molecular biology and pathophysiology, Exp. Physiol. 104 (2019) 605–616. https://doi.org/10.1113/EP086204.

[41] K. Bedard, V. Jaquet, K.-H. Krause, NOX5: from basic biology to signaling and disease, Free Radic. Biol. Med. 52 (2012) 725–734. https://doi.org/10.1016/j.freeradbiomed.2011.11.023.

[42] K.-K. Prior, M.S. Leisegang, I. Josipovic, O. Löwe, A.M. Shah, N. Weissmann, K. Schröder, R.P. Brandes, CRISPR/Cas9-mediated knockout of p22phox leads to loss of Nox1 and Nox4, but not Nox5 activity, Redox Biol. 9 (2016) 287–295. https://doi.org/10.1016/j.redox.2016.08.013.

[43] I. Takac, K. Schröder, L. Zhang, B. Lardy, N. Anilkumar, J.D. Lambeth, A.M. Shah, F. Morel, R.P. Brandes, The E-loop is involved in hydrogen peroxide formation by the NADPH oxidase Nox4, J. Biol. Chem. 286 (2011) 13304–13313. https://doi.org/10.1074/jbc.M110.192138.

[44] A. Sturrock, B. Cahill, K. Norman, T.P. Huecksteadt, K. Hill, K. Sanders, S.V. Karwande, J.C. Stringham, D.A. Bull, M. Gleich, T.P. Kennedy, J.R. Hoidal, Transforming growth factor-beta1 induces Nox4 NAD(P)H oxidase and reactive oxygen species-dependent proliferation in human pulmonary artery smooth muscle cells, Am. J. Physiol. Lung Cell. Mol. Physiol. 290 (2006) L661-L673. https://doi.org/10.1152/ajplung.00269.2005.

[45] C. Goettsch, A. Babelova, O. Trummer, R.G. Erben, M. Rauner, S. Rammelt, N. Weissmann, V. Weinberger, S. Benkhoff, M. Kampschulte, B. Obermayer-Pietsch, L.C. Hofbauer, R.P. Brandes, K. Schröder, NADPH oxidase 4 limits bone mass by promoting osteoclastogenesis, J. Clin. Invest. 123 (2013) 4731–4738. https://doi.org/10.1172/JCI67603.

[46] L. Hecker, R. Vittal, T. Jones, R. Jagirdar, T.R. Luckhardt, J.C. Horowitz, S. Pennathur, F.J. Martinez, V.J. Thannickal, NADPH oxidase-4 mediates myofibroblast activation and fibrogenic responses to lung injury, Nat. Med. 15 (2009) 1077–1081. https://doi.org/10.1038/nm.2005.

[47] J. Li, M. Stouffs, L. Serrander, B. Banfi, E. Bettiol, Y. Charnay, K. Steger, K.-H. Krause, M.E. Jaconi, The NADPH oxidase NOX4 drives cardiac differentiation: Role in regulating cardiac transcription factors and MAP kinase activation, Mol. Biol. Cell 17 (2006) 3978–3988. https://doi.org/10.1091/mbc.E05-06-0532.

[48] R.E. Clempus, D. Sorescu, A.E. Dikalova, L. Pounkova, P. Jo, G.P. Sorescu, H.H.H. Schmidt, B. Lassègue, K.K. Griendling, Nox4 is required for maintenance of the differentiated vascular smooth muscle cell phenotype, ATVB 27 (2007) 42–48. https://doi.org/10.1161/01.ATV.0000251500.94478.18.

[49] J.V. Filina, A.G. Gabdoulkhakova, V.G. Safronova, RhoA/ROCK downregulates FPR2-mediated NADPH oxidase activation in mouse bone marrow granulocytes, Cell. Signal. 26 (2014) 2138–2146. https://doi.org/10.1016/j.cellsig.2014.05.017.

[50] D.A. Callaway, J.X. Jiang, Reactive oxygen species and oxidative stress in osteoclastogenesis, skeletal aging and bone diseases, J. Bone Miner. Metab. 33 (2015) 359–370. https://doi.org/10.1007/s00774-015-0656-4.

[51] A. Nakanishi, M. Hie, N. Iitsuka, I. Tsukamoto, A crucial role for reactive oxygen species in macrophage colony-stimulating factor-induced RANK expression in osteoclastic differentiation, Int. J. Mol. Med. 31 (2013) 874–880. https://doi.org/10.3892/ijmm.2013.1258.

[52] S. Chaubey, G.E. Jones, A.M. Shah, A.C. Cave, C.M. Wells, Nox2 is required for macrophage chemotaxis towards CSF-1, PLoS One 8 (2013) e54869. https://doi.org/10.1371/journal.pone.0054869.

[53] H. Sasaki, H. Yamamoto, K. Tominaga, K. Masuda, T. Kawai, S. Teshima-Kondo, K. Rokutan, NADPH oxidase-derived reactive oxygen species are essential for differentiation of a mouse macrophage cell line (RAW264.7) into osteoclasts, J. Med. Invest. 56 (2009) 33–41. https://doi.org/10.2152/jmi.56.33.

[54] D.M. Knapik, P. Perera, J. Nam, A.D. Blazek, B. Rath, B. Leblebicioglu, H. Das, L.C. Wu, T.E. Hewett, S.K. Agarwal, A.G. Robling, D.C. Flanigan, B.S. Lee, S. Agarwal, Mechanosignaling in bone health, trauma and inflammation, Antioxid. Redox Signal. 20 (2014) 970–985. https://doi.org/10.1089/ars.2013.5467.

[55] R.P. Brandes, N. Weissmann, K. Schröder, Nox family NADPH oxidases in mechano-transduction: mechanisms and consequences: mechanisms and consequences, Antioxid. Redox Signal. 20 (2014) 887–898. https://doi.org/10.1089/ars.2013.5414.

[56] J.S. Lyons, H.C. Joca, R.A. Law, K.M. Williams, J.P. Kerr, G. Shi, R.J. Khairallah, S.S. Martin, K. Konstantopoulos, C.W. Ward, J.P. Stains, Microtubules tune mechanotransduction through NOX2 and TRPV4 to decrease sclerostin abundance in osteocytes, Sci. Signal. 10 (2017). https://doi.org/10.1126/scisignal.aan5748.

[57] H. Zhang, R.A. Clemens, F. Liu, Y. Hu, Y. Baba, P. Theodore, T. Kurosaki, C.A. Lowell, STIM1 calcium sensor is required for activation of the phagocyte oxidase during inflammation and host defense, Blood 123 (2014) 2238–2249. https://doi.org/10.1182/blood-2012-08-450403.

[58] M. Zhang, B.L. Prosser, M.A. Bamboye, A.N.S. Gondim, C.X. Santos, D. Martin, A. Ghigo, A. Perino, A.C. Brewer, C.W. Ward, E. Hirsch, W.J. Lederer, A.M. Shah, Contractile function during angiotensin-II activation: increased Nox2 activity modulates cardiac calcium handling via phospholamban phosphorylation, J. Am. Coll. Cardiol. 66 (2015) 261–272. https://doi.org/10.1016/j.jacc.2015.05.020.

[59] Z.Y. Li, W.Y. Jiang, Z.J. Cui, An essential role of NAD(P)H oxidase 2 in UVA-induced calcium oscillations in mast cells, Photochem. Photobiol. Sci. 14 (2015) 414–428. https://doi.org/10.1039/c4pp00304g.

[60] M. van Driel, J.P.T.M. van Leeuwen, Vitamin D endocrine system and osteoblasts, BoneKEy Reports 3 (2014) 493. https://doi.org/10.1038/bonekey.2013.227.

[61] M.W. Tilyard, G.F. Spears, J. Thomson, S. Dovey, Treatment of postmenopausal osteoporosis with calcitriol or calcium, N Engl J Med 326 (1992) 357–362. https://doi.org/10.1056/NEJM199202063260601.

[62] D. Somjen, S. Katzburg, M. Grafi-Cohen, E. Knoll, O. Sharon, G.H. Posner, Vitamin D metabolites and analogs induce lipoxygenase mRNA expression and activity as well as reactive oxygen species (ROS) production in human bone cell line, J. Steroid Biochem. Mol. Biol. 123 (2011) 85–89. https://doi.org/10.1016/j.jsbmb.2010.11.010.

[63] B.D. Boyan, L.F. Bonewald, V.L. Sylvia, I. Nemere, D. Larsson, A.W. Norman, J. Rosser, D.D. Dean, Z. Schwartz, Evidence for distinct membrane receptors for 1 alpha,25-(OH)(2)D(3) and 24R,25-(OH)(2)D(3) in osteoblasts, Steroids 67 (2002) 235–246.

[64] J.-R. Chen, O.P. Lazarenko, M.L. Blackburn, K.E. Mercer, T.M. Badger, M.J.J. Ronis, p47phox-Nox2-dependent ROS signaling inhibits early bone development in mice but protects against skeletal aging, J. Biol. Chem. 290 (2015) 14692–14704. https://doi.org/10.1074/jbc.M114.633461.

[65] C. Engdahl, C. Lindholm, A. Stubelius, C. Ohlsson, H. Carlsten, M.K. Lagerquist, Periarticular bone loss in antigen-induced arthritis, Arthritis Rheum. 65 (2013) 2857–2865. https://doi.org/10.1002/art.38114.

[66] N.K. Lee, Y.G. Choi, J.Y. Baik, S.Y. Han, D.-W. Jeong, Y.S. Bae, N. Kim, S.Y. Lee, A crucial role for reactive oxygen species in RANKL-induced osteoclast differentiation, Blood 106 (2005) 852–859. https://doi.org/10.1182/blood-2004-09-3662.

[67] H. Sasaki, H. Yamamoto, K. Tominaga, K. Masuda, T. Kawai, S. Teshima-Kondo, K. Matsuno, C. Yabe-Nishimura, K. Rokutan, Receptor activator of nuclear factor-kappaB ligand-induced mouse osteoclast differentiation is associated with switching between NADPH oxidase homologues, Free Radic. Biol. Med. 47 (2009) 189–199. https://doi.org/10.1016/j.freeradbiomed.2009.04.025.

[68] H. Sasaki, H. Yamamoto, K. Tominaga, K. Masuda, T. Kawai, S. Teshima-Kondo, K. Rokutan, NADPH oxidase-derived reactive oxygen species are essential for differentiation of a mouse macrophage cell line (RAW264.7) into osteoclasts, J. Med. Invest. 56 (2009) 33–41.

[69] K. Ambe, H. Watanabe, S. Takahashi, T. Nakagawa, Immunohistochemical localization of Nox1, Nox4 and Mn-SOD in mouse femur during endochondral ossification, Tissue Cell 46 (2014) 433–438. https://doi.org/10.1016/j.tice.2014.07.005.

[70] S. Yang, P. Madyastha, S. Bingel, W. Ries, L. Key, A new superoxide-generating oxidase in murine osteoclasts, J. Biol. Chem. 276 (2001) 5452–5458. https://doi.org/10.1074/jbc.M001004200.

[71] S. Yang, Y. Zhang, W. Ries, L. Key, Expression of Nox4 in osteoclasts, J. Cell. Biochem. 92 (2004) 238–248. https://doi.org/10.1002/jcb.20048.

[72] T. van Phan, O.-J. Sul, K. Ke, M.-H. Lee, W.-K. Kim, Y.-S. Cho, H.-J. Kim, S.-Y. Kim, H.-T. Chung, H.-S. Choi, Carbon monoxide protects against ovariectomy-induced bone loss by inhibiting osteoclastogenesis, Biochem. Pharmacol. 85 (2013) 1145–1152. https://doi.org/10.1016/j.bcp.2013.01.014.

[73] S. Hyeon, H. Lee, Y. Yang, W. Jeong, Nrf2 deficiency induces oxidative stress and promotes RANKL-induced osteoclast differentiation, Free Radic. Biol. Med. 65 (2013) 789–799. https://doi.org/10.1016/j.freeradbiomed.2013.08.005.

[74] K. Schröder, M. Zhang, S. Benkhoff, A. Mieth, R. Pliquett, J. Kosowski, C. Kruse, P. Luedike, U.R. Michaelis, N. Weissmann, S. Dimmeler, A.M. Shah, R.P. Brandes, Nox4 is a protective reactive oxygen species generating vascular NADPH oxidase, Circ. Res. 110 (2012) 1217–1225. https://doi.org/10.1161/CIRCRESAHA.112.267054.

[75] V. Helfinger, K. Palfi, A. Weigert, K. Schröder, The NADPH oxidase Nox4 controls macrophage polarization in an NFκB-Dependent manner, Oxid. Med. Cell. Longev. 2019 (2019) 3264858. https://doi.org/10.1155/2019/3264858.

[76] N. Lampiasi, R. Russo, F. Zito, The alternative faces of macrophage generate osteoclasts, Biomed. Res. Int. 2016 (2016) 9089610. https://doi.org/10.1155/2016/9089610.

[77] C.C. Mandal, S. Ganapathy, Y. Gorin, K. Mahadev, K. Block, H.E. Abboud, S.E. Harris, G. Ghosh-Choudhury, N. Ghosh-Choudhury, Reactive oxygen species derived from Nox4 mediate BMP2 gene transcription and osteoblast differentiation, Biochem. J. 433 (2011) 393–402. https://doi.org/10.1042/BJ20100357.

[78] K. Schröder, K. Wandzioch, I. Helmcke, R.P. Brandes, Nox4 acts as a switch between differentiation and proliferation in preadipocytes, Arterioscler. Thromb. Vasc. Biol. 29 (2009) 239–245. https://doi.org/10.1161/ATVBAHA.108.174219.

[79] J. Watt, A.W. Alund, C.F. Pulliam, K.E. Mercer, L.J. Suva, J.-R. Chen, M.J.J. Ronis, NOX4 deletion in male mice exacerbates the effect of ethanol on trabecular bone and osteoblastogenesis, J. Pharmacol. Exp. Ther. 366 (2018) 46–57. https://doi.org/10.1124/jpet.117.247262.

[80] J. Li, C. He, W. Tong, Y. Zou, D. Li, C. Zhang, W. Xu, Tanshinone IIA blocks dexamethasone-induced apoptosis in osteoblasts through inhibiting Nox4-derived ROS production, Int. J. Clin. Exp. Pathol. 8 (2015) 13695–13706.

Signaling Pathways during Differentiation and Longevity

CHAPTER NINE

Identification of Redox-Regulated Pathways via Redox Proteomics

Gereon Poschmann

CONTENTS

9.1 INTRODUCTION

In recent years, the view on protein oxidation changed from a primarily biomolecular damaging process to a highly dynamical post-translational modification involved in the regulation of the function, structure, and signaling properties of proteins. Along with this, methods evolved to detect and characterize those modifications. Although the pioneering work on the characterization of oxidative post-translational protein modifications (ox-PTM) was initially carried out on a more global level at the beginning of this millennium, there has been a major leap forward in the development of methods and probes in the last 5 to 10 years. This technical development has enabled a more specific, reliable, and sensitive characterization of these modifications. Here especially, mass spectrometry–based methods provided a great contribution. Therefore, this chapter will focus on those kinds of methods. The development of liquid chromatography coupled with tandem mass spectrometry (LC-MS/MS)–based methods was further accelerated by recent progress in instrument development, significantly improving speed, resolution, and sensitivity of mass spectrometers. This beneficial combination enables the identification, quantification, and characterization of thousands of ox-PTM with recent approaches in one experiment.

Especially, oxidative modifications of cysteine residues of proteins are of great interest and in the focus of this chapter. One important reason for this is the great variability of at least 15 different types of modification that are relevant in biological systems [1]. Among them, particularly disulfides, S-nitrosothiols, persulfides, and sulfenic, sulfinic, and sulfonic acids can be investigated and detected by specifically developed methods. Some of those will be described later in this chapter. This diversity of cysteine ox-PTM highlights that they not only act as an "on-off switch" but can also fine-tune processes or enable several different functionalities of one protein by the modification of only one single amino acid. Moreover, most of the mentioned modifications except sulfonic acids (and partially sulfinic acids) are known to be reversible, enabling a dynamic regulation of protein functions and properties. To make the picture even more complex, cysteine modifications might occur only substoichiometrically and might be present only at certain subcellular sites or compartments because ox-PTM can play a crucial role in protein localization.

DOI: 10.4324/9781003204091-13

Although this complexity and partial reversibility are fascinating from a biological perspective, from an analytical point of view they are associated with a lot of challenges. Therefore, there is at the moment no "one-size-fits-all" solution to analyze all kinds of possible modifications at a greater depth. Nevertheless, a lot of different methods have been developed to analyze either reversible cysteine modifications globally or certain types of modifications specifically. Each of the methods has its strengths and weaknesses. In the following paragraphs, general considerations for sample preparation are made and different basic principles and exemplary approaches are summarized to enable successful analysis of cysteine ox-PTM. Furthermore, analytical limitations and the need for a proper experimental design and controls will be discussed.

9.2 GENERAL CONSIDERATIONS FOR SAMPLE PREPARATION

Due to the labile nature of many ox-PTM and potential artifacts due to oxidation during sample preparation, it is important to adapt sample preparation procedures to preserve the native redox status and ox-PTM as well as possible.

Nevertheless, optimal conditions might vary depending on the analyzed type of ox-PTM. For most approaches, working with degassed solution and buffers and/or working under oxygen-free conditions, as well as the addition of enzymes like catalase to remove hydrogen peroxide, proved to be beneficial. Moreover, it might be a good idea to add metal-chelating compounds such as diethylenetriaminepentaacetic acid or neocuproine to minimize metal-catalyzed thiol oxidation or decomposition of certain oxPTM [2].

Often, an initial alkylation step of cysteines' thiols is useful to block further reactions. Here, alkylating agents like N-ethylmaleimide (NEM) and iodoacetamide (IAM) proved to be very useful. Nevertheless, a direct application of NEM to cells before lysis might in some cases be beneficial and in others not, because the effectivity of blocking might vary from site to site and the perturbation of the dynamic redox cycling can also lead to artifacts [3]. Furthermore, it is important to keep in mind that the alkylation with NEM might be reversible, whereas IAM-based alkylation is not reversible but not as fast as alkylation with NEM. Moreover, alkylation using NEM and IAM is not in all cases

100% specific, as reactions with sulfenic acids and other amino acids have also been reported [1].

Generally, more acidic conditions help to prevent thiol-disulfide exchange reactions like the scrambling of disulfide bonds. Therefore, an initial trichloroacetic acid precipitation step and lower pH during sample preparation including protein digestion are widely used [1].

9.3 TYPES OF EXPERIMENTAL APPROACHES

There are two different types of mass spectrometric setups available for the analysis of proteins and their modifications. First, top-down approaches have made big progress in recent years. Here, whole proteins are analyzed on the precursor and fragment levels, making the discrimination of different so-called proteoforms feasible, which might represent different isoforms or differ in their modification pattern. Nevertheless, top-down methods are not routinely applied to highly complex samples and data analysis and interpretation are still complex. Therefore, the so-called bottom-up methods have a wider application for the analysis of ox-PTM. For these, proteins are digested by proteases—in most settings with trypsin—and analyzed on the peptide level. The analysis of peptides with mass spectrometry is straightforward in many cases. However, depending on the cleavage site of the used protease and sequence properties of peptides, some peptides are not suited for the analysis (e.g., because they are too small to include enough sequence information, too large, too hydrophobic, or not well suited for ionization or fragmentation). Furthermore, the information of isoforms and PTM patterns within a protein might get lost. Therefore, a complete picture without taking bigger efforts cannot be expected from bottom-up approaches.

Meanwhile, there is a high diversity in proposed approaches for the analysis of ox-PTM. Nevertheless, most strategies rely on one of three basic principles. In the following paragraphs, a rather subjective selection of available approaches will be described and grouped accordingly.

1. The direct detection of ox-PTM (Figure 9.1A). Here, oxidative modifications of cysteines are directly detected by mass spectrometry. As different ox-PTM will lead to different mass shifts of modified peptides, an unambiguous identification might be straight forward. After fragment-spectrum recording, additional

information about the modified site might be available. Direct detection methods are efficiently usable for the identification of relatively stable ox-PTM. Nevertheless, a global enrichment of relevant modified peptides might not be possible in all cases, which will impact the sensitivity of such approaches. But once an ox-PTM is identified and localized directly, the reliability of its identity and associated quantity is usually very high.

2. The detection of ox-PTM by indirect label–based approaches (Figure 9.1B). Indirect label–based approaches start with the blocking of cysteine thiols. In a second step,

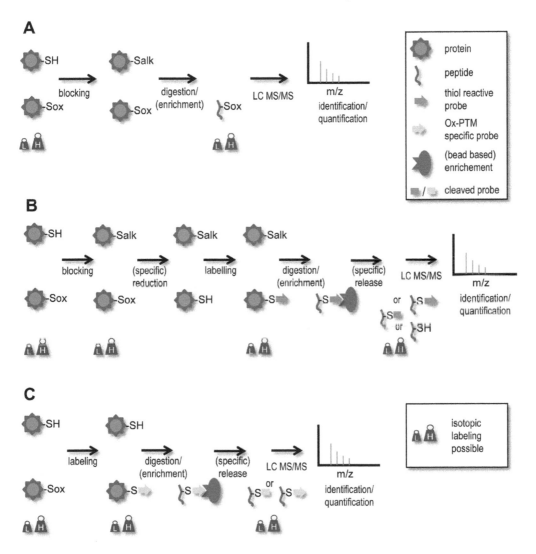

Figure 9.1 Proteomic profiling strategies for the detection of oxidative post-translational modifications (oxPTM).

Direct detection of ox-PTM (Sox, A) can be carried out without or with enrichment. Nevertheless, enrichment is necessary for a sensitive detection of oxPTM from complex samples. In indirectly labeling approaches (B), free thiols are blocked by alkylation (Salk), and after a specific or unspecific reduction step, reversibly oxidized cysteines are alkylated with a different probe as used for blocking. This can be a small molecule with or without enrichment function or directly a bead-based matrix. After an optional enrichment step, digestion of proteins and release of peptides is carried out before analysis. The enrichment can be on the protein level or after digestion on the peptide level. Direct detection approaches (C) apply probes directly reacting with specific ox-PTM like cysteine sulfenic and sulfinic acids. Enrichment is carried on the protein or peptide level before (specific) release and mass spectrometric analysis. In comparative setups, there are several ways to include isotopic labels: by including heavy and light amino acids in cell culture, by including isotopes directly into probes or linkers added by click chemistry, or by labeling peptides after digestion.

reversibly oxidized PTM will (selectively) be reduced and newly built thiols labeled with an alternative alkylating agent. This can be linked to an affinity tag, allowing the enrichment of relevant proteins or peptides and thus a sensitive detection.

3. The detection of ox-PTM by direct label–based approaches (Figure 9.1C). In direct detection approaches, single ox-PTM like sulfenic or sulfinic acids are directly targeted by a certain probe. This is in some cases also feasible in a cellular context *in situ*. Recent approaches include also the possibility of enrichment.

Nearly all approaches can be combined with stable isotope labeling of different conditions or different oxidation states of cysteine residues. There are several ways to introduce isotope labels for analysis. In cells or living animals, isotope-coded amino acids (usually lysine and arginine in light, intermediate, and heavy variants) can be incorporated and used to distinguish different treatment groups, for example. Furthermore, isotope labels can directly be incorporated into the used probes or affinity tags in labeling approaches. Furthermore, it is possible to label proteins or peptides before or after enrichment with isotope-coded probes. Tandem mass tags (TMT) make a high multiplexing capacity feasible, as up to 16 different linker variants are available that can be differentiated and quantified after peptide fragmentation.

9.4 DIRECT DETECTION OF CYSTEINE OXIDATIVE POST-TRANSLATIONAL MODIFICATIONS

Direct detection of cysteine ox-PTM is feasible for relatively stable modifications like glutathione-based disulfides, disulfides, or sulfonic acids (Figure 9.1A). As cysteine ox-PTM occur often sub-stoichiometrically and their total abundance in a complex background is rather low, the direct detection of cysteine ox-PTM is in many cases not an easy task. Well-controlled enrichment procedures might help here for efficient analysis.

The direct detection works well after affinity enrichment of single proteins for the analysis of glutathionylated sites, for example. Nevertheless, global approaches for the detection of glutathionylated sites are more focused on indirect- or label-based approaches.

The detection of intramolecular disulfides has taken a big step forward especially due to the development of efficient analysis software. Recent strategies are capable to identify several hundred peptides linked by disulfides even from complex samples. For example, the pLink-SS approach is based on protein alkylation with N-ethylmaleimide during sample preparation as well as protein digestion at lower pH to prevent disulfide bond scrambling. A subsequent analysis with high-resolution mass spectrometry and HCD fragmentation of di-peptides follows, and crosslink identification is enabled within the pLink2 software environment [4].

But features of mass spectrometers also can be utilized for the detection of disulfide-linked peptides. An interesting method has been developed in the Heck laboratory. Also in this workflow, sample preparation and digestion is performed at acid pH. Precursor ions are sequentially fragmented in a suitable mass spectrometer: first, electron transfer dissociation (ETD) is used to preferentially fragment disulfide linked peptides at the disulfide bond; and second, resulting ions are further fragmented by higher energy collisional dissociation (HCD). The whole procedure is called EThcD and enables a good compromise of preserving the signals of the disulfide bond cleaved precursor peptides as well as proving ETD and HCD ions. The information from all ions is interpreted by the SlinkS software [5]. Both strategies suggest also the use of rather unspecific proteases like pepsin for lower complex samples or the use of different proteases or their combinations to enhance the pool of accessible di-peptides.

It is also possible to directly identify cysteines oxidized to sulfinic and sulfonic acids even from complex biological samples. Here, the unique charge distribution of the acidic modifications can be used for enrichment. For instance, 181 cysteine sulfinic/sulfonic acid–containing peptides could be identified in an approach based on negative selection by strong cation exchange chromatography and positive selection with hydrophilic interaction chromatography [6].

9.5 INDIRECT DETECTION APPROACHES

Whereas a direct detection of ox-PTM offers the advantage of an unambiguous and often site-specific PTM identification, it is still in many cases not an easy task. The main reasons are still the non-trivial interpretation of complex fragmentation patterns of disulfide-linked peptides, the low abundance of many ox-PTM, unfavorable peptide properties in standard positive mode mass spectrometric experiments, and the unstable nature of some ox-PTM. To circumvent at least some of these issues, direct and indirect detection approaches based on chemical probes have been developed for the identification of ox-PTM. Direct detection approaches are based on a direct reaction of a chemical probe with a certain ox-PTM and will be discussed later in this chapter. Indirect labeling approaches rely on a reduction of a reversible ox-PTM and subsequent reaction with a chemical probe. Here, a wide range of different strategies using diverse probes have been developed over the years. Common features of indirect detection methods are an initial blocking step of thiol groups by chemical reagents like N-ethylmaleimide (NEM) or iodoacetamide (IAM), a subsequent reduction of reversible ox-PTM, and finally an alkylation of newly generated cysteine thiols (Figure 9.1B).

In a basic variant, two different alkylating agents (e.g., NEM for blocking and IAM after reduction) can be used, and ratio shifts of NEM- and IAM-labeled peptide intensities point towards altered cysteine oxidation. Alternatively, using a light and a heavy variant of the same alkylating agent (e.g., NEM and NEM-D5 or IAM and IAM-D4) enables a direct ratio determination of reduced and reversibly oxidized peptides.

Nevertheless, a main advantage of indirect detection approaches is the possibility to enrich previously oxidized peptides. Pioneering work (e.g., the OxICAT approach) makes use of the isotope-coded affinity tag (ICAT) reagent [7]. This reagent (which at the moment has been discontinued commercially) consists of an iodoacetamide based warhead, an isotope-coded linker region—available in heavy and light variants—and a biotin moiety for enrichment (Figure 9.2A). In the OxICAT approach, first reduced cysteines of denatured proteins are labeled with the light version of the ICAT reagent, and after reduction of reversibly oxidized cysteines, proteins are labeled with the heavy variant of the probe. After tryptic digestion, an enrichment step of labeled peptides is carried out and finally, peptides are analyzed by LC-MS/MS for peptide identification and quantification. The intensity ratio between heavy and light peptide variants is an approximation of the ratio between reversibly oxidized and reduced cysteine-containing peptide variants [7].

Alternatively, iodoacetyl tandem mass tag (iodoTMT) reagents are alternative probes for OxICAT-like approaches [8]. The reagent is currently available in six different isotopically labeled

variants, enabling multiplexing of six different samples in one experiment. In contrast to ICAT, the nominal masses of all six TMT reagent variants are the same, because the mass difference is encoded by a reporter group which is neutralized by a balancer region in the probe (Figure 9.2B). Quantification and separation of isotopically separated variants are carried out after peptide fragmentation, releasing also the reporter group. For enrichment of labeled peptides, an anti-TMT resin can be used.

Nevertheless, direct isotopic coding of probes is not necessarily required. For instance, the biotinylated iodoacetamide (BIAM) switch assay makes use of iodoacetyl-polyethylene glycol-biotin (Figure 9.2D) to enrich reversibly oxidized proteins solely based on label-free precursor intensity–based quantification [9].

The abovementioned probes are a bit bulky, which might impact their usefulness, especially in a native environment. To enhance the accessibility of sterically buried cysteines, approaches have been developed making use of smaller iodoacetamide-based compounds like iodoacetamide alkyne (IA, Figure 9.2F) or iodopropynylacetamide (IPM, Figure 9.1G). After reaction with thiols, probes and linked proteins/peptides can be coupled via copper-catalyzed alkyne azide cycloaddition reaction to linkers like Azide (AZ)-UV biotin, enabling biotin mediated enrichment and, in the case of AZ-UV-biotin, UV mediated release [10, 11]. If different conditions should be compared, it is possible to include also a heavy AZ-UV-biotin variant to label both conditions independently. Furthermore, those probes offer the possibility to carry out a quantitative reactivity profiling using a lower and higher IA or IPM concentration in the initial reaction with cysteines' thiols [10, 11].

An alternative to the click chemistry–based approaches is the use of cysteine reactive phosphate tags (CPT, Figure 9.2E). Here, iodoacetamide is coupled over a linker to a phosphate group which can be used for enrichment by immobilized metal affinity chromatography, a well-established method for the analysis of phosphopeptides. This method was used recently for the determination of tissue-specific cysteine oxidation in mouse quantifying about 34,000 cysteine sites on about 9,400 proteins [12].

Besides methods relying on chemical probes, thiol-reactive resins like thiopropyl sepharose (Figure 9.2C) have been widely used for the enrichment of oxPTM. In these approaches, thiol groups are first blocked by alkylating agents followed by a

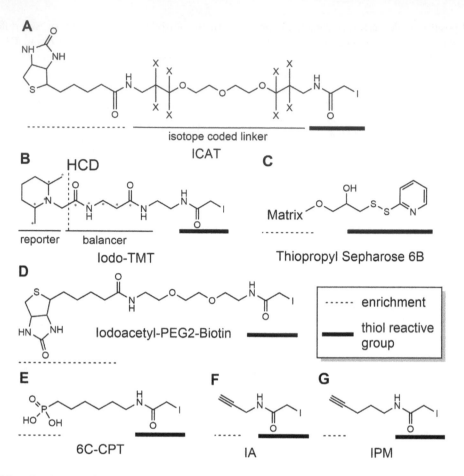

Figure 9.2 Probes (A, B, D–G) and resins (C) used for indirect profiling of cysteine oxPTM.

Most reagents contain an iodoacetamide based thiol-reactive group and a tag enabling enrichment. This tag can be based on biotin (A, D), a phosphate group (E), or an alkyne for further reaction via Copper(I)-catalyzed azide-alkyne cycloaddition (F, G). Some of the probes contain additionally isotope-coded linkers based on heavy/light variants of carbon and nitrogen (positions marked by * in B), for instance in iodo-tandem mass tag (TMT) reagents or hydrogen/deuterium (positions marked by X in A) in the original isotope-coded affinity tag (ICAT) reagent. Deuterium was replaced in later versions of the reagent by coding with ^{13}C. CPT: cysteine reactive phosphate tags (E); IA: iodoacetamide alkyne (F); IPM: iodo-N-propynyl-acetamide (G).

selective or non-selective reduction step. Reversibly oxidized cysteines are subsequently directly coupled to the resin for the enrichment and released by a reducing agent. This method can also be combined with isotope labeling, for instance by using primary amine-reactive tandem mass tag reagents [2, 13].

Indirect detection approaches can be carried out either by targeting a broad range of reversible ox-PTM using an unspecific reduction step (e.g., by dithiothreitol [DTT] or tris(carboxyethyl)phosphine [TCEP]) or by a rather selective reduction step focusing on a single type of oxidative cysteine modification. Here, ascorbate (S-nitrosothiols), arsenite (cysteine sulfenic acids), and enzymatically

approaches with glutaredoxins (S-glutathionylation), for example, have been widely used [13].

9.6 DIRECT DETECTION APPROACHES

Although arsenite-based reduction has been used in indirect labeling approaches for the detection of cysteine sulfenic acids, there are several issues with corresponding methods. On the one hand, are sulfenic acids quite labile and often stabilized by their environment. This can be problematic in denaturing environments and during prolonged sample preparation. On the other hand, several thiol-reactive compounds usually used for blocking as NEM or IAM might cross-react with protein sulfenic acids

[14]. Therefore, great efforts have been made to directly target labile ox-PTM like cysteine sulfenic and sulfonic acid with chemical probes to enable their identification and quantification (Figure 9.1C). Direct detection approaches come with several advantages: labeling with certain probes can be carried out both in *vitro* and in *vivo*, and the direct targeting might help to minimize artifacts during sample preparation. Nevertheless, a usable probe should be specific and show a fast reaction with its target. Moreover, it has to be kept in mind that prolonged labeling times might have an impact on intracellular reaction kinetics by removing protein sulfenic acids from equilibrium or by disturbing redox cycling.

Initial approaches for trapping cysteine sulfenic acids are based on dimedone and its derivatives. The core 1,3-cyclohexadione reactive group of dimedone is known to react with protein sulfenic acids but not with thiols. To enable enrichment of labeled cysteine sulfenic acids, probes containing affinity functionalities have been developed. These include biotin (Figure 9.3B) or click chemistry–compatible moieties like alkynes (Figure 9.3A) or azides [1]. Moreover, probes based on strained alkynes and alkenes like cyclooctynes (Figure 9.3D) and norbornene derivatives (Figure 9.3E) have been shown to react faster than dimedone-based probes with cysteine sulfenic acids. The faster-reaction

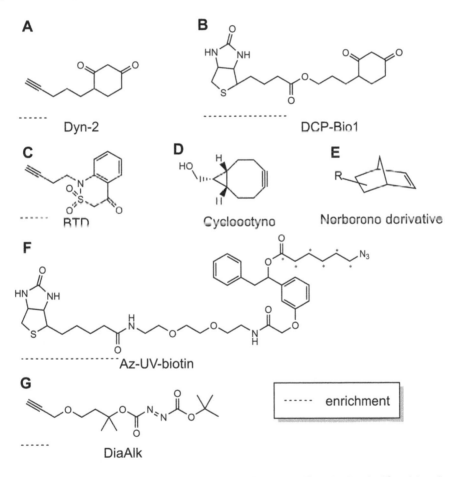

Figure 9.3 Selected chemical probes used for the direct detection of cysteine sulfenic (A–E) and sulfinic (G) acids.

Original probes are based on dimedone derivatives (A, B) and might include a linker for affinity-based enrichment. As the reaction with sulfenic acids is relatively slow for dimedone derivatives, probes based on benzothiazine (BTD, C) and strained alkyne- and alkene-based probes like cyclooctynes (D) and norbornene derivatives (E) have been developed. Furthermore, a clickable electrophilic diazene probe (DiaAlk, G) has been reported for the specific enrichment of cysteine sulfinic acid–containing proteins and peptides. Probes for the direct detection of ox-PTM include either a direct a biotin moiety for enrichment (B) or a click chemistry compatible moiety like an alkyne (A, C, G). R represents either an alkyne-based or biotin tag. Click chemistry–based approaches have been proved useful if (UV)-cleavable linkers like azido-UV-biotin (az-UV-biotin, F) are included. Those linkers might also include isotope labels (*).

kinetics offer the possibility to trap also short-lived sulfenic acids, but these probes might favor also off-target reactions [15].

Furthermore, the Carroll group developed a set of dimedone-inspired carbon-based nucleophiles including a benzothiazine-based (BTD) probe (Figure 9.3C) showing a 170-fold higher reaction rate with cysteine sulfenic acids. Especially in workflows including the click chemistry–based coupling of (UV)-cleavable linkers for enrichment like az-UV-biotin (Figure 9.3F), this probe represents a great improvement in comparison to conventional dimedone-based molecules [16].

For a long time, cysteine sulfinic acids could not efficiently be accessed by chemical proteomics approaches, but new developments make this more and more feasible. Recently, a clickable electrophilic diazene probe (DiaAlk) was developed. This probe enabled the identification of several hundred S-sulfinylated sites and novel potential substrates of cysteine sulfinic acid reductase from cultured cells [17].

In principle, direct labeling approaches offer the possibility to include isotopic labels directly in the probe or a clickable linker molecule, simplifying quantitative comparisons of two or more samples or several different ox-PTM in parallel.

A special case is the direct labeling–based detection of S-glutathionylation sites. Here, an approach has been described relying on a modification of glutathione itself instead of using a chemical probe mediated detection. The idea here is to use a glutathione synthetase mutant to preferentially include azido-alanine instead of glycine into newly synthesized glutathione in living cells [18]. This azido-glutathione derivative can form S-glutathionylated sites in cells and subsequently be enriched by click chemistry-based purification strategies. This approach has been used to identify over 1,000 glutathionylated proteins from cultured cells [18].

9.7 IMPORTANCE OF THE EXPERIMENTAL DESIGN

Before designing an experiment, several questions should be addressed:

1. Which experimental system should I use? Lower complex samples and in vitro–generated samples usually are easier to analyze, as the relevant ox-PTM might be present in higher amounts and better accessibly for downstream analytics by the lower sample complexity. Higher complex samples like living systems might be better models, but "freezing" of ox-PTM can be more difficult and subcellular localization patterns might further impact the analysis.

2. Which ox-PTM is interesting for me? As mentioned, there is no global method providing high analytical sensitivity for all kinds of ox-PTM. In indirect approaches, based on global reduction of reversible ox-PTM, different types of ox-PTM can be accessed in one experiment, but the information of the type of ox-PTM gets lost. In direct labeling and targeted approaches, the focus might lie on a single ox-PTM, which can be accessed with high sensitivity but there will be not much information on other available ox-PTM.

3. Which depth of analysis will I reach? Depending on the analyzed system, question, and used method, many resources in terms of biological material, reagents, and instrument time will be needed. To reach sufficient sensitivity, in some cases a more targeted approach concentrating on a few sites/proteins might be more feasible. Moreover, a wise choice should be made if the analysis should focus on protein or peptide/site level. An analysis on protein level might provide a higher sensitivity as more peptides for one protein are available, but there might be difficulties if more than one ox-PTM is present in one protein. However, an analysis on peptide/site level might provide valuable information about the modified site but usually needs more material, and problems might occur if available peptides are not well suited for LC-MS approaches (e.g., because they are too long, too short, or have unfavorable ionization or liquid chromatography retention properties).

According to the answers to these questions, a method promising the best compromise should be chosen. Furthermore, it might be a good idea to use several methods in parallel to profit from their advantages and complementary information. Moreover, a stepwise approach might be an option in many cases: in a first round, targets of cysteine oxidation on protein or site level could be identified and in targeted follow-up experiments confirmed, and details like stoichiometry, the nature of ox-PTM, and subcellular information retrieved. Nevertheless, in most cases, it is a big advantage if the initial experiment already provides a high level of detailed information about the nature, position, and stoichiometry of ox-PTM.

Highly important for the success of a redox proteomics experiment is also the experimental design. Looking at ox-PTM, the choice of a setup quantitatively comparing two (or more) different conditions is in many cases straightforward and better to control than trying to establish a global catalog of ox-PTM, because artifacts introduced by sample preparation will affect all analyzed samples. Moreover, it is important to control also the total level of a certain protein or the modification ratio of an ox-PTM to exclude that the abundance change of a certain modified peptide is due to an abundance change of the associated protein. Therefore, including a sufficient number of meaningful control samples and/or control proteins/peptides as a spike-in is always a good idea. Especially if complex workflows are used, including click-chemistry, enrichment, and (isotope)-labeling, it is a big advantage if each step could be controlled separately.

Additionally, including fully reduced samples or samples that have been fully reduced and blocked by alkylating agents before enrichment might help to evaluate the results of an experiment. Nevertheless, it has to be noted that in approaches focused on reversibly oxidized cysteines, peptides including sites oxidized to sulfonic acids, for example, might not be captured. Therefore, it is difficult to report values like "percent oxidation" for these cases.

Concerning the experimental design, especially the number and nature of replicates is highly important. Generally, the technical variance of repeated LC-MS measurements is much smaller in comparison to the variance in biological systems or the variance introduced by a complex sample preparation procedure. Therefore, it makes much more sense to make sample replicates on the level of analyzed biological systems, for example, tissue or cells from different animals/culture dishes that subsequently were processed in parallel instead of analyzing the same sample several times. As global approaches are capable to quantify several thousand proteins/peptides in one experiment, statistical analysis of this kind of data requires a correction for multiple testing. This is one of the reasons why it is important to choose an appropriate number of replicates (e.g., by power calculation). Moreover, in some settings including isotope labels, it might be wise to include "label-swaps" to control label-associated variances.

REFERENCES

1. Poole, L.B., C.M. Furdui, and S.B. King, *Introduction to approaches and tools for the evaluation of protein cysteine oxidation.* Essays Biochem, 2020. **64**(1): p. 1–17.
2. Duan, J., M.J. Gaffrey, and W.J. Qian, *Quantitative proteomic characterization of redox-dependent post-translational modifications on protein cysteines.* Mol Biosyst, 2017. **13**(5): p. 816–829.
3. Nagy, P., et al., *Measuring reactive sulfur species and thiol oxidation states: challenges and cautions in relation to alkylation-based protocols.* Antioxid Redox Signal, 2020. **33**(16): p. 1174–1189.
4. Lu, S., et al., *Mapping disulfide bonds from sub-micrograms of purified proteins or micrograms of complex protein mixtures.* Biophys Rep, 2018. **4**(2): p. 68–81.
5. Liu, F., B. van Breukelen, and A.J. Heck, *Facilitating protein disulfide mapping by a combination of pepsin digestion, electron transfer higher energy dissociation (EThcD), and a dedicated search algorithm SlinkS.* Mol Cell Proteomics, 2014. **13**(10): p. 2776–2786.
6. Paulech, J., et al., *Global analysis of myocardial peptides containing cysteines with irreversible sulfinic and sulfonic acid post-translational modifications.* Mol Cell Proteomics, 2015. **14**(3): p. 609–620.
7. Leichert, L.I., et al., *Quantifying changes in the thiol redox proteome upon oxidative stress in vivo.* Proc Natl Acad Sci U S A, 2008. **105**(24): p. 8197–8202.
8. Sievers, S., et al., *Comprehensive redox profiling of the thiol proteome of clostridium difficile.* Mol Cell Proteomics, 2018. **17**(5): p. 1035–1046.
9. Lowe, O., et al., *BIAM switch assay coupled to mass spectrometry identifies novel redox targets of NADPH oxidase 4.* Redox Biol, 2019. **21**: p. 101125.
10. Fu, L., et al., *A quantitative thiol reactivity profiling platform to analyze redox and electrophile reactive cysteine proteomes.* Nat Protoc, 2020. **15**(9): p. 2891–2919.
11. Weerapana, E., et al., *Quantitative reactivity profiling predicts functional cysteines in proteomes.* Nature, 2010. **468**(7325): p. 790–795.
12. Xiao, H., et al., *A quantitative tissue-specific landscape of protein redox regulation during aging.* Cell, 2020. **180**(5): p. 968–983 e24.
13. Guo, J., et al., *Resin-assisted enrichment of thiols as a general strategy for proteomic profiling of cysteine-based reversible modifications.* Nat Protoc, 2014. **9**(1): p. 64–75.
14. Reisz, J.A., et al., *Thiol-blocking electrophiles interfere with labeling and detection of protein sulfenic acids.* FEBS J, 2013. **280**(23): p. 6150–6161.

15. Pople, J.M.M. and J.M. Chalker, *A critical evaluation of probes for cysteine sulfenic acid*. Curr Opin Chem Biol, 2020. **60**: p. 55–65.

16. Gupta, V., et al., *Diverse redoxome reactivity profiles of carbon nucleophiles*. J Am Chem Soc, 2017. **139**(15): p. 5588–5595.

17. Akter, S., et al., *Chemical proteomics reveals new targets of cysteine sulfinic acid reductase*. Nat Chem Biol, 2018. **14**(11): p. 995–1004.

18. Samarasinghe, K.T., et al., *A clickable glutathione approach for identification of protein glutathionylation in response to glucose metabolism*. Mol Biosyst, 2016. **12**(8): p. 2471–2480.

CHAPTER TEN

Glutathione during Development

Jason M. Hansen, Joshua E. Monsivais, and Brandon M. Davies

CONTENTS

DOI: 10.4324/9781003204091-14

10.1 INTRODUCTION

Glutathione (GSH) is an abundant non-protein thiol that promotes antioxidant defense by reducing reactive oxygen species (ROS). In addition to antioxidant defense, GSH participates in several other biological processes: cysteine carrier and storage, cofactors in several enzymatic reactions, maintenance of cellular redox state, protein regulation, and xenobiotic detoxification (Sies 1999; Schafer and Buettner 2001; Pastore, Federici, et al. 2003). Glutathione is found in all mammalian tissues in concentrations between 1 and 10 mM, with the highest concentration typically being found in the liver (Ookhtens and Kaplowitz 1998; Pastore, Federici, et al. 2003). Glutathione is a tripeptide, composed of cysteine (Cys), glutamate (Glu), and glycine (Gly) (γ-glutamyl cysteinylglycine). The active portion of GSH is represented in the thiol (-SH) found on the cysteine. Glutathione can exist in a reduced or oxidized state; the oxidized form primarily occurs in dimer form as a disulfide (GSSG) (Kaplowitz, Tak, and Ookhtens 1985). The ratio of concentrations between the GSSG and GSH is the GSH system, which works as a redox buffer, using GSH to decrease the amount of available ROS. The structure of GSH is unique due to a specialized γ-peptide bond that links glutamate and cysteine; this bond links the amino acid residues through the side chain in place of the normal carboxyl terminus on the amino acid. Due to this unconventional bond configuration, GSH is unrecognized and resistant to intracellular peptidases, allowing GSH to accumulate within cells to a high concentration. Most organisms lack the necessary importers to transport GSH into the cell. Due to the lack of GSH importing, *de novo* synthesis within the cell is required using precursors that originated in the extracellular compartment. There is substantial evidence demonstrating GSH's crucial role in protecting a developing embryo and related roles in embryogenesis (Bock et al. 1987; Hayes, Flanagan, and Jowsey 2005; Ketterer, Coles, and Meyer 1983). This review will focus on the synthesis, function, and role of GSH throughout the developmental process.

10.2 GAMMA GLUTAMYL CYCLE

Although GSH has many roles, a primary role of GSH is to act in antioxidant defense of the cell preventing oxidative damage from ROS and reactive chemicals. Glutathione peroxidase-1 (GPx1) is an enzyme that uses GSH to detoxify hydrogen peroxide (H_2O_2) by catalyzing the reaction necessary to reduce H_2O_2 to H_2O, oxidizing GSH into GSSG in the process (Lu 2009; Zhang and Forman 2012) (Figure 10.1).

After GSH oxidation, GSSG can be recycled back into GSH by glutathione disulfide reductase (GSR) using the electron donor, reduced nicotinamide adenine dinucleotide phosphate (NADPH). Conversely, GSH conjugates can be exported from the cell for eventual excretion (Lu 2009). Recycling restores GSH concentrations, allowing it to function again in conjunction with GPx1 to reduce additional H_2O_2. During exportation, GSH is primarily transported from cells as GSSG or oxidized GSH-conjugates, where the K_m for GSSG promotes rapid export (~100 μmol/L). Reduced GSH has also been shown to be exported but has a higher K_m (~10 mmol/L) and appears to be more cell specific. Plasma membrane transporters in mammals include members of the ATP-binding cassette subfamily C (ABCC) transporters, which more specifically include the multidrug-resistant protein (MRP) subclass of proteins (ABCC1-5) and CFTR (ABCC7) (Oestreicher and Morgan 2019). However, the role of ABCC transporters during embryo development is not well understood but has been suggested to play important roles (Bloise et al. 2016). Clearly, this field requires more research to determine any mechanistic contribution. Once exported outside of the cell, GSSG or oxidized GSH-conjugates can be enzymatically degraded into GSH precursors by γ-glutamyl transpeptidase (GGT) and re-imported into the cell via amino acid transporters for use in intercellular *de novo* synthesis (Meister 1983).

10.3 *DE NOVO* GLUTATHIONE SYNTHESIS

Because the majority of cells lack the components necessary to import GSH directly from extracellular sources, GSH *de novo* synthesis is required. When oxidized, GSSG can be exported into the extracellular space or reduced and reused for detoxification purposes within the cell. The enzyme GGT is found on the exterior portion of the plasma membrane of cells and works by cleaving extracellular GSH, GSSG, and oxidized GSH-conjugates at the unique γ-peptide bond (Pastore, Federici, et al. 2003). This reaction yields free Glu and the Cys-Gly dipeptide, which can be further degraded into Cys and Gly via extracellular peptidases. Free amino acids can

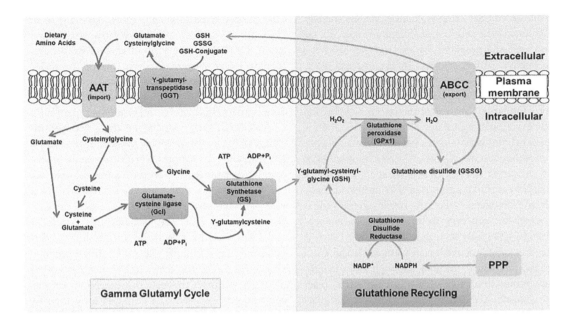

Figure 10.1 The gamma glutamyl cycle (GGC), *de novo* synthesis of GSH, and GSH recycling pathways.

In the GGC, extracellular GSH is broken down via GGT into GSH amino acid precursors that are then transported back into the intracellular compartment. Once these amino acids are present within the cell GSH, *de novo* synthesis occurs through two steps involving ATP and the enzymes gamma glutamate-cysteine ligase (Gcl) and glutathione synthetase (GS). Recycling of GSH/GSSG occurs during the reduction of H_2O_2 to water. Through glutathione peroxidase (GPx1), H_2O_2 is reduced to water and GSH is oxidized to GSSG. GSSG can then be reduced back to GSH through glutathione disulfide reductase (GSR) and the electron donor, NADPH, originating from the pentose phosphate pathway (PPP). GSSG and other GSH conjugates can be exported via ABCC transporters, after which it is broken down via the GGC.

then be transported back into the cell to support *de novo* biosynthesis (Hansen and Harris 2015). Of particular interest is Cys import, as Cys is the rate limiting precursor in GSH synthesis. Liberation of Cys or its oxidized form, cystine (CySS), from GGT-mediated breakdown can be imported back into cells through the xCT transporter (Lewerenz et al. 2013). Preimplantation expression of xCT has not been well studied, but in post-implantation rat embryo (gestational day [GD] 16) brains, xCT was expression was observed. Brain xCT levels continued to increase on post-natal day (PD) 1 through adulthood (La Bella et al. 2007). Interestingly, in muscle the converse was observed, where xCT expression declined into adulthood and GD 16 embryos showed the highest levels. De novo GSH synthesis relies heavily on GGT function to provide free Cys to increase its systemic bioavailability because the only organs and cells with GSH importers with low K_m that would significantly contribute to import are found in the gut and proximal tubule of the kidney (Hagen et al. 1990).

Embryonic Cys availability for *de novo* synthesis comes primarily from dietary amino acids. Due to the rat embryo's nutritional circumstances, the conceptus has a unique feature to compensate for the lack of direct amino acid transport from maternal nutrients (Harris 1993). Through the process of histotrophic nutrition (HN), the rodent conceptus can supply large quantities of amino acids, including those involved in GSH synthesis and other precursors, to support new protein synthesis, nucleic acid synthesis, and many other metabolic needs. The process of HN uses receptor-mediated endocytosis (RME), which captures maternal sources into primary vesicles, fuses with lysosomes, and is degraded by cysteine proteases to liberate amino acids. As such, HN is believed to be the primary mechanism by which amino acids, including Cys, Glu and Gly, are supplied to the embryo.

Synthesis of GSH from its constituent amino acids involves 2 ATP-requiring enzymatic steps (Lu 2013). The first step is the formation of γ-glutamylcysteine using free Glu and Cys, and the second is the formation of GSH using γ-glutamylcysteine and Gly. Once inside the cell, the precursor amino acids are processed and prepared for the first step involving the ligation of Glu and Cys by the enzyme glutamate cysteine ligase (Gcl; Yang, Zeng, et al. 2002). Glutamate cysteine ligase is composed of two subunits; the larger (heavy) of the two is called glutamate cysteine ligase catalytic subunit (Gclc), and the smaller (light) is glutamate cysteine ligase regulatory or modifier subunit (Gclm). When Gclm is knocked out, the larger subunit Gclc is sufficient to produce GSH; however, the amount of GSH produced is significantly reduced. When the larger subunit Gclc is knocked out and only Gclm remains, de novo synthesis of GSH fails to occur. The subunit Gclm is enzymatically inactive but helps to lower the K_m of Gcl and raise the K_i for GSH, allowing for more efficient production of GSH. The Gcl enzymatic reaction requires ATP and Mg^{2+} as cofactors to complete catalysis. The subunits of the enzyme rely heavily on induction, but the holoenzyme activity can be regulated through non-allosteric feedback inhibition by GSH concentrations and the concentration of the rate limiting precursor, Cys (Richman and Meister 1975). The final synthetic step is completed when Gly is added to γ-glutamylcysteine through the enzyme, glutathione synthetase (GS), which also requires ATP and Mg^{2+} (Anderson and Meister 1983). Glutathione synthetase is composed of two identical subunits and unlike the first step is not regulated by feedback inhibition from GSH concentrations. The only limiting factor for this portion of synthesis is the availability of γ-glutamylcysteine and Gcl activity. As such, when GS is overexpressed, there is no significant increase in GSH concentrations. Though GS may not be rate limiting, there is evidence to suggest that it plays an important role in determining overall GSH synthetic capacity in certain tissues and under stressful conditions (Yang, Zeng, et al. 2002).

10.4 METABOLIC SHIFTS DURING DEVELOPMENT

Spatial and temporal fluctuations in ambient O_2 concentrations occur throughout the early stages of development and influence bioenergetic processes, which meet requisite energy demands through central carbon metabolism but also regulate the metabolic signaling functions (Miyazawa and Aulehla 2018) to enable adaptation and protection from potentially harmful metabolic by-products.

Changes to the metabolic processes, especially in an O_2-requiring process such as oxidative phosphorylation (OXPHOS), may support alterations to intracellular redox states, including that of the GSH/GSSG redox couple. As such, understanding metabolic shifts may provide insight into redox-sensitive developmental programs and embryonic periods.

The overall early result of metabolic reprogramming in the embryo is the marked shift from a pre-implant reliance on OXPHOS to supply bioenergetic requirements to an increasing dependence on glycolysis, as embryogenesis proceeds from early cleavage stages to post-implantation organogenesis. These shifts may rework ROS availability and have weighty effects on intracellular redox potentials to support specific developmental processes. Following fertilization and passage of the early embryo through the uterine duct, the conceptus is exposed to a comparatively high O_2 environment, having access to a variety of high-energy substrates, including glucose, pyruvate, lactate, and glutamine (Dumollard et al. 2009). The early cleavage stages, up to the 8-cell stage, primarily rely on OXPHOS, but glucose is not well utilized for energy production, and glycolysis is largely inactive due, in part, to poor glucose transport (Houghton and Leese 2004; Pantaleon et al. 1997; Riley and Moley 2006). In bovine embryos, much of the available glucose at this stage is shunted through the pentose phosphate pathway (PPP; Javed and Wright 1991), which would increase the production of NADPH, to prepare for rapid growth and possibly to reduce the effects of upcoming periods of metabolic shifts and subsequent oxidative stress. NADPH production does not require lamellar cristae but rather would be supported by a large matrix, a notable observation of mitochondria during specific stages of early development, suggesting low, sustained ATP output through upcoming periods of an increased reliance on OXPHOS (Dumollard et al. 2009). Pyruvate and glutamine remain the principle ATP substrates during these early developmental stages (Biggers, Whittingham, and Donahue 1967). Extraembryonic sources of pyruvate and glutamine appear to support mitochondria-mediated ATP synthesis. Together, the observations that lower glycolysis and increased pyruvate-derived ATP synthesis through OXPHOS demonstrate an increasingly important role of the mitochondria early in preimplantation stages of development. Only during later pre-implantation stages (morula and mid-blastocyst) do glycolytic processes increase and result in a decreased ATP/ADP ratio (Quinn and Wales 1973; Barbehenn, Wales, and Lowry 1978).

Mitochondrial dynamics rapidly shift during early development, where in the 8-cell stage metabolically active mitochondria are replaced by immature mitochondria containing few lamellar cristae having low OXPHOS activity, low ATP production, and low O_2 consumption (Trimarchi et al. 2000a, 2000b; Shepard, Muffley, and Smith 1998). In the mouse, from early stages of development to the late blastula, total ATP concentrations continually decrease and synthesis significantly slows (Ginsberg and Hillman 1973). In order to optimize nutrient and O_2 consumption during hypoxia and also to minimize unintended ROS production in mitochondria, activation of O_2-senstive transcription factors, such as hypoxia inducible factors (HIFs), orchestrate a metabolic shift away from high O_2 demand (TCA cycle/OXPHOS) towards anaerobic metabolism with an increased activation of glycolytic and pentose phosphate pathways (Folmes and Terzic 2014; Zheng et al. 2016). During this period, all glycolytic enzymes are significantly upregulated, which coincides with a concomitant limitation to TCA/OXPHOS metabolism and decreases mitochondrial number and volume, maintaining redox homeostasis (Dengler, Galbraith, and Espinosa 2014; Samanta and Semenza 2017).

The shift to glycolytic dependence from early OXPHOS is a hallmark of highly proliferative cells, a prerequisite for proper development and ensuing morphogenesis. The conversion to glycolytic dependence and its consequences is well characterized as the Warburg effect of cancer cells and is shown to be very similar to reprogramming events that occur during development (Krisher and Prather 2012). Under the reducing redox conditions of hypoxia, bioenergetics, and carbon metabolism, glycolysis/aerobic glycolysis can not only meet the needs for energy (ATP) demands but can also address the other metabolic needs for development by providing adequate reducing equivalents (NADPH), substrates for nucleic acid production (serine/glycine), phospholipid biosynthesis (DHAP; dihydroacetone phosphate, a precursory substrate), and substrates for one carbon metabolism (Miyazawa and Aulehla 2018).

During the blastula stage, the embryo implants into the endometrial wall, mitochondria have few lamellar cristae, energy production through OXPHOS is greatly diminished, and a high proportion of available ATP is produced through glycolytic processes (anaerobic glycolysis) as a result of metabolic reprogramming (Houghton et al. 1996; Trimarchi et al. 2000a). As a consequence of implantation, O_2 availability is dramatically decreased, and both glucose consumption and glycolysis are high (Shepard, Tanimura, and Park 1997). In the hypoxic state of rodent gastrulation (approximately 24 hours post-implantation), mitochondrial maturation resumes, although OXPHOS activity remains low. In early somite stages (2–4 somites), embryonic mitochondrial cristae become more numerous but have a blebbed morphology and are not yet lamellar, suggesting a gradual progression to support a more OXPHOS-competent embryo (Dumollard et al. 2009; Shepard, Muffley, and Smith 1998). Essential mitochondrial function during this "switching" period of organogenesis has been demonstrated in numerous studies utilizing various knockout (KO) mouse models involving mitochondrial proteins. For instance, knocking out a subunit of succinate dehydrogenase (complex II in the electron transport chain) resulted in early embryonic death evident at GD 7.5 (Piruat et al. 2004). In another study, cytochrome c, which shuttles electrons to complex IV in the electron transport chain, was deleted and development stalled on GD 8.5 (Li et al. 2000). Thioredoxin 2 (Trx2) is a mitochondrial protein reductase that is involved in the regulation of apoptosis (Yoshioka, Schreiter, and Lee 2006). Loss of Trx2 results in dysmorphogenesis, including exencephaly and embryonic death, and increases ROS production on GD 10.5 (Nonn et al. 2003). Mitochondrial thioredoxin reductase 2 (TR2) reduces Trx2 using NADPH as a cofactor. Loss of TR2 is also embryonic lethal, occurring around GD 13.5, where embryonic heart development was highly affected (Conrad et al. 2004). Together, these studies (and others reviewed in Baker and Ebert, 2013) show that disruption of mitochondrial components tend to largely result in cessation and disruption of development during periods of metabolic reprogramming, where embryos flip from anaerobic glycolysis to OXPHOS for the majority of their ATP needs. As such, it is during these early periods of organogenesis and metabolic switching that affected redox signaling pathways may function to support varying aspects of normal development, such as proliferation, migration, differentiation, and apoptosis. Regulation of redox signaling and control is likely a core outcome of these key metabolic pathways, although relatively little molecular or biochemical detail is currently available to elucidate their mechanisms.

10.5 GLUTATHIONE DURING EARLY AND MID-GESTATION

Much work has been performed to better understand the role of GSH and its redox status during development. Perhaps some of the most supportive evidence for the importance of GSH during development is the generation of mice lacking GSH synthesis enzymes. Generation of homozygous KO Gclc, the enzyme responsible for catalyzing the initial and rate limiting step of *de novo* GSH synthesis (glutamate + cysteine → glutamylcysteine), conceptuses did not produce any viable pups carried to term, showed cessation of development, and produced high levels of lethality prior to implantation. Additionally, Gclc KO embryos failed to undergo proper gastrulation, did not form mesoderm, and failed to develop exocoelomic and amniotic cavities, all occurring concomitantly with increases in apoptosis,

demonstrating a disruption in developmental programs (Shi et al. 2000b). Interestingly, Gclc KO embryos appear normal on GD 6.5 but fail to live beyond GD 8.5, suggesting that events between GD 6.5 and GD 8.5 require GSH for normal development to occur (Dalton et al. 2000).

Prior to fertilization, murine oocytes contain high concentrations of GSH (approximately 1.2 pmoles GSH/oocyte, [6–7 mM]) (Gardiner and Reed 1994) (Figure 10.2). Following fertilization, GSH concentrations decrease slightly (1.0 pmoles GSH/conceptus) but decline sharply beginning at the 2-cell stage (0.6 pmoles/embryo) and continue to decline through the blastocyst stage to a low of ~0.1 pmoles/embryo (approximately 0.7 mM GSH). Interestingly, GSSG levels in preimplantation embryos were very low and were often below the limit of detection. However, when detectable, fertilized oocytes contained 0.03 pmoles GSSG/embryo but decreased to 0.001 pmoles GSSG/embryo at the blastula stage (Gardiner and Reed 1994). While GSH:GSSG ratios increase from the 2-cell stage to the blastula, using calculated concentrations in the Nernst equation to derive a redox potential, these data show a slight oxidation. However, this is likely an overestimation of oxidation, as many samples

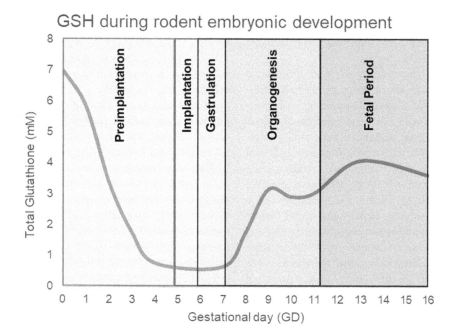

Figure 10.2 Total glutathione concentrations during specific periods of development.

Total GSH (GSH+2XGSSG) are shown. During early preimplantation, total GSH levels are this highest but dramatically drop as the embryo implants. Total GSH levels remain low during gastrulation but increase slightly as cells enter organogenesis and fetal stages of development.

were reported as too low to detect and as such, during pre-implantation, GSH redox potential (E_h) is likely preserved in a relatively reducing state.

Reducing environments during these early preimplantation stages are further supported by observations that bovine zygotic ROS production progressively decreases as cleavage is initiated (Li et al. 2019), which may be, in part, a result of gradual metabolic shifts from OXPHOS to glycolysis. Additionally, supplementation of vitrified murine embryos with reduced, membrane-permeable glutathione ethylester (GSHEE) artificially increases intracellular GSH levels and promotes a reducing intracellular environment while coincidentally increasing fertilization rates and supporting positive developmental outcomes through increased blastocyst viability and decreased ROS availability (Li et al. 2018).

Rapid decline in GSH concentrations in the early preimplantation (2-cell stage) embryo may stem from the inability to sufficiently synthesize GSH and have been suggested to be a result of initially having only maternal GSH available during preimplantation (Dalton et al. 2000). Murine 2-cell stage embryos were treated with GSH depleters (diethyl maleate [DEM]), and during the ensuing recovery periods were unable to properly replenish GSH stores, suggesting poor GSH de novo synthesis at this stage of development (Gardiner and Reed 1995b). However, in later stages, early day 3 embryos were similarly unable to recover GSH concentrations by 5 hours following DEM treatments, but late day 3 embryos fully recovered at 5 hours, demonstrating that prior to day 3 embryos show significantly reduced de novo GSH synthesis capabilities. Inhibition of protein synthesis with cycloheximide on early day 3 prevented GSH recovery on later day 3 (Gardiner and Reed 1995b), suggesting that protein expression of the GSH de novo synthesis pathways are increased during late day 3.

Although decreasing concentrations of GSH during preimplantation periods are not entirely understood, poor Cys intracellular availability may be a contributing factor. Cysteine is the rate limiting substrate to GSH de novo synthesis, and low intracellular concentrations can compromise GSH synthesis (Gardiner and Reed 1995b). In the oviductal and uterine fluids, total Cys (Cys+ 2XCySS) concentrations are 1.4 mM and 0.55 mM, respectively (Elhassan et al. 2001). Three factors may contribute to limited intracellular Cys concentrations: (i) poor Cys transporting capabilities, (ii) highly oxidized

Cys:CySS ratios, and (iii) underdeveloped transsulfuration of methionine-cysteine interconversion via the cystathionine pathway. While initially low, Cys transport into preimplantation embryos does increase from the 1-cell stage to the blastula stage (Takahashi 2012), but overall increase is relatively modest. Additionally, in bovine embryos, adding reductants to preimplantation embryo media resulted in an increase in intracellular Cys concentrations, which promote an increase in GSH levels and further growth (Takahashi 2012; Takahashi et al. 2002). Authors conclude that reduced Cys is more readily transported, whereas CySS is poorly mobilized. Cysteine availability may be further impaired, as murine preimplantation embryos supplemented with methionine, but depleted of Cys, did not demonstrate GSH recovery following depletion (Gardiner and Reed 1995b), suggesting that preimplantation embryos cannot convert methionine into Cys through the cystathionine pathway at this stage of development and may limit usable Cys concentrations. Poor Cys accessibility would constrain GSH de novo synthesis and potentially impair regulation of GSH redox environments.

While concentrations of GSH are dramatically lowered in the preimplantation blastocyst stage embryo, less is known about GSH levels during implantation. This is clearly a critical time for embryonic development, especially in the context of nutrients, O_2 availability, and metabolic reprogramming towards an increasing dependence on OXPHOS. In early, post-implantation mouse embryos, total GSH concentrations are relatively low (~0.4–0.7 mM), but this demarcates the low point in total GSH levels during early/midgestation. On GD 8, organogenesis-stage mouse conceptuses show increased GSH levels to nearly 2 mM and GSSG levels to nearly 30 μM, constituting a moderate shift in GSH E_h to a more oxidized state (−232 mV) compared to values estimated in a murine 2-cell embryo (−242 mV; Gardiner and Reed 1994; Hansen and Harris 2015). Similar levels have been reported in rat embryos of comparable developmental stage (Harris et al. 2013). Increases in post-implantation GSH concentrations may occur, in part, from an increase in de novo GSH synthesis compared to that in preimplantation embryos. Seminal studies demonstrate an increase in de novo GSH synthesis capabilities in rat organogenesis-stage embryos and that GSH turnover may become increasingly dynamic during post-implantation (Harris 1993). Following DEM-mediated GSH depletion, GSH *de novo* synthesis rates rapidly increased until GSH concentrations approached control levels. Interestingly, post-implantation embryonic Cys stores required for GSH *de novo* synthesis did not appear to be a consequence of an increase in transsulfuration through methionine conversion to Cys via the cystathionine pathway (Harris 1993). As such, Cys transport is likely the means by which embryonic Cys supply occurs during post-implantation.

In the organogenesis stage, rat embryos grown in whole embryo culture, the switch to 95% O_2 on GD 11 caused a significant increase in total GSH (Harris et al. 2013), demonstrating a correlation between O_2 availability, ROS production, and GSH content and redox state. Gestational days 8–11 in the mouse encompass early organogenesis-related processes such as neurulation, establishment of a heartbeat, axial rotation/flexion, and increased somitogenesis, and all occur with relatively lower intracellular concentrations of GSH compared to 2-cell stage preimplantation embryos. Lower total GSH pools at these stages, where these developmental milestones occur, may support a more malleable GSH E_h during organogenesis to promote the associated developmental pathway (de)activation. Because there is less GSH, the GSH E_h may be more susceptible to dysregulation via exogenous pressures, including physical or chemical influences. Untimely alterations to the GSH E_h are likely to contribute to disruption of developmental programs and provide rationale as to why the early/mid-organogenesis periods are the most sensitive stages to chemical-induced disruption and dysmorphogenesis (Wilson 1977).

10.6 GLUTATHIONE-RELATED ENZYMES DURING DEVELOPMENT

Various enzymatic and non-enzymatic antioxidants are required for healthy development in humans as well as other model organisms. Many of these endogenous enzymes in the developing embryo have increased activity and presence as development progresses, supporting their participation in redox regulatory pathways during development (Qanungo and Mukherjea 2000).

10.6.1 Glutathione Disulfide Reductase

During early mouse development, specifically from the 2-cell to blastocyst-stage, embryos were treated *in vitro* with tert-butyl hydroperoxide (tBH), an oxidant. In 2-cell stage embryos, tBH treatments caused a significant depletion of GSH pools (by nearly 90% at 15 min) and an overall increase in GSSG concentrations. Also, in 2-cell stage embryos, GSH levels were restored by 60 minute and occurred with a concomitant decrease in GSSG levels arriving at pre-treatment GSSG concentrations. Interestingly, the quick recovery of GSH levels and decreased GSSG concentrations that normally occurs after tBH exposure was abolished when exposed to a pre-treatment of 1,3-bis(2-chloroethyl)-1-nitrosurea (BCNU, carmustine), a known irreversible GSR inhibitor (Gardiner and Reed 1995a). Similarly, in the blastocyst stage embryo, GSH depletion was also observed following tBH treatment, but unlike 2-cell stage embryos only decreased GSH concentrations by approximately 30% at 15 minutes but did also fully recover by 60 minutes. To verify the role of GSR and exclude the involvement of *de novo* GSH synthesis in embryonic recovery, embryos were pretreated with the GSH synthesis inhibitor, buthionine sulphoximine (BSO), before tBH treatments. Pretreatments with BSO did not affect recovery at 60 minutes, suggesting that recovery of GSH stores at 60 minutes is not a consequence of *de novo* synthesis but rather was dependent on recycling through GSR. Additionally, embryos that received the BCNU pretreatment only demonstrated a decrease in GSH, suggesting that GSH/GSSG recycling is a dynamic, ongoing process that occurs even in the absence of exogenous sources of oxidative stress. Another study investigating the dynamic redox regulation of GSH in murine oocytes revealed an increase of GSH in maturing oocytes. Glutathione decreased during development, after fertilization. This agrees with previous studies in that GSH production relies on *de novo* synthesis during oocyte maturation, where *de novo* synthesis halts post-fertilization, using GSR in the GSH regeneration process (Gardiner and Reed 1995a; Dumollard et al. 2007).

In post-implantation embryos, GSR has a constant activity level in the developing rat embryo at GD 9–13, but in the visceral yolk sac (VYS), GSR activity nearly triples during the same developmental period (Choe, Hansen, and Harris 2001; Hiranruengchok and Harris 1993). Increasing VYS activity suggests that the surrounding VYS has a special protective role for the developing embryo during this time frame, as the embryo proper may not be capable of fully responding to changes in metabolic activity and/or oxygen availability. Glutathione disulfide reductase activity in the VYS appears to decrease by 45% between GD 14 and GD 21 while remaining stable in the embryo (Choe, Hansen, and Harris 2001). Glutathione disulfide reductase gene expression shows significant, consistent increases between GD 11 and GD16 in the developing rat embryo (Liu et al. 2018). Reducing GSR expression with all-trans retinoic acid (atRA) in rats (GD 11–16) showed a decrease in the GSH/GSSG ratio and total antioxidant capacity and correlated with an increase in 8-hydroxy-2′-deoxyguanosine, a marker of DNA oxidative damage. In this same study, the occurrence of embryos exhibiting spina bifida aperta also increased, indicating crosstalk between GSR activity and the formation of neural tube defects (NTDs).

Studies using BCNU have shown decreases in viability and lower GSH levels after exposure to the oxidant diamide (Hiranruengchok and Harris 1993). Additionally, co-treatment with 2-nitrosofluorene, a known teratogen and inhibitor of GSH synthesis, and BCNU altered growth parameters such as normal axial rotation (decreased by 62%–97%) and flexion (decreased by 80%–85%; Hiranruengchok and Harris 1993; Berberian et al. 1996). Together, these studies showed that the inability to properly recycle GSH can result in a disruption of developmental programming. Few reports have been published using GSR KO animal models in developmental studies (Han et al. 2017; Pretsch 1999; Rogers et al. 2006; Rogers et al. 2004). However, interestingly, GSR-deficient mice, which exhibit just 2%–10% GSR activity in the liver when compared to normal mice, have been developed with no observable abnormal phenotypes when compared to wild-type mice (Pretsch 1999; Rogers et al. 2004). Altogether, questions remain as to what the role of GSR is during development and to the implications a KO model has in redox signaling, requiring further study.

10.6.2 Glutathione Peroxidase

Glutathione peroxidases (GPx) have been studied using various animal models (Hansen et al. 2001; Cheng et al. 1997; Choe, Hansen, and Harris 2001). Remarkably, when wild-type mouse embryos were treated with mercaptosuccinate, a potent GPx inhibitor, from the 1-cell to the blastocyst stages, no changes in GSH were observed (Nasr-Esfahani and Johnson 1992). Furthermore, with H_2O_2 treatment,

mercaptosuccinate-treated embryos did not have more GSH depletion than with H_2O_2 alone, suggesting H_2O_2 alters GSH levels independent of GPx activity during these early developmental phases. As such, if H_2O_2 detoxification activities act freely from GPx, the GPx system may be mostly absent or non-functional in preimplantation embryos. Using the post-implantation rat embryo model during GD 9–13, GPx, like GSR, remained stable in the embryo but increased in the VYS (Choe, Hansen, and Harris 2001). Interestingly, GD 9–10 showed more GPx activity in the embryo, but in GD 11–13 embryos, GPx activity was higher in the VYS. Glutathione peroxidase continues to show higher activity in the VYS than in the embryo throughout late gestation (GD 14–21). Post-implantation embryonic alterations in activity suggest the embryo is less capable to respond to changes in increasing oxygen tensions during organogenesis stages of development, a period where metabolic reprogramming is being established. Glutathione peroxidase activity was significantly decreased in the spinal cords of GD 11–16 embryos with induced NTDs after atRA exposure (Liu et al. 2018).

Eight mammalian GPx isoforms are known (GPx1–8), with GPx1–4 being selenium dependent and discussed in this review. For all others (GPx5–8), the authors direct the reader to other reviews (Toppo et al. 2008; Ufer and Wang 2011; Hanschmann et al. 2013). Glutathione peroxidase-1 (GPx1) is the predominant isoform, expressed mainly in the cytosol but also detected in the mitochondria to a lesser degree (Toppo et al. 2008). In adult GPx1 KO mice, GSSG concentrations in a variety of tissues are very low compared to wild-type mice, suggesting that while embryonic H_2O_2 metabolism is independent of GPx1 activity, adult mice are very reliant on GPx1 activities (Ho et al. 1997; Cheng et al. 1997). Glutathione peroxidase-1 KO mice and fetal neuronal cultures derived from GD 15 GPx1 KO mouse embryos showed higher susceptibility to paraquat, an oxidant, and H_2O_2, respectively. Glutathione peroxidase-1 KO mice demonstrated a dose-dependent sensitivity to paraquat, where none survived after 24 hours at a dose of 30 mg/kg, but at the same concentration, wild-type mice were unaffected. In vitro, GPx1 KO fetal neurons treated with H_2O_2 also showed a dose-dependent effect, providing evidence of a protective role for GPx1 in the developing nervous system (de Haan et al. 1998). Expression of glutathione peroxidase-2 (GPx2) occurs in the mammalian gastrointestinal system and human livers, offering protection against ROS. Glutathione peroxidase-2 KO mice showed a normal phenotype at birth, suggesting this isoform may be unnecessary during embryonic development, but were more susceptible to carcinogens as adults (Esworthy et al. 2000; Walshe et al. 2007). Glutathione peroxidase-3 (GPx3) is found in the kidney and extracellular fluids. Conflicting accounts using GPx3 KO mice have shown that KO embryos develop normally and there are no observable defects, but other studies report lethality between GD 12.5 and 14.5 due to head hemorrhaging (Olson et al. 2010; Cha et al. 2008). Not much information has been revealed about the regulation and impact of the GPx3 isoform during development and its role in cellular protection.

In mammalian cells, the glutathione peroxidease-4 (GPx4) isoform consists of three different types—cytosolic (c-GPx4), mitochondrial (m-GPx4), and nuclear (n-GPx4)—all of which derive from a single gene and are steadily expressed during early embryogenesis, beginning on GD 7.5, with the nuclear isoform being expressed the least (Borchert et al. 2006; Ufer and Wang 2011; Schneider et al. 2006). Glutathione peroxidase-4 KO mice progressed to GD 7.5–8.5 but died prior to gastrulation (Imai et al. 2003; Yant et al. 2003). Interestingly, when the n-GPx4 or m-GPx4 isoforms are individually knocked down, via siRNA-mediated suppression, mice are viable, suggesting the cytoplasmic isoform is most important for murine embryogenesis (Conrad et al. 2005; Schneider et al. 2009). A study using transgene constructs at an attempt to restore embryonic viability in GPx KO embryos revealed that the reintroduction of c-GPx into fertilized oocytes rescued the KO phenotype, whereas reinserting m-GPx did not, further indicating the requirement of c-GPx during development (Liang et al. 2009). Although embryos are viable, in GD 7.5 embryos knockdown of n-GPx4 led to suppressed growth of heart atrium and overall growth depression, while knockdown of m-GPx4 showed microencephaly and increased apoptosis in the hindbrain region, indicating roles in embryonic heart and brain formation, respectively (Borchert et al. 2006). Additionally, n-GPx4 knockdown mice exhibited male germ cell abnormalities with chromatin instability in spermatozoa, and m-GPx4 knockdown mice exhibited sterility (Conrad et al. 2005; Schneider et al. 2009; Liang et al. 2009).

Glutathione peroxidase-4-deficient mice are viable, although mRNA expression and protein levels are reduced to ~50% when compared to wild-type mice, specifically in the heart, liver, and brain (Yant et al. 2003; Imai et al. 2003). Both mice and mouse embryonic fibroblasts (MEFs) derived from GPx4-deficient mice were more sensitive to oxidative stress caused by tBH, H_2O_2, paraquat, and radiation by showing both decreased growth and colony formation and increased death rates (Yant et al. 2003). Overall, these data demonstrate that GPx4 isoforms are expressed during embryogenesis with complex and vital isoform-specific roles during development and adulthood.

10.6.3 Glutamate Cysteine Ligase

Glutamate cysteine ligase-catalytic and modifier expression levels are independently regulated and show differential expression between the GD 10–11 rat embryo and VYS (twofold to threefold), as specific activities decrease as gestation progresses (Hansen, Lee, and Harris 2004). Glutathione levels were rapidly restored after depletion by DEM in GD 10 rat embryos, where restoration of GSH concentrations occurred first in the VYS followed by repletion in the embryo proper (Harris 1993). Assuming that Cys is the rate-limiting amino acid in GSH synthesis and that Gcl activity is the rate limiting enzyme for GSH de novo synthesis, it can be inferred that Gcl function is active in post-implantation, organogenesis-stage embryos, unlike during preimplantation embryos (Gardiner and Reed 1995a; Gardiner and Reed 1994). Both Gclc and Gclm gene expression was observed in GD 11 embryos (Liu et al. 2018). Expression was not significantly increased until GD 14. These data still demonstrate the potential for de novo synthesis even during early stages of organogenesis but appear to increase during latter gestational periods.

Treatment with methylmercury (MeHg; 10 ppm) prior to and throughout gestation (up to GD 16) showed a significant decrease in fetal GSH/GSSG ratios, specifically a significant decrease in GSH and a simultaneous increase in GSSG levels indicating a more oxidized redox status, but VYS GSH/GSSG ratios remained unchanged (Thompson et al. 2000). Concordantly, following MeHg treatments, Gcl activities were increased in the VYS but remained unchanged in the embryo, indicating the responsive nature of the VYS to oxidants but the poor responsivity to oxidants in the embryo proper. As such, severe embryonic malformations were observed, including hemorrhaging, hydrocephalus, and limb defects. Together, these findings indicate that re-regulation of the GSH pool via de novo synthesis is critical for ensuing development and embryogenesis.

Early in vitro studies using BSO during GD 10–12 showed a 50% decrease in GSH after 8 hours and a 93% decrease after 24 hours, with up to ~80% of embryos showing malformations in head, brain, and limb structures (Slott and Hales 1987). Treatment with BSO in vivo from GD 10–11 revealed similar results with significantly decreased GSH levels, increased mortality rates, and an increased presence of malformations in surviving GD 12 embryos (Hales and Brown 1991). Interestingly, embryos observed on GD 20 appeared normal when compared to untreated embryos, suggesting reregulation of GSH levels and correction of abnormalities during later developmental stages. Embryos from mice treated with BSO during gestation showed a 60% and 51% decrease in GD 21 embryonic brain and liver GSH levels, respectively, and a 14% decrease in the maternal liver GSH level, with no change in maternal brain GSH levels or Gcl activity in either the embryo or mother (Reyes et al. 1995). Further studies implicating the importance of GSH during development have shown that a decrease in GSH synthesis after BSO exposure during GD 5–16 led to decreased levels of GSH in the adult liver, kidney, and brain and also led to social and cognitive defects (Gorny et al. 2019; Gorny et al. 2020). The Gcl mRNA expression levels in murine spinal cords during GD 11–16 show a gradual increase (Liu et al. 2018). After exposure to atRA, the presence of NTDs at GD 14 significantly increases, with the embryos exhibiting decreased Gcl mRNA expression levels and relative protein levels when compared to untreated embryos, indicating a crucial role of the GSH system during embryogenesis to prevent NTDs.

Glutamyl-cysteine ligase catalytic subunit KO mouse embryos do not develop beyond days 6.5–8.5 of gestation (Dalton et al. 2000; Shi et al. 2000a). Mutant embryos lacked formation of mesoderm and showed an enhanced rate of apoptosis beyond programmed cell death during embryonic remodeling (Shi et al. 2000a). Blastocyst-derived cell lines from Gcl KO GD 3.5 murine embryos can be cultured with N-acetylcysteine (NAC) in GSH deprived media, demonstrating that GSH synthesis is required for development but dispensable in cell culture. Interestingly, Gclc-deficient mice exhibit lowered hepatic GSH levels and no increase of Gcl expression levels (Dalton et al. 2000). In another study, cultured Gclc-deficient mouse lymphocytes showed increased ROS and decreased T-helper cell differentiation following differentiation protocols when compared to wild-type cells, while NAC supplementation lowered ROS levels and restored cell differentiation (Lian et al. 2018).

Gclm KO embryos are viable; however, the GSH levels in the post-natal liver, lung, pancreas, erythrocytes, and plasma are decreased to around 10% of the levels in wild-type mice (Yang, Dieter, et al. 2002). Embryonic fibroblasts from GD 14.5 Gclm KO embryos showed an enhanced sensitivity (about tenfold) to H_2O_2. These data show that although Gclm is not essential for viability, the loss of Gclm lowers GSH levels and increases cellular sensitivity to oxidative stress. Another study observed an increased sensitivity to induced apoptosis in ovaries from both GD 13.5 Gclm KO and Gclm-deficient embryos exposed to benzo[a]pyrene (BaP), a known toxicant, indicating a role for GSH synthesis and antioxidant properties in early ovary development and meiosis (Lim and Luderer 2018).

These studies show the expression of Gclc is developmentally essential for early mouse embryogenesis, with Gclm expression being less important developmentally but equally critical during H_2O_2-induced oxidative stress.

10.6.4 Glutathione Synthetase

Glutathione synthetase (GSS) KO mice have been developed and fail to gastrulate at GD 7.5–8.5, indicating a disruption in differentiation and patterning, verifying that GSH is required in early embryogenesis (Winkler et al. 2011). In GSS-deficient embryos, the mice reached parturition and show no distinct phenotypical differences when comparable to WT mice but exhibited ~50% GSS activity and protein expression. While GSH is required for development, it appears that even decreased GSS activity is sufficient to support normal embryogenesis. While mechanistic details regarding the function of GSS in the GSH cycle and redox regulation has been studied, more research is needed to evaluate the developmental ontogeny of GSS in different animal models, the effects of chemical exposures on the redox regulation in terms of GSS activity, and spatiotemporal observations of GSS during development.

10.6.5 Glutaredoxin

Glutaredoxins (Grx) exist in four isoforms: the dithiol Grxs including Grx1 and Grx2, the principal forms, primarily localized to the cytoplasm and mitochondria, respectively; and the monothiol Grxs, Grx3 and Grx5, localized to the cytoplasm/nucleus and mitochondria, respectively (Hanschmann et al. 2013; Matsui et al. 2020). Glutaredoxin localization in the developing mouse (GD 8.5) appears first in the myocardium and neuroepithelium, but later at GD 16.5, Grx1 expression is distributed throughout the entire mouse embryo (Kobayashi et al. 2000). Glutaredoxin-1 KO mice reach full term and exhibit no observable morphological discrepancies or any greater injury after ischemia or hypoxia insults when compared to wild-type mice (Ho et al. 2007). Mouse embryonic fibroblasts derived from Grx1-deficient embryos at GD 12.5–14.5 showed a greater susceptibility to diquat and paraquat, exhibiting increased cell damage and death when compared to MEFs from wild-type embryos. Glutaredoxin-2 KO mice live to parturition but show dysfunctional cardiac hypertrophy later in life and impaired oxidative phosphorylation in cultured neonatal cardiomyocytes (Kanaan et al. 2018; Mailloux et al. 2014). In contrast to Grx1 and Grx2 KO mice, Grx3 KO mice are embryonic lethal at GD 12.5–14.5 with some embryos showing a smaller body size, hemorrhaging in the head, and growth defects such as open neural tubes and pericardial effusion (Cha et al. 2008; Cheng et al. 2011). Research investigating the role of Grx5 in model organisms has shown oxidative damage and iron accumulation in Grx5 KO yeast and inhibition of iron cluster formation in zebrafish, leading to defective hematopoiesis (Wingert et al. 2005). Although the family of Grx enzymes are purported to have similar functions, it is apparent that each isoform has specific roles in development to regulate critical and complex cellular functions. Alterations of protein S-glutathionylation after GSH becomes oxidized may lead to impaired development. Glutaredoxin-related targets are unknown during development and require further study to understand their role during embryogenesis.

10.6.6 Glutathione S-Transferase

The glutathione S-transferase (GST) gene family comprises 16 genes divided into six subfamilies (alpha, mu, omega, pi, theta, and zeta; Nebert and Vasiliou 2004). Levels of GST appear to be detectable in the mouse embryo as early as GD 9, with an increase in the alpha and pi families and a parallel decrease in the mu family of GSTs from GD 14–18 (Di Ilio et al. 1995). However, in the fetal rat liver, GST levels consistently increased up to birth but experienced a large decrease at parturition, with GST subfamilies undergoing different expressional changes (Tee et al. 1992). A study showing the levels of GST-omega 1 and 2 reported an increase in the developing rat embryo (GD 14–16) when compared to earlier timepoints (GD 11–13; Liu et al. 2018). As with GSR and Gcl, GD 11–16 embryos with induced NTDs after atRA exposure showed a significant decrease of GST 1 and 2 mRNA levels. These studies imply that the inhibitory effect of oxidants can lead to the disruption of biological processes during embryogenesis, specifically the roles of enzymes in the GSH defense system, causing major developmental malformations such as NTDs.

Many GST KO mice have been used in research, yet to our knowledge, the ontogeny has not been thoroughly described during development (Board 2007; Henderson and Wolf 2011). Knockout mice have been shown to have a higher sensitivity to various toxicants in adulthood (Ilic et al. 2010; Henderson and Wolf 2011). Although shown not to be evolutionarily related to the 16 GST genes, microsomal GST1 (MGST1) is an enzyme with GST4-like activity, the ability to reduce lipid hydroperoxides in membranes (Nebert and Vasiliou 2004). A study investigating the effects of mice missing MGST1 showed that its deletion is embryonic lethal at GD 7.5–10 (Brautigam et al. 2018). Zebrafish have been proven to be a very useful vertebrate model to observe the ontogeny of GST subtypes (Yamashita et al. 2018; Rastogi et al. 2019; Timme-Laragy et al. 2013). These studies indicate the importance of genes, such as GST, during development, showing that decreased expression levels lead to increased susceptibility to oxidative stress from toxicant exposure. Hematopoiesis in the early developing zebrafish embryo occurs mainly in the intermediate cell mass, which was revealed to have the highest concentration of MGST1 in wild-type zebrafish during peak MGST1 expression at 24 hpf (Brautigam et al. 2018). Knockdown of MGST1 in zebrafish embryos revealed impaired hematopoiesis. Additionally, MGST knockdown significantly reduced differentiated cells of the hematopoietic system, suggesting an important role during embryonic cardiovascular development. Other studies have examined the expression of many GSTs in human fetal tissues including the liver, lungs, and kidneys, in addition to others, showing differing levels of activities and localizations of various GST subtypes (Strange et al. 1989; McCarver and Hines 2002). The studies mentioned above indicate GST subtypes have overlapping activities and functions during development to provide an adequate, capable detoxifying environment required for proper embryogenesis.

10.6.7 Gamma-Glutamyl Transpeptidase

Gamma-glutamyl transpeptidase-deficient mice mature slowly, showing decreased chondrocyte proliferation, and begin to die early, between 10–18 weeks post-birth (Lieberman et al. 1996; Levasseur et al. 2003; Will et al. 2000). Glutathione content in GGT-deficient 6-week-old mice is increased 600% and 245,000% in the plasma and urine, respectively, due to the lack of extracellular GSH metabolism (Lieberman et al. 1996). Due to the lack of GSH cycling, intracellular Cys levels also decreased by 33%, 90%, and 98% in GGT-deficient mice liver, kidney, and lung, respectively (Rojas et al. 2000). Glutathione also decreased by 75%, 10%, 40%, and 50% in the liver, kidney, lung, and pancreas, respectively (Lieberman et al. 1996, Will et al. 2000; Rojas et al. 2000). Accumulation of DNA damage in GGT-deficient mice liver and kidneys were found due to decreased redox capacity with limited Cys and GSH (Rojas et al. 2000). Gamma-glutamyl transpeptidase can by inhibited pharmacologically by acivicin (Stark, Harris, and Juchau 1987). Embryos treated with acivicin showed decreased embryonic lengths and somite numbers, increased total malformations such as cephalic edema, neural tube necrosis, and microphthalmia, higher occurrences of incomplete axial rotations, and higher mortality rates. Acivicin treatment (10.0 uM) for 24 hours reduced GSH concentrations in embryonic tissues to ~60% of untreated embryos, with no changes in the VYS. Gamma-glutamyl transpeptidase also exhibits different activities within the VYS having higher activity from GD 10–11 compared to the embryo, which may lead to higher GSH synthesis and higher rates of detoxification of drugs and reactive chemicals. While not as robustly examined, these studies demonstrate that proper GGT activity in embryonic development promotes organogenesis and growth, while improper activities may lead to deformities and cellular dysfunction from oxidative stress.

10.6.8 Glucose-6-Phosphate Dehydrogenase

Many redox reactions require NADPH as an electron donor, and while NADPH availability is not solely GSH centric, it is important for maintenance of GSH E_h, specifically through the GSR mediated for conversion of GSSG into GSH. NADPH is a product of the pentose phosphate pathway (PPP). Glucose-6-phosphate, a product of glycolytic metabolism, is shunted into the PPP and converted to 6-phosphate-gluconolactone by the PPP rate-limiting enzyme, the X-linked glucose-6-phosphate dehydrogenase (G6PD). Subsequent metabolism within the PPP converts NADP+ to NADPH. Metabolic products of the PPP, such as glyceraldhyde-3-phosphate and fructose-6-phosphate, can be put back into glycolysis and then can be further metabolized. When elevated, NADPH concentrations promote negative feedback on G6PD, and the PPP is inhibited (Stincone et al. 2015).

In male hemizygous G6PD (−) mouse embryos, development proceeds as normal until approximately GD 7.5, where signs of slowed growth are observed (Longo et al. 2002). In latter stages of G6PD (−) embryonic development (GD 8.5), embryo flexion, somitogenesis, neurogenesis, and cardiogenesis are all abnormally affected. In female heterozygous G6PD (+/−) mouse embryos development progressed further, where on GD 8.5, some were largely indistinguishable from wild-type embryos, but in others, abnormalities of the neuroepithelium and the heart were observed. By GD 11.5, heterozygous G6PD (+/−) mouse embryos displayed massive necrosis in the brain and head and trunk mesenchyme, followed by death on GD 12.5. Clearly, this study shows how critical the PPP is during development.

In regards to G6PD activities during mouse pre-implantation, proportional glucose shunting into the PPP was highest shortly after fertilization (2-cell stage) but declined sharply during the blastocyst stage (O'Fallon and Wright 1986). Similar stage-specific PPP patterns are observed in other species as well but may be due to an overall increase in total glucose turnover (Wales and Du 1993). Activation of the PPP during these stages correlate well with GSH recycling during early preimplantation as the primary means to regulate reduced GSH pools (see above). As GSH *de novo* synthesis increases later in development, requirements for NADPH for GSH recycling may also decrease. However, by post-implantation stages, both GSH *de novo* synthesis and GSH recycling appear to be required to promote normal development.

10.7 REDOX THEORY OF DEVELOPMENT

Development is characterized by change. Cells undergo reprogramming to alter their biochemistry, three-dimensional shape and orientation, and physiological function to support the specificities of development. As such, undifferentiated cells undergo differentiation to establish cellular mechanisms to cope with extrauterine life and promote homeostasis as a multicellular organism, where particular cells perform specialized functions. Undifferentiated cells (i.e., embryonic stem cells) have the plasticity to differentiate into a wide variety of phenotypes to accomplish this purpose.

Reduction-oxidation (redox) potentials are defined as the ratio of interconvertible oxidized and reduced forms of a specific redox couple (GSH and glutathione disulfide [GSSG], Cys and CySS, etc.) and are reported as a voltage, where the more negative the value, the more electrons are available (Schafer and Buettner 1999). Thus, by definition, a more negative E_h represents a greater reducing capacity and the more positive E_h, a more oxidizing capacity. Typical intracellular E_h states, specifically focused on the GSH/GSSG redox couple, are known to correlate closely with cellular processes related to development such as proliferation, differentiation, and apoptosis (Schafer and Buettner 1999).

Original definitions of oxidative stress describe global oxidation of reducing equivalents and, at extreme levels, macromolecule damage and cell death (Sies 1985). Consequently, the classical definition of oxidative stress would support oxidizing environments to cause macromolecule damage, such as DNA oxidation, protein oxidation, and lipid peroxidation, and would lead to cell death. This particular view is likely to occur under extreme conditions of oxidation, but under lesser, non-toxic levels that are more likely to occur during physiological signaling, shifts in specific redox couples may promote responses through redox-sensitive pathways to support cell survival until redox homeostasis can be restored. While there are many redox couples that exist within cells, an increasing number of studies have shown that these couples are not in equilibrium but rather are individually controlled and regulated (Hansen, Jones, and Harris 2020; Jones 2006). Newer paradigms and definitions of oxidative stress address couple disequilibrium as a focal point to rationalize fine-tuned redox-mediated signaling through the regulation of specific cellular pathways *within* specific redox couple control (Jones 2006). Because disequilibrium exists between various thiol redox couples, the individual redox couple control of specific redox-sensitive pathways becomes increasingly complex, dynamic, and specialized. As such, redox disequilibrium is a hallmark for proliferation and differentiation, and logically it would support the rationale that redox regulation is an important regulator of development.

10.8 REDOX REGULATION OF PROTEINS

Regulatory control of proteins can occur through a variety of post-translational modifications (PTMs) of specific protein moieties. These PTMs include phosphorylation, acetylation, glycosylation, and methylation, among others. For the purpose of this chapter, the focus of discussion will be on oxidation/reduction of protein cysteine residues as a form of regulatory control.

Albeit one of the less frequent amino acids in proteins (occurrence in proteins in humans is approximately 2.3%; Miseta and Csutora 2000), Cys are unique due to their -SH group, which can convey their redox sensitivity. Thiols are subject, in some cases, to deprotonation to yield a much more reactive thiolate ($-S^-$). Formation of thiol and thiolates are, in part, a consequence of the thiol pK_a (Poole 2015; Roos, Foloppe, and Messens 2013). Cysteines that exhibit lower pK_a can favor the formation of thiolates and thus can be more reactive. Relative alterations to the cysteine pK_a in proteins can be lowered compared to free cysteine, where the pK_a is approximately 8.5. Often, lowered pK_a can be a consequence of vicinal polar group amino acids and/or a determinant of the protein three-dimensional structure. In general, when pK_a values are below 7, these Cys are usually maintained in a thiolate or more reactive form.

The localized pH can directly affect the redox state of these proteins and can be defined for specific subcellular/organellar locations (Poole 2015). For example, human Grx1, which is primarily cytosolic/nuclear (Fernando et al. 1994; Sahlin et al. 2000), has an approximate pK_a of 3.5 (Mieyal et al. 1991), whereas human Grx2, which is localized to both the cytosol and mitochondrial matrix (Hudemann et al. 2009; Lundberg et al. 2001), has a pK$_a$ of 4.8 (Gallogly et al. 2008). At the approximated cytosolic pH of 7.4, Grx1 is more active than Grx2, as the redox sensitive Cys (Cys^{22}) would more readily favor a thiolate form, a difference that, in part, explains biochemical activity measurements where Grx2 activity is much lower, constituting only 10% of the activity measured in Grx1 in the cytosol (Gallogly et al. 2008). However, in cells, mitochondrial Grx2 activity is likely increased compared to cytosolic Grx2 as the mitochondrial matrix pH is considerably higher, approaching a pH of 8.0 (Abad et al. 2004). Thus, mitochondrial Grx2 is likely to have

a higher activity compared to cytosolic Grx2 as the increase in pH drives an increase in the formation of a Cys^{22} thiolate. Accordingly, Grx2 provides an example where reactivity is based on both pK$_a$ and pH and is contingent upon cellular localization. Regardless of compartmentation, regulatory parameters of redox sensitivity, namely pH and pK$_a$, are conveyed through Cys residue modification under physiological conditions and can act as primary drivers (either deactivated or activated) of their function. Under periods of oxidative stress where ROS (i.e., H$_2$O$_2$) can directly modify proteins, susceptible protein thiols (PrSH) can be oxidized to yield a protein sulfenic acid (PrSOH) (Figure 10.3).

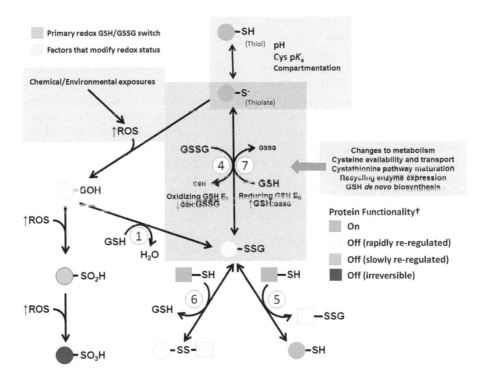

Figure 10.3 Glutathione regulation of protein redox switches.

Thiol residues are susceptible to oxidation/reduction. Glutathione switch reactions (highlighted in blue) can occur where the GSH E$_h$ can regulate the post-translational modification of target proteins. A more oxidizing GSH E$_h$ promotes S-glutathionylation of susceptible thiols (see reaction 10.4), but in a more reducing GSH E$_h$ environment, proteins are de-S-glutathionylated (see reaction 10.7). Proteins oxidized to a sulfenic acid (-SOH) from ROS (i.e., H$_2$O$_2$) can interact with GSH to yield an S-glutathionylated protein (PrSSG; see reaction 10.1) and prevent overoxidation to either sulfinic (-SO$_2$H) or sulfonic (-SO$_3$H) acids. S-glutathionylated proteins can interact with other proteins to promote disulfide switching of the glutathione moiety to other proteins as a post-translational modification (see reaction 10.5) or to liberate the glutathione moiety to free GSH and result in a protein disulfide formation (-SS-; see reaction 10.6). Thiol redox states dictate protein function. Note reactions 10.2 and 10.3 are not shown here.

†While many proteins are deactivated through S-glutathionylation (as denoted here), there are some instances where S-glutathionylation results in activation of proteins. Pink boxes denote factors that can contribute and control protein redox states.

Specific to the GSH redox couple, redox-sensitive proteins can be S-glutathionylated as a type of dynamic, regulatory PTM. For example, PrSOH can be S-glutathionylated through a reaction with reduced GSH to yield H_2O and the modified target protein (PrSSG; see reaction 10.1), an oxidative modification that can change or disrupt protein function. PrSSG can also be formed through a one-electron oxidation with either a protein thiyl radical (PrS$^{\cdot}$) or a GSH radical (GS$^{\cdot}$) (see reactions 10.2 and 10.3), giving superoxide anion ($O_2^{\cdot-}$) as a by-product. During periods of high GSSG concentrations, indicative of oxidative stress, disulfide exchange reactions can occur where GSSG can oxidize PrSH to PrSSG and yield GSH (see Reaction 10.4). Similarly, disulfide exchange reactions can occur between PrSSG and reduced proteins as well, where S-glutathionylated proteins can interact with other proteins to pass on the S-glutathionylation modification (see reaction 10.5) or a disulfide between two proteins to give reduced GSH (see reaction 10.6). The S-glutathionylation adduct can be effectively removed by reduced GSH to give PrSH and GSSG (see reaction 10.7). Restoration of GSH pools can occur when GSSG is effectively reduced to GSH through glutathione disulfide reductase using NADPH as a cofactor.

Reaction 10.1: PrSOH + GSH $\rightarrow$ PrSSG + H_2O

Reaction 10.2: PrS$^{\cdot}$ + GS- + O_2 $\rightarrow$ PrSSG + $O_2^{\cdot-}$

Reaction 10.3: PrS$^-$ + GS$^{\cdot}$ + O_2 $\rightarrow$ PrSSG + $O_2^{\cdot-}$

Reaction 10.4: PrSH + GSSG $\rightarrow$ PrSSG + GSH

Reaction 10.5: PrSH + 'PrSSG $\rightarrow$ PrSSG + 'PrSH

Reaction 10.6: PrSH + 'PrSSG $\rightarrow$ PrSS'Pr + GSH

Reaction 10.7: PrSSG + GSH $\rightarrow$ PrSH + GSSG

An excellent example of a protein affected by S-glutathionylation is protein kinase A (PKA). PKA is primarily activated via increasing concentrations of cyclic AMP (cAMP) and is involved in both a variety of homeostatic responses of the autonomic nervous system and the regulation of numerous cellular functions. Interestingly, under cAMP-activated conditions, the catalytic subunit of PKA is susceptible to S-glutathionylation at Cys[199] and is inactivated (Humphries, Deal, and Taylor 2005; Humphries, Juliano, and Taylor 2002). During chick limb development, PKA activities have been shown involvement in events implicated in limb tissue differentiation (Smales and Biddulph 1985), and in mice, PKA deficiency causes region-specific neural tube defects (Huang, Roelink, and McKnight 2002). Fine-tuning the activity of PKA provides one instance of the importance of GSH availability, redox regulation, and S-glutathionylation in controlling cellular functions and activities during development.

While there are few studies to implicate fluctuating PrSSG levels during development, there are some seminal studies that have modified Grx2 expression as a means to better understand how S-glutathionylation may be connected to proper embryogenesis. In a zebrafish model, morpholino-generated Grx2 KOs demonstrated the loss of neurogenesis, highlighted by an increased level of apoptosis and dysmorphogenesis (Brautigam et al. 2011). In follow-up studies from the same group, zebrafish lacking Grx2 displayed poor cardiogenesis, where proper heart looping failed, which was attributed to cell death and/or disruption of migration of neural crest cells (NCC) to the primary heart field (Berndt et al. 2014). Remarkably, in both studies, dual injection of Grx2 capped mRNA with Grx2 morpholinos to restore Grx2 expression rescued zebrafish neural and heart morphology. Both of these studies highlight the critical nature of the regulation of S-glutathionylation during development.

10.9 GLUTATHIONE-MEDIATED REGULATION OF DEVELOPMENT

10.9.1 Proliferation and Cell Cycle

Proper embryonic development relies not only on temporal and spatial expression of specific genes and protein concentrations, but it also relies heavily on the cellular redox status. Cellular proliferation is regulated by shifts in redox state, where reducing redox states support proliferative activities (Figure 10.4). Oxidant exposure can affect the GSH redox state and alter the cell cycle (dos Santos et al. 2015). Fluctuations of GSH concentrations, ROS availability, and redox states fine-tune phase-specific control of the cell cycle. Untimely onset of oxidative stress can disrupt proliferative control, where proliferation has been shown to be inhibited when the oxidation of intracellular GSH and/or the production of ROS increase (Kviecinski et al. 2012; Panieri, Millia, and Santoro 2017; Hatori et al. 2020).

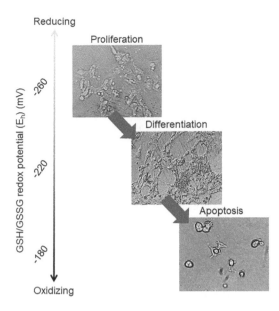

Figure 10.4 Glutathione redox states shift during specific cellular states.

In general, reducing GSH E_h environments support proliferation, but as the GSH E_h shifts towards a more oxidizing state, cells favor differentiation. These shifts are noted in a variety of *in vitro* models of cellular differentiation for many different cell types (see Table 10.1). Shifting of the GSH E_h to very oxidizing conditions can promote apoptosis.

The nucleus changes dramatically during the cell cycle. It has been shown that GSH concentrations in the nucleus increase prior to the start of cell proliferation (Markovic et al. 2009). Glutathione is recruited into the nucleus of plant and animal cells during early cell proliferation, suggesting GSH plays a critical role in the cell cycle and its regulation and may possibly be a key factor for proper DNA replication and gene expression (Vivancos et al. 2010; Pallardo et al. 2009; Markovic et al. 2007). In fibroblasts treated with DEM, cellular GSH levels were depleted, including in the nuclear and cytosolic compartments. Cellular depletion resulted in the inhibition of cellular proliferation. In fibroblasts treated with BSO, GSH depletion in the cytosol was observed but not in the nucleus. BSO treated fibroblast proliferation was unaffected (Markovic et al. 2009). This study highlights the importance of the nuclear GSH pool as a regulator of cellular proliferation and provides rationale to the dynamic, compartmentalized redox shifts and regulation for cell division.

The cellular GSH redox profile during mitosis is crucial for proper proliferation. In an early study, fibroblasts showed a more reduced GSH E_h during active proliferation, but upon confluency and the cessation of proliferation due to contact inhibition, cells shifted towards a more oxidized GSH E_h (Hutter, BG, and Green 1997). In comparison, the GSH E_h of fibrosarcoma cells, which do not exhibit contact inhibition and show uninhibited proliferation, remained reduced as confluency increased. These data show that in normal cells, proliferation is directly linked to a more reducing GSH redox state. Other cell lines have shown similar mechanisms of redox control, including rapidly proliferating colon adenocarcinoma cells, human myeloid leukemia (HL-60) cells, and hybridoma T-cells (Kirlin et al. 1999; Cai and Jones 1998; Jones et al. 1995). All together, these data provide ample evidence that a more reducing GSH redox state promotes cellular proliferation.

These redox changes during mitosis are similar during meiosis. For example, in human chorionic gonadotropin–induced hamster ovulation, oocytes showed a significant increase of GSH concentrations (~70% increase at 16 hours post-ovulation) during meiotic maturation, specifically metaphase I and II, which was associated with nuclear progression (condensation of meiotic chromosomes; Zuelke et al. 2003). Regulation of proper GSH levels is important as improper GSH E_h in pre-implantation embryos has been shown to promote irregular development (Gardiner and Reed 1994; Zuelke et al. 2003). In meiotic metaphase, high GSH levels are believed to support spindle growth and chromatin remodeling, but in pro-nuclear phases GSH levels begin to decrease. Upon fertilization, GSH levels are substantially decreased to less than ~10% of the original concentrations observed in oocytes (Zuelke et al. 2003; Gardiner and Reed 1994). The investigation of GSH concentrations and GSH E_h in cellular compartments during meiosis has provided evidence for regulatory mechanisms of GSH in response to oxidative stress during developmental stages.

As part of the progression through the cell cycle and cellular proliferation, regulation of reactive cysteine residues is a mean by which GSH E_h can control cellular function. Transcription factors, kinases, and elements of the cytoskeleton can ensure correct cell proliferative activities associated with development (Chiu and Dawes 2012; Haddad 2011; Tanaka, Honda, and Takabatake 2001).

10.9.2 Transcription Factor Regulation

The AP-1 transcription factor is made up of Jun and Fos proteins and controls the expression levels of

TABLE 10.1
Redox State Shifts during Cellular Differentiation[a]

Phenotype	Cell type	Undifferentiated cells	Differentiated cells	Ref.
Neuron	Hues9 hESC	↑ GSH/GSSG Ratio	↓ GSH/GSSG Ratio	(Yanes et al. 2010)
	P19	↑ free thiols [-SH]	↓ free thiols [-SH]	(Konopka et al. 2008)
	P19	E_h: −238mV	E_h: −218mV	(Ross et al. 2019)
	Primary chick neuron	↑ GSH/GSSG Ratio at HH 20	↓ GSH/GSSG Ratio at HH 23	(Hatori et al. 2016)
	SH-SY5Y	↓ GSH/GSSG Ratio	↑ GSH/GSSG Ratio	(Hatori et al. 2016)
	IMR-32	↑ GSH/GSSG Ratio	↓ GSH/GSSG Ratio	(Erlejman and Oteiza 2002)
	NSC	↑ GSH/GSSG Ratio	↓ GSH/GSSG Ratio	(Trivedi et al. 2016)
Muscle	C2C12	E_h: −248mV	E_h: −230mV	(Hansen et al. 2007)
	C2C12	E_h: −253mV	E_h: −223mV	(Catani et al. 2004)
	L6C5	E_h: −241mV	E_h: −222mV	(Catani et al. 2004)
	H9c2	E_h: −245mV	E_h: −273mV	(Ding et al. 2008)
Fat	hMSC	E_h: −259mV	E_h: −219mV	(Imhoff and Hansen 2011)
	3T3-L1	↑ GSH/GSSG Ratio	↓ GSH/GSSG Ratio	(Vigilanza et al. 2011)
	3T3-L1	↑ GSH/GSSG Ratio	↓ GSH/GSSG Ratio	(Takahashi and Zeydel 1982)
	3T3-L1	E_h: −238mV	E_h: −216mV	(Imhoff and Hansen 2010)
Bone	hMSC	E_h: −259mV	E_h: −226mV	(Imhoff and Hansen 2011)
	SaOS-2	↑ GSH/GSSG Ratio	↓ GSH/GSSG Ratio	(Romagnoli et al. 2013)
	RAW264.7	↑ GSH/GSSG Ratio	↓ GSH/GSSG Ratio	(Huh et al. 2006)
	BMM	↑ GSH/GSSG Ratio	↓ GSH/GSSG Ratio	(Huh et al. 2006)
Intestine	Caco-2	E_h: −230mV	E_h: −190mV	(Nkabyo et al. 2002)
	HT29	E_h: −258mV	E_h: −198mV	(Kirlin et al. 1999)
Macrophage	BMM	↑ GSH/GSSG Ratio	↓ GSH/GSSG Ratio	(Kim et al. 2004)

[a] Redox shift comparisons between undifferentiated and differentiated cells are shown as changes of E_h (more positive value = more oxidizing environment; more negative value = more reducing environment), free thiol [-SH] groups, or GSH/GSSG ratio (lower ratio = more oxidizing environment; higher ratio = more reducing environment).

E_H: glutathione redox potential in millivolts (mV); GSH: glutathione; GSSG: glutathione disulfide; hESC: human embryonic stem cell; HH: Hamburger-Hamilton stage; NSC: neural stem cell; hMSC: human mesenchymal stem cell; BMM: bone marrow–derived monocyte.

various genes involved in proliferation. It has been known that transcriptional activity of AP-1 is redox-regulated via post-translational modification of a conserved Cys residue in the DNA binding domains of Fos and Jun (Abate et al. 1990). Under normal conditions, ERK1/2 phosphorylate Jun and Fos proteins, thus activating AP-1. Upon activation, AP-1 promotes the transcriptional regulation of cyclin D1, where increased levels of cyclin D1 support cell division (Shaulian and Karin 2001). In cells undergoing oxidative stress, Jun is S-glutathionylated in the DNA-binding site at Cys[269], inhibiting AP-1 activity and causing a subsequent decrease in cyclin D1 gene expression (Tanaka, Honda, and Takabatake 2001; Klatt, Molina, and Lamas 1999; Bakiri et al. 2000). The introduction of a reducing agent, like dithiothreitol (DTT), to restore the redox potential of the cell can reactivate redox-dependent cell proliferative pathways, including that of AP-1 (Klatt et al. 1999; Abate et al. 1990). Redox regulation of cellular proliferation can also be disrupted when Cys[269] is replaced with a serine showing that growth is unaffected by oxidative stress conditions, exhibiting no S-glutathionylation in the DNA-binding region (Klatt et al. 1999; Okuno et al. 1993). Additionally, the exposure of GD 8 mouse embryos to embryotoxic levels of BSO (1 mM) induced prolonged AP-1 DNA binding activity and exhibited an 86% incidence of swollen hindbrains, suggesting improper redox regulation of AP-1 can result in various embryonic malformations (Ozolins et al. 2002).

The transcription factor, nuclear factor κβ (NF-κB) controls the expression of many genes involved in proliferation. S-Glutathionylation can occur on critical DNA binding sites in specific NF-κB subunits, specifically the Cys[62] on the p50 subunit, which must remain in a reduced state for DNA binding to occur (Pineda-Molina et al. 2001). The S-glutathionylation of or sulfenic acid formation on p50 can result in the inhibition of NF-κB activity and downstream transcription of cell proliferation genes (Morgan and Liu 2011; Pineda-Molina et al. 2001). Mutated Cys[62] to a serine shows less DNA binding indicating the redox regulatory mechanism for NF-κB activity (Pineda-Molina et al. 2001). More examples of transcription factors effected by GSH E_h include STAT3 and ER-DBD, which also show a molecular basis connecting oxidative conditions to gene expression during cell division and other cellular processes (Xie et al. 2009; Kushner et al. 2000).

10.9.3 Kinases/Phosphatases

Regulation of kinase activities upstream of transcription factors by oxidants, like H_2O_2, also plays a role in cellular proliferation. Oxidative stress can regulate kinase activity through post-transcription modification of Cys residues in the ATP-binding region. For example, the human MAP kinase kinase, MEKK1, in the ERK pathway can be S-glutathionylated at Cys[1238], prohibiting ATP-binding and subsequent phosphorylation of target substrates, while the mutation of Cys[1238] to valine desensitizes the redox regulation of MEKK1 (Cross and Templeton 2004). Further research shows that S-glutathionylation of redox-sensitive Cys residues in kinases and phosphatases such as PKA, IKKβ, PTP1B, and PKCα, which can result in the inhibition of cellular proliferation (Dominko and Dikic 2018; Reynaert et al. 2006; Humphries, Juliano, and Taylor 2002; Salmeen et al. 2003; Mahadev et al. 2001; Mondol, Tonks, and Kamata 2014; Barrett et al. 1999). In unstimulated cells, NF-κB remains in the cytosol, inhibited by inhibitory κB proteins (IκBs). The phosphorylation of IκBs, and subsequent degradation by the proteasome, thus releasing NF-κB to be translocated to the nucleus, occurs by IκB kinase (IKK), which consists of two subunits, α and β. Negative regulation of IKKβ through modification of Cys[179], such as S-glutathionylation after H_2O_2 exposure, results in inhibited activation of NF-κB (Reynaert et al. 2006). Other oxidative modifications such as disulfide formation in Src, Cdc25C, PTEN, SHP1, SHP2, AKT2, and FGFR1, and sulfenic acid formation in PI3K and AKT1, also alter cell proliferation—all of which can result in the inhibition of cellular proliferation (Kemble and Sun 2009; Leonard, Reddie, and Carroll 2009; Antico Arciuch et al. 2009; Huang et al. 2003; Savitsky and Finkel 2002; Reddy et al. 2008; Kwon et al. 2004; Chen, Willard, and Rudolph 2009). Because of these complex activities, there is a need for GSH redox control during proliferation to promote the correct signaling. Upon restoration of the GSH E_h, the regulation of proliferative processes is corrected, demonstrating redox active switches in related underlying events (Haddad 2011).

10.9.4 Cytoskeleton

Many studies have shown that GSH and ROS levels have an impact on proteins that interact with cytoskeleton rearrangement, a process involved in cell adhesion, spreading, proliferation, differentiation, and migration (Abdelsaid and El-Remessy 2012; Chiarugi et al. 2001). Protein tyrosine phosphatases (PTPs) inhibit kinase substrates through dephosphorylation and disrupt downstream signaling including cell cycle progression, transcriptional regulation, cell growth, differentiation, and apoptosis (Chiarugi et al. 2004). Low molecular weight PTPs (LMW-PTPs) are redox-sensitive proteins involved in cytoskeleton structure rearrangement through the ability to bind and dephosphorylate focal adhesion kinase (FAK), where phosphorylation of FAK promotes activity. Two redox-regulated Cys residues (Cys[15] and Cys[17]) in the catalytic pocket of LMW-PTP must be reduced for it to dephosphorylate FAK, rendering FAK inactive (Chiarugi et al. 2001). Therefore, oxidation or reduction of the molecular switch LMW-PTP can lead to downstream dynamic cytoskeletal changes, including cell migration. Reduced GSH E_h promotes LMW-PTP expression and activity in osteoblastic cell differentiation, which requires cell-cell contact through movement made by changes in the cytoskeleton, while an oxidized GSH E_h promoted decreased LMW-PTP expression and activity, showing a redox regulation of cytoskeletal components as well as differentiation (de Souza Malaspina et al. 2009). In another study, the addition of H_2O_2 to murine fibroblast (NIH-3T3) cells inhibited the activity of LMW-PTPs, with activity returning after H_2O_2 removal, showing redox-regulated function

(Chiarugi et al. 2001; Chiarugi et al. 2003). A co-treatment of DEM or BSO and H_2O_2 revealed no return of activity, indicating GSH E_h regulation plays a key role in LMW-PTPs function.

During oxidative shifts inside the cell, the abundant S-glutathionylation of actin has been shown to impact polymerization resulting in dynamic cytoskeletal changes both in vivo and in vitro (Wang et al. 2001; Fiaschi et al. 2006; West et al. 2006; Rokutan, Johnston Jr, and Kawai 1994; Chai et al. 1994). Modifications to actin can occur with the monomeric G-actin or the polymerized form, F-actin, after changes in the redox environment (Dalle-Donne, Milzani, and Colombo 1999; Dalle-Donne, Milzani, and Colombo 1995). Polymerization of S-glutathionylated actin monomers is slower (decreased by 33%) and less efficient than the polymerization of non-S-glutathionylated actin (Dalle-Donne et al. 2003). Upon incubation with DTT, the capability of actin to polymerize is completely restored within 30 minutes. Another study showed that slower polymerization due to S-glutathionylation of actin filaments led to depressed force development in human cardiac myofibrils, suggesting a redox control mechanism in muscle functionality (Passarelli et al. 2010). The decline of force and contractility was also seen in vivo using trabeculae in the heart of a rat model of ischemia-reperfusion injury (Chen and Ogut 2006). The maximum force fell by 71% during ischemia as the G-actin showed S-glutathionylation and decreased polymerization rates and efficiency compared to native G-actin.

Actin contains six highly conserved Cys residues, however only Cys[374] is the only exposed residue and is the residue that undergoes S-glutathionylation (Stournaras et al. 1990; Dalle-Donne et al. 2003; Wang et al. 2001; Wang et al. 2003; Rokutan, Johnston Jr, and Kawai 1994). The S-glutathionylation of Cys[374] on G-actin in A431 (human epidermal) cells after incubation with H_2O_2 is reversible, with epithelial growth factor (EGF) promoting an increased rate of polymerization to form F-actin (Wang et al. 2001). The involvement of Grx during deglutathionylation was investigated in the same study by preincubating the cells with Cd^{2+}, a known Grx and deglutathionylation inhibitor, where it was revealed that EGF was unable to promote deglutathionylation, indicating a specific mechanism of actin polymerization inhibition or stimulation (Wang et al. 2001; Wang et al. 2003). In another study investigating

actomyosin NIH-3T3 cells, depletion of GSH with BSO significantly impaired actomyosin contractility by promoting disassembly leading to cell dysmorphology, through cell spreading, and cytoskeleton disorganization, both of which were reversed with co-treatment of GSH (Fiaschi et al. 2006). The mutation of Cys[374] to a serine residue showed impeded S-glutathionylation, thus verifying Cys[374] as a redox-regulated GSH target in actin. Further substitutions of Cys[374] with negative amino acids, such as aspartate and glutamate, revealed severe disorganization and higher sensitivity to destabilization upon chemical treatment when compared to wild-type or neutral amino acid substitution, like alanine (Tsapara et al. 1999). The intracellular GSH E_h tightly regulates post-translational modifications of actin leading to impaired cytoskeletal functions that can possibly contribute to the progression of neurodegeneration or loss of muscle function in different diseases, demonstrating the importance of redox-controlled mechanisms in cellular function (Pastore, Tozzi, et al. 2003; Passarelli et al. 2010).

Significant changes were observed in the accessible thiols of proteins associated with cytoskeleton dynamics after ethanol exposure in rat embryos (Jilek et al. 2015). For example, α-actin-4 and profilin showed a decreased abundance after ethanol treatment. Changes to cytoskeleton associated proteins were observed with GSH oxidation and more positive oxidative redox states in the VYS of GD 10–11 rat embryos. The dysregulation of these proteins during development may lead to abnormal structural features and improper cellular function, showing the importance of the GSH E_h and protein modifications, like glutathionylation, due to ROS. Additionally, cytoskeleton defects were presented in zebrafish with altered ROS production during normal development, which correlates with GSH patterns during embryo development, suggesting a possible role of GSH and redox balance to promote cytoskeleton formation (Mendieta-Serrano et al. 2018; Timme-Laragy et al. 2013). Together, these studies display a vital role in the GSH redox state to regulate and promote proper cytoskeleton activity.

10.9.5 Differentiation

Cellular differentiation is required to prepare the embryo for extrauterine life. Before cell fate is determined, an early embryo consists of undifferentiated cells with the potential to differentiate into a range of cell types at specific locations and developmental periods. A major deciding factor of proper cellular differentiation is the intracellular GSH redox state, which can influence epigenetics, metabolism, and various cell signaling pathways. Inherent shifts towards a more oxidative environment support chondrogenic and neuronal differentiation (Kim et al. 2010, Pashkovskaia, Gey, and Rodel 2018). Under hyper-oxidative conditions, differentiation of p19 cells, neuronal precursors, is inhibited, suggesting over-oxidizing cellular environments prevents neurogenic developmental pathways (Pashkovskaia, Gey, and Rodel 2018). Treatment with electron transport chain inhibitors or oxidants, such as paraquat, during p19 neuronal differentiation, decreased the expression of β-III-tubulin, a marker of neuronal differentiation. Together, these data show that redox shifts are important and regulated but are highly susceptible to perturbation.

The specific GSH redox shifts during differentiation are critical to support different cell phenotypes. The GSH redox status changes during osteogenic and adipogenic differentiation, providing details into the role of the GSH redox state during differentiation (Imhoff and Hansen 2011). Human mesenchymal stem cells (hMSCs) can differentiate into many cell types including adipocytes and osteocytes over a 21-day period. Research in this area showed that terminally differentiated cell phenotypes showed a significant shift towards a more oxidizing GSH E_h from undifferentiated cells (undifferentiated, −259mV; differentiated adipocyte, −219 mV; differentiated osteocyte, −226 mV; Imhoff and Hansen 2011). However, the rates at which the GSH E_h shifted during hMSC differentiation were significantly different, where adipocytes demonstrated a rapid GSH oxidation within three days, whereas cells becoming osteocytes did not show an oxidizing shift until day 10. These data suggest the timing of GSH oxidation helps promote a specific terminal differentiation phenotype. Human colon cancer cells treated with sodium butyrate showed a similar oxidizing shift in GSH E_h (+60 mV shift) during differentiation, whereas non-differentiating, proliferating cells showed a constant reduced GSH E_h (Kirlin et al. 1999). Mouse bone marrow–derived monocytes were exposed to macrophage colony-stimulating factor to differentiate cells into macrophages

(Kim et al. 2004). As cells differentiated, the GSH E_h became significantly more oxidized and remained oxidized over the course of three days. Differentiation of embryonic stem cells showed changes in GSH E_h as expressed in a lowered GSH/GSSG ratio, more specifically, a significant oxidation during differentiation and a more reducing environment prior to differentiation and post-terminal differentiation (Yanes et al. 2010). These examples, and others showing similar shifts in GSH E_h (see Table 10.1), imply that differentiation and developmental mechanisms require an oxidizing environment and regulate terminal cell phenotypes.

The data shared above with regard to redox regulation, and more specifically GSH, conclude that embryo development, on an organismal scale and on a cellular, mechanistic level, responds to changes in the redox state and has major implications on growth, gene expression, protein function, and overall cell identity.

10.9.6 Apoptosis

Teratogens are known to cause shifts in the GSH E_h in both in vivo and in vitro models, resulting developmental abnormalities such as NTDs, decreased proliferation, and increased apoptosis rates (Tung and Winn 2011, Cai and Jones 1998; Chen, Wang, et al. 2020; Hansen, Zhang, and Jones 2006). Cysteinyl aspartate–specific proteases (caspases) are responsible for cellular disassembly into apoptotic bodies during apoptosis and have been implicated as targets of S-glutathionylation (Huang et al. 2008). Caspases require a reduced Cys in the active site for proper cleavage and downstream activity (Pan and Berk 2007; Nicholson et al. 1995; Borutaite and Brown 2001). The incubation of HL-60 cells with actinomycin D (an apoptosis inducer) showed high caspase-3 activity after 6 hours (Huang et al. 2008). The addition of GSSG decreased caspase-3 activity by 80% within 2 hours, by means of S-glutathionylation. Interestingly, both caspase-3 and procaspase-3 were S-glutathionylated causing an inhibition of being proteolytically activated by cytochrome c. Furthermore, the inhibition of caspase-3 activity by GSSG was reversed by GSH, DTT, and Cys with capase-3 activity reaching ~80%–100% pre-GSSG treatment. The deglutathionylation of caspase-3 by Grx positively regulates apoptosis (Pan and Berk 2007). These results indicate that the S-glutathionylation of key proteins, such as caspase-3, in important cell processes, like apoptosis, is a molecular switch where cells respond to oxidative stress.

Caspases are responsible for the proteolysis of multiple cellular proteins, including components of cellular defense systems like the GSH synthesis enzyme Gcl, facilitating the cell death process (Franklin et al. 2002). Both HeLa and Jurkat cells showed direct cleavage of Gclc, not Gclm, by caspase-3 and depletion of GSH during apoptosis after caspase-3 was activated by tumor-necrosis factor, cycloheximide, staurosporine, or ultraviolet irradiation. This study revealed a post-translational modification that plays a role in regulating GSH synthesis and redox state during apoptosis. Although Gclm was shown to not be directly cleaved by caspase-3, one study has shown that ovaries from Gclm KO mouse embryos (GD 13.5) exhibited decreased GSH levels and increased BaP-induced apoptosis sensitivity through the activation of caspase-3 when compared to wild-type embryonic ovaries.

HL-60 cells were shown to have a significant oxidizing GSH E_h shift (+60mV over 6 hours) while undergoing apoptosis when exposed to caspase pathway–inducing staurosporine, providing evidence that an oxidizing GSH E_h can lead to the activation of apoptotic pathways and irregular cellular function (Cai and Jones 1998). Another study used zebrafish as an in vivo model to investigate the exposure of nodularin, a cyanobacteria hepatotoxin, and saxitoxin, a neurotoxin, both of which cause ROS (Chen, Jia, et al. 2020; Chen, Wang, et al. 2020). Drug administration led to decreased GSH production (74%), higher pro-apoptotic gene expression, such as BAX and caspase-3, and an increase in DNA damage, all of which were reversed when drugs were co-administered with NAC, implicating the role of GSH in apoptosis and cell survival.

Together these data show that redox potential shifts and endogenous antioxidant systems are important regulators of post-translational modifications in proteins that control cell death pathways in organisms and cells. Many other studies have investigated oxidants, thiol redox switches, ROS, cellular redox systems, specifically involving the GSH/GSSG redox couple, and apoptosis and have been reviewed elsewhere (Circu and Aw 2010; Benhar 2020; Circu and Aw 2008).

10.9.7 Cell Migration

Normal embryogenesis requires cell migration. This recurring developmental event, migration, involves many redox-sensitive cellular changes, including the regulation of cytoskeleton dynamics, the expression of adhesion molecules, and maintenance of signaling pathways to form protrusions used in migration (Arseneault et al. 2013; Chiarugi et al. 2004; Fiaschi et al. 2006). Early investigations in mice provided evidence of redox control during early, crucial developmental process such as cranial NCC migration (Suzuki, Svensson, and Eriksson 1996). Cranial NCCs are precursors for various facial bone structures. The disruption of proper migration has been implicated in underdeveloped mandibles of embryos of diabetic mothers, suggesting high extracellular glucose may be teratogenic (Eriksson 1988; Styrud and Eriksson 1990). Neural crest cells from diabetic embryos (GD 9) in culture showed decreased migration in high glucose conditions compared to normal cells (Suzuki, Svensson, and Eriksson 1996). The high glucose–inhibition of migration was attenuated in the presence of NAC, whereas the addition of superoxide dismutase was not, supporting redox-regulated migration is essential for proper development.

Vascular endothelial growth factor (VEGF) has been shown to cause an oxidizing GSH E_h in human microvascular endothelial (HME) cells (Abdelsaid and El-Remessy 2012). Further investigation revealed that VEGF causes S-glutathionylation and inhibition of LMW-PTP within 30 minutes, stimulating FAK activation and downstream HME cell migration. The LMW-PTP inhibition was reversed after NAC pre-treatment, resulting in a reduced GSH E_h and increased LMW-PTP/FAK interactions. Other studies have also shown that ROS generation in cells also led to the inactivation of LMW-PTPs, likely due to redox changes (Swanson et al. 2011). Redox control of migration through shifts in GSH E_h has also been observed in various cell types (Kviecinski et al. 2012; DeNunzio and Gomez 2014; Mondol, Tonks, and Kamata 2014; Franca et al. 2020; Arseneault et al. 2013). Continued research is needed, as precise mechanisms and proteins targets involved, in addition to important information regarding the GSH E_h, remain to be elucidated.

10.10 GLUTATHIONE DURING TERATOGENESIS

Disturbances in GSH homeostasis are implicated in the etiology and progression of several developmental disorders and are likely a consequence of interferences with important cellular processes such as cell differentiation, proliferation, and apoptosis (see above). There are numerous pharmacological agents that can promote ROS generation, causing a disruption in cellular redox states of an organism leading to improper embryogenesis. Through a chemically induced dysregulation of the GSH system, there are multiple poor developmental outcomes, including neurobehavioral deficits, dysmorphogenesis, and spontaneous abortion.

Initial investigations into understanding GSH function during development focused on GSH protective properties against xenobiotic insults (Lim et al. 2020; Harris, Dixon, and Hansen 2004; Harris, Fantel, and Juchau 1986; Harris et al. 1995; Yan and Hales 2006). However, experiments using whole embryo culture expose conceptuses to environmental pollutants, therapeutic drugs, natural toxins, and other insults and have shown the importance of GSH in the embryo as an antioxidant and protectant. A study treated rat embryos with buthionine-(S,R)-sulfoximine (4 or 8 mmol/kg) on GD 10 to deplete the GSH concentration; the higher dose resulted in the death of 13.2% of total implanted embryos and 21.7% of the surviving embryos being malformed (Hales and Brown 1991). The data demonstrate the crucial protective and developmental role of GSH during development and provides evidence of potentially fatal results via chemically induced disruption of the GSH redox system. Known developmental toxins are acetaminophen (Stark, Harris, and Juchau 1989); valproic acid (Seegmiller et al. 1991; Tung and Winn 2011); retinoic acid (Chen and Juchau 1998; Seegmiller et al. 1991); thalidomide (Hansen and Harris 2013, 2004); methanol (Harris, Dixon, and Hansen 2004); ethanol (Brocardo et al. 2017); lindane (McNutt and Harris 1994); phenytoin (Azarbayjani, Borg, and Danielsson 2006; Parman, Chen, and Wells 1998); MeHg (Robitaille, Mailloux, and Chan 2016); ceramide (Ross et al. 2019), and several other agents. These chemicals can induce oxidative stress to disrupt normal developmental programing, targeting, DNA, lipids, and proteins. Below, we review thalidomide, MeHg, and ethanol in the context of alterations to the GSH system.

10.10.1 Thalidomide

Thalidomide was a popular sedative/antinausea agent introduced in the late 1950s (Vargesson 2015). Because of its supposed ability to treat nausea, it was often prescribed to pregnant women and thought to have little to no side effects. Four years after being initially introduced to the public, thalidomide was discovered to be a source of serious birth defects, which were primarily manifested as a malformation of limbs, termed phocomelia. Because it was so widely prescribed before discovering the teratogenic side effects, an estimated 100,000 babies were born with a form of phocomelia worldwide. The failure to screen out thalidomide as a teratogen is mostly attributed to the thalidomide-resistant rodent models on which it was originally tested (Kim and Scialli 2011).

A leading hypothesis behind the dysmorphogenesis caused by thalidomide is attributed to oxidative stress-induced alteration of the GSH redox state and subsequent disruption of important transcription factors involved in limb development. The transcription factor NF-κB is involved in cellular proliferation, differentiation, and apoptosis and is regulated by the GSH redox state (Hansen and Harris 2004; Knobloch et al. 2008). Misregulation of NF-κB can alter the expression of critical developmental genes within the limb bud, such as Twist and Fgf-10 in the progress zone (PZ) mesenchyme, Fgf-8 in the apical ectodermal ridge (AER), and Shh in the zone of polarizing activity (ZPA; Bushdid et al. 1998; Kanegae et al. 1998). The activity level of NF-κB increases when moderate shifts in GSH E_h occur, where optimal NF-κB activity correlates with a more intermediate shift. However, if dramatic shifts in GSH E_h occur, NF-κB activity will decrease (Galter, Mihm, and Droge 1994). Thalidomide's capability to deplete GSH within the limb bud during development is most likely a contributing factor in the dysregulation of NF-κB and its downstream gene products, leading to phocomelia. To study the oxidative effects of thalidomide, a rodent thalidomide-resistant group (rodents) were compared to a thalidomide-sensitive group (rabbits) (Parman, Wiley, and Wells 1999). In in vivo experiments using mice and rabbits, thalidomide-sensitive rabbit embryos showed an increase in DNA oxidative damage and higher frequencies of birth defects, but mice were unaffected. Using a free radical trap, α-phenyl-N-t-butylnitrone (PBN), rabbit embryos were protected from oxidative damage and a reduction in dysmorphogenesis was observed (Parman, Wiley, and Wells 1999). When rabbits were cotreated with thalidomide and PBN, there was a significant decrease in offspring exhibiting phocomelia, omphalocele, and/or adactyly when compared to the control group.

More specific to GSH, a follow-up study showed that rabbit conceptuses treated in culture with thalidomide showed a significant decrease in GSH concentrations, whereas rat conceptuses in similar stages of development were unaffected (Hansen, Carney, and Harris 1999). In a subsequent study, pregnant rabbits were treated with thalidomide with or without PBN pretreatment and NF-κB genes were evaluated in the limb bud via in situ hybridization (Hansen et al. 2002). In embryos from rabbits treated with thalidomide only, limb dysmorphogenesis and a dysregulation of AER and PZ genes were observed. PBN-pretreated rabbit embryos were protected from thalidomide co-treatments and normal limb gene expression was restored. The data suggests that key factors in the mechanism of thalidomide-induced phocomelia are promotion of oxidative stress, disruption of the GSH redox system, and de-regulation of NF-κB.

10.10.2 Methylmercury

Methylmercury is a global pollutant that affects the health of millions worldwide (Shapiro and Chan 2008). Methylmercury is the organic form of mercury and is a known neurotoxin that binds strongly to thiol residues, such as those found on GSH and other protein and low molecular weight thiols (Robitaille, Mailloux, and Chan 2016). The interaction between GSH and MeHg can lead to the formation of an excretable GS-MeHg complex, but as a result can decrease the amount of available GSH within the cell, causing a shift to a more oxidized redox state (Ballatori and Clarkson 1982). Indeed, in studies where pregnant dams were exposed to MeHg, embryos demonstrated a decrease in GSH content and a concomitant increase in GSSG (Thompson et al. 2000). Though the GSH redox state is a primary protectant against MeHg toxicity, there are various components that are affected as a result of exposure (Ou et al. 1999). Rat CNS cells, when treated with MeHg (2 or 4 μM), exhibited a G_2/M-phase inhibition; however, when the concentration of MeHg increased, any phase of the cell cycle could be inhibited as a result (Ponce et al. 1994). The data expressed suggests that MeHg toxicity can disrupt proper development via various intracellular mechanisms including dysregulation of the GSH redox system. As a result of MeHg's ability to disrupt proper growth of CNS cells, it has been linked to various developmental disorders, especially neurological disorders in embryos, if the mother is exposed to sufficient quantities of MeHg. During the third week of gestation, the human nervous system begins to form and is the most susceptible to oxidative influences such as MeHg. Due to the toxicity of MeHg, disorders such as attention deficit, behavioral, cognitive, and motor skills can occur.

One of the primary effects of MeHg on the GSH redox system is its capacity to disrupt GPx. Glutathione peroxidase isoforms are selenoproteins, which contain critical selenocysteine residues at their catalytic sites (Brigelius-Flohé and Maiorino 2013). Studies have shown that both in vitro and in vivo the activities of different GPx isoforms, such as GPx1, GPx3, and GPx4, are decreased following MeHg exposure (Farina et al. 2009; Branco et al. 2012; Meinerz et al. 2017; Franco et al. 2009; Zemolin et al. 2012; Usuki and Fujimura 2016). An in vitro study conducted with cultured mouse cerebellar neurons reported that MeHg is a direct inhibitor of GPx1, likely due to the interaction of MeHg and the GPx1 selenocysteine. As such, MeHg would disrupt the function of GPx1 in detoxifying peroxides, allowing increased lipid peroxidation and promoting apoptosis (Farina et al. 2009). The ability of MeHg to inhibit GPx1, combined with the affinity that it has for GSH producing GS-MeHg, creates a more oxidizing environment causing disruption in proper neurological development (Ballatori and Clarkson 1982).

10.10.3 Ethanol

Fetal alcohol spectrum disorder continues to be a significant problem throughout our society (Lemoine et al. 2003). Alcohol consumption by pregnant women can cause various birth defects such as facial deformities, growth deficiencies, psychomotor retardation, low IQ, and atypical brain activity. The various mechanisms of ethanol teratogenicity include increased oxidative stress (Ornoy 2007), impaired neurogenesis and increased cellular apoptosis (Heaton et al. 2002), and alterations on gene expression (Wilke et al. 1994). One mechanism by which alcohol can be metabolized involves alcohol dehydrogenase (ADH), which converts ethanol into the reactive metabolite, acetaldehyde. Mitochondrial aldehyde dehydrogenase (ALDH) then converts aldehyde into acetic acid. Alcohol metabolism through ADH and ALDH uses NAD+ as a cofactor, increasing the amount of NADH within the mitochondria (Liang,

Yeligar, and Brown 2012). An increase of NADH, an electron donor to the electron transport chain (ETC), results in the uncoupling of the ETC, electron leakage, and an increase of ROS. An increase in the ethanol-mediated ROS production can alter the GSH E_h shifting the environment into a more oxidizing condition.

Maternal rats who were supplemented with ethanol during pregnancy resulted in offspring with a decreased GSH pool, especially in the embryonic liver, where the decrease of GSH correlated with an increase in ROS (Amini et al. 1996; Colton, Snell-Callanan, and Chernyshev 1998; Devi et al. 1996; Henderson et al. 1995; Ramachandran et al. 2001; Brooks 1997). Similar GSH depletion was observed in chick embryonic brains treated with ethanol (Berning et al. 2013). Interestingly, in rat cultures, embryos treated at 1.5 mg ethanol/ml showed a significant decrease in GSH but not an increase in GSSG (Jilek et al. 2015). In fact, GSSG levels were not shown to significantly increase until doses reached 6 mg ethanol/ml. Alterations to GSH pools resulted in a GSH E_h shift from −216 mV in untreated rat embryos to −186 mV in ethanol treated embryos at 1.5 mg ethanol/ml. As such, ethanol-induced changes to the GSH E_h appear to be largely driven by decreases in GSH, not increases in GSSG concentrations. Ethanol exposure in pregnant mice also showed embryonic depletion of GSH (Addolorato et al. 1997). Still, the resultant decrease in GSH is positively correlated with an increase in apoptosis and is likely to also contribute to ethanol-related effects (Berning et al. 2013).

Additional studies showed alterations within the CNS after treatment with ethanol, resulting in abnormalities with the production and activity of endogenous antioxidants in various organ systems, including the brain and placenta (Ornoy 2007; Heaton et al. 2002; Kay et al. 2006). When alcohol was administered to pregnant mice during gestation, the rat embryo showed significant increases in lipid peroxidation and a decrease in the levels of GSH within various organs in the embryo (Liang, Yeligar, and Brown 2012). Brain GSH homeostasis is quite heterogeneous, where some cell populations have comparatively high concentrations of GSH but others do not. In low GSH–containing fetal neurons, the level of 4-hydroxynonenal (4HNE), a product of lipid peroxidation and oxidative damage, was shown to be 60% higher following ethanol treatments compared to high GSH–containing fetal neurons (Maffi et al. 2008), suggesting that lowered GSH concentrations allow for a greater degree of ethanol-induced damage. Embryonic exposure to ethanol *in utero* can result in disruption of the GSH redox state of the post-natal lungs, where there is an 80% decrease in GSH within the epithelial lining fluid of the lung, promoting a shift in the GSH E_h by ~+30 mV. These pulmonary GSH E_h shifts from prenatal ethanol exposures correlate with an increase in chronic lung disorders, increased risk of bacterial infection, and acute lung injury upon birth (Yeh et al. 2007; Liang, Yeligar, and Brown 2012).

10.11 SUMMARY

Glutathione is now recognized as a molecule that has multiple functions beyond that of an antioxidant preventing oxidative injury, more specifically the regulation of enzyme/protein function and protein folding. Regulatory control of redox-sensitive processes involving GSH are more broadly associated with important developmental events, such as proliferation, differentiation, and apoptosis. As such, GSH is likely an important regulatory node of control that interfaces cellular development with environmental influences, both inherent and exogenous. Identification of GSH-responsive targets, both those that function more directly and permissively, requires further, intense study to better flesh out redox-sensitive pathways that regulate normal development but also are involved in processes that may be perturbed, leading to poor developmental outcomes.

REFERENCES

Abad, M.F., G. Di Benedetto, P.J. Magalhaes, L. Filippin, and T. Pozzan. 2004. "Mitochondrial pH monitored by a new engineered green fluorescent protein mutant." J Biol Chem 279 (12):11521–11529. doi:10.1074/jbc.M306766200.

Abate, C., L. Patel, F. Rauscher 3rd, and T. Curran. 1990. "Redox regulation of fos and jun DNA-binding activity in vitro." Science 249 (4973):1157–1161.

Abdelsaid, M.A., and A.B. El-Remessy. 2012. "S-glutathionylation of LMW-PTP regulates VEGF-mediated FAK activation and endothelial cell migration." J Cell Sci 125 (Pt 20):4751–4760. doi:10.1242/jcs.103481.

Addolorato, G., A. Gasbarrini, S. Marcoccia, M. Simoncini, P. Baccarini, G. Vagni, A. Grieco, A. Sbriccoli, A. Granato, G.F. Stefanini, and G. Gasbarrini.

1997. "Prenatal exposure to ethanol in rats: effects on liver energy level and antioxidant status in mothers, fetuses, and newborns." *Alcohol* 14 (6):569–573. doi:10.1016/s0741-8329(97)00049-9.

Amini, S.A., R.H. Dunstan, P.R. Dunkley, and R.N. Murdoch. 1996. "Oxidative stress and the fetotoxicity of alcohol consumption during pregnancy." *Free Radic Biol Med* 21 (3):357–365. doi:10.1016/0891-5849(96)00027-5.

Anderson, M.E., and A. Meister. 1983. "Transport and direct utilization of gamma-glutamylcyst(E)Ine for glutathione synthesis." *Proc Natl Acad Sci U S A—Biol Sci* 80 (3):707–711. doi:10.1073/pnas.80.3.707.

Antico Arciuch, V.G., S. Galli, M.C. Franco, P.Y. Lam, E. Cadenas, M.C. Carreras, and J.J. Poderoso. 2009. "Akt1 intramitochondrial cycling is a crucial step in the redox modulation of cell cycle progression." *PLoS One* 4 (10):e7523. doi:10.1371/journal.pone.0007523.

Arseneault, R., A. Chien, J.T. Newington, T. Rappon, R. Harris, and R.C. Cumming. 2013. "Attenuation of LDHA expression in cancer cells leads to redox-dependent alterations in cytoskeletal structure and cell migration." *Cancer Lett* 338 (2):255–266. doi:10.1016/j.canlet.2013.03.034.

Azarbayjani, F., L.A. Borg, and B.R. Danielsson. 2006. "Increased susceptibility to phenytoin teratogenicity: excessive generation of reactive oxygen species or impaired antioxidant defense?" *Basic Clin Pharmacol Toxicol* 99 (4):305–311. doi:10.1111/j.1742-7843.2006.pto_416.x.

Baker, C.N., and S.N. Ebert. 2013. "Development of aerobic metabolism in utero: requirement for mitchondrial function during embryonic and fetal periods." *OA Biotechnol* 2 (2):16.

Bakiri, L., D. Lallemand, E. Bossy-Wetzel, and M. Yaniv. 2000. "Cell cycle-dependent variations in c-Jun and JunB phosphorylation: a role in the control of cyclin D1 expression." *EMBO J* 19 (9):2056–2068.

Ballatori, N., and T.W. Clarkson. 1982. "Developmental changes in the biliary excretion of methylmercury and glutathione." *Science* 216 (4541):61–63. doi:10.1126/science.7063871.

Barbehenn, E.K., R.G. Wales, and O.H. Lowry. 1978. "Measurement of metabolites in single preimplantation embryos; a new means to study metabolic control in early embryos." *J Embryol Exp Morphol* 43:29–46.

Barrett, W., J. DeGnore, S. König, H. Fales, Y. Keng, Z. Zhang, M. Yim, and P. Chock. 1999. "Regulation of PTP1B via glutathionylation of the active site cysteine 215." *Biochemistry* 38 (20):6699–6705.

Benhar, M. 2020. "Oxidants, antioxidants and thiol redox switches in the control of regulated cell death pathways." *Antioxidants (Basel)* 9 (4). doi:10.3390/antiox9040309.

Berberian, R.M., G.E. Eurich, G.A. Rios, and C. Harris. 1996. "Formation of glutathione adducts and 2-aminofluorene from 2-nitrosofluorene in postimplantation rat conceptuses in vitro." *Reproduct Toxicol* 10 (4):273–284.

Berndt, C., G. Poschmann, K. Stuhler, A. Holmgren, and L. Brautigam. 2014. "Zebrafish heart development is regulated via glutaredoxin 2 dependent migration and survival of neural crest cells." *Redox Biol* 2:673–678. doi:10.1016/j.redox.2014.04.012.

Berning, E.J., N. Bernhardson, K. Coleman, D.A. Farhat, C.M. Gushrowski, A. Lanctot, B.H. Maddock, K.G. Michels, L.A. Mugge, C.M. Nass, S.M. Yearsley, and R.R. Miller, Jr. 2013. "Ethanol- and/or taurine-induced oxidative stress in chick embryos." *J Amino Acids* 2013:240537. doi:10.1155/2013/240537.

Biggers, J.D., D.G. Whittingham, and R.P. Donahue. 1967. "The pattern of energy metabolism in the mouse oocyte and zygote." *Proc Natl Acad Sci U S A* 58 (2):560–567. doi:10.1073/pnas.58.2.560.

Bloise, E., T.M. Ortiga-Carvalho, F.M. Reis, S.J. Lye, W. Gibb, and S.G. Matthews. 2016. "ATP-binding cassette transporters in reproduction: a new frontier." *Hum Reprod Update* 22 (2):164–181. doi:10.1093/humupd/dmv049.

Board, P.G. 2007. "The use of glutathione transferase-knockout mice as pharmacological and toxicological models." *Expert Opin Drug Metab Toxicol* 3 (3):421–433. doi:10.1517/17425255.3.3.421.

Bock, K.W., W. Lilienblum, G. Fischer, G. Schirmer, and B.S. Bockhennig. 1987. "The role of conjugation reactions in detoxication." *Arch Toxicol* 60 (1–3):22–29. doi:10.1007/Bf00296941.

Borchert, A., C.C. Wang, C. Ufer, H. Schiebel, N.E. Savaskan, and H. Kuhn. 2006. "The role of phospholipid hydroperoxide glutathione peroxidase isoforms in murine embryogenesis." *J Biol Chem* 281 (28):19655–19664. doi:10.1074/jbc.M601195200.

Borutaite, V., and G.C. Brown. 2001. "Caspases are reversibly inactivated by hydrogen peroxide." *FEBS Lett* 500 (3):114–118.

Branco, V., J. Canario, J. Lu, A. Holmgren, and C. Carvalho. 2012. "Mercury and selenium interaction in vivo: effects on thioredoxin reductase and glutathione peroxidase." *Free Radic Biol Med* 52 (4):781–793. doi:10.1016/j.freeradbiomed.2011.12.002.

Brautigam, L., L.D. Schutte, J.R. Godoy, T. Prozorovski, M. Gellert, G. Hauptmann, A. Holmgren, C.H. Lillig, and C. Berndt. 2011. "Vertebrate-specific glutaredoxin is essential for brain development." *Proc Natl*

Acad Sci U S A 108 (51):20532–20537. doi:10.1073/pnas.1110085108.

Brautigam, L., J. Zhang, K. Dreij, L. Spahiu, A. Holmgren, H. Abe, K.D. Tew, D.M. Townsend, M.J. Kelner, R. Morgenstern, and K. Johansson. 2018. "MGST1, a GSH transferase/peroxidase essential for development and hematopoietic stem cell differentiation." Redox Biol 17:171–179. doi:10.1016/j.redox.2018.04.013.

Brigelius-Flohé, R., and M. Maiorino. 2013. "Glutathione peroxidases." Biochim Biophys Acta 1830 (5):3289–3303. doi:10.1016/j.bbagen.2012.11.020.

Brocardo, P.S., J. Gil-Mohapel, R. Wortman, A. Noonan, E. McGinnis, A.R. Patten, and B.R. Christie. 2017. "The effects of ethanol exposure during distinct periods of brain development on oxidative stress in the adult rat brain." Alcohol Clin Exp Res 41 (1):26–37. doi:10.1111/acer.13266.

Brooks, P.J. 1997. "DNA damage, DNA repair, and alcohol toxicity: a review." Alcohol Clin Exp Res 21 (6):1073–1082.

Bushdid, P.B., D.M. Brantley, F.E. Yull, G.L. Blaeuer, L.H. Hoffman, L. Niswander, and L.D. Kerr. 1998. "Inhibition of NF-kappaB activity results in disruption of the apical ectodermal ridge and aberrant limb morphogenesis." Nature 392 (6676):615–68. doi:10.1038/33435.

Cai, J., and D. Jones. 1998. "Superoxide in apoptosis. Mitochondrial generation triggered by cytochrome c loss." J Biol Chem 273 (19):11101–11404.

Catani, M.V., I. Savini, G. Duranti, D. Caporossi, R. Ceci, S. Sabatini, and L. Avigliano. 2004. "Nuclear factor kappaB and activating protein 1 are involved in differentiation-related resistance to oxidative stress in skeletal muscle cells." Free Radic Biol Med 37 (7):1024–1036. doi:10.1016/j.freeradbiomed.2004.06.021.

Cha, H., J.M. Kim, J.G. Oh, M.H. Jeong, C.S. Park, J. Park, H.J. Jeong, B.K. Park, Y.H. Lee, D. Jeong, D.K. Yang, O.Y. Bernecker, D.H. Kim, R.J. Hajjar, and W.J. Park. 2008. "PICOT is a critical regulator of cardiac hypertrophy and cardiomyocyte contractility." J Mol Cell Cardiol 45 (6):796–803. doi:10.1016/j.yjmcc.2008.09.124.

Chai, Y., S. Ashraf, K. Rokutan, R. Johnston Jr, and J. Thomas. 1994. "S-thiolation of individual human neutrophil proteins including actin by stimulation of the respiratory burst: evidence against a role for glutathione disulfide." Arch Biochem Biophys 310 (1):273–281.

Chen, C., D. Willard, and J. Rudolph. 2009. "Redox regulation of SH2-domain-containing protein tyrosine phosphatases by two backdoor cysteines." Biochemistry 48 (6):1399–1409.

Chen, F.C., and O. Ogut. 2006. "Decline of contractility during ischemia-reperfusion injury: actin glutathionylation and its effect on allosteric interaction with tropomyosin." Am J Physiol Cell Physiol 290 (3):C719–727. doi:10.1152/ajpcell.00419.2005.

Chen, G., Z. Jia, L. Wang, and T. Hu. 2020. "Effect of acute exposure of saxitoxin on development of zebrafish embryos (Danio rerio)." Environ Res 185:109432. doi:10.1016/j.envres.2020.109432.

Chen, G., L. Wang, W. Li, Q. Zhang, and T. Hu. 2020. "Nodularin induced oxidative stress contributes to developmental toxicity in zebrafish embryos." Ecotoxicol Environ Saf 194:110444. doi:10.1016/j.ecoenv.2020.110444.

Chen, H., and M.R. Juchau. 1998. "Inhibition of embryonic retinoic acid synthesis by aldehydes of lipid peroxidation and prevention of inhibition by reduced glutathione and glutathione S-transferases." Free Radic Biol Med 24 (3):408–417. doi:10.1016/s0891-5849(97)00272-4.

Cheng, N.H., W. Zhang, W.Q. Chen, J. Jin, X. Cui, N.F. Butte, L. Chan, and K.D. Hirschi. 2011. "A mammalian monothiol glutaredoxin, Grx3, is critical for cell cycle progression during embryogenesis." FEBS J 278 (14):2525–2539. doi:10.1111/j.1742-4658.2011.08178.x.

Cheng, W., Y. Ho, D. Ross, B. Valentine, G. Combs, and X. Lei. 1997. "Cellular glutathione peroxidase knockout mice express normal levels of selenium-dependent plasma and phospholipid hydroperoxide glutathione peroxidases in various tissues." J Nutr 127 (8):1445–1450.

Chiarugi, P., T. Fiaschi, M.L. Taddei, D. Talini, E. Giannoni, G. Raugei, and G. Ramponi. 2001. "Two vicinal cysteines confer a peculiar redox regulation to low molecular weight protein tyrosine phosphatase in response to platelet-derived growth factor receptor stimulation." J Biol Chem 276 (36):33478–33487. doi:10.1074/jbc.M102302200.

Chiarugi, P., G. Pani, E. Giannoni, L. Taddei, R. Colavitti, G. Raugei, M. Symons, S. Borrello, T. Galeotti, and G. Ramponi. 2003. "Reactive oxygen species as essential mediators of cell adhesion: the oxidative inhibition of a FAK tyrosine phosphatase is required for cell adhesion." J Cell Biol 161 (5):933–944. doi:10.1083/jcb.200211118.

Chiarugi, P., M.L. Taddei, N. Schiavone, L. Papucci, E. Giannoni, T. Fiaschi, S. Capaccioli, G. Raugei, and G. Ramponi. 2004. "LMW-PTP is a positive regulator of tumor onset and growth." Oncogene 23 (22):3905–3914. doi:10.1038/sj.onc.1207508.

Chiu, J., and I.W. Dawes. 2012. "Redox control of cell proliferation." *Trends Cell Biol* 22 (11):592–601. doi:10.1016/j.tcb.2012.08.002.

Choe, H., J. Hansen, and C. Harris. 2001. "Spatial and temporal ontogenies of glutathione peroxidase and glutathione disulfide reductase during development of the prenatal rat." *J Biochem Mol Toxicol* 15 (4):197–206.

Circu, M.L., and T.Y. Aw. 2008. "Glutathione and apoptosis." *Free Radic Res* 42 (8):689–706. doi:10.1080/10715760802317663.

Circu, M.L., and T.Y. Aw. 2010. "Reactive oxygen species, cellular redox systems, and apoptosis." *Free Radic Biol Med* 48 (6):749–762. doi:10.1016/j.freeradbiomed.2009.12.022.

Colton, C.A., J. Snell-Callanan, and O.N. Chernyshev. 1998. "Ethanol induced changes in superoxide anion and nitric oxide in cultured microglia." *Alcohol Clin Exp Res* 22 (3):710–716.

Conrad, M., C. Jakupoglu, S.G. Moreno, S. Lippl, A. Banjac, M. Schneider, H. Beck, A.K. Hatzopoulos, U. Just, F. Sinowatz, W. Schmahl, K.R. Chien, W. Wurst, G.W. Bornkamm, and M. Brielmeier. 2004. "Essential role for mitochondrial thioredoxin reductase in hematopoiesis, heart development, and heart function." *Mol Cell Biol* 24 (21):9414–9423. doi:10.1128/MCB.24.21.9414-9423.2004.

Conrad, M., S.G. Moreno, F. Sinowatz, F. Ursini, S. Kolle, A. Roveri, M. Brielmeier, W. Wurst, M. Maiorino, and G.W. Bornkamm. 2005. "The nuclear form of phospholipid hydroperoxide glutathione peroxidase is a protein thiol peroxidase contributing to sperm chromatin stability." *Mol Cell Biol* 25 (17):7637–7644. doi:10.1128/MCB.25.17.7637-7644.2005.

Cross, J., and D. Templeton. 2004. "Oxidative stress inhibits MEKK1 by site-specific glutathionylation in the ATP-binding domain." *Biochem J* 381:675–683.

Dalle-Donne, I., D. Giustarini, R. Rossi, R. Colombo, and A. Milzani. 2003. "Reversible S-glutathionylation of Cys 374 regulates actin filament formation by inducing structural changes in the actin molecule." *Free Radic Biol Med* 34 (1):23–32.

Dalle-Donne, I., A. Milzani, and R. Colombo. 1995. "H_2O_2-treated actin: assembly and polymer interactions with cross-linking proteins." *Biophys J* 69 (6):2710–2719.

Dalle-Donne, I., A. Milzani, and R. Colombo. 1999. "The tert-butyl hydroperoxide-induced oxidation of actin Cys-374 is coupled with structural changes in distant regions of the protein." *Biochemistry* 38 (38):12471–12480.

Dalton, T.P., M.Z. Dieter, Y. Yang, H.G. Shertzer, and D.W. Nebert. 2000. "Knockout of the mouse glutamate cysteine ligase catalytic subunit (Gclc) gene: embryonic lethal when homozygous, and proposed model for moderate glutathione deficiency when heterozygous." *Biochem Biophys Res Commun* 279 (2):324–329. doi:10.1006/bbrc.2000.3930.

de Haan, J., C. Bladier, P. Griffiths, M. Kelner, R. O'Shea, N. Cheung, R. Bronson, M. Silvestro, S. Wild, S. Zheng, P. Beart, P. Hertzog, and I. Kola. 1998. "Mice with a homozygous null mutation for the most abundant glutathione peroxidase, Gpx1, show increased susceptibility to the oxidative stress-inducing agents paraquat and hydrogen peroxide." *J Biol Chem* 273 (35):22528–22536.

de Souza Malaspina, T.S., W.F. Zambuzzi, C.X. dos Santos, A.P. Campanelli, F.R. Laurindo, M.C. Sogayar, and J.M. Granjeiro. 2009. "A possible mechanism of low molecular weight protein tyrosine phosphatase (LMW-PTP) activity modulation by glutathione action during human osteoblast differentiation." *Arch Oral Biol* 54 (7):642–650. doi:10.1016/j.archoralbio.2009.03.011.

Dengler, V.L., M. Galbraith, and J.M. Espinosa. 2014. "Transcriptional regulation by hypoxia inducible factors." *Crit Rev Biochem Mol Biol* 49 (1):1–15. doi:10.3109/10409238.2013.838205.

DeNunzio, M., and G. Gomez. 2014. "Extracellular glutathione promotes migration of hydrogen peroxide-stressed cultured chick embryonic skin cells." *In Vitro Cell Dev Biol Anim* 50 (4):350–357. doi:10.1007/s.

Devi, B.G., S. Schenker, B. Mazloum, and G.I. Henderson. 1996. "Ethanol-induced oxidative stress and enzymatic defenses in cultured fetal rat hepatocytes." *Alcohol* 13 (4):327–332. doi:10.1016/0741-8329 (96)00002-x.

Di Ilio, C., G. Tiboni, P. Sacchetta, S. Angelucci, T. Bucciarelli, U. Bellati, and A. Aceto. 1995. "Time-dependent and tissue-specific variations of glutathione transferase activity during gestation in the mouse." *Mech Ageing Dev* 78:47–62.

Ding, Y., K.J. Choi, J.H. Kim, X. Han, Y. Piao, J.H. Jeong, W. Choe, I. Kang, J. Ha, H.J. Forman, J. Lee, K.S. Yoon, and S.S. Kim. 2008. "Endogenous hydrogen peroxide regulates glutathione redox via nuclear factor erythroid 2-related factor 2 downstream of phosphatidylinositol 3-kinase during muscle differentiation." *Am J Pathol* 172 (6):1529–1541. doi:10.2353/ajpath.2008.070429.

Dominko, K., and D. Dikic. 2018. "Glutathionylation: a regulatory role of glutathione in physiological processes." *Arh Hig Rada Toksikol* 69 (1):1–24. doi:10.2478/aiht-2018-69-2966.

dos Santos, N.V., A.C. Matias, G.S. Higa, A.H. Kihara, and G. Cerchiaro. 2015. "Copper uptake in mammary epithelial cells activates cyclins and triggers antioxidant response." *Oxid Med Cell Longev* 2015:162876. doi:10.1155/2015/162876.

Dumollard, R., J. Carroll, M.R. Duchen, K. Campbell, and K. Swann. 2009. "Mitochondrial function and redox state in mammalian embryos." *Semin Cell Dev Biol* 20 (3):346–353. doi:10.1016/j.semcdb.2008.12.013.

Dumollard, R., Z. Ward, J. Carroll, and M.R. Duchen. 2007. "Regulation of redox metabolism in the mouse oocyte and embryo." *Development* 134 (3):455–465. doi:10.1242/dev.02744.

Elhassan, Y.M., G. Wu, A.C. Leanez, R.J. Tasca, A.J. Watson, and M.E. Westhusin. 2001. "Amino acid concentrations in fluids from the bovine oviduct and uterus and in KSOM-based culture media." *Theriogenology* 55 (9):1907–1918. doi:10.1016/S0093-691x(01)00532-5.

Eriksson, U. 1988. "Importance of genetic predisposition and maternal environment for the occurrence of congenital malformations in offspring of diabetic rats." *Teratology* 37 (4):365–374.

Erlejman, A., and P. Oteiza. 2002. "The oxidant defense system in human neuroblastoma IMR-32 cells predifferentiation and postdifferentiation to neuronal phenotypes." *Neurochem Res* 27 (11):1499–1506.

Esworthy, R., J. Mann, M. Sam, and F. Chu. 2000. "Low glutathione peroxidase activity in Gpx1 knockout mice protects jejunum crypts from gamma-irradiation damage." *Am J Physiol Gastrointest Liver Physiol* 279 (2):G426–436.

Farina, M., F. Campos, I. Vendrell, J. Berenguer, M. Barzi, S. Pons, and C. Sunol. 2009. "Probucol increases glutathione peroxidase-1 activity and displays long-lasting protection against methylmercury toxicity in cerebellar granule cells." *Toxicol Sci* 112 (2):416–426. doi:10.1093/toxsci/kfp219.

Fernando, M.R., H. Sumimoto, H. Nanri, S. Kawabata, S. Iwanaga, S. Minakami, Y. Fukumaki, and K. Takeshige. 1994. "Cloning and sequencing of the cDNA encoding human glutaredoxin." *Biochim Biophys Acta* 1218 (2):229–231. doi:10.1016/0167-4781(94)90019-1.

Fiaschi, T., G. Cozzi, G. Raugei, L. Formigli, G. Ramponi, and P. Chiarugi. 2006. "Redox regulation of beta-actin during integrin-mediated cell adhesion." *J Biol Chem* 281 (32):22983–22991. doi:10.1074/jbc. M603040200.

Folmes, C.D.L., and A. Terzic. 2014. "Metabolic determinants of embryonic development and stem cell fate." *Reprod Fertil Dev* 27 (1):82–88. doi:10.1071/RD14383.

Franca, K.C., P.A. Martinez, M.L. Prado, S.M. Lo, B.E. Borges, S.M. Zanata, A. San Martin, and L.S. Nakao. 2020. "Quiescin/sulfhydryl oxidase 1b (QSOX1b) induces migration and proliferation of vascular smooth muscle cells by distinct redox pathways." *Arch Biochem Biophys* 679:108220. doi:10.1016/j. abb.2019.108220.

Franco, J.L., T. Posser, P.R. Dunkley, P.W. Dickson, J.J. Mattos, R. Martins, A.C. Bainy, M.R. Marques, A.L. Dafre, and M. Farina. 2009. "Methylmercury neurotoxicity is associated with inhibition of the antioxidant enzyme glutathione peroxidase." *Free Radic Biol Med* 47 (4):449–457. doi:10.1016/j. freeradbiomed.2009.05.013.

Franklin, C.C., C.M. Krejsa, R.H. Pierce, C.C. White, N. Fausto, and T.J. Kavanagh. 2002. "Caspase-3-dependent cleavage of the glutamate-l-cysteine ligase catalytic subunit during apoptotic cell death." *Am J Pathol* 160 (5):1887–1894. doi:10.1016/ s0002-9440(10)61135-2.

Gallogly, M.M., D.W. Starke, A.K. Leonberg, S.M. Ospina, and J.J. Mieyal. 2008. "Kinetic and mechanistic characterization and versatile catalytic properties of mammalian glutaredoxin 2: implications for intracellular roles." *Biochemistry* 47 (42):11144–11157. doi:10.1021/bi800966v.

Galter, D., S. Mihm, and W. Droge. 1994. "Distinct effects of glutathione disulphide on the nuclear transcription factor kappa B and the activator protein-1." *Eur J Biochem* 221 (2):639–648. doi:10.1111/j.1432-1033.1994.tb18776.x.

Gardiner, C., and D.J. Reed. 1995a. "Glutathione redox cycle-driven recovery of reduced glutathione after oxidation by tertiary-butyl hydroperoxide in preimplantation mouse embryos." *Arch Biochem Biophys* 321 (1):6–12.

Gardiner, C.S., and D.J. Reed. 1994. "Status of glutathione during oxidant-induced oxidative stress in the preimplantation mouse embryo." *Biol Reprod* 51 (6):1307–1314. doi:10.1095/ biolreprod51.6.1307.

Gardiner, C.S., and D.J. Reed. 1995b. "Synthesis of glutathione in the preimplantation mouse embryo." *Arch Biochem Biophys* 318 (1):30–36. doi:10.1006/ abbi.1995.1200.

Ginsberg, L., and N. Hillman. 1973. "ATP metabolism in cleavage-staged mouse embryos." *J Embryol Exp Morphol* 30 (1):267–282.

Gorny, M., A. Bilska-Wilkosz, M. Iciek, M. Hereta, K. Kaminska, A. Kaminska, G. Chwatko, Z. Rogoz, and E. Lorenc-Koci. 2020. "Alterations in the antioxidant enzyme activities in the neurodevelopmental rat model of schizophrenia induced by glutathione deficiency during early postnatal life." *Antioxidants (Basel)* 9 (6). doi:10.3390/antiox9060538.

Gorny, M., A. Wnuk, A. Kaminska, K. Kaminska, G. Chwatko, A. Bilska-Wilkosz, M. Iciek, M. Kajta, Z. Rogoz, and E. Lorenc-Koci. 2019. "Glutathione deficiency and alterations in the sulfur amino acid homeostasis during early postnatal development as potential triggering factors for schizophrenia-like behavior in adult rats." *Molecules* 24 (23). doi:10.3390/molecules24234253.

Haddad, J.J. 2011. "A redox microenvironment is essential for MAPK-dependent secretion of pro-inflammatory cytokines: modulation by glutathione (GSH/GSSG) biosynthesis and equilibrium in the alveolar epithelium." *Cell Immunol* 270 (1):53–61. doi:10.1016/j.cellimm.2011.04.001.

Hagen, T.M., G.T. Wierzbicka, A.H. Sillau, B.B. Bowman, and D.P. Jones. 1990. "Bioavailability of dietary glutathione: effect on plasma concentration." *Am J Physiol* 259 (4 Pt 1):G524–529. doi:10.1152/ajpgi.1990.259.4.G524.

Hales, B.F., and H. Brown. 1991. "The effect of in vivo glutathione depletion with buthionine sulfoximine on rat embryo development." *Teratology* 44 (3):251–257. doi:10.1002/tera.1420440305.

Han, C., M.J. Kim, D. Ding, H.J. Park, K. White, L. Walker, T. Gu, M. Tanokura, T. Yamasoba, P. Linser, R. Salvi, and S. Someya. 2017. "GSR is not essential for the maintenance of antioxidant defenses in mouse cochlea: possible role of the thioredoxin system as a functional backup for GSR." *PLoS One* 12 (7):e0180817. doi:10.1371/journal.pone.0180817.

Hanschmann, E.M., J.R. Godoy, C. Berndt, C. Hudemann, and C.H. Lillig. 2013. "Thioredoxins, glutaredoxins, and peroxiredoxins—molecular mechanisms and health significance: from cofactors to antioxidants to redox signaling." *Antioxid Redox Signal* 19 (13):1539–1605. doi:10.1089/ars.2012.4599.

Hansen, J.M., E.W. Carney, and C. Harris. 1999. "Differential alteration by thalidomide of the glutathione content of rat vs. rabbit conceptuses in vitro." *Reprod Toxicol* 13 (6):547–554. doi:10.1016/s0890-6238(99)00053-2.

Hansen, J.M., H. Choe, E. Carney, and C. Harris. 2001. "Differential antioxidant enzyme activities and glutathione content between rat and rabbit conceptuses." *Free Radic Biol Med* 30 (10):1078–1088.

Hansen, J.M., S.G. Gong, M. Philbert, and C. Harris. 2002. "Misregulation of gene expression in the redox-sensitive NF-kappaB-dependent limb outgrowth pathway by thalidomide." *Dev Dyn* 225 (2):186–194. doi:10.1002/dvdy.10150.

Hansen, J.M., and C. Harris. 2004. "A novel hypothesis for thalidomide-induced limb teratogenesis: redox misregulation of the NF-kappaB pathway." *Antioxid Redox Signal* 6 (1):1–14. doi:10.1089/152308604771978291.

Hansen, J.M., and C. Harris. 2013. "Redox control of teratogenesis." *Reprod Toxicol* 35:165–179. doi:10.1016/j.reprotox.2012.09.004.

Hansen, J.M., and C. Harris. 2015. "Glutathione during embryonic development." *Biochim Biophys Acta* 1850 (8):1527–1542. doi:10.1016/j.bbagen.2014.12.001.

Hansen, J.M., D.P. Jones, and C. Harris. 2020. "The redox theory of development." *Antioxid Redox Signal* 32 (10):715–740. doi:10.1089/ars.2019.7976.

Hansen, J.M., M. Klass, C. Harris, and M. Csete. 2007. "A reducing redox environment promotes C2C12 myogenesis: implications for regeneration in aged muscle." *Cell Biol Int* 31 (6):546–553. doi:10.1016/j.cellbi.2006.11.027.

Hansen, J.M., E. Lee, and C. Harris. 2004. "Spatial activities and induction of glutamate-cysteine ligase (GCL) in the postimplantation rat embryo and visceral yolk sac." *Toxicol Sci* 81 (2):371–378. doi:10.1093/toxsci/kfh154.

Hansen, J.M., H. Zhang, and D.P. Jones. 2006. "Differential oxidation of thioredoxin-1, thioredoxin-2, and glutathione by metal ions." *Free Radic Biol Med* 40 (1):138–1345. doi:10.1016/j.freeradbiomed.2005.09.023.

Harris, C. 1993. "Glutathione biosynthesis in the postimplantation rat conceptus in vitro." *Toxicol Appl Pharmacol* 120 (2):247–256. doi:10.1006/taap.1993.1109.

Harris, C., M. Dixon, and J.M. Hansen. 2004. "Glutathione depletion modulates methanol, formaldehyde and formate toxicity in cultured rat conceptuses." *Cell Biol Toxicol* 20 (3):133–145. doi:10.1023/b:cbto.0000029466.08607.86.

Harris, C., A.G. Fantel, and M.R. Juchau. 1986. "Differential glutathione depletion by L-buthionine-S,R-sulfoximine in rat embryo versus visceral yolk sac in vivo and in vitro." *Biochem Pharmacol* 35 (24):4437–4441. doi:10.1016/0006-2952(86)90760-4.

Harris, C., R. Hiranruengchok, E. Lee, R.M. Berberian, and G.E. Eurich. 1995. "Glutathione status in chemical embryotoxicity: synthesis, turnover and adduct formation." *Toxicol In Vitro* 9 (5):623–631. doi:10.1016/0887-2333(95)00072-g.

Harris, C., D.Z. Shuster, R. Roman Gomez, K.E. Sant, M.S. Reed, J. Pohl, and J.M. Hansen. 2013. "Inhibition of glutathione biosynthesis alters compartmental redox status and the thiol proteome in organogenesis-stage rat conceptuses." *Free Radic Biol Med* 63:325–337. doi:10.1016/j.freeradbiomed.2013.05.040.

Hatori, Y., T. Kubo, Y. Sato, S. Inouye, R. Akagi, and T. Seyama. 2020. "Visualization of the redox status of cytosolic glutathione using the organelle- and cytoskeleton-targeted redox sensors." *Antioxidants (Basel)* 9 (2). doi:10.3390/antiox9020129.

Hatori, Y., Y. Yan, K. Schmidt, E. Furukawa, N.M. Hasan, N. Yang, C.N. Liu, S. Sockanathan, and S. Lutsenko. 2016. "Neuronal differentiation is associated with a redox-regulated increase of copper flow to the secretory pathway." *Nat Commun* 7:10640. doi:10.1038/ncomms10640.

Hayes, J.D., J.U. Flanagan, and I.R. Jowsey. 2005. "Glutathione transferases." *Ann Rev Pharmacol Toxicol* 45:51–88. doi:10.1146/annurev.pharmtox.45.120403.095857.

Heaton, M.B., M. Paiva, J. Mayer, and R. Miller. 2002. "Ethanol-mediated generation of reactive oxygen species in developing rat cerebellum." *Neurosci Lett* 334 (2):83–86. doi:10.1016/s0304-3940(02)01123-0.

Henderson, C.J., and C.R. Wolf. 2011. "Knockout and transgenic mice in glutathione transferase research." *Drug Metab Rev* 43 (2):152–164. doi:10.3109/03602532.2011.562900.

Henderson, G.I., B.G. Devi, A. Perez, and S. Schenker. 1995. "In utero ethanol exposure elicits oxidative stress in the rat fetus." *Alcohol Clin Exp Res* 19 (3):714–720. doi:10.1111/j.1530-0277.1995.tb01572.x.

Hiranruengchok, R., and C. Harris. 1993. "Glutathione oxidation and embryotoxicity elicited by diamide in the developing rat conceptus in vitro." *Toxicol Appl Pharmacol* 120 (1):62–71. doi:10.1006/taap.1993.1087.

Ho, Y., J. Magnenat, R. Bronson, J. Cao, M. Gargano, M. Sugawara, and C. Funk. 1997. "Mice deficient in cellular glutathione peroxidase develop normally and show no increased sensitivity to hyperoxia." *J Biol Chem* 272 (26):16644–16651.

Ho, Y.S., Y. Xiong, D.S. Ho, J. Gao, B.H. Chua, H. Pai, and J.J. Mieyal. 2007. "Targeted disruption of the glutaredoxin 1 gene does not sensitize adult mice to tissue injury induced by ischemia/reperfusion and hyperoxia." *Free Radic Biol Med* 43 (9):1299–1312. doi:10.1016/j.freeradbiomed.2007.07.025.

Houghton, F.D., and H.J. Leese. 2004. "Metabolism and developmental competence of the preimplantation embryo." *Eur J Obstetr Gynecol Reproduct Biol* 115:S92-S96. doi:10.1016/j.ejogrb.2004.01.019.

Houghton, F.D., J.G. Thompson, C.J. Kennedy, and H.J. Leese. 1996. "Oxygen consumption and energy metabolism of the early mouse embryo." *Mol Reprod Dev* 44 (4):476–485. doi:10.1002/(SICI)1098-2795(199608)44:4<476::AID-MRD7>3.0.CO;2-I.

Huang, X., M. Begley, K. Morgenstern, Y. Gu, P. Rose, H. Zhao, and X. Zhu. 2003. "Crystal structure of an inactive Akt2 kinase domain." *Structure* 11 (1):21–30.

Huang, Y., H. Roelink, and G.S. McKnight. 2002. "Protein kinase A deficiency causes axially localized neural tube defects in mice." *J Biol Chem* 277 (22):19889–19896. doi:10.1074/jbc.M111412200.

Huang, Z., J.T. Pinto, H. Deng, and J.P. Richie, Jr. 2008. "Inhibition of caspase-3 activity and activation by protein glutathionylation." *Biochem Pharmacol* 75 (11):2234–2244. doi:10.1016/j.bcp.2008.02.026.

Hudemann, C., M.E. Lonn, J.R. Godoy, F. Zahedi Avval, F. Capani, A. Holmgren, and C.H. Lillig. 2009. "Identification, expression pattern, and characterization of mouse glutaredoxin 2 isoforms." *Antioxid Redox Signal* 11 (1):1–14. doi:10.1089/ars.2008.2068.

Huh, Y.J., I.M. Kim, H. Kim, H. Song, H. So, S.T. Lee, S.B. Kwon, H.J. Kim, H.H. Kim, S.H. Lee, Y. Choi, S.C. Chung, D.W. Jeong, and B.M. Min. 2006. "Regulation of osteoclast differentiation by the redox-dependent modulation of nuclear import of transcription factors." *Cell Death Differ* 13 (7):1138–1146. doi:10.1038/sj.cdd.4401793.

Humphries, K.M., M.S. Deal, and S.S. Taylor. 2005. "Enhanced dephosphorylation of cAMP-dependent protein kinase by oxidation and thiol modification." *J Biol Chem* 280 (4):2750–2758. doi:10.1074/jbc.M410242200.

Humphries, K.M., C. Juliano, and S.S. Taylor. 2002. "Regulation of cAMP-dependent protein kinase activity by glutathionylation." *J Biol Chem* 277 (45):43505–43511. doi:10.1074/jbc.M207088200.

Hutter, D., T. BG, and J. Green. 1997. "Redox state changes in density-dependent regulation of proliferation." *Exp Cell Res* 232 (2):435–438.

Ilic, Z., D. Crawford, D. Vakharia, P.A. Egner, and S. Sell. 2010. "Glutathione-S-transferase A3 knockout mice are sensitive to acute cytotoxic and genotoxic effects of aflatoxin B1." *Toxicol Appl Pharmacol* 242 (3):241–246. doi:10.1016/j.taap.2009.10.008.

Imai, H., F. Hirao, T. Sakamoto, K. Sekine, Y. Mizukura, M. Saito, T. Kitamoto, M. Hayasaka, K. Hanaoka, and Y. Nakagawa. 2003. "Early embryonic lethality caused by targeted disruption of the mouse PHGPx gene." *Biochemical and Biophysical Research Communications* 305 (2):278–286. doi:10.1016/s0006-291x(03)00734-4.

Imhoff, B.R., and J.M. Hansen. 2010. "Extracellular redox environments regulate adipocyte differentiation." *Differentiation* 80 (1):31–39. doi:10.1016/j.diff.2010.04.005.

Imhoff, B.R., and J.M. Hansen. 2011. "Differential redox potential profiles during adipogenesis and osteogenesis." *Cell Mol Biol Lett* 16 (1):149–161. doi:10.2478/s11658-010-0042-0.

Javed, M.H., and R.W. Wright. 1991. "Determination of pentose-phosphate and Embden-Meyerhof pathway activities in bovine embryos." *Theriogenology* 35 (5):1029–1037. doi:10.1016/0093-691x(91)90312-2.

Jilek, J.L., K.E. Sant, K.H. Cho, M.S. Reed, J. Pohl, J.M. Hansen, and C. Harris. 2015. "Ethanol attenuates histiotrophic nutrition pathways and alters the intracellular redox environment and thiol proteome during rat organogenesis." *Toxicol Sci* 147 (2):475–489. doi:10.1093/toxsci/kfv145.

Jones, D., E. Maellaro, S. Jiang, A. Slater, and S. Orrenius. 1995. "Effects of N-acetyl-l-cysteine on T-cell apoptosis are not mediated by increased cellular glutathione." *Immunology Letters* 45 (3):205–209.

Jones, D.P. 2006. "Redefining oxidative stress." *Antioxid Redox Signal* 8 (9–10):1865–1879. doi:10.1089/ars.2006.8.1865.

Kanaan, G.N., B. Ichim, L. Gharibeh, W. Maharsy, D.A. Patten, J.Y. Xuan, A. Reunov, P. Marshall, J. Veinot, K. Menzies, M. Nemer, and M.E. Harper. 2018. "Glutaredoxin-2 controls cardiac mitochondrial dynamics and energetics in mice, and protects against human cardiac pathologies." *Redox Biol* 14:509–521. doi:10.1016/j.redox.2017.10.019.

Kanegae, Y., A.T. Tavares, J.C. Izpisua Belmonte, and I.M. Verma. 1998. "Role of Rel/NF-kappaB transcription factors during the outgrowth of the vertebrate limb." *Nature* 392 (6676):611–614. doi:10.1038/33429.

Kaplowitz, N., Y.A. Tak, and M. Ookhtens. 1985. "The regulation of hepatic glutathione." *Ann Rev Pharmacol Toxicol* 25:715–744. doi:10.1146/annurev.pharmtox.25.1.715.

Kay, H.H., S. Tsoi, K. Grindle, and R.R. Magness. 2006. "Markers of oxidative stress in placental villi exposed to ethanol." *J Soc Gynecol Investig* 13 (2):118–121. doi:10.1016/j.jsgi.2005.11.007.

Kemble, D., and G. Sun. 2009. "Direct and specific inactivation of protein tyrosine kinases in the Src and FGFR families by reversible cysteine oxidation." *Proc Natl Acad Sci U S A* 106 (13):5070–5075.

Ketterer, B., B. Coles, and D.J. Meyer. 1983. "The Role of Glutathione in Detoxication." *Environ Health Perspect* 49 (Mar):59–69. doi:10.2307/3429581.

Kim, J.H., and A.R. Scialli. 2011. "Thalidomide: the tragedy of birth defects and the effective treatment of disease." *Toxicol Sci* 122 (1):1–6. doi:10.1093/toxsci/kfr088.

Kim, J.M., H. Kim, S.B. Kwon, S.Y. Lee, S.C. Chung, D.W. Jeong, and B.M. Min. 2004. "Intracellular glutathione status regulates mouse bone marrow monocyte-derived macrophage differentiation and phagocytic activity." *Biochem Biophys Res Commun* 325 (1):101–108. doi:10.1016/j.bbrc.2004.09.220.

Kim, K.S., H.W. Choi, H.E. Yoon, and I.Y. Kim. 2010. "Reactive oxygen species generated by NADPH oxidase 2 and 4 are required for chondrogenic differentiation." *J Biol Chem* 285 (51):40294–40302. doi:10.1074/jbc.M110.126821.

Kirlin, W., J. Cai, S. Thompson, D. Diaz, T.J. Kavanagh, and D. Jones. 1999. "Glutathione redox potential in response to differentiation and enzyme inducers." *Free Radic Biol Med* 27 (11–12):1208–1218.

Klatt, P., E. Molina, M. De Lacoba, C. Padilla, E. Martinez-Galesteo, J. Barcena, and S. Lamas. 1999. "Redox regulation of c-Jun DNA binding by reversible S-glutathiolation." *FASEB J* 13 (12):1481–1490.

Klatt, P., E. Molina, and S. Lamas. 1999. "Nitric oxide inhibits c-Jun DNA binding by specifically targeted S-glutathionylation." *J Biol Chem* 274 (22):15857–15864.

Knobloch, J., K. Reimann, L.O. Klotz, and U. Ruther. 2008. "Thalidomide resistance is based on the capacity of the glutathione-dependent antioxidant defense." *Mol Pharm* 5 (6):1138–1144. doi:10.1021/mp8001232.

Kobayashi, M., H. Nakamura, J. Yodoi, and K. Shiota. 2000. "Immunohistochemical localization of thioredoxin and glutaredoxin in mouse embryos and fetuses." *Antioxid Redox Signal* 2 (4):653–663.

Konopka, R., L. Kubala, A. Lojek, and J. Pachernik. 2008. "Alternation of retinoic acid induced neural differentiation of P19 embryonal carcinoma cells by reduction of reactive oxygen species intracellular production." *Neuro Endocrinol Lett* 29 (5):770–774.

Krisher, R.L., and R.S. Prather. 2012. "A role for the Warburg effect in preimplantation embryo development: metabolic modification to support rapid cell proliferation." *Molecular Reproduction and Development* 79 (5):311–320. doi:10.1002/mrd.22037.

Kushner, P., D. Agard, G. Greene, T. Scanlan, A. Shiau, R. Uht, and P. Webb. 2000. "Estrogen receptor pathways to AP-1." *J Steroid Biochem Mol Biol* 74 (5).

Kviecinski, M.R., R.C. Pedrosa, K.B. Felipe, M.S. Farias, C. Glorieux, M. Valenzuela, B. Sid, J. Benites, J.A. Valderrama, J. Verrax, and P. Buc Calderon. 2012. "Inhibition of cell proliferation and migration by oxidative stress from ascorbate-driven juglone redox cycling in human bladder-derived T24 cells." *Biochem Biophys Res Commun* 421 (2):268–273. doi:10.1016/j.bbrc.2012.03.150.

Kwon, J., S. Lee, K. Yang, Y. Ahn, Y. Kim, E. Stadtman, and S. Rhee. 2004. "Reversible oxidation and inactivation of the tumor suppressor PTEN in cells stimulated with peptide growth factors." *Proc Natl Acad Sci U S A* 101 (47):16419–16424.

La Bella, V., F. Valentino, T. Piccoli, and F. Piccoli. 2007. "Expression and developmental regulation of the cystine/glutamate exchanger (xc-) in the rat." *Neurochem Res* 32 (6):1081–1090. doi:10.1007/s11064-006-9277-6.

Lemoine, P., H. Harousseau, J.P. Borteyru, and J.C. Menuet. 2003. "Children of alcoholic parents-observed anomalies: discussion of 127 cases." *Ther Drug Monit* 25 (2):132–136. doi:10.1097/00007691-200304000-00002.

Leonard, S., K. Reddie, and K. Carroll. 2009. "Mining the thiol proteome for sulfenic acid modifications reveals new targets for oxidation in cells." *ACS Chem Biol* 4 (9):783–799.

Levasseur, R., R. Barrios, F. Elefteriou, D.A. Glass, 2nd, M.W. Lieberman, and G. Karsenty. 2003. "Reversible skeletal abnormalities in gamma-glutamyl transpeptidase-deficient mice." *Endocrinology* 144 (7): 2761–2764. doi:10.1210/en.2002-0071.

Lewerenz, J., S.J. Hewett, Y. Huang, M. Lambros, P.W. Gout, P.W. Kalivas, A. Massie, I. Smolders, A. Methner, M. Pergande, S.B. Smith, V. Ganapathy, and P. Maher. 2013. "The cystine/glutamate antiporter system x(c)(−) in health and disease: from molecular mechanisms to novel therapeutic opportunities." *Antioxid Redox Signal* 18 (5):522–555. doi:10.1089/ars.2011.4391.

Li, F., L.X. Cui, D.W. Yu, H.S. Hao, Y. Liu, X.M. Zhao, Y.W. Pang, H.B. Zhu, and W.H. Du. 2019. "Exogenous glutathione improves intracellular glutathione synthesis via the gamma-glutamyl cycle in bovine zygotes and cleavage embryos." *J Cell Physiol* 234 (5):7384–7394. doi:10.1002/jcp.27497.

Li, K., Y. Li, J.M. Shelton, J.A. Richardson, E. Spencer, Z.J. Chen, X. Wang, and R.S. Williams. 2000. "Cytochrome c deficiency causes embryonic lethality and attenuates stress-induced apoptosis." *Cell* 101 (4):389–399.

Li, Z.C., R.H. Gu, X.W. Lu, S. Zhao, Y. Feng, and Y.J. Sun. 2018. "Preincubation with glutathione ethyl ester improves the developmental competence of vitrified mouse oocytes." *J Assist Reprod Genet* 35 (7):1169–1178. doi:10.1007/s10815-018-1215-4.

Lian, G., J.R. Gnanaprakasam, T. Wang, R. Wu, X. Chen, L. Liu, Y. Shen, M. Yang, J. Yang, Y. Chen, V. Vasiliou, T.A. Cassel, D.R. Green, Y. Liu, T.W. Fan, and R. Wang. 2018. "Glutathione de novo synthesis but not recycling process coordinates with glutamine catabolism to control redox homeostasis and directs murine T cell differentiation." *Elife* 7. doi:10.7554/eLife.36158.

Liang, H., S.E. Yoo, R. Na, C.A. Walter, A. Richardson, and Q. Ran. 2009. "Short form glutathione peroxidase 4 is the essential isoform required for survival and somatic mitochondrial functions." *J Biol Chem* 284 (45):30836–30844. doi:10.1074/jbc.M109.032839.

Liang, Y., S.M. Yeligar, and L.A. Brown. 2012. "Chronic-alcohol-abuse-induced oxidative stress in the development of acute respiratory distress syndrome." *Scientific World Journal* 2012:740308. doi:10.1100/2012/740308.

Lieberman, M., A. Wiseman, Z. Shi, B. Carter, R. Barrios, C. Ou, P. Chevez-Barrios, Y. Wang, G. Habib, J. Goodman, S. Huang, R. Lebovitz, and M. Matzuk. 1996. "Growth retardation and cysteine deficiency in gamma-glutamyl transpeptidase-deficient mice." *PNAS* 93 (15):7923–7926.

Lim, J., S. Ali, L.S. Liao, E.S. Nguyen, L. Ortiz, S. Reshel, and U. Luderer. 2020. "Antioxidant supplementation partially rescues accelerated ovarian follicle loss, but not oocyte quality, of glutathione-deficient micedagger." *Biol Reprod* 102 (5):1065–1079. doi:10.1093/biolre/ioaa009.

Lim, J., and U. Luderer. 2018. "Glutathione deficiency sensitizes cultured embryonic mouse ovaries to benzo[a]pyrene-induced germ cell apoptosis." *Toxicol Appl Pharmacol* 352:38–45. doi:10.1016/j.taap.2018.05.024.

Liu, D., J. Xue, Y. Liu, H. Gu, X. Wei, W. Ma, W. Luo, L. Ma, S. Jia, N. Dong, J. Huang, Y. Wang, and Z. Yuan. 2018. "Inhibition of NRF2 signaling and increased reactive oxygen species during embryogenesis in a rat model of retinoic acid-induced neural tube defects." *Neurotoxicology* 69:84–92. doi:10.1016/j.neuro.2018.09.005.

Longo, L., O.C. Vanegas, M. Patel, V. Rosti, H. Li, J. Waka, T. Merghoub, P.P. Pandolfi, R. Notaro, K. Manova, and L. Luzzatto. 2002. "Maternally transmitted severe glucose 6-phosphate dehydrogenase deficiency is an embryonic lethal." *EMBO J* 21 (16):4229–4239. doi:10.1093/emboj/cdf426.

Lu, S.C. 2009. "Regulation of glutathione synthesis." *Mol Aspects Med* 30 (1–2):42–59. doi:10.1016/j.mam.2008.05.005.

Lu, S.C. 2013. "Glutathione synthesis." *Biochim Biophys Acta* 1830 (5):3143–3153. doi:10.1016/j.bbagen.2012.09.008.

Lundberg, M., C. Johansson, J. Chandra, M. Enoksson, G. Jacobsson, J. Ljung, M. Johansson, and A. Holmgren. 2001. "Cloning and expression of a novel human glutaredoxin (Grx2) with mitochondrial and nuclear isoforms." *J Biol Chem* 276 (28):26269–26275. doi:10.1074/jbc.M011605200.

Maffi, S.K., M.L. Rathinam, P.P. Cherian, W. Pate, R. Hamby-Mason, S. Schenker, and G.I. Henderson. 2008. "Glutathione content as a potential mediator of the vulnerability of cultured fetal cortical neurons to ethanol-induced apoptosis." *J Neurosci Res* 86 (5):1064–1076. doi:10.1002/jnr.21562.

Mahadev, K., A. Zilbering, L. Zhu, and B.J. Goldstein. 2001. "Insulin-stimulated hydrogen peroxide reversibly inhibits protein-tyrosine phosphatase 1b in vivo and enhances the early insulin action cascade." *J Biol Chem* 276 (24):21938–21942. doi:10.1074/jbc.C100109200.

Mailloux, R.J., J.Y. Xuan, S. McBride, W. Maharsy, S. Thorn, C.E. Holterman, C.R. Kennedy, P. Rippstein, R. deKemp, J. da Silva, M. Nemer, M. Lou, and M.E. Harper. 2014. "Glutaredoxin-2 is required to control oxidative phosphorylation in cardiac muscle by mediating deglutathionylation reactions." *J Biol Chem* 289 (21):14812–14828. doi:10.1074/jbc.M114.550574.

Markovic, J., C. Borras, A. Ortega, J. Sastre, J. Vina, and F.V. Pallardo. 2007. "Glutathione is recruited into the nucleus in early phases of cell proliferation." *J Biol Chem* 282 (28):20416–20424. doi:10.1074/jbc.M609582200.

Markovic, J., N.J. Mora, A.M. Broseta, A. Gimeno, N. de-la-Concepcion, J. Vina, and F.V. Pallardo. 2009. "The depletion of nuclear glutathione impairs cell proliferation in 3t3 fibroblasts." *PLoS One* 4 (7):e6413. doi:10.1371/journal.pone.0006413.

Matsui, R., B. Ferran, A. Oh, D. Croteau, D. Shao, J. Han, D.R. Pimentel, and M.M. Bachschmid. 2020. "Redox regulation via glutaredoxin-1 and protein s-glutathionylation." *Antioxid Redox Signal* 32 (10):677–700. doi:10.1089/ars.2019.7963.

McCarver, D.G., and R.N. Hines. 2002. "The ontogeny of human drug-metabolizing enzymes: phase II conjugation enzymes and regulatory mechanisms."

J Pharmacol Exp Ther 300 (2):361–366. doi:10.1124/jpet.300.2.361.

McNutt, T.L., and C. Harris. 1994. "Lindane embryotoxicity and differential alteration of cysteine and glutathione levels in rat embryos and visceral yolk sacs." *Reprod Toxicol* 8 (4):351–362. doi:10.1016/0890-6238(94)90051-5.

Meinerz, D.F., V. Branco, M. Aschner, C. Carvalho, and J.B.T. Rocha. 2017. "Diphenyl diselenide protects against methylmercury-induced inhibition of thioredoxin reductase and glutathione peroxidase in human neuroblastoma cells: a comparison with ebselen." *J Appl Toxicol* 37 (9):1073–1081. doi:10.1002/jat.3458.

Meister, A. 1983. "Transport and metabolism of glutathione and gamma-glutamyl-transferase aminoacids." *Biochem Soc Transact* 11 (6):793–794. doi:10.1042/bst0110793.

Mendieta-Serrano, M.A., F.J. Mendez-Cruz, M. Antunez-Mojica, D. Schnabel, L. Alvarez, L. Cardenas, H. Lomeli, J.A. Ruiz-Santiesteban, and E. Salas-Vidal. 2018. "NADPH-Oxidase-derived reactive oxygen species are required for cytoskeletal organization, proper localization of E-cadherin and cell motility during zebrafish epiboly." *Free Radic Biol Med* 130:82–98. doi:10.1016/j.freeradbiomed.2018.10.416.

Mieyal, J.J., D.W. Starke, S.A. Gravina, and B.A. Hocevar. 1991. "Thioltransferase in human red blood cells: kinetics and equilibrium." *Biochemistry* 30 (36):8883–8891. doi:10.1021/bi00100a023.

Miseta, A., and P. Csutora. 2000. "Relationship between the occurrence of cysteine in proteins and the complexity of organisms." *Mol Biol Evol* 17 (8):1232–1239. doi:10.1093/oxfordjournals.molbev.a026406.

Miyazawa, H., and A. Aulehla. 2018. "Revisiting the role of metabolism during development." *Development* 145 (19). doi:10.1242/dev.131110.

Mondol, A.S., N.K. Tonks, and T. Kamata. 2014. "Nox4 redox regulation of PTP1B contributes to the proliferation and migration of glioblastoma cells by modulating tyrosine phosphorylation of coronin-1C." *Free Radic Biol Med* 67:285–291. doi:10.1016/j.freeradbiomed.2013.11.005.

Morgan, M.J., and Z.G. Liu. 2011. "Crosstalk of reactive oxygen species and NF-kappaB signaling." *Cell Res* 21 (1):103–115. doi:10.1038/cr.2010.178.

Nasr-Esfahani, M., and M. Johnson. 1992. "Quantitative analysis of cellular glutathione in early preimplantation mouse embryos developing in vivo and in vitro." *Hum Reprod* 7 (9):1281–1290.

Nebert, D., and V. Vasiliou. 2004. "Analysis of the glutathione S-transferase (GST) gene family." *Hum Genomics* 1 (6):460–464.

Nicholson, D., A. Ali, N. Thornberry, J. Vaillancourt, C. Ding, M. Gallant, Y. Gareau, P. Griffin, M. Labelle, Y. Lazebnik, N. Munday, S. Raju, M. Smulson, T. Yamin, V. Yu, and D. Miller. 1995. "Identification and inhibition of the ICE/CED-3 protease necessary for mammalian apoptosis." *Nature* 376 (6535):37–43.

Nkabyo, Y., T.R. Ziegler, L. Gu, W. Watson, and D. Jones. 2002. "Glutathione and thioredoxin redox during differentiation in human colon epithelial (Caco-2) cells." *Am J Physiol Gastrointest Liver Physiol* 283 (6):G1352–1359.

Nonn, L., R.R. Williams, R.P. Erickson, and G. Powis. 2003. "The absence of mitochondrial thioredoxin 2 causes massive apoptosis, exencephaly, and early embryonic lethality in homozygous mice." *Mol Cell Biol* 23 (3):916–922. doi:10.1128/mcb.23.3.916-922.2003.

O'Fallon, J.V., and R.W. Wright, Jr. 1986. "Quantitative determination of the pentose phosphate pathway in preimplantation mouse embryos." *Biol Reprod* 34 (1):58–64. doi:10.1095/biolreprod34.1.58.

Oestreicher, J., and B. Morgan. 2019. "Glutathione: subcellular distribution and membrane transport (1)." *Biochem Cell Biol* 97 (3):270–289. doi:10.1139/bcb-2018-0189.

Okuno, H., A. Akahori, H. Sato, S. Xanthoudakis, T. Curran, and H. Iba. 1993. "Escape from redox regulation enhances the transforming activity of Fos." *Oncogene* 8 (3):695–701.

Olson, G.E., J.C. Whitin, K.E. Hill, V.P. Winfrey, A.K. Motley, L.M. Austin, J. Deal, H.J. Cohen, and R.F. Burk. 2010. "Extracellular glutathione peroxidase (Gpx3) binds specifically to basement membranes of mouse renal cortex tubule cells." *Am J Physiol Renal Physiol* 298 (5):F1244–1253. doi:10.1152/ajprenal.00662.2009.

Ookhtens, M., and N. Kaplowitz. 1998. "Role of the liver in interorgan homeostasis of glutathione and cyst(e)ine." *Semin Liver Dis* 18 (4):313–329. doi:10.1055/s-2007-1007167.

Ornoy, A. 2007. "Embryonic oxidative stress as a mechanism of teratogenesis with special emphasis on diabetic embryopathy." *Reprod Toxicol* 24 (1):31–41. doi:10.1016/j.reprotox.2007.04.004.

Ou, Y.C., C.C. White, C.M. Krejsa, R.A. Ponce, T.J. Kavanagh, and E.M. Faustman. 1999. "The role of intracellular glutathione in methylmercury-induced toxicity in embryonic neuronal cells." *Neurotoxicology* 20 (5):793–804.

Ozolins, T.R., W. Harrouk, T. Doerksen, J.M. Trasler, and B.F. Hales. 2002. "Buthionine sulfoximine embryotoxicity is associated with prolonged AP-1 activation." *Teratology* 66 (4):192–200. doi:10.1002/tera.10084.

Pallardo, F.V., J. Markovic, J.L. Garcia, and J. Vina. 2009. "Role of nuclear glutathione as a key regulator of cell proliferation." *Mol Aspects Med* 30 (1–2):77–85. doi:10.1016/j.mam.2009.01.001.

Pan, S., and B.C. Berk. 2007. "Glutathiolation regulates tumor necrosis factor-alpha-induced caspase-3 cleavage and apoptosis: key role for glutaredoxin in the death pathway." *Circ Res* 100 (2):213–219. doi:10.1161/01.RES.0000256089.30318.20.

Panieri, E., C. Millia, and M.M. Santoro. 2017. "Real-time quantification of subcellular H_2O_2 and glutathione redox potential in living cardiovascular tissues." *Free Radic Biol Med* 109:189–200. doi:10.1016/j.freeradbiomed.2017.02.022.

Pantaleon, M., M.B. Harvey, W.S. Pascoe, D.E. James, and P.L. Kaye. 1997. "Glucose transporter GLUT3: ontogeny, targeting, and role in the mouse blastocyst." *Proc Nat Acad Sci U S A* 94 (8):3795–3800. doi:10.1073/pnas.94.8.3795.

Parman, T., G. Chen, and P.G. Wells. 1998. "Free radical intermediates of phenytoin and related teratogens. Prostaglandin H synthase-catalyzed bioactivation, electron paramagnetic resonance spectrometry, and photochemical product analysis." *J Biol Chem* 273 (39):25079–25088. doi:10.1074/jbc.273.39.25079.

Parman, T., M.J. Wiley, and P.G. Wells. 1999. "Free radical-mediated oxidative DNA damage in the mechanism of thalidomide teratogenicity." *Nat Med* 5 (5):582–585. doi:10.1038/8466.

Pashkovskaia, N., U. Gey, and G. Rodel. 2018. "Mitochondrial ROS direct the differentiation of murine pluripotent P19 cells." *Stem Cell Res* 30:180–191. doi:10.1016/j.scr.2018.06.007.

Passarelli, C., A. Di Venere, N. Piroddi, A. Pastore, B. Scellini, C. Tesi, S. Petrini, P. Sale, E. Bertini, C. Poggesi, and F. Piemonte. 2010. "Susceptibility of isolated myofibrils to in vitro glutathionylation: potential relevance to muscle functions." *Cytoskeleton (Hoboken)* 67 (2):81–89. doi:10.1002/cm.20425.

Pastore, A., G. Federici, E. Bertini, and F. Piemonte. 2003. "Analysis of glutathione: implication in redox and detoxification." *Clin Chim Acta* 333 (1):19–39. doi:10.1016/s0009-8981(03)00200-6.

Pastore, A., G. Tozzi, L.M. Gaeta, E. Bertini, V. Serafini, S. Di Cesare, V. Bonetto, F. Casoni, R. Carrozzo, G. Federici, and F. Piemonte. 2003. "Actin glutathionylation increases in fibroblasts of patients with Friedreich's ataxia: a potential role in the pathogenesis of the disease." *J Biol Chem* 278 (43):42588–42595. doi:10.1074/jbc.M301872200.

Pineda-Molina, E., P. Platt, J. Vazquez, A. Marina, M. Garcia de Lacoba, D. Perez-Sala, and S. Lamas.

2001. "Glutathionylation of the p50 subunit of NF-kappaB: a mechanism for redox-induced inhibition of DNA binding." *Biochemistry* 40 (47):14134–14142.

Piruat, J.I., C.O. Pintado, P. Ortega-Saenz, M. Roche, and J. Lopez-Barneo. 2004. "The mitochondrial SDHD gene is required for early embryogenesis, and its partial deficiency results in persistent carotid body glomus cell activation with full responsiveness to hypoxia." *Mol Cell Biol* 24 (24):10933–10940.doi:10.1128/MCB.24.24.10933-10940.2004.

Ponce, R.A., T.J. Kavanagh, N.K. Mottet, S.G. Whittaker, and E.M. Faustman. 1994. "Effects of methyl mercury on the cell cycle of primary rat CNS cells in vitro." *Toxicol Appl Pharmacol* 127 (1):83–90. doi:10.1006/taap.1994.1142.

Poole, L.B. 2015. "The basics of thiols and cysteines in redox biology and chemistry." *Free Radic Biol Med* 80: 148–157. doi:10.1016/j.freeradbiomed.2014.11.013.

Pretsch, W. 1999. "Glutathione reductase activity deficiency in homozygous Gr1a1Neu mice does not cause haemolytic anaemia." *Genet Res* 73 (1):1–5. doi:10.1017/s0016672398003590.

Qanungo, S., and M. Mukherjea. 2000. "Ontogenic profile of some antioxidants and lipid peroxidation in human placental and fetal tissues." *Mol Cell Biochem* 215 (1–2):11–19.

Quinn, P., and R.G. Wales. 1973. "The relationships between the ATP content of preimplantation mouse embryos and their development in vitro during culture." *J Reprod Fertil* 35 (2):301–309.

Ramachandran, V., A. Perez, J. Chen, D. Senthil, S. Schenker, and G.I. Henderson. 2001. "In utero ethanol exposure causes mitochondrial dysfunction, which can result in apoptotic cell death in fetal brain: a potential role for 4-hydroxynonenal." *Alcohol Clin Exp Res* 25 (6):862–871.

Rastogi, A., C.W. Clark, S.M. Conlin, S.E. Brown, and A.R. Timme-Laragy. 2019. "Mapping glutathione utilization in the developing zebrafish (*Danio rerio*) embryo." *Redox Biol* 26:101235. doi:10.1016/j.redox.2019.101235.

Reddy, N.M., S.R. Kleeberger, J.H. Bream, P.G. Fallon, T.W. Kensler, M. Yamamoto, and S.P. Reddy. 2008. "Genetic disruption of the Nrf2 compromises cell-cycle progression by impairing GSH-induced redox signaling." *Oncogene* 27 (44):5821–5832. doi:10.1038/onc.2008.188.

Reyes, E., S. Ott, B. Robinson, and R. Contreras. 1995. "The effect of in utero administration of buthionine sulfoximine on rat development." *Pharmacol Biochem Behav* 50 (4):491–497.

Reynaert, N., A. van der Vliet, A. Guala, T. McGovern, M. Hristova, C. Pantano, N. Heintz, J. Heim, Y. Ho, D. Matthews, E. Wouters, and Y. Janssen-Heininger. 2006. "Dynamic redox control of NF-kappaB through glutaredoxin-regulated S-glutathionylation of inhibitory kappaB kinase beta." *Proc Natl Acad Sci U S A* 103 (35):13086–13091.

Richman, P.G., and A. Meister. 1975. "Regulation of gamma-glutamyl-cysteine synthetase by nonallosteric feedback inhibition by glutathione." *J Biol Chem* 250 (4):1422–1426.

Riley, J.K., and K.H. Moley. 2006. "Glucose utilization and the P13-K pathway: mechanisms for cell survival in preimplantation embryos." *Reproduction* 131 (5):823–835. doi:10.1530/rep.1.00645.

Robitaille, S., R.J. Mailloux, and H.M. Chan. 2016. "Methylmercury alters glutathione homeostasis by inhibiting glutaredoxin 1 and enhancing glutathione biosynthesis in cultured human astrocytoma cells." *Toxicol Lett* 256:1–10. doi:10.1016/j.toxlet.2016.05.013.

Rogers, L.K., C.M. Bates, S.E. Welty, and C.V. Smith. 2006. "Diquat induces renal proximal tubule injury in glutathione reductase-deficient mice." *Toxicol Appl Pharmacol* 217 (3):289–298. doi:10.1016/j.taap.2006.08.012.

Rogers, L.K., T. Tamura, B.J. Rogers, S.E. Welty, T.N. Hansen, and C.V. Smith. 2004. "Analyses of glutathione reductase hypomorphic mice indicate a genetic knockout." *Toxicol Sci* 82 (2):367–373. doi:10.1093/toxsci/kfh268.

Rojas, E., M.Valverde, S.Kala, G.Kala, and M.W.Lieberman. 2000. "Accumulation of DNA damage in the organs of mice deficient in γ-glutamyltranspeptidase." *Mutat Res* 447 (2):305–316.

Rokutan, K., R. Johnston Jr, and K. Kawai. 1994. "Oxidative stress induces S-thiolation of specific proteins in cultured gastric mucosal cells." *Am J Physiol* 266 (2):G247–254.

Romagnoli, C., G. Marcucci, F. Favilli, R. Zonefrati, C. Mavilia, G. Galli, A. Tanini, T. Iantomasi, M.L. Brandi, and M.T. Vincenzini. 2013. "Role of GSH/GSSG redox couple in osteogenic activity and osteoclastogenic markers of human osteoblast-like SaOS-2 cells." *FEBS J* 280 (3):867–879. doi:10.1111/febs.12075.

Roos, G., N. Foloppe, and J. Messens. 2013. "Understanding the pK(a) of redox cysteines: the key role of hydrogen bonding." *Antioxid Redox Signal* 18 (1):94–127. doi:10.1089/ars.2012.4521.

Ross, M.M., T.B. Piorczynski, J. Harvey, T.S. Burnham, M. Francis, M.W. Larsen, K. Roe, J.M. Hansen, and M.R. Stark. 2019. "Ceramide: a novel inducer for neural tube defects." *Dev Dyn* 248 (10):979–996. doi:10.1002/dvdy.93.

Sahlin, L., H. Wang, Y. Stjernholm, M. Lundberg, G. Ekman, A. Holmgren, and H. Eriksson. 2000. "The expression of glutaredoxin is increased in the human cervix in term pregnancy and immediately postpartum, particularly after prostaglandin-induced delivery." *Mol Hum Reprod* 6 (12):1147–1153. doi:10.1093/molehr/6.12.1147.

Salmeen, A., J. Andersen, M. Myers, T. Meng, J. Hinks, N.K. Tonks, and D. Barford. 2003. "Redox regulation of protein tyrosine phosphatase 1B involves a sulphenyl-amide intermediate." *Nature* 423 (6941):769–773.

Samanta, D., and G.L. Semenza. 2017. "Maintenance of redox homeostasis by hypoxia-inducible factors." *Redox Biology* 13:331–335. doi:https://doi.org/10.1016/j.redox.2017.05.022.

Savitsky, P.A., and T. Finkel. 2002. "Redox regulation of Cdc25C." *J Biol Chem* 277 (23):20535–20540. doi:10.1074/jbc.M201589200.

Schafer, F.Q., and G.R. Buettner. 1999. "Singlet oxygen toxicity is cell line-dependent: a study of lipid peroxidation in nine leukemia cell lines." *Photochem Photobiol* 70 (6):858–867. doi:10.1111/j.1751-1097.1999.tb08294.x.

Schafer, F.Q., and G.R. Buettner. 2001. "Redox environment of the cell as viewed through the redox state of the glutathione disulfide/glutathione couple." *Free Radic Biol Med* 30 (11):1191–1212.

Schneider, M., H. Forster, A. Boersma, A. Seiler, H. Wehnes, F. Sinowatz, C. Neumuller, M.J. Deutsch, A. Walch, M. Hrabe de Angelis, W. Wurst, F. Ursini, A. Roveri, M. Maleszewski, M. Maiorino, and M. Conrad. 2009. "Mitochondrial glutathione peroxidase 4 disruption causes male infertility." *FASEB J* 23 (9):3233–3242. doi:10.1096/fj.09-132795.

Schneider, M., D.M. Vogt Weisenhorn, A. Seiler, G.W. Bornkamm, M. Brielmeier, and M. Conrad. 2006. "Embryonic expression profile of phospholipid hydroperoxide glutathione peroxidase." *Gene Expr Patterns* 6 (5):489–494. doi:10.1016/j.modgep.2005.11.002.

Seegmiller, R.E., C. Harris, D.L. Luchtel, and M.R. Juchau. 1991. "Morphological differences elicited by two weak acids, retinoic and valproic, in rat embryos grown in vitro." *Teratology* 43 (2):133–150. doi:10.1002/tera.1420430206.

Shapiro, A.M., and H.M. Chan. 2008. "Characterization of demethylation of methylmercury in cultured astrocytes." *Chemosphere* 74 (1):112–118. doi:10.1016/j.chemosphere.2008.09.019.

Shaulian, E., and M. Karin. 2001. "AP-1 in cell proliferation and survival." *Oncogene* 20 (19):2390–2400.

Shepard, T.H., L.A. Muffley, and L.T. Smith. 1998. "Ultrastructural study of mitochondria and their cristae in embryonic rats and primate (N. nemistrina)." *Anat Rec* 252 (3):383–392. doi:10.1002/(SICI)1097-0185(199811)252:3<383::AID-AR6>3.0.CO;2-Z.

Shepard, T.H., T. Tanimura, and H.W. Park. 1997. "Glucose absorption and utilization by rat embryos." *Int J Dev Biol* 41 (2):307–314.

Shi, Z., J. Osei-Frimpong, G. Kala, S. Kala, R. Barrios, G. Habib, D. Lukin, C. Danney, M. Matzuk, and M.W. Lieberman. 2000a. "Glutathione synthesis is essential for mouse development but not for cell growth in culture." *Proc Natl Acad Sci U S A* 97 (10):5101–5106.

Shi, Z.Z., J. Osei-Frimpong, G. Kala, S.V. Kala, R. Barrios, G.M. Habib, D.J. Lukin, C.M. Danney, M.M. Matzuk, and M.W. Lieberman. 2000b. "Glutathione-deficiency causes embryonic lethality in mice, but is compatible with cell growth in vitro." *FASEB J* 14 (8):A1401–A1401.

Sies, H. 1985. "Oxidative stress: introductory remarks." In *Oxidative stress*, edited by H. Sies, 1–7. London: Academic Press.

Sies, H. 1999. "Glutathione and its role in cellular functions." *Free Radic Biol Med* 27 (9–10):916–921. doi:10.1016/S0891-5849(99)00177-X.

Slott, V., and B. Hales. 1987. "Effect of glutathione depletion by buthionine sulfoximine on rat embryonic development in vitro." *Biochem Pharmacol* 36 (5):683–688.

Smales, W.P., and D.M. Biddulph. 1985. "Limb development in chick embryos: cyclic AMP-dependent protein kinase activity, cyclic AMP, and prostaglandin concentrations during cytodifferentiation and morphogenesis." *J Cell Physiol* 122 (2):259–265. doi:10.1002/jcp.1041220215.

Stark, K., C. Harris, and M. Juchau. 1987. "Embryotoxicity elicited by inhibition of γ-glutamyltransferase by Acivicin and transferase antibodies in cultured rat embryos." *Toxicol Appl Pharmacol* 89 (1):88–96.

Stark, K.L., C. Harris, and M.R. Juchau. 1989. "Modulation of the embryotoxicity and cytotoxicity elicited by 7-hydroxy-2-acetylaminofluorene and acetaminophen via deacetylation." *Toxicol Appl Pharmacol* 97 (3):548–560. doi:10.1016/0041-008x(89)90260-3.

Stincone, A., A. Prigione, T. Cramer, M.M. Wamelink, K. Campbell, E. Cheung, V. Olin-Sandoval, N.M. Gruning, A. Kruger, M. Tauqeer Alam, M.A. Keller, M. Breitenbach, K.M. Brindle, J.D. Rabinowitz, and M. Ralser. 2015. "The return of metabolism: biochemistry and physiology of the pentose phosphate pathway." *Biol Rev Camb Philos Soc* 90 (3):927–963. doi:10.1111/brv.12140.

Stournaras, C., G. Drewes, H. Blackholm, I. Merkler, and H. Faulstich. 1990. "Glutathionyl(cysteine-374) actin forms filaments of low mechanical stability." *Biochim Biophys Acta* 1037 (1):86–91.

Strange, R., A. Howie, R. Hume, B. Matharoo, J. Bell, C. Hiley, P. Jones, and G. Beckett. 1989. "The developmental expression of alpha-, mu- and pi-class glutathione S-transferases in human liver." *Biochim Biophys Acta* 993 (2–3):186–190.

Styrud, J., and U. Eriksson. 1990. "Effects of D-glucose and beta-hydroxybutyric acid on the in vitro development of (pre)chondrocytes from embryos of normal and diabetic rats." *Acta Endocrinol (Copenh)* 122 (4):487–498.

Suzuki, N., K. Svensson, and U. Eriksson. 1996. "High glucose concentration inhibits migration of rat cranial neural crest cells in vitro." *Diabetologia* 39 (4):401–411.

Swanson, P.A., 2nd, A. Kumar, S. Samarin, M. Vijay-Kumar, K. Kundu, N. Murthy, J. Hansen, A. Nusrat, and A.S. Neish. 2011. "Enteric commensal bacteria potentiate epithelial restitution via reactive oxygen species-mediated inactivation of focal adhesion kinase phosphatases." *Proc Natl Acad Sci U S A* 108 (21):8803–8808. doi:10.1073/pnas.1010042108.

Takahashi, M. 2012. "Oxidative stress and redox regulation on in vitro development of mammalian embryos." *J Reprod Dev* 58 (1):1–9. doi:10.1262/jrd.11-138N.

Takahashi, M., T. Nagai, N. Okamura, H. Takahashi, and A. Okano. 2002. "Promoting effect of beta-mercaptoethanol on in vitro development under oxidative stress and cystine uptake of bovine embryos." *Biol Reprod* 66 (3):562–567. doi:10.1095/biolreprod66.3.562.

Takahashi, S., and M. Zeydel. 1982. "gamma-Glutamyl transpeptidase and glutathione in aging IMR-90 fibroblasts and in differentiating 3T3 L1 preadipocytes." *Arch Biochem Biophys* 214 (1):260–267.

Tanaka, K., M. Honda, and T. Takabatake. 2001. "Redox regulation of MAPK pathways and cardiac hypertrophy in adult rat cardiac myocyte." *J Am Col Cardiol* 37 (2):676–685. doi:10.1016/s0735-1097(00)01123-2.

Tee, L., K. Gilmore, D. Meyer, B. Ketterer, Y. Vandenberghe, and G. Yeoh. 1992. "Expression of glutathione S-transferase during rat liver development." *Biochem J* 282:209–218.

Thompson, S.A., C.C. White, C.M. Krejsa, D.L. Eaton, and T.J. Kavanagh. 2000. "Modulation of glutathione and glutamate-L-cysteine ligase by methylmercury during mouse development." *Toxicol Sci* 57 (1):141–146. doi:10.1093/toxsci/57.1.141.

Timme-Laragy, A.R., J.V. Goldstone, B.R. Imhoff, J.J. Stegeman, M.E. Hahn, and J.M. Hansen. 2013. "Glutathione redox dynamics and expression of glutathione-related genes in the developing embryo." *Free Radic Biol Med* 65:89–101. doi:10.1016/j.freeradbiomed.2013.06.011.

Toppo, S., S. Vanin, V. Bosello, and S.C. Tosatto. 2008. "Evolutionary and structural insights into the multifaceted glutathione peroxidase (Gpx) superfamily." *Antioxid Redox Signal* 10 (9):1501–1514. doi:10.1089/ars.2008.2057.

Trimarchi, J.R., L. Liu, D.M. Porterfield, P.J. Smith, and D.L. Keefe. 2000a. "A non-invasive method for measuring preimplantation embryo physiology." *Zygote* 8 (1):15–24.

Trimarchi, J.R., L. Liu, D.M. Porterfield, P.J. Smith, and D.L. Keefe. 2000b. "Oxidative phosphorylation-dependent and -independent oxygen consumption by individual preimplantation mouse embryos." *Biol Reprod* 62 (6):1866–1874. doi:10.1095/biolreprod 62.6.1866.

Trivedi, M., Y. Zhang, M. Lopez-Toledano, A. Clarke, and R. Deth. 2016. "Differential neurogenic effects of casein-derived opioid peptides on neuronal stem cells: implications for redox-based epigenetic changes." *J Nutr Biochem* 37:39–46. doi:10.1016/j.jnutbio.2015.10.012.

Tsapara, A., D. Kardassis, A. Moustakas, A. Gravanis, and C. Stournaras. 1999. "Expression and characterization of Cys374 mutated human beta-actin in two different mammalian cell lines: impaired microfilament organization and stability." *FEBS Lett* 455 (1–2):117–122.

Tung, E.W., and L.M. Winn. 2011. "Valproic acid increases formation of reactive oxygen species and induces apoptosis in postimplantation embryos: a role for oxidative stress in valproic acid-induced neural tube defects." *Mol Pharmacol* 80 (6):979–987. doi:10.1124/mol.111.072314.

Ufer, C., and C.C. Wang. 2011. "The Roles of Glutathione Peroxidases during Embryo Development." *Front Mol Neurosci* 4:12. doi:10.3389/fnmol.2011.00012.

Usuki, F., and M. Fujimura. 2016. "Decreased plasma thiol antioxidant barrier and selenoproteins as potential biomarkers for ongoing methylmercury intoxication and an individual protective capacity." *Arch Toxicol* 90 (4):917–926. doi:10.1007/s00204-015-1528-3.

Vargesson, N. 2015. "Thalidomide-induced teratogenesis: history and mechanisms." *Birth Defects Res C Embryo Today* 105 (2):140–156. doi:10.1002/bdrc.21096.

Vigilanza, P., K. Aquilano, S. Baldelli, G. Rotilio, and M.R. Ciriolo. 2011. "Modulation of intracellular glutathione affects adipogenesis in 3T3-L1 cells." *J Cell Physiol* 226 (8):2016–2024. doi:10.1002/jcp.22542.

Vivancos, P.D., Y. Dong, K. Ziegler, J. Markovic, F.V. Pallardo, T.K. Pellny, P.J. Verrier, and C.H. Foyer. 2010. "Recruitment of glutathione into the nucleus during cell proliferation adjusts whole-cell redox homeostasis in *Arabidopsis thaliana* and lowers the oxidative defence shield." *Plant J* 64 (5):825–838. doi:10.1111/j.1365-313X.2010.04371.x.

Wales, R.G., and Z.F. Du. 1993. "Contribution of the pentose phosphate pathway to glucose utilization by preimplantation sheep embryos." *Reprod Fertil Dev* 5 (3):329–340. doi:10.1071/rd9930329.

Walshe, J., M.M. Serewko-Auret, N. Teakle, S. Cameron, K. Minto, L. Smith, P.C. Burcham, T. Russell, G. Strutton, A. Griffin, F.F. Chu, S. Esworthy, V. Reeve, and N.A. Saunders. 2007. "Inactivation of glutathione peroxidase activity contributes to UV-induced squamous cell carcinoma formation." *Cancer Res* 67 (10):4751–4758. doi:10.1158/0008-5472.CAN-06-4192.

Wang, J., E.S. Boja, W. Tan, E. Tekle, H.M. Fales, S. English, J.J. Mieyal, and P.B. Chock. 2001. "Reversible glutathionylation regulates actin polymerization in A431 cells." *J Biol Chem* 276 (51):47763–47766. doi:10.1074/jbc.C100415200.

Wang, J., E. Tekle, H. Oubrahim, J.J. Mieyal, E. Stadtman, and P.B. Chock. 2003. "Stable and controllable RNA interference: investigating the physiological function of glutathionylated actin." *PNAS* 100 (9):5103–5106.

West, M.B., B.G. Hill, Y.T. Xuan, and A. Bhatnagar. 2006. "Protein glutathiolation by nitric oxide: an intracellular mechanism regulating redox protein modification." *FASEB J* 20 (10):1715–1717. doi:10.1096/fj.06-5843fje.

Wilke, N., M. Sganga, S. Barhite, and M.F. Miles. 1994. "Effects of alcohol on gene expression in neural cells." *EXS* 71:49–59. doi:10.1007/978-3-0348-7330-7_6.

Will, Y., K.A. Fischer, R.A. Horton, R.S. Kaetzel, M.K. Brown, O. Hedstrom, M.W. Lieberman, and D.J. Reed. 2000. "gamma-glutamyltranspeptidase-deficient knockout mice as a model to study the relationship between glutathione status, mitochondrial function, and cellular function." *Hepatology* 32 (4 Pt 1):740–749. doi:10.1053/jhep.2000.17913.

Wilson, J.G. 1977. "Teratogenic effects of environmental chemicals." *Fed Proc* 36 (5):1698–1703.

Wingert, R.A., J.L. Galloway, B. Barut, H. Foott, P. Fraenkel, J.L. Axe, G.J. Weber, K. Dooley, A.J. Davidson, B. Schmid, B.H. Paw, G.C. Shaw, P. Kingsley, J. Palis, H. Schubert, O. Chen, J. Kaplan, L.I. Zon, and C. Tubingen Screen. 2005. "Deficiency of glutaredoxin 5 reveals Fe-S clusters are required for vertebrate haem synthesis." *Nature* 436 (7053):1035–1039. doi:10.1038/nature03887.

Winkler, A., R. Njalsson, K. Carlsson, A. Elgadi, B. Rozell, L. Abraham, N. Ercal, Z.Z. Shi, M.W. Lieberman, A. Larsson, and S. Norgren. 2011. "Glutathione is essential for early embryogenesis: analysis of a glutathione synthetase knockout mouse." *Biochem Biophys Res Commun* 412 (1):121–126. doi:10.1016/j.bbrc.2011.07.056.

Xie, Y., S. Kole, P. Precht, M.J. Pazin, and M. Bernier. 2009. "S-glutathionylation impairs signal transducer and activator of transcription 3 activation and signaling." *Endocrinology* 150 (3):1122–1131. doi:10.1210/en.2008-1241.

Yamashita, A., J. Deguchi, Y. Honda, T. Yamada, I. Miyawaki, Y. Nishimura, and T. Tanaka. 2018. "Increased susceptibility to oxidative stress-induced toxicological evaluation by genetically modified nrf2a-deficient zebrafish." *J Pharmacol Toxicol Methods*. doi:10.1016/j.vascn.2018.12.006.

Yan, J., and B.F. Hales. 2006. "Depletion of glutathione induces 4-hydroxynonenal protein adducts and hydroxyurea teratogenicity in the organogenesis stage mouse embryo." *J Pharmacol Exp Ther* 319 (2):613–621. doi:10.1124/jpet.106.109850.

Yanes, O., J. Clark, D.M. Wong, G.J. Patti, A. Sanchez-Ruiz, H.P. Benton, S.A. Trauger, C. Desponts, S. Ding, and G. Siuzdak. 2010. "Metabolic oxidation regulates embryonic stem cell differentiation." *Nat Chem Biol* 6 (6):411–417. doi:10.1038/nchembio.364.

Yang, H., Y. Zeng, T.D. Lee, Y. Yang, X. Ou, L. Chen, M. Haque, R. Rippe, and S.C. Lu. 2002. "Role of AP-1 in the coordinate induction of rat glutamate-cysteine ligase and glutathione synthetase by tert-butylhydroquinone." *J Biol Chem* 277 (38):35232–35239. doi:10.1074/jbc.M203812200.

Yang, Y., M.Z. Dieter, Y. Chen, H.G. Shertzer, D.W. Nebert, and T.P. Dalton. 2002. "Initial characterization of the glutamate-cysteine ligase modifier subunit Gclm(-/-) knockout mouse. Novel model system for a severely compromised oxidative stress response." *J Biol Chem* 277 (51):49446–49452. doi:10.1074/jbc.M209372200.

Yant, L.J., Q. Ran, L. Rao, H. Van Remmen, T. Shibatani, J.G. Belter, L. Motta, A. Richardson, and T.A. Prolla. 2003. "The selenoprotein GPX4 is essential for mouse development and protects from radiation and oxidative damage insults." *Free Radic Biol Med* 34 (4):496–502. doi:10.1016/s0891-5849(02)01360-6.

Yeh, M.Y., E.L. Burnham, M. Moss, and L.A. Brown. 2007. "Chronic alcoholism alters systemic and pulmonary glutathione redox status." *Am J Respir Crit Care Med* 176 (3):270–276. doi:10.1164/rccm.200611-1722OC.

Yoshioka, J., E.R. Schreiter, and R.T. Lee. 2006. "Role of thioredoxin in cell growth through interactions with signaling molecules." *Antioxid Redox Signal* 8 (11–12):2143–2151. doi:10.1089/ars.2006.8.2143.

Zemolin, A.P., D.F. Meinerz, M.T. de Paula, D.O. Mariano, J.B. Rocha, A.B. Pereira, T. Posser, and J.L. Franco. 2012. "Evidences for a role of glutathione peroxidase 4 (GPx4) in methylmercury induced neurotoxicity in vivo." *Toxicology* 302 (1):60–67. doi:10.1016/j.tox.2012.07.013.

Zhang, H., and H.J. Forman. 2012. "Glutathione synthesis and its role in redox signaling." *Semin Cell Dev Biol* 23 (7):722–728. doi:10.1016/j.semcdb.2012.03.017.

Zheng, X., L. Boyer, M. Jin, J. Mertens, Y. Kim, L. Ma, L. Ma, M. Hamm, F.H. Gage, and T. Hunter. 2016. "Metabolic reprogramming during neuronal differentiation from aerobic glycolysis to neuronal oxidative phosphorylation." *eLife* 5:e13374. doi:10.7554/eLife.13374.

Zuelke, K.A., S.C. Jeffay, R.M. Zucker, and S.D. Perreault. 2003. "Glutathione (GSH) concentrations vary with the cell cycle in maturing hamster oocytes, zygotes, and pre-implantation stage embryos." *Mol Reprod Dev* 64 (1):106–112. doi:10.1002/mrd.10214.

CHAPTER ELEVEN

Roles of NRF2 in Quiescence and Differentiation

Shohei Murakami and Hozumi Motohashi

CONTENTS

11.1 INTRODUCTION

Oxidative stress was previously considered a cytotoxic factor that should be removed to support homeostasis. However, emerging studies have demonstrated that cellular redox balance is an indispensable factor for the regulation of diverse cellular functions, such as signal transduction, cell proliferation and differentiation, and apoptosis (Chaudhari et al., 2014; Sart et al., 2015). Among the interesting targets in the study of redox signaling, stem cells (SCs) are one of the highlighted topics, considering their promising applications in clinical therapies. SCs draw high attention in terms of the replacement/regeneration of damaged tissues and rejuvenation of aged tissues, and the discovery of induced pluripotent stem cells (iPSCs) has greatly extended the possibilities of SC therapies (Shi et al., 2017; Takahashi and Yamanaka, 2006). Currently, approaches to develop intervention strategies for the maintenance of healthy SCs in vivo and techniques for the generation and expansion of functional SCs ex vivo are critical issues that remain to be solved. Accumulation of

oxidative stress is highly related to this problem; a moderate level of oxidative stress, represented by reactive oxygen species (ROS), is necessary for the proliferation and differentiation of SCs, but excessive oxidative stress decreases the quality of SCs and disrupts their self-renewal capacity and multipotency (Chaudhari et al., 2014; Sart et al., 2015). Thus, finely tuned regulation of the oxidative stress level is indispensable for stem cell therapies, and a better understanding of molecular mechanisms underlying the loss of stemness and the development of novel techniques for SC maintenance is required for improving the engraftment of transplanted SCs and, thereby, designing effective therapies.

Various molecular factors have been reported to be involved in redox regulation in SCs, depending on the organism and tissue. Identifying transcription factors regulating these factors is one of the most effective and powerful means to control intracellular redox balance. NF-E2-related factor 2 (NRF2) is an inducible transcription factor that coordinately activates genes essential for

DOI: 10.4324/9781003204091-15

detoxification and antioxidant functions by binding to a cis-acting sequence, the antioxidant response element (ARE), in its target genes (Chaudhari et al., 2014; Yamamoto et al., 2018). Recent studies have reported that NRF2 activity substantially impacts SC functions, namely, their self-renewal capacity and balance between proliferation and differentiation (Dai et al., 2020; Murakami and Motohashi, 2015). These studies have involved hematopoietic stem cells (HSCs), mesenchymal stem cells (MSCs), embryonic stem cells (ESCs), iPSCs, and so on. This chapter focuses on the function of NRF2 and its contribution to several types of SCs and discusses the application of these findings to clinical therapies.

11.2 THE KEAP1-NRF2 SYSTEM

NRF2 has been shown to be a homolog of the transcription factor NF-E2 p45 and identified to bind to AREs (Itoh et al., 1997; Robledinos-Antón et al., 2019; Yamamoto et al., 2018). Its inducibility is attributed to KEAP1 (Itoh et al., 1999), which forms a ubiquitin E3 ligase complex with CULLIN3 and polyubiquitinates NRF2 in the cytosol, leading to proteasomal degradation of NRF2 under steady-state conditions (Figure 11.1A; Suzuki and Yamamoto, 2017; Yamamoto et al., 2018). KEAP1

contains highly reactive thiol groups, which sense electrophiles and oxidative stress (Figure 11.2A). Modification of the thiol groups in KEAP1 perturbs its interaction with NRF2 or CULLIN3 or induces conformational changes in KEAP1, resulting in stabilization and nuclear translocation of de novo translated NRF2 (Figure 11.1B; Suzuki and Yamamoto, 2017; Yamamoto et al., 2018). Thus, upon exposure to electrophiles or oxidative stress, NRF2 is translocated into the nucleus, forms heterodimers with small MAFs, and induces the transcription of target genes. In response to xenobiotic metabolites or oxidative stress, NRF2 activation enhances the transcription of genes encoding detoxification enzymes and antioxidant proteins and maintains cellular homeostasis.

NRF2 is composed of six NRF2-ECH homology (Neh) domains (Figure 11.2B; Robledinos-Antón et al., 2019; Yamamoto et al., 2018). Neh1 comprises the cap'n'collar (CNC) and bZIP domain, which are necessary for both the formation of dimers with small MAFs and DNA binding. Neh2 is the domain responsible for KEAP1-mediated degradation (Itoh et al., 1999), which binds to the double glycine repeat and carboxy-terminal region (DC) domain of KEAP1 homodimers (Figure 11.2B) through ETGE and DLG motifs; the former has higher affinity and the latter has lower affinity for

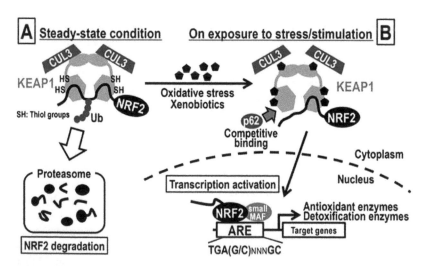

Figure 11.1 The KEAP1-NRF2 system.

(A) Under steady-state conditions, NRF2 is trapped and polyubiquitinated by the KEAP1-CULLIN3 complex and is then degraded via the proteasome. (B) Upon exposure to stress/stimulation, KEAP1 thiol groups sense the signals, and KEAP1 fails to form a complex with CULLIN3 or to bind to NRF2, which allows NRF2 to translocate to the nucleus and to promote transcription of its target genes by binding to their AREs. Typical NRF2 target genes encode antioxidant enzymes and detoxification enzymes. In addition, the autophagy-related protein p62 can activate NRF2 by competing with NRF2 for binding to KEAP1. CUL3, CULLIN3.

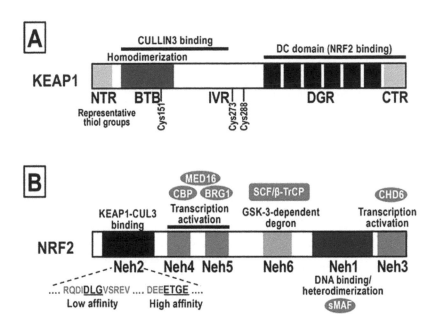

Figure 11.2 Molecular structures of KEAP1 and NRF2.

(A) KEAP1 structure and three representative reactive cysteine residues. KEAP1 is composed of an N-terminal region (NTR), a BTB domain, an intervening region (IVR), a double glycine repeat domain (DGR), and a C-terminal region domain (CTR). The BTB domain and IVR are essential for KEAP1 interaction with CULLIN3, and the DGR and CTR domains (DC) are responsible for binding to the Neh2 domain of NRF2. (B) NRF2 structure and cofactors. Neh1 recognizes AREs and interacts with small MAFs (sMAFs). Neh2 binds to the KEAP1 DC domain through the DLG and EGTG motifs. Neh3 is a transcriptional activation domain that recruits CHD6. Neh4 and Neh5 are required for full NRF2 transcriptional activation by interacting with transcriptional activators or mediators, including CBP, BRG1 and MED16. Neh6 is a KEAP1-independent degron that is phosphorylated by GSK-3 and leads to SCF/β-TrCP-mediated degradation of NRF2.

the DC domain (Figure 11.2B; Robledinos-Antón et al., 2019; Yamamoto et al., 2018). Neh3, Neh4, and Neh5 act as transactivation domains required for full transcriptional activation of NRF2. Chromodomain helicase DNA binding protein 6 (CHD6) can bind to Neh3 (Nioi et al., 2005), and CBP (histone acetyl-transferase cAMP responsive element binding protein), BRG1 (a catalytic subunit of SWI2/SNF2-like chromatin-remodeling complexes), and MED16 (a component of the Mediator complex) are recruited to Neh4 and Neh5 (Katoh et al., 2001; Sekine et al., 2016; Zhang et al., 2006). Neh6 is another degron in NRF2 that is independent of KEAP1. Glycogen synthase kinase-3 (GSK-3) phosphorylates a serine residue in Neh6, and NRF2 is subsequently degraded in an SCF/β-TrCP-dependent manner (Rada et al., 2011).

Because the KEAP1-NRF2 system responds to and eliminates oxidative stress, it has been considered a major regulator of cellular redox balance. However, recent studies have shown that NRF2 can be activated by abnormal accumulation of metabolites or inflammatory mediators (Adam et al., 2011; Bollong et al., 2018; Kobayashi et al., 2009; Suzuki and Yamamoto, 2017; Yamamoto et al., 2018). Concomitantly, other studies have shown that NRF2 targets genes encoding not only detoxification and antioxidant enzymes but also metabolic enzymes, especially genes related to the pentose phosphate pathway and glutamine metabolism (Bollong et al., 2018; Mitsuishi et al., 2012). Furthermore, NRF2 activation exerts suppressive effects on inflammatory responses at least partially due to direct suppression of proinflammatory cytokine production (Kobayashi et al., 2016). These new findings regarding NRF2 target genes, independent of positive or negative regulation, have gradually unveiled functions of NRF2 that had been obscured by the complexity of biological phenomena. Currently, this diversity of the KEAP1-NRF2 system is at the forefront of SC research, especially research on SC therapies and regenerative medicine.

11.3 COMMON FEATURES OF STEM CELLS

In the last decade, SCs started to attract attention for use in regenerative medicine and repair of damaged/hypoactive tissues. Considerable effort in SC studies has been directed towards characterizing SCs in several tissues in terms of their function and specific microenvironment. Adult tissue SCs have some common features beyond the types of SCs (Figure 11.3). SCs are localized in special microenvironments called "niches," which secrete essential factors and provide direct interactions with SCs for their maintenance; SCs are then maintained in a quiescent state to avoid replication-mediated senescence or exhaustion (Figure 11.3A; Chaudhari et al., 2014; Ito and Suda, 2014; Mohyeldin et al., 2010). In addition, because the microenvironment of SCs is hypoxic, SCs rely on glycolysis for energy production (Ito and Suda, 2014; Sart et al., 2015; Zhang et al., 2012). Accordingly, mitochondrial oxidative phosphorylation (OXPHOS) is suppressed in SCs, which protects SCs against the accumulation of ROS as by-products of OXPHOS and thereby against ROS-mediated exhaustion and dysfunction (Figure 11.3A; Ito and Suda, 2014; Sart et al., 2015). However, a moderate increase in the ROS level is required for the proliferation of certain SCs, although this increase is accompanied by a risk of stemness loss (Figure 11.3B). Therefore, the redox balance must be finely tuned to maintain SCs with intact stemness and to allow SCs to proliferate and differentiate in response to stimulation.

Considering that the KEAP1-NRF2 system is a major modifier of cellular redox status, one can expect that NRF2 activity is involved in SC regulation. Indeed, NRF2 affects the function and action of some SCs through ROS regulation, but interestingly, in other SCs, NRF2 directly regulates some genes essential for cell proliferation and differentiation (Dai et al., 2020; Murakami and Motohashi, 2015). The specific roles of NRF2 in different SCs are discussed below.

11.4 HEMATOPOIETIC STEM CELLS

HSCs are one of the most well characterized types of SCs because they are easily identified and assessed

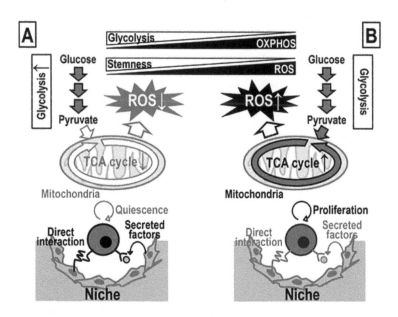

Figure 11.3 Common features of SCs.

(A) When SCs are maintained in the niche, SCs receive signals indispensable for their maintenance/quiescence from the niche, which secretes humoral factors, including cytokines and chemokines, and directly interacts with SCs. The niche is maintained under hypoxic conditions; therefore, SC metabolism relies on glycolysis, which suppresses mitochondrial activity, the tricarboxylic acid (TCA) cycle, and OXPHOS. Low mitochondrial activity results in reduced ROS production and subsequently protects SCs against ROS-mediated loss of stemness. (B) During differentiation, SCs are released from the niche due to attenuated signaling from the niche and start to proliferate for self-renewal or the production of progenitor cells. In addition, SCs undergo a metabolic shift from glycolysis to the TCA cycle and OXPHOS, which allows mitochondrial ROS production. The produced ROS enhance SC proliferation or differentiation.

through flow cytometry. HSCs are maintained in a quiescent state and in hypoxic bone marrow (BM) niches, where they interact with MSCs, osteoblasts, and so on (Ito and Suda, 2014; Mendelson and Frenette, 2014; Mohyeldin et al., 2010). HSCs do not participate in hematopoietic cell production under steady-state conditions (Sun et al., 2014), and their energy production is highly dependent on glycolysis (Suda et al., 2011). Many studies have demonstrated that HSCs are highly sensitive to oxidative stress; excessive accumulation of ROS attenuates HSC stemness (Suda et al., 2011). Based on the ROS-dependent dysfunction of HSCs, NRF2 was expected to offer strong buffering capability against oxidative stress in HSCs. However, a study using Nrf2 knockout mice showed that Nrf2 deficiency diminished the survival rate of hematopoietic stem and progenitor

cells (HSPCs) upon exposure to oxidative stress in vitro, but the dysfunction of HSPCs was not rescued by treatment with the ROS scavenger N-acetyl cysteine (Merchant et al., 2011). In another study, NRF2 was shown to directly regulate the expression of Cxcr4, which is important for retention and/ or homing of HSCs in BM niches (Figure 11.4A-i; Tsai et al., 2013). Downregulation of Nrf2 leads to the release of HSCs from the niches, which is followed by hyperproliferation at the expense of quiescence and self-renewal. After transplantation, Nrf2-deficient HSCs cannot efficiently migrate to BM niches, consistent with downregulation of Cxcr4. Therefore, NRF2 supports the maintenance and protection of HSCs, even though its activity is maintained at a low level under physiological conditions due to KEAP1-mediated regulation.

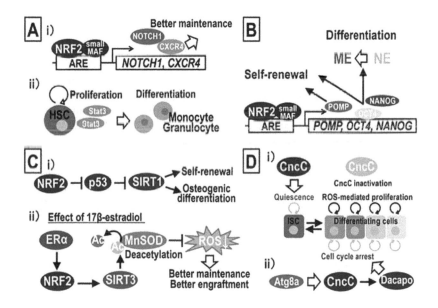

Figure 11.4 Effects of NRF2 activation on SCs.

(A) Effects on HSCs. (i) NRF2 activation induces CXCR4 or NOTCH1 gene expression by direct upregulation. (ii) Constitutive NRF2 activation causes HSCs to exit quiescence with enhanced STAT3 signaling and simultaneously skews their differentiation potential towards monocyte-granulocyte lineages. (B) Effects on ESCs. NRF2 directly enhances the expression of POMP and the transcription factor pair OCT4 and NANOG. The former augments the self-renewal capacity by enhancing proteasome activity, and the latter also modulates the self-renewal capacity and fate determination. (C) Effects on MSCs. (i) NRF2 suppresses p53 expression, which results in derepression of SIRT1 expression. SIRT1 expression enhances self-renewal under stress conditions and facilitates osteogenic differentiation. (ii) 17β-Estradiol treatment leads to promotion of NRF2 expression under hyperglycemia, which supports SIRT3 expression and thereby deacetylation of MnSOD by SIRT3 for its activation. Through this pathway, ROS levels are decreased, resulting in improved maintenance and engraftment of MSCs. (D) Effects on CncC, an NRF2 homolog in Drosophila. (i) CncC is constitutively activated in Drosophila ISCs and maintains ISC quiescence by suppressing ROS production. For ISC proliferation, CncC must be inhibited so that ROS can accumulate to induce ISC proliferation. (ii) In the setting of proteasome dysfunction, accumulation of Atg8a, which can inhibit the function of Keap1, promotes CncC activity. Activated CncC induces the expression of Dacapo, a homolog of the p21^{CIP}/p27^{KIP} family, which arrests the cell cycle until the protein aggregation is eliminated.

On the other hand, enforced NRF2 activation by *Keap1* deficiency is detrimental to HSC maintenance. Constitutive activation of NRF2 reinforces exit from quiescence accompanied by activation of Stat3 signaling, resulting in reduced engraftment of HSCs after transplantation (Figure 11.4A-ii; Murakami et al., 2017). Because pharmacological activation of NRF2 also enhances the exit of HSCs from quiescence, NRF2 activation can stimulate a switch in the HSC status from dormancy to proliferation and differentiation (Murakami et al., 2017), which induces a risk of replication-dependent HSC exhaustion. In addition, NRF2 activation skews the cell fate decision of HSCs towards granulocyte-monocyte lineages at the expense of lymphoid and erythrocyte-megakaryocyte lineages (Figure 11.4A-ii; Murakami et al., 2014). These negative aspects of NRF2 activity should be considered in the manipulation of HSCs.

However, provided that NRF2 activity is carefully controlled in terms of the stimulation period and/ or the timing of its activation, NRF2 activation can be beneficial for HSCs, especially upon exposure to irradiation. Total body irradiation induces chronic oxidative stress in HSCs and thereby disrupts their self-renewal capacity, causing myelosuppression or BM failure. In the setting of irradiation-induced HSC injury, some reagents/drugs, such as theaflavin, Vam3, and hydrogen-rich water, seem to exert protective effects on HSCs by activating NRF2 (Han et al., 2017; Zhang et al., 2016, 2017). Consistent with this notion, NRF2 inducers can ameliorate myelosuppression after irradiation, increasing HSPC proliferation, decreasing ROS levels and inducing NOTCH1 expression (Figure 11.4A-i; Kim et al., 2014).

Taken together, these findings indicate that NRF2 greatly impacts HSC function; under steady-state conditions, NRF2 activity must be maintained at an optimal level for HSCs to function appropriately without losing stemness, but under stressed conditions, such as radiation exposure, intense activation of NRF2 can prevent HSC exhaustion in both ROS-dependent and ROS-independent manners.

11.5 EMBRYONIC STEM CELLS

ESCs originate from the inner cell mass of blastocysts and possess self-renewal capacity and multipotency, which allows them to differentiate into the three germ layers: the ectoderm, endoderm, and mesoderm. Accordingly, ESCs have been considered a promising source for tissue repair and regeneration. Consistent with their important roles in development, ESCs are highly protected against oxidative stress via their high expression of antioxidant genes and low dependence on mitochondria for energy production, and these characteristics are gradually lost during differentiation (Ito and Suda, 2014; Saretzki, 2004; Saretzki et al., 2008; Zhang et al., 2012).

NRF2 is abundantly expressed in human ESCs, but its level decreases during differentiation (Jang et al., 2014). NRF2 dysfunction in ESCs attenuates the expression of OCT4, NANOG, SOX2, and TRET, which are indicators of pluripotency, whereas pharmacological activation of NRF2 blocks induced differentiation of ESCs by maintaining the expression of OCT4 and NANOG (Jang et al., 2014). One mechanism underlying the NRF2-dependent maintenance of the undifferentiated state is upregulation of a proteasome chaperone (POMP), whose dysregulation reduces the expression of pluripotency marker genes (Figure 11.4B; Jang et al., 2014). As another explanation for the NRF2-mediated undifferentiated state, NRF2 seems to directly upregulate OCT4 and NANOG gene expression by binding to elements upstream of these genes (Figure 11.4B; Jang et al., 2016). Through these pathways, NRF2 is likely to regulate the self-renewal capacity and pluripotency maintenance of ESCs.

Because OCT4 and NANOG induce differentiation towards mesendoderm (ME), NRF2 activity also alters the lineage specification of ESCs by enhancing differentiation towards ME and suppressing differentiation towards neuroectoderm (NE; Jang et al., 2016). During ESC differentiation, the segregation between the ME and NE is the earliest event in the fate decision. During differentiation towards NE, p62, which can activate NRF2 by competing for binding to KEAP1 (Figure 11.1B), is degraded through autophagy, which is likely to restore KEAP1-mediated degradation of NRF2 (Jang et al., 2016). This downregulation of NRF2 allows cells to differentiate into NE; conversely, autophagy inhibition or NRF2 activation strongly suppresses the expression of NE markers. Under oxidative stress induced by paraquat treatment, enhanced accumulation of ROS by NRF2 knockdown accelerates neuronal differentiation of ESC-like cells (Hu et al., 2018), which is an interesting example demonstrating that an altered redox balance appears to be involved in the NRF2-mediated NE specification of ESCs.

These observations suggest that NRF2 impacts the self-renewal capacity and fate determination of ESCs, but these findings should be interpreted with caution because they are based only on *in vitro* experiments with manipulation of ESCs under culture conditions. To understand the function of NRF2 in ESCs during development, appropriate *in vivo* investigation is necessary.

11.6 INDUCED PLURIPOTENT STEM CELLS

ESCs are useful tools to obtain better insights into the activity and features of SCs and are likely a promising source of regenerative medicine. However, the necessity of blastocyst removal for ESC establishment raises controversial ethical issues regarding the manipulation of human ESCs. Furthermore, before ESCs can be applied extensively for regenerative medicine, the problem of immune rejection must be solved. Because of these problems, the application of ESCs to regenerative medicine is restricted and limited. By contrast, these problems do not apply to iPSCs because these cells are produced from the somatic cells of the subjects themselves and do not induce immune rejection (Shi et al., 2017). Therefore, since the discovery of iPSC induction by OCT3/4, SOX2, c-MYC, and KLF4, iPSCs have attracted great interest for application in regenerative medicine.

iPSCs overcome the problems of ESCs, and their characteristics are similar to those of ESCs (Takahashi and Yamanaka, 2006). Like ESCs, iPSCs possess renewal capacity and can differentiate into all cell lineages. Regarding the relation of SCs with oxidative stress, iPSCs express high levels of antioxidant genes, and their intracellular ROS level remains low, a characteristic associated with low mitochondrial biogenesis as represented by cristae-poor structures (Armstrong et al., 2010; Folmes et al., 2011; Prigione et al., 2010; Zhang et al., 2012). Mitochondria are dramatically reorganized during reprogramming into iPSCs by a decrease in their number and alterations in their morphology, which simultaneously shifts energy production from OXPHOS to glycolysis, leading to low production of ROS. However, a moderate increase in the ROS level is required for appropriate reprogramming of somatic cells; loss of ROS via antioxidant treatment or excessive ROS production interrupts efficient reprogramming

(Zhou et al., 2016). Accordingly, maintenance of a proper ROS level is required for efficient and sufficient reprogramming and maintenance of iPSCs.

Only a limited number of reports have described the contribution of Nrf2 activity to the reprogramming or function of iPSCs. During the reprogramming of human dermal fibroblasts into iPSCs, increasing ROS levels appear to activate NRF2 prior to upregulation of HIF1α, a factor responsible for metabolic switching, by enhancing the expression of glycolysis-related genes (Hawkins et al., 2016). Inhibition of NRF2 via KEAP1 overexpression resulted in failure to upregulate HIF1α and to promote a metabolic shift towards glycolysis, and consistently reduced iPSC colony formation. Interestingly, because NRF2 activates genes related to the pentose phosphate pathway (Mitsuishi et al., 2012), which is activated in the early phase of reprogramming at the peak of NRF2 activity, NRF2 activation may also contribute to the metabolic shift before HIF1α-mediated metabolic switching (Hawkins et al., 2016). Therefore, NRF2 likely plays an essential role in reprogramming as a downstream factor of ROS produced in the early stage of reprogramming.

Similar to the observations in ESCs, a role of NRF2 as a determinant of NE/ME differentiation bias in iPSCs is implied: the lower the NRF2 activity in iPSCs, the more abundantly are NE markers expressed, and NRF2 downregulation facilitates NE differentiation of iPSCs (refer to Figure 11.4B; Jang et al., 2016). Although this pattern implies that NRF2 plays similar roles in lineage specification in both iPSCs and ESCs, whether NRF2 activation impacts the self-renewal capacity of iPSCs remains to be clarified.

These observations suggest that NRF2 is involved in efficient reprogramming and proper differentiation of iPSCs. However, the knowledge regarding the effects of NRF2 activity on iPSCs is still unconnected and insufficient, and many questions need to be addressed: for example, whether NRF2 is abundantly expressed in iPSCs, as observed in hESCs, and the effect of NRF2 activation, if any, on the self-renewal capacity and maintenance of iPSCs. Considering that iPSCs must be manipulated *in vitro*—under exposure to 21% oxygen—and that iPSCs are sensitive to oxidative stress (Shi et al., 2017; Wu et al., 2013), investigating whether NRF2 regulates the redox balance in iPSCs and guarantees their quality is important.

11.7 MESENCHYMAL STEM CELLS

MSCs are well investigated and are the most thoroughly clinically tested SCs; these cells are a valuable source of diverse mesodermal cells, including adipocytes, chondrocytes, and osteoblasts (Squillaro et al., 2016). MSCs are highly available and exhibit immunomodulatory effects attenuating immune rejection, making them a promising tool for regenerative medicine (Mushahary et al., 2018; Squillaro et al., 2016). However, even though MSCs have these considerable advantages, their clinical application is still limited because of issues with low engraftment or survival after transplantation.

Like other SCs, MSCs are also maintained in a hypoxic microenvironment in vivo, and their ROS levels are low in the undifferentiated state, a characteristic attributed to low mitochondrial biogenesis (Figure 11.3A; Ito and Suda, 2014; Mohyeldin et al., 2010). In undifferentiated MSCs, metabolism relies heavily on glycolysis, but during differentiation, a metabolic shift from glycolysis to OXPHOS occurs, accompanied by a drastic alteration in mitochondrial morphology (Figure 11.3B; Ito and Suda, 2014; Sart et al., 2015). Owing to the dominance of glycolysis, electron leakage from the respiratory chain is minimized, and undifferentiated MSCs are protected from mitochondrial ROS in vivo. Although MSCs have a highly developed antioxidant system, they are susceptible to oxidative stress. Dysfunction or low engraftment of MSCs upon transplantation seems to be at least partially due to exposure to 21% oxygen, accompanied by a metabolic shift towards OXPHOS (Ito and Suda, 2014).

Substantial evidence shows that NRF2 activation improves the self-renewal capacity and engraftment efficiency of MSCs. In human BM-derived MSCs (BM-MSCs), NRF2 activation confers resistance to oxidative stress- and hypoxia-induced cell death in vitro (Mohammadzadeh et al., 2012), and consistently, inhibition of NRF2 by triclosan treatment impairs the proliferation of human BM-MSCs in vitro, accompanied by reduced expression of antioxidant enzymes (Yoon et al., 2017). In human umbilical cord MSCs (UC-MSCs), NRF2 overexpression leads to increased expression of the stemness markers NANOG and BMI-1 and decreased expression of proapoptotic factors, suggesting that NRF2 protects MSCs against exhaustion during in vitro culture (Yuan et al., 2017).

Considering the application of MSCs to regenerative medicine, pharmacological enhancement of their self-renewal capacity is an ideal method for their stimulation. Pharmacological NRF2 activation in human BM-MSCs has also been shown to evoke SIRT1 expression and simultaneously suppress p53 activity, by which induced osteogenic differentiation is blocked and self-renewal is accelerated (Figure 11.4C-i) (Yoon et al., 2016). Upon stimulation by an NRF2 activator, these BM-MSCs maintain normal differentiation potential towards osteoblasts. As another NRF2-mediated protective pathway in MSCs, 17β-estradiol (E2) treatment of UC-MSCs under high glucose conditions has been reported (Oh et al., 2019). High-glucose conditions increase mitochondrial ROS production by decreasing nuclear NRF2 protein expression, SIRT3 expression and MnSOD activity, causing autophagy-mediated cell death (Figure 11.4C-ii). In contrast, E2 treatment induces NRF2 expression through upregulation of estrogen receptor α followed by increased expression of SIRT3 and activation of MnSOD, which ultimately reduces the ROS level in UC-MSCs. Importantly, E2-stimulated UC-MSCs showed enhanced wound healing ability in ovariectomized diabetic mice with hyperglycemia, suggesting that the E2-induced NRF2/SIRT3/MnSOD axis improves the in vitro maintenance of MSCs and enhances their engraftment. Consistent with this observation, NRF2-overexpressing human amniotic MSCs were found to mediate more efficient protection against or recovery from lipopolysaccharide-induced lung injury than control amniotic MSCs (Zhang et al., 2018).

These observations suggest that in regenerative medicine, NRF2 is a promising target for preconditioning MSCs prior to transplantation in order to strengthen and maintain their function.

11.8 INTESTINAL STEM CELLS

An interesting effect of NRF2 activity on the quiescence and proliferation of intestinal stem cells (ISCs) has been reported. In Drosophila, whereas the NRF2 homolog CncC is constitutively active in ISCs to inhibit their proliferation, KEAP1-mediated suppression of CncC allows ISCs to proliferate in response to mitogenic stimulation induced by accumulation of ROS (Figure 11.4D-i) (Hochmuth et al., 2011). Interestingly, overproliferation of intestinal epithelium in aged flies is caused by loss of CncC expression, which leads to elevated ROS levels. A similar phenomenon was also reproduced in the midgut of the mosquito Aedes aegypti

(Bottino-Rojas et al., 2018). Furthermore, this CncC-mediated suppression of proliferation is likely to function as a "proteostatic checkpoint" in *Drosophila* ISCs (Rodriguez-Fernandez et al., 2019); proteasome dysfunction imposes a risk of cell death because of impaired cell function, and the cell cycle is therefore arrested in ISCs until protein aggregation is completely eliminated. To induce cell cycle arrest upon exposure to protein aggregation, Atg8a/LC3 upregulation is promoted, which results in CncC activation via disruption of KEAP1-mediated suppression, and the upregulated CncC enhances the expression of a homolog of the p21/p27 family, *Dacapo*, leading to cell cycle arrest (Figure 11.4D-ii). Overgrowth of the midgut epithelium in aged flies can also be explained by attenuation of Atg8a/CncC/Dacapo pathway activity (Rodriguez-Fernandez et al., 2019).

The important question is whether these CncC-mediated pathways found in *Drosophila* and *Aedes aegypti* are conserved in mammals; indeed, a corresponding enhancement in the proliferation of ISC-like cells was observed in Nrf2-deficient mice after abdominal irradiation (Yang et al., 2017). However, this observation must be interpreted carefully because the NRF2 status differs significantly between *Drosophila* and mammals. Unlike in *Drosophila*, NRF2 in mammals is constantly controlled by KEAP1 through the Neh2 domain under steady state conditions, and Nrf2-deficient mice exhibited disadvantages in the intestine for glutathione synthesis in response to treatment with the phenolic antioxidant butylated hydroxyanisole (Itoh et al., 1997) and for the repair of damage induced by burn-induced intestinal injury (Chen et al., 2016), implying that NRF2 has positive stress-protective effects on the intestine. These observations suggest that under certain conditions, NRF2 can function as a protective factor against stress in intestinal cells, as observed in other SCs. Thus, the roles played by NRF2 in mammalian ISCs should be further examined.

11.9 OTHER STEM CELLS

Another contribution of NRF2 has been demonstrated in airway basal stem cells (ABSCs) in a ROS-dependent manner (Paul et al., 2014). In ABSCs, ROS have been shown to act upstream of NRF2 as a stimulus to induce NRF2 activation; in the setting of tracheobronchial epithelial injury induced by polidocanol treatment, high levels of ROS are produced in response to the stimulus, which induces NRF2 activation and sequentially activates the NOTCH1 pathway via direct binding of NRF2 to ARE sequences in the *Notch1* promoter region. Because the direct regulation of *Notch1* by NRF2 has also been observed in HSCs (Kim et al., 2014), the NRF2-NOTCH pathway may modulate the self-renewal capacity and/or differentiation of some types of SCs.

Considering that Nrf2 deficiency guides the differentiation potential of ESCs and iPSCs towards a neuronal fate (Hu et al., 2018; Jang et al., 2016), it is not surprising that neuronal stem cells (NSCs) are influenced by NRF2 activity. However, the impacts of NRF2 activation on both the self-renewal and differentiation of NSCs are controversial. Regarding the contribution of NRF2 to NSC self-renewal, NRF2 activation abolishes the self-renewal capacity during neuronal differentiation induced by mitochondrial dynamics (Khacho et al., 2016). In contrast, other reports showed that NRF2 overexpression can restore the regenerative potential of NSCs derived from middle-aged rats (Ray et al., 2018) and that in Nrf2-deficient mice, the self-renewal capacity of NSCs in the subgranular zone is compromised (Robledinos-Antón et al., 2017). Regarding NSC fate determination, NRF2-activated NSCs preferentially differentiate towards a neuronal fate; exposure to hyperglycemia elevates the ROS level, which promotes NRF2-mediated neuronal differentiation of NSCs instead of reducing glial differentiation (Si et al., 2019). In a rat model, neuronal differentiation was also found to be favored over glial differentiation when NRF2 expression was enforced in neuronal stem and progenitor cells (Ray et al., 2018). Consistent with this finding, Nrf2-deficient mice showed impaired neuronal differentiation in the subgranular zone, accompanied by abnormal production of astrocytes and oligodendrocytes (Robledinos-Antón et al., 2017). However, another report conversely stated that NRF2 must be downregulated by epigenetic repression during neuronal differentiation from an undifferentiated state, which induces loss of antioxidant defenses with concomitant activation of Wnt signaling, and that enforced NRF2 activation blocks proper acquisition of neuronal features (Bell et al., 2015). Although these findings support the contributions of NRF2 to NSC self-renewal, proliferation, and differentiation, further studies are required for a precise understanding of NRF2-mediated regulation of NSCs.

Other SCs that have been examined in relation to NRF2 activity include muscle SCs (Yamaguchi et al., 2015) and pilosebaceous SCs in skin (Muzumdar et al., 2019). However, evidence regarding these SCs is limited. Elucidating the roles played by NRF2 in these SCs will help us to better understand the applicability of NRF2 modulation for regenerative medicine. Moreover, this elucidation will provide integrated knowledge regarding the contribution of NRF2 to SC functions—quiescence, self-renewal, and differentiation—under steady-state conditions, as well as knowledge regarding their survival, engraftment, and pluripotency for their application in regenerative medicine.

11.10 SUMMARY

Because novel mechanisms of NRF2 regulation and new NRF2 target genes have been identified and verified in various tissues and different organisms under diverse conditions, the range of physiological actions covered by NRF2 activity has been extended. We have reviewed the mechanisms by which NRF2 contributes to SC function by focusing on several types of SCs. Among the types of SCs addressed herein, our knowledge about the effects of NRF2 in iPSCs is still limited and is insufficient to determine the applicability of NRF2 manipulation to their reprogramming, maintenance, differentiation induction, and engraftment. Based on their availability and lack of controversy regarding ethics and graft-versus-host disease, studying the impacts of NRF2 activation on the function of iPSCs will be useful. Questions such as whether NRF2 activation facilitates the reprogramming efficiency of iPSCs and protects them against undesirable dysfunction during culture, whether NRF2 activation promotes differentiation specifically towards certain lineages or disrupts the induction of differentiation towards cells of the target tissue, and whether NRF2 maintains the undifferentiated state of iPSCs or induces the proper differentiated state before and after transplantation, are especially important to answer.

Manipulation SCs to exploit their advantages for regenerative medicine requires their exposure to an environment containing 21% oxygen during incubation or transfer, which induces SC exhaustion/dysfunction due to oxidative stress. Whereas the enhanced reducing conditions induced by NRF2 activation can be considered helpful for protecting SCs against these adverse conditions, NRF2 also likely has more desirable impacts on SCs in a redox-independent manner. As the need for regenerative medicine is increasing, it is imperative to establish new manipulation techniques and/or improve the already established methods for SC therapies.

REFERENCES

Adam, J., Hatipoglu, E., O'Flaherty, L., Ternette, N., Sahgal, N., Lockstone, H., Baban, D., Nye, E., Stamp, G.W., Wolhuter, K., et al. (2011). Renal cyst formation in Fh1-deficient mice is independent of the Hif/Phd pathway: roles for fumarate in KEAP1 succination and Nrf2 signaling. Cancer Cell 20, 524–537.

Armstrong, L., Tilgner, K., Saretzki, G., Atkinson, S.P., Stojkovic, M., Moreno, R., Przyborski, S., and Lako, M. (2010). Human induced pluripotent stem cell lines show stress defense mechanisms and mitochondrial regulation similar to those of human embryonic stem cells. Stem Cells 28, 661–673.

Bell, K.F.S., Al-Mubarak, B., Martel, M.-A., McKay, S., Wheelan, N., Hasel, P., Márkus, N.M., Baxter, P., Deighton, R.F., Serio, A., et al. (2015). Neuronal development is promoted by weakened intrinsic antioxidant defences due to epigenetic repression of Nrf2. Nature Communications 6, 7066.

Bollong, M.J., Lee, G., Coukos, J.S., Yun, H., Zambaldo, C., Chang, J.W., Chin, E.N., Ahmad, I., Chatterjee, A.K., Lairson, L.L., et al. (2018). A metabolite-derived protein modification integrates glycolysis with KEAP1–NRF2 signalling. Nature 562, 600–604.

Bottino-Rojas, V., Talyuli, O.A.C., Carrara, L., Martins, A.J., James, A.A., Oliveira, P.L., and Paiva-Silva, G.O. (2018). The redox-sensing gene Nrf2 affects intestinal homeostasis, insecticide resistance, and Zika virus susceptibility in the mosquito Aedes aegypti. Journal of Biological Chemistry 293, 9053–9063.

Chaudhari, P., Ye, Z., and Jang, Y.-Y. (2014). Roles of reactive oxygen species in the fate of stem cells. Antioxidants & Redox Signaling 20, 1881–1890.

Chen, Z., Zhang, Y., Ma, L., Ni, Y., and Zhao, H. (2016). Nrf2 plays a pivotal role in protection against burn trauma-induced intestinal injury and death. Oncotarget 7, 19272–19283.

Dai, X., Yan, X., Wintergerst, K.A., Cai, L., Keller, B.B., and Tan, Y. (2020). Nrf2: redox and metabolic regulator of stem cell state and function. Trends in Molecular Medicine 26, 185–200.

Folmes, C.D.L., Nelson, T.J., Martinez-Fernandez, A., Arrell, D.K., Lindor, J.Z., Dzeja, P.P., Ikeda, Y., Perez-Terzic, C., and Terzic, A. (2011). Somatic oxidative bioenergetics transitions into pluripotency-dependent glycolysis to facilitate nuclear reprogramming. Cell Metabolism 14, 264–271.

Han, X., Zhang, J., Xue, X., Zhao, Y., Lu, L., Cui, M., Miao, W., and Fan, S. (2017). Theaflavin ameliorates ionizing radiation-induced hematopoietic injury via the NRF2 pathway. Free Radical Biology and Medicine 113, 59–70.

Hawkins, K.E., Joy, S., Delhove, J.M.K.M., Kotiadis, V.N., Fernandez, E., Fitzpatrick, L.M., Whiteford, J.R., King, P.J., Bolanos, J.P., Duchen, M.R., et al. (2016). NRF2 orchestrates the metabolic shift during induced pluripotent stem cell reprogramming. Cell Reports 14, 1883–1891.

Hochmuth, C.E., Biteau, B., Bohmann, D., and Jasper, H. (2011). Redox regulation by Keap1 and Nrf2 controls intestinal stem cell proliferation in Drosophila. Cell Stem Cell 8, 188–199.

Hu, Q., Khanna, P., Ee Wong, B.S., Lin Heng, Z.S., Subhramanyam, C.S., Thanga, L.Z., Sing Tan, S.W., and Baeg, G.H. (2018). Oxidative stress promotes exit from the stem cell state and spontaneous neuronal differentiation. Oncotarget 9, 4223–4238.

Ito, K., and Suda, T. (2014). Metabolic requirements for the maintenance of self renewing stem cells. Nature Reviews Molecular Cell Biology 15, 243–256.

Itoh, K., Chiba, T., Takahashi, S., Ishii, T., Igarashi, K., Katoh, Y., Oyake, T., Hayashi, N., Satoh, K., Hatayama, I., et al. (1997). An Nrf2/small Maf heterodimer mediates the induction of phase II detoxifying enzyme genes through antioxidant response elements. Biochemical and Biophysical Research Communications 236, 313–322.

Itoh, K., Wakabayashi, N., Katoh, Y., Ishii, T., Igarashi, K., Engel, J.D., and Yamamoto, M. (1999). Keap1 represses nuclear activation of antioxidant responsive elements by Nrf2 through binding to the amino-terminal Neh2 domain. Genes & Development 13, 76–86.

Jang, J., Wang, Y., Kim, H.-S., Lalli, M.A., and Kosik, K.S. (2014). Nrf2, a regulator of the proteasome, controls self-renewal and pluripotency in human embryonic stem cells: Nrf2-proteasome pathway controls stemness in hESCs. Stem Cells 32, 2616–2625.

Jang, J., Wang, Y., Lalli, M.A., Guzman, E., Godshalk, S.E., Zhou, H., and Kosik, K.S. (2016). Primary cilium-autophagy-Nrf2 (PAN) axis activation commits human embryonic stem cells to a neuroectoderm fate. Cell 165, 410–420.

Katoh, Y., Itoh, K., Yoshida, E., Miyagishi, M., Fukamizu, A., and Yamamoto, M. (2001). Two domains of Nrf2 cooperatively bind CBP, a CREB binding protein, and synergistically activate transcription. Genes Cells 6, 857–868.

Khacho, M., Clark, A., Svoboda, D.S., Azzi, J., MacLaurin, J.G., Meghaizel, C., Sesaki, H., Lagace, D.C., Germain, M., Harper, M.-E., et al. (2016). Mitochondrial dynamics impacts stem cell identity and fate decisions by regulating a nuclear transcriptional program. Cell Stem Cell 19, 232–247.

Kim, J.-H., Thimmulappa, R.K., Kumar, V., Cui, W., Kumar, S., Kombairaju, P., Zhang, H., Margolick, J., Matsui, W., Macvittie, T., et al. (2014). NRF2-mediated Notch pathway activation enhances hematopoietic reconstitution following myelosuppressive radiation. Journal of Clinical Investigation 124, 730–741.

Kobayashi, E.H., Suzuki, T., Funayama, R., Nagashima, T., Hayashi, M., Sekine, H., Tanaka, N., Moriguchi, T., Motohashi, H., Nakayama, K., et al. (2016). Nrf2 suppresses macrophage inflammatory response by blocking proinflammatory cytokine transcription. Nature Communications 7, 11624.

Kobayashi, M., Li, L., Iwamoto, N., Nakajima-Takagi, Y., Kaneko, H., Nakayama, Y., Eguchi, M., Wada, Y., Kumagai, Y., and Yamamoto, M. (2009). The antioxidant defense system Keap1-Nrf2 comprises a multiple sensing mechanism for responding to a wide range of chemical compounds. MCB 29, 493–502.

Mendelson, A., and Frenette, P.S. (2014). Hematopoietic stem cell niche maintenance during homeostasis and regeneration. Nature Medicine 20, 833–846.

Merchant, A.A., Singh, A., Matsui, W., and Biswal, S. (2011). The redox-sensitive transcription factor Nrf2 regulates murine hematopoietic stem cell survival independently of ROS levels. Blood 118, 6572–6579.

Mitsuishi, Y., Taguchi, K., Kawatani, Y., Shibata, T., Nukiwa, T., Aburatani, H., Yamamoto, M., and Motohashi, H. (2012). Nrf2 redirects glucose and glutamine into anabolic pathways in metabolic reprogramming. Cancer Cell 22, 66–79.

Mohammadzadeh, M., Halabian, R., Gharehbaghian, A., Amirizadeh, N., Jahanian-Najafabadi, A., Roushandeh, A.M., and Roudkenar, M.H. (2012). Nrf-2 overexpression in mesenchymal stem cells reduces oxidative stress-induced apoptosis and cytotoxicity. Cell Stress and Chaperones 17, 553–565.

Mohyeldin, A., Garzón-Muvdi, T., and Quiñones-Hinojosa, A. (2010). Oxygen in stem cell biology: a critical component of the stem cell niche. Cell Stem Cell 7, 150–161.

Murakami, S., and Motohashi, H. (2015). Roles of Nrf2 in cell proliferation and differentiation. Free Radical Biology and Medicine 88, 168–178.

Murakami, S., Shimizu, R., Romeo, P.-H., Yamamoto, M., and Motohashi, H. (2014). Keap1-Nrf2 system regulates cell fate determination of hematopoietic stem cells. Genes Cells 19, 239–253.

Murakami, S., Suzuki, T., Harigae, H., Romeo, P.-H., Yamamoto, M., and Motohashi, H. (2017). NRF2 activation impairs quiescence and bone marrow reconstitution capacity of hematopoietic stem cells. Molecular and Cellular Biology 37, e00086-17, e00086-17.

Mushahary, D., Spittler, A., Kasper, C., Weber, V., and Charwat, V. (2018). Isolation, cultivation, and characterization of human mesenchymal stem cells: hMSC. Cytometry 93, 19–31.

Muzumdar, S., Hiebert, H., Haertel, E., Ben-Yehuda Greenwald, M., Bloch, W., Werner, S., and Schäfer, M. (2019). Nrf2-mediated expansion of pilosebaceous cells accelerates cutaneous wound healing. The American Journal of Pathology 189, 568–579.

Nioi, P., Nguyen, T., Sherratt, P.J., and Pickett, C.B. (2005). The carboxy-terminal Neh3 domain of Nrf2 is required for transcriptional activation. MCB 25, 10895–10906.

Oh, J.Y., Choi, G.E., Lee, H.J., Jung, Y.H., Chae, C.W., Kim, J.S., Lee, C.-K., and Han, H.J. (2019). 17β-Estradiol protects mesenchymal stem cells against high glucose-induced mitochondrial oxidants production via Nrf2/Sirt3/MnSOD signaling. Free Radical Biology and Medicine 130, 328–342.

Paul, M.K., Bisht, B., Darmawan, D.O., Chiou, R., Ha, V.L., Wallace, W.D., Chon, A.T., Hegab, A.E., Grogan, T., Elashoff, D.A., et al. (2014). Dynamic changes in intracellular ROS levels regulate airway basal stem cell homeostasis through Nrf2-dependent notch signaling. Cell Stem Cell 15, 199–214.

Prigione, A., Fauler, B., Lurz, R., Lehrach, H., and Adjaye, J. (2010). The senescence-related mitochondrial/oxidative stress pathway is repressed in human induced pluripotent stem cells. STEM CELLS 28, 721–733.

Rada, P., Rojo, A.I., Chowdhry, S., McMahon, M., Hayes, J.D., and Cuadrado, A. (2011). SCF/-TrCP promotes glycogen synthase Kinase 3-dependent degradation of the Nrf2 transcription factor in a Keap1-independent manner. Molecular and Cellular Biology 31, 1121–1133.

Ray, S., Corenblum, M.J., Anandhan, A., Reed, A., Ortiz, F.O., Zhang, D.D., Barnes, C.A., and Madhavan, L. (2018). A role for Nrf2 expression in defining the aging of hippocampal neural stem cells. Cell Transplant 27, 589–606.

Robledinos-Antón, N., Fernández-Ginés, R., Manda, G., and Cuadrado, A. (2019). Activators and inhibitors of NRF2: A review of their potential for clinical development. Oxidative Medicine and Cellular Longevity 2019, 1–20.

Robledinos-Antón, N., Rojo, A.I., Ferreiro, E., Núñez, Á., Krause, K.-H., Jaquet, V., and Cuadrado, A. (2017). Transcription factor NRF2 controls the fate of neural stem cells in the subgranular zone of the hippocampus. Redox Biology 13, 393–401.

Rodriguez-Fernandez, I.A., Qi, Y., and Jasper, H. (2019). Loss of a proteostatic checkpoint in intestinal stem cells contributes to age-related epithelial dysfunction. Nature Communications 10, 1050.

Saretzki, G. (2004). Stress defense in murine embryonic stem cells is superior to that of various differentiated murine cells. Stem Cells 22, 962–971.

Saretzki, G., Walter, T., Atkinson, S., Passos, J.F., Bareth, B., Keith, W.N., Stewart, R., Hoare, S., Stojkovic, M., Armstrong, L., et al. (2008). Downregulation of multiple stress defense mechanisms during differentiation of human embryonic stem cells. Stem Cells 26, 455–464.

Sart, S., Song, L., and Li, Y. (2015). Controlling redox status for stem cell survival, expansion, and differentiation. Oxidative Medicine and Cellular Longevity 2015, 1–14.

Sekine, H., Okazaki, K., Ota, N., Shima, H., Katoh, Y., Suzuki, N., Igarashi, K., Ito, M., Motohashi, H., and Yamamoto, M. (2016). The mediator subunit MED16 transduces NRF2-activating signals into antioxidant gene expression. Molecular and Cellular Biology 36, 407–420.

Shi, Y., Inoue, H., Wu, J.C., and Yamanaka, S. (2017). Induced pluripotent stem cell technology: a decade of progress. Nature Reviews Drug Discovery 16, 115–130.

Si, Z.-P., Wang, G., Han, S.-S., Jin, Y., Hu, Y.-X., He, M.-Y., Brand-Saberi, B., Yang, X., and Liu, G.-S. (2019). CNTF and Nrf2 are coordinately involved in regulating self-renewal and differentiation of neural stem cell during embryonic development. IScience 19, 303–315.

Squillaro, T., Peluso, G., and Galderisi, U. (2016). Clinical trials with mesenchymal stem cells: an update. Cell Transplant 25, 829–848.

Suda, T., Takubo, K., and Semenza, G.L. (2011). Metabolic regulation of hematopoietic stem cells in the hypoxic niche. Cell Stem Cell 9, 298–310.

Sun, J., Ramos, A., Chapman, B., Johnnidis, J.B., Le, L., Ho, Y.-J., Klein, A., Hofmann, O., and Camargo, F.D. (2014). Clonal dynamics of native haematopoiesis. Nature 514, 322–327.

Suzuki, T., and Yamamoto, M. (2017). Stress-sensing mechanisms and the physiological roles of the

Keap1–Nrf2 system during cellular stress. Journal of Biological Chemistry 292, 16817–16824.

Takahashi, K., and Yamanaka, S. (2006). Induction of pluripotent stem cells from mouse embryonic and adult fibroblast cultures by defined factors. Cell 126, 663–676.

Tsai, J.J., Dudakov, J.A., Takahashi, K., Shieh, J.-H., Velardi, E., Holland, A.M., Singer, N.V., West, M.L., Smith, O.M., Young, L.F., et al. (2013). Nrf2 regulates haematopoietic stem cell function. Nature Cell Biology 15, 309–316.

Wu, Y., Zhang, X., Kang, X., Li, N., Wang, R., Hu, T., Xiang, M., Wang, X., Yuan, W., Chen, A., et al. (2013). Oxidative stress inhibits adhesion and transendothelial migration, and induces apoptosis and senescence of induced pluripotent stem cells. Clinical and Experimental Pharmacology and Physiology 40, 626–634.

Yamaguchi, M., Murakami, S., Yoneda, T., Nakamura, M., Zhang, L., Uezumi, A., Fukuda, S., Kokubo, H., Tsujikawa, K., and Fukada, S. (2015). Evidence of Notch-Hesr-Nrf2 axis in muscle stem cells, but absence of Nrf2 has no effect on their quiescent and undifferentiated state. PLoS ONE 10, e0138517.

Yamamoto, M., Kensler, T.W., and Motohashi, H. (2018). The KEAP1-NRF2 system: a thiol-based sensor-effector apparatus for maintaining redox homeostasis. Physiological Reviews 98, 1169–1203.

Yang, W., Sun, Z., Yang, B., and Wang, Q. (2017). Nrf2-knockout protects from intestinal injuries in C57BL/6J mice following abdominal irradiation with γ rays. IJMS 18, 1656.

Yoon, D.S., Choi, Y., Cha, D.S., Zhang, P., Choi, S.M., Alfhili, M.A., Polli, J.R., Pendergrass, D., Taki, F.A., Kapalavavi, B., et al. (2017). Triclosan disrupts SKN-1/Nrf2-mediated oxidative stress response in C. elegans and human mesenchymal stem cells. Scientific Reports 7, 12592.

Yoon, D.S., Choi, Y., and Lee, J.W. (2016). Cellular localization of NRF2 determines the self-renewal and osteogenic differentiation potential of human MSCs via the P53–SIRT1 axis. Cell Death & Disease 7, e2093–e2093.

Yuan, Z., Zhang, J., Huang, Y., Zhang, Y., Liu, W., Wang, G., Zhang, Q., Wang, G., Yang, Y., Li, H., et al. (2017). NRF2 overexpression in mesenchymal stem cells induces stem-cell marker expression and enhances osteoblastic differentiation. Biochemical and Biophysical Research Communications 491, 228–235.

Zhang, J., Nuebel, E., Daley, G.Q., Koehler, C.M., and Teitell, M.A. (2012). Metabolic regulation in pluripotent stem cells during reprogramming and self-renewal. Cell Stem Cell 11, 589–595.

Zhang, J., Ohta, T., Maruyama, A., Hosoya, T., Nishikawa, K., Maher, J.M., Shibahara, S., Itoh, K., and Yamamoto, M. (2006). BRG1 interacts with Nrf2 to selectively mediate HO-1 Induction in response to oxidative stress. MCB 26, 7942–7952.

Zhang, J., Xue, X., Han, X., Li, Y., Lu, L., Li, D., and Fan, S. (2017). Hydrogen-rich water ameliorates total body irradiation-induced hematopoietic stem cell injury by reducing hydroxyl radical. Oxidative Medicine and Cellular Longevity 2017, 1–16.

Zhang, J., Xue, X., Han, X., Yao, C., Lu, L., Li, D., Hou, Q., Miao, W., Meng, A., and Fan, S. (2016). Vam3 ameliorates total body irradiation-induced hematopoietic system injury partly by regulating the expression of Nrf2-targeted genes. Free Radical Biology and Medicine 101, 455–464.

Zhang, S., Jiang, W., Ma, L., Liu, Y., Zhang, X., and Wang, S. (2018). Nrf2 transfection enhances the efficacy of human amniotic mesenchymal stem cells to repair lung injury induced by lipopolysaccharide. Journal of Cellular Biochemistry 119, 1627–1636.

Zhou, G., Meng, S., Li, Y., Ghebre, Y.T., and Cooke, J.P. (2016). Optimal ROS signaling is critical for nuclear reprogramming. Cell Reports 15, 919–925.

CHAPTER TWELVE

Role of Iron in Cell Differentiation

Chinmay K. Mukhopadhyay, Sameeksha Yadav, Diksha Kulshreshtha, and Ilora Ghosh

CONTENTS

12.1 INTRODUCTION

Cell differentiation is the process through which a cell undergoes changes in gene expression to become a more specific type to reach its mature form and function. Development, growth, reproduction, and longevity of all multicellular organisms depend on this essential process. The detailed understanding of cell differentiation has been the focus of intense investigation for the past several decades. Cells gradually become committed towards maturing into a particular cell type with specialized functions. Properties of these committed cells are not similar to fully differentiated cells. Differentiation process alters the cell significantly in its shape, size, and energy requirements.

It has now been established that both gene structure and environmental factors are important for cell differentiation. Both mechanical and chemical stimuli are part of the environmental factors. Iron acts as a chemical stimulus for the cell differentiation process. It is found in heme-containing enzymes, in iron-sulphur clusters (ISCs), or as mono- or dinuclear iron-containing proteins for carrying out essential reactions in almost every cell type (Ponka, 1997; Rouault and Tong, 2005; Sheftel et al., 2012). Iron plays a vital role in carrying oxygen by RBCs. The generation and the final destruction of RBCs are carried out by macrophages. Specifically differentiated macrophages are needed to generate and destroy RBCs for recycling iron. Macrophages are also one of the most important players of the immune system. They are differentiated into pro-inflammatory and anti-inflammatory categories, and iron plays a decisive role in the polarization of macrophages. They also carry out very specialized functions in different tissues as resident cells, including handling of local iron homeostasis and inflammation. To produce ATP continuously, muscle cells need an adequate supply of iron and oxygen. For performing this function, skeletal muscle myoblasts are

DOI: 10.4324/9781003204091-16

differentiated into myotubes. Smooth muscle cells are important components of vasculature in different tissues. They are remarkably plastic, as they even have the ability to differentiate into osteoblasts in which iron can contribute significantly. Adipose tissue is an endocrine organ and forms an energy storage depot (McGown et al., 2014) in the form of triglycerides, which are packaged into lipid droplets and are crucial for systemic metabolic homeostasis (Rosen et al., 2006; Rosen and Spiegelman, 2014). This tissue consists of adipocytes or fat cells, interstitial fibroblastic cells, and progenitor cells. Among them, adipocyte is remarkably known for its plasticity (Farmer, 2006). This chapter will focus on the understanding of the role of iron and iron homeostasis components on the differentiation of these three important cell types that play a vital role in the life process.

12.2 IRON HOMOEOSTASIS

Iron, a d-block transition metal, played a pivotal role in the development of the earliest forms of life on the Earth (Russell et al., 1993). It readily interconverts into its oxidation states (ferrous, Fe^{2+}, and ferric, Fe^{3+}) in the biological system, making it an indispensable component for almost all living forms. Though highly abundant, its bioavailability is poor because under aerobic conditions, the relatively soluble Fe^{2+} form gets oxidized to insoluble Fe^{3+} form (Papanikolaou and Pantopoulos, 2005). It is central to various essential biological processes including photosynthesis, respiration, oxygen transport, TCA cycle, cell cycle maintenance, and DNA synthesis (Aisen et al., 2001; Gao et al., 1999; Gros et al., 2010; Kakar et al., 2010). Although vital for survival, it is extremely reactive and potentially toxic, which is why iron homeostasis is tightly regulated at the genetic, cellular, and systemic levels and by a set of proteins including receptors, transcription factors, and RNA-binding proteins.

Only 5%–35% of total dietary iron is absorbed, depending on physiology and the form in which the iron is ingested (McDowell, 2003). On average, about 10% of the 10–20 mg of ingested iron is absorbed each day to balance the 1–2 mg iron that is lost daily. Efficacy of cellular iron absorption may increase up to 20% to support growth in children, for maintenance of pregnancy, and during haemorrhages and menstruation. There are two major sources of dietary iron: heme and non-heme. Heme-derived iron is more readily absorbed than non-heme iron (Wang and Pantopoulos, 2011).

Ferric reductases present in the gastrointestinal tract aid in reduction of dietary iron (Fe^{3+}) to the soluble Fe^{2+} form to enhance iron uptake. Heme, being hydrophobic, enters through the intestinal heme transporter, heme carrier protein 1 (HCP1), present on the membrane of enterocytes in the duodenum (Shayeghi et al., 2005). Non-heme iron is actively transported by the divalent metal transporter 1 (DMT1/DCT1/NRAMP2) of the enterocytes in the duodenum and jejunum (Gunshin et al., 1997; Mims et al., 2005; Muir and Hopfer, 1985) (Figure 12.1). Expression of DMT1 depends on the amount of ingested iron. Normally, DMT1 expression is relatively low; however, dietary iron deficiency induces DMT1 expression in the enterocytes (Canonne-Hergaux et al., 1999). Non-heme iron (Fe^{3+}), being insoluble, is first reduced by ascorbate-dependent ferric reductase DCytb1, to be then transported by DMT1 (McKie et al., 2001). Iron enters the bloodstream once it crosses the duodenal mucosa, where transferrin (Tf) mediates its transport to the bone marrow and other parts of the body while keeping its toxicity under control (Frazer and Anderson, 2005) (Figure 12.1). At the physiological serum iron level (20 μmol/L) and pH, Tf saturation remains only 20%–35% so that sufficient unsaturated Tf is available for scavenging free iron. The level of non-transferrin-bound iron (NTBI) (1 μmol/L) often remains undetectable in a healthy individual (Anderson, 1999; Cazzola et al., 1985). During iron overload, as in hemochromatosis and thalassemia, the level of NTBI is elevated due to increased Tf saturation (Batey et al., 1980; Hershko et al., 1978). Iron bound Tf readily binds to its cell surface receptor, transferrin receptor 1 (TfR1), leading to endocytosis of the iron-Tf-TfR1 complex and releasing bound iron in the endosome and recycling TfR1 back to the plasma membrane (Figure 12.2; Cheng et al., 2004; Kim and Ponka, 2000; Klausner et al., 1983; Ponka et al., 1998). Apart from being crucial for iron absorption in the intestine, DMT1 is also involved in exporting released iron from the iron-Tf-TfR1 complex in the acidic endosome to the cytosol (Figure 12.2; Gunshin et al., 1997; Kakhlon and Cabantchik, 2002; Picard et al., 2000). An endosomal ferric reductase, Steap3, reduces iron to facilitate its release into the cytosol (Ohgami et al., 2005). Release of iron from the endosome to the cytosol leads to an increase in the cytosolic labile iron pool (LIP), a transit pool of iron (Figure 12.2; Kakhlon and Cabantchik, 2002; Picard et al., 2000). Iron may

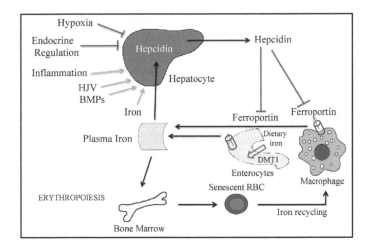

Figure 12.1 Components of systemic iron homeostasis.

Hepcidin released from hepatocytes degrades ferroportin in macrophages and enterocytes of the duodenum and jejunum to control the plasma iron level. Erythropoiesis happens in the bone marrow as per iron availability. Iron is recycled from senescent RBCs by splenic macrophages.

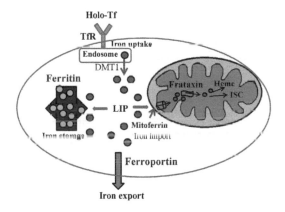

Figure 12.2 Overview of cellular iron homeostasis.

Holo-transferrin (Hol-Tf) binds to transferrin receptor 1 (TfR) and the complex is endocytosed. Released iron in endosome is transported to the cytosol by DMT1 to enrich the labile iron pool (LIP). Iron is utilized for heme and iron sulphur cluster (ISC) formation in mitochondria in which frataxin plays a role. Excess iron is stored in ferritin and released by ferroportin.

be channelized from LIP for various cellular metabolic processes (e.g., heme synthesis, ISC biogenesis in the mitochondrion) or for storage into major iron storage protein ferritin (Ft; Figure 12.2). Ft is a globular heteropolymer (~475 kDa) composed of 24 heavy (H) and light (L) subunits, the ratio of which varies from cell to cell type (Harrison and Arosio, 1996). H-subunit acts as a ferroxidase whereas the L-subunit acts as a nucleation centre for stabilization of Ft nanocage (Arosio and Levi, 2002; Lawson et al., 1989). Excess iron needs to be exported out of the cell for avoiding toxicity. The transmembrane protein ferroportin (Fpn or Iregl or MTP1) is the only identified iron exporter in mammals (Donovan et al., 2000) (Figure 12.2). Fpn associates with multi-copper oxidase ceruloplasmin (Cp) in astrocytes and macrophages. Fpn exports iron in Fe^{2+} form to the plasma that must be oxidized by Cp to incorporate into Tf (Harris et al., 1999; Jeong and David, 2003; Osaki, 1966; Sato and Gitlin, 1991; Zaitsev et al., 1999). In the intestine, Fpn couples with another multicopper oxidase, hephaestin (Heph), which has about 50% sequence similarity with Cp, for iron transport in the body fluid (Figure 12.1; Abboud and Haile, 2000; Vulpe et al., 1999; Yeh et al., 2009).

Most proteins involved in iron metabolism are regulated post-transcriptionally by the interaction between iron regulatory proteins (IRPs) and iron responsive elements (IREs) present in the UTR of their mRNAs. IRPs bind IREs present in the 5′ UTR of Ft and Fpn mRNAs and the 3′ UTR of DMT1 and TfR1 mRNAs (Casey et al., 1988; Gunshin et al., 1997; McKie et al., 2000; Pantopoulos, 2004; Rouault et al., 1988). During iron deficiency, binding of IRPs to their cognate IRE sequences on DMT1 and TfR1 mRNA leads to stabilization of respective transcripts for increased cellular iron uptake, whereas binding of IRPs to IREs in Ft and Fpn mRNA leads to their translational repression (Hentze et al., 2004;

Liu et al., 2002; Pantopoulos, 2004; Rouault and Klausner, 1996). Conversely, during iron overload, IRP1 functions as a cytosolic aconitase (ACO1) for the conversion of citrate to isocitrate and is not available to bind to IREs. TfR1 and DMT1 mRNAs are thus destabilized, whereas translation of Ft and Fpn mRNAs is increased (Rouault and Klausner, 1996; Theil and Eisenstein, 2000).

Fpn is post-translationally regulated by an anti-microbial peptide hormone, hepcidin, which is the major systemic regulator of iron homeostasis (Figure 12.1; Nemeth et al., 2004b). Hepcidin is mainly synthesized in the liver as an 84 amino acid precursor, which is finally secreted as a 25 amino acid peptide (Park et al., 2001). Iron and inflammation induce its expression, whereas anaemia and hypoxia reduce it (Nicolas et al., 2002). Hepcidin binds Fpn to promote ubiquitination and proteasomal degradation of the latter, thus negatively regulating iron efflux (Nemeth et al., 2004b; Ganz and Nemeth, 2006; Nemeth and Ganz, 2006). Fpn is abundantly expressed in duodenal enterocytes and reticuloendothelial macrophages in the liver, marrow, and spleen (Papanikolaou et al., 2005). Hepcidin inhibits the release of iron at the level of absorption in the intestine as well as during recycling of iron from senescent erythrocytes, thus controlling both iron absorption and recycling (Figure 12.1). Iron promotes induction of hepcidin mRNA in hepatocytes (Pigeon et al., 2001). It is transcriptionally regulated by the CCAAT/enhancer binding protein (C/EBP) in response to iron levels (Courselaud et al., 2002). During chronic inflammation, IL-6 induces hepcidin expression through the STAT3 signalling pathway, leading to hypoferremia (Nemeth et al., 2004a; Wrighting and Andrews, 2006). Hepcidin expression can also be regulated by bone morphogenetic proteins (BMPs) and hemojuvelin (HJV or HFE2; Figure 12.1; Babitt et al., 2006). BMPs control cell growth, differentiation, and apoptosis by binding to their respective serine/threonine kinase receptors, leading to phosphorylation and activation of downstream signalling pathways involving Smad proteins and mitogen-activated protein (MAP) kinases (Nohe et al., 2004). HJV is a BMP receptor, and overexpression of both HJV and BMP upregulates hepcidin expression in hepatocytes due to downstream activation of Smad4 (Figure 12.1; Babitt et al., 2006). During hypoxia, HIF-1 functions as a negative regulator of hepcidin expression (Figure 12.1; Peyssonnaux et al., 2007).

12.2.1 Recycling of Body Iron

Our body contains about 4,000 mg of iron, most of which is present within erythrocytes (~2,500 mg) and splenic and hepatic macrophages (~1,000 mg). Remaining iron is distributed throughout the body as numerous iron-containing proteins (e.g., cytochromes, myoglobin). Plasma Tf remains bound to only about 3 mg of iron, constituting the ambulatory pool of iron for supplying to intracellular iron stores (Ft and hemosiderin). Dietary iron absorption balances only the daily iron losses, thus iron recycling accounts for most of the maintenance of iron homeostasis. Reticuloendothelial macrophages release 20–25 mg of iron from senescent erythrocytes each day (Collins et al., 2008). The inducible isoform of heme oxygenase (HO), heme oxygenase 1 (HO-1), binds heme from senescent RBCs in a 1:1 molar ratio and catalyses its oxidative degradation to release carbon monoxide, iron, and biliverdin (Gozzelino et al., 2010). HO-1 thus plays a crucial role in heme-iron recycling and protecting cells from iron-mediated damage. HO-1 also aids in metabolizing dietary heme to liberate iron in the enterocytes (Wang and Pantopoulos, 2011).

12.2.2 Role of Iron in Erythropoiesis

A crosstalk exists between erythropoiesis and iron homeostasis, where each regulates the other. Erythropoietin (EPO), secreted from the kidney, is the major driver of erythropoiesis, which binds to its receptors and stimulates the proliferation and terminal differentiation of erythroid precursors (Elliott et al., 2008). Expression of EPO gets altered in response to iron status through the involvement of IRP1 (Anderson et al., 2013). During iron deficiency, IRP1 reduces the synthesis of EPO by repressing hypoxia-inducible factor 2 alpha (HIF2α). Iron deficiency also inhibits the activity of erythroid aconitase, leading to overexpression of transcription factor PU.1, which inhibits erythropoiesis (Richardson et al., 2013). In turn, erythropoiesis regulates iron homeostasis by modulating hepcidin expression (Pak et al., 2006). Erythroferrone (ERFE) affects hepcidin suppression, thus increasing erythroid activity and erythropoiesis (Kautz et al., 2014).

12.2.3 Iron Metabolism and Mitochondria

Mitochondrial iron metabolism is crucial for synthesis of heme and Fe-S clusters (Figure 12.2; Ajioka et al., 2006; Lill, 2009). The mitochondrion is particularly more susceptible to oxidative damage due to coexistence of ROS (derived from the electron transport chain) and iron. The major sources of iron delivery into mitochondria are iron-loaded endosomes and intracellular Ft (Sheftel et al., 2007; Vaisman et al., 1997). Iron importers mitoferrin 1 (Mfrn1) and mitoferrin 2 (Mfrn2) located on the mitochondrial inner membrane play essential roles in supplying iron to mitochondria and synthesizing heme and Fe-S clusters (Figure 12.2; Shaw et al., 2006). Fe-S clusters are prosthetic groups involved in mitochondrial electron transport and catalysis, the defects in assembly of which are associated with mitochondrial iron overload, abnormal heme synthesis, and blood disorders (Lill, 2009; Rouault and Tong, 2008). Fe-S clusters assemble in mitochondria on a protein scaffold consisting of an ISCU homodimer, where iron (Fe^{2+}) is donated by frataxin (FXN) and sulphur from cysteine via Nfs1/ISD11 complex-mediated de-sulphuration (Sheftel et al., 2010). Fe-S clusters regulate the expression of ALAS2, which is the rate-limiting enzyme for heme biosynthesis (Hentze et al., 2004). Where there is a low level of Fe-S clusters, IRP1 associates with IRE in the 5′ UTR of ALAS2 mRNA, blocking its translation and heme synthesis.

12.3 IRON AND MACROPHAGE DIFFERENTIATION

Macrophages play a wide range of biological activities, including scavenging and recycling, tissue repair, host defence, and embryonic development (Ganz, 2012). It is in the centre stage of mammalian iron homeostasis, as macrophages recycle iron from senescent erythrocytes and other used-up cells for erythropoiesis (hemoglobin [Hb] synthesis for new erythrocytes) in bone marrow. This process is also needed for the synthesis of numerous iron-containing essential proteins. During inflammation and infection, macrophages retain iron to limit iron availability in tissues that can alter inflammatory conditions and affect the growth of the invading extracellular microbes as an important means of host defence. In these conditions, two distinct macrophage activation states have been recognized (Cairo et al., 2011): M1 (classically activated) and M2 (alternatively activated; Biswas and Mantovani, 2012). When exposed to Toll-like receptor (TLR) ligands and/or interferon-γ (IFN-γ), macrophages become polarized into the pro-inflammatory

M1 phenotype, which is characterized by the expression of inflammatory cytokines, inducible nitric oxide synthase (NOS2) with the generation of reactive nitrogen species (RNS), generation of reactive oxygen species (ROS), promotion of Th1 responses, and significant microbicidal and tumouricidal activity. IL-4 and IL-13 exposure polarizes macrophages to M2 phenotype, thus dampening inflammation, helping in parasite clearance, promoting tissue remodelling, and tumour progression (Biswas and Mantovani, 2012). Incidentally, IL-10 and immune complexes also can polarize macrophages to an M2-like phenotype with pro-tumoural and immunoregulatory functions (Biswas and Mantovani, 2012).

Accumulating evidence suggests that tissue-resident macrophages also play a significant and specialized role in local iron availability and in modulating the tissue microenvironment to contribute to cellular and tissue function (Winn et al., 2020). These specialized functions depend on the remarkable plasticity of macrophages to be present and function as per micro-environmental demand. Thus, macrophages are differentiated into various phenotypes to meet wide array of micro-environmental demand as described below. An overview of iron and transcription factors at the crossroads of macrophage differentiation is presented in Table 12.1.

12.3.1 Iron and Differentiation of Erythrophagocytic Macrophages

Senescent or damaged RBCs are sensed and phagocytosed by erythrophagocytic macrophages to digest the Hb content of RBCs and recycle the iron back to erythroid progenitors, for heme synthesis and Hb production (Hamza and Dailey, 2012; Korolnek and Hamza, 2015). The lineage of erythrophagocytic macrophage is developed from bone marrow–derived monocyte progenitor cells by the heme-responsive transcription factor Spi-C (Kohyama et al., 2009; Haldar et al., 2014). Transcription factors such as PU.1 and C/EBPα have general roles in myelomonocytic development (Ye and Graf, 2007; Friedman, 2007); however, Spi-C controls the development of the highly specialized erythrophagocytic macrophage (red pulp macrophages [RPM]). Spi-C is highly expressed in erythrophagocytic macrophages but not in monocytes, dendritic cells, or most other tissue macrophages, as Spi-C-/- mice have a defect

TABLE 12.2

Overview of Iron and Transcription Factors at the Crossroads of Macrophage Differentiation

Transcription factor	Associated cell type	Progenitor	Cell markers	Signaling involved	Functional crosstalk with iron	References
BACH 1	Bone marrow macrophages	BM-derived monocytes	F4/80$^+$, VCAM1$^+$	GSK3β-mediated signaling	Heme-regulated transcription factor essential for phenotypic transition of macrophages; regulates osteoclastogenesis	Haldar et al., 2014
C/EBPα	M1 macrophages	Monocytes/ Microglia	CD16, CD32, CD80, CD86	NF-κB-mediated signaling	Functions in synchrony with PU.1 to direct LMP to GMP stage; also is a key regulator of hepatic metabolism and controls transcription of hepcidin	Courselaud et al., 2002; Ye and Graf, 2007; Friedman, 2007
Nfatc1	Osteoclasts	BM-derived monocytes → Preosteoclasts → Osteoclasts	CSF-1	RANKL-induced signalling	Increase in cellular iron level due to Fpn downregulation promotes osteoclast differentiation	Wang et al., 2018
NRF-2	Kupffer cells	Yolk sac and foetal HSCs	CD11b^{lo}, CD80^{lo}, CD169$^+$, F4/80^h, Galectin-3$^+$	Keap-1-dependent signalling	Key regulator of redox homeostasis; plays major role during anti-inflammation along with HO-1	Scott et al., 2016
Pgc-1β	Osteoclasts	BM-derived monocytes → Preosteoclasts → Osteoclasts	CSF-1	RANKL-induced signalling	Increase in cellular iron level due to Fpn downregulation promotes osteoclast differentiation	Wang et al., 2018
PU.1	Bone marrow macrophages	BM-derived monocytes	F4/80$^+$, VCAM1$^+$	PKCβ/ NF-κB-mediated signalling	Crucial for monocyte to macrophage differentiation and macrophage proliferation (directs HSCs to LMP); upregulates in iron-deficient condition, thus inhibiting erythropoiesis	Richardson et al., 2013; Friedman, 2007
Spi-C	Splenic red pulp macrophages	Yolk sac and BM-derived monocytes	CD206$^+$, Dectin-2$^+$, F4/80$^+$, Spi-C	P38- and PI3K-mediated signalling	Central to differentiation of monocytes to iron-recycling macrophages; also important for osteoclastogenesis	Kohyama et al., 2009; Haldar et al., 2014

in the development of red pulp macrophages in bone marrow but have normal monocytes and other subset of macrophages (Kohyama et al., 2009). These macrophages highly express genes involved in capturing Hb and in iron regulation. In Spi-C/ mice RBCs are captured in the spleen but not efficiently phagocytized, resulting to an iron overload condition selectively to splenic red pulp (Kohyama et al., 2009). The Spi-C also regulates the development of F4/80⁺VCAM1⁺ bone marrow macrophages (BMM; Haldar et al., 2014). Heme regulates both BMM and RPM (Haldar et al., 2014). This was the first report of a metabolite-driven differentiation of a tissue-resident macrophage subset. A transcription repressor BACH1 inhibits Spi-C expression in monocytes. Heme promotes proteasomal degradation of BACH1 for rapid derepression of Spi-C resulting in differentiation of BMM and RPM (Halder et al., 2014). Heme is usually derived from Hb and subsequently translocated from phagolysosome to cytoplasm by the heme responsive gene-1 (HRG1; Rajagopal et al., 2008) and degraded by HO-1. HO-1 deletion is detected with progressive depletion of erythrophagocytic macrophages, suggesting heme degradation by HO-1 is required to maintain the integrity of this particular macrophage lineage (Kovtunovych et al., 2010).

12.3.2 Iron and Macrophage Polarization

Macrophages in response to inflammatory condition can polarize into two different functional phenotypes like M1 and M2. M1-like macrophages function as key effector cells for eliminating pathogens and cancer cells, while M2-like macrophages maintain tissue integrity by eliminating/repairing damaged cells and tissue matrices. Macrophages can switch their metabolic functions from a heal/growth promotional set up (M2) to a killing/inhibitory type (M1; Mills, 2012; Mills et al., 2000). One of the major differences between M1 and M2 macrophages is arginine metabolism. In M1 cells it is shifted to nitric oxide and citrulline, but in M2 it is shifted to ornithine and polyamine (Mills et al., 2000). Tissue macrophages basically behave like the M2 phenotype, while infiltrating recruited monocytes can differentiate either into M1 or M2 depending on the microenvironment (Italiani and Boraschi, 2014). Interestingly, available evidence suggests that M2 macrophages can switch to M1 macrophages, but the reverse does not occur in general (Italiani and Boraschi, 2014). M1 and M2 macrophages also

have distinct features in terms of iron and glucose metabolism (Biswas and Mantovani, 2012) and chemokine synthesis profiles (Mantovani et al., 2004).

Interestingly, a number of studies revealed that polarization of macrophages could alter the systemic iron homeostasis components hepcidin and Fpn (Soares and Hamza, 2016; Recalcati et al., 2010). M1 macrophages showed an increased level of FtH and repression of Fpn-favouring iron sequestration in the reticuloendothelial system, whereas Fpn expression was increased with simultaneous downregulation of FtH in M2 macrophages (Recalcati et al., 2010). Thus, M2 macrophages transcriptional level by M1 polarization show iron release capacity. The condition media of M2 macrophages favour cell proliferation potentially due to increased iron release, as conditioned media of macrophages of patient with Fpn mutation do not show similar cell proliferation ability (Recalcati et al., 2010). Tumour-associated macrophages (TAM) show M2 characteristics and their iron release tendency thus become helpful for proliferation of tumour cells. In complementary, Fpn transcription is modulated in M2 macrophages, while hepcidin expression is increased at the transcriptional level by M1 polarization (Agoro and Mura, 2016; Corna et al., 2010). An increased load of cellular iron triggers the expression of an M2-like phenotype in resting macrophages and affects pro-inflammatory immune responses, whereas iron depletion reduces M2 markers (Agoro et al., 2018). The precise cellular and biochemical reasons are not defined so far. Another study reported that heme and iron could activate macrophages to differentiate towards an M1-like pro-inflammatory phenotype, as detected by increased production of inflammatory cytokines and ROS (Vinchi et al., 2016). This is critically dependent on ROS generation and activation of TLR-4. Hemopexin, the scavenger of free heme, could reverse this pro-inflammatory activation of macrophages and ROS generation both in vivo and in vitro (Vinchi et al., 2016). The study further provided evidence that the iron moiety of the heme actually could promote the M1-like pro-inflammatory phenotype. Evidences in mice and patients suggest that iron overloading polarizes macrophages towards a pro-inflammatory M1-like phenotype in chronic venous leg ulcers (Sindrilaru et al., 2011). In genetically determined iron overload disease hereditary hemochromatosis, the hemochromatosis gene Hfe is found defective. In macrophages isolated from

Hfe knockout mice, a depleted intracellular iron level correlates with a decreased inflammatory cytokine response to LPS along-with an impaired innate immune response (Cairo et al., 2011). A similar observation was reported in patients with defects in iron metabolism (Cairo et al., 2011).

12.3.3 Iron and Differentiation of Other Tissue Resident Macrophages

Tissue resident macrophages (TRM) sense and react to wide range of environmental signals to help parenchymal cells for contributing to tissue repair and regeneration in case of a tissue injury (Soares and Hamza, 2016). Their differentiation is driven by specific genetic programs mostly influenced by micro-environmental cues generated from parenchyma cells of different tissues (Davies et al., 2013). TRMs have the ability to self-sustain the lineage throughout adulthood independent of monocyte recruitment (Hashimoto et al., 2013).

Kupffer cells (KC) are present in the liver and are the largest pool of TRMs in the body (Bouwens et al., 1986). They handle many iron-homeostasis and inflammatory functions of liver. They are derived from the yolk sac (Hoeffel et al., 2015; Gomez et al., 2015) or arise from foetal-hematopoietic stem cells and are found within liver sinusoids (Winn et al., 2020). Transcription factors Spi-C and NRF-2 are involved in the generation of KCs (Scott et al., 2016). They are primary cells to recycle iron released from senescent erythrocytes around sinusoids. Aberrant KC activation may lead to inflammatory cytokine release to influence hepatic hepcidin production (Winn et al., 2020). KCs are also reported in inhibiting hepcidin production (Theurl et al., 2008).

Multipotent hematopoietic stem cells (HSCs) in bone marrow determine the regulation of haematopoiesis and subsequent erythropoiesis in which iron is involved as the fundamental component. A specialized pool of resident erythroid island macrophages help erythropoiesis by serving as iron-rich nurse cells (Winn et al., 2020). Like splenic red pulp macrophages, the differentiation of these macrophages also depends on the transcription factor Spi-C (Haldar et al., 2014). These resident macrophages are also essential for bone remodelling by differentiating into osteoclasts to drive bone resorption. Iron released by Fpn is necessary for normal osteoclastogenesis and skeletal homeostasis in mice (Wang et al., 2018). The specific deletion of the iron exporter Fpn in myeloid osteoclast precursors increased osteoclastogenesis and decreased bone mass in mice. Interestingly, these phenotypes were more pronounced in female mice. Further, it was found that the elevated intracellular iron pool due to Fpn deletion increased expression of the nuclear factor of activated T cells 1 (Nfatc1) and PPARG coactivator 1β (Pgc-1β; Wang et al., 2018). These transcription factors play critical role in osteoclast differentiation. A population of non-osteoclast resident bone macrophages has been speculated as iron-handling regulatory cells (Winn et al., 2020).

Microglia are the resident macrophages of the central nervous system (CNS) and express iron transport and storage proteins like any other cell types. They can modulate cellular iron transport in accordance with their polarization state and extracellular milieu. They exhibit an enhanced preference for NTBI uptake and storage in the pro-inflammatory M1-like stage (Nnah and Wessling-Resnick, 2018, McCarthy et al., 2018). There are reports of shifting of pro-inflammatory to anti-inflammatory state by releasing Ft from M2-like microglial cells to help neuronal remyelination and repair after tissue injury (Nnah and Wessling-Resnick, 2018, McCarthy et al., 2018).

An iron-rich macrophage pool termed MFe^{hi} has been detected in adipose tissue of lean mice; the remaining pool is termed as MFe^{lo} (Orr et al., 2014). The MFe^{hi} pool accumulates excess iron in adipose tissue and maintains an M2-like character that is disturbed during diet-induced obesity (Hubler et al., 2018). Usually, these macrophages accumulate excess iron during high-iron diet to keep stable iron concentrations in adipocytes (Hubler et al., 2018). Intriguingly, Spi-C transcription factor is elevated in the MFe^{hi} but not in the MFe^{lo} macrophage pool isolated from mice subjected to high-iron diets (Hubler et al., 2018). These findings suggest that iron plays an instrumental role in adipocyte differentiation process. The control of this local iron homeostasis in adipose tissue is important for iron availability for normal adipogenesis (Gabrielsen et al., 2012) and to avoid iron-induced lipid peroxidation, particularly in the context of obesity (Winn et al., 2020).

Resident macrophages play an essential role in the maintenance of intestinal homeostasis as part of the immune system. However, they are also implicated in pathologies of the gastrointestinal tract, such as inflammatory bowel disease (IBD).

There is substantial advancement in the understanding of intestinal macrophage heterogeneity, their ontogeny, and factors that regulate their origin (Bain and Schridde, 2018). The influence of the local environment and its alteration during inflammation and infection on the phenotypic and functional identity of the macrophages is now better understood. Although the body iron uptake happens mainly through intestine, the role of iron in intestinal macrophage differentiation remains an area to be explored.

12.4 IRON IN MUSCLE CELL DIFFERENTIATION

Muscle tissue is a high consumer of iron because of the presence of heme-containing myoglobin and the need to generate ATP for contraction (Kaplan and Ward, 2015). Iron availability is an essential requirement for proper functioning of skeletal, cardiac, and smooth muscle. Among these, skeletal and cardiac muscle have more similar functioning mechanisms than smooth muscles.

The skeletal muscle, the most abundant tissue in the body, is essential for mobility and movement. Moreover, it also contributes significantly to glucose and lipid metabolism, both of which require adequate iron availability. Usually, the myogenic precursor satellite cells initially proliferate and then differentiate into myoblasts that further differentiate to form muscle cells. The mononucleated myoblasts proliferate, differentiate, and fuse with each other as well as pre-existing myofibers to form multinucleated myotubes and myofibers. The fusion of myoblasts is specific to skeletal muscle (e.g., biceps brachii), not cardiac or smooth muscle. Like proliferation, the differentiation of muscle cells also happens during development and post-natal myogenesis (White et al., 2010).

Like proliferative muscle cells, myotubes also require Tf-bound iron (Fe-Tf) for growth and maintenance of the normal differentiated state (Hagiwara et al., 1987). Cells did not mature or differentiate when lacking supplementation of Fe-Tf in chick myotubes (Ozawa and Hagiwara, 1982; Gerstenfeld et al., 1984); instead they degenerated when incubated with basal culture medium. The presence of more TfR on the surfaces of myotubes than myoblasts suggests higher requirement of iron in differentiated muscle cells (Hasegawa and Ozawa, 1982). However, a later study reported that presumptive myoblasts in its exponential growth phase had $3.78 \pm 0.24 \times 10^{10}$ TfR/µg DNA, while

myotubes had $3.80 \pm 0.26 \times 10^{10}$ TfR/µg DNA (Sorokin et al., 1987). Interestingly, the same study found that the maximum iron uptake was significantly higher in myotubes than in any growth phase of presumptive myoblasts. This might be attributed to higher TfR recycling in myotubes. RNA and global protein syntheses in myotubes were affected in absence of supplementation of Fe-Tf (Shoji and Ozawa, 1985). Further experiments provided evidence that actually iron was indispensable to the myotube differentiation but not the Tf protein molecule (Hagiwara et al., 1987). Myoglobin synthesis happens when cells are differentiated into multinucleated myotubes (Graber and Woodsworth, 1986). Myotubes have a high abundance of mitochondria with heme and non-heme iron–containing enzymes. So, myotubes need more iron than myoblasts, and that may be reflected by the higher capacity of Fe-Tf binding and iron uptake in differentiated muscle cells. Interestingly, in several other cell types, the terminally differentiated cells are found with little or no significant Tf binding, unlike myotubes (Chitambar et al., 1983; Iacopetta et al., 1982; Panet al., 1983).

The differentiation of muscle cells is regulated by the myogenic regulatory factor family (MRF) of transcription factors: MyoD, Myogenin, MYF5, and MRF4 (Zammit, 2017). The discovery of these factors was a seminal step in understanding specification of the skeletal muscle lineage and control of myogenic differentiation during development. They also play crucial roles in guiding satellite cell function for regenerating skeletal muscle and to link the genetic control of developmental and regenerative myogenesis (Zammit, 2017). MyoD was the first among the family of MRF found to convert a small proportion of fibroblasts and other differentiated cell types into the skeletal muscle lineage (Davis et al., 1987). In fact, this was the first report of a single transcription factor to drive transdifferentiation for reprogramming of a differentiated cell type. Myogenin, which was discovered in 1989, could force myoblasts to undergo myogenic differentiation and result in the exit of myoblasts from the cell cycle to fuse to form multinucleated myofibers (Wright et al., 1989). Myogenic factor 5 (MYF5) and myogenic regulatory factor 4 (MRF4, also known as myogenic factor-6 or Myf6) were later identified using sequence homology to MyoD and Myogenin (Braun et al., 1989; Rhodes and Konieczny, 1989).

Lactoferrin (Lf), a multifunctional non-heme binding glycoprotein, has a role in proliferation of myoblasts as well as in differentiation of myoblasts (Kitakaze et al., 2018). Lf was found to increase myotube specific structural protein, myosin heavy chain (MyHC), and myotube formation in C2C12 myoblasts. Lf augments mRNA expression of MyoD and Myogenin by binding with low-density lipoprotein receptor-related protein (LRP1) mediated by ERK1/2 phosphorylation (Figure 12.3). Another study also reported the role of Lf in promoting osteoblastic and chondroblastic differentiation of C2C12 myoblasts (Yagi et al., 2009).

The overexpression of HO-1 was suggested as a strategy for improving survival of transplanted muscle precursors because of a speculative role of HO-1 on MRF. HO-1 overexpression blocked C2C12 myoblasts to myotube differentiation (Kozakowska et al., 2012). Further experimentation provided evidence of abrogation of MyoD, myogenin, and MyHC in HO-1 overexpressed cells. Because MyoD could induce myogenin and Myf6, the role of MyoD was further scrutinized. Blocking of HO-1 activity by pharmaceutical inhibitor or silencing of the enzyme by siRNA increased MyoD expression and subsequent myotube formation. Intriguingly, the effect of HO-1 could be mimicked by iron, one of the products of HO-1 activity but not by antioxidants, suggesting the iron-releasing capacity of HO-1 as the

crucial factor in controlling MyoD but not its antioxidant capacity. CO-dependent inhibition of C/EBPδ binding to MyoD promoter was found to be one of the mechanisms of this process. Further investigation revealed that HO-1 could affect microRNA transcriptome by downregulating the miRNA processing enzymes Lin28 and DGCR8 for reducing the total pool of miRNA for myotube differentiation. This study provided strong evidence of HO-1 as an important regulator of muscle differentiation (Figure 12.3). The HO-1 activation also reported to inhibit differentiation of osteoclasts (Lee et al., 2010) and adipocytes (Vanella et al., 2010).

Fe-S cofactors are formed on the scaffold protein ISCU in mitochondria. ISCU works in conjunction with the cysteine desulfurases NFS1, ISD11, and FXN to assemble [2Fe-2S] and [4Fe-4S] clusters (Schmucker et al., 2011); those are subsequently transferred to recipients by a chaperone complex (Uhrigshardt et al., 2010). ISCU myopathy is a disease occurring due to an intronic mutation leading to abnormally spliced ISCU mRNA affecting the formation of Fe-S clusters as well as mitochondrial iron overload in affected myofibers (Mochel et al., 2008). The affected Fe-S cluster formation causes striking deficiencies in aconitase and succinate dehydrogenase, and lesser deficiencies of mitochondrial complex I and III and the Rieske protein (Hall et al., 1993). Overexpression of MyoD in myoblasts derived from patients decreased the abundance of normally spliced ISCU mRNA and subsequent protein expression; however, this increased the abnormally spliced ISCU mRNA level. It is still not clear how MyoD affects the balance of the splicing of these ISCU mRNAs (Crooks et al., 2012).

Hepcidin is the master regulator of systemic iron metabolism (Muckenthaler et al., 2017). HJV or repulsive guidance molecule c (RGMc), a secreted glycoprotein, is expressed in striated muscle cells and is one of the regulators of hepcidin synthesis in hepatocytes (Babitt et al., 2006). Mutations in the HJV gene were detected in juvenile hemochromatosis (Lanzara et al., 2004; Papanikolaou et al., 2004), and a targeted knockout mouse yielded a similar disease phenotype (Huang et al., 2005). Synthesis of RGMc occurs during muscle cell differentiation (Kuninger et al., 2004). In situ hybridization showed that RGMc mRNA was expressed during early embryonic development in committed muscle precursor cells as well as in foetal muscle, liver, and heart (Kuninger et al., 2004). Very rapid expression of mRNA and RGMc was detected

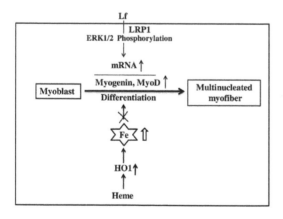

Figure 12.3 Role of heme and non-heme iron protein on muscle cell differentiation.

Lactoferrin (Lf) binds to low-density lipoprotein receptor-related protein (LRP1) to induce the transcript level of MyoD and Myogenin mediated by ERK1/2 phosphorylation. Heme induces HO-1 expression, resulting in the release of iron that blocks muscle cell differentiation by increasing ferritin-H level and decreasing extracellular inorganic phosphate. Overexpression of HO-1 can also block the differentiation via CO-dependent inhibition of C/EBPδ binding to MyoD promoter.

during incubation of myoblasts with differentiation media. RGMc was also reported to be secreted in the extracellular fluid (Kuninger et al., 2006). Although RGMc synthesis is dependent on muscle cell differentiation, overexpression of the molecule does not influence the muscle differentiation process. This may be depicted in juvenile hemochromatosis or knockout mice, as they display normal muscle (Kuninger et al., 2006). HJV functions as a co-receptor for BMPs to enhance BMP-mediated signalling to lead hepcidin gene expression mediated by Smads (Babitt et al., 2006). These observations implied a global role of differentiated muscle cells in systemic iron homeostasis in mammals (Figure 12.4). This may be supported by the observation that disturbance of muscle cell iron homeostasis by specifically knocking out TfR1 results in systemic damage in the liver and adipocytes (Barrientos et al., 2015). However, the molecular mechanism of HJV regulation and secretion during muscle cell differentiation still remains largely unknown.

Smooth muscle cells (SMC) are also well studied for their ability to be differentiated during normal development and maturation of the vasculature as well as during vascular injury leading to different pathogenesis. The understanding of differentiation of SMC is very challenging, as this cell exhibits a wide range of different phenotypes at different developmental stages, and more intriguingly, it is not terminally differentiated unlike skeletal or cardiac muscle cell (Owens et al., 2004). Fully matured SMC exhibit remarkable plasticity apparently to provide a better survival advantage. Another important feature of SMC is that unlike other muscle types, they express multiple markers indicative of their relative state of differentiation but not any exclusive one. However, smooth muscle myosin heavy chain (SM MHC) and smoothelin are the most widely attributed differentiating markers for SMCs. Interestingly, the role of ROS is well appreciated in SMC differentiation, but the role of iron is far less explored.

Human smooth muscle cells (HSMC) have the ability for osteoblastic differentiation. High extracellular inorganic phosphate (Pi) may enter vascular cells to induce the calcification of vascular cells leading to the transition of HSMCs to osteoblast-like cells accompanied by increased expression of core binding factor α-1 (Cbfa-1), the osteoblast-specific transcription factor for osteoblast differentiation (Otto et al., 1997). Heme is found to decrease the calcification and enhance the osteoblastic differentiation (Zarjou et al., 2009). Heme induces HO-1, which releases iron. Further experiments revealed that iron regardless of its oxidized state could block HSMC to osteoblastic differentiation by two distinct mechanisms: (i) by increasing FtH and (ii) by decreasing extracellular Pi level. FtH blocks this differentiation process by virtue of its ferroxidase activity. Another ferroxidase Cp also showed similar effect. The decreased expression of Cbfa-1 was found as the mechanism of iron-induced inhibition of HSMC differentiation to osteoblast (Zarjou et al., 2009). Vascular calcification also contributes to the pathogenesis of chronic kidney disease, diabetes, and atherosclerosis. Thus, the HO-1/Ft system plays an essential role in maintaining homeostasis of vascular function in different tissues (Zarjou et al., 2009).

12.5 IRON AND ADIPOCYTE DIFFERENTIATION

Adipocytes play a crucial role in the energy balance of the body (Gregoire et al., 1998). Pre-adipocytes undergo growth arrest and are subsequently differentiated into adipocytes. In higher eukaryotes, differentiated adipocytes serve as the major energy reservoir for the purpose of storing triglycerides during the period of energy excess and mobilization during energy deprivation. Cells lose their fibroblastic appearance during adipose differentiation to acquire the characteristic morphology of mature

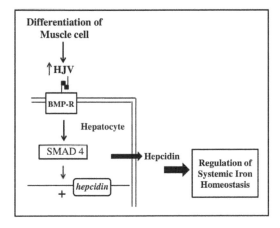

Figure 12.4 A model implicating the role of differentiated muscle cells on regulating systemic iron homeostasis.

Differentiated muscle cells release hemojuvelin (HJV), which binds to the BMP receptor (BMP-R) for SMAD4-induced hepcidin synthesis and release from hepatocytes.

adipocytes (Novikoff et al., 1980). Simultaneously, activities of enzymes involved in the biosynthesis of fatty acids and triglycerides are increased (Chen and London, 1981) due to a dramatic increase in expression of adipocytic genes, including those responsible for expressing fatty acid binding proteins and lipid metabolizing enzymes. Several decades of research characterized regulatory regions of adipogenic genes and identified transcription factors like peroxisome proliferator-activated receptor-γ (PPAR-γ) and C/EBP as key players in the complex transcriptional cascade during adipocyte differentiation. A potential role of the basic helix-loop-helix (bHLH) family of proteins in activating adipocyte differentiation was proposed, as overexpression of inhibitor of DNA binding 3 (Id3) could prevent the differentiation of adipocytes (Moldes et al., 1997). Ids are dominant negative members of the bHLH family. The differentiation of pre-adipocytes is controlled by communication between individual cells as well as between cells and the extracellular environment that is influenced by various hormones, growth factors, cytokines, and nutritional components (Gregoire et al., 1998).

Circulating markers of iron overload were reported to be positively associated with visceral and subcutaneous fat depots (Iwasaki et al., 2005) and with adipocyte insulin resistance (IR; Wlazlo et al., 2013). An iron-enriched diet induces iron accumulation and insulin resistance in visceral adipose tissue (VAT) in mice (Dongiovanni et al., 2013). The iron overload in adipocytes affected adiponectin synthesis simultaneously with induction of IR (Gabrielsen et al., 2012). Adipose tissues from obese and insulin-resistant individuals were detected with increased Ft subunits (Moreno-Navarrete et al., 2014), suggesting a role of iron in the adipocyte differentiation process. The role of iron homeostasis components on adipogenesis is summarized in Table 12.2 and described below.

The heme-induced increase in adipocyte differentiation was perhaps one of the earliest findings suggesting involvement of iron in adipocyte differentiation. Heme alone as well as in presence of insulin was found to induce this process, although mechanisms appeared to be different (Chen and London, 1981). Increased synthesis of iron storage protein Ft is associated with development, cellular differentiation, and inflammation (Harrison and Arosio, 1996). During differentiation of 3T3-L1 adipocytes, protein and mRNA levels of both subunits of Ft were increased (Festa et al., 2000). Interestingly, the same report also found increased levels of IRP1 mRNA and interaction between IRP with IRE of FtH, suggesting a decrease in Ft translation—apparently a contradictory phenomenon of increased FtH transcript expression. However, the gross increase in FtH level was for sequestering iron, probably to avoid lipid peroxidation of abundant fat depots present in adipocytes (Festa et al., 2000). The IRP1 functions as cytosolic ACO1 in iron-saturated cells. A later study further confirmed an increased ACO1 gene expression and activity in parallel with adipogenic genes during differentiation of 3T3L1 cells as well

TABLE 12.2

Role of Iron Homeostasis Components on Adipogenesis

Iron homeostasis components	Cellular localization	Effect on adipogenesis	References
Heme	Extracellular	Promote differentiation	Chen & London, 1981
Ferritin	Cytosol	Promote proliferation and differentiation	Harrison & Arosio, 1996; Festa et al., 2000; Moreno et al., 2015
Mitoferrin 1, Mitoferrin 2	Mitochondria (inner membrane)	Promote differentiation	Chen et al., 2015
MitoNEET	Mitochondria (outer membrane)	Promote differentiation	Kusminski et al., 2012
Rieske's iron-sulphur protein	Mitochondria (inner membrane)	Promote differentiation	Tormos et al., 2011
Aconitase 1	Cytosol	Sustain adipogenic capacity	Moreno et al., 2015; Moreno-Navarrete et al., 2014a
Lactoferrin	Extracellular	Promote differentiation	Moreno-Navarrete et al., 2011, 2014b

REDOX REGULATION OF DIFFERENTIATION AND DE-DIFFERENTIATION

as in human preadipocytes (Moreno et al., 2015). The knocking-down of ACO1 impaired adipogenesis with a simultaneous decrease in lipogenic, isocitrate dehydrogenase 1, adiponectin, and Glut4 gene expression. The iron uptake component TfR1 was affected due to ACO1 knockdown disrupting iron uptake in differentiated adipocytes (Moreno et al., 2015). Results from two human cohort studies also found a positive association between ACO1 gene expression with adipogenic markers in sub-cutaneous and visceral adipose tissue. These findings suggest that ACO1 activity is required for normal adipogenic capacity of adipose tissue linking iron, adipogenesis, and energy metabolism (Moreno-Navarrete et al., 2014a).

Iron by its capacity in regulating mitochondrial biogenesis (Rensvold et al., 2013) may also control adipocytic differentiation because the mitochondrial biogenesis programme is defined as a key metabolic process in adipocyte differentiation (Medina-Gómez, 2012). Others have also suggested that mitochondrial function is important during adipogenesis and in mature adipocytes (Kusminski and Scherer, 2012). Mitochondrial iron availability depends on two importers: Mfrn1 and Mfrn2 (Chen et al., 2015). Their expressions were found elevated during adipogenic induction (Chen et al., 2015). Simultaneous knocking-down of both these importers in 3T3 L1 preadipocytes reduced expressions of adipogenic genes and lipid production during adipogenic differentiation (Chen et al., 2015). Interestingly, insulin-induced glucose uptake and signalling pathways are affected in adipocytes derived from preadipocytes deficient in Mfrns (Chen et al., 2015). These observations suggest that mitochondrial iron availability may control the adipogenic process and even insulin sensitivity in adipocytes. The loss of function of a mitochondrial iron-sulphur protein mitoNEET in adipocytes resulted into reduced mitochondrial iron content and heme synthesis leading to extreme obesity in ob/ob mice (Kusminski et al., 2012). The loss of function of Rieske's iron-sulphur protein in Complex III of mitochondrial respiratory chain resulted in impairment of adipogenic differentiation (Tormos et al., 2011). Experimental iron deficiency affected adipocyte differentiation severely but was reversed in a dose-dependent iron supplementation, suggesting a threshold of intracellular iron availability is required for optimal adipocyte differentiation (Moreno-Navarrete et al., 2014a). The knocking-down of adipocytic

Tf affected expressions of adipogenic markers like PPARγ, adiponectin, Glut4, CEBP/α, and others with a simultaneous increase in expressions of inflammatory genes like IL6 and monocyte chemoattractant protein-1 (MCP1) (Moreno-Navarrete et al., 2014a).

Lf, another iron-related component, has been reported to be associated with increased adipogenesis and decreased inflammatory markers in mouse and human adipose tissue. The knocking-down of Lf in human subcutaneous and visceral pre-adipocytes led to decreased expression of adipogenic, lipogenic, and insulin-signalling related genes with a simultaneous increase in gene expression of inflammatory mediators (Moreno-Navarrete et al., 2014b). While addition of purified Lf recovers the adipocyte differentiation in a similar condition, addition of human Lf to pre-human adipocytes led to enhanced PPAR-γ expression and insulin-signalling activity (Moreno-Navarrete et al., 2011). These observations underscore the significance of increased Lf mRNA and protein levels during adipocyte differentiation (Moreno-Navarrete et al., 2013). It supports the view of autocrine production of Lf in adipocyte tissue for adipogenesis (Moreno-Navarrete et al., 2014b).

12.6 SUMMARY AND FUTURE PERSPECTIVES

This chapter summarizes the understanding of the influence of iron and iron homeostasis components on differentiation of three cell types: macrophages, muscle cells, and adipocytes. Research over the last four decades has established that several key iron homeostasis components, such as HO-1, Ft, TfR, and Fpn; Fe-S cluster proteins like ISCU and iron sensor IRP1; and mitochondrial iron importers (Mfrn1 and Mfrn2) have key roles in cell differentiation. Interestingly, the involvement of soluble secretory iron homeostasis proteins like Lf and HJV suggests cells can influence the differentiation of other cell types. All cell types need iron, and they differentiate to perform specialized functions. However, less is known about most other cell types like oligodendrocytes, which utilize iron for specialized functions. Iron deficiency or anaemia during pregnancy is well-known to affect the growth and development of the embryo as well as the new-born. So, it will be intriguing to know how cells within the human embryo differentiate into complex tissues and organ systems during anaemia or in iron-overload conditions. Differentiated cells like adipocytes also can be de-differentiated; however, the role of iron in de-differentiation has not yet been well understood. Better understanding of these areas may be helpful to allow "to direct" or "not to direct" cell differentiation for improved health care in the future.

12.7 ACKNOWLEDGMENT

C.K.M. acknowledges financial support received from the Department of Biotechnology and Science and Engineering Research Board (SERB) of the Department of Science and Technology, India. S.Y. is a recipient of a fellowship from the Council of Scientific and Industrial Research, India. D.K. acknowledges support as a fellowship from Jawaharlal Nehru University.

REFERENCES

Abboud S., Haile D.J. 2000. A novel mammalian iron-regulated protein involved in intracellular iron metabolism. J Biol Chem 275: 19906–19912.

Agoro R., Mura C. 2016. Inflammation-induced up-regulation of hepcidin and down-regulation of ferroportin transcription are dependent on macrophage polarization. Blood Cells Mol Dis 61: 16–25.

Agoro R., Taleb M., Quesniaux V.F.J., Mura C. 2018. Cell iron status influences macrophage polarization. PLoS ONE 13: e0196921.

Aisen P., Enns C., Wessling R.M. 2001. Chemistry and biology of eukaryotic iron metabolism. Int J Biochem Cell Biol 33: 940–959.

Ajioka R.S., Phillips J.D., Kushner J.P. Biosynthesis of heme in mammals. 2006. Biochim Biophys Acta 1763: 723–736.

Anderson G.J. 1999. Non-transferrin-bound iron and cellular toxicity. J Gastroenterol Hepatol 14: 105–108.

Anderson S.A., Nizzi C.P., Chang Y.I., Kathryn M.D., Schmidt P.J., Galy B., Damnernsawad A., Broman A.T., Kendziorski C., Hentze M.W., Fleming M.D., Zhang J. Eisenstein R.S. 2013. The IRP1-HIF-2alpha axis coordinates iron and oxygen sensing with erythropoiesis and iron absorption. Cell Metab 17: 282–290.

Arosio P., Levi S. 2002. Ferritin, iron homeostasis, and oxidative damage. Free Radic Biol Med 33: 457–463.

Babitt J.L., Huang F.W., Wrighting D.M., Xia Y., Sidis Y., Samad T.A., Campagna J.A., Chung R.T., Schneyer A.L., Woolf C.J., Andrews N.C., Lin H.Y. 2006. Bone morphogenetic protein signaling by hemojuvelin regulates hepcidin expression. Nat Genet 38: 531–539.

Bain C.C., Schridde A. 2018. Origin, differentiation, and function of intestinal macrophages. Front Immunol 9: 2733.

Barrientos T., Laothamatas I., Koves T.R., Soderblum M., Muoio M.A., Andrew D.M. 2015. Metabolic catastrophe in mice lacking transferrin receptor in muscle. EBioMedicine 2: 1705–1717.

Batey R.G., Lai Chung Fong P., Shamir S., Sherlock S. 1980. A nontransferrin-bound serum iron in idiopathic hemochromatosis. Dig Dis Sci 25: 340–346.

Biswas S.K., Mantovani A. 2012. Orchestration of metabolism by macrophages. Cell Metabol 15: 432–437.

Bouwens L., Baekeland M., De Zanger R., Wisse E. 1986. Quantitation, tissue distribution and proliferation kinetics of Kupffer cells in normal rat liver. Hepatology 6: 718–722.

Braun T., Buschhausen-Denker G., Bober E. Tannich E., Arnold H.H. 1989. A novel human muscle factor related to but distinct from MyoD1 induces myogenic conversion in 10T1/2 fibroblasts. EMBO J 8: 701–709.

Cairo G., Recalcati S., Mantovani A., Locati M. 2011. Iron trafficking and metabolism in macrophages: contribution to the polarized phenotype. Trends Immunol 32: 241–247.

Canonne-Hergaux F., Gruenheid S., Ponka P., Gros P. 1999. Cellular and subcellular localization of the Nramp2 iron transporter in the intestinal brush border and regulation by dietary iron. Blood 93: 4406–4417.

Casey J.L., Hentze M.W., Koeller D.M., Caughman S.W., Rouault T.A. 1988. Iron-responsive elements: regulatory RNA sequences that control mRNA levels and translation. Science 240: 924–928.

Cazzola M., Arosio P., Bellotti V. 1985. Use of a monoclonal antibody against human heart ferritin for evaluating acidic ferritin concentration in human serum. Br J Haematol 61: 445–453.

Chen J.J., London I.M. 1981. Hemin enhances the differentiation of mouse 3T3 cells to adipocytes. Cell 26: 117–122.

Chen Y.C., Wu Y.T., Wei Y.H. 2015. Depletion of mitoferrins leads to mitochondrial dysfunction and impairment of adipogenic differentiation in 3T3-L1 preadipocytes. Free Radic Res 49: 1285–1295.

Cheng Y., Zak O., Aisen P., Harrison S.C., Walz T. 2004. Structure of the human transferrin receptor-transferrin complex. Cell 116: 565 576.

Chitambar C.R., Massey E.J., Seligman P.A. 1983. Regulation of transferrin receptor expression on human leukemic cells during proliferation and induction of differentiation. Effects of gallium and dimethylsulfoxide. *J Clin Invest* 72: 1314–1325.

Collins J.F., Wessling-Resnick M., Knutson M.D. 2008. Hepcidin regulation of iron transport. *J Nutr.* 2008. 138: 2284–2288.

Corna G., Campana L., Pignatti E., Castiglioni A., Tagliafico E., Bosurgi L., et al. 2010. Polarization dictates iron handling by inflammatory and alternatively activated macrophages. *Haematologica* 95: 1814–1822.

Courselaud B., Pigeon C., Inoue Y., Inoue J., Gonzalez F.J., Leroyer P., Gilot D., Boudjema K., Guguen-Guillouzo C., Brissot P., Loreal O., Ilyin G. 2002. C/EBPalpha regulates hepatic transcription of hepcidin, an antimicrobial peptide and regulator of iron metabolism: cross-talk between C/EBP pathway and iron metabolism. *J Biol Chem* 277: 41163–41170.

Crooks D.R., Jeong S.Y., Tong W.-H., Ghosh M.C., Olivierre H., Haller R.G., Rouault T.A. 2012.Tissue specificity of a human mitochondrial disease: differentiation enhanced mis-splicing of the Fe-S scaffold gene ISCU renders patient cells more sensitive to oxidative stress in ISCU myopathy. *J Biol Chem* 287: 40119–40130.

Davies L.C., Jenkins S.J., Allen J.E., Taylor P.R. 2013. Tissue-resident macrophages. *Nat Immunol* 14: 986–995.

Davis R.L., Weintraub H., Lassar A.B. 1987. Expression of a single transfected cDNA converts fibroblasts to myoblasts. *Cell* 51: 987–1000.

Dongiovanni P., Ruscica M., Rametta R., Recalcati S., Steffani L., Gatti S., Girelli D., Cairo G., Magni P., Fargion S., Valenti L. 2013. Dietary iron overload induces visceral adipose tissue insulin resistance. *Am J Pathol* 182: 2254–2263.

Donovan A., Brownlie A., Zhou Y., Shepard J., Pratt S.J., Moynihan J., Paw B.H., Drejer A., Barut B., Zapata A., Law T.C., Brugnara C., Lux S.E., Pinkus G.S., Pinkus J.L., Kingsley P.D., Palis J., Fleming M.D., Andrews N.C., Zon L.I. 2000. Positional cloning of zebrafish ferroportin1 identifies a conserved vertebrate iron exporter. *Nature* 403: 776–781.

Elliott S., Pham E., Macdougall I.C. 2008. Erythropoietins: a common mechanism of action. *Exp Hematol* 36: 1573–1584.

Farmer S.R. 2006. Transcriptional control of adipocyte formation. *Cell Metab* 4: 263–273

Festa M., Ricciardelli G., Mele G., Pietropaolo C., Ruffo A., Colonna A. 2000. Overexpression of H ferritin and up-regulation of iron regulatory protein genes during differentiation of 3T3-L1 pre-adipocytes. *J Biol Chem* 275: 36708–36712.

Frazer D.M., Anderson G.J. 2005. Iron imports. I. Intestinal iron absorption and its regulation. *Am J Physiol Gastrointest Liver Physiol* 289:G631–G635.

Friedman A.D. 2007. Transcriptional control of granulocyte and monocyte development. *Oncogene* 26, 6816–6828.

Gabrielsen J.S., Gao Y., Simcox J.A., Huang J., Thorup D., Jones D., Cooksey R.C., Gabrielsen D., Adams T.D., Hunt S.C., Hopkins P.N., Cefalu W.T., McClain D.A. 2012. Adipocyte iron regulates adiponectin and insulin sensitivity. *J Clin Invest* 122: 3529–3540.

Ganz T. 2012. Macrophages and Systemic Iron Homeostasis. *J Innate Immun* 4: 446–453

Ganz T., Nemeth E. 2006. Iron imports. IV. Hepcidin and regulation of body iron metabolism. *Am J Physiol Gastrointest Liver Physiol* 290: G199–G203.

Gao F.B., Brenman J.E., Jan L.Y., Jan Y.N. 1999. Genes regulating dendritic outgrowth, branching, and routing in *Drosophila*. *Genes Dev* 13: 2549–2561.

Gerstenfeld C., Crawfordd R., Boedtker H., Doty P. 1984. Expression of type I and III collagen genes during differentiation of embryonic chicken myoblasts in culture. *Mol Cell Biol.* 4: 1483–1492.

Gomez P.E., Klapproth K., Schulz C., Busch K., Azzoni E., Crozet L., Garner H., Trouillet C., de Bruijn M.F., Geissmann F., Rodewald H.R. 2015. Tissue-resident macrophages originate from yolk-sac-derived erythro-myeloid progenitors. *Nature* 518: 547–551.

Gozzelino R., Jeney V., Soares M.P. 2010. Mechanisms of cell protection by heme oxygenase-1. *Annu Rev Pharmacol Toxicol* 50: 323–354.

Graber S.G., Woodworth R.C. 1986. Myoglobin expression in L6 muscle cells: role of differentiation and heme. *J Biol Chem* 261: 9150–9154.

Gregoire F.M., Smas C.M., Sul H.S. 1998. Understanding adipocyte differentiation. *Physiol Rev* 78: 783–809.

Gros G., Wittenberg B.A., Jue T. 2010. Myoglobin's old and new clothes: from molecular structure to function in living cells. *J Exp Biol* 213: 2713–2725.

Gunshin H., Mackenzie B., Berger U.V., Gunshin Y., Romero M.F., Boron W.F., NussbergerS., Gollan J.L., Hediger M.A. 1997. Cloning and characterization of a mammalian proton-coupled metal-ion transporter. *Nature* 388: 482–488.

Hagiwara Y., Saito K., Atsumi S., Ozawa E. 1987. Iron supports myogenic cell differentiation to the same degree as does iron-bound transferrin. *Dev Biol* 120: 236–244.

Haldar M., Kohyama M., So A.Y., Kc W., Wu X., Briseno C.G., Satpathy A.T., Kretzer N.M., Arase H., Rajasekaran N.S., et al. 2014. Heme-mediated SPI-C induction promotes monocyte differentiation into iron-recycling macrophages. *Cell* 156: 1223–1234

Hall R.E., Henriksson K.G., Lewis S.F., Haller R.G., Kennaway N.G. 1993. Mitochondrial myopathy with succinate dehydrogenase and aconitase deficiency. Abnormalities of several iron-sulfur proteins. *J Clin Invest* 92: 2660–2666.

Hamza I., Dailey H.A. 2012. One ring to rule them all: trafficking of heme and heme synthesis intermediates in the metazoans. *Biochim Biophys Acta* 1823: 1617–1632.

Harris Z.L., Durley A.P., Man T.K., Gitlin J.D. 1999. Targeted gene disruption reveals an essential role for ceruloplasmin in cellular iron efflux. *Proc Natl Acad Sci USA* 96: 10812–10817.

Harrison P.M., Arosio P. 1996. The ferritins: molecular properties, iron storage function and cellular regulation. *Biochim Biophys Acta* 1275: 161–203.

Hasegawa T., Ozawa E. 1982. Transferrin receptor on chick fibroblast cell surface and the binding affinity in relevance to the growth promoting activity of transferrin. *Dev Growth Diff* 24: 581–587.

Hashimoto D., Chow A., Noizat C., Teo P., Beasley M.B., Leboeuf M., Becker C.D., See P., Price J., Lucas D., Greter M., Mortha A., Boyer S.W., Forsberg E.C., Tanaka M., van Rooijen N., García-Sastre A., Stanley E.R., Ginhoux F., Frenette P.S., Merad M. 2013. Tissue-resident macrophages self-maintain locally throughout adult life with minimal contribution from circulating monocytes. *Immunity* 38: 792–804.

Hentze M.W., Muckenthaler M.U., Andrews N.C. 2004. Balancing acts: molecular control of mammalian iron metabolism. *Cell* 117: 285–297.

Hershko H., Graham G., Bates G.W., Rachmilewitz E. 1978. Nonspecific serum iron in thalassaemia: an abnormal serum iron fraction of potential toxicity. *Br J Haematol* 40: 255–263.

Hoeffel G., Chen J., Lavin Y., Low D., Almeida F.F., See P., Beaudin A.E., Lum J., Low I., Forsberg E.C., Poidinger M., Zolezzi F., Larbi A., Ng L.G., Chan J.K., Greter M., Becher B., Samokhvalov I.M., Merad M., Ginhoux F. 2015. C-Myb(+) erythro-myeloid progenitor-derived fetal monocytes give rise to adult tissue-resident macrophages. *Immunity* 42: 665–678.

Huang F.W., Pinkus J.L., Pinkus G.S., Fleming M.D., Andrews N.C. 2005. A mouse model of juvenile hemochromatosis. *J Clin Invest* 115: 2187–2191.

Hubler M.J., Erikson K.M., Kennedy A.J., Hasty A.H. 2018. MFe[hi] adipose tissue macrophages compensate for tissue iron perturbations in mice. *Am J Physiol, Cell Physiol* 315:C319–C329.

Iacopetta B.J., Morgan E.H., Yeoh G.C.T. 1982. Transferrin receptors and iron uptake during erythroid cell development. *Biochim Biophys Acta* 687: 204–210.

Italiani P., Boraschi D. 2014. From Monocytes to M1/M2 Macrophages: phenotypical vs. Functional Differentiation. *Front Immunol* 5: 514. eCollection.

Iwasaki T., Nakajima A., Yoneda M., Yamada Y., Mukasa K., Fujita K., Fujisawa N., Wada K., Terauchi Y. 2005. Serum ferritin is associated with visceral fat area and subcutaneous fat area. *Diabetes Care* 28: 2486–2491.

Jeong S.Y., David S. 2003. Glycosylphosphatidylinositol-anchored ceruloplasmin is required for iron efflux from cells in the central nervous system. *J Biol Chem* 278: 27144–27148.

Kakar S., Hoffman F.G., Storz J.F., Fabian M., Hargrove M.S. 2010. Structure and reactivity of hexacoordinate hemoglobins. *Biophys Chem* 152: 1–14.

Kakhlon O., Cabantchik Z.I. 2002. The labile iron pool: characterization, measurement, and participation in cellular processes. *Free Radic Biol Med* 33: 1037–1046.

Kaplan J., Ward D.M. 2015. Muscle specific iron deficiency has systemic consequences. *EBioMedicine* 2: 1582–1583.

Kautz L., Jung G., Valore E.V., Rivella S., Nemeth E., Ganz T. 2014. Identification of erythroferrone as an erythroid regulator of iron metabolism. *Nat Genet* 46: 678–684.

Kim S., Ponka P. 2000. Effects of interferon-gamma and lipopolysaccharide on macrophage iron metabolism are mediated by nitric oxide-induced degradation of iron regulatory protein 2. *J Biol Chem* 275: 6220–6226.

Kitakaze T., Oshimo M., Kobayashi Y., Mizuyukiryu M., Suzuki Y.A., Inui H., Harada N., Yamaji R. 2018. Lactoferrin promotes murine C2C12 myoblast proliferation and differentiation and myotube hypertrophy. *Mol Med Rep* 17: 5912–5920.

Klausner R.D., Ashwell G., Renswoude J., Harford J.B., Bridges K.R. 1983. Binding of apotransferrin to K562 cells: explanation of the transferrin cycle. *Proc Natl Acad Sci* 80: 2263–2266.

Kohyama M., Ise W., Edelson B.T., Wilker P.R., Hildner K., Mejia C., Frazier W.A., Murphy T.L., Murphy K.M. 2009. Role for Spi-C in the development of red pulp macrophages and splenic iron homeostasis. *Nature* 457: 318–321.

Korolnek T., Hamza I. 2015. Macrophages and iron trafficking at the birth and death of red cells. *Blood* 125: 2893–2897.

Kovtunovych G., Eckhaus M.A., Ghosh M.C., Ollivierre-Wilson H., Rouault T.A. 2010. Dysfunction of the heme recycling system in heme oxygenase 1-deficient mice: effects on macrophage viability and tissue iron distribution. *Blood* 116: 6054–6062.

Kozakowska M., Ciesla M., Stefanska A., Skrzypek K., Was H., Jazwa A., et al. 2012. Heme oxygenase-1 inhibits myoblast differentiation by targeting myomirs. *Antioxid Redox Signal* 16: 113–127.

Kuninger D., Kuns-Hashimoto R., Kuzmickas R., Rotwein P. 2006. Complex biosynthesis of the muscle-enriched iron regulator RGMc. *J Cell Sci* 119: 3273–3283.

Kuninger D., Kuzmickas R., Peng B., Pintar J.E., Rotwein P. 2004. Gene discovery by microarray: identification of novel genes induced during growth factor mediated muscle cell survival and differentiation. *Genomics* 84: 876–889.

Kusminski C.M., Holland W.L., Sun K., Park J., Spurgin S.B., Lin Y., et al. 2012. MitoNEET-driven alterations in adipocyte mitochondrial activity reveal a crucial adaptive process that preserves insulin sensitivity in obesity. *Nat Med* 18: 1539–1549.

Kusminski C.M., Scherer P.E. Mitochondrial dysfunction in white adipose tissue. 2012. *Trends Endocrinol Metab* 23: 435–443.

Lanzara C., Roetto A., Daraio F., Rivard S., Ficarella R., Simard H., Cox T.M., Cazzola M., Piperno A., Gimenez-Roqueplo A.P. et al. 2004. Spectrum of hemojuvelin gene mutations in 1q-linked juvenile hemochromatosis. *Blood* 103: 4317–4321.

Lawson D.M., Treffry A., Artymiuk P.J., Harrison P.M., Yewdall S.J., Luzzago A., Cesareni G., Levi S., Arosio P. 1989. Identification of the ferroxidase centre in ferritin. *FEBS Lett* 254: 207–210.

Lee S.K., Lee C.Y., Kook Y.A., Lee S.K., Kim E.C. 2010. Mechanical stress promotes odontoblastic differentiation via the heme oxygenase-1 pathway in human dental pulp cell line. *Life Sci* 86: 107–114.

Lill R. 2009. Function and biogenesis of iron-sulphur proteins. *Nature* 460: 831–838.

Liu X.B., Hill P., Haile D.J. 2002. Role of the ferroportin iron-responsive element in iron and nitric oxide dependent gene regulation. *Blood Cells Mol Dis* 29: 315–326.

Mantovani A., Sica A., Sozzani S., Allavena P., Vecchi A., Locati M. 2004. The chemokine system in diverse forms of macrophage activation and polarization. *Trends Immunol* 25: 677–686.

McCarthy R.C., Sosa J.C., Gardeck A.M., Baez A.S., Lee C.H., Wessling-Resnick M. 2018. Inflammation-induced iron transport and metabolism by brain microglia. *J Biol Chem* 293: 7853–7863.

McDowell L.R. 2003. *Minerals in Animal And Human Nutrition.* 2nd ed. Elsevier Science Amsterdam.

McGown C., Birerdinc A., Younossi Z.M. 2014. Adipose tissue as an endocrine organ. *Clinics In Liver Dis* 18: 41–58.

McKie A.T., Marciani P., Rolfs A., Brennan K., Wehr K., Barrow D., Miret S., Bomford A., Peters T.J., Farzaneh F., Hediger M.A., Hentze M.W., Simpson R.J. 2000. A Novel Duodenal Iron-Regulated Transporter, IREG1, Implicated in the Basolateral Transfer of Iron to the Circulation. *Mol Cell* 5: 299–309.

McKie A.T., Barrow D., Latunde-Dada G.O., Rolfs A., Sager G., Mudaly E., Mudaly M., Richardson C., Barlow D., Bomford A., Peters T.J., Raja K.B., Shirali S., Hediger M.A., Farzaneh F., Simpson R.J. 2001. An iron-regulated ferric reductase associated with the absorption of dietary iron. *Science* 291: 1755–1759.

Medina-Gómez G. 2012. Mitochondria and endocrine function of adipose tissue. *Best Pract Res Clin Endocrinol Metab* 26: 791–804.

Mills C.D. 2012. M1 and M2 macrophages: oracles of health and disease. *Crit Rev Immunol* 32: 463–488.

Mills C.D., Kincaid K., Alt J.M., Heilman M.J., Hill A.M. 2000. M-1/M-2 macrophages and the Th1/Th2 paradigm. *J Immunol* 164: 6166–6173.

Mims M.P., Guan Y., Pospisilova D., Priwitzerova M., Indrak K., Ponka P., Divoky V., Prchal J.T. 2005. Identification of a human mutation of DMT1 in a patient with microcytic anemia and iron overload. *Blood* 105: 1337–1342.

Mochel F., Knight M.A., Tong W.H., Hernandez D., Ayyad K., Taivassalo T., Andersen P.M., Singleton A., Rouault T.A., Fischbeck K.H., Haller R.G. 2008. Splice mutation in the iron-sulfur cluster scaffold protein ISCU causes myopathy with exercise intolerance. *Am J Hum Genet* 82: 652–660.

Moldes M., Lasnier F., Fève B., Pairault J., Djian P. 1997. Id3 prevents differentiation of preadipose cells. *Mol Cell Biol* 17: 1796–1804.

Moreno M., Ortega F., Xifra G., Ricart W., Fernández-Real J.M., Moreno-Navarrete J.M. 2015. Cytosolic aconitase activity sustains adipogenic capacity of adipose tissue connecting iron metabolism and adipogenesis. *FASEB J* 29: 1529–1539.

Moreno-Navarrete J.M., Novelle M.G., Catalán V., Ortega F., Moreno M., Gomez-Ambrosi J., Xifra G., Serrano

M., Guerra E., Ricart W., Frühbeck G., Diéguez C., Fernández-Real J.M. 2014. Insulin resistance modulates iron-related proteins in adipose tissue. *Diabetes Care* 37: 1092–1100.

Moreno-Navarrete J.M., Ortega F., Moreno M., Ricart W., Fernández-Real JM. 2014a. Fine-tuned iron availability is essential to achieve optimal adipocyte differentiation and mitochondrial biogenesis. *Diabetologia* 57: 1957–1967.

Moreno-Navarrete J.M., Ortega F., Moreno M., Serrano M., Ricart W., Fernández-Real J.M. 2014b. Lactoferrin gene knockdown leads to similar effects to iron chelation in human adipocytes. *J Cell Mol Med* 18: 391–395.

Moreno-Navarrete J.M., Ortega F., Sabater M., Ricart W., Fernández-Real J.M. 2011. Proadipogenic effects of lactoferrin in human subcutaneous and visceral preadipocytes. *J Nutr Biochem* 22: 1143–1149.

Moreno-Navarrete J.M., Serrano M., Sabater M., Ortega F., Serino M., Pueyo N., Luche E., Waget A., Rodriguez-Hermosa J.I., Ricart W., Burcelin R., Fernández-Real J.M. 2013. Study of lactoferrin gene expression in human and mouse adipose tissue, human preadipocytes and mouse 3T3-L1 fibroblasts. Association with adipogenic and inflammatory markers. *J Nutr Biochem* 24: 1266–1275.

Muckenthaler M.U., Rivella S., Hentze M.W., Galy B. 2017. A Red Carpet for Iron Metabolism. *Cell* 168(3): 344–361.

Muir A., Hopfer U. 1985. Regional specificity of iron uptake by small intestinal brush-border membranes from normal and iron deficient mice. *Am J Physiol* 248:G376-G379.

Nemeth E., Ganz T. 2006. Regulation of iron metabolism by hepcidin. *Annu Rev Nutr* 266: 323–342.

Nemeth E., Rivera S., Gabayan V., Keller C., Taudorf S., Pedersen B.K., Ganz T. 2004a. IL-6 mediates hypoferremia of inflammation by inducing the synthesis of the iron regulatory hormone hepcidin. *J Clin Invest* 113: 1271–1276.

Nemeth E., Tuttle M.S., Powelson J., Vaughn M.B., Donovan A., Ward D.M., Ganz T., Kaplan J. 2004b. Hepcidin regulates cellular iron efflux by binding to ferroportin and inducing its internalization. *Science* 306: 2090–2093.

Nicolas G., Chauvet C., Viatte L., Danan J.L., Bigard X., Devaux I., Beaumont C., Kahn A., Vaulont S. 2002. The gene encoding the iron regulatory peptide hepcidin is regulated by anemia, hypoxia, and inflammation. *J Clin Invest* 110. 1037 1044.

Nnah I.C., Wessling-Resnick M. 2018. Brain iron homeostasis: a focus on microglial iron. *Pharmaceuticals (Basel)*.11: E129.

Nohe A., Keating E., Knaus P., Petersen N.O. 2004. Signal transduction of bone morphogenetic protein receptors. *Cell Signal* 16: 291–299.

Novikoff A.B., Novikoff P.M., Rosen O.M., Rubin C.S. 1980. Organelle relationships in cultured 3T3-L1 preadipocytes. *J Cell Biol* 87: 180–196.

Ohgami R.S., Campagna D.R., Greer E.L., Antiochos B., McDonald A., Chen J., Sharp J.J., Fujiwara Y., Barker J.E., Fleming M.D. 2005. Identification of a ferrireductase required for efficient transferrin dependent iron uptake in erythroid cells. *Nat Genet* 37: 1264–1269.

Orr J.S., Kennedy A., Anderson-Baucum E.K., Webb C.D., Fordahl S.C., Erikson K.M., Zhang Y., Etzerodt A., Moestrup S.K., Hasty A.H. 2014. Obesity alters adipose tissue macrophage iron content and tissue iron distribution. *Diabetes* 63: 421–432.

Osaki S. 1966. Kinetic studies of ferrous ion oxidation with crystalline human ferroxidase (ceruloplasmin). *J Biol Chem* 241: 5053–5059.

Otto F., Thornell A.P., Crompton T., Denzel A., Gilmour K.C., Rosewell I.R., Stamp G.W., Beddington R.S., Mundlos S., Olsen B.R., Selby P.B., Owen M.J. 1997. Cbfa1, a candidate gene for cleidocranial dysplasia syndrome, is essential for osteoblast differentiation and bone development. *Cell* 89: 765–771.

Owens G.K., Kumar M.S., Wamhoff B.R. 2004. Molecular Regulation of Vascular Smooth Muscle Cell Differentiation in Development and Disease. *Physiol Rev* 84: 767–801.

Ozawa E., Hagiwara Y. 1982. Degeneration of large myotubes following removal of transferrin from culture medium. *Biomed Res* 3: 16–23.

Pak M., Lopez M.A., Gabayan V., Ganz T., Rivera S. 2006. Suppression of hepcidin during anemia requires erythropoietic activity. *Blood* 108: 3730–3735.

Pan B.T., Blostein R., Johnstone R.M. 1983. Loss of the transferrin receptor during maturation of sheep reticulocytes in vitro. *Biochem J* 210: 37–47.

Pantopoulos K. 2004. Iron metabolism and the IRE/IRP regulatory system: an update. *Ann N Y Acad Sci* 1012: 1–13.

Papanikolaou G., Pantopoulos K. 2005. Iron metabolism and toxicity. *Toxicol Appl Pharmacol* 202: 199 211.

Papanikolaou G., Samuels M.E., Ludwig E.H., MacDonald M.L., Franchini P.L., Dube M.P., Andres L., MacFarlane J., Sakellaropoulos N., Politou M. et al.

2004. Mutations in HFE2 cause iron overload in chromosome 1q-linked juvenile hemochromatosis. *Nat Genet* 36: 77–82.

Papanikolaou G., Tzilianos M., Christakis J.I., Bogdanos D., Tsimirika K., MacFarlane J., Goldberg Y.P., Sakellaropoulos N., Ganz T., Nemeth E. 2005. Hepcidin in iron overload disorders. *Blood* 105: 4103–4105.

Park C.H., Valore E.V., Waring A.J., Ganz T. 2001. Hepcidin, a urinary antimicrobial peptide synthesized in the liver. *J Biol Chem* 276: 7806–7810.

Peyssonnaux C., Zinkernagel A.S., Schuepbach R.A., Rankin E., Vaulont S., Haase V.H., Nizet V., Johnson R.S. 2007. Regulation of iron homeostasis by the hypoxia-inducible transcription factors (HIFs). *J Clin Invest* 117: 1926–1932.

Picard V., Govoni G., Jabado N., Gros P. 2000. Nramp 2 (DCT1/DMT1) expressed at the plasma membrane transports iron and other divalent cations into a calcein-accessible cytoplasmic pool. *J Biol Chem* 275: 35738–35745.

Pigeon C., Ilyin G., Courselaud B., Leroyer P., Turlin B., Brissot P., Loreal O. 2001. A new mouse liver-specific gene, encoding a protein homologous to human antimicrobial peptide hepcidin, is overexpressed during iron overload. *J Biol Chem* 276: 7811–7819.

Ponka P. 1997. Tissue-specific regulation of iron metabolism and heme synthesis: distinct control mechanisms in erythroid cells. *Blood* 89: 1–25.

Ponka P., Beaumont C., Richardson D.R. 1998. Function and regulation of transferrin and ferritin. *Semin Hematol* 35: 35–54.

Rajagopal A., Rao A.U., Amigo J., Tian M., Upadhyay S.K., Hall C., Uhm S., Mathew M.K., Fleming M.D., Paw B.H., et al. 2008 Haem homeostasis is regulated by the conserved and concerted functions of HRG-1 proteins. *Nature* 453: 1127–1131.

Recalcati S., Locati M., Marini A., Santambrogio P., Zaninotto F., De Pizzol M., et al. 2010. Differential regulation of iron homeostasis during human macrophage polarized activation. *Eur J Immunol* 40: 824–835.

Rensvold J.W., Ong S.E., Jeevananthan A., Carr S.A., Mootha V.K., Pagliarini D.J. 2013. Complementary RNA and protein profiling identifies iron as a key regulator of mitochondrial biogenesis. *Cell Rep* 3: 237–245.

Rhodes S.J., Konieczny S.F. 1989. Identification of MRF4: a new member of the muscle regulatory factor gene family. *Genes Dev.* 3 (12B): 2050–2061.

Richardson C.L., Delehanty L.L., Bullock G.C., Rival C.M., Tung K.S., Kimpel D.L., Gardenghi S., Rivella S., Goldfarb A.N. 2013. Isocitrate ameliorates anemia by suppressing the erythroid iron restriction response. *J Clin Invest* 123: 3614–3623.

Rosen E.D., MacDougald O.A. 2006. Adipocyte differentiation from the inside out. *Nat Rev Mol Cell Biol* 7: 885–896.

Rosen E.D., Spiegelman B.M. 2014. What We Talk About When We Talk About Fat. *Cell* 156: 20–44.

Rouault T.A., Hentze M.W., Caughman S.W., Harford J.B., Klausner R.D. 1988. Binding of a cytosolic protein to the iron-responsive element of human ferritin messenger RNA. *Science* 241: 1207–1210.

Rouault T.A., Klausner R.D. 1996. Iron-sulfur clusters as biosensors of oxidants and iron. *Trends Biochem Sci* 21: 174–177.

Rouault T.A., Tong W.H. 2005. Iron-sulphur cluster biogenesis and mitochondrial iron homeostasis. *Nat Rev Mol Cell Biol* 6: 345–351

Rouault T.A., Tong W.H. 2008. Iron-sulfur cluster biogenesis and human disease. *Trends Genet* 24: 398–407.

Russell M.J., Roy M.D., Allan J.H. 1993. On the emergence of life via catalytic iron-sulphide membranes. *Terra Nova* 5: 343–347.

Sato M., Gitlin J.D. 1991. Mechanisms of copper incorporation during the biosynthesis of human ceruloplasmin. *J Biol Chem* 266. 5128–5134.

Schmucker S., Martelli A., Colin F., Page A., Wattenhofer-Donzé M., Reutenauer L., Puccio H. 2011. Mammalian frataxin: an essential function for cellular viability through an interaction with a preformed ISCU/NFS1/ISD11 iron-sulfur assembly complex. *PloS One* 6: e16199

Scott C.L., Zheng F., De Baetselier P., Martens L., Saeys Y., De Prijck S., Lippens S., Abels C., Schoonooghe S., Raes G., Devoogdt N., Lambrecht B.N., Beschin A., Guilliams M. 2016. Bone marrow-derived monocytes give rise to self-renewing and fully differentiated Kupffer cells. *Nat Commun* 7: 10321.

Shaw G.C., Cope J.J., Li L., Corson K., Hersey C., Ackermann G.E., Gwynn B., Lambert A.J., Wingert R.A., Traver D., Trede N.S., Barut B.A., Zhou Y., Minet E., Donovan A., Brownlie A., Balzan R., Weiss M.J., Peters L.L., Kaplan J., Zon L.I., Paw B.H. 2006. Mitoferrin is essential for erythroid iron assimilation. *Nature* 440: 96–100.

Shayeghi M., Latunde-Dada G.O., Oakhill J.S., Laftah A.H., Takeuchi K., Halliday N., Khan Y., Warley A., McCann F.E., Hider R.C., Frazer D.M., Anderson

G.J., Vulpe C.D., Simpson R.J., McKie A.T. 2005. Identification of an intestinal heme transporter. *Cell* 122: 789–801.

Sheftel A.D., Mason A.B., Ponka P. 2012. The long history of iron in the Universe and in health and disease. *Biochim Biophys Acta* 1820: 161–187.

Sheftel A.D., Stehling O., Lill R. 2010. Iron-sulfur proteins in health and disease. *Trends Endocrinol Metab* 21: 302–314.

Sheftel A.D., Zhang A.S., Brown C., Shirihai O.S., Ponka P. 2007. Direct interorganellar transfer of iron from endosome to mitochondrion. *Blood* 110: 125–132.

Shoji A., Ozawa E. 1985. Requirement of Fe ion for activation of RNA polymerase. *Proc Japan Acad B.* 61, 494–496.

Sindrilaru A., Peters T., Wieschalka S., Baican C., Baican A., Peter H., Hainzl A., Schatz S., Qi Y., Schlecht A., et al. 2011. An unrestrained proinflammatory M1 macrophage population induced by iron impairs wound healing in humans and mice. *J Clin Invest* 121, 985–997.

Soares M.P., Hamza I. 2016. Macrophages and iron metabolism. *Immunity* 44: 492–504.

Sorokin L.M., Morgan E.H., Yeoh C.C.T. 1987. Transferrin receptor numbers and transferrin and iron uptake in cultured chick muscle cells at different stages of development. *J Cell Physiol* 131: 342–353.

Theil E.C., Eisenstein R.S. 2000. Combinatorial mRNA regulation: iron regulatory proteins and iso-iron-responsive elements (Iso-IREs). *J Biol Chem* 275: 40659–40662.

Theurl M., Theurl I., Hochegger K., Obrist P., Subramaniam N., van Rooijen N., Schuemann K., Weiss G. 2008. Kupffer cells modulate iron homeostasis in mice via regulation of hepcidin expression. *J Mol Med* 86: 825–835.

Tormos K.V., Anso E., Hamanaka R.B., Eisenbart J., Joseph J., Kalyanaraman B., et al. 2011. Mitochondrial complex III ROS regulate adipocyte differentiation. *Cell Metab* 14: 537–544.

Uhrigshardt H., Singh A., Kovtunovych G., Ghosh M., Rouault T.A. 2010. Characterization of the human HSC20, an unusual DnaJ type III protein, involved in iron sulfur cluster biogenesis. *Hum Mol Genet* 19: 3816–3834.

Vaisman B., Fibach E., Konijn A.M. 1997. Utilization of intracellular ferritin iron for hemoglobin synthesis in developing human erythroid precursors. *Blood* 90: 831–838.

Vanella L., Kim D.H., Asprinio D., Peterson S.J., Barbagallo I., Vanella A., Goldstein D., Ikehara S.,

Kappas A., Abraham N.G. 2010. HMOX1 expression increases mesenchymal stem cell derived osteoblasts but decreases adipocyte lineage. *Bone* 46: 236–243.

Vinchi F., Costa da Silva M., Ingoglia G., Petrillo S., Brinkman N., Zuercher A., Cerwenka A., Tolosano E., Muckenthaler M.U. 2016. Hemopexin therapy reverts heme-induced proinflammatory phenotypic switching of macrophages in a mouse model of sickle cell disease. *Blood* 127: 473–486.

Vulpe C.D., Kuo Y.M., Murphy T.L., Cowley L., Askwith C., Libina N., Gitschier J., Anderson G.J. 1999. Hephaestin, a ceruloplasmin homologue implicated in intestinal iron transport, is defective in the sla mouse. *Nat Genet* 21: 195–199.

Wang J., Pantopoulos K. 2011. Regulation of cellular iron metabolism. *Biochem J* 434: 365–381.

Wang L., Fang B., Fujiwara T., Krager K., Gorantla A., Li C., Feng J.Q., Jennings M.L., Zhou J., Aykin-Burns N., Zhao H. 2018. Deletion of ferroportin in murine myeloid cells increases iron accumulation and stimulates osteoclastogenesis in vitro and in vivo. *J Biol Chem* 293: 9248–9264.

White R.B., Bièrinx A.S., Gnocchi V.F., Zammit P.S. 2010. Dynamics of muscle fibre growth during postnatal mouse development. *BMC Dev Biol* 10: 21.

Winn N.C., Katrina M., Volk K.M., Hasty A.H. 2020. Regulation of tissue iron homeostasis: the macrophage "ferrostat." *JCI Insight* 5: e132964.

Wlazlo N., van Greevenbroek M.M., Ferreira I Jansen E.H., Feskens E.J., van der Kallen C.J., Schalkwijk C.G., Bravenboer B., Stehouwer C.D. 2013. Iron metabolism is associated with adipocyte insulin resistance and plasma adiponectin: the Cohort on Diabetes and Atherosclerosis Maastricht (CODAM) study. *Diabetes Care* 36: 309–315.

Wright W.E., Sassoon D.A., Lin V.K. 1989. Myogenin, a factor regulating myogenesis, has a domain homologous to MyoD. *Cell* 56: 607–617.

Wrighting D.M., Andrews N.C. 2006. Interleukin-6 induces hepcidin expression through STAT3. *Blood* 108: 3204–3209.

Yagi M., Suzuki N., Takayama T., Arisue M., Kodama T., Yoda Y., Otsuka K., Ito K. 2009. Effects of lactoferrin on the differentiation of pluripotent mesenchymal cells. *Cell Biol Int* 33: 283–289.

Ye M., Graf T. 2007. Early decisions in lymphoid development. *Curr Opin Immunol* 19: 123–128.

Yeh K.Y., Yeh M., Mims L., Glass J. 2009. Iron feeding induces ferroportin 1 and hephaestin migration and interaction in rat duodenal epithelium. *Am J Physiol Gastrointest Liver Physiol* 296: 55–65.

Zaitsev V.N., Zaitseva I., Papiz M., Lindley P.F. 1999. An X-ray crystallographic study of the binding sites of the azide inhibitor and organic substrates to ceruloplasmin, a multi-copper oxidase in the plasma. *J Biol Inorg Chem* 4: 579–587.

Zammit P.S. 2017. Function of the myogenic regulatory factors Myf5, MyoD, Myogenin and MRF4 in skeletal muscle, satellite cells and regenerative myogenesis. *Semin Cell Dev Biol* 72: 19–32.

Zarjou A., Jeney V., Arosio P., Poli M., Antal-Szalmás, P., Agarwal A., Balla G., Balla J. 2009. Ferritin prevents calcification and osteoblastic differentiation of vascular smooth muscle cells. *J Am Soc Nephrol* 20: 1254–1263.

CHAPTER THIRTEEN

Redox Regulation of Cytoskeletal Dynamics

Clara Ortegón Salas, Manuela Gellert, and Christopher Horst Lillig

CONTENTS

13.1 REDOX SIGNALLING

Cells constantly exchange and process information with and from their environment. Appropriate responses to such signals and information are essential for the functions of cells, organs, and the whole organism. Errors, overreactions, lack of responses, or misinterpretations contribute to numerous pathological conditions (e.g., malignant transformation, inflammation, or degenerative disorders). In general, signal transduction is initiated by signalling molecules that are sensed by intra- or extracellular receptors that in turn become activated. The information is usually passed on to intracellular transducer proteins, mostly enzymes that produce or release second messenger molecules or introduce reversible post-translational modifications in other proteins. Eventually, signal transduction will reach effector molecules that trigger an appropriate biological response and often also terminate the original signal. Signal transduction pathways can branch and communicate with other information transmitting pathways, thus allowing for the amplification, modulation, interconnection, and adaptation of signals and responses. Redox signalling refers to information transduction through the (mostly reversible) redox modification of proteins. Previously, redox modifications were considered to be the mere consequence of undefined "stress" conditions. In recent years, however, reversible redox modifications of protein side chains has emerged as a rapid, reversible, and highly specific key mechanism in cell signalling (for more in-depth discussions see, e.g., [1–5]). Today, we learn more and more that redox switches may not be toggled randomly but through the action of specific enzymes that act as receptor, transducer, or effector molecules. The major targets of redox signalling are the side chains of the sulphur-containing amino acids cysteine and methionine (Figure 13.1).

13.2 REDOX MODIFICATIONS OF CYSTEINYL AND METHIONYL SIDE CHAINS

Two thiol groups of cysteinyl side chains can be oxidized to an intra- or intermolecular disulfide (Figure 13.1). Disulfides may also form between

DOI: 10.4324/9781003204091-17

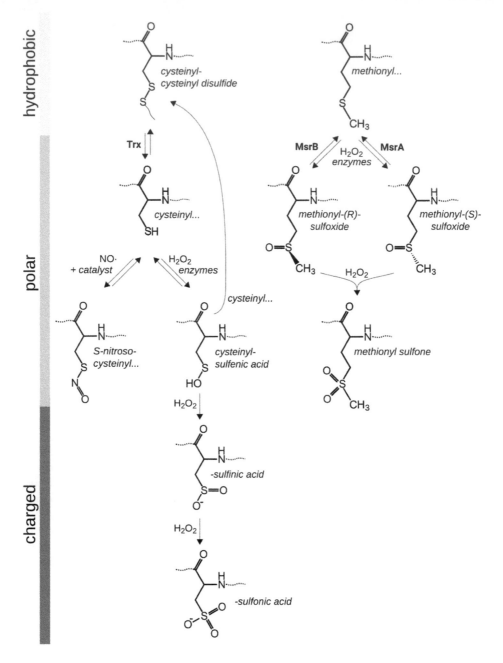

Figure 13.1 Redox modifications of protein cysteinyl and methionyl residues.

In the presence of another thiol (SH), the cysteinyl residue (left side) can be modified to a protein (cysteinyl-cysteinyl) disulfide, which can be reduced by Trxs and Grxs or cysteinyl glutathione–mixed disulfides. Protein cysteinyl residues can be oxidized to sulfenic acid (R-SOH) by peroxides or (at least in some cases) specific enzymes. In the presence of excessive peroxides, this may be irreversibly "over"-oxidized to sulfinic (R-SO$_2$H) and sulfonic acid (R-SO$_3$H). Cysteinyl-glutathione disulfides may also be formed through thiol-disulfide exchange reactions with glutathione disulfide or by specific enzymes (e.g., Grxs) that also specifically catalyze the reduction of these disulfides. Nitric oxide (NO·) in general can only lead to the nitrosylation of cysteinyl residues through the catalysis during which one electron is transferred from the NO· to a recipient (e.g., a metal cofactor). S-nitrosylation can be reversed by transnitrosylation to another protein thiol (e.g., to the active site of Trxs). Methionyl residues (right side) are oxidized stereoselective to R- or S-methionyl sulfoxides. These are specific substrates for methionine sulfoxide reductases (Msr) B and A, respectively. Further oxidation of methionyl sulfoxides results in methionyl sulfone, a step that has to be considered irreversible.

protein thiols and small molecular weight thiol-containing compounds, the most prominent being the ubiquitous tripeptide glutathione. This process is also known as thiolation or more specifically protein S-glutathionylation (see, e.g., [6,7]). Disulfides may form through sulfenic acid intermediates (see discussion below); however, we see a growing number of examples where disulfides are introduced specifically through thiol-disulfide exchange reactions in relay reactions (i.e., redox signal transduction cascades). The key enzymes/proteins for these exchange reactions belong to the thioredoxin (Trx) family of proteins, namely Trxs, glutaredoxins (Grxs), peroxiredoxins (Prxs), and glutathione peroxidases (Gpxs) [8,9]. Isoforms of these proteins are ubiquitously expressed and targeted to all subcellular compartments, including the extracellular space. The thiol-disulfide exchange reaction is initiated by the nucleophilic attack of a thiolate on a given disulfide. Trxs and Grxs contain a conserved Cys-x-x-Cys motif, where the more N-terminal cysteinyl residue displays an unusual low pK_a value and is responsible for the initial attack. The trigonal bipyramidal transition state of the reaction requires the attack to occur in line with the disulfide (i.e., with an angle of 180° relative to the disulfide bond) (for a more detailed discussion on the geometrical constrains, see [10]). This reaction results in a mixed disulfide intermediate between the attacking protein and its target. Subsequently, this disulfide is reduced by the second thiol of the attacking protein, yielding a disulfide bond in the attacking protein and the reduced target (i.e., the exchange of the disulfide). A disulfide in the active site of Trxs can be reduced by NADPH-dependent thioredoxin reductase [11]. Disulfides in the active sites of Grxs are reduced by two molecules of glutathione [12], and the resulting glutathione disulfide can be reduced by NADPH-dependent glutathione reductase. In addition, Grxs can specifically catalyse the reversible glutathionylation of protein thiols in a similar mechanism, where the more N-terminal thiol attacks a mixed disulfide with glutathione, resulting in the transfer of the glutathionyl group to the Grx. The structural requirements for this reaction have recently been unrevealed in molecular detail (see [13,14]). We want to emphasize that all these reactions are reversible. Dependent on the external constrains, disulfides may be either introduced or reduced by Trxs and Grxs. Through the reaction of protein cysteinyl side chains with hydrogen peroxide or peroxynitrite, their thiol groups may also be oxidized to

sulfenic acids (Figure 13.1) [15]. Sulfenic cysteinyl residues may react with a second reduced cysteinyl residue or a low molecular weight thiol, resulting in a disulfide (Figure 13.1). Through the reaction with further peroxides, they may also be oxidized to sulfinic and sulfonic acids (Figure 13.1), reactions that are considered to be irreversible under physiological conditions, although they do occur at specific target sites as well [16]. The significance of sulfenylation for a broad range of proteins is still under debate. While the modifications can be trapped in biological samples, the low rate constants of any given thiol with hydrogen peroxide may not allow the reaction to take place in competition with dedicated peroxidases whose rate constants are 10^5 to 10^7 higher (for detailed discussions and a mathematical analysis, see [10,17,18]). The function of peroxidases such as Prxs and Gpxs as receptors and transducers of signal transmitted via the second messenger H_2O_2 is an emerging concept ([8,17]; for an example, see [19]). S-nitrosylation is often reported to be the product of a reaction of a reduced protein thiol with a nitric oxide molecule produced by nitric oxide synthases. The reaction of nitric oxide ($\cdot N=O$ or $\cdot NO$) with a protein thiol, however, is a redox reaction: cysteinyl-SH + $\cdot N=O$ $\rightleftharpoons$ cysteinyl-S-N=O + 1 e^- + H^+. The surplus electron requires a second reactant to be reduced. It is commonly assumed that this reaction must thus be catalyzed by metal (iron)-containing enzymes. Once formed, S-nitroso groups may also be transferred to other thiols in transnitrosylation reactions, which may also be a critical function of Trx family proteins [20,21].

The second sulphur-containing amino acid methionine can be reversibly oxidized to methionyl/methionine sulfoxide (Figure. 13.1). Unlike the cysteinyl thiol group, the methionyl thioether group can be oxidized to a mixture of two diastereomers: methionine-S-sulfoxide and methionine-R-sulfoxide (Figure 13.1). Similar to the oxidation of cysteinyl residues to sulfenic acids, methionine sulfoxidation was suggested to be the product of the reaction of free hydrogen peroxide with methionyl residues or free methionine. However, as discussed above, the rate constant for this reaction is in the order of 10^{-2} $M^{-1} \cdot s^{-1}$ [22], and thus 10^7 lower compared to the dedicated peroxidases. The discovery of enzymes that specifically oxidize methionyl residues with high specificity and at sufficient rates does not only solve this dilemma, but it also emphasizes the importance of this reaction for the redox regulation of cellular

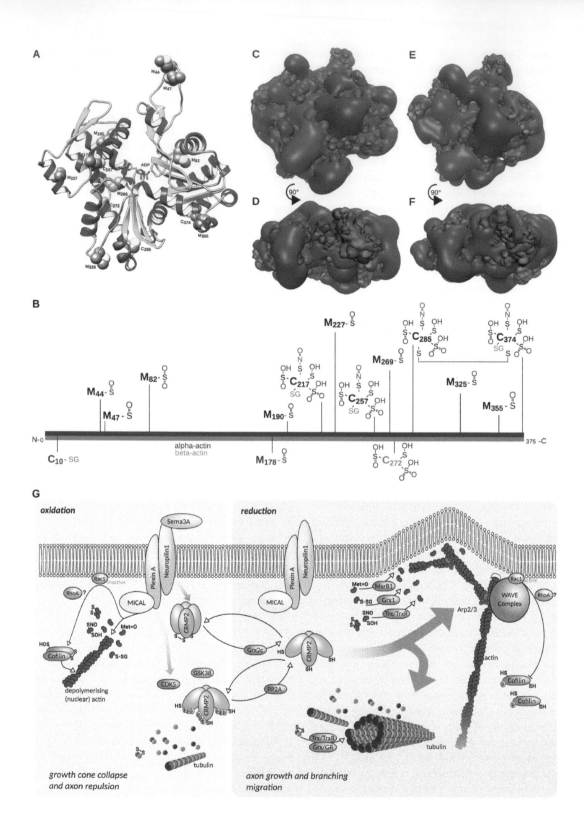

REDOX REGULATION OF DIFFERENTIATION AND DE-DIFFERENTIATION

functions [23]; more details are discussed below. Methionine-S-sulfoxide is specifically reduced by methionine sulfoxide reductase (Msr)A, the methionine-R-sulfoxide by MsrB. Due to alternative splicing and transcription initiation, human MsrA is expressed in multiple isoforms localised in mitochondria, cytosol, and the nucleus [24–26]. The reaction mechanism of MsrA enzymes requires three cysteinyl residues in its catalytic centre [27]. The first catalytic cysteinyl residue of MsrA reacts with methionine-S-sulfoxide, yielding a sulfenic acid intermediate, hence this residue is also named the sulfenic cysteinyl residue. This sulfenic cysteinyl residue is reduced by the second cysteinyl residue yielding a disulfide between them. This disulfide is attacked by the third cysteinyl residue, evincing a disulfide between the second and third cysteinyl residues. This disulfide is reduced by Trxs or Grxs as outlined above, regenerating the enzyme [27,28]. The reaction mechanism of MsrBs requires two cysteinyl (or selenol) residues [27]. Most mammalian genomes encode three MsrB variants. MsrB1 is characterized by a selenocysteinyl residue in its active site instead of the sulfenic catalytic cysteinyl residue and exhibits the highest activity. MsrB2 and 3 contain only cysteinyl residues and are less efficient [29,30]. The reaction of MsrB with methionine-R-sulfoxide yields a sulfenic or selenic acid intermediate in the first active site residue that is reduced by the second cysteinyl yielding a disulfide or selenosulfide bond. Unlike for the MsrA type enzymes, this disulfide is a direct substrate for Trxs [29,27]. As an alternative mechanism, the sulfenic/selenic intermediate may also be directly reduced by reaction with glutathione, yielding a glutathione-mixed disulfide that can be subsequently reduced by Grxs [28].

How do the abovementioned redox modifications effect the function of the suspected transducer and effector proteins? From a structural point of view, the formation of a disulfide may significantly alter the conformation of any given protein, and there has been much speculation as to how frequent larger structural changes may be. As of today, we are not aware of any comprehensive study that has addressed this question. If a thiol group was involved in the reaction mechanisms of a protein or enzyme, any redox modification would inhibit its function. However, even without significant structural rearrangements or mechanistic functions, all of the abovementioned modifications significantly alter the protein's physiochemical properties (Figure 13.1). The formation of a disulfide, for instance, turns the polar thiol group into a rather hydrophobic disulfide. For methionyl residues this is exactly the opposite: oxidation turns the hydrophobic side chain into a polar group (see also Figure 13.2C–F). The

---

Figure 13.2 Redox modifications of actin and redox switches the regulation of cytoskeletal dynamics.

(A) Structure of human beta-actin (PDB code 5onv) in cartoon representation with all redox susceptible cysteinyl and methionyl residues in space-filled representation. ATP is highlighted in its central binding pocket. (B) Schematic representation of the cysteinyl and methionyl residues of alpha- (blue) or beta-actin (green) that are susceptible to S-glutathionylation (orange/-SG), S-nitrosylation (red/-NO), and oxidation (dark red/disulfide: S-S, sulfenic acid: -SOH, sulfinic acid: -SO$_2$H, sulfonic acid: -SO$_3$H, sulfoxide: -SO, sulfone: -SO$_2$). All modifications described for both alpha and beta actin are depicted above, the ones specific for either of them below in their respective colours (modified from reference [175]). (C–F) Electrostatic isosurfaces of unmodified ADP:actin from rabbit (pdb code 5onv) (C–D) and methionine sulfoxide (M44, M47)-modified ADP:actin from rabbit (pdb code 6avb) at ±25 mV (E–F); red negative potential, blue positive potential). Electrostatic features were computed as outlined in [176]. The orientation of the molecule in (D) and (F) has been tilted by 90° towards the viewer. (G) Sema3A binding to the NP1/PlexA receptor activates the signalling cascade. MICAL activation via PlexA leads to actin-depolymerization through sulfoxidation of Met44 and Met47 of actin and/or cofilin activation. Without ligand binding, MICALs are not activated by the NP1/PlexA receptor. CRMP2, also a mediator of the Sema3A signalling cascade, induces growth cone collapse and axon repulsion in its oxidized or phosphorylated state (see left side: "oxidation"). Other redox modifications that lead to actin depolymerization are S-nitrosylation, S-glutathionylation, sulfenic acid, and disulfide bond formation, which is also observed in tubulin dimers. MICALs are not activated without ligand binding to the NP1/PlexA receptor. The reduction of CRMP2 by cytosolic Grx2 (Grx2c) leads to a conformational change of the homo-tetrameric structure. Polymerization of actin and tubulin is caused (directly or indirectly) by this conformational change. Also in this reduced state, the phosphorylation sites of CRMP2 are accessible for phosphorylation by CDK5 and GSK3β (de-phosphorylation by PP2A), that may allow a more fine-tuned regulation of growth cone collapse and axon repulsion. Close to the membrane, the WAVE complex is activated by active Rac1 and binds Arp2/3, leading to actin polymerization and branching. Activated Rac1 also leads to the inactivation of cofilin by phosphorylation. The redox modifications of actin can be reduced by MsrB1 (sulfoxidation of methionyl residues), Grx1 (S-glutathionylation), or the Trx/TrxR system (S-nitrosylation, sulfenic acids, disulfides). Disulfides occurring in tubulin dimers can be reduced by either the Grx/GR or the Trx/TrxR system. The absence of Sema3A leads to axon outgrowth and branching as well as enabling active migration via changes in the cytoskeletal dynamics (right side: "reduction").

formation of sulfenic acids increases the polarity of cysteinyl residues further; higher oxidation states may even add charge as the pK_a of sulfinic and sulfonic is <2. Protein glutathionylation results in geometric alterations and also the addition of two negatively charged carboxyl and one positively charged amino groups. These changes in polarity or charge are in line with the effects caused by other post-translational modifications. Acetylation, for instance, removes a positive charge, and phosphorylation adds a negative charge.

13.3 REDOX MODIFICATIONS OF OTHER AMINO ACID SIDE CHAINS

In addition to the modifications discussed above, redox modifications of non-sulphur-containing amino acid side chains have also been reported, all of which are irreversible under physiological conditions. They can thus not be part of reversible signalling cascades; however, they may still serve as signals recognised by other receptor molecules. Most of these reactions can be traced back to the reactivity of hydroxyl radicals as product of the Fenton reaction. Due to its high reactivity, the hydroxyl radical likely abstracts a hydrogen atom of the first biomolecule in its diffusion way. Aromatic amino acid side chains are particularly susceptible to modifications, including hydroxy derivatives, dityrosine, and others [31,32]. The most common non-specific irreversible oxidative modifications caused by hydroxy radicals are protein carbonylation and fragmentation. They have been widely used as a marker of "oxidative stress," aging, and numerous pathologies [31–35]. As an example for the potential role of these modifications as signals themselves, we refer to a study that demonstrated that the endothelin-1-mediated carbonylation and subsequent degradation of annexin A1 promotes the growth of smooth muscle cells [36]. Tyrosine residues are also susceptible to reacting with peroxynitrite or nitrogen dioxide, yielding 3-nitrotyrosin [37–39], a reaction that was also discussed to require catalysis [40].

The hydroxylation of prolyl side chains by prolyl hydroxylases is used as a signalling mechanism sensing the amount and presence of molecular oxygen, as these monooxygenases require molecular oxygen and α-ketoglutarate as substrates. Prolyl hydroxylation of hypoxia-inducible factors (HIFs) targets them for degradation through the proteasome. During hypoxia, when oxygen is lacking, the proteins cannot be hydroxylated and degraded, and the proteins act as transcription factors promoting the transcription of hypoxic response genes [41–43].

13.4 REDOX SIGNALLING IN ACTIN DYNAMICS AND CELL MOVEMENTS

The most dynamic cytoskeletal structures are likely the actin filaments. Actin polymerization and depolymerization is controlled by numerous actin-binding proteins that stabilize filaments, promote elongation, severing, or nucleation of de novo filaments [44]. Actin filaments provide mechanical support for cells and participate in cell movements, all of which are controlled by numerous signalling pathways, including phosphorylation cascades, calcium signalling, and phospholipid signalling. These pathways involve numerous proteins that allow the spatiotemporal control of actin dynamics in specific regions of the cytosol, for instance at the plasma membrane at the leading edge of the moving cell. One of the mechanisms that initiates new actin filaments involves the actin-related protein (ARP) 2/3 complex. This complex is active at the leading edge of motile cells, where it initiates branches on the sides of existing filaments. The growing filaments push the membrane forward to protrude (see [45–47] for more detailed discussions). The subunits ARP2 and ARP3 are too far apart to form an active complex. The activation of the ARP2/3 complex is regulated by Wiskott-Aldrich syndrome proteins (WASP) and WASP family verprolin-homologous protein (WAVE) family proteins that bind to the subunits inducing conformational changes that locks the complex in its active state [48,49]. The WASP/WAVE complexes are recruited to the membrane where they are activated by the Ras (rat sarcoma) family GTPases Rho (Ras homology gene family member), Rac (Ras-related C3 botulinum toxin substrate), and CDC42 (cell division control protein 42) in response to various signalling pathways [50]. These small GTPases are probably the key signal integrators and mediators controlling the dynamic changes that enable cell motility.

13.5 REDOX MODIFICATIONS AND REGULATION OF ACTIN

Actin was most likely isolated for the first time by W. D. Halliburton (1860–1931) in 1887 while studying the properties of myosin in warm-blooded animals [51]. However, the discovery and initial characterization of actin took place in 1942 in the laboratory of Albert Szent-Györgyi by Brúnó F. Straub (1914–1996), who was able to isolate actin from acetone-dried muscle. It is remarkable that Straub's protocol is still the basis of the method to purify actin from muscle sources. Actin is a well-characterized, abundant, and essential protein in eukaryotic cells. Its dynamic properties allow it to shift between monomeric, globular (G)-actin and polymeric, filamentous (F)-actin. The latter forms microfilaments that are crucial for the cytoskeleton in all eukaryotic cells and thereby for many cellular functions such as apoptosis, cell division, cell-cell interaction, cell polarity, and migration.

Actin dynamics are regulated by reversible post-translational modifications (PTMs) on various levels [52], including acetylation, phosphorylation, methylation, and thiol redox modifications [53]. Except for one, all cysteinyl residues and almost half of the methionyl residues of actin are susceptible to redox modifications (Figure 13.2A–F). These sulphur-containing amino acids are distributed all over the primary structure of the protein (Figure 13.2B), Cys374 may be the most vulnerable. Cysteinyl residues at positions 10 (only in α-actin), 217, 257, and 374 are targets for S-glutathionylation [54–56]. Cysteines 217, 257, 272 (not in α-actin), 285, and 374 can be oxidized to sulfenic, sulfinic, and sulfonic acids [57–59]. Cysteines 217, 257, 285, and 374 can be S-nitrosylated [60–62]. Methionyl residues at positions 44, 47, 176, 190, 227, 269, and 355 were found to be oxidized to sulfoxide derivatives, whereas methionine 82 can be further oxidized to the sulfone [57,63]. Additionally, histidyl residues 40, 87, and 173, and all four tryptophan residues present in actin (79, 86, 340, 356) can be irreversibly oxidized [57]. Numerous variables determine which residues are more susceptible and how actin responds to oxidation. For example, at high Ca^{2+} concentrations, the highly reactive Cys374 is protected from H_2O_2 oxidation [58], or in case of the not very reactive Cys10, it becomes more susceptible to oxidation when myosin II is bound to actin [64]. All these variables and the different outcome of the oxidation suggest that the effect of this oxidation is specific to the cell type, the redox environment, and crosstalk to other signalling pathways and mechanisms. Likewise, the effects of redox PTMs cannot be grouped together with respect to their effect on actin properties and

dynamics. Most of them have been associated with conformational changes that affect the polymerization/de-polymerization dynamics as well as the interaction with binding partners. Although in many studies redox modifications of actin are usually connected to a variety of diseases (see below), redox-regulated actin dynamics are also important in physiological conditions (e.g., during neurite outgrowth or cell migration). Hence, redox modifications of actin must be regulated by specific enzymes. Oxidative cysteine modifications are enzymatically reversed by the glutaredoxin (Grx), thioredoxin (Trx), and peroxiredoxin (Prx) systems, among others [65–67]. S-glutathionylation of Cys374 in actin is known to decrease the polymerization dynamics, shifting the ratio of F-actin to G-actin. Grx1 catalyses specifically and efficiently the deglutathionylation at the expense of NADPH, being able to restore the F-actin levels in vivo [68,69]. Additionally, loss of Grx1 function leads to increased levels of glutathionylated actin and decreased F-actin dynamics [70]. Trx1 has been identified as an interaction partner of actin, restoring the actin polymerization by reducing disulfide bonded cysteinyl residues [58,71]. Moreover, by Trx1, actin can be kept reduced and protected from further oxidation at high concentrations of H_2O_2. Interestingly, the interaction with actin was reported to be mediated by Cys62 of Trx1, although this residue is not part of the active site involved in catalysis. The same cysteinyl residue has been shown to play an important role in protecting cells from apoptosis, which suggests that Trx1, by binding to actin and regulating its dynamics, could protect cells from apoptosis [72]. The Trx system also mediates actin de-nitrosylation [73,74]. Oxidized methionyl residues in actin are reduced by Msrs (see above). The requirement for the oxidation reaction to be catalyzed as well, discussed above, was finally met with the identification of specific oxidases. Hung et al. [75] identified a flavin monooxygenase named MICAL (see below) able to bind F-actin and selectively oxidize Met44 and Met47 residues into the Met-R-SOs, resulting in actin filaments destabilization and disassembly in vitro and in vivo. Conformational changes of F-actin increase the susceptibility for severing by cofilin. As outlined above, the oxidative modification of methionyl residues turns a hydrophobic side chain into a polar. In Figure 13.2C–F we have depicted the effect of Met44 and Met47 sulfoxidation on the electrostatic potential (i.e., the polarity) of the protein. These modifications clearly increase the negative surface potential of the protein (compare Figure 13.2 C–D, wild type actin, to Figure 13.2 E–F, M44 and M47, oxidized actin). Undoubtedly, this will have severe consequences on the interaction of the G-actin with other proteins, including its own kind in F-actin filaments. After all, complementary electrostatics are a major determinant of protein-protein interactions.

13.6 REDOX REGULATION OF COFILIN ACTIVITY

The cofilin or actin de-polymerizing factor (ADF, hereafter referred to as cofilin) protein family consists of small (13–20 kDa) actin binding proteins ubiquitously expressed among all eukaryotes. Cofilin binds to both G-actin and F-actin, with a 40-fold preference for the ADP-bound subunits, thus enhancing the rate of monomer dissociation from the pointed end of actin filaments; in addition, cofilin can also sever actin filaments causing the formation of free barbed ends, as cofilin is an essential protein [76] and its activity is tightly regulated. Phosphorylated cofilin is inactive and represents the main fraction of the cofilin pool in the cytoplasm. The Ser/Thr LIM kinases (LIMK1 and LIMK2) and related testicular protein kinases (TESK1 and TESK2), specifically phosphorylate cofilin at the seryl residue at position 3 (P-cofilin), inhibit binding to actin and thus enhanced severing and depolymerization. Chronopin (CIN) and protein phosphatase 1 (PPI1) and 2 (PPI2) have been suggested to control cofilin phosphorylation as well. Slingshot family protein phosphatases (SSHs) specifically dephosphorylate Ser3-P-cofilin and switch back the activity of cofilin [77–79]. Cofilin contains four cysteinyl residues (at positions 39, 80, 139, and 147) that are potential targets for redox modifications. In the folded cofilin molecule, Cys39 and 80 are buried inside, whereas Cys139 and 147 are solvent exposed [80]. Oxidation of cofilin leads to the formation of sulfonic acid in Cys139, preventing the participation of this residue in disulfide formation. As Cys147 is not in proximity to either Cys39 nor Cys80, the most likely intramolecular disulfide bond is formed between Cys39 and Cys80. An intermolecular disulfide bridge between Cys39 and Cys147 of two adjacent subunits has been shown to form in vitro [81] and in vivo [82]. Although oxidized cofilin is still able to bind F-actin, it does not induce F-actin de-polymerization, resulting in enhanced F-actin stability, actin polymerization, and cell motility. Consistent with this, oxidized cofilin is more prone to become phosphorylated [83,84].

13.7 SEMAPHORIN SIGNALLING, CRMP2, AND MICALS

Semaphorins are extracellular signalling molecules that play a crucial role in cytoskeleton function, as they induce alterations in the organization of actin filaments and microtubules. Semaphorins were initially characterized in the nervous system, where they control axon guidance and neuronal polarization [85–87], but the expression of semaphorins has been also described in most (if not all) other tissues [88–90], where they mediate a plethora of functions. Semaphorins comprise a large and diverse family of secreted, membrane-bound, and transmembrane proteins widely conserved among animal phyla. They are grouped into eight classes, 1–7 and V. Classes 1 and 2 are restricted to invertebrates, classes 3–7 to vertebrates, and class V is only found in viri. Each of these classes includes subgroups with different members (see [91] for an overview). All members of the family contain a conserved extracellular domain called the semaphorin (Sema) domain. Semaphorin 3A (Sema3A), a prototypical class 3 secreted semaphorin, is involved in axon repulsion, dendritic branching, and synapse formation in neurons [92–94]. The receptors that specifically mediate class 3 semaphorin signals consists of a heterodimeric transmembrane receptor composed of neuropilins (NRP)-1 or (NRP)-2, and plexin class A (PlexA) proteins that act as specific ligand-binding partners and signal transducers, respectively. For Sema3A in particular, NRP1 is necessary to mediate the signal transduction [95–97]. Upon Sema3A binding to the NRP1/PlexA complex, PlexA directly interacts with Rho family GTPases (e.g., Rho). GTP-bound Rho activates Rho kinase (ROCK), responsible for the disruption of actin polymerization causing the collapse of axonal growth cones and axon retraction. ROCK is additionally responsible for the regulation of the motor protein myosin II. Phosphorylation of Ser19 of the regulatory myosin light chains (rMLC) activates myosin II, which in turn interacts with F-actin generating contractile forces that contribute to axon retraction [98]. Furthermore, phosphorylation of cofilin by LIMK1 has been shown to be necessary for Sema3A induced growth cone collapse [99].

Other effectors of the semaphorin signalling cascade are the Ser/Thr kinases cyclin-dependent kinase 5 (Cdk5) and glycogen synthase kinase 3 (GSK3) [100,101]. These kinases have one substrate in common: collapsin response mediator protein 2 (CRMP2 or DPYL2, dihydropyrimidinase like 2), one of the five members of the CRMP family. Phosphorylated CRMP2 has been identified as a mediator of the axon repulsion in Sema3A signalling, hence this protein has a strong impact on F-actin and microtubule dynamics as well. Additionally, at least some of the CRMP2 functions are controlled by redox modifications. CRMP2 can form an inter-molecular disulfide between two Cys504 residues in the homo-tetrameric quaternary complex [102,103]. This reversible intermolecular thiol switch determines two conformations of the complex, resulting in a positive or negative effect on axon outgrowth (Figure 13.2). Oxidized CRMP2 has been identified as a target of cytosolic glutaredoxin 2 (Grx2c), which is responsible for the reduction of the thiol switch. This regulation is crucial for normal axon outgrowth during embryonic brain development [104]. However, Grx2c is also specifically induced in many cancer cells, and the Grx2c-mediated CRMP2 regulation was implied in cell motility and malignant transformation [105,106]. The mechanism(s) that leads to the oxidation of the CRMP2 thiol switch are still unclear.

MICAL proteins have been identified as interaction partners of CRMP2, leading to the oxidation of CRMP2 through the production of H_2O_2 [107]. The human genome encodes three MICAL genes (MICAL1–3) [108]. All members of the MICAL family genes contain an N-terminal flavo-monooxygenase (MO) domain, with FAD as prosthetic group. This is followed by calponin homology (CH) and LIM domains. The MO domain is responsible for the enzymatic activity, utilizing oxygen and NADPH to oxidize substrates; the CH domain is required for cellular localization and actin binding [109]; the LIM domain mediates further specific protein-protein interactions [110]. For human MICAL1 and MICAL3, an additional C-terminal Rab-binding domain (RBD) has been identified, which appears to autoregulate the oxidase activity, as the C-terminal-truncated MICAL1 resulted in constitutively active enzyme in vivo, in contrast to full-length MICAL1 [111]. MICALs directly interact with PlexA [112], supporting the idea that MICALs function in the semaphorin-mediated axon repulsion through

the regulation of cytoskeletal dynamics. The actin-binding CH domain in MICALs is a single type 2 CH domain, which by itself is not sufficient to bind F-actin [113]. However, this domain enhances the binding of the MO domain to actin binding as well as its catalytic activity [114]. Several studies have identified MICALs as oxidases of F-actin filaments, specifically methionyl residues located in the D-loop at positions 44 and 47, causing actin de-polymerization [75,115,116]. MICAL proteins may act through the local production of H_2O_2 upon activation [117]. For a timely overview on the functions of MICALs as specific oxidases and their relevance in the redox regulation of cellular events, we refer to reference [118].

13.8 REDOX REGULATION OF MICROTUBULE DYNAMICS

Microtubules consist of small globular tubulin proteins, hetero-dimers of α- and β-tubulin, and microtubule-associated proteins (MAPs). They are a key component of the cytoskeleton, being especially important for neuronal functions. Tubulin hetero-dimers assemble to create polarized and length variable microtubules, whose dynamics and interaction with motor proteins are essential for mitosis, cell motility, cell shape, and the long-distance transport of cargo within cells [119]. Microtubules, as well as actin, are dynamically redox regulated in both physiological and pathological contexts. α- and β-tubulin contain 12 and 8 cysteinyl residues, respectively. These 20 cysteinyl residues make neuronal microtubules, due to their high concentration, particularly susceptible to oxidation. Landino et al. demonstrated that peroxy-nitrite treatment in bovine brain led to intermolecular disulfides between α- and β-subunits, resulting in the inhibition of microtubule polymerization, which was restored only after addition of reducing agents [120]. This disulfide is a target of the thioredoxin reductase system [121]. Peroxy-nitrite had the same effect in porcine brains, but here the disulfide bond was reported to be a target for the glutaredoxin system [122]. In general, oxidation of microtubules reduces their ability to polymerize and causes severing of tubules. Interestingly, tubulin oxidation also impairs translocation of the transcription factor NFκB to the nucleus, suggesting a link between redox state and microtubule-dependent trafficking in cells [123], a mechanism in which MICAL1 has been implied as well [124–126]. CRMP2 is also a regulator of microtubule dynamics [127]. It can bind directly to tubulin hetero-dimers [128], increasing the stability of microtubules and promoting their polymerization, respectively (Figure 13.2).

13.9 PATHOLOGICAL CONDITIONS CAUSED BY DISTURBED REDOX REGULATION

Cell migration is regulated by cytoskeletal dynamics. Development of all organs and tissues during embryogenesis depends on the accurate migration of countless cells. Regulation of cell migration is essential for various processes in adults such as regeneration, immune response, and tissue repair. Dysregulation of cytoskeletal dynamics and cell migration contributes to various pathological conditions (e.g., cancer metastasis and age-related disorders).

Redox regulation of cytoskeletal dynamics and cell migration during embryogenesis remains to be elucidated. One protein that has been identified to be crucial for vertebrate development is the cytosolic isoform of glutaredoxin (Grx) 2, Grx2c [105,129]. Grx2c regulates reversible thiol redox modifications of CRMP2 [103,104], actin [56], and SIRT1 [130] that are essential for the control of axon formation, migration of cardiac neural crest cells, and vessel formation, respectively. Thereby Grx2c is crucially involved in the formation of the cardiovascular system and the brain during vertebrate development.

In adults, however, Grx2c can also promote cancer progression and metastasis; patients suffering from clear cell renal cell carcinoma (ccRCC) with high Grx2 expression show a decreased cancer-specific survival [106]. Already a moderate expression of Grx2c is sufficient to increase cell motility and invasiveness by at least twofold on a plain surface and a 3D collagen matrix using different cell culture models. Expression of Grx2c in ccRCC was significantly higher in more advanced tumour stages [106]. Metastasis, the spreading of cancer cells, is a form of pathological cell migration and is linked to cellular nucleoside diphosphate kinase (NDPK) and SIRT1 [131–133]. Brain metastatic tissues of non-small cell lung cancers show high expression of SIRT1 [134]. The activities of NDPK and SIRT1 are regulated by redox modification of a single cysteinyl residue. Depending on the redox state of Cys109, which is reduced by the Trx system, NDPK loses its ability to suppress

tumour metastasis [135]. The activity of SIRT1, on the other hand, is modulated by reversible S-glutathionylation of its Cys204, a reaction catalyzed by Grx2c [130]. Both, SIRT1 and NDPK lead to the release of vascular endothelial growth factor-C (VEGF-C) via the activation of the transcription factor FOXO-1. T-cell migration and metastasis are positively regulated by the enhanced release of VEGF-C [136]. These findings demonstrate the connection of redox-regulated pathways and cancer cell migration that may be interesting targets in cancer therapies.

Persistent deficits in neurodegenerative disorders (e.g., Alzheimer's disease, multiple sclerosis [MS], or following a cerebral stroke) are also caused by the low regenerative capacity of our central nervous system [137–139]. In the neuroinflammatory disorder MS, neuronal cell loss and thereby neuronal dysfunctions are caused by the de-myelination due to oligodendroglial cells death and disturbed re-myelination [140]. Migration of oligodendroglial progenitor cells (OPCs) into de-myelinated lesion areas for myelin regeneration is thought to improve symptoms and delay the progression, while altered migration capacities are a major cause for the failing regeneration of the myelin [141]. OPCs migration and differentiation are inhibited by changes in the chemotactic milieu [142]. Some of the molecules responsible for these effects are Semas (see above) [142–144]. The NAD$^+$-dependent protein deacetylase SIRT1 was identified as a target for Grx2-specific deglutathionylation in the regulation of vascular development [130]. Recently, it was suggested that SIRT1 inhibition may help to expand the endogenous OPCs pool without affecting their differentiation in mouse experimental autoimmune encephalomyelitis (EAE), a model for MS [145]. Increased S-nitrosylation and carbonylation of β-actin as well as α/β-tubulin were demonstrated in EAE animals [146,147]. Taken together, these findings illustrate the importance of functioning redox-regulated OPCs migration in the progression of MS [148]. A better understanding of these processes may also benefit other neuro-inflammatory diseases (e.g., traumatic brain injury, stroke, and other oligo-/neuro-degenerative diseases) [149,150].

Redox modifications of the cytoskeleton play a major role in axonal de- and regeneration in both Alzheimer's and Parkinson's disease [151,152]. Increased carbonylation of β-actin and oxidation of cysteinyl residues in microtubule-associated proteins directly influence the degeneration rates and the regeneration capacity of axonal loss [33,153]. In Alzheimer's disease brain extracts, β-tubulin carbonylation was not enhanced [152]. A recent study suggests that reduced S-glutathionylation due to decreased Grx1 levels may contribute to synaptic dysfunction via direct disruption of the F-actin in spines [154]. Some patients suffering from amyotrophic lateral sclerosis (ALS) carry mutations in a protein disulfide isomerase (PDI) that lacks redox activity. The redox function of the PDI was shown to be protective against, for example, protein misfolding and cytoplasmic mislocalization of TAR DNA-binding protein 43 (TPD-43) or superoxide dismutase 1 (SOD1) [155].

Cell migration is also essential at the onset of atherosclerosis and impaired wound healing as age-related diseases [156]. CD4+ T cells, endothelial cells, smooth muscle cells, and dermal fibroblasts are among the cell types known for alterations in their migration potential with the progression of ageing [157–160]. The decreased function of integrin α1β2, a collagen-binding protein, and the disorganized β-actin are the basis for age-related impaired migration in dermal fibroblasts [161]. One indication for an important role of specific redox switches in cytoskeletal dynamics during ageing is the redox regulation of β-actin's Cys347 during integrin-mediated cell adhesion [162]. The beneficial effects of hypoxia on wound healing, that depends on increased TGFβ-1-mediated migration of dermal fibroblasts, are also lost during ageing [158]. The underlying signalling pathway depends on the amount of free thiols, however, the exact type of redox modifications and switches remains unclear as of today [163,164].

The number of pharmacological substances being evaluated that potentially affect the redox regulation of cytoskeletal dynamics in the described pathologies is very low. One is U-83836E, a scavenger of lipid peroxyl radicals, thereby attenuating cytoskeletal damage after traumatic brain injury [165]. Other includes the long known nitroxyl (HNO) and HNO donors that emerged as a potential pharmacological agent for cancer and cardiovascular diseases [166]. HNO can react as nucleophile and electrophile targeting predominantly protein thiols and iron haem proteins [166]. Cardiac contractility is increased by HNO due to the specific promotion of disulfides between Cys81 and Cys37 in the myosin light chain, Cys 257 of actin, and Cys190 of tropomyosin. Drugs used for anticancer

treatment targeting microtubules either destabilize or stabilize microtubules resulting in the inhibition of mitosis and cell proliferation. Targeting of microtubules dynamics may also be a promising strategy in the treatment of neurodegenerative diseases [167,168]. More specific pharmacological inhibitors of actin assembly and nucleation may be used in the future to treat cancer development and metastasis, but so far they are not implemented in the clinics [169]. Another promising target is CRMP2 (see above), which is involved in the regulation of both actin and tubulin dynamics. The anti-epileptic drug lacosamide, also known as VIMPAT, inhibits CRMP2 although the mode of action is not understood at the molecular level. However, it was suggested that the aberrant neurite outgrowth during epilepsy is inhibited by the lacosamide-mediated CRMP2 inactivation [170]. Besides epilepsy, CRMP2 is considered a promising target in numerous neurological diseases including ALS, Parkinson's, Alzheimer's disease, and spinal cord injury [171,172]. Other compounds under investigation to modulate CRMP2's function in different models are lanthionine ketimine (LK) and lanthionine ketimine ester (LKE). In a model for Parkinson's disease, data suggest a modification of CRMP2's phosphorylation state, thereby suppressing microglial activation [172]. It was suggested before that the redox regulation and the regulation via phosphorylation of CRMP2 are tightly connected, if not dependent on each other [173]. The reduction of CRMP2 Ser522 phosphorylation by LKE appears to be beneficial in EAE mice [174].

13.10 CONCLUSION

The constant remodelling of both the actin and tubulin cytoskeleton is essential for normal cell functions. It is spatiotemporally controlled by various regulatory mechanisms including posttranslational redox modifications: thiol and methionyl switches. These redox switches are not operated by random modifications by some unspecified oxidants. They are controlled by specific enzymes, oxidases and reductases, that catalyse the oxidation and reduction of these modifications comparable to the kinase/phosphatase antagonists in phosphorylation signalling. The emerging importance of reversible cysteinyl and methionyl oxidation in cytoskeletal dynamics may offer new strategies of various diseases, including cancer and neuro-degenerative disorders.

REFERENCES

[1] P. Ghezzi, V. Bonetto, M. Fratelli, Thiol disulfide balance: from the concept of oxidative stress to that of redox regulation, Antioxid. Redox Signal. 7 (2005) 964–972.

[2] D.P. Jones, Redefining oxidative stress, Antioxid. Redox Signal. 8 (2006) 1865–1879. https://doi.org/10.1089/ars.2006.8.1865.

[3] P.J. Halvey, W.H. Watson, J.M. Hansen, Y.-M. Go, A. Samali, D.P. Jones, Compartmental oxidation of thiol-disulphide redox couples during epidermal growth factor signalling, Biochem. J. 386 (2005) 215–219. https://doi.org/BJ20041829.

[4] Y.-M. Go, D.P. Jones, Redox compartmentalization in eukaryotic cells, Biochim. Biophys. Acta BBA—Gen. Subj. 1780 (2008) 1273–1290.

[5] C. Berndt, C.H. Lillig, L. Flohé, Redox regulation by glutathione needs enzymes, Front. Pharmacol. 5 (2014). https://doi.org/10.3389/fphar.2014.00168.

[6] M.M. Gallogly, J.J. Mieyal, Mechanisms of reversible protein glutathionylation in redox signaling and oxidative stress, Curr. Opin. Pharmacol. 7 (2007) 381–391. https://doi.org/S1471-4892(07)00103-8.

[7] M. Imber, A.J. Pietrzyk-Brzezinska, H. Antelmann, Redox regulation by reversible protein S-thiolation in Gram-positive bacteria, Redox Biol. 20 (2018) 130–145. https://doi.org/10.1016/j.redox.2018.08.017.

[8] L. Flohé, Changing paradigms in thiology from antioxidant defense toward redox regulation, Methods Enzymol. 473 (2010) 1–39. https://doi.org/10.1016/S0076-6879(10)73001-9.

[9] E.-M. Hanschmann, J.R. Godoy, C. Berndt, C. Hudemann, C.H. Lillig, Thioredoxins, glutaredoxins, and peroxiredoxins: molecular mechanisms and health significance: from cofactors to antioxidants to redox signaling, Antioxid. Redox Signal. 19 (2013) 1539–1605. https://doi.org/10.1089/ars.2012.4599.

[10] M. Deponte, C.H. Lillig, Enzymatic control of cysteinyl thiol switches in proteins, Biol. Chem. 396 (2015) 401–413. https://doi.org/10.1515/hsz-2014-0280.

[11] A. Holmgren, M. Björnstedt, Thioredoxin and thioredoxin reductase, Methods Enzymol. 252 (1995) 199–208.

[12] A. Holmgren, F. Aslund, Glutaredoxin, Methods Enzymol. 252 (1995) 283–292. https://doi.org/7476363.

[13] D. Trnka, A.D. Engelke, M. Gellert, A. Moseler, M.F. Hossain, T.T. Lindenberg, L. Pedroletti, B. Odermatt, J.V. de Souza, A.K. Bronowska, T.P. Dick, U. Mühlenhoff, A.J. Meyer, C. Berndt, C.H. Lillig, Molecular basis for the distinct functions of redox-active and FeS-transfering glutaredoxins, Nat. Commun. 11 (2020) 3445. https://doi.org/10.1038/s41467-020-17323-0.

[14] L. Liedgens, J. Zimmermann, L. Wäschenbach, F. Geissel, H. Laporte, H. Gohlke, B. Morgan, M. Deponte, Quantitative assessment of the determinant structural differences between redox-active and inactive glutaredoxins, Nat. Commun. 11 (2020) 1725. https://doi.org/10.1038/s41467-020-15441-3.

[15] L.B. Poole, Formation and functions of protein sulfenic acids, Curr. Protoc. Toxicol. Editor. Board Mahin Maines Ed.—Chief Al. Chapter 17 (2004) Unit 17.1. https://doi.org/10.1002/0471140856.tx1701s18.

[16] Y.-C. Chang, C.-N. Huang, C.-H. Lin, H.-C. Chang, C.-C. Wu, Mapping protein cysteine sulfonic acid modifications with specific enrichment and mass spectrometry: an integrated approach to explore the cysteine oxidation, Proteomics. 10 (2010) 2961–2971. https://doi.org/10.1002/pmic.200900850.

[17] L. Flohé, The impact of thiol peroxidases on redox regulation, Free Radic. Res. 50 (2016) 126–142. https://doi.org/10.3109/10715762.2015.1046858.

[18] F. Antunes, P.M. Brito, Quantitative biology of hydrogen peroxide signaling, Redox Biol. 13 (2017) 1–7. https://doi.org/10.1016/j.redox.2017.04.039.

[19] M.C. Sobotta, W. Liou, S. Stöcker, D. Talwar, M. Oehler, T. Ruppert, A.N.D. Scharf, T.P. Dick, Peroxiredoxin-2 and STAT3 form a redox relay for H_2O_2 signaling, Nat. Chem. Biol. 11 (2015) 64–70. https://doi.org/10.1038/nchembio.1695.

[20] D.T. Hess, A. Matsumoto, S.-O. Kim, H.E. Marshall, J.S. Stamler, Protein S-nitrosylation: purview and parameters, Nat. Rev. Mol. Cell Biol. 6 (2005) 150–166. https://doi.org/10.1038/nrm1569.

[21] M.W. Foster, D.T. Hess, J.S. Stamler, Protein S-nitrosylation in health and disease: a current perspective, Trends Mol. Med. 15 (2009) 391–404. https://doi.org/10.1016/j.molmed.2009.06.007.

[22] B. Sjöberg, S. Foley, B. Cardey, M. Fromm, M. Enescu, Methionine oxidation by hydrogen peroxide in peptides and proteins: a theoretical and Raman spectroscopy study, J. Photochem. Photobiol. B. 188 (2018) 95–99. https://doi.org/10.1016/j.jphotobiol.2018.09.009.

[23] A. Drazic, J. Winter, The physiological role of reversible methionine oxidation, Biochim. Biophys. Acta. 1844 (2014) 1367–1382. https://doi.org/10.1016/j.bbapap.2014.01.001.

[24] A. Hansel, L. Kuschel, S. Hehl, C. Lemke, H.-J. Agricola, T. Hoshi, S.H. Heinemann, Mitochondrial targeting of the human peptide methionine sulfoxide reductase (MSRA), an enzyme involved in the repair of oxidized proteins, FASEB J. Off. Publ. Fed. Am. Soc. Exp. Biol. 16 (2002) 911–913. https://doi.org/10.1096/fj.01-0737fje.

[25] A. Hansel, S.H. Heinemann, T. Hoshi, Heterogeneity and function of mammalian MSRs: enzymes for repair, protection and regulation, Biochim. Biophys. Acta. 1703 (2005) 239–247. https://doi.org/10.1016/j.bbapap.2004.09.010.

[26] J.W. Lee, N.V. Gordiyenko, M. Marchetti, N. Tserentsoodol, D. Sagher, S. Alam, H. Weissbach, M. Kantorow, I.R. Rodriguez, Gene structure, localization and role in oxidative stress of methionine sulfoxide reductase A (MSRA) in the monkey retina, Exp. Eye Res. 82 (2006) 816–827. https://doi.org/10.1016/j.exer.2005.10.003.

[27] S. Boschi Muller, A. Gand, G. Branlant, The methionine sulfoxide reductases: catalysis and substrate specificities, Arch. Biochem. Biophys. 474 (2008) 266–273. https://doi.org/10.1016/j.abb.2008.02.007.

[28] H.-Y. Kim, Glutaredoxin serves as a reductant for methionine sulfoxide reductases with or without resolving cysteine, Acta Biochim. Biophys. Sin. 44 (2012) 623–627. https://doi.org/10.1093/abbs/gms038.

[29] H.-Y. Kim, V.N. Gladyshev, Methionine sulfoxide reduction in mammals: characterization of methionine-R-sulfoxide reductases, Mol. Biol. Cell. 15 (2004) 1055–1064. https://doi.org/10.1091/mbc.E03-08-0629.

[30] B.C. Lee, A. Dikiy, H.-Y. Kim, V.N. Gladyshev, Functions and evolution of selenoprotein methionine sulfoxide reductases, Biochim. Biophys. Acta. 1790 (2009) 1471–1477. https://doi.org/10.1016/j.bbagen.2009.04.014.

[31] R.T. Dean, S. Fu, R. Stocker, M.J. Davies, Biochemistry and pathology of radical-mediated protein oxidation, Biochem. J. 324 (Pt 1) (1997) 1–18.

[32] E.R. Stadtman, Free radical mediated oxidation of proteins, in: T. Özben (Ed.), Free Radicals, Oxidative Stress, and Antioxidants, Springer US (1998), pp. 51–64. http://doi.org/10.1007/978-1-4757-2907-8_5.

[33] I. Dalle-Donne, G. Aldini, M. Carini, R. Colombo, R. Rossi, A. Milzani, Protein carbonylation, cellular dysfunction, and disease progression, J. Cell. Mol. Med. 10 (2006) 389–406.

[34] W.M. Garrison, Reaction mechanisms in the radiolysis of peptides, polypeptides, and proteins, Chem. Rev. 87 (1987) 381–398. https://doi.org/10.1021/cr00078a006.

[35] E.R. Stadtman, Metal ion-catalyzed oxidation of proteins: biochemical mechanism and biological consequences, Free Radic. Biol. Med. 9 (1990) 315–325.

[36] C.M. Wong, L. Marcocci, L. Liu, Y.J. Suzuki, Cell signaling by protein carbonylation and decarbonylation, Antioxid. Redox Signal. 12 (2010) 393–404. https://doi.org/10.1089/ars.2009.2805.

[37] J.S. Beckman, Oxidative damage and tyrosine nitration from peroxynitrite, Chem. Res. Toxicol. 9 (1996) 836–844. https://doi.org/10.1021/tx9501445.

[38] O. Augusto, M.G. Bonini, A.M. Amanso, E. Linares, C.C.X. Santos, S.L. De Menezes, Nitrogen dioxide and carbonate radical anion: two emerging radicals in biology, Free Radic. Biol. Med. 32 (2002) 841–859.

[39] R. Radi, Peroxynitrite, a stealthy biological oxidant, J. Biol. Chem. 288 (2013) 26464–26472. https://doi.org/10.1074/jbc.R113.472936.

[40] M. Zou, A. Yesilkaya, V. Ullrich, Peroxynitrite inactivates prostacyclin synthase by heme-thiolate-catalyzed tyrosine nitration, Drug Metab. Rev. 31 (1999) 343–349. https://doi.org/10.1081/DMR-100101922.

[41] D.R. Mole, P.H. Maxwell, C.W. Pugh, P.J. Ratcliffe, Regulation of HIF by the von Hippel-Lindau tumour suppressor: implications for cellular oxygen sensing, IUBMB Life. 52 (2001) 43–47. https://doi.org/10.1080/15216540252774757.

[42] M. Safran, W.G. Kaelin, HIF hydroxylation and the mammalian oxygen-sensing pathway, J. Clin. Invest. 111 (2003) 779–783. https://doi.org/10.1172/JCI18181.

[43] P.J. Ratcliffe, Oxygen sensing and hypoxia signalling pathways in animals: the implications of physiology for cancer, J. Physiol. 591 (2013) 2027–2042. https://doi.org/10.1113/physiol.2013.251470.

[44] T.D. Pollard, L. Blanchoin, R.D. Mullins, Molecular mechanisms controlling actin filament dynamics in nonmuscle cells, Annu. Rev. Biophys. Biomol. Struct. 29 (2000) 545–576. https://doi.org/10.1146/annurev.biophys.29.1.545.

[45] F. Merino, S. Pospich, S. Raunser, Towards a structural understanding of the remodeling of the actin cytoskeleton, Semin. Cell Dev. Biol. 102 (2020) 51–64. https://doi.org/10.1016/j.semcdb.2019.11.018.

[46] T.D. Pollard, Regulation of actin filament assembly by Arp2/3 complex and formins, Annu. Rev. Biophys. Biomol. Struct. 36 (2007) 451–477. https://doi.org/10.1146/annurev.biophys.35.040405.101936.

[47] V.I. Risca, E.B. Wang, O. Chaudhuri, J.J. Chia, P.L. Geissler, D.A. Fletcher, Actin filament curvature biases branching direction, Proc. Natl. Acad. Sci. 109 (2012) 2913–2918. https://doi.org/10.1073/pnas.1114292109.

[48] A.Y. Pollitt, R.H. Insall, WASP and SCAR/WAVE proteins: the drivers of actin assembly, J. Cell Sci. 122 (2009) 2575–2578. https://doi.org/10.1242/jcs.023879.

[49] S. Kurisu, T. Takenawa, WASP and WAVE family proteins: friends or foes in cancer invasion?, Cancer Sci. 101 (2010) 2093–2104. https://doi.org/10.1111/j.1349-7006.2010.01654.x.

[50] T.E.B. Stradal, K. Rottner, A. Disanza, S. Confalonieri, M. Innocenti, G. Scita, Regulation of actin dynamics by WASP and WAVE family proteins, Trends Cell Biol. 14 (2004) 303–311. https://doi.org/10.1016/j.tcb.2004.04.007.

[51] W.D. Halliburton, On muscle-plasma, J. Physiol. 8 (1887) 133–202.

[52] T.D. Pollard, J.A. Cooper, Actin, a central player in cell shape and movement, Science. 326 (2009) 1208–1212. https://doi.org/10.1126/science.1175862.

[53] J.R. Terman, A. Kashina, Post-translational modification and regulation of actin, Curr. Opin. Cell Biol. 25 (2013) 30–38. https://doi.org/10.1016/j.ceb.2012.10.009.

[54] J. Wang, E.S. Boja, W. Tan, E. Tekle, H.M. Fales, S. English, J.J. Mieyal, P.B. Chock, Reversible glutathionylation regulates actin polymerization in A431 cells, J. Biol. Chem. 276 (2001) 47763–47766. https://doi.org/11684673.

[55] Y. Hamnell-Pamment, C. Lind, C. Palmberg, T. Bergman, I.A. Cotgreave, Determination of site-specificity of S-glutathionylated cellular proteins, Biochem. Biophys. Res. Commun. 332 (2005) 362–369. https://doi.org/10.1016/j.bbrc.2005.04.130.

[56] C. Berndt, G. Poschmann, K. Stühler, A. Holmgren, L. Bräutigam, Zebrafish heart development is regulated via glutaredoxin 2 dependent migration and survival of neural crest cells, Redox Biol. 2 (2014) 673–678. https://doi.org/10.1016/j.redox.2014.04.012.

[57] M. Fedorova, N. Kuleva, R. Hoffmann, Identification of cysteine, methionine and tryptophan residues of actin oxidized in vivo during oxidative stress, J. Proteome Res. 9 (2010) 1598–1609. https://doi.org/10.1021/pr901099e.

[58] I. Lassing, F. Schmitzberger, M. Björnstedt, A. Holmgren, P. Nordlund, C.E. Schutt, U. Lindberg, Molecular and structural basis for redox regulation of beta-actin, J. Mol. Biol. 370 (2007) 331–348. https://doi.org/S0022-2836(07)00555-4.

[59] M.E. Farah, V. Sirotkin, B. Haarer, D. Kakhniashvili, D.C. Amberg, Diverse protective roles of the actin cytoskeleton during oxidative stress, Cytoskelet. 68 (2011) 340–354. https://doi.org/10.1002/cm.20516.

[60] D. Su, A.K. Shukla, B. Chen, J.-S. Kim, E. Nakayasu, Y. Qu, U. Aryal, K. Weitz, T.R.W. Clauss, M.E. Monroe, D.G. Camp, D.J. Bigelow, R.D. Smith, R.N. Kulkarni, W.-J. Qian, Quantitative site-specific reactivity profiling of S-nitrosylation in mouse skeletal muscle using cysteinyl peptide enrichment coupled with mass spectrometry, Free Radic. Biol. Med. 57 (2013) 68–78. https://doi.org/10.1016/j.freeradbiomed.2012.12.010.

[61] S.R. Thom, V.M. Bhopale, D.J. Mancini, T.N. Milovanova, Actin S-nitrosylation inhibits neutrophil beta2 integrin function, J. Biol. Chem. 283 (2008) 10822–10834. https://doi.org/10.1074/jbc.M709200200.

[62] I. Dalle-Donne, A. Milzani, D. Giustarini, P. Di Simplicio, R. Colombo, R. Rossi, S-NO-actin: S-nitrosylation kinetics and the effect on isolated vascular smooth muscle, J. Muscle Res. Cell Motil. 21 (2000) 171–181. https://doi.org/10.1023/a:1005671319604.

[63] I. Dalle-Donne, R. Rossi, D. Giustarini, N. Gagliano, P. Di Simplicio, R. Colombo, A. Milzani, Methionine oxidation as a major cause of the functional impairment of oxidized actin, Free Radic. Biol. Med. 32 (2002) 927–937. https://doi.org/10.1016/S0891-5849(02)00799-2.

[64] J. Duke, R. Takashi, K. Ue, M.F. Morales, Reciprocal reactivities of specific thiols when actin binds to myosin, Proc. Natl. Acad. Sci. U. S. A. 73 (1976) 302–306.

[65] M.D. Shelton, P.B. Chock, J.J. Mieyal, Glutaredoxin: role in reversible protein S-glutathionylation and regulation of redox

signal transduction and protein translocation, Antioxid. Redox Signal. 7 (2005) 348–366.

[66] P. Ghezzi, Review: Regulation of protein function by glutathionylation, Free Radic. Res. 39 (2005) 573–580. https://doi.org/10.1080/10715760500072172.

[67] Y.M.W. Janssen-Heininger, B.T. Mossman, N.H. Heintz, H.J. Forman, B. Kalyanaraman, T. Finkel, J.S. Stamler, S.G. Rhee, A. van der Vliet, Redox-based regulation of signal transduction: principles, pitfalls, and promises, Free Radic. Biol. Med. 45 (2008) 1–17. https://doi.org/10.1016/j.freeradbiomed.2008.03.011.

[68] R.P. Kommaddi, D.S. Tomar, S. Karunakaran, D. Bapat, S. Nanguneri, A. Ray, B.L. Schneider, D. Nair, V. Ravindranath, Glutaredoxin1 diminishes amyloid beta-mediated oxidation of F-actin and reverses cognitive deficits in an Alzheimer's disease mouse model, Antioxid. Redox Signal. 31 (2019) 1321–1338. https://doi.org/10.1089/ars.2019.7754.

[69] C. Wilson, C. González-Billault, Regulation of cytoskeletal dynamics by redox signaling and oxidative stress: implications for neuronal development and trafficking, Front. Cell. Neurosci. 9 (2015). https://doi.org/10.3389/fncel.2015.00381.

[70] J. Sakai, J. Li, K.K. Subramanian, S. Mondal, B. Bajrami, H. Hattori, Y. Jia, B.C. Dickinson, J. Zhong, K. Ye, C.J. Chang, Y.-S. Ho, J. Zhou, H.R. Luo, Reactive oxygen species-induced actin glutathionylation controls actin dynamics in neutrophils, Immunity. 37 (2012) 1037–1049. https://doi.org/10.1016/j.immuni.2012.08.017.

[71] Zschauer Tim-Christian, Kunze Kerstin, Jakob Sascha, Haendeler Judith, Altschmied Joachim, Oxidative stress-induced degradation of thioredoxin-1 and apoptosis is inhibited by thioredoxin-1-actin interaction in endothelial cells, Arterioscler. Thromb. Vasc. Biol. 31 (2011) 650–656. https://doi.org/10.1161/ATVBAHA.110.218982.

[72] X. Wang, S. Ling, D. Zhao, Q. Sun, Q. Li, F. Wu, J. Nie, L. Qu, B. Wang, X. Shen, Y. Bai, Y. Li, Y. Li, Redox regulation of actin by thioredoxin-1 is mediated by the interaction of the proteins via cysteine 62, Antioxid. Redox Signal. 13 (2010) 565–573. https://doi.org/10.1089/ars.2009.2833.

[73] M. Benhar, M.T. Forrester, J.S. Stamler, Protein denitrosylation: enzymatic mechanisms and cellular functions, Nat. Rev. Mol. Cell Biol. 10 (2009) 721–732. https://doi.org/10.1038/nrm2764.

[74] X. Ren, L. Zou, X. Zhang, V. Branco, J. Wang, C. Carvalho, A. Holmgren, J. Lu, Redox signaling mediated by thioredoxin and glutathione systems in the central nervous system, Antioxid. Redox Signal. 27 (2017) 989–1010. https://doi.org/10.1089/ars.2016.6925.

[75] R.-J. Hung, U. Yazdani, J. Yoon, H. Wu, T. Yang, N. Gupta, Z. Huang, W.J.H. van Berkel, J.R. Terman, Mical links semaphorins to F-actin disassembly, Nature. 463 (2010) 823–827. https://doi.org/10.1038/nature08724.

[76] C.B. Gurniak, E. Perlas, W. Witke, The actin depolymerizing factor n-cofilin is essential for neural tube morphogenesis and neural crest cell migration, Dev. Biol. 278 (2005) 231–241. https://doi.org/10.1016/j.ydbio.2004.11.010.

[77] K. Mizuno, Signaling mechanisms and functional roles of cofilin phosphorylation and dephosphorylation, Cell. Signal. 25 (2013) 457–469. https://doi.org/10.1016/j.cellsig.2012.11.001.

[78] B.J. Agnew, L.S. Minamide, J.R. Bamburg, Reactivation of phosphorylated actin depolymerizing factor and identification of the regulatory site, J. Biol. Chem. 270 (1995) 17582–17587. https://doi.org/10.1074/jbc.270.29.17582.

[79] K. Moriyama, K. Iida, I. Yahara, Phosphorylation of Ser-3 of cofilin regulates its essential function on actin, Genes Cells Devoted Mol. Cell. Mech. 1 (1996) 73–86. https://doi.org/10.1046/j.1365-2443.1996.05005.x.

[80] B.J. Pope, K.M. Zierler-Gould, R. Kühne, A.G. Weeds, L.J. Ball, Solution structure of human cofilin: actin binding, pH sensitivity, and relationship to actin-depolymerizing factor, J. Biol. Chem. 279 (2004) 4840–4848. https://doi.org/10.1074/jbc.M310148200.

[81] J. Pfannstiel, M. Cyrklaff, A. Habermann, S. Stoeva, G. Griffiths, R. Shoeman, H. Faulstich, Human cofilin forms oligomers exhibiting actin bundling activity, J. Biol. Chem. 276 (2001) 49476–49484. https://doi.org/10.1074/jbc.M104760200.

[82] B.W. Bernstein, A.E. Shaw, L.S. Minamide, C.W. Pak, J.R. Bamburg, Incorporation of cofilin into rods depends on disulfide intermolecular bonds: implications for actin regulation and neurodegenerative disease, J. Neurosci. 32 (2012) 6670–6681. https://doi.org/10.1523/JNEUROSCI.6020-11.2012.

[83] M. Klemke, G.H. Wabnitz, F. Funke, B. Funk, H. Kirchgessner, Y. Samstag, Oxidation of cofilin mediates T cell hyporesponsiveness under oxidative stress conditions, Immunity. 29 (2008) 404–413. https://doi.org/10.1016/j.immuni.2008.06.016.

[84] J.M. Cameron, M. Gabrielsen, Y.H. Chim, J. Munro, E.J. McGhee, D. Sumpton, P. Eaton, K.I. Anderson, H. Yin, M.F. Olson, Polarized cell motility induces hydrogen peroxide to inhibit cofilin via cysteine oxidation, Curr. Biol. CB. 25 (2015) 1520–1525. https://doi.org/10.1016/j.cub.2015.04.020.

[85] A.L. Kolodkin, D.J. Matthes, C.S. Goodman, The semaphorin genes encode a family of trans-membrane and secreted growth cone guidance molecules, Cell. 75 (1993) 1389–1399. https://doi.org/10.1016/0092-8674(93)90625-z.

[86] M. Tessier-Lavigne, C.S. Goodman, the molecular biology of axon guidance, Science. 274 (1996) 1123–1133. https://doi.org/10.1126/science.274.5290.1123.

[87] Y. Luo, D. Raible, J.A. Raper, Collapsin: a protein in brain that induces the collapse and paralysis of neuronal growth cones, Cell. 75 (1993) 217–227. https://doi.org/10.1016/0092-8674(93)80064-L.

[88] L. Roth, E. Koncina, S. Satkauskas, G. Crémel, D. Aunis, D. Bagnard, The many faces of semaphorins: from development to pathology, Cell. Mol. Life Sci. 66 (2008) 649. https://doi.org/10.1007/s00018-008-8518-z.

[89] J.A. Epstein, H. Aghajanian, M.K. Singh, Semaphorin signaling in cardiovascular development, Cell Metab. 21 (2015) 163–173. https://doi.org/10.1016/j.cmet.2014.12.015.

[90] Q. Lu, L. Zhu, the role of semaphorins in metabolic disorders, Int. J. Mol. Sci. 21 (2020). https://doi.org/10.3390/ijms21165641.

[91] U. Yazdani, J.R. Terman, The semaphorins, Genome Biol. 7 (2006) 211. https://doi.org/10.1186/gb-2006-7-3-211.

[92] R.J. Pasterkamp, Getting neural circuits into shape with semaphorins, Nat. Rev. Neurosci. 13 (2012) 605–618. https://doi.org/10.1038/nrn3302.

[93] A. Morita, N. Yamashita, Y. Sasaki, Y. Uchida, O. Nakajima, F. Nakamura, T. Yagi, M. Taniguchi, H. Usui, R. Katoh-Semba, K. Takei, Y. Goshima, Regulation of dendritic branching and spine maturation by Semaphorin3A-Fyn signaling, J. Neurosci. 26 (2006) 2971–2980. https://doi.org/10.1523/JNEUROSCI.5453-05.2006.

[94] V. Fenstermaker, Y. Chen, A. Ghosh, R. Yuste, Regulation of dendritic length and branching by semaphorin 3A, J. Neurobiol. 58 (2004) 403–412. https://doi.org/10.1002/neu.10304.

[95] H. Fujisawa, Discovery of semaphorin receptors, neuropilin and plexin, and their functions in neural development, J. Neurobiol. 59 (2004) 24–33. https://doi.org/10.1002/neu.10337.

[96] Z. He, M. Tessier-Lavigne, Neuropilin is a receptor for the axonal chemorepellent semaphorin III, Cell. 90 (1997) 739–751. https://doi.org/10.1016/S0092-8674(00)80534-6.

[97] T. Takahashi, A. Fournier, F. Nakamura, L.-H. Wang, Y. Murakami, R.G. Kalb, H. Fujisawa, S.M. Strittmatter, Plexin-Neuropilin-1 complexes form functional Semaphorin-3A receptors, Cell. 99 (1999) 59–69. https://doi.org/10.1016/S0092-8674(00)80062-8.

[98] G. Gallo, RhoA-kinase coordinates F-actin organization and myosin II activity during semaphorin-3A-induced axon retraction, J. Cell Sci. 119 (2006) 3413. https://doi.org/10.1242/jcs.03084.

[99] H. Aizawa, S. Wakatsuki, A. Ishii, K. Moriyama, Y. Sasaki, K. Ohashi, Y. Sekine-Aizawa, A. Sehara-Fujisawa, K. Mizuno, Y. Goshima, I. Yahara, Phosphorylation of cofilin by LIM-kinase is necessary for semaphorin 3A-induced growth cone collapse, Nat. Neurosci. 4 (2001) 367–373. https://doi.org/10.1038/86011.

[100] Y. Sasaki, C. Cheng, Y. Uchida, O. Nakajima, T. Ohshima, T. Yagi, M. Taniguchi, T. Nakayama, R. Kishida, Y. Kudo, S. Ohno, F. Nakamura, Y. Goshima, Fyn and Cdk5 mediate semaphorin-3A signaling, which is involved in regulation of dendrite orientation in cerebral cortex, Neuron. 35 (2002) 907–920. https://doi.org/10.1016/s0896-6273(02)00857-7.

[101] B.J. Eickholt, F.S. Walsh, P. Doherty, An inactive pool of GSK-3 at the leading edge of growth cones is implicated in Semaphorin 3A signaling, J. Cell Biol. 157 (2002) 211–217. https://doi.org/10.1083/jcb.200201098.

[102] L.H. Wang, S.M. Strittmatter, Brain CRMP forms heterotetramers similar to liver dihydropyrimidinase, J. Neurochem. 69 (1997) 2261–2269.

[103] M. Gellert, S. Venz, J. Mitlöhner, C. Cott, E.-M. Hanschmann, C.H. Lillig, Identification of a dithiol-disulfide switch in collapsin response mediator protein 2 (CRMP2) that is toggled in a model of neuronal differentiation, J. Biol. Chem. 288 (2013) 35117–35125. https://doi.org/10.1074/jbc.M113.521443.

[104] L. Bräutigam, L.D. Schütte, J.R. Godoy, T. Prozorovski, M. Gellert, G. Hauptmann, A. Holmgren, C.H. Lillig, C. Berndt, Vertebrate-specific glutaredoxin is essential for brain development, Proc. Natl. Acad. Sci. U. S. A. 108 (2011) 20532–20537. https://doi.org/10.1073/pnas.1110085108.

[105] M.E. Lönn, C. Hudemann, C. Berndt, V. Cherkasov, F. Capani, A. Holmgren, C.H. Lillig, Expression pattern of human glutaredoxin 2 isoforms: identification and characterization of two testis/cancer cell-specific isoforms, Antioxid. Redox Signal. 10 (2008) 547–557. https://doi.org/10.1089/ars.2007.1821.

[106] M. Gellert, E. Richter, J. Mostertz, L. Kantz, K. Masur, E.-M. Hanschmann, S. Ribback, N. Kroeger, E. Schaeffeler, S. Winter, F. Hochgräfe, M. Schwab, C.H. Lillig, The cytosolic isoform of glutaredoxin 2 promotes cell migration and invasion, Biochim. Biophys. Acta BBA— Gen. Subj. 1864 (2020) 129599. https://doi.org/10.1016/j.bbagen.2020.129599.

[107] S.S.P. Giridharan, S. Caplan, MICAL-family proteins: complex regulators of the actin cytoskeleton, Antioxid. Redox Signal. 20 (2014) 2059–2073. https://doi.org/10.1089/ars.2013.5487.

[108] T. Suzuki, T. Nakamoto, S. Ogawa, S. Seo, T. Matsumura, K. Tachibana, C. Morimoto, H. Hirai, MICAL, a Novel CasL Interacting molecule, associates with vimentin, J. Biol. Chem. 277 (2002) 14933–14941 https://doi.org/10.1074/jbc.M111842200.

[109] J.H. Hartwig, Actin-binding proteins. 1: spectrin super family, Protein Profile. 2 (1995) 703–800.

[110] I. Bach, The LIM domain: regulation by association, Mech. Dev. 91 (2000) 5–17. https://doi.org/10.1016/S0925-4773(99)00314-7.

[111] E.F. Schmidt, S.-O. Shim, S.M. Strittmatter, Release of MICAL autoinhibition by semaphorin-plexin signaling promotes interaction with collapsin response mediator protein, J. Neurosci. Off. J. Soc. Neurosci. 28 (2008) 2287–2297. https://doi.org/10.1523/JNEUROSCI.5646-07.2008.

[112] J.R. Terman, T. Mao, R.J. Pasterkamp, H.-H. Yu, A.L. Kolodkin, MICALs, a Family of conserved flavoprotein oxidoreductases, function in plexin-mediated axonal repulsion, Cell. 109 (2002) 887–900. https://doi.org/10.1016/S0092-8674(02)00794-8.

[113] M. Gimona, R. Mital, The single CH domain of calponin is neither sufficient nor necessary for F-actin binding, J. Cell Sci. 111 (Pt 13) (1998) 1813–1821.

[114] S.S. Alqassim, M. Urquiza, E. Borgnia, M. Nagib, L.M. Amzel, M.A. Bianchet, Modulation of MICAL monooxygenase activity by its calponin homology domain: structural and mechanistic insights, Sci. Rep. 6 (2016) 22176. https://doi.org/10.1038/srep22176.

[115] B. Manta, V.N. Gladyshev, Regulated methionine oxidation by monooxygenases, Free Radic. Biol. Med. 109 (2017) 141–155. https://doi.org/10.1016/j.freeradbiomed.2017.02.010.

[116] R.-J. Hung, C.W. Pak, J.R. Terman, Direct redox regulation of F-Actin assembly and disassembly by Mical, Science. 334 (2011) 1710–1713. https://doi.org/10.1126/science.1211956.

[117] M. Nadella, M.A. Bianchet, S.B. Gabelli, J. Barrila, L.M. Amzel, Structure and activity of the axon guidance protein MICAL, Proc. Natl. Acad. Sci. U. S. A. 102 (2005) 16830–16835. https://doi.org/10.1073/pnas.0504838102.

[118] C. Ortegón Salas, K. Schneider, C.H. Lillig, M. Gellert, Signal-regulated oxidation of proteins via MICAL, Biochem. Soc. Trans. 48 (2020) 613–620. https://doi.org/10.1042/BST20190866.

[119] G.M. Cooper, Microtubules, Cell Mol. Approach 2nd Ed. (2000). www.ncbi.nlm.nih.gov/books/NBK9932/.

[120] L.M. Landino, R. Hasan, A. McGaw, S. Cooley, A.W. Smith, K. Masselam, G. Kim, Peroxynitrite oxidation of tubulin sulfhydryls inhibits microtubule polymerization, Arch. Biochem. Biophys. 398 (2002) 213–220. https://doi.org/10.1006/abbi.2001.2729.

[121] L.M. Landino, J.S. Iwig, K.L. Kennett, K.L. Moynihan, Repair of peroxynitrite damage to tubulin by the thioredoxin reductase system, Free Radic. Biol. Med. 36 (2004) 497–506. https://doi.org/10.1016/j.freeradbiomed.2003.11.026.

[122] L.M. Landino, K.L. Moynihan, J.V. Todd, K.L. Kennett, Modulation of the redox state of tubulin by the glutathione/glutaredoxin reductase system, Biochem. Biophys. Res. Commun. 314 (2004) 555–560. https://doi.org/14733943.

[123] G.G. Mackenzie, G.A. Salvador, C. Romero, C.L. Keen, P.I. Oteiza, A deficit in zinc availability can cause alterations in tubulin thiol redox status in cultured neurons and in the developing fetal rat brain, Free Radic. Biol. Med. 51 (2011) 480–489. https://doi.org/10.1016/j.freeradbiomed.2011.04.028.

[124] J. Rahajeng, S.S.P. Giridharan, B. Cai, N. Naslavsky, S. Caplan, Important relationships between Rab and MICAL proteins in endocytic trafficking, World J. Biol. Chem. 1 (2010) 254–264. https://doi.org/10.4331/wjbc.v1.i8.254.

[125] S.S.P. Giridharan, B. Cai, N. Naslavsky, S. Caplan, Trafficking cascades mediated by Rab35 and its membrane hub effector, MICAL-L1, Commun.

Integr. Biol. 5 (2012) 384–387. https://doi. org/10.4161/cib.20064.

[126] M. Fukuda, E. Kanno, K. Ishibashi, T. Itoh, Large scale screening for novel rab effectors reveals unexpected broad Rab binding specificity, Mol. Cell. Proteomics MCP. 7 (2008) 1031–1042. https://doi.org/10.1074/mcp.M700569-MCP200.

[127] Y. Gu, Y. Ihara, Evidence that collapsin response mediator Protein-2 is involved in the dynamics of microtubules, J. Biol. Chem. 275 (2000) 17917–17920. https://doi.org/10.1074/jbc. C000179200.

[128] Y. Fukata, T.J. Itoh, T. Kimura, C. Ménager, T. Nishimura, T. Shiromizu, H. Watanabe, N. Inagaki, A. Iwamatsu, H. Hotani, K. Kaibuchi, CRMP-2 binds to tubulin heterodimers to promote microtubule assembly, Nat. Cell Biol. 4 (2002) 583–591. https://doi.org/10.1038/ncb825.

[129] C. Hudemann, M.E. Lönn, J.R. Godoy, F. Zahedi Avval, F. Capani, A. Holmgren, C.H. Lillig, Identification, expression pattern, and characterization of mouse glutaredoxin 2 isoforms, Antioxid. Redox Signal. 11 (2009) 1–14. https://doi.org/10.1089/ars.2008.2068.

[130] L. Bräutigam, L.D.E. Jensen, G. Poschmann, S. Nyström, S. Bannenberg, K. Dreij, K. Lepka, T. Prozorovski, S.J. Montano, O. Aktas, P. Uhlén, K. Stühler, Y. Cao, A. Holmgren, C. Berndt, Glutaredoxin regulates vascular development by reversible glutathionylation of sirtuin 1, Proc. Natl. Acad. Sci. U. S. A. 110 (2013) 20057–20062. https://doi.org/10.1073/pnas.1313753110.

[131] V. Byles, L. Zhu, J.D. Lovaas, L.K. Chmilewski, J. Wang, D.V. Faller, Y. Dai, SIRT1 induces EMT by cooperating with EMT transcription factors and enhances prostate cancer cell migration and metastasis, Oncogene. 31 (2012) 4619–4629. https://doi.org/10.1038/onc.2011.612.

[132] R. Kunimoto, K. Jimbow, A. Tanimura, M. Sato, K. Horimoto, T. Hayashi, S. Hisahara, T. Sugino, T. Hirobe, T. Yamashita, Y. Horio, SIRT1 regulates lamellipodium extension and migration of melanoma cells, J. Invest. Dermatol. 134 (2014) 1693–1700. https://doi.org/10.1038/jid.2014.50.

[133] M.A.E. Almgren, K.C.E. Henriksson, J. Fujimoto, C.L. Chang, Nucleoside diphosphate kinase A/ nm23-H1 promotes metastasis of NB69-derived human neuroblastoma, Mol. Cancer Res. MCR. 2 (2004) 387–394.

[134] L. Han, X.-H. Liang, L.-X. Chen, S.-M. Bao, Z.-Q. Yan, SIRT1 is highly expressed in brain metastasis tissues of non-small cell lung cancer

(NSCLC) and in positive regulation of NSCLC cell migration, Int. J. Clin. Exp. Pathol. 6 (2013) 2357–2365.

[135] E. Lee, J. Jeong, S.E. Kim, E.J. Song, S.W. Kang, K.-J. Lee, Multiple functions of Nm23-H1 are regulated by oxido-reduction system, PLoS One. 4 (2009) e7949. https://doi.org/10.1371/journal. pone.0007949.

[136] J. Li, E. Wang, F. Rinaldo, K. Datta, Upregulation of VEGF-C by androgen depletion: the involvement of IGF-IR-FOXO pathway, Oncogene. 24 (2005) 5510–5520. https://doi.org/10.1038/ sj.onc.1208693.

[137] J.W. Fawcett, R.A. Asher, The glial scar and central nervous system repair, Brain Res. Bull. 49 (1999) 377–391.

[138] G. Yiu, Z. He, Glial inhibition of CNS axon regeneration, Nat. Rev. Neurosci. 7 (2006) 617–627. https://doi.org/10.1038/nrn1956.

[139] J.W. Prineas, F. Connell, Remyelination in multiple sclerosis, Ann. Neurol. 5 (1979) 22–31. https://doi.org/10.1002/ana.410050105.

[140] B. Ferguson, M.K. Matyszak, M.M. Esiri, V.H. Perry, Axonal damage in acute multiple sclerosis lesions, Brain J. Neurol. 120 (Pt 3) (1997) 393–399.

[141] A. Boyd, H. Zhang, A. Williams, Insufficient OPC migration into demyelinated lesions is a cause of poor remyelination in MS and mouse models, Acta Neuropathol. (Berl.). 125 (2013) 841–859. https://doi.org/10.1007/s00401-013-1112-y.

[142] T. Koda, T. Okuno, K. Takata, J.A. Honorat, M. Kinoshita, S. Tada, M. Moriya, S. Sakoda, H. Mochizuki, A. Kumanogoh, Y. Nakatsuji, Sema4A inhibits the therapeutic effect of IFN-β in EAE, J. Neuroimmunol. 268 (2014) 43–49. https://doi.org/10.1016/j.jneuroim.2013.12.014.

[143] Y.A. Syed, E. Hand, W. Möbius, C. Zhao, M. Hofer, K.A. Nave, M.R. Kotter, Inhibition of CNS remyelination by the presence of semaphorin 3A, J. Neurosci. Off. J. Soc. Neurosci. 31 (2011) 3719–3728. https://doi.org/10.1523/ JNEUROSCI.4930-10.2011.

[144] G. Piaton, M.-S. Aigrot, A. Williams, S. Moyon, V. Tepavcevic, I. Moutkine, J. Gras, K.S. Matho, A. Schmitt, H. Soellner, A.B. Huber, P. Ravassard, C. Lubetzki, Class 3 semaphorins influence oligodendrocyte precursor recruitment and remyelination in adult central nervous system, Brain J. Neurol. 134 (2011) 1156–1167. https://doi.org/10.1093/brain/awr022.

[145] T. Prozorovski, J. Ingwersen, D. Lukas, P. Göttle, B. Koop, J. Graf, R. Schneider, K. Franke, S.

Schumacher, S. Britsch, H.-P. Hartung, P. Küry, C. Berndt, O. Aktas, Regulation of sirtuin expression in autoimmune neuroinflammation: induction of SIRT1 in oligodendrocyte progenitor cells, Neurosci. Lett. 704 (2019) 116–125. https://doi.org/10.1016/j.neulet.2019.04.007.

[146] O.A. Bizzozero, J. Zheng, Identification of major S-nitrosylated proteins in murine experimental autoimmune encephalomyelitis, J. Neurosci. Res. 87 (2009) 2881–2889. https://doi.org/10.1002/jnr.22113.

[147] S.M. Smerjac, O.A. Bizzozero, Cytoskeletal protein carbonylation and degradation in experimental autoimmune encephalomyelitis, J. Neurochem. 105 (2008) 763–772. https://doi.org/10.1111/j.1471-4159.2007.05178.x.

[148] I. de la Pena, M. Pabon, S. Acosta, P.R. Sanberg, N. Tajiri, Y. Kaneko, C.V. Borlongan, Oligodendrocytes engineered with migratory proteins as effective graft source for cell transplantation in multiple sclerosis, Cell Med. 6 (2014) 123–127. https://doi.org/10.3727/215517913X674144.

[149] A. Arvidsson, T. Collin, D. Kirik, Z. Kokaia, O. Lindvall, Neuronal replacement from endogenous precursors in the adult brain after stroke, Nat. Med. 8 (2002) 963–970. https://doi.org/10.1038/nm747.

[150] K. Jin, M. Minami, J.Q. Lan, X.O. Mao, S. Batteur, R.P. Simon, D.A. Greenberg, Neurogenesis in dentate subgranular zone and rostral subventricular zone after focal cerebral ischemia in the rat, Proc. Natl. Acad. Sci. U. S. A. 98 (2001) 4710–4715. https://doi.org/10.1073/pnas.081011098.

[151] M. Sparaco, L.M. Gaeta, G. Tozzi, E. Bertini, A. Pastore, A. Simonati, F.M. Santorelli, F. Piemonte, Protein glutathionylation in human central nervous system: potential role in redox regulation of neuronal defense against free radicals, J. Neurosci. Res. 83 (2006) 256–263. https://doi.org/10.1002/jnr.20729.

[152] M.Y. Aksenov, M.V. Aksenova, D.A. Butterfield, J.W. Geddes, W.R. Markesbery, Protein oxidation in the brain in Alzheimer's disease, Neuroscience. 103 (2001) 373–383.

[153] L.M. Landino, S.H. Robinson, T.E. Skreslet, D.M. Cabral, Redox modulation of tau and microtubule-associated protein-2 by the glutathione/glutaredoxin reductase system, Biochem. Biophys. Res. Commun. 323 (2004) 112–117. https://doi.org/15351709.

[154] R.P. Kommaddi, D.S. Tomar, S. Karunakaran, D. Bapat, S. Nanguneri, A. Ray, B.L. Schneider, D. Nair, V. Ravindranath, Glutaredoxin1 diminishes amyloid beta-mediated oxidation of F-Actin and reverses cognitive deficits in an Alzheimer's disease mouse model, Antioxid. Redox Signal. 31 (2019) 1321–1338. https://doi.org/10.1089/ars.2019.7754.

[155] S. Parakh, S. Shadfar, E.R. Perri, A.M.G. Ragagnin, C.V. Piattoni, M.B. Fogolín, K.C. Yuan, H. Shahheydari, E.K. Don, C.J. Thomas, Y. Hong, M.A. Comini, A.S. Laird, D.M. Spencer, J.D. Atkin, The redox activity of protein disulfide isomerase inhibits ALS phenotypes in cellular and zebrafish models, IScience. 23 (2020) 101097. https://doi.org/10.1016/j.isci.2020.101097.

[156] H.S. Kim, S.L. Ullevig, H.N. Nguyen, D. Vanegas, R. Asmis, Redox regulation of 14-3-3ζ controls monocyte migration, Arterioscler. Thromb. Vasc. Biol. 34 (2014) 1514–1521. https://doi.org/10.1161/ATVBAHA.114.303746.

[157] M.J. Reed, A.C. Corsa, S.A. Kudravi, R.S. McCormick, W.T. Arthur, A deficit in collagenase activity contributes to impaired migration of aged microvascular endothelial cells, J. Cell. Biochem. 77 (2000) 116–126.

[158] J.E. Mogford, N. Tawil, A. Chen, D. Gies, Y. Xia, T.A. Mustoe, Effect of age and hypoxia on TGFbeta1 receptor expression and signal transduction in human dermal fibroblasts: impact on cell migration, J. Cell. Physiol. 190 (2002) 259–265. https://doi.org/10.1002/jcp.10060.

[159] S. Cané, S. Ponnappan, U. Ponnappan, Altered regulation of CXCR4 expression during aging contributes to increased CXCL12-dependent chemotactic migration of CD4(+) T cells, Aging Cell. 11 (2012) 651–658. https://doi.org/10.1111/j.1474-9726.2012.00830.x.

[160] S.T. Nikkari, J. Koistinaho, O. Jaakkola, Changes in the composition of cytoskeletal and cytocontractile proteins of rat aortic smooth muscle cells during aging, Differ. Res. Biol. Divers. 44 (1990) 216–221.

[161] M.J. Reed, N.S. Ferara, R.B. Vernon, Impaired migration, integrin function, and actin cytoskeletal organization in dermal fibroblasts from a subset of aged human donors, Mech. Ageing Dev. 122 (2001) 1203–1220.

[162] T. Fiaschi, G. Cozzi, G. Raugei, L. Formigli, G. Ramponi, P. Chiarugi, Redox regulation of beta-actin during integrin-mediated cell adhesion, J.

Biol. Chem. 281 (2006) 22983–22991. https://doi.org/10.1074/jbc.M603040200.

[163] R. Blakytny, L.J. Erkell, G. Brunner, Inactivation of active and latent transforming growth factor beta by free thiols: potential redox regulation of biological action, Int. J. Biochem. Cell Biol. 38 (2006) 1363–1373. https://doi.org/10.1016/j.biocel.2006.01.017.

[164] M.H. Barcellos-Hoff, T.A. Dix, Redox-mediated activation of latent transforming growth factor-beta 1, Mol. Endocrinol. Baltim. Md. 10 (1996) 1077–1083. https://doi.org/10.1210/mend.10.9.8885242.

[165] A.G. Mustafa, J.A. Wang, K.M. Carrico, E.D. Hall, Pharmacological inhibition of lipid peroxidation attenuates calpain-mediated cytoskeletal degradation after traumatic brain injury, J. Neurochem. 117 (2011) 579–588. https://doi.org/10.1111/j.1471-4159.2011.07228.x.

[166] C.H. Switzer, W. Flores-Santana, D. Mancardi, S. Donzelli, D. Basudhar, L.A. Ridnour, K.M. Miranda, J.M. Fukuto, N. Paolocci, D.A. Wink, The emergence of nitroxyl (HNO) as a pharmacological agent, Biochim. Biophys. Acta. 1787 (2009) 835–840. https://doi.org/10.1016/j.bbabio.2009.04.015.

[167] C.C. Rohena, S.L. Mooberry, Recent progress with microtubule stabilizers: new compounds, binding modes and cellular activities, Nat. Prod. Rep. 31 (2014) 335–355. https://doi.org/10.1039/c3np70092e.

[168] J.J. Field, A. Kanakkanthara, J.H. Miller, Microtubule-targeting agents are clinically successful due to both mitotic and interphase impairment of microtubule function, Bioorg. Med. Chem. 22 (2014) 5050–5059. https://doi.org/10.1016/j.bmc.2014.02.035.

[169] A. Nürnberg, A. Kollmannsperger, R. Grosse, Pharmacological inhibition of actin assembly to target tumor cell motility, Rev. Physiol. Biochem. Pharmacol. 166 (2014) 23–42. https://doi.org/10.1007/112_2013_16.

[170] S.M. Wilson, R. Khanna, Specific binding of lacosamide to Collapsin Response Mediator Protein 2 (CRMP2) and direct impairment of its canonical function: implications for the therapeutic potential of lacosamide, Mol. Neurobiol. (2014). https://doi.org/10.1007/s12035-014-8775-9.

[171] K. Hensley, K. Venkova, A. Christov, W. Gunning, J. Park, Collapsin response mediator protein-2: an emerging pathologic feature and therapeutic target for neurodisease indications, Mol. Neurobiol. 43 (2011) 180–191. https://doi.org/10.1007/s12035-011-8166-4.

[172] K. Togashi, M. Hasegawa, J. Nagai, K. Kotaka, A. Yazawa, M. Takahashi, D. Masukawa, Y. Goshima, K. Hensley, T. Ohshima, Lanthionine ketimine ester improves outcome in an MPTP-induced mouse model of Parkinson's disease via suppressions of CRMP2 phosphorylation and microglial activation, J. Neurol. Sci. 413 (2020) 116802. https://doi.org/10.1016/j.jns.2020.116802.

[173] D. Möller, M. Gellert, W. Langel, C.H. Lillig, Molecular dynamics simulations and in vitro analysis of the CRMP2 thiol switch, Mol. Biosyst. 13 (2017) 1744–1753. https://doi.org/10.1039/c7mb00160f.

[174] A. Moutal, S. Kalinin, K. Kowal, N. Marangoni, J. Dupree, S.X. Lin, K. Lis, L. Lisi, K. Hensley, R. Khanna, D.L. Feinstein, Neuronal conditional knockout of collapsin response mediator protein 2 ameliorates disease severity in a mouse model of multiple sclerosis, ASN Neuro. 11 (2019) 1759091419892090. https://doi.org/10.1177/1759091419892090.

[175] M. Gellert, E.-M. Hanschmann, K. Lepka, C. Berndt, C.H. Lillig, Redox regulation of cytoskeletal dynamics during differentiation and de-differentiation, Biochim. Biophys. Acta. 1850 (2015) 1575–1587. https://doi.org/10.1016/j.bbagen.2014.10.030.

[176] M. Gellert, M.F. Hossain, F.J.F. Berens, L.W. Bruhn, C. Urbainsky, V. Liebscher, C.H. Lillig, Substrate specificity of thioredoxins and glutaredoxins—towards a functional classification, Heliyon. 5 (2019) e02943. https://doi.org/10.1016/j.heliyon.2019.e02943.

CHAPTER FOURTEEN

At the Interface between Metabolism and Redox Regulation

SIRTUIN 1 AS CHECKPOINT FOR NEUROGENESIS AND LONGEVITY

Tim Prozorovski, Christian Kroll, Carsten Berndt, and Orhan Aktas

CONTENTS

14.1 INTRODUCTION

Chromatin undergoes dynamic changes in structure and flexibility to permit regulation of transcription and DNA replication during the transition from the stem cell stage towards lineage-specific progenitors and terminal differentiation within the target tissue. Sensing the intracellular and environmental metabolic, energy and redox states during replication and transition is necessary for proper generation of valuable and functional cells and their integration in pre-existing circuits. Growing evidence suggests that deployment of chromatin-remodelling enzymes, such as Sirtuin 1 (SIRT1), is coordinated by metabolic reprogramming and emphasizes an essential role of specific metabolites, multiple redox pairs, and reactive oxygen species (ROS)-induced signalling pathways in nucleosome re-arrangement and stem cell transcription. Interference with these pathways disturbs developmental processes and disrupts maintenance of the stem cell reservoir throughout life, while a proper adaptation to metabolic heterogeneity is required for stem cell resistance to a wide range of stress conditions, the way by which SIRT1 may contribute to its positive effect on tissue homeostasis and longevity. A better understanding of these processes should offer novel therapeutic targets for regulation of pathological processes, succeeding tissue repair and extension in life span.

SIRT1 (the mammalian orthologue for yeast Sir2) is a redox-sensitive NAD^+-dependent deacetylase that plays a key role in the establishment and maintenance of repressive chromatin (Figure 14.1). Particularly, SIRT1 has a role in chromatin organization and epigenetic control of bivalent gene

DOI: 10.4324/9781003204091-18

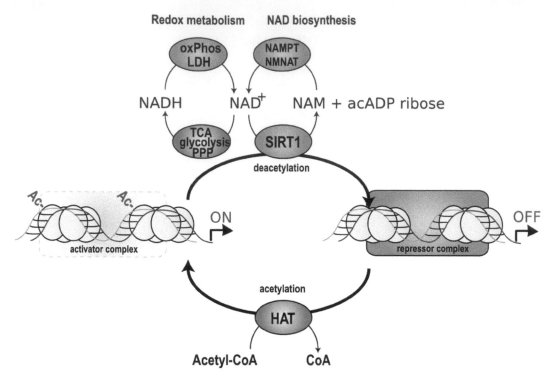

Figure 14.1 Overview of NAD⁺-dependent regulation of SIRT1 deacetylating activity.

SIRT1-mediated deacetylation of histone and non-histone substrates at gene promoters is dependent on NAD⁺ availability. NAD⁺ biosynthesis is exerted by NAMPTs (nicotinamide phosphoribosyl transferases) and NMNATs (nicotinamide mono-nucleotide adenylyltransferases), while NAD⁺/NADH ratio is regulated by the activities of PPP (pentose phosphate pathway), oxPhos (oxidative phosphorylation), glycolysis, tricarboxylic acid cycle (TCA) and lactate dehydrogenase (LDH). Using NAD⁺ as a co-substrate, SIRT1 catalyses the transfer of the protein-bound acetyl group onto the ADP-ribose moiety, thereby forming O-acetyl-ADP ribose (acADP ribose) and nicotinamide (NAM). Histone de-acetylation leads to formation of heterochromatin via an attraction of transcriptional repressor complexes. Acetylation of histone tails by histone acetyltransferases (HAT) favours active chromatin formation and an assembly of transcriptional enhancer/activator complexes.

expression [1] of the key developmental transcriptional factors and promotes effective cell differentiation and lineage commitment during exit from pluripotency [2–5].

SIRT1 mediates transcriptional repression by deacetylation of the lysines at key histones, such as histone H4 lysine K16 (H4K16) and H3K9 [6], and a number of non-histone proteins [7], including transcriptional factors, histone methyl- and demethyltransferases [8–11], arginine methyltransferases [12], and DNA methyltransferases [13, 14]. Furthermore, SIRT1 may influence acetylation state of histones by regulating the activity of histone acetyltransferases (HAT) [15, 16].

In parallel, due to its ability to sense NAD⁺ concentrations, SIRT1 constitutes an obvious link between metabolism and chromatin modifications thus transmitting the transcriptional response

to physiological and pathological alterations in the cell [17]. Here, we will focus on SIRT1 functions during pluripotency, self-renewal, and differentiation of stem cells, as well as longevity and regulation of SIRT1 activity via redox-dependent posttranslational modifications.

14.2 SIRT1 IS CRITICALLY REQUIRED DURING EARLY STEPS OF NEURAL DEVELOPMENT

Using in vitro models for recapitulation of neural development from embryonic stem cells (ESCs) [18], several independent studies depict the essential role of SIRT1 in the maintenance of pluripotency and ESC differentiation towards neuroectodermal lineage [19, 14, 3]. These data helped to unravel the mechanism of pro-neuronal action of mammalian SIRT1 and are consistent with neurodevelopmental abnormalities, such as neural tube closure defects (exencephaly) and disturbed neuroretinal morphogenesis, described in three independently generated mutant mouse lineages carrying null SIRT1 alleles [20–22].

SIRT1 contributes to the maintenance of ESC pluripotency, self-renewal, and survival through regulation of the expression and/or deacetylation of key components of the core pluripotency network, such as Nanog, Oct4, Sox2 [23, 24], and the inhibition of the p53 signalling pathway [25, 26].

Epigenetically, SIRT1 mediates transcriptional repression as a component of a transformation-specific PcG complex, PRC4 [27, 3], which is required to retain promoters in a "poised" state permissive for subsequent activation [28]. In ESCs, PRC4 silences the expression of a range of inducible developmental genes including DLL4, TBX3, PAX6, and NeuroD1, which are important for the neuroectodermal lineage and morphogenesis of the nervous tissue [19]. In addition to histone deacetylation, SIRT1 facilitates the formation of a repressive H3K27me3 histone mark, which is catalysed by another component of the complex, methyltransferase EZH2 [29]. Bivalent epigenetic state of genes is supported not only at the level of chromatin modifications, but also requires the maintenance of cytosine guanine repeats (CpG) to reach DNA promoter regions in unmethylated state. To this, it was demonstrated that SIRT1 inhibits DNMT3l transcriptional expression leading to sustained DNA hypomethylated profile of imprinted and germline developmental genes [14]. Genetic ablation of SIRT1 is associated with an abnormal DNA methylation and inability to induce differentiation factors important for pluripotency exit in vitro and in vivo. Further, SIRT1 plays a role in lineage commitment via regulation of retinoic acid pathway, an important mechanism promoting neuroectodermal differentiation by suppressing mesodermal fate [30, 31].

The SIRT1 expression in mouse and human ESCs is controlled at the level of post-transcriptional regulation and by enzymes of the ubiquitin-proteasome system (UPS). The redox-sensitive ubiquitin-specific peptidase 22 (USP22) has been identified as a bona fide deubiquitinase that promotes stabilization of SIRT1 in ESCs and protects from early embryonic lethality [32]. USP22 is a component of the chromatin-modifying transcription co-activator complex SAGA [33] that reverses the polycomb-catalysed ubiquitination of histones and non-histone proteins [34]. Upon or following differentiation, the USP22-SIRT1 complex is re-located to inhibit transcription of pluripotency gene Sox2 [35].

Generally, mechanisms that direct re-location of repressor complexes upon differentiation are largely unknown but findings from the molecular biology of diseases offer certain insights. PCR4 re-positioning has been recently described in cancer cells exposed to genotoxic concentrations of ROS. Treatment with hydrogen peroxide shifted PRC4-SIRT1 away from CG-poor transcriptionally inactive regions to DNA damage areas encompassing high levels of CGs and leading to increased DNA methylation and transcriptional repression [36]. Binding of SIRT1 to chromatin under oxidative conditions was mediated by DNMT1 [37]. Similarly, re-location of PRC4 occurs in vivo under inflammatory conditions in the mouse model of colitis [36]. SIRT1 re-location along with massive epigenetic changes and alteration in gene transcription has been shown in other settings associated with oxidative stress [38–40] including ROS-dependent epigenetic reprogramming during malignant transformation [41]. Whether recruitment of silencing repressor complexes to pluripotent genes upon differentiation occurs in response to oxidative-mediated CpG damage is an intriguing question in stem cell biology [42, 43].

Neuronal lineage commitment and differentiation of ESCs is paralleled by a rapid decline in SIRT1 levels [19, 47] and diminishing of the PcG proteins at neuronal lineage genes [44–46]. Induction of pro-neuronal genes in ESCs is tightly coupled with E3 ubiquitin ligase SMURF2 that targets EZH2 for proteasomal degradation [48]. SIRT1 is also known as a substrate of SMURF2 during neuronal differentiation [49], suggesting the major role of the UPS in disassembly of repressive transcriptional complexes. SMURF2 is a redox-sensitive enzyme, and its developmental role has been shown during establishment of neural polarity [50], dorsal ventral patterning, and neural fold formation [50, 51].

CHFR, another E3 ubiquitin ligase, has been demonstrated to target SIRT1 for degradation in a redox-dependent manner in response to JNK activation [52]. JNK signalling is important in chromatin reprogramming and differentiation of neural stem and progenitor cells (hereinafter NSCs) [53] and thereby might be potentially involved in downregulation of SIRT1 levels. CHFR targets other chromatin modifiers such as HDAC1 [54] and PARP-1 [55] for nuclear export and UPS-dependent degradation. Similar to ESC, targeted proteolysis via the UPS has a role in the strict control of temporal and spatial gene expression during embryonic and adult neurogenesis [56, 57].

Decline in SIRT1 levels upon ESC differentiation is also regulated at post-transcriptional level through microRNA [58–61] and HuR-mediated RNA stability [62]. Thus, upon differentiation of ESCs, the SIRT1 transcripts are downregulated due to diminished interaction with ubiquitous RNA-binding protein HuR [19]. This interaction is affected by PRMT4 (protein arginine N-methyltransferase 4)-dependent methylation of HuR proteins.

14.3 SIRT1 FUNCTIONS IN EMBRYONIC NEUROGENESIS

During all stages of the CNS development SIRT1 is highly expressed in NSCs, such as neuroepithelial, radial glia, and radial glia-like adult NSCs. Of note, in embryonic NSCs (eNSCs) SIRT1 has an instructive role on lineage specification rather than decisive role in the self-renewal, multipotency, or cell differentiation. SIRT1 depletion is associated with late-developmental defects, such as neuronal DNA damage [63], neuronal survival [64, 65], axonal survival [66], neurite outgrowth and branching [67, 68], axonogenesis [69], and brain senescence [70], while its deficiency in NSCs per se does not preclude differentiation into neuronal or glial lineages.

Mice with conditional deletion of SIRT1 catalytic domain (exon 4), which is mediated by Cre recombinase expressed under nestin promoter, are characterized by elevated generation of neurons [71–73], indicating that enzymatic activity of SIRT1 may negatively influence neurogenesis.

Similarly, SIRT1 inhibition enhances the differentiation of neural stem cells derived from inducible pluripotent stem cells (iPS) [74]. In line with these findings, the overexpression of SIRT1 decreases neurogenesis in the neural tube of chick embryos [75].

Several studies depict the role of SIRT1 in inhibition of neuronal fate and favouring glial lineage under oxidative conditions. ROS critically contribute to multipotency, self-renewal, and differentiation in the nervous system [76, 77], and even mild alterations towards more oxidative state direct eNSCs differentiation to astrocytic lineage in SIRT1-dependent manner. SIRT1 silencing by in utero electroporation abolishes oxidation-dependent shift of NSCs towards astrocytes in early postnatal pups [78]. NSCs deficient for the mitochondrial 8-oxoguanine DNA glycosylase (OGG1) accumulate mitochondrial DNA damage, ROS and differentiate predominantly into astrocytes in SIRT1-dependent manner [79–81].

Helix-loop-helix (HLH) proteins orchestrate developmental-specific gene programs specifying the neural identity of progenitors [82–84]. Basic HLH (bHLH) transcription factors neurogenin 1 and 2 (Ngn1, Ngn2) initiate a cascade of bHLH gene activation events that promote differentiation and NSC migration and eventually lead to the expression of terminal neuronal differentiation genes [85, 84]. SIRT1 interacts with bHLH proteins through its N-terminal sirtuin homology domain [86, 87]. bHLHs recruit SIRT1 to a promoter region resulting in deacetylation of H3 and/or H4 histones and transcriptional repression [88].

In mouse eNSCs, SIRT1 forms repressor transcriptional complex with bHLH Hes1, the one of the known markers or determinants of NSC identity [90], at the promoter of the pro-neuronal factor Mash1 (Ascl1 in humans) [78]. Differentiation stimuli promote dissociation of repressor complex and recruitment of activator complex via action of another NAD+-dependent enzyme, PARP1, resulting in Mash1 transcription [89]. This mechanism operates in redox-dependent manner, where mild oxidative conditions abrogate dissociation of SIRT1 from Hes1, limiting Mash1 expression and favouring astrocytic differentiation. Such non-toxic oxidative conditions do not induce global changes in histone acetylation pattern of H3K9 or H4K16, and thus aberrant SIRT1 activity is likely restricted to dysregulation of specific genes. This research illustrates the necessity of a switch in NAD+-consuming enzymes playing opposite roles in gene transcription and the requirement of a proper redox/metabolic state underlying the neuronal fate of NSCs.

Interestingly, while SIRT1 helps to maintain the pluripotent state of ESCs in part through epigenetic inhibition of paired-box 3 Pax3 [19], in committed NSCs SIRT1 suppresses Pax3 via deacetylation, leading to induction of Hes1 [75]. Furthermore, SIRT1-mediated interaction of Pax3 with Ngn2 impairs its activity to initiate differentiation [75].

SIRT1-mediated chromatin remodelling is also required to "switch off" a stemness program of NSCs. SIRT1 forms a transcriptional repressor complex with nuclear receptor co-repressor N-CoR and bHLH Bcl6 to epigenetically repress various stemness factors, including Hes1 and Hes5 [91, 92].

In contrast to mice lacking SIRT1 enzymatic domain, blockade of SIRT1 activity with specific inhibitors (e.g., Ex527, sirtinol, splitomicin) impairs cortical [93] and cerebellar granular neuron differentiation [94].

How is the SIRT1 action directed from an inhibition of the developmental program to suppression of stemness coordinated? Recent studies have emphasized the ability of SIRT1 to sense cellular changes in redox and energetic states during stem cell differentiation. Obviously, SIRT1 directs the differentiation program of mammalian NSCs downstream from postmitotic events. In daughter cells, commitment to differentiate is associated with mitochondria fission, whereas the original self-renewal cell undergoes mitochondria fusion following mitosis [95]. Biogenesis of mitochondria is ultimately linked to the differentiation process that allows metabolic reprogramming of NSCs from aerobic glycolysis (high lactate dehydrogenase activity; hexokinase activity; high glucose consumption) [96, 97] towards high-energy production via oxidative phosphorylation (oxPhos) [98, 99]. In line, increased and compartmentalized mitochondrial ROS production promotes neuronal fate commitment [100]. Treatment with the proton ionophore CCCP that overactivates the mitochondrial oxidative complexes and enhances NAD$^+$/NADH ratios has been demonstrated to promote embryonic cortical neurogenesis in a SIRT1-dependent manner [95].

In parallel to metabolic reprogramming, massive morphological alterations during neurogenesis, such as recently described structural changes in shape and polarity of the progenitor cell, comprise remodelling of organelles and chromatin compaction, thus processes driven by SIRT1 as well [101, 102].

Oxygen availability is an important factor for the regulation of NSC fate. Thus, gradient oxygen tension during neural tube development differentially modulates the activities of the CtBP, an oxygen-sensitive transcriptional repressor of SIRT1 [103]. The modulation of CtBP activities may illustrate how the pattern and anatomical architecture of metabolic microenvironmental signals direct developmental processes. High oxygen concentrations favour the formation of the CtBP/

Hes1 repressor complex at the promoter of pro-neuronal factor Math1 and thereby the inhibition of neuronal fate [104]. For comparison, in other parts of the dorsal neural tube a more hypoxic environment permits neurogenesis [104]. In these hypoxic niches, CtBP promotes cell-cycle exit and neuronal differentiation through repression of transcriptional activities of Wnt-target genes by forming a repressor complex with β-catenin/TCF [105]. Lowering NAD$^+$/NADH ratio in cultured NSCs by decreasing oxygen tension enhances CtBP-mediated transcriptional repression of Hes1 [104]. In vitro, hypoxic conditions affect the binding of the CtBP/HIC1 complex to the Sirt1 promoter and leads to de-repression of SIRT1 transcription [103]. Given that SIRT1 deficiency is associated with generation of neural tube defects, the instructive role of microenvironment (such as gradient oxygen tension) directing SIRT1 activity and expression can be proposed.

14.4 SIRT1 IN ADULT NEUROGENESIS

In the adult brain, neurogenesis is restricted to specialized regions commonly referred to as neural stem cell niches. Adult stem cells are largely existing in dormant quiescent state, which is actively controlled by multiple extrinsic cues [106–108]. Neurogenic niches play a supportive role by assisting maintenance of the delicate balance between self-renewal and commitment to differentiation into neuronal, astrocytic, and oligodendrocytic lineages [109, 110]. NSC niches share a common architecture that allows close contact between NSCs and the extensive blood vessel network [111]. Nevertheless, they possess heterogeneity that defines specific NSC functions [112]. For example, NSCs derived from embryonic cortices mostly rely on glycolysis that permits generation of glial cells, which is consistent with more glycolytic nature of differentiated astrocytes and oligodendrocytes [113]. Notably, NSCs transplanted into other areas of the brain predominantly differentiate into glial cells indicating the high degree of neural plasticity and the pro-neuronal role of microenvironment in adult stem niches. Neurogenic niches have been also identified in the adult human [114] and rodent hypothalamus [115] and olfactory bulb [116]. In addition to local factors produced by the microenvironment, growing evidence suggests that neurogenesis is influenced by various factors and nutritional cues derived via blood capillary and cerebral spinal fluid.

The continuous generation and integration of new neurons into existing circuitries is considered as a precisely regulated process under control of brain activity [117], circadian molecular clock [118], physical activity [119], and body metabolism [120]. Furthermore, neurogenesis is influenced by a subtype of hypothalamic neurons, the main area of metabolic regulation [117]. On the other hand, various pathological conditions, such as hyperglycemia and dyslipidemia affect neurogenesis by reducing the number of generated neurons and their function [121].

SIRT1 is increasingly recognized for its role in brain homeostasis and adult neurogenesis in response to energy state and nutrient availability. In response to stress conditions, SIRT1 limits proliferation and generation of new neurons that can be considered in the context of genomic stability maintenance. In adult subventricular zone (SVZ) and hippocampal subgranular zone (SGZ) niches, SIRT1 is expressed in the nucleus of NSCs. Lentivirus-mediated SIRT1 knockdown in the SGZ enhances the number of new-born neurons [72]. These data corroborate to results showing enhanced neurogenesis in the SVZ and SGZ of mice with conditional deletion of SIRT1's catalytic domain [71, 73]. SIRT1 deficiency in the adult brain is associated with increased proliferation of NSCs [71, 122].

Interestingly, despite elevated neuronal generation in the hippocampus, SIRT1-deficient mice exhibit behavioural defects associated with hippocampal functions like learning and memory. This is consistent with the notion that SIRT1 has a role in synaptic plasticity [123–125]. Recent studies indicate accumulated DNA damage in SIRT1-deficient neurons that may further contribute to neurological dysfunction. Whether DNA damage appears as a result of abolished SIRT1-mediated DNA repair [126, 22] during early steps of neurogenesis needs further investigation.

Consistent with an effect of SIRT1 deficiency on adult neurogenesis, lentiviral-based overexpression of SIRT1 decreases neuronal differentiation in vivo and in vitro [72]. Factors that promote activation of SIRT1, such as inhibition of glycolysis by 2-deoxy-D-glucose or treatment with resveratrol, block NSCs proliferation and neurogenesis in a SIRT1 dependent manner [71] Taken together, these data suggest that SIRT1 is a negative regulator of adult neurogenesis.

Reduced hippocampal NSC proliferation, or a transiently increased progenitor expansion followed by defective maturation into adult neurons, has been reported in rodent models of diabetes [127], whereas calorie restriction (CR) improves brain function and cognitive performance in mice [128, 121]. CR is accompanied in mice by lower blood glucose levels, SIRT1 induction in the brain, and preserved or even enhanced adult neurogenesis [129, 130].

Recent findings shed light on the epigenetic mechanisms by which SIRT regulates proliferation in response to nutrient availability. It was shown that energy balance and nutrients modulate Hes1 expression through coordinated action of CREB and SIRT1 proteins. High glucose levels directly modulate SIRT1 signalling leading to inactivation of transcriptional factor CREB and gene silencing of Hes1 and thereby inhibiting cell cycle progression of NSCs [122]. CR has an opposite effect promoting displacement of SIRT1 from Hes1 promoter and association with transcriptional activator CREB [122]. SIRT1 regulates the balance between proliferation and differentiation downstream from AMPK and serine/threonine kinase mTOR [131–134], major nutrient sensors [135]. In Xenopus, CR induces reversible cell cycle (G2) arrest and inhibits proliferation of NSCs through the mechanism involving GADD34 activation [136]. Re-feeding restores neurogenesis in mTOR-dependent manner, promoting G2 progression into mitosis. These studies illustrate an adaptive mechanism to control neurogenic processes in response to nutrient availability. In mammalian stem cells, GADD34 and mTOR control SIRT1 activity via phosphorylation of Ser47. Under oxidative stress conditions, de-phosphorylation of Ser47 through the GADD34/PP1α complex activates SIRT1 [137]. This mechanism is involved in cell cycle regulation and DNA damage repair and arrest in S and G2/M phases of the cell cycle [138]. Nutrient availability activates mTOR signalling that phosphorylates the Ser47 resulting in SIRT1 inhibition [132] and enhanced mitotic activity. The physiological role of SIRT1-mediated inhibition on replication is likely linked with preservation of genome integrity under stress conditions [22, 139] and maintenance of the stem cell pool, a way by which SIRT1 may contribute to mammalian longevity. Intriguingly, the lack of neurogenesis in adult brain was postulated as a result of stem cell pool depletion rather than inability of adult NSCs to generate new neurons [140].

14.5 SIRT1 IN LONGEVITY

SIRT1 is often labelled as "the" longevity gene as it has been associated [141] and disputed [142] with the epigenetic regulation of aging and longevity in a variety of organisms. At birth, genetic contribution has only a minor impact on longevity of an organism. Therefore, lifestyle and thus epigenetic regulations are considered to play an important role for the determination of the final life span [143]. Involved mechanisms on these epigenetic regulations are DNA-methylation, posttranslational histone modifications, and non-coding RNA interference [144, 145], all linked to SIRT1 activation [146, 147].

In yeast, worms, and flies, increased activity of the SIRT1 homologue Sir2 results in an extended life span, whereas inactivation of Sir2 shortens it [148]. Inactivation of SIRT1/Sir2 happens for example upon S-glutathionylation (see below). Expression of Sir2 mutants lacking cysteines targeted by S-glutathionylation increased longevity in yeast [149]. Knockout of SIRT1 in mice causes genomic instability, as well as developmental and metabolic defects, resulting in a reduction of life span [150, 151], whereas transgenic mice with increased SIRT1 expression display beneficial phenotypes on health and longevity [152].

SIRT1 is a major signalling module of the FOXO transcription factor family [153]. The FOXO family is known as a key factor on human longevity [154, 155] and associated with multiple cellular pathways, which regulate growth, stress resistance, metabolism, cellular differentiation, and apoptosis in mammals [156]. FOXO proteins have a strong influence on the insulin and IGF-signalling pathway, which also affects longevity [157].

A well-described method to prolong life span in mammals, as well as in nematode and yeast, is CR [158]. CR has an upregulating effect on SIRT1 levels, through the interdependent regulation of ChREBP, which promotes SIRT1 transcription [159]. Studies in mice suggest a meditative role of SIRT1 on the beneficial effects of CR, as SIRT1 knockout mice do not show the beneficial metabolic responses of low nutritional intake [160].

14.6 MODIFICATION OF SIRT1 ACTIVITY

The energy carrier molecule NAD^+ is a principal cofactor of SIRT1 activity [161, 133, 162], by which SIRT1 senses the redox/energy state of the cell.

The subnuclear dynamics of SIRT1 activity have been recently demonstrated [163], suggesting that SIRT1 molecules are organized in nuclear niches and their activity correlates with the local pool of free NAD^+. Although the biosynthesis of NAD^+ occurred in all compartments [164], compartmentalization and different levels of the redox couple NAD+/NADH as well as a competition among NAD-consuming enzymes [165] emerged as a powerful regulatory mechanism of SIRT1 activity. Restoration of SIRT1 activity is also dependent on the recycling of nicotinamide, a product of NAD^+ catalysis and its natural inhibitor. Nicotinamide is used in NAD^+ biosynthesis by enzymes of salvage pathway [166, 167]. Deletion of the key enzymes of NAD pathway such as NAMPT has been shown to abrogate self-renewal and differentiation of NSCs and affect oligodendrocyte lineage progression via SIRT1/SIRT2-dependent mechanism [168]. NAD^+ has emerged as a crucial factor for pluripotency, and its availability underlies the reprogramming efficiency of somatic cells [169].

Depletion of NAD^+ availability in satellite cells (i.e., muscle stem cells) upon switching from oxPhos towards glycolysis is associated with SIRT1 inactivation, transcriptional de-repression of muscle developmental genes, and differentiation into muscle cells [170].

SIRT1-mediated NAD^+ catalysis generates ADP-ribosyl moiety, which is used for deacetylation or ADP-ribosylation of proteins and may serve as a precursor of cyclic ADP-ribose mediating mobilization of intracellular calcium. Transfer of an ADP-ribosyl group to an acceptor protein (reversible mono- and poly-ADP-ribosylation) is considered as another NAD^+-dependent signal transduction event modifying chromatin structure [171–173]. Although SIRT1 may not possess ADP-ribosyl transferase activity [174], ADP-transferases such as PARP-1 and other nuclear sirtuins with proved ADP-ribosylation activity such as SIRT4 [175], SIRT6 [176], and SIRT7 [177] likely cooperate with SIRT1 in order to direct ADP-ribosylation signalling. For example, histone desuccinylase SIRT7, which also functions in NSCs, moonlights as a regulator of the switch in NAD^+-consuming enzymes. SIRT7 interaction with SIRT1 inhibits SIRT1 self-activation (auto-deacetylation) and promotes formation of less-active SIRT1 oligomeric complexes [178, 179]. On the other hand, SIRT7 is known to activate PARP-1 and exerts its function on chromatin remodelling [180].

Similar to SIRT1 expression, NAD$^+$ biosynthesis is regulated in a circadian manner and defines the dynamic activity of SIRT1 for high-magnitude circadian expression of several core clock genes [181]. Consistently, circadian rhythm is a driving force of neurogenesis [118, 182], and depletion of circadian proteins is associated with neurological deficits in SIRT1-dependent manner [183].

Cellular NAD$^+$ levels are differentially regulated by hypoxia. In a condition of insufficient oxygen supply the redox state of eukaryotic cells and consecutively the NAD$^+$/NADH ratio changes. For example, perinatal hypoxia in the brain leads to elevated NAD$^+$ levels and increased SIRT1 activity in neural progenitor cells [184]. HIF1α is a direct target of SIRT1 in cellular adaptation to environmental stress orchestrated by hypoxia [185, 186]. In other settings, such as myocardial ischemia/reperfusion (IR) injury, depletion of NAD$^+$ occurs because of hyperactivation of the NAD$^+$-consuming enzyme PARP1 [187] and/or downregulation of NAMPT [188]. Exogenous NAD$^+$ supplementation after hypoxic stress restores SIRT1 activity and inhibits apoptotic events [189]. Numerous studies demonstrate the protective effects of SIRT1 during hypoxic conditions [190, 191]. As described above, SIRT1 mediates a wide range of cellular responses through its deacetylating activity, targeting several transcription factors, including HIF1α, NF-κB, FOXO1 and p53 [192]. The SIRT1-dependent modulation of these pathways promotes cell survival under hypoxic conditions through induction of DNA repair, resistance to oxidative stress, and reduction of inflammatory response [192]. Activation of SIRT1 under low oxygen enhances energy homeostasis through PGC1α-dependent induction of mitochondrial biogenesis [193]. In addition to SIRT1 activity, hypoxia controls transcription and localization of SIRT1. Expression of SIRT1 is regulated by many feedback-controlled mechanisms, including the HIF1α/HIF2α–SIRT1 pathway [194]. SIRT1 harbours a HIF-responsible element (HRE) in its promotor region [195], leading to increased expression under hypoxic conditions [196]. Ischemia/reperfusion has been demonstrated to impact SIRT1 sumoylation and translocation to the cytoplasm in cancer cells [197]. Nuclear-cytoplasmic shuttling of SIRT1 has been reported upon differentiation of embryonic NSCs [198]. NSCs reside in hypoxic niches, and the fate of these cells is influenced by oxygen availability [100, 199, 200]. Therefore, the connection between SIRT1 and hypoxia/reoxygenation may represent a central point in SIRT1's role during self-renewal and neural differentiation.

Next to the NAD$^+$/NADH, the GSSG/GSH ratio also changes upon different redox states. Depletion of GSH synthesis in NSCs favours the astrocytic fate in a SIRT1-dependent manner [78]. In mammals, zebrafish, and yeast, regulation of SIRT1/Sir2 activity upon S-glutathionylation was described [201–203]. Several cysteine residues were identified as S-glutathionylated in these publications. Exchange of these cysteines to serines abolished oxidative inhibition of SIRT1 activity [202–204]. Both glutaredoxins 1 and 2 (GRX1 and GRX2) are able to reduce S-glutathionylated SIRT1 [202, 204]. GRX1-deficient mice develop fatty liver under normal diet because of decreased activity of S-glutathionylated SIRT1 [205]. Hepatic steatosis was inhibited in GRX1-/- mice after expression of a non-glutathionylated cysteine mutant of SIRT1 [205]. In zebrafish lacking GRX2, increased amounts of S-glutathionylated SIRT1 inhibited formation of the vasculature [202]. Next to this reversible posttranslational redox modification, SIRT1 undergoes various other posttranslational modifications of which many are also redox dependent (Figure 14.2).

14.7 REGULATION OF THE CATALYTIC DOMAIN

The catalytic domain of SIRT1 is predominantly sensitive to oxidation and contains several sites targeted by redox-sensitive signalling cascades. Two zinc finger domains of SIRT1 are subjected to sulfhydration mediated by CSE that enhances zinc ion binding activity, stabilization of alpha-helix structure, and reduces degradation [206]. Signalling properties of hydrogen sulphide (H$_2$S) [207], a gaseous molecule, were considered in adult neurogenesis. Supplementation with H$_2$S enhances proliferation in SVZ of adult mice [208].

Phosphorylation of SIRT1 at Tyr280 and Tyr301 by tyrosine kinase JAK1 was demonstrated in context of the negative feedback mechanism for the JAK1-STAT3 signalling pathway [209, 210]. Phosphorylation of tyrosines does not affect deacetylase activity but is required for SIRT1 interaction with transcription factor STAT3. SIRT1-STAT3 binding maintains the de-acetylated state of STAT3 and inhibits its transcriptional activity towards induction of gluconeogenic genes. This negative

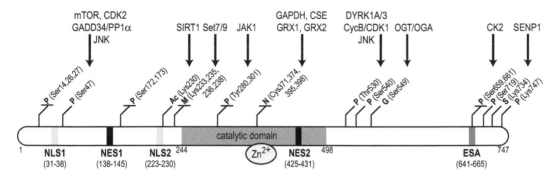

Figure 14.2 Schematic overview and summary of posttranslational modifications of human SIRT1.

Sites of post-translational modifications are indicated as **P**, phosphorylation; **M**, methylation; **N**, the sites for nitrosylation, sulfhydration and glutathionylation that coordinate Zn^{2+} binding; **S**, sumoylation; **Ac**, acetylation; **G**, N-acetyl-β-D-glucosamination. The **ESA** (essential for SIRT1 activity), **NLS** (nuclear localization sequences), **NES** (nuclear export sequences), and the Zn^{2+} binding site are mentioned. The arrows show the sites of action of the key modifiers of SIRT1 activity, such as JNK, JUN N-terminal kinase; Cyc, cyclin; CDK, cyclin-dependent kinase, mTOR, mammalian target of rapamycin; Set7/9, methyltransferase Set7/9; JAK, Janus kinase; GADD34, growth arrest and DNA damage inducible protein 34; PP1a, serine/threonine protein phosphatase 1α; GRX, glutaredoxin; GAPDH, glycerinaldehyde 3-phosphate dehydrogenase; SENP1, sentrin-specific protease 1; CSE, cystathionine gamma lyase; DYRK, dual-specificity tyrosine phosphorylation-regulated kinase; CK2, casein kinase 2; OGT, O-GlcNAc transferase; OGA, O-GlcNAcase. The numbers below indicate amino acid numbers.

feedback on STAT3 signalling may play an essential role in NSCs. In neocortical development, STAT3 is important for the maintenance of neural progenitors via induction of DLL1 and subsequent upregulation of DLL1-NOTCH signalling [211], thereby providing the proper timing of neuronal commitment. SIRT1 seemed to be a good candidate to limit STAT3 action in self-renewal cells to initiate differentiation towards neuronal lineage. In postnatal neocortex SIRT3 signalling is switched off in cells committed to neuronal lineage [211–213], whereas overexpression of STAT3 or stimulation of STAT signalling upon differentiation directs NSCs to an astrocytic fate [214, 215]. In turn, JAK2-STAT3 pathway induces SIRT1 expression through engaging redox-sensitive transcription factor NRF2 to ARE elements in the Sirt1 promoter [216].

14.8 ATP-SENSITIVE DOMAIN OF SIRT1

The ATP-sensitive domain of SIRT1 is located in the N-terminal part and is responsible for boosting SIRT1's catalytic activity [217, 218]. The ATP-sensitive domain is assumed to play a role as a region of action of sirtuin-activating compounds, such as resveratrol [219].

Phosphorylation of Ser47 has a unique role in SIRT1-mediated regulation of cell cycle progression and integrates signals downstream from redox-sensitive pathways. Under oxidative conditions, de-phosphorylation of Ser47 through GADD34 activates SIRT1 and induces cell cycle arrest [137, 138]. Of note, GADD34 triggers reversible G2 arrest and inhibits proliferation of NSCs in response to nutrient deprivation [136]. In conditions associated with nutrients and energy excess, mTOR signalling phosphorylates Ser47 resulting in SIRT1 inhibition [132]. mTOR is active in NSCs and promotes proliferation and differentiation in response to nutrient abundance [220, 221].

Additionally, Ser47 (and Ser27) match minimal consensus sequence of CDKs phosphorylation substrates. Under hypoxic conditions, SIRT1 binds CDK2 and became phosphorylated at Ser47. This complex formation is associated with CDK2 deacetylation and enhanced proliferation of oligodendrocyte progenitors in response to low oxygen levels in vitro and in vivo [222].

14.9 INSULIN-SENSITIVE DOMAIN OF SIRT1

Insulin/IGF signalling is the driving force of cell growth during development and in adult stem cells [226]. The N-terminal region of SIRT1 encompass a hidden "insulin-response sensor" (IRS), comprising an acidic cluster and a 3-helix bundle. Results of several works illustrate the mechanism by which insulin/IGF receptor signalling impairs SIRT1 function. In the presence of insulin, IRS engages natural SIRT1 inhibitor DBC1 [223, 224]. DBC1 sequesters both acidic cluster and 3-helix bundle for interaction with phosphorylated PACS-2 and inhibits SIRT1 activity via destabilization of the 3-helix bundle [225]. By this mechanism, DBC1/PACS-2 axis triggers the expression of genes negatively regulated by SIRT1 in response to insulin signalling in metabolically active tissues. This is in line with increased hepatic SIRT1 activity in PACS-2 deficient mice [225]. CK2-dependent phosphorylation of Ser164 in obese mice causes re-distribution of SIRT1 from the nucleus to the cytoplasm [227].

Lys230 residue located in a 3-helix bundle is a site of auto-deacetylation, which is regulated by another member of the sirtuin family, SIRT7. While auto-deacetylation of Lys230 is required for efficient deacetylation of p53 and histone substrates (H3K9ac and H4K16ac), SIRT7 antagonizes SIRT1 through a direct interaction and inhibition of autocatalytic activation [179]. Increased SIRT1 activity in SIRT7 deficient mice affect adipocyte differentiation and maintenance of white adipose tissue, indicating the antagonistic actions of SIRT7 and SIRT1 during adipogenesis. Moreover, SIRT7 promotes SIRT1 oligomerization and its binding to Set7/9 that methylates lysines 233, 235, 236, and 238 abrogating SIRT1's substrate binding [228]. SIRT1 autoregulation seems to be a powerful mechanism to amplify signal responses (discussed in [179]).

Interestingly, in addition to methylation of H3K4 (as a permissive mark), Set7/9 targets many proteins involved in transcriptional regulation and being targets of SIRT1, such as p53, TAF10, P65, STAT3, SOX2, pRb, DNMT1, SUV39H1, and FOXO3. In contrast to SIRT1, Set7/9-induced methylation of PARP1 leads to its activation and increases PARP-dependent poly-ADP ribosylation of multiple substrates. Thus, Set7/9 can be viewed as a molecule promoting a switch in the redox sensors in order to preserve the NAD$^+$ pool. The mechanism of the crosstalk between SIRT1 and PARP1 has been described in other cellular contexts [229].

14.10 THE C-TERMINAL DOMAIN OF SIRT1

The reversible covalent attachment of O-linked N-acetyl-β-D-glucosamine (O-GlcNAc) that is regulated by two opposing enzymes, O-GlcNAc transferase (OGT) and O-GlcNAcase (OGA) has emerged as a novel redox-sensitive signalling at the interface of metabolism, epigenetics [230], and development [231]. In many studied proteins, O-GlcNacylation and phosphorylation are reciprocal [232]. O-GlcNAc modification of SIRT1 at Ser549 increases deacetylase activity. Mutation of Ser549 to alanine abrogates SIRT1-dependent deacetylation of p53 and affects cell survival under stress conditions [233]. O-GlcNAc modification occurs in vitro and in vivo following nutrition (glucose) deprivation or oxidative stress. O-GlcNacylation, is considered as a sensor of cellular state that regulates a broad spectrum of homeostasis processes, including transcription and cell signalling [230]. In eNSCs, OGT targets Pax3 and impact the function of the Polycomb group (PcG). Its deficiency is associated with neural-tube defects [234]. In turn, PRC2 is necessary to maintain normal level of OGT and cellular distribution of O-GlcNAc modification [235]. The relation of O-GlcNacylation of Ser549 and Ser47 de-phosphorylation by GADD43 in response to nutritional restriction or oxidative stress has not been studied so far.

While effect of serine kinases on Ser549 has not been described, phosphorylation of several serine residues in the C-terminal domain, such as serines 649, 651, and 683 by CK2, were linked to enhanced enzymatic activity [236]. Of note, Ser659 and Ser661 lie within a region of SIRT1 that is referred to the ESA (essential for SIRT1 activity) motif and may affect binding to DBC1. Enzymatic activity of SIRT1 is also enhanced upon Tyr522 phosphorylation by DYRK1A and DYRK3 [237]. DYRK1A is increasingly recognized as a chromatin modifying kinase and gene-specific regulator of proteins involved in cell growth [238]. Both enzymes, DYRK1A and DYRK3, play a role in neuronal development [238].

14.11 CONCLUSION

SIRT1 serves as a critical redox-sensitive regulatory element for transcriptional programs by directing neural development, homeostasis, and longevity. NAD$^+$-dependent enzymatic activity of SIRT1 is required for the early development and maintenance of stem cells, and for regulation of their differentiation into specific cell types. SIRT1 is embedded in a network of complex interactions of chromatin remodellers that adapts stem cell activity according to the metabolic and redox states of the cell as well as the given physiological or pathological tissue environment. This emerging association highlights the potential role of SIRT1 in preservation and limiting exhaustion of the stem and progenitor pool throughout life, suggesting a pivotal role in delaying ageing processes and promoting longevity. Understanding the mechanisms of instructive redox changes influencing SIRT1 activity and subsequently the transcriptional switch between developmental and stemness genes will open exciting new perspectives for stem cell biology and stem cell therapy.

14.12 ABBREVIATIONS

AMP-activated protein kinase (AMPK); antioxidant response element (ARE); adenosine triphosphate (ATP); B-cell lymphoma 6 (BCL6); basic helix-loop-helix (bHLH); cyclin-dependent kinase (CDK); carbohydrate response-element-binding protein (ChREB); calorie restriction (CR); CAMP responsive element binding protein (CREB); cystathionine gamma lyase (CSE); C-terminal binding protein (CtBP); deleted in breast cancer 1 (DBC1); delta-like protein 4 (DLL4); deoxyribonucleic acid (DNA); DNA methyltransferases (Dnmt); dual specificity tyrosine phosphorylation-regulated kinase (DYRK); embryonic stem cell (ESC); enhancer of zeste homolog 2 (EZH2); forkhead Box O (FOXO); growth arrest and DNA damage-inducible gene 34 (GADD34); glycerinaldehyd-3-phosphat-dehydrogenase (GAPDH); glial fibrillary acidic protein (GFAP); glutaredoxin (GRX); glutathione (GSH); glutathione disulfide (GSSG); histone acetyltransferase (HAT); histone deacetylase (HDAC); hairy and enhancer of split homolog (HES); hypermethylated in cancer 1 (HIC1); hypoxia inducible factor-1 (HIF-1); human antigen R (HuR); insulin-like growth factor (IGF); c-Jun N-terminal kinase (JNK); mouse achaete-scute complex homolog 1 (MASH1); mouse atonal homolog (Math); mammalian target of rapamycin (mTOR); nicotinamide adenine dinucleotide (NAD); nicotinamide phosphoribosyltransferase (NAMPT); nuclear receptor co-repressor (N-CoR); neurogenic differentiation 1 (NeuroD1); neurogenin (Ngn); nicotinamide mononucleotide adenylyltransferase (NMNAT); NADPH oxidase (NOX); nuclear factor erythroid 2-related factor 2 (NRF2); neural stem and progenitor cells (NSCs); octamer-binding transcription factor 4 (OCT4); O-GlcNAc hydrolase (OGA); 8-oxoguanine DNA glycosylase (OGG); O-linked N-acetyl-β-D-glucosamine (O-GlcNAc); O-GlcNAc transferase (OGT); phosphofurin acidic cluster sorting protein (PACS); ADP-ribosyltransferase diphtheria toxin-like 1 (PARP-1); paired box gene (PAX); polycomb group (PcG); peroxisome proliferator-activated receptor-gamma coactivator (PGC); phosphoinositide-3-kinase (PI3K); protein phosphatase 1 (PP1); polycomb repressor complex (PRC); PR domain containing (PRDM); reactive oxygen species (ROS); Spt-Ada-Gcn5-Acetyl transferase (SAGA); subgranular zone (SGZ); sirtuin (SIRT); SMA ("small" worm phenotype) and MAD family protein (SMAD); SMAD specific E3 ubiquitin protein ligase 2 (Smurf2); sex determining region Y-box 2 (SOX2); signal transducers and activators of transcription (STAT); subventricular zone (SVZ); T-box transcription factor (TBX); T-cell factor (TCF); ubiquitin-proteasome system (UPS); ubiquitin-specific peptidase 22 (USP22); Wingless and Int-1 (Wnt); vascular cell adhesion molecule 1 (VCAM1).

REFERENCES

[1] T. A. Boyer, D. Mathur, and R. Jaenisch, "Molecular control of pluripotency," Curr. Opin. Genet. Dev., vol. 16, no. 5, pp. 455–462, Oct. 2006, doi:10.1016/j.gde.2006.08.009.

[2] L. Guarente, "Sir2 links chromatin silencing, metabolism, and aging," Genes Dev., vol. 14, no. 9, pp. 1021–1026, May 2000.

[3] A. Kuzmichev et al., "Composition and histone substrates of polycomb repressive group complexes change during cellular differentiation," Proc. Natl. Acad. Sci. U. S. A., vol. 102, no. 6, pp. 1859–1864, Feb. 2005, doi:10.1073/pnas.0409875102.

[4] C. Mantel and H. E. Broxmeyer, "Sirtuin 1, stem cells, aging, and stem cell aging," Curr. Opin. Hematol., vol. 15, no. 4, pp. 326–331, Jul. 2008, doi:10.1097/MOH.0b013e3283043819.

[5] S. Michan and D. Sinclair, "Sirtuins in mammals: insights into their biological function," Biochem. J., vol. 404, no. 1, pp. 1–13, May 2007, doi:10.1042/BJ20070140.

[6] A. Vaquero, M. Scher, D. Lee, H. Erdjument-Bromage, P. Tempst, and D. Reinberg, "Human SirT1 interacts with histone H1 and promotes formation of facultative heterochromatin," Mol. Cell, vol. 16, no. 1, pp. 93–105, Oct. 2004, doi:10.1016/j.molcel.2004.08.031.

[7] C. L. Brooks and W. Gu, "How does SIRT1 affect metabolism, senescence and cancer?," Nat. Rev.

Cancer, vol. 9, no. 2, pp. 123–128, Feb. 2009, doi:10.1038/nrc2562.

[8] A. Vaquero, M. Scher, H. Erdjument-Bromage, P. Tempst, L. Serrano, and D. Reinberg, "SIRT1 regulates the histone methyl-transferase SUV39H1 during heterochromatin formation," *Nature*, vol. 450, no. 7168, pp. 440–444, Nov. 2007, doi:10.1038/nature06268.

[9] P. Mulligan *et al.*, "A SIRT1-LSD1 corepressor complex regulates Notch target gene expression and development," *Mol. Cell*, vol. 42, no. 5, pp. 689–699, Jun. 2011, doi:10.1016/j.molcel. 2011.04.020.

[10] J.-Y. Kang *et al.*, "KDM2B is a histone H3K79 demethylase and induces transcriptional repression via sirtuin-1-mediated chromatin silencing," *FASEB J. Off. Publ. Fed. Am. Soc. Exp. Biol.*, vol. 32, no. 10, pp. 5737–5750, Oct. 2018, doi:10.1096/fj.201800242R.

[11] L. Bosch-Presegué *et al.*, "Stabilization of Suv39H1 by SirT1 is part of oxidative stress response and ensures genome protection," *Mol. Cell*, vol. 42, no. 2, pp. 210–223, Apr. 2011, doi:10.1016/j.molcel.2011.02.034.

[12] Y. Lai, J. Li, X. Li, and C. Zou, "Lipopolysaccharide modulates p300 and Sirt1 to promote PRMT1 stability via an SCFFbxl17-recognized acetyldegron," *J. Cell Sci.*, vol. 130, no. 20, pp. 3578–3587, Oct. 2017, doi:10.1242/jcs.206904.

[13] L. Peng *et al.*, "SIRT1 deacetylates the DNA methyltransferase 1 (DNMT1) protein and alters its activities," *Mol. Cell. Biol.*, vol. 31, no. 23, pp. 4720–4734, Dec. 2011, doi:10.1128/MCB.06147-11.

[14] J. Heo *et al.*, "Sirt1 regulates DNA methylation and differentiation potential of embryonic stem cells by antagonizing Dnmt3l," *Cell Rep.*, vol. 18, no. 8, pp. 1930–1945, Feb. 2017, doi:10.1016/j. celrep.2017.01.074.

[15] T. Bouras *et al.*, "SIRT1 deacetylation and repression of p300 involves lysine residues 1020/1024 within the cell cycle regulatory domain 1," *J. Biol. Chem.*, vol. 280, no. 11, pp. 10264–10276, Mar. 2005, doi:10.1074/jbc.M408748200.

[16] W. C. Hallows, S. Lee, and J. M. Denu, "Sirtuins deacetylate and activate mammalian acetyl-CoA synthetases," *Proc. Natl. Acad. Sci. U. S. A.*, vol. 103, no. 27, pp. 10230–10235, Jul. 2006, doi:10.1073/ pnas.0604392103.

[17] T. Zhang and W. L. Kraus, "SIRT1-dependent regulation of chromatin and transcription: linking NAD(+) metabolism and signaling to the control of cellular functions," *Biochim. Biophys.*

Acta, vol. 1804, no. 8, pp. 1666–1675, Aug. 2010, doi:10.1016/j.bbapap.2009.10.022.

[18] T. Dvash and N. Benvenisty, "Human embryonic stem cells as a model for early human development," *Best Pract. Res. Clin. Obstet. Gynaecol.*, vol. 18, no. 6, pp. 929–940, Dec. 2004, doi:10.1016/j. bpobgyn.2004.06.005.

[19] V. Calvanese *et al.*, "Sirtuin 1 regulation of developmental genes during differentiation of stem cells," *Proc. Natl. Acad. Sci. U. S. A.*, vol. 107, no. 31, pp. 13736–13741, Aug. 2010, doi:10.1073/pnas. 1001399107.

[20] H.-L. Cheng *et al.*, "Developmental defects and p53 hyperacetylation in Sir2 homolog (SIRT1)-deficient mice," *Proc. Natl. Acad. Sci. U. S. A.*, vol. 100, no. 19, pp. 10794–10799, Sep. 2003, doi:10.1073/pnas.1934713100.

[21] M. W. McBurney *et al.*, "The mammalian SIR2alpha protein has a role in embryogenesis and gametogenesis," *Mol. Cell. Biol.*, vol. 23, no. 1, pp. 38–54, Jan. 2003, doi:10.1128/mcb. 23.1.38-54.2003.

[22] R.-H. Wang *et al.*, "Impaired DNA damage response, genome instability, and tumorigenesis in SIRT1 mutant mice," *Cancer Cell*, vol. 14, no. 4, pp. 312–323, Oct. 2008, doi:10.1016/j. ccr.2008.09.001.

[23] D. S. Yoon *et al.*, "SIRT1 directly regulates SOX2 to maintain self-renewal and multipotency in bone marrow-derived mesenchymal stem cells," *Stem Cells Dayt. Ohio*, vol. 32, no. 12, pp. 3219–3231, Dec. 2014, doi:10.1002/stem.1811.

[24] X. Ou *et al.*, "SIRT1 deficiency compromises mouse embryonic stem cell hematopoietic differentiation, and embryonic and adult hematopoiesis in the mouse," *Blood*, vol. 117, no. 2, pp. 440–450, Jan. 2011, doi:10.1182/ blood-2010-03-273011.

[25] M.-K. Han, E.-K. Song, Y. Guo, X. Ou, C. Mantel, and H. E. Broxmeyer, "SIRT1 regulates apoptosis and Nanog expression in mouse embryonic stem cells by controlling p53 subcellular localization," *Cell Stem Cell*, vol. 2, no. 3, pp. 241–251, Mar. 2008, doi:10.1016/j.stem.2008.01.002.

[26] Z.-N. Zhang, S.-K. Chung, Z. Xu, and Y. Xu, "Oct4 maintains the pluripotency of human embryonic stem cells by inactivating p53 through Sirt1-mediated deacetylation," *Stem Cells Dayt. Ohio*, vol. 32, no. 1, pp. 157–165, Jan. 2014, doi:10.1002/stem.1532.

[27] T. Furuyama, R. Banerjee, T. R. Breen, and P. J. Harte, "SIR2 is required for polycomb silencing and

is associated with an E(Z) histone methyltransferase complex," *Curr. Biol. CB*, vol. 14, no. 20, pp. 1812–1821, Oct. 2004, doi:10.1016/j.cub.2004.09.060.

[28] P. Voigt, W.-W. Tee, and D. Reinberg, "A double take on bivalent promoters," *Genes Dev.*, vol. 27, no. 12, pp. 1318–1338, Jun. 2013, doi:10.1101/gad.219626.113.

[29] J. E. Ohm et al., "A stem cell-like chromatin pattern may predispose tumor suppressor genes to DNA hypermethylation and heritable silencing," *Nat. Genet.*, vol. 39, no. 2, pp. 237–242, Feb. 2007, doi:10.1038/ng1972.

[30] M.-R. Kang et al., "Reciprocal roles of SIRT1 and SKIP in the regulation of RAR activity: implication in the retinoic acid-induced neuronal differentiation of P19 cells," *Nucleic Acids Res.*, vol. 38, no. 3, pp. 822–831, Jan. 2010, doi:10.1093/nar/gkp1056.

[31] S. Tang et al., "SIRT1-mediated deacetylation of CRABPII regulates cellular retinoic acid signaling and modulates embryonic stem cell differentiation," *Mol. Cell*, vol. 55, no. 6, pp. 843–855, Sep. 2014, doi:10.1016/j.molcel.2014.07.011.

[32] Z. Lin et al., "USP22 antagonizes p53 transcriptional activation by deubiquitinating Sirt1 to suppress cell apoptosis and is required for mouse embryonic development," *Mol. Cell*, vol. 46, no. 4, pp. 484–494, May 2012, doi:10.1016/j.molcel.2012.03.024.

[33] N. L. Samara et al., "Structural insights into the assembly and function of the SAGA deubiquitinating module," *Science*, vol. 328, no. 5981, pp. 1025–1029, May 2010, doi:10.1126/science.1190049.

[34] B. S. Atanassov and S. Y. R. Dent, "USP22 regulates cell proliferation by deubiquitinating the transcriptional regulator FBP1," *EMBO Rep.*, vol. 12, no. 9, pp. 924–930, Sep. 2011, doi:10.1038/embor.2011.140.

[35] R. T. Sussman, T. J. Stanek, P. Esteso, J. D. Gearhart, K. E. Knudsen, and S. B. McMahon, "The epigenetic modifier ubiquitin-specific protease 22 (USP22) regulates embryonic stem cell differentiation via transcriptional repression of sex-determining region Y-box 2 (SOX2)," *J. Biol. Chem.*, vol. 288, no. 33, pp. 24234–24246, Aug. 2013, doi:10.1074/jbc.M113.469783.

[36] H. M. O'Hagan, H. P. Mohammad, and S. B. Baylin, "Double strand breaks can initiate gene silencing and SIRT1-dependent onset of DNA methylation in an exogenous promoter CpG island," *PLoS Genet.*, vol. 4, no. 8, p. e1000155, Aug. 2008, doi:10.1371/journal.pgen.1000155.

[37] G. L. Sen, J. A. Reuter, D. E. Webster, L. Zhu, and P. A. Khavari, "DNMT1 maintains progenitor function in self-renewing somatic tissue," *Nature*, vol. 463, no. 7280, pp. 563–567, Jan. 2010, doi:10.1038/nature08683.

[38] K. D. Mills, D. A. Sinclair, and L. Guarente, "MEC1-dependent redistribution of the Sir3 silencing protein from telomeres to DNA double-strand breaks," *Cell*, vol. 97, no. 5, pp. 609–620, May 1999, doi:10.1016/s0092-8674(00)80772-2.

[39] P. Oberdoerffer et al., "SIRT1 redistribution on chromatin promotes genomic stability but alters gene expression during aging," *Cell*, vol. 135, no. 5, pp. 907–918, Nov. 2008, doi:10.1016/j.cell.2008.10.025.

[40] F. M. Meliso et al., "SIRT1 regulates Mxd1 during malignant melanoma progression," *Oncotarget*, vol. 8, no. 70, pp. 114540–114553, Dec. 2017, doi:10.18632/oncotarget.21457.

[41] H. M. O'Hagan et al., "Oxidative damage targets complexes containing DNA methyltransferases, SIRT1, and polycomb members to promoter CpG Islands," *Cancer Cell*, vol. 20, no. 5, pp. 606–619, Nov. 2011, doi:10.1016/j.ccr.2011.09.012.

[42] M. H. Sherman, C. H. Bassing, and M. A. Teitell, "Regulation of cell differentiation by the DNA damage response," *Trends Cell Biol.*, vol. 21, no. 5, pp. 312–319, May 2011, doi:10.1016/j.tcb.2011.01.004.

[43] N. Gu et al., "DNA damage triggers reprogramming of differentiated cells into stem cells in Physcomitrella," *Nat. Plants*, vol. 6, no. 9, pp. 1098–1105, Sep. 2020, doi:10.1038/s41477-020-0745-9.

[44] L. A. Boyer et al., "Polycomb complexes repress developmental regulators in murine embryonic stem cells," *Nature*, vol. 441, no. 7091, pp. 349–353, May 2006, doi:10.1038/nature04733.

[45] T. I. Lee et al., "Control of developmental regulators by polycomb in human embryonic stem cells," *Cell*, vol. 125, no. 2, pp. 301–313, Apr. 2006, doi:10.1016/j.cell.2006.02.043.

[46] T. S. Mikkelsen et al., "Genome-wide maps of chromatin state in pluripotent and lineage-committed cells," *Nature*, vol. 448, no. 7153, pp. 553–560, Aug. 2007, doi:10.1038/nature06008.

[47] J. Sakamoto, T. Miura, K. Shimamoto, and Y. Horio, "Predominant expression of Sir2alpha, an NAD-dependent histone deacetylase, in the embryonic mouse heart and brain," *FEBS Lett.*, vol. 556, no. 1–3, pp. 281–286, Jan. 2004, doi:10.1016/s0014-5793(03)01444-3.

[48] Y.-L. Yu et al., "Smurf2-mediated degradation of EZH2 enhances neuron differentiation and improves functional recovery after ischaemic stroke," EMBO Mol. Med., vol. 5, no. 4, pp. 531–547, Apr. 2013, doi:10.1002/emmm.201201783.

[49] L. Yu et al., "Ubiquitination-mediated degradation of SIRT1 by SMURF2 suppresses CRC cell proliferation and tumorigenesis," Oncogene, vol. 39, no. 22, pp. 4450–4464, May 2020, doi:10.1038/s41388-020-1298-0.

[50] M. Narimatsu et al., "Regulation of planar cell polarity by Smurf ubiquitin ligases," Cell, vol. 137, no. 2, pp. 295–307, Apr. 2009, doi:10.1016/j.cell.2009.02.025.

[51] H. Zhu, P. Kavsak, S. Abdollah, J. L. Wrana, and G. H. Thomsen, "A SMAD ubiquitin ligase targets the BMP pathway and affects embryonic pattern formation," Nature, vol. 400, no. 6745, pp. 687–693, Aug. 1999, doi:10.1038/23293.

[52] M. Kim, Y. E. Kwon, J. O. Song, S. J. Bae, and J. H. Seol, "CHFR negatively regulates SIRT1 activity upon oxidative stress," Sci. Rep., vol. 6, p. 37578, Nov. 2016, doi:10.1038/srep37578.

[53] V. K. Tiwari, M. B. Stadler, C. Wirbelauer, R. Paro, D. Schübeler, and C. Beisel, "A chromatin-modifying function of JNK during stem cell differentiation," Nat. Genet., vol. 44, no. 1, pp. 94–100, Dec. 2011, doi:10.1038/ng.1036.

[54] Y. M. Oh et al., "Chfr is linked to tumour metastasis through the downregulation of HDAC1," Nat. Cell Biol., vol. 11, no. 3, pp. 295–302, Mar. 2009, doi:10.1038/ncb1837.

[55] L. Kashima et al., "CHFR protein regulates mitotic checkpoint by targeting PARP-1 protein for ubiquitination and degradation," J. Biol. Chem., vol. 287, no. 16, pp. 12975–12984, Apr. 2012, doi:10.1074/jbc.M111.321828.

[56] T. C. Tuoc and A. Stoykova, "Roles of the ubiquitin-proteosome system in neurogenesis," Cell Cycle Georget. Tex, vol. 9, no. 16, pp. 3174–3180, Aug. 2010, doi:10.4161/cc.9.16.12551.

[57] A. M. Hamilton and K. Zito, "Breaking it down: the ubiquitin proteasome system in neuronal morphogenesis," Neural Plast., vol. 2013, p. 196848, 2013, doi:10.1155/2013/196848.

[58] C. Delaloy et al., "MicroRNA-9 coordinates proliferation and migration of human embryonic stem cell-derived neural progenitors," Cell Stem Cell, vol. 6, no. 4, pp. 323–335, Apr. 2010, doi:10.1016/j.stem.2010.02.015.

[59] Z. Xu, L. Zhang, X. Fei, X. Yi, W. Li, and Q. Wang, "The miR-29b-Sirt1 axis regulates self-renewal of mouse embryonic stem cells in response to reactive oxygen species," Cell. Signal., vol. 26, no. 7, pp. 1500–1505, Jul. 2014, doi:10.1016/j.cellsig.2014.03.010.

[60] M. Yamakuchi, "MicroRNA regulation of SIRT1," Front. Physiol., vol. 3, p. 68, 2012, doi:10.3389/fphys.2012.00068.

[61] L. R. Saunders et al., "miRNAs regulate SIRT1 expression during mouse embryonic stem cell differentiation and in adult mouse tissues," Aging, vol. 2, no. 7, pp. 415–431, Jul. 2010, doi:10.18632/aging.100176.

[62] K. Abdelmohsen et al., "Phosphorylation of HuR by Chk2 regulates SIRT1 expression," Mol. Cell, vol. 25, no. 4, pp. 543–557, Feb. 2007, doi:10.1016/j.molcel.2007.01.011.

[63] M. M. Dobbin et al., "SIRT1 collaborates with ATM and HDAC1 to maintain genomic stability in neurons," Nat. Neurosci., vol. 16, no. 8, pp. 1008–1015, Aug. 2013, doi:10.1038/nn.3460.

[64] Y. Li, W. Xu, M. W. McBurney, and V. D. Longo, "SirT1 inhibition reduces IGF-I/IRS-2/Ras/ERK1/2 signaling and protects neurons," Cell Metab., vol. 8, no. 1, pp. 38–48, Jul. 2008, doi:10.1016/j.cmet.2008.05.004.

[65] K. Hasegawa and K. Yoshikawa, "Necdin regulates p53 acetylation via Sirtuin1 to modulate DNA damage response in cortical neurons," J. Neurosci. Off. J. Soc. Neurosci., vol. 28, no. 35, pp. 8772–8784, Aug. 2008, doi:10.1523/JNEUROSCI.3052-08.2008.

[66] T. Araki, Y. Sasaki, and J. Milbrandt, "Increased nuclear NAD biosynthesis and SIRT1 activation prevent axonal degeneration," Science, vol. 305, no. 5686, pp. 1010–1013, Aug. 2004, doi:10.1126/science.1098014.

[67] J. F. Codocedo, C. Allard, J. A. Godoy, L. Varela-Nallar, and N. C. Inestrosa, "SIRT1 regulates dendritic development in hippocampal neurons," PloS One, vol. 7, no. 10, p. e47073, 2012, doi:10.1371/journal.pone.0047073.

[68] W. Guo et al., "Sirt1 overexpression in neurons promotes neurite outgrowth and cell survival through inhibition of the mTOR signaling," J. Neurosci. Res., vol. 89, no. 11, pp. 1723–1736, Nov. 2011, doi:10.1002/jnr.22725.

[69] X.-H. Li et al., "Sirt1 promotes axonogenesis by deacetylation of Akt and inactivation of GSK3," Mol. Neurobiol., vol. 48, no. 3, pp. 490–499, Dec. 2013, doi:10.1007/s12035-013-8437-3.

[70] A. Z. Herskovits and L. Guarente, "SIRT1 in neurodevelopment and brain senescence," Neuron,

vol. 81, no. 3, pp. 471–483, Feb. 2014, doi:10. 1016/j.neuron.2014.01.028.

[71] C.-Y. Ma, M. Yao, Q. Zhai, J. Jiao, X. Yuan, and M. Poo, "SIRT1 suppresses self-renewal of adult hippocampal neural stem cells," *Dev. Camb. Engl.*, vol. 141, no. 24, pp. 4697–4709, Dec. 2014, doi:10.1242/dev.117937.

[72] S. Saharan, D. J. Jhaveri, and P. F. Bartlett, "SIRT1 regulates the neurogenic potential of neural precursors in the adult subventricular zone and hippocampus," *J. Neurosci. Res.*, vol. 91, no. 5, pp. 642–659, May 2013, doi:10.1002/jnr.23199.

[73] V. A. Rafalski et al., "Expansion of oligodendrocyte progenitor cells following SIRT1 inactivation in the adult brain," *Nat. Cell Biol.*, vol. 15, no. 6, pp. 614–624, Jun. 2013, doi:10.1038/ncb2735.

[74] B. Hu et al., "Repression of SIRT1 promotes the differentiation of mouse induced pluripotent stem cells into neural stem cells," *Cell. Mol. Neurobiol.*, vol. 34, no. 6, pp. 905–912, Aug. 2014, doi:10.1007/s10571-014-0071-8.

[75] S. Ichi et al., "Role of Pax3 acetylation in the regulation of Hes1 and Neurog2," *Mol. Biol. Cell*, vol. 22, no. 4, pp. 503–512, Feb. 2011, doi:10.1091/mbc.E10-06-0541.

[76] J. E. Le Belle et al., "Proliferative neural stem cells have high endogenous ROS levels that regulate self-renewal and neurogenesis in a PI3K/Akt-dependant manner," *Cell Stem Cell*, vol. 8, no. 1, pp. 59–71, Jan. 2011, doi:10.1016/j. stem.2010.11.028.

[77] K. Forsberg, A. Wuttke, G. Quadrato, P. M. Chumakov, A. Wizenmann, and S. Di Giovanni, "The tumor suppressor p53 fine-tunes reactive oxygen species levels and neurogenesis via PI3 kinase signaling," *J. Neurosci. Off. J. Soc. Neurosci.*, vol. 33, no. 36, pp. 14318–14330, Sep. 2013, doi:10.1523/JNEUROSCI.1056-13.2013.

[78] T. Prozorovski et al., "Sirt1 contributes critically to the redox-dependent fate of neural progenitors," *Nat. Cell Biol.*, vol. 10, no. 4, pp. 385–394, Apr. 2008, doi:10.1038/ncb1700.

[79] R. Beckervordersandforth et al., "Role of mitochondrial metabolism in the control of early lineage progression and aging phenotypes in adult hippocampal neurogenesis," *Neuron*, vol. 93, no. 3, pp. 560-573.e6, Feb. 2017, doi:10.1016/j.neuron.2016.12.017.

[80] W. Wang et al., "Mitochondrial DNA damage level determines neural stem cell differentiation fate," *J. Neurosci. Off. J. Soc. Neurosci.*, vol. 31,

no. 26, pp. 9746–9751, Jun. 2011, doi:10.1523/JNEUROSCI.0852-11.2011.

[81] W. Wang, P. Osenbroch, R. Skinnes, Y. Esbensen, M. Bjørås, and L. Eide, "Mitochondrial DNA integrity is essential for mitochondrial maturation during differentiation of neural stem cells," *Stem Cells Dayt. Ohio*, vol. 28, no. 12, pp. 2195–2204, Dec. 2010, doi:10.1002/stem.542.

[82] C. Murre, "Helix-loop-helix proteins and the advent of cellular diversity: 30 years of discovery," *Genes Dev.*, vol. 33, no. 1–2, pp. 6–25, Jan. 2019, doi:10.1101/gad.320663.118.

[83] F. Guillemot and B. A. Hassan, "Beyond proneural: emerging functions and regulations of proneural proteins," *Curr. Opin. Neurobiol.*, vol. 42, pp. 93–101, Feb. 2017, doi:10.1016/j. conb.2016.11.011.

[84] N. Bertrand, D. S. Castro, and F. Guillemot, "Proneural genes and the specification of neural cell types," *Nat. Rev. Neurosci.*, vol. 3, no. 7, pp. 517–530, Jul. 2002, doi:10.1038/nrn874.

[85] W. Ge et al., "Coupling of cell migration with neurogenesis by proneural bHLH factors," *Proc. Natl. Acad. Sci. U. S. A.*, vol. 103, no. 5, pp. 1319–1324, Jan. 2006, doi:10.1073/pnas.0510419103.

[86] T. Senawong, V. J. Peterson, and M. Leid, "BCL11A-dependent recruitment of SIRT1 to a promoter template in mammalian cells results in histone deacetylation and transcriptional repression," *Arch. Biochem. Biophys.*, vol. 434, no. 2, pp. 316–325, Feb. 2005, doi:10.1016/j. abb.2004.10.028.

[87] T. Takata and F. Ishikawa, "Human Sir2-related protein SIRT1 associates with the bHLH repressors HES1 and HEY2 and is involved in HES1- and HEY2-mediated transcriptional repression," *Biochem. Biophys. Res. Commun.*, vol. 301, no. 1, pp. 250–257, Jan. 2003, doi:10.1016/s0006-291x(02)03020-6.

[88] T. Senawong et al., "Involvement of the histone deacetylase SIRT1 in chicken ovalbumin upstream promoter transcription factor (COUP-TF)-interacting protein 2-mediated transcriptional repression," *J. Biol. Chem.*, vol. 278, no. 44, pp. 43041–43050, Oct. 2003, doi:10.1074/jbc.M307477200.

[89] B.-G. Ju et al., "Activating the PARP-1 sensor component of the groucho/TLE1 corepressor complex mediates a CaMKinase IIdelta-dependent neurogenic gene activation pathway," *Cell*, vol. 119, no. 6, pp. 815–829, Dec. 2004, doi:10.1016/j.cell.2004.11.017.

[90] T. Ohtsuka, M. Sakamoto, F. Guillemot, and R. Kageyama, "Roles of the basic helix-loop-helix genes Hes1 and Hes5 in expansion of neural stem cells of the developing brain," *J. Biol. Chem.*, vol. 276, no. 32, pp. 30467–30474, Aug. 2001, doi:10.1074/jbc.M102420200.

[91] S. Hisahara *et al.*, "Histone deacetylase SIRT1 modulates neuronal differentiation by its nuclear translocation," *Proc. Natl. Acad. Sci. U. S. A.*, vol. 105, no. 40, pp. 15599–15604, Oct. 2008, doi:10.1073/pnas.0800612105.

[92] L. Tiberi *et al.*, "BCL6 controls neurogenesis through Sirt1-dependent epigenetic repression of selective Notch targets," *Nat. Neurosci.*, vol. 15, no. 12, pp. 1627–1635, Dec. 2012, doi:10.1038/nn.3264.

[93] J. Bonnefont *et al.*, "Cortical neurogenesis requires Bcl6-mediated transcriptional repression of multiple self-renewal-promoting extrinsic pathways," *Neuron*, vol. 103, no. 6, pp. 1096–1108.e4, Sep. 2019, doi:10.1016/j.neuron.2019.06.027.

[94] L. Tiberi *et al.*, "A BCL6/BCOR/SIRT1 complex triggers neurogenesis and suppresses medulloblastoma by repressing Sonic Hedgehog signaling," *Cancer Cell*, vol. 26, no. 6, pp. 797–812, Dec. 2014, doi:10.1016/j.ccell.2014.10.021.

[95] R. Iwata, P. Casimir, and P. Vanderhaeghen, "Mitochondrial dynamics in postmitotic cells regulate neurogenesis," *Science*, vol. 369, no. 6505, pp. 858–862, Aug. 2020, doi:10.1126/science.aba9760.

[96] D.-Y. Kim, I. Rhee, and J. Paik, "Metabolic circuits in neural stem cells," *Cell. Mol. Life Sci. CMLS*, vol. 71, no. 21, pp. 4221–4241, Nov. 2014, doi:10.1007/s00018-014-1686-0.

[97] V. A. Rafalski and A. Brunet, "Energy metabolism in adult neural stem cell fate," *Prog. Neurobiol.*, vol. 93, no. 2, pp. 182–203, Feb. 2011, doi:10.1016/j.pneurobio.2010.10.007.

[98] X. Zheng *et al.*, "Metabolic reprogramming during neuronal differentiation from aerobic glycolysis to neuronal oxidative phosphorylation," *eLife*, vol. 5, Jun. 2016, doi:10.7554/eLife.13374.

[99] M. Agostini *et al.*, "Metabolic reprogramming during neuronal differentiation," *Cell Death Differ.*, vol. 23, no. 9, pp. 1502–1514, Sep. 2016, doi:10.1038/cdd.2016.36.

[100] T. Prozorovski, R. Schneider, C. Berndt, H.-P. Hartung, and O. Aktas, "Redox-regulated fate of neural stem progenitor cells," *Biochim. Biophys. Acta*, vol. 1850, no. 8, pp. 1543–1554, Aug. 2015, doi:10.1016/j.bbagen.2015.01.022.

[101] M. A. Le Gros *et al.*, "Soft X-ray tomography reveals gradual chromatin compaction and reorganization during neurogenesis in vivo," *Cell Rep.*, vol. 17, no. 8, pp. 2125–2136, Nov. 2016, doi:10.1016/j.celrep.2016.10.060.

[102] M. Götz and W. B. Huttner, "The cell biology of neurogenesis," *Nat. Rev. Mol. Cell Biol.*, vol. 6, no. 10, pp. 777–788, Oct. 2005, doi:10.1038/nrm1739.

[103] Q. Zhang *et al.*, "Metabolic regulation of SIRT1 transcription via a HIC1: CtBP corepressor complex," *Proc. Natl. Acad. Sci. U. S. A.*, vol. 104, no. 3, pp. 829–833, Jan. 2007, doi:10.1073/pnas.0610590104.

[104] J. M. Dias *et al.*, "CtBPs sense microenvironmental oxygen levels to regulate neural stem cell state," *Cell Rep.*, vol. 8, no. 3, pp. 665–670, Aug. 2014, doi:10.1016/j.celrep.2014.06.057.

[105] Z. Xie *et al.*, "Smad6 promotes neuronal differentiation in the intermediate zone of the dorsal neural tube by inhibition of the Wnt/beta-catenin pathway," *Proc. Natl. Acad. Sci. U. S. A.*, vol. 108, no. 29, pp. 12119–12124, Jul. 2011, doi:10.1073/pnas.1100160108.

[106] E. Pastrana, L.-C. Cheng, and F. Doetsch, "Simultaneous prospective purification of adult subventricular zone neural stem cells and their progeny," *Proc. Natl. Acad. Sci. U. S. A.*, vol. 106, no. 15, pp. 6387–6392, Apr. 2009, doi:10.1073/pnas.0810407106.

[107] S. Ahn and A. L. Joyner, "In vivo analysis of quiescent adult neural stem cells responding to Sonic hedgehog," *Nature*, vol. 437, no. 7060, pp. 894–897, Oct. 2005, doi:10.1038/nature03994.

[108] P. Codega *et al.*, "Prospective identification and purification of quiescent adult neural stem cells from their in vivo niche," *Neuron*, vol. 82, no. 3, pp. 545–559, May 2014, doi:10.1016/j.neuron.2014.02.039.

[109] S. Temple, "The development of neural stem cells," *Nature*, vol. 414, no. 6859, pp. 112–117, Nov. 2001, doi:10.1038/35102174.

[110] F. H. Gage and S. Temple, "Neural stem cells: generating and regenerating the brain," *Neuron*, vol. 80, no. 3, pp. 588–601, Oct. 2013, doi:10.1016/j.neuron.2013.10.037.

[111] F. D. Miller and A. Gauthier-Fisher, "Home at last: neural stem cell niches defined," *Cell Stem Cell*, vol. 4, no. 6, pp. 507–510, Jun. 2009, doi:10.1016/j.stem.2009.05.008.

[112] J. P. Andreotti *et al.*, "Neural stem cell niche heterogeneity," *Semin. Cell Dev. Biol.*, vol. 95, pp. 42–53, Nov. 2019, doi:10.1016/j.semcdb.2019.01.005.

[113] R. Beckervordersandforth, "Mitochondrial metabolism-mediated regulation of adult neurogenesis," *Brain Plast. Amst. Neth.*, vol. 3, no. 1, pp. 73–87, Nov. 2017, doi:10.3233/BPL-170044.

[114] G. Pellegrino *et al.*, "A comparative study of the neural stem cell niche in the adult hypothalamus of human, mouse, rat and gray mouse lemur (Microcebus murinus)," *J. Comp. Neurol.*, vol. 526, no. 9, pp. 1419–1443, Jun. 2018, doi:10.1002/cne.24376.

[115] S. C. Robins *et al.*, "α-Tanycytes of the adult hypothalamic third ventricle include distinct populations of FGF-responsive neural progenitors," *Nat. Commun.*, vol. 4, p. 2049, 2013, doi:10.1038/ncomms3049.

[116] D. A. Lim and A. Alvarez-Buylla, "The adult Ventricular-Subventricular Zone (V-SVZ) and Olfactory Bulb (OB) neurogenesis," *Cold Spring Harb. Perspect. Biol.*, vol. 8, no. 5, May 2016, doi:10.1101/cshperspect.a018820.

[117] A. Paul, Z. Chaker, and F. Doetsch, "Hypothalamic regulation of regionally distinct adult neural stem cells and neurogenesis," *Science*, vol. 356, no. 6345, pp. 1383–1386, Jun. 2017, doi:10.1126/science.aal3839.

[118] P. Bouchard-Cannon, L. Mendoza-Viveros, A. Yuen, M. Kaern, and H.-Y. M. Cheng, "The circadian molecular clock regulates adult hippocampal neurogenesis by controlling the timing of cell-cycle entry and exit," *Cell Rep.*, vol. 5, no. 4, pp. 961–973, Nov. 2013, doi:10.1016/j.celrep.2013.10.037.

[119] G. Kempermann, "Activity dependency and aging in the regulation of adult neurogenesis," *Cold Spring Harb. Perspect. Biol.*, vol. 7, no. 11, Nov. 2015, doi:10.1101/cshperspect.a018929.

[120] D. Stangl and S. Thuret, "Impact of diet on adult hippocampal neurogenesis," *Genes Nutr.*, vol. 4, no. 4, pp. 271–282, Dec. 2009, doi:10.1007/s12263-009-0134-5.

[121] H. R. Park and J. Lee, "Neurogenic contributions made by dietary regulation to hippocampal neurogenesis," *Ann. N. Y. Acad. Sci.*, vol. 1229, pp. 23–28, Jul. 2011, doi:10.1111/j.1749-6632.2011.06089.x.

[122] S. Fusco *et al.*, "A CREB-Sirt1-Hes1 circuitry mediates neural stem cell response to glucose availability," *Cell Rep.*, vol. 14, no. 5, pp. 1195–1205, Feb. 2016, doi:10.1016/j.celrep.2015.12.092.

[123] S. Michán *et al.*, "SIRT1 is essential for normal cognitive function and synaptic plasticity," *J. Neurosci. Off. J. Soc. Neurosci.*, vol. 30, no. 29, pp. 9695–9707, Jul. 2010, doi:10.1523/JNEUROSCI.0027-10.2010.

[124] J. Gao *et al.*, "A novel pathway regulates memory and plasticity via SIRT1 and miR-134," *Nature*, vol. 466, no. 7310, pp. 1105–1109, Aug. 2010, doi:10.1038/nature09271.

[125] M. O. Dietrich *et al.*, "Agrp neurons mediate Sirt1's action on the melanocortin system and energy balance: roles for Sirt1 in neuronal firing and synaptic plasticity," *J. Neurosci. Off. J. Soc. Neurosci.*, vol. 30, no. 35, pp. 11815–11825, Sep. 2010, doi:10.1523/JNEUROSCI.2234-10.2010.

[126] J. Jang *et al.*, "SIRT1 enhances the survival of human embryonic stem cells by promoting DNA repair," *Stem Cell Rep.*, vol. 9, no. 2, pp. 629–641, Aug. 2017, doi:10.1016/j.stemcr.2017.06.001.

[127] B. T. Lang, Y. Yan, R. J. Dempsey, and R. Vemuganti, "Impaired neurogenesis in adult type-2 diabetic rats," *Brain Res.*, vol. 1258, pp. 25–33, Mar. 2009, doi:10.1016/j.brainres.2008.12.026.

[128] I. Casetta *et al.*, "Epidemiology of Parkinson's disease in Italy. A descriptive survey in the U.S.L. of Cento, province of Ferrara, Emilia-Romagna," *Acta Neurol. (Napoli)*, vol. 12, no. 4, pp. 284–291, Aug. 1990.

[129] H. Y. Cohen *et al.*, "Calorie restriction promotes mammalian cell survival by inducing the SIRT1 deacetylase," *Science*, vol. 305, no. 5682, pp. 390–392, Jul. 2004, doi:10.1126/science.1099196.

[130] M. P. Mattson, "Neuroprotective signaling and the aging brain: take away my food and let me run," *Brain Res.*, vol. 886, no. 1–2, pp. 47–53, Dec. 2000, doi:10.1016/s0006-8993(00)02790-6.

[131] H. S. Ghosh, B. Reizis, and P. D. Robbins, "SIRT1 associates with eIF2-alpha and regulates the cellular stress response," *Sci. Rep.*, vol. 1, p. 150, 2011, doi:10.1038/srep00150.

[132] J. H. Back *et al.*, "Cancer cell survival following DNA damage-mediated premature senescence is regulated by mammalian target of rapamycin (mTOR)-dependent Inhibition of sirtuin 1," *J. Biol. Chem.*, vol. 286, no. 21, pp. 19100–19108, May 2011, doi:10.1074/jbc.M111.240598.

[133] C. Cantó *et al.*, "AMPK regulates energy expenditure by modulating NAD+ metabolism and SIRT1 activity," *Nature*, vol. 458, no. 7241, pp. 1056–1060, Apr. 2009, doi:10.1038/nature07813.

[134] C. Cantó *et al.*, "Interdependence of AMPK and SIRT1 for metabolic adaptation to fasting and exercise in skeletal muscle," *Cell Metab.*, vol. 11, no. 3, pp. 213–219, Mar. 2010, doi:10.1016/j.cmet.2010.02.006.

[135] S. Cetrullo, S. D'Adamo, B. Tantini, R. M. Borzi, and F. Flamigni, "mTOR, AMPK, and Sirt1: key

players in metabolic stress management," *Crit. Rev. Eukaryot. Gene Expr.*, vol. 25, no. 1, pp. 59–75, 2015, doi:10.1615/critreveukaryotgeneexpr.2015012975.

[136] C. R. McKeown and H. T. Cline, "Correction: nutrient restriction causes reversible G2 arrest in Xenopus neural progenitors," *Dev. Camb. Engl.*, vol. 147, no. 15, Aug. 2020, doi:10.1242/dev.195479.

[137] I. C. Lee *et al.*, "Oxidative stress promotes SIRT1 recruitment to the GADD34/PP1α complex to activate its deacetylase function," *Cell Death Differ.*, vol. 25, no. 2, pp. 255–267, Feb. 2018, doi:10.1038/cdd.2017.152.

[138] M. C. Hollander, S. Poola-Kella, and A. J. Fornace, "Gadd34 functional domains involved in growth suppression and apoptosis," *Oncogene*, vol. 22, no. 25, pp. 3827–3832, Jun. 2003, doi:10.1038/sj.onc.1206567.

[139] K. Utani and M. I. Aladjem, "Extra view: Sirt1 acts as a gatekeeper of replication initiation to preserve genomic stability," *Nucl. Austin Tex*, vol. 9, no. 1, pp. 261–267, Jan. 2018, doi:10.1080/19491034.2018.1456218.

[140] A. Lazutkin, O. Podgorny, and G. Enikolopov, "Modes of division and differentiation of neural stem cells," *Behav. Brain Res.*, vol. 374, p. 112118, Nov. 2019, doi:10.1016/j.bbr.2019.112118.

[141] H.-C. Chang and L. Guarente, "SIRT1 and other sirtuins in metabolism," *Trends Endocrinol Metab*, vol. 25, no. 3, pp. 138–145, Mar. 2014, doi:10.1016/j.tem.2013.12.001.

[142] W. Dang, "The controversial world of sirtuins," *Drug Discov Today Technol*, vol. 12, pp. e9–e17, Jun. 2014, doi:10.1016/j.ddtec.2012.08.003.

[143] G. Taormina and M. G. Mirisola, "Longevity: epigenetic and biomolecular aspects," *Biomol. Concepts*, vol. 6, no. 2, pp. 105–117, Apr. 2015, doi:10.1515/bmc-2014-0038.

[144] A. D. Goldberg, C. D. Allis, and E. Bernstein, "Epigenetics: a landscape takes shape," *Cell*, vol. 128, no. 4, pp. 635–638, Feb. 2007, doi:10.1016/j.cell.2007.02.006.

[145] L. A. Baker, C. D. Allis, and G. G. Wang, "PHD fingers in human diseases: disorders arising from misinterpreting epigenetic marks," *Mutat. Res.*, vol. 647, no. 1–2, pp. 3–12, Dec. 2008, doi:10.1016/j.mrfmmm.2008.07.004.

[146] K. R. Patel, E. Scott, V. A. Brown, A. J. Gescher, W. P. Steward, and K. Brown, "Clinical trials of resveratrol," *Ann. N. Y. Acad. Sci.*, vol. 1215, pp. 161–169, Jan. 2011, doi:10.1111/j.1749-6632.2010.05853.x.

[147] L. Subramanian, S. Youssef, S. Bhattacharya, J. Kenealey, A. S. Polans, and P. R. van Ginkel,

"Resveratrol: challenges in translation to the clinic: a critical discussion," *Clin. Cancer Res. Off. J. Am. Assoc. Cancer Res.*, vol. 16, no. 24, pp. 5942–5948, Dec. 2010, doi:10.1158/1078-0432.CCR-10-1486.

[148] V. D. Longo and B. K. Kennedy, "Sirtuins in aging and age-related disease," *Cell*, vol. 126, no. 2, pp. 257–268, Jul. 2006.

[149] N. Vall-Llaura, N. Mir, L. Garrido, C. Vived, and E. Cabiscol, "Redox control of yeast Sir2 activity is involved in acetic acid resistance and longevity," *Redox Biol.*, vol. 24, p. 101229, Jun. 2019, doi:10.1016/j.redox.2019.101229.

[150] H.-S. Kim *et al.*, "SIRT2 maintains genome integrity and suppresses tumorigenesis through regulating APC/C activity," *Cancer Cell*, vol. 20, no. 4, pp. 487–499, Oct. 2011, doi:10.1016/j.ccr.2011.09.004.

[151] T. Finkel, C.-X. Deng, and R. Mostoslavsky, "Recent progress in the biology and physiology of sirtuins," *Nature*, vol. 460, no. 7255, pp. 587–591, Jul. 2009, doi:10.1038/nature08197.

[152] L. Bordone *et al.*, "SIRT1 transgenic mice show phenotypes resembling calorie restriction," *Aging Cell*, vol. 6, no. 6, pp. 759–767, Dec. 2007, doi:10.1111/j.1474-9726.2007.00335.x.

[153] A. Brunet *et al.*, "Stress-dependent regulation of FOXO transcription factors by the SIRT1 deacetylase," *Science*, vol. 303, no. 5666, pp. 2011–2015, Mar. 2004, doi:10.1126/science.1094637.

[154] C. V. Anselmi *et al.*, "Association of the FOXO3A locus with extreme longevity in a southern Italian centenarian study," *Rejuvenation Res.*, vol. 12, no. 2, pp. 95–104, Apr. 2009, doi:10.1089/rej.2008.0827.

[155] F. Flachsbart *et al.*, "Association of FOXO3A variation with human longevity confirmed in German centenarians," *Proc. Natl. Acad. Sci. U. S. A.*, vol. 106, no. 8, pp. 2700–2705, Feb. 2009, doi:10.1073/pnas.0809594106.

[156] G. Murtaza, A. K. Khan, R. Rashid, S. Muneer, S. M. F. Hasan, and J. Chen, "FOXO transcriptional factors and long-term living," *Oxid. Med. Cell. Longev.*, vol. 2017, no. 3494289, p. 8, Aug. 2017.

[157] C. Kenyon, "The plasticity of aging: insights from long-lived mutants," *Cell*, vol. 120, no. 4, pp. 449–460, Feb. 2005, doi:10.1016/j.cell.2005.02.002.

[158] G. Taormina and M. G. Mirisola, "Calorie restriction in mammals and simple model organisms," *BioMed Res. Int.*, vol. 2014, p. 308690, 2014, doi:10.1155/2014/308690.

[159] L. G. Noriega et al., "CREB and ChREBP oppositely regulate SIRT1 expression in response to energy availability," EMBO Rep., vol. 12, no. 10, pp. 1069–1076, Sep. 2011, doi:10.1038/embor.2011.151.

[160] D. Chen, A. D. Steele, S. Lindquist, and L. Guarente, "Increase in activity during calorie restriction requires Sirt1," Science, vol. 310, no. 5754, p. 1641, Dec. 2005, doi:10.1126/science.1118357.

[161] L. Aguilar-Arnal, S. Katada, R. Orozco-Solis, and P. Sassone-Corsi, "NAD(+)-SIRT1 control of H3K4 trimethylation through circadian deacetylation of MLL1," Nat. Struct. Mol. Biol., vol. 22, no. 4, pp. 312–318, Apr. 2015, doi:10.1038/nsmb.2990.

[162] A. P. Gomes et al., "Declining NAD(+) induces a pseudohypoxic state disrupting nuclear-mitochondrial communication during aging," Cell, vol. 155, no. 7, pp. 1624–1638, Dec. 2013, doi:10.1016/j.cell.2013.11.037.

[163] L. Aguilar-Arnal, S. Ranjit, C. Stringari, R. Orozco-Solis, E. Gratton, and P. Sassone-Corsi, "Spatial dynamics of SIRT1 and the subnuclear distribution of NADH species," Proc. Natl. Acad. Sci. U. S. A., vol. 113, no. 45, pp. 12715–12720, Nov. 2016, doi:10.1073/pnas.1609227113.

[164] A. Nikiforov, V. Kulikova, and M. Ziegler, "The human NAD metabolome: functions, metabolism and compartmentalization," Crit. Rev. Biochem. Mol. Biol., vol. 50, no 4, pp 284–297, 2015, doi:10.3109/10409238.2015.1028612.

[165] M. S. Cohen, "Interplay between compartmentalized NAD+ synthesis and consumption: a focus on the PARP family," Genes Dev., vol. 34, no. 5–6, pp. 254–262, Mar. 2020, doi:10.1101/gad.335109.119.

[166] Y. Yang and A. A. Sauve, "NAD(+) metabolism: bioenergetics, signaling and manipulation for therapy," Biochim. Biophys. Acta, vol. 1864, no. 12, pp. 1787–1800, Dec. 2016, doi:10.1016/j.bbapap.2016.06.014.

[167] E. M. Kropp et al., "Inhibition of an NAD$^+$ salvage pathway provides efficient and selective toxicity to human pluripotent stem cells," Stem Cells Transl. Med., vol. 4, no. 5, pp. 483–493, May 2015, doi:10.5966/sctm.2014-0163.

[168] L. R. Stein and S. Imai, "Specific ablation of Nampt in adult neural stem cells recapitulates their functional defects during aging," EMBO J., vol. 33, no. 12, pp. 1321–1340, Jun. 2014, doi:10.1002/embj.201386917.

[169] M. J. Son et al., "Nicotinamide overcomes pluripotency deficits and reprogramming barriers," Stem Cells Dayt. Ohio, vol. 31, no. 6, pp. 1121–1135, Jun. 2013, doi:10.1002/stem.1368.

[170] J. G. Ryall et al., "The NAD(+)-dependent SIRT1 deacetylase translates a metabolic switch into regulatory epigenetics in skeletal muscle stem cells," Cell Stem Cell, vol. 16, no. 2, pp. 171–183, Feb. 2015, doi:10.1016/j.stem.2014.12.004.

[171] J. Landry et al., "The silencing protein SIR2 and its homologs are NAD-dependent protein deacetylases," Proc. Natl. Acad. Sci. U. S. A., vol. 97, no. 11, pp. 5807–5811, May 2000, doi:10.1073/pnas.110148297.

[172] D. Corda and M. Di Girolamo, "Functional aspects of protein mono-ADP-ribosylation," EMBO J., vol. 22, no. 9, pp. 1953–1958, May 2003, doi:10.1093/emboj/cdg209.

[173] M. O. Hottiger, "Nuclear ADP-ribosylation and its role in chromatin plasticity, cell differentiation, and epigenetics," Annu. Rev. Biochem., vol. 84, pp. 227–263, 2015, doi:10.1146/annurev-biochem-060614-034506.

[174] J. Du, H. Jiang, and H. Lin, "Investigating the ADP-ribosyltransferase activity of sirtuins with NAD analogues and 32P-NAD," Biochemistry, vol. 48, no. 13, pp. 2878–2890, Apr. 2009, doi:10.1021/bi802093g.

[175] M. C. Haigis et al., "SIRT4 inhibits glutamate dehydrogenase and opposes the effects of calorie restriction in pancreatic beta cells," Cell, vol. 126, no. 5, pp. 941–954, Sep. 2006, doi:10.1016/j.cell.2006.06.057.

[176] G. Liszt, E. Ford, M. Kurtev, and L. Guarente, "Mouse Sir2 homolog SIRT6 is a nuclear ADP-ribosyltransferase," J. Biol. Chem., vol. 280, no. 22, pp. 21313–21320, Jun. 2005, doi:10.1074/jbc.M413296200.

[177] N. G. Simonet et al., "SirT7 auto-ADP-ribosylation regulates glucose starvation response through mH2A1," Sci. Adv., vol. 6, no. 30, p. eaaz2590, Jul. 2020, doi:10.1126/sciadv.aaz2590.

[178] X. Guo et al., "The NAD(+)-dependent protein deacetylase activity of SIRT1 is regulated by its oligomeric status," Sci. Rep., vol. 2, p. 640, 2012, doi:10.1038/srep00640.

[179] J. Fang et al., "Sirt7 promotes adipogenesis in the mouse by inhibiting autocatalytic activation of Sirt1," Proc. Natl. Acad. Sci. U. S. A., vol. 114, no. 40, pp. E8352–E8361, Oct. 2017, doi:10.1073/pnas.1706945114.

[180] L. Li et al., "SIRT7 is a histone desuccinylase that functionally links to chromatin compaction and genome stability," *Nat. Commun.*, vol. 7, p. 12235, Jul. 2016, doi:10.1038/ncomms12235.

[181] Y. Nakahata, S. Sahar, G. Astarita, M. Kaluzova, and P. Sassone-Corsi, "Circadian control of the NAD+ salvage pathway by CLOCK-SIRT1," *Science*, vol. 324, no. 5927, pp. 654–657, May 2009, doi:10.1126/science.1170803.

[182] A. Malik, R. V. Kondratov, R. J. Jamasbi, and M. E. Geusz, "Circadian clock genes are essential for normal adult neurogenesis, differentiation, and fate determination," *PloS One*, vol. 10, no. 10, p. e0139655, 2015, doi:10.1371/journal.pone.0139655.

[183] A. A. H. Ali, B. Schwarz-Herzke, A. Stahr, T. Prozorovski, O. Aktas, and C. von Gall, "Premature aging of the hippocampal neurogenic niche in adult Bmal1-deficient mice," *Aging*, vol. 7, no. 6, pp. 435–449, Jun. 2015, doi:10.18632/aging.100764.

[184] B. Jablonska et al., "Sirt1 regulates glial progenitor proliferation and regeneration in white matter after neonatal brain injury," *Nat. Commun.*, vol. 7, p. 13866, Dec. 2016, doi:10.1038/ncomms13866.

[185] S. F. Leiser and M. Kaeberlein, "A role for SIRT1 in the hypoxic response," *Mol. Cell*, vol. 38, no. 6, pp. 779–780, Jun. 2010, doi:10.1016/j.molcel.2010.06.015.

[186] J.-H. Lim, Y.-M. Lee, Y.-S. Chun, J. Chen, J.-E. Kim, and J.-W. Park, "Sirtuin 1 modulates cellular responses to hypoxia by deacetylating hypoxia-inducible factor 1α," *Mol. Cell*, vol. 38, no. 6, pp. 864–878, Jun. 2010, doi:10.1016/j.molcel.2010.05.023.

[187] P. O. Hassa and M. O. Hottiger, "The diverse biological roles of mammalian PARPS, a small but powerful family of poly-ADP-ribose polymerases," *Front. Biosci. J. Virtual Libr.*, vol. 13, pp. 3046–3082, Jan. 2008, doi:10.2741/2909.

[188] C.-P. Hsu, S. Oka, D. Shao, N. Hariharan, and J. Sadoshima, "Nicotinamide phosphoribosyltransferase regulates cell survival through NAD+ synthesis in cardiac myocytes," *Circ. Res.*, vol. 105, no. 5, pp. 481–491, Aug. 2009, doi:10.1161/CIRCRESAHA.109.203703.

[189] L. Liu, P. Wang, X. Liu, D. He, C. Liang, and Y. Yu, "Exogenous NAD+ supplementation protects H9c2 cardiac myoblasts against hypoxia/reoxygenation injury via Sirt1-p53 pathway," *Fundam. Clin. Pharmacol.*, vol. 28, no. 2, pp. 180–189, Apr. 2014, doi:10.1111/fcp.12016.

[190] Z. Meng et al., "Resveratrol relieves ischemia-induced oxidative stress in the hippocampus by activating SIRT1," *Exp. Ther. Med.*, Jun. 2015, doi:10.3892/etm.2015.2555.

[191] E. Pantazi et al., "Silent information regulator 1 protects the liver against ischemia-reperfusion injury: implications in steatotic liver ischemic preconditioning," *Transpl. Int. Off. J. Eur. Soc. Organ Transplant.*, vol. 27, no. 5, pp. 493–503, May 2014, doi:10.1111/tri.12276.

[192] X. Meng, J. Tan, M. Li, S. Song, Y. Miao, and Q. Zhang, "Sirt1: role under the condition of ischemia/hypoxia," *Cell Mol Neurobiol*, vol. 37, no. 1, pp. 17–28, Mar. 2016.

[193] P. Li, Y. Liu, N. Burns, K.-S. Zhao, and R. Song, "SIRT1 is required for mitochondrial biogenesis reprogramming in hypoxic human pulmonary arteriolar smooth muscle cells," *Int. J. Mol. Med.*, vol. 39, no. 5, pp. 1127–1136, May 2017, doi:10.3892/ijmm.2017.2932.

[194] E.-J. Yeo, "Hypoxia and aging," *Exp. Mol. Med.*, vol. 51, no. 6, pp. 1–15, Jun. 2019, doi:10.1038/s12276-019-0233-3.

[195] R. Chen, E. M. Dioum, R. T. Hogg, R. D. Gerard, and J. A. Garcia, "Hypoxia increases sirtuin 1 expression in a hypoxia-inducible factor-dependent manner," *J. Biol. Chem.*, vol. 286, no. 16, pp. 13869–13878, Apr. 2011, doi:10.1074/jbc.M110.175414.

[196] Y. Yuan, V. F. Cruzat, P. Newsholme, J. Cheng, Y. Chen, and Y. Lu, "Regulation of SIRT1 in aging: roles in mitochondrial function and biogenesis," *Mech. Ageing Dev.*, vol. 155, pp. 10–21, Apr. 2016, doi:10.1016/j.mad.2016.02.003.

[197] C. Tong et al., "Impaired SIRT1 nucleocytoplasmic shuttling in the senescent heart during ischemic stress," *FASEB J.*, vol. 27, no. 11, pp. 4332–4342, Sep. 2012.

[198] M. Tanno, J. Sakamoto, T. Miura, K. Shimamoto, and Y. Horio, "Nucleocytoplasmic shuttling of the NAD+-dependent histone deacetylase SIRT1," *J. Biol. Chem.*, vol. 282, no. 9, pp. 6823–6832, Mar. 2007, doi:10.1074/jbc.M609554200.

[199] L. De Filippis and D. Delia, "Hypoxia in the regulation of neural stem cells," *Cell. Mol. Life Sci. CMLS*, vol. 68, no. 17, pp. 2831–2844, Sep. 2011, doi:10.1007/s00018-011-0723-5.

[200] D. M. Panchision, "The role of oxygen in regulating neural stem cells in development and disease," *J. Cell. Physiol.*, vol. 220, no. 3, pp. 562–568, Sep. 2009, doi:10.1002/jcp.21812.

[201] R. S. Zee et al., "Redox regulation of sirtuin-1 by S-glutathiolation," *Antioxid. Redox Signal.*, vol. 13, no. 7, pp. 1023–1032, Oct. 2010, doi:10.1089/ars.2010.3251.

[202] L. Bräutigam et al., "Glutaredoxin regulates vascular development by reversible glutathionylation of sirtuin 1," *Proc. Natl. Acad. Sci. U. S. A.*, vol. 110, no. 50, pp. 20057–20062, Dec. 2013, doi:10.1073/pnas.1313753110.

[203] N. Vall-Llaura, G. Reverter-Branchat, C. Vived, N. Weertman, M. J. Rodríguez-Colman, and E. Cabiscol, "Reversible glutathionylation of Sir2 by monothiol glutaredoxins Grx3/4 regulates stress resistance," *Free Radic. Biol. Med.*, vol. 96, pp. 45–56, Jul. 2016, doi:10.1016/j.freeradbiomed.2016.04.008.

[204] D. Shao et al., "A redox-resistant sirtuin-1 mutant protects against hepatic metabolic and oxidant stress," *J. Biol. Chem.*, vol. 289, no. 11, pp. 7293–7306, Mar. 2014, doi:10.1074/jbc.M113.520403.

[205] D. Shao et al., "Glutaredoxin-1 deficiency causes fatty liver and dyslipidemia by inhibiting Sirtuin-1," *Antioxid. Redox Signal.*, vol. 27, no. 6, pp. 313–327, Aug. 2017, doi:10.1089/ars.2016.6716.

[206] C. Du et al., "Sulfhydrated Sirtuin-1 increasing its deacetylation activity is an essential epigenetics mechanism of anti-atherogenesis by hydrogen sulfide," *Antioxid. Redox Signal.*, vol. 30, no. 2, pp. 184–197, Jan. 2019, doi:10.1089/ars.2017.7195.

[207] R. Wang, "Hydrogen sulfide: the third gasotransmitter in biology and medicine," *Antioxid. Redox Signal.*, vol. 12, no. 9, pp. 1061–1064, May 2010, doi:10.1089/ars.2009.2938.

[208] M. Wang, J.-J. Tang, L.-X. Wang, J. Yu, L. Zhang, and C. Qiao, "Hydrogen sulfide enhances adult neurogenesis in a mouse model of Parkinson's disease," *Neural Regen. Res.*, vol. 16, no. 7, pp. 1353–1358, Jul. 2021, doi:10.4103/1673-5374.301026.

[209] Y. Nie et al., "STAT3 inhibition of gluconeogenesis is downregulated by SirT1," *Nat. Cell Biol.*, vol. 11, no. 4, pp. 492–500, Apr. 2009, doi:10.1038/ncb1857.

[210] W. Wang et al., "JAK1-mediated Sirt1 phosphorylation functions as a negative feedback of the JAK1-STAT3 pathway," *J. Biol. Chem.*, vol. 293, no. 28, pp. 11067–11075, Jul. 2018, doi:10.1074/jbc.RA117.001387.

[211] T. Yoshimatsu et al., "Non-cell-autonomous action of STAT3 in maintenance of neural precursor cells in the mouse neocortex," *Dev. Camb. Engl.*, vol. 133, no. 13, pp. 2553–2563, Jul. 2006, doi:10.1242/dev.02419.

[212] F. Gu et al., "Suppression of Stat3 promotes neurogenesis in cultured neural stem cells," *J. Neurosci. Res.*, vol. 81, no. 2, pp. 163–171, Jul. 2005, doi:10.1002/jnr.20561.

[213] X. Kong et al., "JAK2/STAT3 signaling mediates IL-6-inhibited neurogenesis of neural stem cells through DNA demethylation/methylation," *Brain. Behav. Immun.*, vol. 79, pp. 159–173, Jul. 2019, doi:10.1016/j.bbi.2019.01.027.

[214] A. Bonni et al., "Regulation of gliogenesis in the central nervous system by the JAK-STAT signaling pathway," *Science*, vol. 278, no. 5337, pp. 477–483, Oct. 1997, doi:10.1126/science.278.5337.477.

[215] P. Rajan and R. D. McKay, "Multiple routes to astrocytic differentiation in the CNS," *J. Neurosci. Off. J. Soc. Neurosci.*, vol. 18, no. 10, pp. 3620–3629, May 1998.

[216] N.-Y. Song, Y.-H. Lee, H.-K. Na, J.-H. Baek, and Y.-J. Surh, "Leptin induces SIRT1 expression through activation of NF-E2-related factor 2: implications for obesity-associated colon carcinogenesis," *Biochem. Pharmacol.*, vol. 153, pp. 282–291, Jul. 2018, doi:10.1016/j.bcp.2018.02.001.

[217] F. Ghisays et al., "The N-terminal domain of SIRT1 is a positive regulator of endogenous SIRT1-dependent deacetylation and transcriptional outputs," *Cell Rep.*, vol. 10, no. 10, pp. 1665–1673, Mar. 2015, doi:10.1016/j.celrep.2015.02.036.

[218] H. Kang et al., "Sirt1 carboxyl domain is an ATP-repressible domain that is transferrable to other proteins," *Nat. Commun.*, vol. 8, p. 15560, May 2017, doi:10.1038/ncomms15560.

[219] B. P. Hubbard et al., "Evidence for a common mechanism of SIRT1 regulation by allosteric activators," *Science*, vol. 339, no. 6124, pp. 1216–1219, Mar. 2013, doi:10.1126/science.1231097.

[220] D. Meng, A. R. Frank, and J. L. Jewell, "mTOR signaling in stem and progenitor cells," *Dev. Camb. Engl.*, vol. 145, no. 1, Jan. 2018, doi:10.1242/dev.152595.

[221] N. K. Love, N. Keshavan, R. Lewis, W. A. Harris, and M. Agathocleous, "A nutrient-sensitive restriction point is active during retinal progenitor cell differentiation," *Dev. Camb. Engl.*, vol. 141, no. 3, pp. 697–706, Feb. 2014, doi:10.1242/dev.103978.

[222] B. Jablonska et al., "Oligodendrocyte regeneration after neonatal hypoxia requires FoxO1-mediated p27Kip1 expression," *J. Neurosci. Off. J. Soc. Neurosci.*, vol. 32, no. 42, pp. 14775–14793, Oct. 2012, doi:10.1523/JNEUROSCI.2060-12.2012.

[223] J.-E. Kim, J. Chen, and Z. Lou, "DBC1 is a negative regulator of SIRT1," *Nature*, vol. 451, no. 7178, pp. 583–586, Jan. 2008, doi:10.1038/nature06500.

[224] W. Zhao, J.-P. Kruse, Y. Tang, S. Y. Jung, J. Qin, and W. Gu, "Negative regulation of the deacetylase SIRT1 by DBC1," *Nature*, vol. 451, no. 7178, pp. 587–590, Jan. 2008, doi:10.1038/nature06515.

[225] T. C. Krzysiak *et al.*, "An insulin-responsive sensor in the SIRT1 disordered region binds DBC1 and PACS-2 to control enzyme activity," *Mol. Cell*, vol. 72, no. 6, pp. 985–998.e7, Dec. 2018, doi:10.1016/j.molcel.2018.10.007.

[226] A. N. Ziegler, S. W. Levison, and T. L. Wood, "Insulin and IGF receptor signalling in neural-stem-cell homeostasis," *Nat. Rev. Endocrinol.*, vol. 11, no. 3, pp. 161–170, Mar. 2015, doi:10.1038/nrendo.2014.208.

[227] S. E. Choi *et al.*, "Obesity-linked phosphorylation of SIRT1 by Casein Kinase 2 inhibits its nuclear localization and promotes fatty liver," *Mol. Cell. Biol.*, vol. 37, no. 15, Aug. 2017, doi:10.1128/MCB.00006-17.

[228] X. Liu *et al.*, "Methyltransferase Set7/9 regulates p53 activity by interacting with Sirtuin 1 (SIRT1)," *Proc. Natl. Acad. Sci. U. S. A.*, vol. 108, no. 5, pp. 1925–1930, Feb. 2011, doi:10.1073/pnas.1019619108.

[229] A. Luna, M. I. Aladjem, and K. W. Kohn, "SIRT1/PARP1 crosstalk: connecting DNA damage and metabolism," *Genome Integr.*, vol. 4, no. 1, p. 6, Dec. 2013, doi:10.1186/2041-9414-4-6.

[230] J. P. Singh, K. Zhang, J. Wu, and X. Yang, "O-GlcNAc signaling in cancer metabolism and epigenetics," *Cancer Lett.*, vol. 356, no. 2 Pt A, pp. 244–250, Jan. 2015, doi:10.1016/j.canlet.2014.04.014.

[231] D. C. Love, M. W. Krause, and J. A. Hanover, "O-GlcNAc cycling: emerging roles in development and epigenetics," *Semin. Cell Dev. Biol.*, vol. 21, no. 6, pp. 646–654, Aug. 2010, doi:10.1016/j.semcdb.2010.05.001.

[232] N. E. Zachara and G. W. Hart, "O-GlcNAc a sensor of cellular state: the role of nucleocytoplasmic glycosylation in modulating cellular function in response to nutrition and stress," *Biochim. Biophys. Acta*, vol. 1673, no. 1–2, pp. 13–28, Jul. 2004, doi:10.1016/j.bbagen.2004.03.016.

[233] C. Han *et al.*, "O-GlcNAcylation of SIRT1 enhances its deacetylase activity and promotes cytoprotection under stress," *Nat. Commun.*, vol. 8, no. 1, p. 1491, Nov. 2017, doi:10.1038/s41467-017-01654-6.

[234] R.-R. Tan *et al.*, "Abnormal O-GlcNAcylation of Pax3 occurring from hyperglycemia-induced neural tube defects is ameliorated by carnosine but not folic acid in chicken embryos," *Mol. Neurobiol.*, vol. 54, no. 1, pp. 281–294, Jan. 2017, doi:10.1007/s12035-015-9581-8.

[235] S. A. Myers, B. Panning, and A. L. Burlingame, "Polycomb repressive complex 2 is necessary for the normal site-specific O-GlcNAc distribution in mouse embryonic stem cells," *Proc. Natl. Acad. Sci. U. S. A.*, vol. 108, no. 23, pp. 9490–9495, Jun. 2011, doi:10.1073/pnas.1019289108.

[236] H. Kang, J.-W. Jung, M. K. Kim, and J. H. Chung, "CK2 is the regulator of SIRT1 substrate-binding affinity, deacetylase activity and cellular response to DNA-damage," *PLoS One*, vol. 4, no. 8, p. e6611, Aug. 2009, doi:10.1371/journal.pone.0006611.

[237] X. Guo, J. G. Williams, T. T. Schug, and X. Li, "DYRK1A and DYRK3 promote cell survival through phosphorylation and activation of SIRT1," *J. Biol. Chem.*, vol. 285, no. 17, pp. 13223–13232, Apr. 2010, doi:10.1074/jbc.M110.102574.

[238] C. Di Vona *et al.*, "Chromatin-wide profiling of DYRK1A reveals a role as a gene-specific RNA polymerase II CTD kinase," *Mol. Cell*, vol. 57, no. 3, pp. 506–520, Feb. 2015, doi:10.1016/j.molcel.2014.12.026.

CHAPTER FIFTEEN

Roles of Hydrogen Peroxide and Peroxiredoxin in the Yeast Replicative Aging Model of Aging and Age-Related Disease

Mikael Molin

CONTENTS

15.1 YEAST AGING MODELS—REPLICATIVE VERSUS CHRONOLOGICAL

Because of its short life span, facile genetics, and the availability of several genome-wide mutant and epitope-tagged gene collections, the unicellular model organism, budding yeast, has become an important tool to understand molecular mechanisms underlying the apparent universally conserved process of aging. Two features of the yeast life cycle have been used to model aging and age-related disease, namely (1) the number of daughter cells a mother cell can produce (replicative aging, Figure 15.1A–B), a property Mortimer and Johnston showed was finite about 60 years ago[1]; and (2) the time cells remain viable upon nutrient starvation (chronological aging, Figure 15.1C). Whereas the former resembles human aging in that it is thought to be an emergent property, not extensively modulated by evolutionary selection[2],

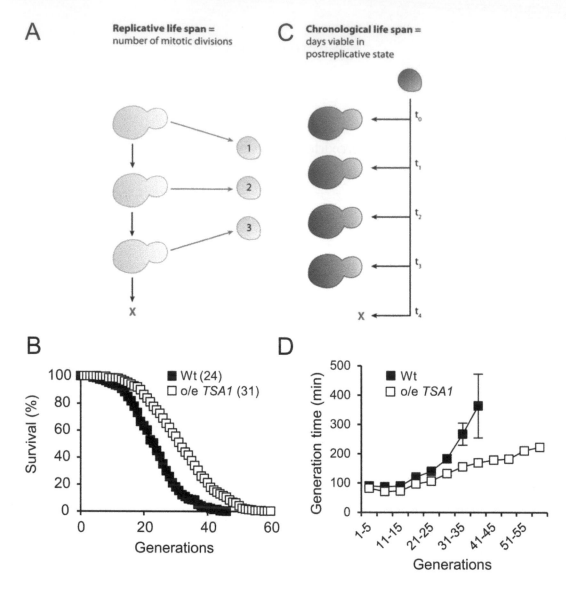

A Replicative life span =
number of mitotic divisions

C Chronological life span =
days viable in
postreplicative state

B

Survival (%) vs Generations

■ Wt (24)
□ o/e *TSA1* (31)

D

Generation time (min) vs Generations

■ Wt
□ o/e *TSA1*

Figure 15.1 Replicative aging in yeast can model aging and age-related deterioration in multicellular organisms.

Replicative life span in yeast equals the number of mitotic divisions (A) whereas chronological life span denotes the time cells remain in a non-dividing state due to nutrient starvation (C). Reproduced from reference[11] with permission. (B) Replicative life span of wild-type cells and cells expressing an extra copy of the *TSA1* gene. (D) Replicative fitness (i.e., the generation times at different replicative age) of wild-type cells and cells expressing an extra copy of the *TSA1* gene. (B) and (D) are reproduced with permission from reference[7].

the latter may better seize the fact that most cells in the human body do not divide to a large extent following development and early differentiation[3].

Studies of yeast replicative aging in the 1980s indicated that a plot of the fraction of surviving cells in a population as a function of the number of generations, rather than the standard chronological time[4], resulted in a survival curve closely resembling the Gompertz curve used in studies of aging

of multicellular organisms[5] (Figure 15.1B). In addition, a plot of the average time it takes cells at different ages to complete cell division (Figure 15.1D) gives a measure of the deterioration of fitness with increased replicative age in cells[6–9].

Yeast replicative aging is a tractable model system in the study of two phenomena closely connected to molecular mechanisms underlying aging. First, the accumulation of molecular aging factors with age; and second, rejuvenation (i.e., the resetting of cellular age in newborn organisms, a process thought to be carried out via mother-cell specific retention of aging factors)[3,10,11]. In fact, the separation of daughter cells and mother cells as damage-free and damage-accumulating cell lineages may resemble the separation of the soma and germ cells in multicellular organisms[12]. Nevertheless, for an aging factor to cause replicative aging, it has to both (1) accumulate with age in a manner causing cellular dysfunction and death and (2) be reset in daughter cells. In this chapter, I will discuss the roles of H_2O_2 and peroxiredoxin in yeast replicative aging.

15.2 EVIDENCE FOR A ROLE OF REACTIVE OXYGEN SPECIES IN SHAPING YEAST REPLICATIVE AGING

Early evidence for the relevance of ROS in yeast replicative aging consisted of observations that cells lacking the yeast genes encoding the superoxide dismutases (SODs)[13,14] or the catalases[15] sustained

shortened life-spans. Furthermore, increased dihydroethidium (DHE)-staining, suggesting increased superoxide levels, was observed as early as in cells 5–9 generations of age[16].

Problems in gathering evidence for the accumulation of specifically H_2O_2 have, as in many other model systems, mirrored the lack of sensors specific and sensitive enough to detect low levels of endogenous H_2O_2[17]. Nevertheless, increased dihydrorhodamine (DHR) staining in middle-aged yeast mother cells[18,19] support its accumulation during aging. In addition, the increased H_2O_2 levels are completely reset in daughter cells via a mechanism involving mother-cell specific retention of damaged antioxidant enzymes (e.g., SOD and catalase)[18]. Together the data support excessive H_2O_2 as a cause of replicative aging. More recently, hyperoxidation of the 2-Cys peroxiredoxin Tsa1, a process specifically occurring upon increased H_2O_2 (Figure 15.2A) was noted in cells from around 5 generations of age[20] and a moderate increase in the H_2O_2 specific oxidation of the genetic sensor

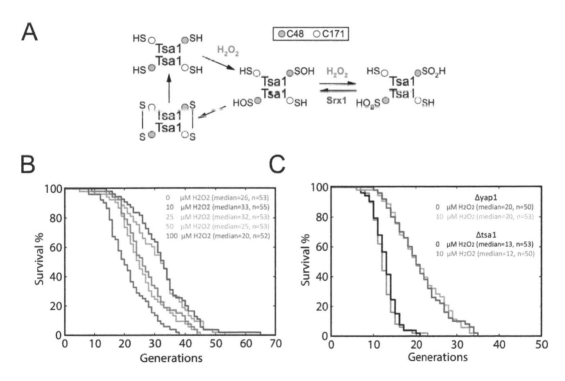

Figure 15.2 Low levels of H_2O_2 extend yeast replicative life span in a Tsa1-dependent manner.

(A) Overview of interactions of the most highly expressed yeast peroxiredoxin Tsa1 with H_2O_2. Reproduced with permission from reference[7]. (B) Replicative life span of wild-type cells measured in a microfluidic device with constant flow of medium containing the indicated levels of H_2O_2. (C) Replicative life span of cells deficient in Tsa1 (tsa1Δ) or Yap1 (yap1Δ) cells measured in a microfluidic device with constant flow of medium containing the indicated levels of H_2O_2. (B) and (C) are reproduced from reference[31] with permission.

HyPer3 in middle aged yeast mother cells (10–12 generations of age) reinforces this notion[21].

Several observations on key roles of the cysteine-dependent H_2O_2-scavengers and regulators of redox signaling, peroxiredoxins, in aging support an integral role of cytosolic H_2O_2 in yeast replicative aging. This since the affinities of typical 2-cysteine peroxiredoxins for H_2O_2 and their abundant expression suggest that they will react with a great majority of the H_2O_2 produced in cells[22], at least until they, upon encountering excess H_2O_2, are inactivated by hyperoxidation[23] (Figure 15.2A). Notably, the most highly expressed cytosolic 2-Cys peroxiredoxin in yeast, Tsa1, is a potent regulator of the rate of yeast replicative aging; its deletion causes ~30% reduced life span[20,24] whereas a mere twofold increased levels leads to 40% extended life span[7] (Figure 15.1B). The key role of the major cytosolic 2-Cys peroxiredoxin in yeast replicative aging, in particular the observation that peroxiredoxin activity limits the life span of wild type cells[7], further documents the importance of H_2O_2 in this process.

15.2.1 H_2O_2 Toxicity versus Signaling (and Mitohormesis)

Early work on roles of ROS, including H_2O_2, in aging focused extensively on their roles in causing irreparable damage in proteins, for example, (i.e., protein carbonylation)[25]. Protein carbonylation results from oxygen radical species reacting with amino acid side chains or the peptide backbone leading to its cleavage and increases gradually with age in several organisms, with more dramatic increases in the last third of their life span. Supporting the idea that oxidative damage is a replicative aging factor in yeast, carbonylated proteins accumulate during yeast replicative aging and their levels are reset in daughter cells[26].

In parallel with examples of low levels of H_2O_2 executing roles in signaling, seminal discoveries on a beneficial role of low levels of ROS in counteracting aging were made, initially in worms[27,28]. The phenomenon was termed "mitohormesis" because it involved moderate levels of the redox-cycling drug paraquat, acting in mitochondria to increase the levels of ROS. Subsequent studies suggest that also other drugs (e.g., the type-2 diabetes drug metformin) may act via mitohormesis to reduce age-related deterioration[29]. However, it has remained unclear which of the ROS superoxide anion or H_2O_2 is most relevant in mitohormetic

signaling responses[30]. However, through a microfluidic device maintaining yeast mother cells in a continuous flow of medium[6] constant levels of H_2O_2 could be maintained and their effects on replicative aging ascertained. Interestingly, whereas low levels of exogenous H_2O_2 (10–25 μM) extend life span in a Tsa1-dependent manner, higher levels (>100 μM) shorten life span[31] (Figure 15.2B–C). Thus, increased levels of H_2O_2 and peroxiredoxin-dependent signaling may explain the beneficial effects of mitohormesis on age-related decline also in multicellular organisms[29,32]. The absolute requirement for a 2-Cys Prx for metformin to extend life span in worms supports this notion[30].

15.3 HALLMARKS OF AGING RELEVANT IN MANY ORGANISMS

A review article from a couple of years back listed nine hallmarks of aging conserved in several organisms[33]. Not all of these are relevant in yeast replicative aging[34], but compelling evidence for age-related genomic instability, epigenetic alterations, loss of proteostasis, mitochondrial dysfunction, and deregulated nutrient sensing have been presented.

In addition, altered cytosolic and vacuolar pH homeostasis and their influence on mitochondrial function have both been identified as early onset causes of replicative aging, and to be reset in daughter cells[35,36]. Furthermore, it has become clear that many of the aging factors are extensively interconnected. A more modern theory embracing such sometimes rather distant interconnections has been put forward (the "integrative theory") and will be discussed below.

In the following sections I will in turn discuss the different specific hallmark deficiencies relevant in yeast replicative aging with particular attention to those implicating H_2O_2 and/or peroxiredoxin in their manifestation.

15.4 GENOMIC INSTABILITY

While the accumulation of different types of mutations with age has indeed been documented in yeast[37,38], and deficiencies in the progeria-related Sgs1 RecQ helicase homologue[39] or components of the DNA damage response[40] result in shortened replicative life span, it is generally accepted that any mutation in the mother will be passed on to its daughters[34]. Thus genomic instability violates the rejuvenation criteria discussed for aging factors above.

Furthermore, strong evidence against genome instability causing replicative aging has been reported through next-generation sequencing of DNA isolated from yeast mother cells of various age[41]. Gradual accumulation of mutations with age occurred at a rate far lower than age-related decline, and mutations correlated neither with age-induced growth arrest nor death. Similarly, long-lived cells mildly overproducing the peroxiredoxin Tsa1 neither display reduced mutation rates in young nor middle-aged cells, and conversely, reducing the eightfold increased mutation rates in young or middle-aged cells lacking Tsa1 by half did not improve their life span[7,20]. In fact, even in a strain lacking all eight functional thioredoxin-dependent peroxidases (peroxiredoxins and glutathione-peroxidase homologs) that accumulates mutations at a 12-fold higher rate than the wild-type, accumulation of mutations is too slow to cause aging[41]. Thus the role of peroxiredoxins in regulating aging seems to be distinct from their role in maintaining genome stability.

Nevertheless, the ribosomal DNA locus has been identified as a regulator of replicative aging through the production of extrachromosomal ribosomal circles (ERCs). ERCs are independently replicating circular DNA containing origins of replication and ribosomal DNA segments. ERCs result from homologous recombination during attempts to repair double-strand breaks near the rDNA locus and are selectively retained in mother cells[42,43]. However, a subsequent scrutinization of the role of the unstable and recombination-prone rDNA region in replicative aging using mutants with different replication potential suggested that replicative aging correlated with rDNA instability rather than the levels of ERCs[44]. Nevertheless, the peroxiredoxin Tsa1 and its dedicated desulfinylase enzyme Srx1 appear to control life span mostly independently of the major sirtuin, Sir2, and ERCs/rDNA stability[20].

15.5 EPIGENETIC ALTERATIONS

In yeast, the most well-studied example of epigenetics affecting replicative aging involves the protein and histone deacetylase and the first sirtuin to be discovered, Sir2. Sir2 appears to play a key role in regulating replicative aging via at least three different mechanisms: (1) through reducing the levels of rDNA instability/ERCs[42], (2) via maintaining silencing at heterochromatic regions at telomeres[45], and (3) through mother cell–specific retention of aggregated proteins[46,47].

Nevertheless, despite extensive interactions between the mammalian Sir2-homologue SIRT-1 and redox metabolism[48], life span control via Tsa1 and Srx1 seems to be mostly independent of Sir2-dependent epigenetic alterations[20]. Conversely, cells lacking Sir2 sustain normal resistance to different forms of oxidative stress[49–52].

15.6 LOSS OF PROTEOSTASIS

Deficiencies in the synthesis, folding, disaggregation, and degradation of proteins (protein homeostasis or proteostasis; Figure 15.3) have been identified as key factors in aging and the dramatically increased incidence of neurodegenerative diseases with age[53]. In particular, these diseases are characterized by the accumulation of misfolded and aggregated proteins in the cytosol of neurons.

Both protein synthesis and folding in the crowded intracellular environment requires an intricate network of helper proteins, so-called molecular chaperones[54]. Molecular chaperones typically bind to exposed hydrophobic patches on the surface and can be divided into foldases and holdases. Foldases are able to exercise molecular work through ATP-driven conformational changes, whereas holdases only bind hydrophobic patches in client proteins to prevent them from aggregating[53,55–>57]. Many chaperones are named heat-shock proteins (Hsps), after their increased synthesis and importance during heat stress, with a suffix relating to their approximate molecular weight. Of them, Hsp70 ATPases, with their co-chaperones the Hsp40s, are involved in all four aspects of proteostasis (Figure 15.3). These will be discussed below along with the protein disaggregase Hsp104, which is endowed with the unique function of binding to protein aggregates and facilitating their disaggregation[55,58].

15.6.1 Deficient Protein Synthesis

The rate of protein synthesis has been noted to decrease with age in both model organisms and human cells[59]. In addition, downregulation of translation by different means has been proposed to extend life span.

Replicatively aged yeast displays a dramatical reduction in protein synthesis at ~12 generations of age[60,61]. In fact, using a method involving the

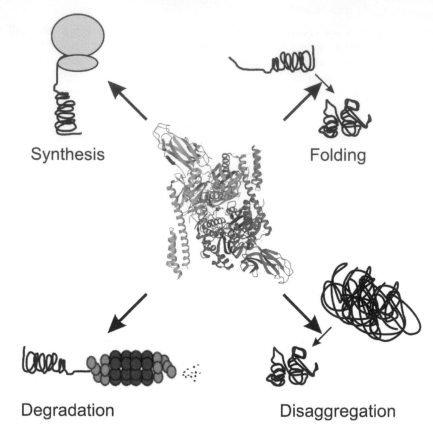

Synthesis

Folding

Degradation

Disaggregation

Figure 15.3 Overview of chaperone-facilitated proteostatic (protein homeostatic) mechanisms contributing to replicative aging in yeast.

At the center is the structure of an Hsp70 (PDB ID: 5TKY[156]), a group of proteins that have been found to facilitate protein synthesis, folding, disaggregation, and degradation in yeast.

incorporation of radioactive leucine, protein synthesis was noted to decrease gradually starting already at 3–4 generations of age[60], supporting its early onset. Interestingly a multi-omics investigation in replicatively aged cells pointed out proteins in the protein synthesis machinery as those that, first, increased the most relative to transcript levels and, second, in dynamic network reconstructions best predicted the behavior of the other genes[62]. Thus dysfunctional protein biogenesis may be a driver of replicative aging in yeast.

Furthermore, ribosomes accumulate in the regulatory regions of genes, encoding ribosomal proteins and components of the translation machinery at increased age, in agreement with the decreased protein synthesis[61]. In addition, translational elongation appears to be reduced with age[60], and aggregates of the mRNA processing body component Lsm1 become evident at 20 generations[61]. In agreement with repressed translational initiation upon replicative aging, the eIF2

kinase Gcn2 is activated but not its downstream transcription factor, Gcn4, best known for its role in amino acid metabolism[61]. Furthermore, roles were proposed for both a translational inhibitor, Ssd1, the levels of which increase in replicatively aged cells[61,63]; and Gcn2, in the repression of translational initiation. However, aged cells lacking both Ssd1 and Gcn2 still repress translational initiation to a large extent. Nevertheless, increased levels of SSD1 were noted to extend life span by 20% in a manner dependent on GCN2. Also, based on the ability of overexpression of the tRNA Methionine gene (IMT4) to extend life span (by 33%), a model was proposed in which uncharged tRNAs, induced by stress and activation of Gcn2, reduce translational initiation.

In addition, the deletion of genes encoding ribosomal proteins, predominantly in the large subunit (60S), was noted to reduce translational initiation and to extend life span[64]. A mechanism was proposed in which 60S subunit ribosomal deficiencies activate the transcription factor Gcn4 independently of Gcn2. A more recent study clarified that Gcn4 overexpression represses ribosomal gene expression via Gcn4 binding directly to ribosomal protein gene promoters, resulting in translational repression and a ~40% extended life span[65].

In summary, several studies support a drastic reduction of translational initiation during yeast replicative aging, although it is still obscure which molecular players are involved and to which aging factor(s) they relate. Another study suggested more complex translational deficiencies, including reduced translational elongation. Nevertheless, repression of translation via different means extends life span, documenting the importance of translational regulation in regulating yeast replicative aging.

15.6.2 Aggregated Proteins

Aggregated proteins have been found to accumulate with age in yeast and to be specifically retained in mother cells via their attachment to the actin cytoskeleton in a mechanism dependent on Sir2, chaperonin, Hsp104, and polarisomes[46,47].

However, the impact of the disaggregase Hsp104 on yeast replicative aging is controversial, with two studies supporting a short life span of Hsp104-deficient cells[16,66], whereas two other report a wild-type life span[7,67]. Nevertheless, increasing Hsp104 levels to levels reducing age-related protein aggregation is not sufficient to extend life span[68].

However, the molecular type A cytosolic Hsp70 chaperones (Ssa family) function in slowing down yeast aging[7,69,70]. There are four functionally related Ssa proteins in yeast and the expression of only one is essential for viability[71]. Ssa1 and Ssa2 are most highly expressed, 97% of their sequences are identical and they display the most pronounced functional similarities, such that cells lacking either Ssa1 or Ssa2 display few obvious phenotypes[72], whereas cells lacking both are short-lived, display temperature sensitive growth, and fail to recruit Hsp104 to aggregated proteins[69,73,74]. A recent study found that the short life span of cells lacking Ssa1 and Ssa2 could be counteracted by increased levels of the slightly more divergent type A Hsp70 Ssa4— whereas the recruitment of Hsp104 to aggregated proteins was not—and rapid degradation of a proteasomal model substrate (ΔssCPY*-LEU) only partially[74], indicating that still other Hsp70 functions

(e.g., in protein synthesis or translocation[75–78]; Figure 15.3) may contribute to life span control through the Hsp70 molecular chaperones.

On a final note, cells lacking the yeast metacaspase Mca1 suffer deficient proteostasis[79] and accumulate aggregated proteins[80], and Mca1 was noted to co-accumulate with Hsp104 in protein aggregates following heat shock. Whereas Mca-deficient cells sustain a wild-type life span, simultaneous deficiency in a major Hsp40 protein, Ydj1, and Mca1 dramatically shortens life span (65% reduction) and is accompanied by hyperaccumulation of protein aggregates[80]. Moreover, Mca1 overexpression extends life span by 50% in an Hsp104-dependent manner and reduces the accumulation of age-related protein aggregates. However, Mca1 has an N-terminal poly Q-/N-rich prion-like domain that modulates its aggregation[81]; therefore it is unclear if Mca1 remains fully functional in the absence of Hsp104[55,79,82,83].

15.6.3 Protein Turnover

Protein turnover may play a key role in aging because cells lacking the transcription factor Rpn4, controlling the expression of proteasome subunits and thus displaying reduced proteasome activity, are short-lived; conversely, cells lacking the ubiquitin ligase, Ubr2, displaying increased Rpn4 levels and proteasome activity, are long-lived[84]. However, the proteasome is highly multifunctional and targets both aggregated and non-aggregated proteins[85–87]. Therefore, the mechanisms by which proteasome activity impinges on yeast replicative aging are still unclear. In fact, one study suggests complex crosstalk with nutrient signaling in life span control[88]. Nevertheless, data from several organisms suggest that protein half-life increases with age, supporting decreased protein degradation[59,68]. In yeast, the accumulation of aggregated proteins seems to cause proteasomal decline because overexpression of the Hsp104 disaggregase in aged cells restored proteasome activity[68]. Interestingly, in worms the half-lives of ribosomal proteins and proteins in the translation apparatus were noted to increase most strongly with age, supporting a specific relaxation in the degradation of the protein synthesis machinery by the proteasome during aging[89].

In summary, data support a complex involvement of proteostatic chaperone-mediated functions in yeast replicative aging and tentative roles for protein aggregates as aging factors. More specifically, multiple lines of evidence support the involvement of one or more Hsp70-, proteasome-, and metacaspase-regulated functions in regulating the rate of replicative aging in yeast.

15.6.4 Roles of H_2O_2 in Age-Related Proteostasis

Life span control via Tsa1 requires Ssa1/2 but largely not Hsp104[7], suggesting the importance of one or more Hsp70 regulated proteostatic functions (Figure 15.3) in peroxiredoxin life span control.

15.6.4.1 Protein Synthesis

Previous studies suggest that the addition of H_2O_2 to yeast cells reduces translation through both Gcn2-dependent and independent manners[90]. Ribosomal sequencing shows that H_2O_2 leads to complex alterations in translation including increased ribosome occupancy of short upstream ORFs, elevated stop codon read-through, and ribosomes accumulating in the beginning of ORFs and in the translation elongation step[91]. Furthermore, the peroxiredoxin Tsa1 has been noted to associate with translating ribosomes, and both Tsa1 and Tsa2 appear to play a role in the proper decoding of STOP codons[92,93]. In addition, cells lacking Tsa1 were noted to be hypersensitive to translational inhibitors and to more strongly arrest translational initiation following H_2O_2 addition[93].

The amino acid starvation-unrelated activation and role of Gcn2 during replicative aging, distinct from Gcn4[61], is interesting given that the peroxiredoxin-sulfinic acid reductase Srx1 was identified as a novel target for translational control by Gcn2 upon H_2O_2 addition[20,91]. Furthermore, Srx1 is required for life span extension by CR or reduced PKA activity. However, whether Srx1 has any role in life span interventions more directly impinging on Gcn2 activity and translational initiation is not known at the moment, and the impact of the peroxiredoxin Tsa1 on these processes is poorly understood.

15.6.4.2 Protein Aggregates

The observation that the peroxiredoxin Tsa1 is critical for age-related protein aggregate control supports a key role of H_2O_2 in this process[7]. Interestingly, the ability of Hsp104 and the Hsp70s Ssa1 and Ssa2 to bind age- and H_2O_2-related aggregates is entirely dependent on Tsa1 hyperoxidation. In addition, its reduction through the sulfinic acid reductase Srx1 is necessary for efficient disaggregation of protein aggregates forming upon H_2O_2 stress but not heat shock[7].

15.6.4.3 Ubiquitin Proteasome Function

Reduced proteasome levels through Rpn4 deficiency impairs the ability of increased Tsa1 levels to extend life span[7]. Thus a ubiquitin-proteasome system (UPS)-regulated function appears to play a key role in peroxiredoxin-mediated longevity. Cells lacking Tsa1 hyperaccumulate ubiquitinated protein following H_2O_2 stress, whereas cells overproducing Tsa1 display reduced levels[7].

Interestingly, linkage-specific ubiquitination (in the form of ubiquitin linked to target proteins via K63) has recently been shown to maintain the translation machinery upon H_2O_2 stress[94]. Notably, K63-linked ubiquitinated proteins accumulate transiently following the addition of H_2O_2, and an inability to form such linkages (cells expressing a single K63R ubiquitin gene) destabilizes polysomes.

Furthermore, a recent comprehensive large-scale screen identified extensive oxidation of cysteines in ribosomal proteins and translation factors upon increased H_2O_2[95]. Notably, the protein synthesis machinery was noted as a major target of H_2O_2 because ribosomal proteins were targeted by both endogenous and exogenous H_2O_2 (1 mM) and required for full repression of protein synthesis upon H_2O_2 addition[95].

In conclusion, current data suggest a key role of H_2O_2 and peroxiredoxin in age-related protein aggregate control, whereas its impact on other aspects of age-related proteostasis in yeast is still largely unknown. Nevertheless, H_2O_2 appears to exert complex control over both protein biogenesis and the ubiquitin-proteasome system.

15.7 MITOCHONDRIAL DYSFUNCTION

Mitochondria are considered the main source of ROS in living cells[96,97], and their roles in aging are intricately intertwined with the roles of ROS in this process. Nevertheless, several other aspects of mitochondrial biogenesis, organization, and metabolism have been implicated in aging as well[98]. Notably, mitochondria play key roles in energy metabolism, anabolic pathways, metal metabolism including the biosynthesis of iron-sulfur clusters, calcium ion homeostasis, and signaling, and also contribute extensively to cellular stress responses such as autophagy and apoptosis[98,99]. In addition to ROS accumulating at increased age, mitochondrial dysfunction is manifested also as decreased efficiency of the respiratory chain and mitochondrial membrane potential, accumulation of mitochondrial DNA mutations and damage, reduced mitochondrial biogenesis, and higher-order tubular organization into filamentous structures (fusion), leading to their more singular appearance (mitochondrial fission). Furthermore, the function of human lysosomes and their homologues in yeast, vacuoles, have been increasingly linked to mitochondrial dysfunction, aging, and age-related disease in recent years[100–104].

15.7.1 Role of Mitochondrial Dysfunction in Yeast Replicative Aging

Mitochondrial structure and function are altered early on during yeast replicative aging[34]. Among the earliest deficiencies detected are reduced membrane potential and mitochondrial fission[36]. The importance of the former process in aging is somewhat cast into doubt, however, by recent observations on cells aged in microfluidic devices showing age-independent and heritable loss of mitochondrial membrane potential leading to clonal senescence in only a fraction of the cells[6]. Nevertheless, the involvement of mitochondrial fission in aging is supported by observations that deletion of genes required for fission (DNM1 or FIS1) extends life span[8].

Interestingly, an overexpression screen of ~250 genes regulating mitochondrial structure for delayed mitochondrial fission during aging identified two genes controlling vacuolar H+-ATPase activity (VMA1 and VPH2[36]). Notably, the functions of the vacuole (and the functionally homologous lysosome in mammals) are closely connected to mitochondria[105]. Strikingly, vacuolar acidity declines already during the first four divisions of yeast mother cells, and VMA1 or VPH2 overexpression restores mitochondrial structure and membrane potential and extends life span by 40%[36]. vATPase function is closely connected to the plasma membrane H+-ATPase Pma1[106]. In fact, vacuolar acidity was found to be suppressed in yeast mother cells through an asymmetric increase in Pma1 protein levels[107] and activity with replicative age[35], suggesting that both cytosolic and vacuolar pH homeostasis exert a strong influence on yeast replicative aging.

Why is it essential for cells to maintain a high vacuolar acidity to sustain longevity? Reduced cytosolic amino acids levels via the vacuolar neutral amino acid transporter, AVT1, were both noted to stimulate longevity and to be required for life span extension by vATPase overproduction[36]. A follow-up study indicated that the accumulation of in particular cysteine has a strong negative impact on the fitness in vATPase-deficient cells[103]. Furthermore, the fitness and mitochondrial function of vATPase deficient cells is tightly linked to iron homeostasis because lack of the iron-starvation-responsive transcription factor Aft1 exacerbates these phenotypes, whereas exogenous iron supplementation restores both of them[103,106,108]. In addition, H_2O_2 levels appear to increase moderately in vATPase-deficient cells, and cells lacking the antioxidant genes SOD1, TSA1, or TSA2 are hypersensitive to vATPase deficiency[103,108]. The authors proposed that ROS produced upon reduced vacuolar acidity induces iron starvation resulting in mitochondrial deficiency. However, it is presently unclear how these data can be reconciled with the key role of H_2O_2 signaling in mechanisms by which Tsa1 slows down aging[21,31]. In particular, the mechanism by which oxidants are produced upon attenuation of vacuolar acidity are unknown. In addition, because cells lacking Tsa1 may not accumulate toxic H_2O_2 levels[21,109], the mechanisms by which peroxiredoxin- or SOD-deficiency exacerbates fitness defects upon reduced vATPase function are still unclear. Nevertheless, the vacuolar ATPase appears to interact closely with the peroxiredoxin Tsa1 and iron homeostasis in controlling yeast replicative longevity, but future studies need to clarify, for example, the role of H_2O_2 signaling via peroxiredoxins in amino acid and iron homeostasis.

The mechanism underlying cytosolic cysteine inducing an iron-starvation response is also not completely understood. The transcription factor Aft1 senses mitochondrial iron levels via an iron-sulfur cluster[110–112], and iron-sulfur cluster deficiency in replicatively aged cells has previously been linked to genome instability in yeast[113], suggesting potential coordination with this essential process.

15.8 DEREGULATED NUTRIENT SENSING

The most successful intervention to extend life span to date, caloric restriction (CR), is a testament to the large importance of nutrient signaling pathways in controlling the rate of aging in most organisms[20,114–116]

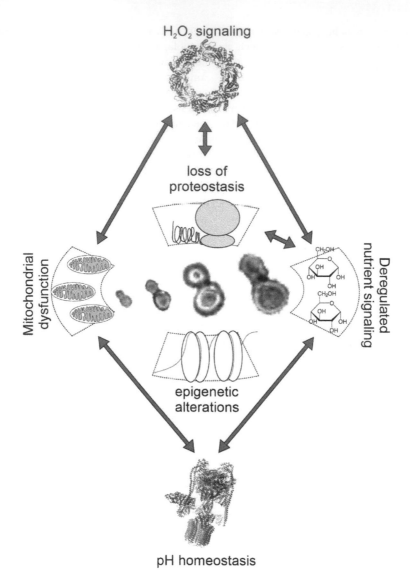

H_2O_2 signaling

loss of
proteostasis

Mitochondrial
dysfunction

Deregulated
nutrient signaling

epigenetic
alterations

pH homeostasis

Figure 15.4 Overview of mechanisms of replicative aging in yeast and their interconnections with H_2O_2 and peroxiredoxin.

(Figure 15.4). Although initial studies in mice suggested that the ability of different dietary restriction regimens to extend life span correlated with a decrease in caloric intake[117], it has become more and more evident that mechanisms by which specific nutrients modulate life span are not identical[118,119].

The most extensively studied intervention in yeast is glucose restriction, whereby lowering medium glucose from 2% to 0.5% or lower extends replicative life span by 20% to 30%[20,120,121]. Furthermore, the amino acid methionine has received particular attention in aging research lately because methionine restriction extends life span in organisms from yeast to mice, potentially by mimicking CR[118,119]. Nevertheless, metabolomic analyses in flies support extensive remodeling of methionine metabolism upon increased age[122].

There is a large consensus in the aging field that reduced nutrient and/or growth factor signaling through less than a handful of conserved pathways (e.g., insulin/IGF-1, target-of-rapamycin [TOR], and cAMP-dependent protein kinase [PKA]) is instrumental for caloric restriction to slow down aging. Upon glucose restriction, PKA plays a key role both in yeast and mice[121,123]. Furthermore, several mechanisms by which TOR senses amino

REDOX REGULATION OF DIFFERENTIATION AND DE-DIFFERENTIATION

acids have recently been identified in mammals and yeast[124].

15.8.1 Protein Kinase A

In yeast, a mild repression of protein kinase A (PKA) signaling (e.g., through gpr1Δ, gpa2Δ, cdc25-10, or cdc35-1 alleles) has been noted to extend replicative life span by ~25%–45%[20,67,121]. Conversely, hyperactivating PKA (e.g., through lack of the high-affinity cAMP-phosphodiesterase PDE2) shortens life span by 25%–45%[21,121], suggesting that PKA is a critical determinant of the rate of replicative aging in yeast. The sensitivity of PKA even to moderate glucose restriction (0.5%) is supported by the PKA-dependent increase in H_2O_2 resistance under these conditions[20].

The PKA pathway monitors glucose via at least two mechanisms. First, a principal pathway is dependent on the GTPase Ras[125,126], monitoring glucose metabolization in the glycolytic pathway via a mechanism in which the Ras GTP-exchange factor (GEF) Cdc25 is thought to directly probe the levels of the glycolytic intermediate fructose-1,6-bisphosphate[127]. GTP loaded Ras, in turn, stimulates production of the second messenger cAMP and PKA activity. Second, a more auxiliary branch of the Ras-cAMP-PKA pathway is composed of a G-PCR receptor system involving the receptor Gpr1 and the alpha subunit Gpa2 stimulating cAMP production[128].

In addition, pH homeostasis is intimately linked to glucose utilization and nutrient signaling in yeast[129,130]. Alkalinization of the cytosol through the H^+-ATPase activities of the plasma membrane, H^+-ATPase Pma1[131], and the vATPase[132] upon the addition of glucose to starved cells stimulates both resumed growth[129] and PKA signaling through the GTPase Arf1[133,134]. This suggests that altered cytosolic pH homeostasis during replicative aging may directly impinge on PKA activity.

15.8.2 TORC1

As in mammals, yeast express two target-of-rapamycin (TOR) protein kinase complexes with distinct functions, but only one, TORC1, has been implicated in nutrient sensing[135]. Moderately reduced TORC1 activity through the deletion of the TOR1 gene or the immunosuppressant drug rapamycin extends yeast replicative life span by 21%[136] or 15%[137], respectively. TORC1 controls ribosomal biogenesis, translation initiation and entry into the G0 phase of the cell cycle, but not nitrogen metabolism, through the AGC kinase Sch9[138]. In support of the importance of the former processes in regulating aging, Sch9 ablation extends life span by 42%[136].

The mechanisms by which TORC1 senses glucose are poorly understood. In general, molecular studies on TORC1 functions in relation to glucose have focused on complete glucose removal (e.g.,[138,139]). In replicative aging, restricting glucose from 2% down to 0.05% has been argued to act through TORC1 because this regimen fails to further extend the life span of Tor1- or Sch9-deficient cells. Notably, TORC1 activity is inhibited upon glucose removal through its redistribution from a uniform vacuolar membrane association into a single vacuole-associated hollow, cylindrical tube (TOROID[139]). This redistribution is dependent on the perivacuolar Gtr1/2 GTPases, which are essential both for TORC1 responses to amino acids[124] and their response to cytosolic alkalinization upon glucose addition[134].

As in other studies of TORC1 function, data on mutant alleles hyperactivating TORC1 function are lacking in replicative aging. However, the human ataxin-2 homologue and stress granule component, Pbp1, sequesters Kog1 in stress granula upon heat shock and inhibits TORC1 function[9,140]. Furthermore, Pbp1 is required for both TORC1 repression and growth fitness of moderately replicatively aged cells (age 5–10 generations) under respiratory conditions (carbon-source lactate)[9]. Interestingly, with increased H_2O_2, a methionine-rich low complexity region in the Pbp1 C-terminus is oxidized, leading the protein to form a phase-separated condensate inhibiting TORC1[141]. Thus Pbp1 acts as a redox sensor of TORC1. However, cells lacking Pbp1 display wild-type growth fitness during replicative aging in 2% glucose, suggesting that the Pbp1 function maintaining age-related fitness is not active under fermentative conditions[9].

Methionine restriction extends the replicative life span of yeast[142,143], possibly through a mechanism resembling glucose restriction (to 0.05%)[143]. Indeed, methionine addition to cells grown under respiratory conditions was found to induce TORC1 activity via a mechanism involving its metabolization to the methyl donor S-adenosylmethionine, the protein phosphatase 2A, and the Gtr1/2 GTPase[144]. Furthermore, methionine biosynthesis is a known activator of PKA[145] and, in fact, the deletion of seven of the eight genes in methionine biosynthesis extends life span. This supports the view that reduced methionine levels extend yeast replicative aging. Whether more modest glucose restriction also impacts on methionine metabolism is presently unknown.

15.9 INTEGRATIVE THEORY OF AGING

Following the description of multiple mechanisms contributing to aging, it may be questioned which are the earliest and/or most important factors leading to organismal decline. A modern integrative theory of aging, however, stresses that different pathways and/or organelles are intricately interconnected into so-called integrons that impinge on aging[146]. In this model, interconnectivity allows the function of a subsystem failing to be carried out by an interconnected node, thus providing robustness to the system[147]. However, when individual subsystems fail, their extensive interconnection will eventually cause a progressive decline of all systems through sequential collapse of homeostasis. Striking examples of the latter have been mentioned in the earlier discussions of the role of loss of vacuolar acidity in mitochondrial dysfunction, iron homeostasis, and genome instability[36,103,113].

Several examples of interconnected cellular subsystems providing buffering functions impinging on aging have also been provided in yeast. Such efforts have been facilitated by the unique ability to perform ordered, comprehensive synthetic genetic array (SGA) screens for backup functions of a gene or specific function in a gene of interest[148]. These efforts have also resulted in one of the most comprehensive cellular maps of function and functional connections in any organism[149,150]. Through this approach, the major Hsp40 Ydj1 protein, for example, was identified to be required for the fitness of cells lacking the yeast metacaspase, Mca1, implicated in proteostasis (see above), as well as its ability to maintain a normal life span[80]. Furthermore, through a clever modification of the SGA approach, backup functions of the sirtuin Sir2 identified a total of 122 genes with functions in (1) genomic silencing and chromatin structure, (2) nuclear transport, and (3) actin cytoskeleton and cell polarity, as required for growth of cells lacking Sir2[47]. Following up on the latter category, the polarisome and myosin-motor machinery involved in establishing cell polarity was found to be required for the fitness of Sir2-deficient cells and in themselves for both the segregation of aggregated proteins and normal life span[47].

Through which cellular components could such wide-ranging integration be achieved? Focus on the PQC network has pointed out different organelles to share PQC factors (e.g., specific chaperones), as well as to communicate via interorganelle trafficking, contact sites, and small ions and molecules to achieve spatial proteostasis[147]. More generally, highly interconnected nodes in the cellular network (hub genes) may be expected to be more important in cell homeostasis and possibly also in aging[151]. Strikingly, among the 40 most highly pleiotropic genes in the global cell map[150], five are involved in the protein kinase A and TORC1 nutrient signaling pathways, supporting the idea that these pathways carry out a multitude of functions, are well situated to integrate information from many different cellular processes and may link the different processes underlying aging together (Figure 15.4). In agreement with a key role of PKA in cellular homeostasis, a genome-wide screen for modifiers of PKA activity identified a striking number of known PKA target functions—involved in glycogen accumulation, filamentous growth, and amino acid biosynthesis, for example—supporting complex feedback control of these processes by the PKA enzyme[145]. On a similar note, vacuolar pH control and proteasome function have been noted to feedback control nutrient signaling[88,106,152]. Other highly interconnected genes in the cell map carry out functions in the CCT chaperonin that folds e.g. actin (4 genes), in plasma membrane potential/pH homeostasis (2 genes), in histone deacetylation (2 genes), in ribosomal biogenesis (2 genes), in mRNA degradation and processing (2 genes), in RNA polymerase II transcription (2 genes) revealing a striking overlap with functions discussed above.

15.9.1 Role of H_2O_2 in the Integrative Theory of Aging

Then what could be the role of H_2O_2 in the integrative theory of aging? In general, H_2O_2 cannot diffuse over long distances, but its extremely high reactivity with peroxiredoxins, the elevated expression levels of peroxiredoxins, and multiple stable redox states and conformations[54] suggest that H_2O_2 may affect many processes through reacting with peroxiredoxins and altering their conformation and function. Indeed, the peroxiredoxin Tsa1 is highly connected in the cellular network, displaying synthetic lethal or sick interactions with >400 other genes[7,150].

Furthermore, a recent study identified a key role of nutrient signaling through PKA in the anti-aging function of the major cytosolic peroxiredoxin Tsa1[21]. A novel mechanism of second messenger-independent kinase control was proposed in which H_2O_2 via peroxiredoxin oxidizes the kinase catalytic subunit. Supporting this, Tsa1 and H_2O_2 targeted a cysteine conserved in all eukaryotic PKA enzymes and 11.5% of all eukaryotic protein kinases[153,154]. This cysteine is situated adjacent to a threonine residue that when phosphorylated maintains the activation loop in a conformation more efficiently binding substrates[155], and redox-modification of this residue in certain isoforms of PKA inhibits kinase activity[21,155].

Overall, redox-regulation of protein kinases and the impact of H_2O_2 on both vacuolar function and the proteostatic machinery, provide potential explanations for the pleiotropic functions of H_2O_2 and peroxiredoxins and their importance in aging (Figure 15.4).

15.10 CONCLUDING REMARKS

Yeast will continue to serve as a model organism for aging research due to its facile genetics and manipulation, technical microfluidic inventions enabling continuous tracking of yeast's replicative life span, the large potential for genome-wide screens, and the available highly interconnected maps of cellular function. It will be especially interesting to clarify the close interconnection of H_2O_2 and peroxiredoxins with cytosolic and vacuolar pH homeostasis, protein synthesis, and ubiquitin-proteasome function for improved understanding of their key importance in the yeast replicative aging model. With the increased possibilities to follow age-related deterioration through microfluidic techniques, we should soon expect a more detailed picture of the homeostatic collapse characterizing age-related deterioration and clues on how to postpone it.

REFERENCES

1 Mortimer, R. K. & Johnston, J. R. Life span of individual yeast cells. *Nature* **183**, 1751–1752, doi:10.1038/1831751a0 (1959).

2 Kowald, A. & Kirkwood, T. B. L. Can aging be programmed? A critical literature review. *Aging Cell* **15**, 986–998, doi:10.1111/acel.12510 (2016).

3 Longo, V. D., Shadel, G. S., Kaeberlein, M. & Kennedy, B. Replicative and chronological aging in *Saccharomyces cerevisiae*. *Cell Metab* **16**, 18–31, doi:10.1016/j.cmet.2012.06.002 (2012).

4 Kuo, P. L. *et al.* A roadmap to build a phenotypic metric of ageing: insights from the Baltimore Longitudinal Study of Aging. *J Intern Med* **287**, 373–394, doi:10.1111/joim.13024 (2020).

5 Muller, I., Zimmermann, M., Becker, D. & Flomer, M. Calendar life span versus budding life span of *Saccharomyces cerevisiae*. *Mech Ageing Dev* **12**, 47–52, doi:10.1016/0047-6374(80)90028-7 (1980).

6 Fehrmann, S. *et al.* Aging yeast cells undergo a sharp entry into senescence unrelated to the loss of mitochondrial membrane potential. *Cell Rep* **5**, 1589–1599, doi:10.1016/j.celrep.2013.11.013 (2013).

7 Hanzen, S. *et al.* Lifespan control by redox-dependent recruitment of chaperones to misfolded proteins. *Cell* **166**, 140–151, doi:10.1016/j.cell.2016.05.006 (2016).

8 Scheckhuber, C. Q. *et al.* Reducing mitochondrial fission results in increased life span and fitness of two fungal ageing models. *Nat Cell Biol* **9**, 99–105 (2007).

9 Yang, Y. S. *et al.* Yeast Ataxin-2 forms an intracellular condensate required for the inhibition of TORC1 signaling during respiratory growth. *Cell* **177**, 697–710 e617, doi:10.1016/j.cell.2019.02.043 (2019).

10 Nystrom, T. & Liu, B. The mystery of aging and rejuvenation: a budding topic. *Curr Opin Microbiol* **18**, 61–67, doi:10.1016/j.mib.2014.02.003 (2014).

11 Steinkraus, K. A., Kaeberlein, M. & Kennedy, B. K. Replicative aging in yeast: the means to the end. *Annu Rev Cell Dev Biol* **24**, 29–54 (2008).

12 Monaghan, P. & Metcalfe, N. B. The deteriorating soma and the indispensable germline: gamete senescence and offspring fitness. *Proc Biol Sci* **286**, 20192187, doi:10.1098/rspb.2019.2187 (2019).

13 Barker, M. G., Brimage, L. J. & Smart, K. A. Effect of Cu, Zn superoxide dismutase disruption mutation on replicative senescence in *Saccharomyces cerevisiae*. *FEMS Microbiol Lett* **177**, 199–204, doi:10.1111/j.1574-6968.1999.tb13732.x (1999).

14 Wawryn, J., Krzepilko, A., Myszka, A. & Bilinski, T. Deficiency in superoxide dismutases shortens life span of yeast cells. *Acta Biochim Pol* **46**, 249–253 (1999).

15 Nestelbacher, R. *et al.* The influence of oxygen toxicity on yeast mother cell-specific aging. *Exp Gerontol* **35**, 63–70 (2000).

16 Lam, Y. T., Aung-Htut, M. T., Lim, Y. L., Yang, H. & Dawes, I. W. Changes in reactive oxygen species begin early during replicative aging of

Saccharomyces cerevisiae cells. *Free Radic Biol Med* **50**, 963–970, doi:10.1016/j.freeradbiomed.2011.01.013 (2011).

17 Murphy, M. P. *et al.* Unraveling the biological roles of reactive oxygen species. *Cell Metab* **13**, 361–366, doi:10.1016/j.cmet.2011.03.010 (2011).

18 Erjavec, N. & Nystrom, T. Sir2p-dependent protein segregation gives rise to a superior reactive oxygen species management in the progeny of *Saccharomyces cerevisiae*. *Proc Natl Acad Sci U S A* **104**, 10877–10881, doi:10.1073/pnas.0701634104 (2007).

19 Laun, P. *et al.* Aged mother cells of *Saccharomyces cerevisiae* show markers of oxidative stress and apoptosis. *Mol Microbiol* **39**, 1166–1173 (2001).

20 Molin, M. *et al.* Life span extension and H_2O_2-resistance elicited by caloric restriction require the peroxiredoxin Tsa1 in *Saccharomyces cerevisiae*. *Mol Cell* **43**, 823–833, doi:10.1016/j.molcel.2011.07.027 (2011).

21 Roger, F. *et al.* Peroxiredoxin promotes longevity and H_2O_2-resistance in yeast through redox-modulation of PKA. *eLife* **9**, e60346, doi:10.7554/eLife.60346 (2020).

22 Winterbourn, C. C. & Hampton, M. B. Thiol chemistry and specificity in redox signaling. *Free Radic Biol Med* **45**, 549–561, doi:10.1016/j.freeradbiomed.2008.05.004 (2008).

23 Peskin, A. V. *et al.* Hyperoxidation of peroxiredoxins 2 and 3: rate constants for the reactions of the sulfenic acid of the peroxidatic cysteine. *J Biol Chem* **288**, 14170–14177, doi:10.1074/jbc.M113.460881 (2013).

24 Fomenko, D. E. *et al.* Thiol peroxidases mediate specific genome-wide regulation of gene expression in response to hydrogen peroxide. *Proc Natl Acad Sci U S A* **108**, 2729–2734, doi:10.1073/pnas.1010721108 (2011).

25 Levine, R. L. Carbonyl modified proteins in cellular regulation, aging, and disease. *Free Radic Biol Med* **32**, 790–796 (2002).

26 Aguilaniu, H., Gustafsson, L., Rigoulet, M. & Nystrom, T. Asymmetric inheritance of oxidatively damaged proteins during cytokinesis. *Science* **299**, 1751–1753 (2003).

27 Ristow, M. & Schmeisser, S. Extending life span by increasing oxidative stress. *Free Radic Biol Med* **51**, 327–336, doi:10.1016/j.freeradbiomed.2011.05.010 (2011).

28 Schulz, T. J. *et al.* Glucose restriction extends *Caenorhabditis elegans* life span by inducing mitochondrial respiration and increasing oxidative stress. *Cell Metab* **6**, 280–293, doi:10.1016/j.cmet.2007.08.011 (2007).

29 De Haes, W. *et al.* Metformin promotes lifespan through mitohormesis via the peroxiredoxin PRDX-2. *Proc Natl Acad Sci U S A* **111**, E2501–2509, doi:10.1073/pnas.1321776111 (2014).

30 Holmstrom, K. M. & Finkel, T. Cellular mechanisms and physiological consequences of redox-dependent signalling. *Nat Rev Mol Cell Biol* **15**, 411–421, doi:10.1038/nrm3801 (2014).

31 Goulev, Y. *et al.* Nonlinear feedback drives homeostatic plasticity in H_2O_2 stress response. *eLife* **6**, 23791, doi:10.7554/eLife.23971 (2017).

32 Matsumura, T. *et al.* N-acetyl-l-tyrosine is an intrinsic triggering factor of mitohormesis in stressed animals. *EMBO Rep* **21**, e49211, doi:10.15252/embr.201949211 (2020).

33 Lopez-Otin, C., Blasco, M. A., Partridge, L., Serrano, M. & Kroemer, G. The hallmarks of aging. *Cell* **153**, 1194–1217, doi:10.1016/j.cell.2013.05.039 (2013).

34 Janssens, G. E. & Veenhoff, L. M. Evidence for the hallmarks of human aging in replicatively aging yeast. *Microb Cell* **3**, 263–274, doi:10.15698/mic2016.07.510 (2016).

35 Henderson, K. A., Hughes, A. L. & Gottschling, D. E. Mother-daughter asymmetry of pH underlies aging and rejuvenation in yeast. *eLife* **3**, e03504, doi:10.7554/eLife.03504 (2014).

36 Hughes, A. L. & Gottschling, D. E. An early age increase in vacuolar pH limits mitochondrial function and lifespan in yeast. *Nature* **492**, 261–265, doi:10.1038/nature11654 (2012).

37 McMurray, M. A. & Gottschling, D. E. An age-induced switch to a hyper-recombinational state. *Science* **301**, 1908–1911 (2003).

38 Lesur, I. & Campbell, J. L. The transcriptome of prematurely aging yeast cells is similar to that of telomerase-deficient cells. *Mol Biol Cell* **15**, 1297–1312, doi:10.1091/mbc.e03-10-0742 (2004).

39 Sinclair, D. A., Mills, K. & Guarente, L. Accelerated aging and nucleolar fragmentation in yeast sgs1 mutants. *Science* **277**, 1313–1316, doi:10.1126/science.277.5330.1313 (1997).

40 Hoopes, L. L., Budd, M., Choe, W., Weitao, T. & Campbell, J. L. Mutations in DNA replication genes reduce yeast life span. *Mol Cell Biol* **22**, 4136–4146, doi:10.1128/mcb.22.12.4136-4146.2002 (2002).

41 Kaya, A., Lobanov, A. V. & Gladyshev, V. N. Evidence that mutation accumulation does not cause aging in *Saccharomyces cerevisiae*. *Aging Cell* **14**, 366–371, doi:10.1111/acel.12290 (2015).

42 Sinclair, D. A. & Guarente, L. Extrachromosomal rDNA circles—a cause of aging in yeast. *Cell* **91**, 1033–1042 (1997).

43 Defossez, P. A. *et al.* Elimination of replication block protein Fob1 extends the life span of yeast mother cells. *Mol Cell* **3**, 447–455, doi:10.1016/s1097-2765(00)80472-4 (1999).

44 Ganley, A. R., Ide, S., Saka, K. & Kobayashi, T. The effect of replication initiation on gene amplification in the rDNA and its relationship to aging. *Mol Cell* **35**, 683–693, doi:10.1016/j.molcel.2009.07.012 (2009).

45 Dang, W. *et al.* Histone H4 lysine 16 acetylation regulates cellular lifespan. *Nature* **459**, 802–807, doi:10.1038/nature08085 (2009).

46 Erjavec, N., Larsson, L., Grantham, J. & Nystrom, T. Accelerated aging and failure to segregate damaged proteins in Sir2 mutants can be suppressed by overproducing the protein aggregation-remodeling factor Hsp104p. *Genes Dev* **21**, 2410–2421 (2007).

47 Liu, B. *et al.* The polarisome is required for segregation and retrograde transport of protein aggregates. *Cell* **140**, 257–267 (2010).

48 Singh, C. K. *et al.* The role of sirtuins in antioxidant and redox signaling. *Antioxid Redox Signal* **28**, 643–661, doi:10.1089/ars.2017.7290 (2018).

49 Ando, A., Nakamura, T., Murata, Y., Takagi, H. & Shima, J. Identification and classification of genes required for tolerance to freeze-thaw stress revealed by genome-wide screening of *Saccharomyces cerevisiae* deletion strains. *FEMS Yeast Res* **7**, 244–253, doi:10.1111/j.1567-1364.2006.00162.x (2007).

50 Brown, J. A. *et al.* Global analysis of gene function in yeast by quantitative phenotypic profiling. *Mol Syst Biol* **2**, 2006 0001, doi:10.1038/msb4100043 (2006).

51 Outten, C. E., Falk, R. L. & Culotta, V. C. Cellular factors required for protection from hyperoxia toxicity in *Saccharomyces cerevisiae*. *Biochem J* **388**, 93–101, doi:10.1042/BJ20041914 (2005).

52 Thorpe, G. W., Fong, C. S., Alic, N., Higgins, V. J. & Dawes, I. W. Cells have distinct mechanisms to maintain protection against different reactive oxygen species: oxidative-stress-response genes. *Proc Natl Acad Sci U S A* **101**, 6564–6569, doi:10.1073/pnas.0305888101 (2004).

53 Hartl, F. U. Protein misfolding diseases. *Annu Rev Biochem* **86**, 21–26, doi:10.1146/annurev-biochem-061516-044518 (2017).

54 Troussicot, L., Burmann, B. M. & Molin, M. Structural determinants of multimerization and dissociation in 2-Cys peroxiredoxin chaperone function. *Structure*, https://doi.org/10.1016/j.str.2021.04.007 (2021).

55 Shorter, J. & Southworth, D. R. Spiraling in control: structures and mechanisms of the Hsp104 disaggregase. *Cold Spring Harb Perspect Biol* **11**, doi:10.1101/cshperspect.a034033 (2019).

56 Rosenzweig, R., Nillegoda, N. B., Mayer, M. P. & Bukau, B. The Hsp70 chaperone network. *Nat Rev Mol Cell Biol* **20**, 665–680, doi:10.1038/s41580-019-0133-3 (2019).

57 Voth, W. & Jakob, U. Stress-activated chaperones: a first line of defense. *Trends Biochem Sci* **42**, 899–913, doi:10.1016/j.tibs.2017.08.006 (2017).

58 Duran, E. C., Weaver, C. L. & Lucius, A. L. Comparative analysis of the structure and function of AAA+ Motors ClpA, ClpB, and Hsp104: common threads and disparate functions. *Frontiers in Molecular Biosciences* **4**, 54, doi:10.3389/fmolb.2017.00054 (2017).

59 Anisimova, A. S., Alexandrov, A. I., Makarova, N. E., Gladyshev, V. N. & Dmitriev, S. E. Protein synthesis and quality control in aging. *Aging (Albany NY)* **10**, 4269–4288, doi:10.18632/aging.101721 (2018).

60 Motizuki, M. & Tsurugi, K. The effect of aging on protein synthesis in the yeast *Saccharomyces cerevisiae*. *Mech Ageing Dev* **64**, 235–245, doi:10.1016/0047-6374(92)90081-n (1992).

61 Hu, Z. *et al.* Ssd1 and Gcn2 suppress global translation efficiency in replicatively aged yeast while their activation extends lifespan. *eLife* **7**, e35551, doi:10.7554/eLife.35551 (2018).

62 Janssens, G. E. *et al.* Protein biogenesis machinery is a driver of replicative aging in yeast. *eLife* **4**, e08527, doi:10.7554/eLife.08527 (2015).

63 Feser, J. *et al.* Elevated histone expression promotes life span extension. *Mol Cell* **39**, 724–735, doi:10.1016/j.molcel.2010.08.015 (2010).

64 Steffen, K. K. *et al.* Yeast life span extension by depletion of 60s ribosomal subunits is mediated by Gcn4. *Cell* **133**, 292–302 (2008).

65 Mittal, N. *et al.* The Gcn4 transcription factor reduces protein synthesis capacity and extends yeast lifespan. *Nat Commun* **8**, 457, doi:10.1038/s41467-017-00539-y (2017).

66 Saarikangas, J. & Barral, Y. Protein aggregates are associated with replicative aging without compromising protein quality control. *eLife* **4**, doi:10.7554/eLife.06197 (2015).

67 Kaeberlein, M., Kirkland, K. T., Fields, S. & Kennedy, B. K. Genes determining yeast replicative life span in a long-lived genetic background. *Mech Ageing Dev* **126**, 491–504, doi:10.1016/j.mad.2004.10.007 (2005).

68 Andersson, V., Hanzen, S., Liu, B., Molin, M. & Nystrom, T. Enhancing protein disaggregation restores proteasome activity in aged cells. *Aging (Albany, NY)* **5**, 802–812 (2013).

69 Craig, E. A. & Jacobsen, K. Mutations of the heat inducible 70 kilodalton genes of yeast confer temperature sensitive growth. *Cell* **38**, 841–849, doi:10.1016/0092-8674(84)90279-4 (1984).

70 Oling, D., Eisele, F., Kvint, K. & Nystrom, T. Opposing roles of Ubp3-dependent deubiquitination regulate replicative life span and heat resistance. *EMBO J* **33**, 747–761, doi:10.1002/embj.201386822 (2014).

71 Werner-Washburne, M., Stone, D. E. & Craig, E. A. Complex interactions among members of an essential subfamily of hsp70 genes in *Saccharomyces cerevisiae*. *Mol Cell Biol* **7**, 2568–2577, doi:10.1128/mcb.7.7.2568 (1987).

72 Schwimmer, C. & Masison, D. C. Antagonistic interactions between yeast [PSI(+)] and [URE3] prions and curing of [URE3] by Hsp70 protein chaperone Ssa1p but not by Ssa2p. *Mol Cell Biol* **22**, 3590–3598, doi:10.1128/mcb.22.11.3590-3598.2002 (2002).

73 Winkler, J., Tyedmers, J., Bukau, B. & Mogk, A. Hsp70 targets Hsp100 chaperones to substrates for protein disaggregation and prion fragmentation. *J Cell Biol* **198**, 387–404, doi:10.1083/jcb.201201074 (2012).

74 Andersson, R. *et al*. Differential role of cytosolic Hsp70s in longevity assurance and protein quality control. *PLOS Genet* **17** (1): e1008951, https://doi.org/10.1371/journal.pgen.1008951 (2021).

75 Horton, L. E., James, P., Craig, E. A. & Hensold, J. O. The yeast hsp70 homologue Ssa is required for translation and interacts with Sis1 and Pab1 on translating ribosomes. *J Biol Chem* **276**, 14426–14433, doi:10.1074/jbc.M100266200 (2001).

76 Walters, R. W. & Parker, R. Coupling of ribostasis and proteostasis: Hsp70 proteins in mRNA metabolism. *Trends Biochem Sci* **40**, 552–559, doi:10.1016/j.tibs.2015.08.004 (2015).

77 Zhong, T. & Arndt, K. T. The yeast SIS1 protein, a DnaJ homolog, is required for the initiation of translation. *Cell* **73**, 1175–1186 (1993).

78 Craig, E. A. Hsp70 at the membrane: driving protein translocation. *BMC Biol* **16**, 11, doi:10.1186/s12915-017-0474-3 (2018).

79 Lee, R. E., Brunette, S., Puente, L. G. & Megeney, L. A. Metacaspase Yca1 is required for clearance of insoluble protein aggregates. *Proc Natl Acad Sci U S A* **107**, 13348–13353, doi:10.1073/pnas.1006610107 (2010).

80 Hill, S. M., Hao, X., Liu, B. & Nystrom, T. Lifespan extension by a metacaspase in the yeast *Saccharomyces cerevisiae*. *Science*, doi:10.1126/science.1252634 (2014).

81 Erhardt, M., Wegrzyn, R. D. & Deuerling, E. Extra N-terminal residues have a profound effect on the aggregation properties of the potential yeast prion protein Mca1. *PLoS One* **5**, e9929, doi:10.1371/journal.pone.0009929 (2010).

82 Kaganovich, D., Kopito, R. & Frydman, J. Misfolded proteins partition between two distinct quality control compartments. *Nature* **454**, 1088–1095, doi:10.1038/nature07195 (2008).

83 Wickner, R. B. Anti-prion systems in yeast. *J Biol Chem* **294**, 1729–1738, doi:10.1074/jbc.TM118.004168 (2019).

84 Kruegel, U. *et al*. Elevated proteasome capacity extends replicative lifespan in *Saccharomyces cerevisiae*. *PLoS Genet* **7**, e1002253, doi:10.1371/journal.pgen.1002253 (2011).

85 Dikic, I. Proteasomal and autophagic degradation systems. *Annu Rev Biochem* **86**, 193–224, doi:10.1146/annurev-biochem-061516-044908 (2017).

86 Varshavsky, A. The ubiquitin system, autophagy, and regulated protein degradation. *Annu Rev Biochem* **86**, 123–128, doi:10.1146/annurev-biochem-061516-044859 (2017).

87 Zheng, N. & Shabek, N. Ubiquitin ligases: structure, function, and regulation. *Annu Rev Biochem* **86**, 129–157, doi:10.1146/annurev-biochem-060815-014922 (2017).

88 Yao, Y. *et al*. Proteasomes, Sir2, and Hxk2 form an interconnected aging network that impinges on the AMPK/Snf1-regulated transcriptional repressor Mig1. *PLoS Genet* **11**, e1004968, doi:10.1371/journal.pgen.1004968 (2015).

89 Dhondt, I. *et al*. Changes of protein turnover in aging *Caenorhabditis elegans*. *Mol Cell Proteomics* **16**, 1621–1633, doi:10.1074/mcp.RA117.000049 (2017).

90 Shenton, D. *et al*. Global translational responses to oxidative stress impact upon multiple levels of

protein synthesis. *J Biol Chem* **281**, 29011–29021 (2006).

91 Gerashchenko, M. V., Lobanov, A. V. & Gladyshev, V. N. Genome-wide ribosome profiling reveals complex translational regulation in response to oxidative stress. *Proc Natl Acad Sci U S A* **109**, 17394–17399, doi:10.1073/pnas.1120799109 (2012).

92 Sideri, T. C., Stojanovski, K., Tuite, M. F. & Grant, C. M. Ribosome-associated peroxiredoxins suppress oxidative stress-induced de novo formation of the [PSI+] prion in yeast. *Proc Natl Acad Sci U S A* **107**, 6394–6399, doi:10.1073/pnas.1000347107 (2010).

93 Trotter, E. W., Rand, J. D., Vickerstaff, J. & Grant, C. M. The yeast Tsa1 peroxiredoxin is a ribosome-associated antioxidant. *Biochem J* **412**, 73–80, doi:10.1042/BJ20071634 (2008).

94 Silva, G. M., Finley, D. & Vogel, C. K63 polyubiquitination is a new modulator of the oxidative stress response. *Nat Struct Mol Biol* **22**, 116–123, doi:10.1038/nsmb.2955 (2015).

95 Topf, U. *et al.* Quantitative proteomics identifies redox switches for global translation modulation by mitochondrially produced reactive oxygen species. *Nat Commun* **9**, 324, doi:10.1038/s41467-017-02694-8 (2018).

96 Longo, V. D., Gralla, E. B. & Valentine, J. S. Superoxide dismutase activity is essential for stationary phase survival in *Saccharomyces cerevisiae*. Mitochondrial production of toxic oxygen species in vivo. *J Biol Chem* **271**, 12275–12280 (1996).

97 Perrone, G. G., Tan, S. X. & Dawes, I. W. Reactive oxygen species and yeast apoptosis. *Biochim Biophys Acta* **1783**, 1354–1368, doi:10.1016/j.bbamcr.2008.01.023 (2008).

98 Wallace, D. C. A mitochondrial paradigm of metabolic and degenerative diseases, aging, and cancer: a dawn for evolutionary medicine. *Annu Rev Genet* **39**, 359–407, doi:10.1146/annurev.genet.39.110304.095751 (2005).

99 Nunnari, J. & Suomalainen, A. Mitochondria: in sickness and in health. *Cell* **148**, 1145–1159, doi:10.1016/j.cell.2012.02.035 (2012).

100 Audano, M., Schneider, A. & Mitro, N. Mitochondria, lysosomes, and dysfunction: their meaning in neurodegeneration. *J Neurochem* **147**, 291–309, doi:10.1111/jnc.14471 (2018).

101 Colacurcio, D. J. & Nixon, R. A. Disorders of lysosomal acidification: the emerging role of v-ATPase in aging and neurodegenerative disease.

Ageing Res Rev **32**, 75–88, doi:10.1016/j.arr.2016.05.004 (2016).

102 Ferguson, S. M. Beyond indigestion: emerging roles for lysosome-based signaling in human disease. *Curr Opin Cell Biol* **35**, 59–68, doi:10.1016/j.ceb.2015.04.014 (2015).

103 Hughes, C. E. *et al.* Cysteine toxicity drives age-related mitochondrial decline by altering iron homeostasis. *Cell* **180**, 296–310 e218, doi:10.1016/j.cell.2019.12.035 (2020).

104 Plotegher, N. & Duchen, M. R. Crosstalk between Lysosomes and Mitochondria in Parkinson's Disease. *Front Cell Dev Biol* **5**, 110, doi:10.3389/fcell.2017.00110 (2017).

105 Rutter, J. & Hughes, A. L. Power(2): the power of yeast genetics applied to the powerhouse of the cell. *Trends Endocrinol Metab* **26**, 59–68, doi:10.1016/j.tem.2014.12.002 (2015).

106 Molin, M. & Demir, A. B. Linking peroxiredoxin and vacuolar-ATPase functions in calorie restriction-mediated life span extension. *Int J Cell Biol* **2014**, 12, doi:10.1155/2014/913071 (2014).

107 Thayer, N. H. *et al.* Identification of long-lived proteins retained in cells undergoing repeated asymmetric divisions. *Proc Natl Acad Sci U S A* **111**, 14019–14026, doi:10.1073/pnas.1416079111 (2014).

108 Diab, H. I. & Kane, P. M. Loss of vacuolar H+-ATPase (V-ATPase) activity in yeast generates an iron deprivation signal that is moderated by induction of the peroxiredoxin TSA2. *J Biol Chem* **288**, 11366–11377, doi:M112.419259 [pii] 10.1074/jbc.M112.419259 (2013).

109 Ogusucu, R., Rettori, D., Netto, L. E. & Augusto, O. Superoxide dismutase 1-mediated production of ethanol- and DNA-derived radicals in yeasts challenged with hydrogen peroxide: molecular insights into the genome instability of peroxiredoxin-null strains. *J Biol Chem* **284**, 5546–5556 (2009).

110 Kispal, G., Csere, P., Guiard, B. & Lill, R. The ABC transporter Atm1p is required for mitochondrial iron homeostasis. *FEBS Lett* **418**, 346–350 (1997).

111 Li, H. *et al.* Histidine 103 in Fra2 is an iron-sulfur cluster ligand in the [2Fe-2S] Fra2-Grx3 complex and is required for in vivo iron signaling in yeast. *J Biol Chem* **286**, 867–876, doi:10.1074/jbc.M110.184176 (2011).

112 Rutherford, J. C. *et al.* Activation of the iron regulon by the yeast Aft1/Aft2 transcription factors depends on mitochondrial but not cytosolic iron-sulfur protein biogenesis. *J Biol Chem*

280, 10135–10140, doi:10.1074/jbc.M413731200 (2005).

113 Veatch, J. R., McMurray, M. A., Nelson, Z. W. & Gottschling, D. E. Mitochondrial dysfunction leads to nuclear genome instability via an iron-sulfur cluster defect. *Cell* **137**, 1247–1258, doi:10.1016/j.cell.2009.04.014 (2009).

114 Fontana, L., Partridge, L., and Longo, V. D. Dietary restriction, growth factors and aging: from yeast to humans. *Science* **328**, 321–326 (2010).

115 Mattison, J. A. *et al.* Caloric restriction improves health and survival of rhesus monkeys. *Nat Commun* **8**, 14063, doi:10.1038/ncomms14063 (2017).

116 Minor, R. K., Allard, J. S., Younts, C. M., Ward, T. M. & de Cabo, R. Dietary interventions to extend life span and health span based on calorie restriction. *J Gerontol A Biol Sci Med Sci* **65**, 695–703, doi:10.1093/gerona/glq042 (2010).

117 Walford, R. L., Harris, S. B. & Weindruch, R. Dietary restriction and aging: historical phases, mechanisms and current directions. *J Nutr* **117**, 1650–1654, doi:10.1093/jn/117.10.1650 (1987).

118 Lee, B. C., Kaya, A. & Gladyshev, V. N. Methionine restriction and life-span control. *Ann N Y Acad Sci* **1363**, 116–124, doi:10.1111/nyas.12973 (2016).

119 McIsaac, R. S., Lewis, K. N., Gibney, P. A. & Buffenstein, R. From yeast to human: exploring the comparative biology of methionine restriction in extending eukaryotic life span. *Ann N Y Acad Sci* **1363**, 155–170, doi:10.1111/nyas.13032 (2016).

120 Jo, M. C., Liu, W., Gu, L., Dang, W. & Qin, L. High-throughput analysis of yeast replicative aging using a microfluidic system. *Proc Natl Acad Sci U S A* **112**, 9364–9369, doi:10.1073/pnas.1510328112 (2015).

121 Lin, S. J., Defossez, P. A. & Guarente, L. Requirement of NAD and SIR2 for life-span extension by calorie restriction in *Saccharomyces cerevisiae*. *Science* **289**, 2126–2128 (2000).

122 Avanesov, A. S. *et al.* Age- and diet-associated metabolome remodeling characterizes the aging process driven by damage accumulation. *eLife* **3**, e02077, doi:10.7554/eLife.02077 (2014).

123 Enns, L. C. *et al.* Disruption of protein kinase A in mice enhances healthy aging. *PLoS One* **4**, e5963, doi:10.1371/journal.pone.0005963 (2009).

124 Gonzalez, A. & Hall, M. N. Nutrient sensing and TOR signaling in yeast and mammals. *EMBO J* **36**, 397–408, doi:10.15252/embj.201696010 (2017).

125 Wang, Y. *et al.* Ras and Gpa2 mediate one branch of a redundant glucose signaling pathway in yeast. *PLoS Biol* **2**, E128 (2004).

126 Santangelo, G. M. Glucose signaling in *Saccharomyces cerevisiae*. *Microbiol Mol Biol Rev* **70**, 253–282 (2006).

127 Peeters, K. *et al.* Fructose-1,6-bisphosphate couples glycolytic flux to activation of Ras. *Nat Commun* **8**, 922, doi:10.1038/s41467-017-01019-z (2017).

128 Deprez, M. A., Eskes, E., Wilms, T., Ludovico, P. & Winderickx, J. pH homeostasis links the nutrient sensing PKA/TORC1/Sch9 menage-a-trois to stress tolerance and longevity. *Microb Cell* **5**, 119–136, doi:10.15698/mic2018.03.618 (2018).

129 Orij, R., Postmus, J., Ter Beek, A., Brul, S. & Smits, G. J. In vivo measurement of cytosolic and mitochondrial pH using a pH-sensitive GFP derivative in *Saccharomyces cerevisiae* reveals a relation between intracellular pH and growth. *Microbiology* **155**, 268–278, doi:10.1099/mic.0.022038-0 (2009).

130 Thevelein, J. M. *et al.* Regulation of the cAMP level in the yeast *Saccharomyces cerevisiae*: intracellular pH and the effect of membrane depolarizing compounds. *J Gen Microbiol* **133**, 2191–2196 (1987).

131 Goossens, A., de La Fuente, N., Forment, J., Serrano, R. & Portillo, F. Regulation of yeast H(+)-ATPase by protein kinases belonging to a family dedicated to activation of plasma membrane transporters. *Mol Cell Biol* **20**, 7654–7661 (2000).

132 Martinez-Munoz, G. A. & Kane, P. Vacuolar and plasma membrane proton pumps collaborate to achieve cytosolic pH homeostasis in yeast. *J Biol Chem* **283**, 20309–20319, doi:10.1074/jbc.M710470200 (2008).

133 Dechant, R. *et al.* Cytosolic pH is a second messenger for glucose and regulates the PKA pathway through V-ATPase. *EMBO J* **29**, 2515–2526, doi:10.1038/emboj.2010.138 (2010).

134 Dechant, R., Saad, S., Ibanez, A. J. & Peter, M. Cytosolic pH regulates cell growth through distinct GTPases, Arf1 and Gtr1, to promote Ras/PKA and TORC1 activity. *Mol Cell* **55**, 409–421, doi:10.1016/j.molcel.2014.06.002 (2014).

135 Loewith, R. *et al.* Two TOR complexes, only one of which is rapamycin sensitive, have distinct roles in cell growth control. *Mol Cell* **10**, 457–468 (2002).

136 Kaeberlein, M. *et al.* Regulation of yeast replicative life span by TOR and Sch9 in response to nutrients. *Science* **310**, 1193–1196 (2005).

137 Medvedik, O., Lamming, D. W., Kim, K. D. & Sinclair, D. A. MSN2 and MSN4 link calorie restriction and TOR to sirtuin-mediated lifespan extension in *Saccharomyces cerevisiae*. *PLoS Biol* **5**, e261 (2007).

138 Urban, J. *et al.* Sch9 is a major target of TORC1 in *Saccharomyces cerevisiae*. *Mol Cell* **26**, 663–674, doi:10.1016/j.molcel.2007.04.020 (2007).

139 Prouteau, M. *et al.* TORC1 organized in inhibited domains (TOROIDs) regulate TORC1 activity. *Nature* **550**, 265–269, doi:10.1038/nature24021 (2017).

140 Takahara, T. & Maeda, T. Transient sequestration of TORC1 into stress granules during heat stress. *Mol Cell* **47**, 242–252, doi:10.1016/j.molcel.2012.05.019 (2012).

141 Kato, M. *et al.* Redox state controls phase separation of the yeast Ataxin-2 protein via reversible oxidation of its methionine-rich low-complexity domain. *Cell* **177**, 711–721 e718, doi:10.1016/j.cell.2019.02.044 (2019).

142 Lee, B. C. *et al.* Methionine restriction extends lifespan of *Drosophila melanogaster* under conditions of low amino-acid status. *Nat Commun* **5**, 3592, doi:10.1038/ncomms4592 (2014).

143 Zou, K. *et al.* Life span extension by glucose restriction is abrogated by methionine supplementation: cross-talk between glucose and methionine and implication of methionine as a key regulator of life span. *Sci Adv* **6**, eaba1306, doi:10.1126/sciadv.aba1306 (2020).

144 Sutter, B. M., Wu, X., Laxman, S. & Tu, B. P. Methionine inhibits autophagy and promotes growth by inducing the SAM-responsive methylation of PP2A. *Cell* **154**, 403–415, doi:10.1016/j.cell.2013.06.041 (2013).

145 Filteau, M. *et al.* Systematic identification of signal integration by protein kinase A. *Proc Natl Acad Sci U S A* **112**, 4501–4506, doi:10.1073/pnas.1409938112 (2015).

146 Dillin, A., Gottschling, D. E. & Nystrom, T. The good and the bad of being connected: the integrons of aging. *Curr Opin Cell Biol* **26**, 107–112, doi:10.1016/j.ceb.2013.12.003 (2014).

147 Gottschling, D. E. & Nystrom, T. The upsides and downsides of organelle interconnectivity. *Cell* **169**, 24–34, doi:10.1016/j.cell.2017.02.030 (2017).

148 Tong, A. H. & Boone, C. Synthetic genetic array analysis in *Saccharomyces cerevisiae*. *Methods Mol Biol* **313**, 171–192 (2006).

149 Costanzo, M. *et al.* The genetic landscape of a cell. *Science* **327**, 425–431, doi:10.1126/science.1180823 (2010).

150 Costanzo, M. *et al.* A global genetic interaction network maps a wiring diagram of cellular function. *Science* **353**, doi:10.1126/science.aaf1420 (2016).

151 Tong, A. H. *et al.* Global mapping of the yeast genetic interaction network. *Science* **303**, 808–813 (2004).

152 Zhang, N., Quan, Z., Rash, B. & Oliver, S. G. Synergistic effects of TOR and proteasome pathways on the yeast transcriptome and cell growth. *Open Biol* **3**, 120137, doi:10.1098/rsob.120137 (2013).

153 Byrne, D. P. *et al.* Aurora A regulation by reversible cysteine oxidation reveals evolutionarily conserved redox control of Ser/Thr protein kinase activity. *Sci Signal* **13**, doi:10.1126/scisignal.aax2713 (2020).

154 Humphries, K. M., Juliano, C. & Taylor, S. S. Regulation of cAMP-dependent protein kinase activity by glutathionylation. *J Biol Chem* **277**, 43505–43511, doi:10.1074/jbc.M207088200 (2002).

155 Humphries, K. M., Deal, M. S. & Taylor, S. S. Enhanced dephosphorylation of cAMP-dependent protein kinase by oxidation and thiol modification. *J Biol Chem* **280**, 2750–2758, doi:10.1074/jbc.M410242200 (2005).

156 Gumiero, A. *et al.* Interaction of the cotranslational Hsp70 Ssb with ribosomal proteins and rRNA depends on its lid domain. *Nat Commun* **7**, 13563, doi:10.1038/ncomms13563 (2016).

Supersulfide-Mediated Signaling during Differentiation and De-Differentiation

CHAPTER SIXTEEN

Supersulfide-Mediated Signaling during Differentiation and De-Differentiation

Tsuyoshi Takata, Masanobu Morita, Tetsuro Matsunaga,
Hozumi Motohashi, and Takaaki Akaike

CONTENTS

16.1 INTRODUCTION

Supersulfide species, represented typically by reactive sulfur-containing persulfides/polysulfides with sulfur catenation (S_n, n > 1), are endogenous metabolites abundantly formed in cells and tissues of mammals and humans (Ida et al. 2014; Akaike et al. 2017; Fukuto et al. 2018). The most typical supersulfides widely distributed among different organisms include various reactive persulfides/polysulfides, such as cysteine hydropersulfide/polysulfide ($CysS_nH$), glutathione hydropersulfide/polysulfide (GS_nH), and trisulfide/polysulfide (GS_nSG), that are known to be more redox active than other simple thiols and disulfides (Toohey 2011; Fukuto et al. 2012; Ida et al. 2014; Shimizu, Fukushima, et al. 2017; Peng et al. 2017; Akaike et al. 2017). Hydrogen sulfide (H_2S) is recently suggested to be a small molecule signaling species, however, it actually acts as a marker for the functionally active supersulfides, because the reported biological activities of H_2S are mediated entirely by supersulfides, and because H_2S is a major degraded metabolite of supersulfides and most likely their

artefactual product (Ida et al. 2014; Akaike et al. 2017; Fukuto et al. 2018). In fact, the reactive persulfides can act as strong antioxidant and redox signaling molecules (Ida et al. 2014; Akaike et al. 2017; Fukuto et al. 2018), which may thereby ameliorate, for example, chronic heart failure by attenuating oxidative or electrophilic stress-induced cellular senescence (Nishida et al. 2012). Persulfides have a unique redox property that differs from that of simple thiols because of the additional sulfur atoms. $CysS_nH$ behaves as a strong nucleophile and an antioxidant and plays an important role in regulating a cellular redox balance and redox signaling (de Beus et al. 2004; Mustafa et al. 2009; Vandiver et al. 2013; Ida et al. 2014; Gao et al. 2015; Yang et al. 2015; Kasamatsu et al. 2016). Excessively oxidative conditions, where reactive oxygen species (ROS) are produced more than physiological levels, are deleterious due to unfavorable oxidation of cellular components including lipids, proteins, and DNA, whereas mild and appropriately controlled production of ROS are beneficial and utilized as essential messengers of signal transduction

DOI: 10.4324/9781003204091-21

(Atashi et al. 2015; Coso et al. 2012; Bedard et al. 2007). Maintenance of cellular homeostasis relies on a delicate balance between ROS generation and neutralization/utilization.

Somatic stem cells localize to specific sites, or "niches," in the tissues, where the stem cells are maintained as undifferentiated state for self-renewal and pluripotency by various environmental factors such as surrounding cells, proteins, and ions (Scadden 2006; Morrison et al. 2008; Li et al. 2005). In order to maintain the functions of hematopoietic stem cells and mesenchymal stem cells present in bone marrow, adhesion molecules such as cadherins and cytokines are involved in a complex manner. In addition to such biological factors, the physical environment of a high degree of metaphyseal oxygen decline is also important for maintaining hematopoietic stem cell function (Takubo et al. 2010). Stem cells existing in low oxygen conditions are unlikely to produce ATP by an aerobic energy production system (oxygen respiration) that uses oxygen, and excess ROS loses its function as a stem cell (Tan et al. 2018). It is inevitable that the production of ROS and the generation of oxidative stress in oxygen respiration—and therefore the energy production depending on the glycolytic system such as glycolytic enzymes and mitochondrial synthesis/decomposition—are beneficial for the cells as suppression of the ROS generation. On the other hand, because mutant mice with impaired energy metabolism in mitochondria also show the metabolic characteristics of hematopoietic stem cells and their activity as stem cells, it has been suggested to be important for proper maintenance of energy metabolism in mitochondria (Nakada et al. 2010; Gurumurthy et al. 2010; Gan et al. 2010; Sahin et al. 2011).

Interestingly, we have recently revealed the existence of sulfur respiration, which uses sulfur rather than oxygen to generate energy in mitochondria (Akaike et al. 2017). Additionally, stem cells may use sulfur instead of oxygen in the mitochondria to generate energy by sulfur respiration and escape oxidative stress through the antioxidant effects of produced supersulfides. This article provides an overview of recent advances in the supersulfide research and discusses our understanding of several biological processes such as cell survival/death, proliferation and differentiation of stem cells, and cellular metabolism.

16.2 IDENTIFICATION OF A NOVEL SULFUR METABOLISM PATHWAY

About four billion years ago, when primitive cells that are the origin of life were born, the earth was anoxic. Instead of oxygen that is required for a lot of aerobic organisms in modern earth, sulfur is presumed to be used for energy production as an electron accepter (Olson 2020). During the long history of life, sulfur-dependent energy metabolism has been making a major contribution to the prosperity of anaerobic organisms.

Humans ingest a wide variety of sulfur-containing molecules that are contained in foods, such as leeks and garlic. Previous studies on sulfur metabolism focused on biosynthesis and metabolism of sulfur-containing amino acids and proteins, mainly dealing with thiols and their oxidized derivatives. A new trend of studies on sulfur metabolism emerged, and physiological roles of H_2S were frequently discussed. The exact mechanism of H_2S biosynthesis and its regulation were totally unclear, however. In fact, H_2S is a major product decomposed either physiologically and artefactually from supersulfides, and more importantly much of its reported biological activities is apparently that of supersulfides (Ida et al. 2014; Akaike et al. 2017; Fukuto et al. 2018).

In our efforts to develop a new quantitative analytical system for reactive sulfur metabolites, we successfully detected a substantial amount of supersulfide species in biological samples derived from various species, from bacteria to humans (Ida et al. 2014). Persulfides, such as $CysS_nH$, have unique redox properties that differ from those of simple thiols because of the presence of additional sulfur atoms. Such supersulfide species possess both high nucleophilicity and electrophilicity compared with ordinary thiol compounds and may play important roles in regulating oxidative stress and redox signaling in vivo (Fukuto et al. 2018; Hamid et al. 2019; Takata et al. 2019; Doka et al. 2020; Khan et al. 2018; Rudyk et al. 2019). It was also found that appreciable amounts of formation of protein persulfides, referred to as persulfidation, exist in different cellular proteins (Akaike et al. 2017; Khan et al. 2018; Rudyk et al. 2019; Doka et al. 2020; Takata et al. 2019; Ida et al. 2014; Takata et al. 2017; Takata et al. 2020). Amazingly, it turned out that cysteinyl-tRNA synthetase (CARS) is a bifunctional enzyme and serves as a major enzyme for supersulfide synthesis. In addition, CARS produces CysSSH-tRNA by

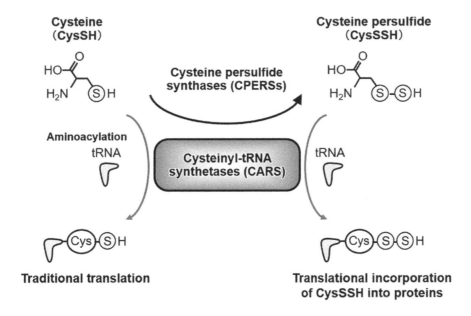

Figure 16.1 Sulfur metabolic pathway coupled to translation by the novel CysSSH-producing enzyme CARS. Cysteinyl-tRNA synthetases (CARS) produce cysteine persulfide (CysSSH) from cysteine (CysSH). CARS are capable of converting CysSH to CysSSH via a pyridoxal phosphate-dependent process using a second CysSH as the sulfur atom donor (independent of ATP and tRNA). The CARS-synthesized CysSSH can then form a tRNA-bound CysSSH adduct (also via CARS catalysis) resulting in the incorporation of CysSSH into proteins, thus generating a protein containing a hydropersulfide functionality. CysSSH producing activity of CARS is critically involved in translation-coupled protein persulfidation.

using cysteine (CysSH) as a substrate, resulting in the production of nascent polypeptide chains containing CysSSH and eventually formation of protein persulfidation (Figure 16.1; Akaike et al. 2017). Importantly, the catalytic center for cysteinyl-tRNA synthetase activity and the one for persulfide synthase activity are separate, and the two enzymatic activities are independent of each other.

Cysteine persulfide synthase (CPERS) activity is observed not only in bacterial CARS but also in mammalian CARS (Akaike et al. 2017). Two different mammalian CARS have been found, one located in the cytosol, CARS1, and the other, CARS2, in the mitochondria (Hallmann et al. 2014; Coughlin et al. 2015). We now know that both CARS (mouse CARS1 and human CARS2) show strong persulfide-producing activities that are dependent on pyridoxal phosphate. Using the gene editing technology CRISPR-Cas9 system, we mutated the motif responsible for CPERS activity, which is conserved in all species. In cells expressing mutant CARS with deficient CPERS activity but with intact aminoacyl-tRNA synthesis activity, the abundance of supersulfides and related sulfur metabolites was decreased. These data suggest that

CARS functions as a major CPERS *in vivo* in all species, including humans (Ida et al. 2014; Akaike et al 2017)

16.3 DISCOVERY OF SULFUR RESPIRATION IN MAMMALS

The inner mitochondrial membrane is a place where electrons are transported from NADH to the final electron acceptor, the oxygen molecule. The proton gradient and ATP production are coupled with electron transport. CARS2, a mitochondrial isoform of CARS, was found to be responsible for the generation and maintenance of mitochondrial membrane potential (Akaike et al. 2017). It is highly plausible that $CysS_nH$ that is generated by CARS2 in mitochondria contributes to mitochondrial energy metabolism via membrane potential formation. $CysS_nH$ acts as an electron acceptor being reduced to H_2S in an electron transport-dependent manner, and the resultant H_2S acts as an electron donor for the electron transport chain (ETC; Figure 16.2). This is the rediscovery of sulfur respiration in the living world and the discovery of sulfur respiration in humans and mammals. In the newly discovered

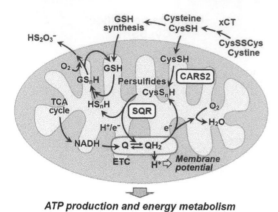

ATP production and energy metabolism

Figure 16.2 Mitochondrial sulfur respiration: energy generation by electron transfer conjugation of CARS2-derived supersulfides.

CysS$_n$H produced by CARS2 in mitochondria may be reductively metabolized to sulfides, and further oxidized by the sulfide:quinone oxidoreductase (SQR), in a manner linked to ETC in mitochondria. The CysS$_n$H-dependent sulfur metabolism is coupled with formation of glutathione polysulfide (GS$_n$H), which is controlled by the mitochondrial ETC. Q/QH$_2$, ubiquinone/ubiquinol; TCA, tricarboxylic acid; xCT, cystine/glutamic acid transporter.

sulfur respiration, the final electron acceptor of the electron transport system is the supersulfide, not the oxygen molecule as in normal oxygen respiration. In the case of oxygen respiration, electrons from the ETC are transferred to oxygen to make water, while in the case of sulfur respiration, electrons are transferred to CysS$_n$H, resulting in H$_2$S generation.

The sulfide:quinone oxidoreductase (SQR) protein is localized in mitochondria and donates protons and electrons derived from sulfides to the mitochondrial ubiquinone (Q: coenzyme Q10 in mammals) cycle. Most recently, SQR-mediated oxidation of sulfides was reported to drive reverse electron transport at mitochondrial complex I (Jia et al. 2020). It is conceivable, therefore, that CysS$_n$H produced by CARS2 in mitochondria is reductively metabolized to CysSH and HS$^-$, which may be further oxidized by the SQR in a manner linked to ETC in mitochondria.

Some photosynthetic bacteria generate reducing power from sulfur compounds such as H$_2$S (Shimizu, Shen, et al. 2017). Sulfur-oxidizing bacteria (e.g., *Acidithiobacillus thiooxidans*) utilize sulfur as an electron donor for energy production (Yin et al. 2014). In *Staphylococcus aureus*, reactive persulfides have been reported not only in CysSH but also in CoA, as CoA-SSH (Peng et al. 2017). The genome of

budding yeast (*Saccharomyces cerevisiae*) contains only a single cysteinyl-tRNA synthetase gene (*YNL247W*, also known as CRS1). Energy-dependent alteration of the transcriptional start site of CRS1 generates two isoforms of CRS1, cytosolic and mitochondrial isoforms, the latter of which is involved in the mitochondrial energy metabolism (Nishimura, Nasuno, et al. 2019). This is the first demonstration of sulfur respiration in eukaryotes, as shown in Figure 16.2 (Akaike et al. 2017). The discovery of sulfur respiration in mammals may have major implications not only for fundamental biology but also for disease pathogenesis related to energy metabolism.

16.4 REGULATION OF PROTEIN FUNCTION BY PERSULFIDATION

Notably, the chemical reactivity of reactive persulfides is thought to depend on the catalytic activity of particular proteins regulated by protein persulfidation (i.e., forming CysS$_n$H on various proteins; Ono et al. 2014). Several research groups have developed methods for detecting persulfidation, and increasing number of reports have been published on the identification of polysulfided proteins and their functional analysis (Jung et al. 2016; Doka et al. 2016). For example, protein kinase G type Iα (PKGIα) promotes pulmonary vasodilation by its persulfide oxidation, which alleviates pulmonary hypertension caused by chronic hypoxia (Rudyk et al. 2019; Feelisch et al. 2020). The activity of Ca^{2+}/calmodulin-dependent protein kinases (CaMKs) was directly regulated by persulfidation (Takata et al. 2020) in the regulatory N-terminal kinase domain of CaMKII (Araki et al. 2019) and in the activation loop of CaMKI (Takata et al. 2019) and CaMKIV (Takata et al. 2017). Members of the CaMK cascade—including CaMKK, CaMKI, and CaMKIV—are present in most mammalian tissues but are predominantly expressed in the brain, and they play pivotal roles in various cellular processes such as outgrowth of axons and dendrites (CaMKI; Ageta-Ishihara et al. 2009), the potentiation of synaptic transmission during learning and memory, and the activation of T lymphocytes (CaMKIV; Kasahara et al. 2001; Westphal et al. 1998). Supersulfides promote neurite outgrowth and differentiation through accelerated intracellular Ca^{2+} influx in mouse neuroblastoma cells (Koike et al. 2015). The CaMKK-CaMKI cascade might be fine-tuned by polysulfide induced Ca^{2+} influx and/or CaMKI modification in neuronal cells. CaMKIV is expressed in the hematopoietic progenitor population that contains stem cells. In the Camk4-null mouse, hematopoietic stem cells and cerebellar granule cells are decreased in number due to the acceleration of apoptosis (Kitsos et al. 2005). The CaMKIV-cAMp-response element-binding protein (CREB) pathway regulates osteoclast differentiation through induction of nuclear factor expression in activated T cells (NFAT; Sato et al. 2006) and also regulates the expression of genes such as insulin and prolactin in endocrine cells, such as pancreatic β cells and pituitary tumor cell lines (Ban et al. 2000; Murao et al. 2004; Yu et al. 2004). Thus, the regulation of CaMK activity through persulfidation may be widely responsible for the regulation of gene expression and differentiation associated with intracellular Ca^{2+} mobilization in vivo.

Living organisms strongly rely on a reaction (redox reaction) in which electrons are transferred between various molecules in order to synthesize and metabolize molecules and maintain vital activity, and cysteine in proteins is important in this redox reaction. However, when the cysteine thiol (R-SH) is irreversibly excessively oxidized to sulfinic acid (R-SO$_2$H) and sulfonic acid (R-SO$_3$H), the function of the oxidized protein is significantly reduced. CysS$_n$H formed on proteins are easily oxidized than their parental cysteine residues. The stepwise oxidation of a persulfide group leads to the consecutive formation of perthiosulfenic acid (R-S$_n$OH), perthiosulfinic acid (R-S$_n$O$_2$H), and perthiosulfonic acid (R-S$_n$O$_3$H; Ono et al. 2014).

Recently, a mechanism by which cells protect proteins from irreparable damage by oxygen was demonstrated (Doka et al. 2020). Unlike R-SO$_2$H and R-SO$_3$H, which are irreversibly oxidized forms, the hydrosulfide group (-SH) can be regenerated from R-S$_n$O$_2$H and R-S$_n$O$_3$H by reductive cleavage of the disulfide bond, which is interpreted as escaping from oxidative damage and preventing functional deterioration of the protein (Figure 16.3). In other words, persulfides confer protection on proteins from irreversible modification by electrophiles (Takata et al. 2019).

Oxidized cysteines are usually reduced and repaired by the glutathione-thioredoxin system, which is the main reducing system in the body. Analysis of protein cysteine residues in mice with a defective glutathione-thioredoxin system showed a decrease in cysteine and an increase in R-S$_n$O$_3$H compared with wild-type mice. This indicates that R-S$_n$O$_2$H and R-S$_n$O$_3$H are abundantly present in the living bodies of mice and protect proteins by preventing their oxidation. Namely, the protein persulfidation plays an extremely important role in living organisms. We demonstrated that the novel mechanism not only protects proteins, but also regulates and converts their functions. Phosphatase and tensin homolog deleted from chromosome 10 (PTEN), protein-tyrosine phosphatase 1B (PTP1B), peroxiredoxins, and heat-shock protein 90 (HSP90) are good examples (Doka et al. 2020). These pathways are generally believed to decline in oxidative conditions, including aging and aging-related diseases.

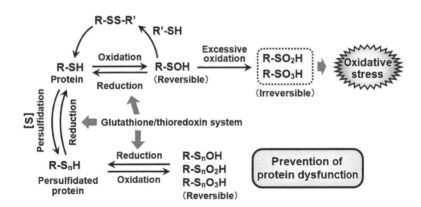

Figure 16.3 Mechanisms for avoiding oxidative damage to proteins due to supersulfides.

The thiol of protein (R-SH) is irreversibly oxidized by being excessively oxidized, but the persulfide species has a reducing ability and provides a functionally reversible oxidation state, protein persulfidation (R-S$_n$H). The glutathione-thioredoxin system contributes to the reduction of excessively oxidized persulfide species, perthiosulfinic acid (R-S$_n$O$_2$H) and perthiosulfonic acid (R-S$_n$O$_3$H).

16.5 STEM CELL REGULATION AND AGING BY SULFUR RESPIRATION

Appropriate maintenance of tissue homeostasis by somatic stem cells is considered to be involved in the progression of organismal aging, and it is postulated that reduced stem cell function and number may contribute to aging-related diseases. Various types of somatic stem cells are commonly thought to be in the quiescent state with minimal production of ROS, which is favorable for retaining a lifetime ability to self-renew (Suda et al. 2011). ROS often causes protein carbonylation and thiol oxidation, resulting in the alterations in various biological processes (Dansen et al. 2009; Guo et al. 2010; Velu et al. 2007). Accumulation of ROS causes cytotoxicity and impairs self-renewal activity of stem cells, leading to senescence and exhaustion of stem cells (Sahin et al. 2011; Rossi et al. 2008). Thus, glutathione, which is known as an antioxidant, is required in high concentrations to maintain stem cell function (Jeong et al. 2018).

CysS$_n$H and GS$_n$H are much more reactive than their parental cysteine and glutathione, respectively, and serve as excellent antioxidants in cells (Ida et al. 2014). Cells with increased level of CysSSH are resistant to hydrogen peroxide–induced cytotoxicity. Intriguingly, supersulfides exert highly anti-inflammatory functions in mice and thus protect mice from lethal shock induced by endotoxin (Zhang et al. 2019; Sawa et al. 2020). In patients with asthma and chronic obstructive pulmonary disease overlap (ACO), the amounts of reactive persulfide and polysulfide were significantly decreased, which skews the redox balance towards oxidative conditions and exacerbates inflammation (Kyogoku et al. 2019). Thus, reactive persulfides provide a main defense system from oxidative stress. Reactive persulfides are likely to protect somatic stem cells from disadvantages such as oxidative stress caused by oxygen respiration and to contribute to the energy production of somatic stem cells (Figure 16.4).

Mitochondrial fission and fusion have significant impacts on tumor and stem cell phenotypes. Our recent study demonstrated novel physiological roles of CARS as a CPERS during regulation of mitochondrial functions (Figure 16.2; Akaike et al. 2017). CARS2-deficient cells show remarkably altered mitochondrial morphology (i.e., shrunken or fragmented appearance), which was improved when mutant CARS2 possessed CPERS activity but lacked cysteinyl tRNA synthesizing activity, indicating that mitochondrial production of supersulfides is required for normal regulation of mitochondrial morphology. Deficiency of CARS2-mediated supersulfide production activated dynamin-related

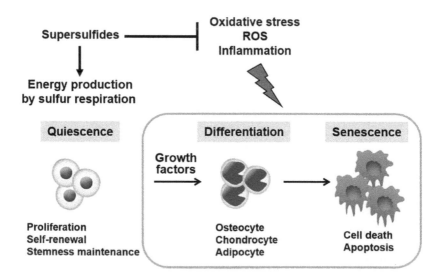

Figure 16.4 A model of the effects of sulfur respiration on stem cells.

Several triggers including oxidative stress, reactive oxygen species (ROS), inflammation, and growth factors affect stem cells and their destinations. High ROS levels cause stem cell senescence and death. Quiescent and/or self-renewing stem cells could display low ROS levels due to their strong antioxidant machinery by supersulfides and protein persulfidation. Sulfur respiration may contribute to also energy production under hypoxia in niche.

protein 1 (Drp1), a primary mitochondrial fission mediator (Akhtar et al. 2016). Under normal culture conditions, Drp1 is extensively persulfidated, whereas defective supersulfide production caused by CARS1/CARS2 deficiency markedly suppressed Drp1 persulfidation. Methylmercury also induces mitochondrial hyperfission in cardiomyocytes through consuming polysulfides in Drp1. As mitochondrial hyperfission increases mechanical stress-induced fragility of cardiomyocytes (Nishimura, Shimoda, et al. 2019), sufficient polysulfides of Drp1 are important for healthy status of cardiomyocytes. Drp1 is an important signal effector molecule that is regulated in a reversible manner via a unique process involving persulfidation and depersulfidation.

Stem cell functionality decreases during aging, contributing to age-associated pathologies and the overall aging process of the organism (Behrens et al. 2014). Diet is also an emerging important regulator of adult stem cell function. For instance, it has been reported that dietary restriction enhances stem cell functionality in muscle (Cerletti et al. 2012) and intestinal epithelium (Yilmaz et al. 2012) and improves repopulation capacity of hematopoietic stem cells in early mouse aging (Tang et al. 2016). Furthermore, various nutrient-sensitive signal transduction systems, such as insulin-Akt/protein kinase B (Tothova et al. 2007; Miyamoto et al. 2007), AMP-activated protein kinase (AMPK; Nakada et al. 2010; Gurumurthy et al. 2010; Gan et al. 2010), and mechanistic target of rapamycin (mTOR; Chen et al. 2008; Chen, Liu, et al. 2009) are known to control individual aging through balancing the self-renewal and quiescent state of stem cells. Interestingly, a few studies suggest that dietary restriction is associated with multiple benefits including extended life span in different species by promoting sulfur metabolism (Hine et al. 2015; Mitchell et al. 2016). The addition of thiosulfate to *Caenorhabditis elegans* shows an increase in supersulfides and a prolongation of life (Zivanovic et al. 2019). These observations suggest that the regulation of stem cell metabolism by supersulfides may be able to collectively manipulate quiescence, self-renewal ability, and pluripotency to delay aging process.

16.6 SULFUR RESPIRATION AND CANCER STEM CELLS

Cancer stem cells are defined as those that have strong self-renewal ability and cancer forming ability in cancer tissues. Cancer stem cells, like normal tissue stem cells, remain dormant in a very hypoxic niche. Breast and liver cancer stem cells tend to have low ROS levels owing to the increased expression of ROS-scavenging systems (Douglass 1989; Kim et al. 2012). The high ROS-scavenging ability in cancer stem cells confers therapeutic resistance (Diehn et al. 2009). Low ROS levels in cancer stem cells are also likely to be attributable to low mitochondrial respiratory activity.

Recent studies described increased expression of various sulfur-metabolizing enzymes and new roles of sulfur-containing metabolites in cancer cells (Hellmich et al. 2015). This is particularly evident in cancers of the colon and ovaries, where the malignant cells overexpress cystathionine β-synthase (CBS) and produce increased amounts of H_2S. The increased H_2S enhances tumor growth and spread by stimulating cellular bioenergetics, activating proliferative, migratory, and invasive signaling pathways, and enhancing tumor angiogenesis (Szabo et al. 2013; Bhattacharyya et al. 2013). Cystathionine γ-lyase (CSE) expression is upregulated in both breast cancers and breast cancer cell lines and results in proliferation and migration of breast cancer cells (You et al. 2017). Additionally, numerous studies revealed that endogenous H_2S produced by CSE promotes proliferation of human hepatoma and colon cells (Yin et al. 2012; Pan et al. 2014). Of note, we found that CSE and CBS produces CysSSH via C-S cleavage of cystine (Nishida et al. 2012; Ida et al. 2014). Inhibition of CSE and CBS significantly decreased intracellular cysteine levels, which in turn reduced supersulfide production. Because CSE and CBS knockdown in the absence of CARS-mediated supersulfide production lowers cysteine levels but not $CysS_nH$ levels (Akaike et al. 2017), we consider that CSE and CBS are not likely to participate directly in the persulfide production but instead may support cysteine biosynthesis and supply to CARS, at least under physiological conditions.

Cystine/glutamic acid transporter (xCT), a cell membrane amino acid transporter, has been implicated in the proliferation and multidrug resistance of several types of cancer cells (Chen, Song, et al. 2009; Huang et al. 2005; Lo et al. 2008). Increased xCT expression in tumor cells maintains intracellular glutathione levels by supplying cystine to tumor cells. This increases the resistance of tumor cells to oxidative stress, leading to chemo- and radio-resistance, especially in cancer stem cells (Lo et al.

2008; Ishimoto et al. 2011). It is possible that the supersulfide generation by CARS may function in this context. In fact, we showed that persulfide is reduced by inhibitor of xCT and suppression of its expression (Ida et al. 2014). Thus, highly malignant carcinomas such as pancreatic cancer that can proliferate vigorously even under hypoxic stress are assumed to utilize sulfur respiration. Supersulfide regulation may be a promising therapeutic target for cancer treatment.

16.7 CONCLUSION: TOWARD SULFUR DRUG DISCOVERY

An anaerobic environment is required for stem cells to maintain self-renewal and pluripotency. In contrast, ROS generation is inevitable in aerobic energy production system using oxygen (oxygen respiration). Reactive sulfur species like $CysS_nH$ and GS_nH are identified as new endogenous antioxidants that efficiently eliminate ROS. We also demonstrated a novel role of supersulfides for protecting proteins from excessive oxidation causing functional deterioration. Searching for supersulfide-producing enzymes, our research led to the discovery of sulfur respiration by supersulfides derived from mitochondrial CARS2. The promotion of sulfur respiration would be beneficial for the maintenance of healthy condition of tissues and organs with relatively high demand of energy production, such as skeletal muscle, liver, and possibly neural tissues. Cancer stem cells that remain dormant in a hypoxic niche can survive by actively utilizing sulfur respiration in starvation and hypoxic environments, and may contribute to highly malignant cancer cells. In the future, mechanistic elucidation of energy metabolism by sulfur respiration in hypoxia, starvation, and hibernation, for example, will help to understand the mechanism of metabolic regulation in stem cells. Novel preventive and therapeutic approaches for various diseases might allow us to control aging and longevity (Motohashi et al. 2019).

REFERENCES

Ageta-Ishihara, N., S. Takemoto-Kimura, M. Nonaka, et al. 2009. Control of cortical axon elongation by a GABA-driven Ca^{2+}/calmodulin-dependent protein kinase cascade. J. Neurosci. 29 (43):13720–13729.

Akaike, T., T. Ida, F. Y. Wei, et al. 2017. Cysteinyl-tRNA synthetase governs cysteine polysulfidation

and mitochondrial bioenergetics. Nat. Commun. 8 (1):1177.

Akhtar, M. W., S. Sanz-Blasco, N. Dolatabadi, et al. 2016. Elevated glucose and oligomeric β-amyloid disrupt synapses via a common pathway of aberrant protein S-nitrosylation. Nat. Commun. 7:10242.

Araki, S., T. Takata, Y. Tsuchiya, and Y. Watanabe. 2019. Reactive sulfur species impair Ca^{2+}/calmodulin-dependent protein kinase II via polysulfidation. Biochem. Biophys. Res. Commun. 508 (2):550–555.

Atashi, F., A. Modarressi, and M. S. Pepper. 2015. The role of reactive oxygen species in mesenchymal stem cell adipogenic and osteogenic differentiation: a review. Stem Cells Dev. 24 (10):1150–1163.

Ban, N., Y. Yamada, Y. Someya, et al. 2000. Activating transcription factor-2 is a positive regulator in CaM kinase IV-induced human insulin gene expression. Diabetes 49 (7):1142–1148.

Bedard, K., and K. H. Krause. 2007. The NOX family of ROS-generating NADPH oxidases: physiology and pathophysiology. Physiol. Rev. 87 (1):245–313.

Behrens, Axel, Jan M. van Deursen, K. Lenhard Rudolph, and Björn Schumacher. 2014. Impact of genomic damage and ageing on stem cell function. Nat. Cell Biol. 16 (3):201–207.

Bhattacharyya, S., S. Saha, K. Giri, et al. 2013. Cystathionine beta-synthase (CBS) contributes to advanced ovarian cancer progression and drug resistance. PLoS One 8 (11):e79167.

Cerletti, M., Y. C. Jang, L. W. Finley, M. C. Haigis, and A. J. Wagers. 2012. Short-term calorie restriction enhances skeletal muscle stem cell function. Cell Stem Cell 10 (5):515–519.

Chen, C., Y. Liu, R. Liu, et al. 2008. TSC-mTOR maintains quiescence and function of hematopoietic stem cells by repressing mitochondrial biogenesis and reactive oxygen species. J. Exp. Med. 205 (10):2397–2408.

Chen, C., Y. Liu, Y. Liu, and P. Zheng. 2009. mTOR regulation and therapeutic rejuvenation of aging hematopoietic stem cells. Sci. Signal. 2 (98):ra75.

Chen, R. S., Y. M. Song, Z. Y. Zhou, et al. 2009. Disruption of xCT inhibits cancer cell metastasis via the caveolin-1/β-catenin pathway. Oncogene 28 (4):599–609.

Coso, S., I. Harrison, C. B. Harrison, et al. 2012. NADPH oxidases as regulators of tumor angiogenesis: current and emerging concepts. Antioxid. Redox Signal. 16 (11):1229–1247.

Coughlin, C. R., 2nd, G. H. Scharer, M. W. Friederich, et al. 2015. Mutations in the mitochondrial cysteinyl-tRNA synthase gene, CARS2, lead to a severe epileptic encephalopathy and complex movement disorder. *J. Med. Genet.* 52 (8):532–540.

Dansen, T. B., L. M. Smits, M. H. van Triest, et al. 2009. Redox-sensitive cysteines bridge p300/CBP-mediated acetylation and FoxO4 activity. *Nat. Chem. Biol.* 5 (9):664–672.

de Beus, M. D., J. Chung, and W. Colón. 2004. Modification of cysteine 111 in Cu/Zn superoxide dismutase results in altered spectroscopic and biophysical properties. *Protein Sci.* 13 (5):1347–1355.

Diehn, M., R. W. Cho, N. A. Lobo, et al. 2009. Association of reactive oxygen species levels and radioresistance in cancer stem cells. *Nature* 458 (7239):780–783.

Doka, E., T. Ida, M. Dagnell, et al. 2020. Control of protein function through oxidation and reduction of persulfidated states. *Sci. Adv.* 6 (1):eaax8358.

Doka, E., I. Pader, A. Biro, et al. 2016. A novel persulfide detection method reveals protein persulfide- and polysulfide-reducing functions of thioredoxin and glutathione systems. *Sci. Adv.* 2 (1):e1500968.

Douglass, C. W. 1989. Estimating periodontal treatment needs from epidemiological data. *J. Periodontol.* 60 (7):417–419.

Fan, K., N. Li, J. Qi, et al. 2014. Wnt/β-catenin signaling induces the transcription of cystathionine-γ-lyase, a stimulator of tumor in colon cancer. *Cell. Signal.* 26 (12):2801–2808.

Feelisch, M., T. Akaike, K. Griffiths, et al. 2020. Long-lasting blood pressure lowering effects of nitrite are NO-independent and mediated by hydrogen peroxide, persulfides, and oxidation of protein kinase G1alpha redox signalling. *Cardiovasc. Res.* 116 (1):51–62.

Fukuto, J. M., S. J. Carrington, D. J. Tantillo, et al. 2012. Small molecule signaling agents: the integrated chemistry and biochemistry of nitrogen oxides, oxides of carbon, dioxygen, hydrogen sulfide, and their derived species. *Chem. Res. Toxicol.* 25 (4):769–793.

Fukuto, J. M., L. J. Ignarro, P. Nagy, et al. 2018. Biological hydropersulfides and related polysulfides: a new concept and perspective in redox biology. *FEBS Lett.* 592 (12):2140–2152.

Gan, B., J. Hu, S. Jiang, et al. 2010. Lkb1 regulates quiescence and metabolic homeostasis of haematopoietic stem cells. *Nature* 468 (7324):701–704.

Gao, X. H., D. Krokowski, B. J. Guan, et al. 2015. Quantitative H$_2$S-mediated protein sulfhydration reveals metabolic reprogramming during the integrated stress response. *Elife* 4:e10067.

Guo, Zhi, Sergei Kozlov, Martin F. Lavin, Maria D. Person, and Tanya T. Paull. 2010. ATM activation by oxidative stress. *Science* 330 (6003):517–521.

Gurumurthy, S., S. Z. Xie, B. Alagesan, et al. 2010. The Lkb1 metabolic sensor maintains haematopoietic stem cell survival. *Nature* 468 (7324):659–663.

Hallmann, K., G. Zsurka, S. Moskau-Hartmann, et al. 2014. A homozygous splice-site mutation in CARS2 is associated with progressive myoclonic epilepsy. *Neurology* 83 (23):2183–2187.

Hamid, H. A., A. Tanaka, T. Ida, et al. 2019. Polysulfide stabilization by tyrosine and hydroxyphenyl-containing derivatives that is important for a reactive sulfur metabolomics analysis. *Redox Biol.* 21: 101096.

Hellmich, M. R., and C. Szabo. 2015. Hydrogen sulfide and cancer. *Handb. Exp. Pharmacol.* 230:233–241.

Hine, C., E. Harputlugil, Y. Zhang, et al. 2015. Endogenous hydrogen sulfide production is essential for dietary restriction benefits. *Cell* 160 (1–2): 132–144.

Huang, Y., Z. Dai, C. Barbacioru, and W. Sadée. 2005. Cystine-glutamate transporter SLC7A11 in cancer chemosensitivity and chemoresistance. *Cancer Res.* 65 (16):7446–7454.

Ida, T., T. Sawa, H. Ihara, et al. 2014. Reactive cysteine persulfides and S-polythiolation regulate oxidative stress and redox signaling. *Proc. Natl. Acad. Sci. U. S. A.* 111 (21):7606–7611.

Ishimoto, T., O. Nagano, T. Yae, et al. 2011. CD44 variant regulates redox status in cancer cells by stabilizing the xCT subunit of system xc– and thereby promotes tumor growth. *Cancer Cell* 19 (3):387–400.

Jeong, E. M., J. H. Yoon, J. Lim, et al. 2018. Real-time monitoring of glutathione in living cells reveals that high glutathione levels are required to maintain stem cell function. *Stem Cell Reports* 10 (2): 600–614.

Jia, J., Z. Wang, M. Zhang, et al. 2020. SQR mediates therapeutic effects of H$_2$S by targeting mitochondrial electron transport to induce mitochondrial uncoupling. *Sci. Adv.* 6 (35):eaaz5752.

Jung, M., S. Kasamatsu, T. Matsunaga, et al. 2016. Protein polysulfidation-dependent persulfide dioxygenase activity of ethylmalonic encephalopathy protein 1. *Biochem. Biophys. Res. Commun.* 480 (2):180–186.

Kasahara, J., K. Fukunaga, and E. Miyamoto. 2001. Activation of calcium/calmodulin-dependent protein kinase IV in long term potentiation in the rat hippocampal CA1 region. *J. Biol. Chem.* 276 (26): 24044–24050.

Kasamatsu, S., A. Nishimura, M. Morita, T. Matsunaga, H. Abdul Hamid, and T. Akaike. 2016. Redox signaling regulated by cysteine persulfide and protein polysulfidation. *Molecules* 21 (12):1721.

Khan, S., S. Fujii, T. Matsunaga, et al. 2018. Reactive persulfides from *Salmonella typhimurium* downregulate autophagy-mediated innate immunity in macrophages by inhibiting electrophilic signaling. *Cell Chem. Biol.* 25 (11):1403–1413 e4.

Kim, Ho Min, Naotsugu Haraguchi, Hideshi Ishii, et al. 2012. Increased CD13 expression reduces reactive oxygen species, promoting survival of liver cancer stem cells via an epithelial–mesenchymal transition-like phenomenon. *Ann. Surg. Oncol.* 19 (3):539–548.

Kitsos, C. M., U. Sankar, M. Illario, et al. 2005. Calmodulin-dependent protein kinase IV regulates hematopoietic stem cell maintenance. *J. Biol. Chem.* 280 (39):33101–33108.

Koike, S., N. Shibuya, H. Kimura, K. Ishii, and Y. Ogasawara. 2015. Polysulfide promotes neuroblastoma cell differentiation by accelerating calcium influx. *Biochem. Biophys. Res. Commun.* 459 (3):488–492.

Kyogoku, Y., H. Sugiura, T. Ichikawa, et al. 2019. Nitrosative stress in patients with asthma-chronic obstructive pulmonary disease overlap. *J. Allergy Clin. Immunol.* 144 (4):972–983 e14.

Li, L., and T. Xie. 2005. Stem cell niche: structure and function. *Annu. Rev. Cell Dev. Biol.* 21:605–631.

Lo, M., Y. Z. Wang, and P. W. Gout. 2008. The x_c^- cystine/glutamate antiporter: a potential target for therapy of cancer and other diseases. *J. Cell. Physiol.* 215 (3):593–602.

Mitchell, S. J., J. Madrigal-Matute, M. Scheibye-Knudsen, et al. 2016. Effects of sex, strain, and energy intake on hallmarks of aging in mice. *Cell Metab.* 23 (6):1093–1112.

Miyamoto, K., K. Y. Araki, K. Naka, et al. 2007. Foxo3a is essential for maintenance of the hematopoietic stem cell pool. *Cell Stem Cell* 1 (1):101–112.

Morrison, S. J., and A. C. Spradling. 2008. Stem cells and niches: mechanisms that promote stem cell maintenance throughout life. *Cell* 132 (4):598–611.

Motohashi, H., and T. Akaike. 2019. Sulfur-utilizing cytoprotection and energy metabolism. *Current Opinion in Physiology* 9:1–8.

Murao, K., H. Imachi, W. M. Cao, et al. 2004. Role of calcium-calmodulin-dependent protein kinase cascade in thyrotropin (TSH)-releasing hormone induction of TSH and prolactin gene expression. *Endocrinology* 145 (11):4846–4852.

Mustafa, A. K., M. M. Gadalla, N. Sen, et al. 2009. H_2S signals through protein S-sulfhydration. *Sci. Signal.* 2 (96):ra72.

Nakada, D., I. L. Saunders, and S. J. Morrison. 2010. Lkb1 regulates cell cycle and energy metabolism in haematopoietic stem cells. *Nature* 468 (7324): 653–658.

Nishida, M., T. Sawa, N. Kitajima, et al. 2012. Hydrogen sulfide anion regulates redox signaling via electrophile sulfhydration. *Nat. Chem. Biol.* 8 (8):714–724.

Nishimura, A., R. Nasuno, Y. Yoshikawa, et al. 2019. Mitochondrial cysteinyl-tRNA synthetase is expressed via alternative transcriptional initiation regulated by energy metabolism in yeast cells. *J. Biol. Chem.* 294 (37):13781–13788.

Nishimura, A., K. Shimoda, T. Tanaka, et al. 2019. Depolysulfidation of Drp1 induced by low-dose methylmercury exposure increases cardiac vulnerability to hemodynamic overload. *Sci. Signal.* 12 (587):eaaw1920.

Olson, K. R. 2020. Are reactive sulfur species the new reactive oxygen species? *Antioxid. Redox Signal.* 33 (16):1125–1142.

Ono, K., T. Akaike, T. Sawa, et al. 2014. Redox chemistry and chemical biology of H_2S, hydropersulfides, and derived species: implications of their possible biological activity and utility. *Free Radic. Biol. Med.* 77:82–94.

Peng, H., J. Shen, K. A. Edmonds, et al. 2017. Sulfide homeostasis and nitroxyl intersect via formation of reactive sulfur species in *Staphylococcus aureus*. *mSphere* 2 (3):e00082–17.

Rossi, D. J., C. H. Jamieson, and I. L. Weissman. 2008. Stems cells and the pathways to aging and cancer. *Cell* 132 (4):681–696.

Rudyk, O., A. Rowan, O. Prysyazhna, et al. 2019. Oxidation of PKGIα mediates an endogenous adaptation to pulmonary hypertension. *Proc. Natl. Acad. Sci. U. S. A.* 116 (26):13016–13025.

Sahin, E., S. Colla, M. Liesa, et al. 2011. Telomere dysfunction induces metabolic and mitochondrial compromise. *Nature* 470 (7334):359–365.

Sato, K., A. Suematsu, T. Nakashima, et al. 2006. Regulation of osteoclast differentiation and function by the CaMK-CREB pathway. *Nat. Med.* 12 (12):1410–1416.

Sawa, T., H. Motohashi, H. Ihara, and T. Akaike. 2020. Enzymatic regulation and biological functions of reactive cysteine persulfides and polysulfides. *Biomolecules* 10 (9):1245.

Scadden, David T. 2006. The stem-cell niche as an entity of action. *Nature* 441 (7097):1075–1079.

Shimizu, K., H. Fukushima, K. Ogura, et al. 2017. The SCFβ^{-TRCP} E3 ubiquitin ligase complex targets Lipin1 for ubiquitination and degradation to promote hepatic lipogenesis. *Sci. Signal.* 10 (160): eaah4117.

Shimizu, T., J. Shen, M. Fang, et al. 2017. Sulfide-responsive transcriptional repressor SqrR functions as a master regulator of sulfide-dependent photosynthesis. *Proc. Natl. Acad. Sci. U. S. A.* 114 (9):2355–2360.

Suda, T., K. Takubo, and G. L. Semenza. 2011. Metabolic regulation of hematopoietic stem cells in the hypoxic niche. *Cell Stem Cell* 9 (4):298–310.

Szabo, C., C. Coletta, C. Chao, et al. 2013. Tumor-derived hydrogen sulfide, produced by cystathionine-β-synthase, stimulates bioenergetics, cell proliferation, and angiogenesis in colon cancer. *Proc. Natl. Acad. Sci. U. S. A.* 110 (30):12474–12479.

Takata, T., S. Araki, Y. Tsuchiya, and Y. Watanabe. 2020. Persulfide signaling in stress-initiated calmodulin kinase response. *Antioxid. Redox Signal.* 33 (18):1308–1319.

Takata, T., H. Ihara, N. Hatano, Y. Tsuchiya, T. Akaike, and Y. Watanabe. 2017. Reactive sulfur species inactivate Ca^{2+}/calmodulin-dependent protein kinase IV via S-polysulfidation of its active-site cysteine residue. *Biochem. J.* 474 (15):2547–2562.

Takata, T., A. Tsukuda, Y. Tsuchiya, T. Akaike, and Y. Watanabe. 2019. The active-site cysteine residue of Ca^{2+}/calmodulin-dependent protein kinase I is pro tected from irreversible modification via generation of polysulfidation. *Nitric Oxide* 86:68–75.

Takubo, K., N. Goda, W. Yamada, et al. 2010. Regulation of the HIF-1α level is essential for hematopoietic stem cells. *Cell Stem Cell* 7 (3):391–402.

Tan, Darren Q., and Toshio Suda. 2018. Reactive oxygen species and mitochondrial homeostasis as regulators of stem cell fate and function. *Antioxid. Redox Signal.* 29 (2):149–168.

Tang, Duozhuang, Si Tao, Zhiyang Chen, et al. 2016. Dietary restriction improves repopulation but impairs lymphoid differentiation capacity of hematopoietic stem cells in early aging. *J. Exp. Med.* 213 (4):535–553.

Toohey, J. I. 2011. Sulfur signaling: is the agent sulfide or sulfane? *Anal. Biochem.* 413 (1):1–7.

Tothova, Zuzana, Ramya Kollipara, Brian J. Huntly, et al. 2007. FoxOs are critical mediators of hematopoietic stem cell resistance to physiologic oxidative stress. *Cell* 128 (2):325–339.

Vandiver, M. S., B. D. Paul, R. Xu, et al. 2013. Sulfhydration mediates neuroprotective actions of parkin. *Nat. Commun.* 4:1626.

Velu, C. S., S. K. Niture, C. E. Doneanu, N. Pattabiraman, and K. S. Srivenugopal. 2007. Human p53 is inhibited by glutathionylation of cysteines present in the proximal DNA-binding domain during oxidative stress. *Biochemistry* 46 (26):7765–7780.

Westphal, R. S., K. A. Anderson, A. R. Means, and B. E. Wadzinski. 1998. A signaling complex of Ca^{2+}-calmodulin-dependent protein kinase IV and protein phosphatase 2A. *Science* 280 (5367):1258–1261.

Yang, R., C. Qu, Y. Zhou, et al. 2015. Hydrogen sulfide promotes Tet1- and Tet2-mediated Foxp3 demethylation to drive regulatory T cell differentiation and maintain immune homeostasis. *Immunity* 43 (2):251–263.

Yilmaz, Ömer H., Pekka Katajisto, Dudley W. Lamming, et al. 2012. mTORC1 in the paneth cell niche couples intestinal stem-cell function to calorie intake. *Nature* 486 (7404):490–495.

Yin, Huaqun, Xian Zhang, Xiaoqi Li, et al. 2014. Whole-genome sequencing reveals novel insights into sulfur oxidation in the extremophile *Acidithiobacillus thiooxidans*. *BMC Microbiol.* 14 (1):179.

Yin, Peng, Chao Zhao, Zengxia Li, et al. 2012. Sp1 is involved in regulation of cystathionine γ-lyase gene expression and biological function by PI3K/Akt pathway in human hepatocellular carcinoma cell lines. *Cell Signal.* 24 (6):1229–1240.

You, J., X. Shi, H. Liang, et al. 2017. Cystathionine-γ-lyase promotes process of breast cancer in association with STAT3 signaling pathway. *Oncotarget* 8 (39):65677–65686.

Yu, X., K. Murao, Y. Sayo, et al. 2004. The role of calcium/calmodulin-dependent protein kinase cascade in glucose upregulation of insulin gene expression. *Diabetes* 53 (6):1475–1481.

Zhang, T., K. Ono, H. Tsutsuki, et al. 2019. Enhanced cellular polysulfides negatively regulate TLR4 signaling and mitigate lethal endotoxin shock. *Cell Chem. Biol.* 26 (5):686–698 e4.

Zivanovic, J., E. Kouroussis, J. B. Kohl, et al. 2019. Selective persulfide detection reveals evolutionarily conserved antiaging effects of S-sulfhydration. *Cell Metab.* 30 (6):1152–1170 e13.

CHAPTER SEVENTEEN

Selenoproteins during Cancer Development and Progression

Anna P. Kipp

CONTENTS

17.1 INTRODUCTION

Cancer is a multifactorial disease driven by various modifications of healthy cells, out of which initial mutations of driver genes that escape the cellular repair machinery have first been identified to be of relevance [1]. Accordingly, tumor cells acquire novel properties over time resulting in unrestricted proliferation and invasion, which have been summarized in the so-called hallmarks of cancer [2, 3]. It is still not entirely clear where cancer cells reside. According to the cancer stem cell hypothesis, endogenous healthy stem cells that can be found in almost every adult tissue develop into cancer stem cells [4, 5]. Besides this, a second source of tumor origin arises from reprogramming or de-differentiation of somatic cells [6]. Every "hallmark of cancer" has been discussed to be affected by redox signaling processes, indicating that the cellular redox state is a very important factor contributing to cancer development and progression [7]. Reactive oxygen species (ROS) originate from cellular aerobic processes.

On the one hand, ROS act as signaling molecules regulating proliferation, among others, but on the other hand, they can also act in a cytotoxic manner and damage nucleic acids, proteins, and lipids. Therefore, cells depend on the maintenance of a tight balance between ROS generation and elimination. The main cancer-relevant, redox-sensitive pathways modulate growth factor signaling. Besides that, transcription factors or their binding proteins are most often modified by oxidation (overview in [8]), including transcription factors such as Nrf2, NF-κB, p53, FOXOs, Wnt, and many more. Via these mechanisms, cells are able to adapt to changes in the redox balance by upregulating antioxidant enzymes. If this adaptation fails and high levels of H_2O_2 or lipid peroxides accumulate, induction of cell death occurs [9, 10]. Thus, the cellular redox state plays a critical role in all stages of carcinogenesis [11].

Reports emerging in the 1970s indicated an inverse association between the dietary availability of the essential trace element selenium and cancer

DOI: 10.4324/9781003204091-22

incidence [12], raising interest in using selenium for chemoprevention. Mechanism-wise, selenium is very interesting because it is incorporated into selenoproteins that are mostly oxidoreductases and thus are essentially involved in regulating the cellular redox status. Besides this, selenoproteins are important mediators of the immune response, thyroid hormone metabolism, and protein folding in the endoplasmic reticulum (ER) [13]. In Europe, the selenium supply via the food chain is rather limited compared to elsewhere, for example, the United States. The selenium status of an organism is defined by the expression of selenoproteins, which contain a selenocysteine moiety in their active center. In humans, 25 genes encode for seleno-proteins. Selenoprotein P (SELENOP) and glutathi-one peroxidase (GPX) 3 are the only extracellular selenoproteins and, therefore, serve as plasma bio-markers for the selenium status [14]. The analy-sis of samples from the EPIC cohort, conducted in Europe, revealed that low plasma selenium as well as low SELENOP levels were strongly associated with an increased risk to develop colorectal can-cer (CRC) [15] and hepatobiliary cancer [16]. This indicates that selenium plays a role in the primary prevention of cancer development. In addition, intervention studies such as the NPC (Nutritional Prevention of Cancer Trial) und SELECT (Selenium and Vitamin E Cancer Prevention Trial) indicate that increasing the systemic selenium status up to 120 µg Se/L plasma lowers the risk to develop cancer. However, above this border, as indicated by the tertile of participants with the highest ini-tial selenium status, no further protective effects of selenium supplementation could be detected [17, 18]. A selenium status of 120 µg Se/L plasma is supposed to be necessary to maximize plasma SELENOP levels [14], which gives a rationale to hypothesize that anticarcinogenic effects of sele-nium are likely mediated by selenoproteins and not directly by selenium itself. As most selenopro-teins act in an antioxidant manner, they are sup-posed to have anticarcinogenic properties during tumor initiation as they are part of the cellular sys-tems protecting from oxidative DNA damage. For example, a knockdown of GPX1 indeed resulted in increased DNA damage and micronuclei forma-tion after treating cells with UV irradiation [19], whereas GPX1 overexpression avoided DNA dam-age [20].

Besides the effects of selenoproteins on limit-ing tumor initiation, established tumor cells may benefit from an enhanced selenium supply and the concomitant upregulation of selenoproteins, as they have a higher need for antioxidant sys-tems than normal cells [21]. So far, there is only one study addressing the time-selective role of selenium (in combination with vitamin E) on carcinogenesis using an esophageal carcinogen-esis rat model induced by application of nitroso-methylbenzylamine (NMBzA). Protective effects were only observed when selenium and vitamin E supplementation directly started during initiation by NMBzA application. At a later time point, when 70% of the rats already developed hyperplasia in the esophagus, supplementation effects could no longer be detected [22]. Based on this, the chap-ter will discuss recent knowledge about the role of individual selenoproteins during cancer develop-ment and progression considering both protective and detrimental effects.

17.2 SELENOPROTEINS IN CANCER CELLS

Cancer cells are characterized by a higher metabolic turnover than healthy cells, and thus they produce more ROS [21] that needs to be counteracted by antioxidant molecules/enzymes. Selenoproteins are important players in this scenario. Whereas sev-eral selenoproteins are upregulated in cancer cells, others are substantially downregulated, resulting in specific selenoprotein expression patterns for dif-ferent types of cancer. This has been systematically reviewed elsewhere for GPXs [23]. Regarding the GPX family, the isoform GPX2 becomes most often upregulated, which is also the case for thioredoxin reductase 1 (TXNRD1) [24]. In contrast, GPX3 is most consistently downregulated in several types of cancer, but there are also examples in which GPX3 is upregulated [25]. The same is the case for GPX4, which is also downregulated in many tumors [23] but upregulated in others (e.g., steroid-producing adrenocortical carcinomas) [26, 27]. Also, tumor type-specific expression changes have been described for deiodinases [28]. In a recent study on CRC, transcript levels of 15 selenoprotein genes were analyzed in parallel to characterize the sele-noprotein expression pattern of CRC. GPX2 and TXNRD3 were consistently upregulated in both adenoma and carcinoma, which can be explained by the high Wnt signaling activity in almost all CRC [29]. In parallel, GPX3, SELENOP, SELENOS, and selenophosphate synthase 2 (SEPHS2) were downregulated [30]. However, there was no correlation of selenoprotein mRNA levels with the systemic selenium status (serum selenium levels or SELENOP) of the patients, which might not be surprising, as selenoprotein transcripts are only marginally modified by selenium availability. Based on the so-called concept of selenoprotein hierarchy, selenoproteins are differentially sensitive to a limited selenium supply, which mainly results in differences on the protein expression level. For example, GPX4 and the TXNRDs are relatively insensitive towards selenium restrictions, whereas others (e.g., GPX1, SELENOH, SELENOW) decline very fast. So far, different mechanisms have been proposed to underlie this phenomenon including selenoprotein mRNA stability, the efficiency of their translation and incorporation of selenocysteine in the growing peptide chain next to specific tRNA subpools, and differences in binding affinity for translation supporting or stabilizing proteins (overview in [13, 31]).

Furthermore, selenoprotein levels are not just modified in cancer cells but correlate with the tumor grade. For example, GPX4 expression negatively correlates with the tumor grade of breast invasive ductal carcinoma, for example [32]. Also, GPX2 upregulation appears to depend on the tumor stage [33, 34]. A growing number of studies indicates that selenoprotein mRNA and protein expression levels correlate with the relapse-free survival time of cancer patients and, thus, selenoproteins are discussed to be potentially useful predictors for survival probability. However, these relationships appear to be rather unique to the type of cancer, indicating that upregulation of some selenoproteins but not all of them might be enough to maintain redox balance. For example, low tumor-resident GPX3 expression predicts worse overall patient survival in low-grade glioma and lung adenocarcinoma, whereas low GPX3 levels correlate with better survival rates of patients with stomach and lung squamous cell carcinoma [25]. Also upregulation of GPX1 within tumor tissue was either correlating with a worse prognosis in case of low-grade glioma or acute myeloid leukemia or a better prognosis in kidney renal papillary cell carcinoma, thus clearly depending on the type of cancer [35].

17.3 GLUTATHIONE PEROXIDASES

GPXs reduce hydroperoxides to their corresponding alcohols and mainly depend on glutathione (GSH) as an electron donor to become recycled. While GPX4 is able to reduce complex fatty acids, phospholipids, and cholesterol hydroperoxides located in membranes, GPX1, 2, and 3 are supposed to react with less complex, soluble fatty acid hydroperoxides (overview in [36]). Besides protecting healthy cells from oxidative DNA damage, GPXs modulate proliferation rates and growth properties of tumor cells by changing the hydroperoxide tone. In line with the anticarcinogenic effects during initiation, Ras-positive pancreatic cancer cells showed impaired growth properties and developed into smaller tumors upon **GPX1** overexpression [37]. In contrast, cell proliferation was reduced in GPX1 knockdown breast cancer [38] and in renal cell carcinoma cells [39]. Mice overexpressing GPx1 developed more tumors than wild-type mice in a chemically induced skin tumor model [40]. Thus, high GPX1 levels can support tumor cell survival but also inhibit tumor growth. GPX1 has been shown to act downstream of glutamate dehydrogenase 1, which is most often upregulated in cancer cells, to maintain the high needs of α-ketoglutarate and fumarate. Fumarate itself is able to bind to and activate GPX1, supporting tumor growth [38].

During the last several years, we have been involved in studying the role of **GPX2** during colorectal carcinogenesis using different in vivo and in vitro approaches. In line with what has been described for GPX1, GPX2 is clearly shifting from an anticarcinogenic function during initiation towards a growth-promoting effect in already established tumor cells. GPX2 knockdown cells develop into smaller tumors using a xenograft model [34, 41], and GPX2 knockout mice developed less aberrant crypts in a chemically induced colon cancer model (azoxymethane (AOM) model) as compared to wild-type mice [42]. Comparable results were consistently obtained in different types of cancer cells including prostate [43], bladder [44], cervical [45], and lung [46] cancer (Table 17.1). However, if the promotion is mainly driven by inflammation induced by dextran sodium sulfate (DSS), GPX2 has anti-inflammatory and thus anti-proliferative effects [23, 47]. This is supposed to be mainly mediated by the inhibition of NF-κB signaling by GPX2 [48]. In line with this, anti-inflammatory effects through NF-κB inhibition have also been described for GPX1 [48] and GPX4 [49]. However, NF-κB does not only play a role during immune response, but it also has direct effects on cancer growth [50].

TABLE 17.1

*Modulation of Selenoprotein Expression (KO = Knockout; KD = Knockdown; OE = Overexpression) in Tumor Cell Lines or Mice Resulting in a Predominant **Tumor Promoter** Function Using In Vitro (e.g., Proliferation and Growth Rate, Soft Agar) and In Vivo Approaches (e.g., Xenografts, Chemically Induced or Genetic Carcinogenesis Models)*

Tumor promoter	Selenoprotein	Tumor suppressor
KD: inhibited cell growth in colon cancer cells and liver cells [89, 95] but not in lung carcinoma LLC1 cells [89] KO: less ACF formation in the colon of mice [92]	SELENOF	?
KD: less proliferation, migration, and tumor formation of renal cell carcinoma cells [86]	SELENOM	?
KO: reduced proliferation and tumor formation of cancer cell lines of the kidney, liver, breast, brain, skin, and bone [109]	SEPHS2	?
KD: reduced proliferation of lung cancer cells [79]	TXNRD2	?
OE: enhanced cell survival of liver cancer cells [81]	mtTXNRD3	?
KD: reduced proliferation and tumor formation of liver [73, 75], colon [93], and lung cancer cells [74] KO: fewer tumors in a chemically induced liver tumor model [76]	TXNRD1	KO: more liver tumors in a chronic, chemically induced stress cancer model [76]
KD: less proliferation and tumor formation of colorectal [34, 41], prostate [43], bladder [44], and cervical cancer [45] KO: fewer adenomas in a chemically induced colorectal cancer model [42]	GPX2	KO: more squamous cell carcinoma in response to UV radiation [113] KO: higher number of inflammation-driven colonic adenomas [47]
OE: more tumors in a chemically induced skin tumor model [40] KD: reduced proliferation of breast cancer [38] and renal cell carcinoma cells [39]	GPX1	OE: reduced cell growth and tumor size of pancreatic cancer cells [37]

For example, overexpression of GPX4 resulted in the suppression of NF-κB and accordingly limited induction of matrix metalloprotease 1 [51]. Besides NF-κB, additional redox-sensitive pathways involving kinases such as extracellular signal-regulated kinase (ERK), and c-Jun N-terminal kinase (JNK) have been described to be blocked by lowering H_2O_2 levels catalyzed by GPXs [52, 53]. In case of GPX1, the dual specificity phosphatase (DUSP) 4 which is known to be an antagonist of mitogen-activated protein kinase (MAPK) signaling, was discussed to act downstream of GPX1 [52]. In triple-negative breast cancer cells, GPX1 interacts with focal adhesion kinase (FAK) and prevents its inhibition by H_2O_2. Downstream of FAK, the regulation of adhesion signaling is affected by GPX1 [54]. These results indicate that GPX1 and 2 modulate redox signaling pathways and in this way bi-directionally affect tumor cell growth.

So far, **GPX3** has been mainly characterized as tumor suppressor (Table 17.2). Overexpression of GPX3 in lung cancer cell lines suppressed proliferation, migration, and invasion, which again involved the inhibition of H_2O_2-mediated activation of NF-κB signaling, subsequently resulting in cell cycle arrest [55]. Further models with GPX3 overexpression showed less proliferation and in general fewer characteristics of tumor cells [56–58], whereas models with reduced GPX3 expression were associated with enhanced proliferation resulting in increased tumor size using mouse models for prostate cancer or inflammation-driven colon carcinogenesis [59, 60]. Only in one colon cancer cell line, a GPX3 knockdown impaired colony formation, and increased susceptibility of the knockdown cells towards apoptosis [60], indicating that under specific circumstances GPX3 might also support tumor cell survival and growth.

In line with GPX3, also **GPX4** has for long been mainly described to act as a tumor suppressor as GPX4 overexpression limited growth properties of different tumor models (Table 17.2) [61, 62]. However, GPX4 has a different substrate spectrum than all other GPX family members and is specifically localized at cellular membranes [36]. This

TABLE 17.2

*Modulation of Selenoprotein Expression (KO = Knockout; KD = Knockdown; OE = Overexpression; TG = Transgenic) in Tumor Cell Lines of Mice Resulting in a Predominant **Tumor Suppressor** Function Using In Vitro (e.g., Proliferation and Growth Rate, Soft Agar) and In Vivo Approaches (e.g., Xenografts, Chemically Induced or Genetic Carcinogenesis Models).*

Tumor promoter	Selenoprotein	Tumor suppressor
KO: less tumor growth and metastasis in a genetic melanoma mouse model [99]	SELENOK	OE: less tumor characteristics of choriocarcinoma cells [97] OE: less cell adhesion and migration in gastric cancer cells [98]
KO: no in vivo tumor growth of melanoma xenografts [67] KO: no tumor relapse after chemotherapy [68]	GPX4	OE: less tumor growth of fibrosarcoma and reduced lung metastasis of melanoma cells [61] OE: reduced tumor growth of pancreatic cancer cells [62]
KD: decreased colony formation of colon cancer cells [60]	GPX3	OE: less colony and tumor formation of prostate cancer cells [56] OE: less lung metastasis of liver cancer cells [58] OE: less proliferation and cell cycle arrest of lung cancer cells [55] OE: less migration, invasiveness and metastasis of liver cancer cells [58] KD: increased tumor size and lymph node metastasis of thyroid cancer cells [59] KO: higher prostate cancer incidence in a genetic mouse model [57] KO: less tumors in a model of colitis-associated colorectal cancer [60]
?	SELENOP	OE: less proliferation in liver cancer cells [107] TG for loss of antioxidant function: more colon tumors [108]
?	SELENOH	KD: more tumors with colon cancer cells [102] KD: more gastrointestinal and liver tumor formation in a zebrafish cancer model [103]

explains why GPX4 also has unique functions in signaling pathways. Ferroptosis is a regulated cell death pathway initiated by iron-dependent lipid peroxidation [63], a process that can be antagonized by GPX4 [64, 65]. Ferroptosis is also important for cancer development as several types of tumors acquire resistance against ferroptosis (e.g., by upregulating GPX4) [26, 27]. Indeed, induction of lipid peroxidation by treatment with docosahexaenoic acid (DHA) killed cancer cells with GPX4 knockdown more efficiently than control cells [66]. De-differentiation of several types of tumors is accompanied by higher expression levels of enzymes catalyzing the synthesis of polyunsaturated fatty acids, increasing their dependency on GPX4

[67]. So-called drug-tolerant, quiescent persister cells develop in a wide range of tumor types during chemotherapy. Drug-resistant tumors may emerge from these cells. In contrast to the parental cells, this cell population has been shown to be very prone to GPX4 inhibition and ferroptosis induction. Based on this, the authors of this study propose that co-treatment of cancer cells with a targeted chemotherapy together with a GPX4 inhibitor would effectively reduce the persister cell pool [68]. Also, in hepatocellular carcinoma, induction of ferroptosis is discussed as new therapeutic strategy [69]. Aside from ferroptosis, GPX4 has recently been shown to modulate stimulator-of-interferon genes (STING), a central receptor of innate immunity that is also involved

in tumor immunity. GPX4 inhibition blocked the STING pathway, [70] and this could impact on cancer development in an indirect manner. These novel GPX4 functions and their role during cancer-acquired drug resistance and immune evasion have been recently discussed more extensively in another review [71].

17.4 THIOREDOXIN REDUCTASES

The family of thioredoxin reductases consists of three members, out of which **TXNRD1** has been mainly studied for its role in cancer. Next to the glutathione/glutathione reductase system, the thioredoxin/TXNRD system is very important for maintaining the cellular redox balance as it is able to reduced peroxiredoxins, for example (overview in [72]). Comparable to GPXs, TXNRDs thus also affect redox signaling involved in cancer cell proliferation, metastasis, and apoptosis.

TXNRD1 is mostly upregulated in tumor cells/tissue, and its downregulation results in growth arrest, impaired anchorage-independent growth, and less xenograft formation, [73–75] resulting in its primary categorization as tumor promoter (Table 17.1). In a diethylnitrosamine (DEN)-induced hepatocarcinoma model, liver-specific TXNRD1 knockout mice developed fewer tumors than wild-type mice, [76] which has been described before in an opposite direction [77]. When co-treated with phenobarbital (PB) to induce oxidative stress, TXNRD1 knockout mice exhibited a very high tumor load compared to wild type-mice. The authors discuss activation of Nrf2 in the first scenario to be of relevance for protecting from tumor development upon loss of TXNRD1, while Nrf2 could not be further activated by co-treatment with PB [76]. Treating mice of a xenograft experiment with TXNRD1 inhibitors retarded tumor growth and increased apoptosis rates within the tumor tissue [78]. Together, these results offer strong evidence that TXNRD1 functions as a tumor promoter in malignant cells but under specific circumstances shifts to a tumor suppressor function as well.

Less data is available for the role of **TXNRD2** and **3** during cancer development. In lung cancer cells, a knockdown of TXNRD2 inhibited proliferation and induced apoptosis while overexpression of TXNRD2 had opposite effects [79]. Multiple myeloma is a malignancy of plasma cells which is frequently treated with proteasome inhibitors such as bortezomib. Upon treatment, TXNRD2 expression becomes downregulated. Overexpression of TXNRD2 in myeloma does not affect tumor formation in a xenograft model under untreated conditions but it reduced tumor-suppressive effects of bortezomib [80]. Recently, a sorafenib-resistant leukemia cell line was used to study a mitochondrial form of TXNRD3 that was highly upregulated under these conditions. Overexpression of mtTXNRD3 in liver cancer cells induced a metabolic shift from oxidative phosphorylation towards more active glycolysis, which was further accompanied by an upregulation of anti-apoptotic proteins and enhanced cell survival [81]. Thus, the family of TXNRDs mainly has been described to exhibit tumor-promoting effects so far.

17.5 SELENOPROTEINS OF THE ER WITH FOCUS ON SELENOF, SELENOK, AND SELENOM

To date, it is known that there are eight mammalian selenoproteins localized in the ER. These include deiodinase 2 (DIO2), selenoprotein K (SELENOK), N (SELENON), M (SELENOM), F (SELENOF), T (SELENOT), S (SELENOS), and I (SELENOI) [82, 83]. Based on their localization, these selenoproteins are involved in regulating protein degradation, ER stress, and redox metabolism. In gastric cancer tissue, the gene expression of SELENOK, S, T, F, and M is downregulated [84]. However, this again essentially depends on the tumor type as, for example, SELENOM is upregulated in human hepatocellular carcinoma tissue [85] and renal cell carcinoma [86]. SELENOM and F are structural homologs, and both contain a CxxU or a CxU motif that is common in proteins involved in cellular redox homeostasis [87]. A SELENOM knockdown reduced proliferation, migration, and xenograft formation of renal carcinoma cells, which appears to be modulated via inhibition of PI3K/AKT/mTOR signaling [86]. In addition, SELENOM has been described to bind to actin 1 and 2, which might play a mechanistic role during migration and thus metastasis of cancer cells [88]. Cancer characteristics such as colony formation, tumor growth, and lung metastasis were reduced upon knockdown of SELENOF in colon cancer cells [89, 90]. Additionally, knockdown of SELENOF decreased cell growth in mesothelioma cells [91] while it had no effect in lung cancer cells [89]. SELENOF knockout mice developed less aberrant crypts in a chemically induced

(AOM) colon tumor model [92], which appears to be modulated by the signal transducer and activator of transcription (STAT) 1 known to increase antitumor immunity [93]. Comparable to SELENOM, SELENOF-deficient cells were characterized by reorganization of F-actin and α-tubulin resulting in membrane blebbing [94], which impaired the cell's ability for migration and invasion [95]. Thus, cell motility and cell cycle regulation are supposed to be the main molecular targets for the tumor-promoting effects of SELENOF, which needs to be studied in more detail for SELENOM. Unexpectedly, a knockdown of the tumor-promoting SELENOF and TXNRD1 in colon cancer cells abrogated the growth-inhibiting effects observed for the single selenoproteins. Instead, components of the Wnt pathway were significantly upregulated, which might be an attempt to compensate for the loss of these two selenoproteins [93].

SELENOK and S are transmembrane proteins being localized in the ER membrane interacting with the ER-associated protein degradation pathway. Both are important for a proper immune response, in the case of SELENOK mainly by modulating calcium-dependent signaling (overview in [96]). Thus, both can potentially affect tumor growth via modulating the immune system, but direct effects in cancer cells are rarely studied and so far only observed for SELENOK. For example, in human choriocarcinoma cells SELENOK downregulation increased the proliferative, migratory, and invasive capacity of these cells while overexpressing SELENOK had opposite effects [97]. Similarly, cell adhesion and migration were inhibited in gastric cancer cells with SELENOK overexpression [98]. However, in melanoma cells, a SELENOK knockout resulted in reduced proliferation and migration, which was accompanied by low levels of a Ca^{2+} channel protein and thus lower calcium flux. Using a transgenic mouse model for spontaneous melanoma development, a SELENOK knockout inhibited primary tumor growth and reduced metastasis to lymph nodes in comparison to wild-type mice [99]. This indicates that SELENOK so far has only been characterized as tumor suppressor but can obviously also promote tumor development as shown for melanoma (Table 17.2).

17.6 SELENOPROTEIN H

SELENOH belongs to the group of Se-sensitive selenoproteins, indicating that it is relatively sensitive towards changes in the Se supply. It is a member of the high mobility group DNA-binding protein family [100] and belongs to the redox-active selenoproteins containing a CxxU motif [101]. SELENOH is expressed in several cancer cell lines, in intestinal tumors from APC[min/+] mice, and in human colon tumors. A SELENOH knockdown in CRC cells resulted in increased migration and proliferation as well as a higher capability to build tumors in a xenograft model [102]. Using a zebrafish model with loss of p53 and carcinogen-induced tumorigenesis showed that downregulation of SELENOH resulted in increased tumor numbers compared to wild-type fishes in gastrointestinal tract and liver [103]. This indicates that SELENOH acts as a tumor suppressor at least in colon cancer. Whether this is also the case for other cancer types warrants further investigation.

17.7 SELENOPROTEIN P

SELENOP is essential to distribute Se from the liver to peripheral tissues, preferentially to brain, testes, and bone, by binding to the apolipoprotein E receptor 2 (APOER2) [104]. Aside from this, SELENOP is supposed to exhibit GPX activity which has been shown only in vitro so far [105]. This could play a role in serum but also intracellularly as SELENOP has been shown to modulate redox signaling in a variety of cell types (overview in [106]). However, modulating SELENOP expression could partially be attributed to changes in Se availability, which would affect the expression of additional selenoproteins. For example, SELENOP overexpression in HepG2 cells inhibited cell proliferation, which was accompanied by GPX1 upregulation, and thus effects cannot be clearly attributed to SELENOP only [107]. Applying an inflammation-driven colon cancer model resulted in less tumors in homozygote SELENOP knockout mice, whereas tumor size increased in heterozygote SELENOP mice in comparison to wild-type mice [108]. Under heterozygote conditions, tumors were highly proliferative and exhibited stem cell characteristics, whereas SELENOP knockout tumors were small and characterized by many apoptotic cells. Using mice with either SELENOP deletion of the antioxidant function or the Se transport function indicated that both transgenic mice developed more tumors compared to wild-type mice, but effects on proliferation and redox status were more pronounced in transgenic mice with loss of SELENOP's antioxidant function

[108]. Based on this, it can be concluded that moderate and not complete loss of SELENOP as well as its antioxidant capability, are mostly driving tumor development.

17.8 SELENOPHOSPHATE SYNTHASE 2

SEPHS2 is a selenoprotein itself but is at the same time essentially involved in selenoprotein synthesis. A SEPHS2 knockout resulted in reduced proliferation, colony-forming capacity, and xenograft formation in several cancer cell types (Table 17.1). Accordingly, the authors could show that SEPHS2 is required to detoxify selenide, which otherwise is able to limit tumor growth or eventually to even kill tumor cells [109]. Very high, supranutritional doses of selenite are indeed used as antineoplastic or adjuvant agents as part of cancer therapy. Selenite in itself or in combination with a chemotherapy or irradiation is described to be selectively cytotoxic to tumor cells and to induce tumor growth inhibition [110]. This was thought to be based on the observation that selenite is able to accumulate in tumor cells [111, 112]. Recently it has been shown that the cystine glutamate exchanger xCT, which is also frequently upregulated in cancer cells, is important for selenite uptake although in an indirect manner. xCT shuttles cystine into cells and accordingly increases both intra- and extracellular cysteine levels that are used as free thiols to reduce selenite to volatile selenide, which can than easily enter the cells. Thus, tumor cells essentially depend on SEPHS2 to shuttle selenide towards selenoprotein synthesis, increasing their antioxidant capacity. Via the same mechanism, SEPHS2 protects cancer cells from direct selenide toxicity [109].

17.9 CONCLUDING REMARKS

It is getting more and more evident that effects of selenoproteins on cancer development are tissue specific. Out of the 25 human selenoproteins, 13 have been mechanistically studied for their role in cancer development so far. From these, eight selenoproteins act mainly as tumor promoters (Table 17.1) and five can be categorized as tumor suppressors (Table 17.2). From the group of tumor promoters, TXNRD1 and GPX2 have been most extensively studied. Loss of both selenoproteins diminished cancer cell characteristics, resulting in the development of fewer or smaller tumors. In contrast, tumor-suppressive functions have been described as well, which can be observed for GPX2 when inflammation is the main driver of tumorigenesis. In case of TXNRD1, it appears that there is a certain threshold of oxidative stress that limits the capability of tumor cells to survive without TXNRD1. One of the best studied tumor suppressor selenoproteins is GPX3, which is frequently downregulated in tumor cells, and its overexpression limits cancer cell characteristics. Although prevention of oxidative DNA damage clearly has anticarcinogenic effects, modulation of redox signaling differs depending on the specific circumstances, such as tumor stage or inflammatory conditions. Accordingly, more research is needed to understand how selenoprotein profiles are establishing in different types of tumors and which properties these selenoproteins provide to the tumor.

17.10 ABBREVIATIONS

Azoxymethane (AOM); apolipoprotein E receptor 2 (APOER2); colorectal cancer (CRC); deiodinase (DIO); diethylnitrosamine (DEN); docosahexaenoic acid (DHA); dextran sodium sulfate (DSS); dual specificity phosphatase (DUSP); endoplasmic reticulum (ER); extracellular signal-regulated kinase (ERK); focal adhesion kinase (FAK); glutathione peroxidase (GPX); glutathione (GSH); c-Jun N-terminal kinase (JNK); mitogen-activated protein kinase (MAPK); nitrosomethylbenzylamine (NMBzA); Nutritional Prevention of Cancer Trial (NPC); phenobarbital (PB); reactive oxygen species (ROS); Selenium and Vitamin E Cancer Prevention Trial (SELECT); selenoprotein (SELENO); selenophosphate synthase 2 (SEPHS2); signal transducer and activator of transcription (STAT); stimulator-of-interferon genes (STING); thioredoxin reductase (TXNRD).

17.11 ACKNOWLEDGMENTS

The support of Kristina Lossow in proofreading the manuscript is highly acknowledged.

REFERENCES

1. Fearon ER and B Vogelstein (1990) A genetic model for colorectal tumorigenesis. Cell 61: 759–767
2. Hanahan D and RA Weinberg (2011) Hallmarks of cancer: the next generation. Cell 144:646–674

3. Hanahan D and RA Weinberg (2000) The hall-marks of cancer. Cell 100:57–70

4. Clevers H (2011) The cancer stem cell: premises, promises and challenges. Nat Med 17:313–319

5. Batlle E and H Clevers (2017) Cancer stem cells revisited. Nat Med 23:1124–1134

6. Liu J (2018) The dualistic origin of human tumors. Semin Cancer Biol 53:1–16

7. Hornsveld M and TB Dansen (2016) The hall-marks of cancer from a redox perspective. Antioxid Redox Signal 25:300–325

8. Brigelius-Flohé R and L Flohé (2011) Basic principles and emerging concepts in the redox control of transcription factors. Antioxid Redox Signal 15:2335–2381

9. Galadari S, A Rahman, S Pallichankandy, and F Thayyullathil (2017) Reactive oxygen species and cancer paradox: to promote or to suppress? Free Radic Biol Med 104:144–164

10. Gaschler MM and BR Stockwell (2017) Lipid peroxidation in cell death. Biochem Biophys Res Commun 482:419–425

11. Kohan R, A Collin, S Guizzardi, N Tolosa de Talamoni, and G Picotto (2020) Reactive oxygen species in cancer: a paradox between pro- and anti-tumour activities. Cancer Chemother Pharmacol 86:1–13

12. Schrauzer GN, DA White, and CJ Schneider (1977) Cancer mortality correlation studies--III: statistical associations with dietary selenium intakes. Bioinorg Chem 7:23–31

13. Labunskyy VM, DL Hatfield, and VN Gladyshev (2014) Selenoproteins: molecular pathways and physiological roles. Physiol Rev 94:739–777

14. Rayman MP (2012) Selenium and human health. Lancet 379:1256–1268

15. Hughes DJ, V Fedirko, M Jenab, L Schomburg, C Meplan, H Freisling, HB Bueno-de-Mesquita, S Hybsier, NP Becker, M Czuban, A Tjonneland, M Outzen, MC Boutron-Ruault, A Racine, N Bastide, T Kuhn, R Kaaks, D Trichopoulos, A Trichopoulou, P Lagiou, S Panico, PH Peeters, E Weiderpass, G Skeie, E Dagrun, MD Chirlaque, MJ Sanchez, E Ardanaz, I Ljuslinder, M Wennberg, KE Bradbury, P Vineis, A Naccarati, D Palli, H Boeing, K Overvad, M Dorronsoro, P Jakszyn, AJ Cross, JR Quiros, M Stepien, SY Kong, T Duarte-Salles, E Riboli, and JE Hesketh (2015) Selenium status is associated with colorectal cancer risk in the European prospective investigation of cancer and nutrition cohort. Int J Cancer 136:1149–1161

16. Hughes DJ, T Duarte-Salles, S Hybsier, A Trichopoulou, M Stepien, K Aleksandrova, K Overvad, A Tjonneland, A Olsen, A Affret, G Fagherazzi, MC Boutron-Ruault, V Katzke, R Kaaks, H Boeing, C Bamia, P Lagiou, E Peppa, D Palli, V Krogh, S Panico, R Tumino, C Sacerdote, HB Bueno-de-Mesquita, PH Peeters, D Engeset, E Weiderpass, C Lasheras, A Agudo, MJ Sanchez, C Navarro, E Ardanaz, M Dorronsoro, O Hemmingsson, NJ Wareham, KT Khaw, KE Bradbury, AJ Cross, M Gunter, E Riboli, I Romieu, L Schomburg, and M Jenab (2016) Prediagnostic selenium status and hepatobiliary cancer risk in the European Prospective Investigation into Cancer and Nutrition cohort. Am J Clin Nutr 104:406–414

17. Lü J and C Jiang (2005) Selenium and cancer chemoprevention: hypotheses integrating the actions of selenoproteins and selenium metabo-lites in epithelial and non-epithelial target cells. Antioxid Redox Signal 7:1715–27

18. Lu J, J Zhang, C Jiang, Y Deng, N Ozten, and MC Bosland (2016) Cancer chemoprevention research with selenium in the post-SELECT era: Promises and challenges. Nutr Cancer 68:1–17

19. Baliga MS, V Diwadkar-Navsariwala, T Koh, R Fayad, G Fantuzzi, and AM Diamond (2008) Selenoprotein deficiency enhances radiation-induced micronuclei formation. Mol Nutr Food Res 52:1300–1304

20. Baliga MS, H Wang, P Zhuo, JL Schwartz, and AM Diamond (2007) Selenium and GPx-1 over-expression protect mammalian cells against UV-induced DNA damage. Biol Trace Elem Res 115:227–242

21. Trachootham D, J Alexandre, and P Huang (2009) Targeting cancer cells by ROS-mediated mechanisms: a radical therapeutic approach? Nat Rev Drug Discov 8:579–591

22. Yang H, X Jia, X Chen, CS Yang, and N Li (2012) Time-selective chemoprevention of vitamin E and selenium on esophageal carcinogenesis in rats: the possible role of nuclear factor kappaB signaling pathway. Int J Cancer 131:1517–1527

23. Kipp AP (2017) Selenium-dependent glutathione peroxidases during tumor development. Adv Cancer Res 136:109–138

24. Selenius M, AK Rundlof, E Olm, AP Fernandes, and M Björnstedt (2010) Selenium and the selenoprotein thioredoxin reductase in the pre-vention, treatment and diagnostics of cancer. Antioxid Redox Signal 12:867–880

25. Chang C, BL Worley, R Phaeton, and N Hempel (2020) Extracellular glutathione peroxidase GPx3 and its role in cancer. Cancers (Basel) 12

26. Belavgeni A, SR Bornstein, A von Massenhausen, W Tonnus, J Stumpf, C Meyer, E Othmar, M Latk, W Kanczkowski, M Kroiss, C Hantel, C Hugo, M Fassnacht, CG Ziegler, AV Schally, NP Krone, and A Linkermann (2019) Exquisite sensitivity of adrenocortical carcinomas to induction of ferroptosis. Proc Natl Acad Sci U S A 116:22269–22274

27. Weigand I, J Schreiner, F Rohrig, N Sun, LS Landwehr, H Urlaub, S Kendl, K Kiseljak-Vassiliades, ME Wierman, JPF Angeli, A Walch, S Sbiera, M Fassnacht, and M Kroiss (2020) Active steroid hormone synthesis renders adrenocortical cells highly susceptible to type II ferroptosis induction. Cell Death Dis 11:192

28. Casula S and AC Bianco (2012) Thyroid hormone deiodinases and cancer. Front Endocrinol (Lausanne) 3:74

29. Kipp AP, MF Müller, EM Göken, S Deubel, and R Brigelius-Flohé (2012) The selenoproteins GPx2, TrxR2 and TrxR3 are regulated by Wnt signalling in the intestinal epithelium. Biochim Biophys Acta 1820:1588–1596

30. Hughes DJ, T Kunicka, L Schomburg, V Liska, N Swan, and P Soucek (2018) Expression of selenoprotein genes and association with selenium status in colorectal adenoma and colorectal cancer. Nutrients 10

31. Bulteau AL and L Chavatte (2015) Update on selenoprotein biosynthesis. Antioxid Redox Signal 23:775–794

32. Cejas P, MA Garcia-Cabezas, E Casado, C Belda-Iniesta, J De Castro, JA Fresno, M Sereno, J Barriuso, E Espinosa, P Zamora, J Feliu, A Redondo, DA Hardisson, J Renart, and M Gonzalez-Baron (2007) Phospholipid hydroperoxide glutathione peroxidase (PHGPx) expression is downregulated in poorly differentiated breast invasive ductal carcinoma. Free Radic Res 41:681–687

33. De Sousa EMF, X Wang, M Jansen, E Fessler, A Trinh, LP de Rooij, JH de Jong, OJ de Boer, R van Leersum, MF Bijlsma, H Rodermond, M van der Heijden, CJ van Noesel, JB Tuynman, E Dekker, F Markowetz, JP Medema, and L Vermeulen (2013) Poor-prognosis colon cancer is defined by a molecularly distinct subtype and develops from serrated precursor lesions. Nat Med 19:614–618

34. Emmink BL, J Laoukili, AP Kipp, J Koster, KM Govaert, S Fatrai, A Verheem, EJ Steller, R Brigelius-Flohé, CR Jimenez, IH Borel Rinkes, and O Kranenburg (2014) GPx2 suppression of H_2O_2 stress links the formation of differentiated tumor mass to metastatic capacity in colorectal cancer. Cancer Res 74:6717–6730

35. Wei R, H Qiu, J Xu, J Mo, Y Liu, Y Gui, G Huang, S Zhang, H Yao, X Huang, and Z Gan (2020) Expression and prognostic potential of GPX1 in human cancers based on data mining. Ann Transl Med 8:124

36. Brigelius-Flohé R and M Maiorino (2013) Glutathione peroxidases. Biochim Biophys Acta 1830:3289–3303

37. Liu J, MM Hinkhouse, W Sun, CJ Weydert, JM Ritchie, LW Oberley, and JJ Cullen (2004) Redox regulation of pancreatic cancer cell growth: role of glutathione peroxidase in the suppression of the malignant phenotype. Hum Gene Ther 15:239–250

38. Jin L, D Li, GN Alesi, J Fan, HB Kang, Z Lu, TJ Boggon, P Jin, H Yi, ER Wright, D Duong, NT Seyfried, R Egnatchik, RJ DeBerardinis, KR Magliocca, C He, ML Arellano, HJ Khoury, DM Shin, FR Khuri, and S Kang (2015) Glutamate dehydrogenase 1 signals through antioxidant glutathione peroxidase 1 to regulate redox homeostasis and tumor growth. Cancer Cell 27:257–270

39. Cheng Y, T Xu, S Li, and H Ruan (2019) GPX1, a biomarker for the diagnosis and prognosis of kidney cancer, promotes the progression of kidney cancer. Aging (Albany NY) 11:12165–12176

40. Lu YP, YR Lou, P Yen, HL Newmark, OI Mirochnitchenko, M Inouye, and MT Huang (1997) Enhanced skin carcinogenesis in transgenic mice with high expression of glutathione peroxidase or both glutathione peroxidase and superoxide dismutase. Cancer Res 57:1468–1474

41. Banning A, A Kipp, S Schmitmeier, M Löwinger, S Florian, S Krehl, S Thalmann, R Thierbach, P Steinberg, and R Brigelius-Flohé (2008) Glutathione peroxidase 2 inhibits cyclooxygenase-2-mediated migration and invasion of HT-29 adenocarcinoma cells but supports their growth as tumors in nude mice. Cancer Res 68:9746–9753

42. Müller MF, S Florian, S Pommer, M Osterhoff, RS Esworthy, FF Chu, R Brigelius-Flohé, and AP Kipp (2013) Deletion of glutathione peroxidase-2 inhibits azoxymethane-induced colon cancer development. PLoS One 8:e72055

43. Naiki T, A Naiki-Ito, M Asamoto, N Kawai, K Tozawa, T Etani, S Sato, S Suzuki, T Shirai, K Kohri, and S Takahashi (2014) GPX2 overexpression is involved in cell proliferation and prognosis of castration-resistant prostate cancer. Carcinogenesis 35:1962–1967

44. Naiki T, A Naiki-Ito, K Iida, T Etani, H Kato, S Suzuki, Y Yamashita, N Kawai, T Yasui, and S Takahashi (2018) GPX2 promotes development of bladder cancer with squamous cell differentiation through the control of apoptosis. Oncotarget 9:15847–15859

45. Wang Y, P Cao, M Alshwmi, N Jiang, Z Xiao, F Jiang, J Gu, X Wang, X Sun, and S Li (2019) GPX2 suppression of H_2O_2 stress regulates cervical cancer metastasis and apoptosis via activation of the beta-catenin-WNT pathway. Onco Targets Ther 12:6639–6651

46. Du H, B Chen, NL Jiao, YH Liu, SY Sun, and YW Zhang (2020) Elevated glutathione peroxidase 2 expression promotes cisplatin resistance in lung adenocarcinoma. Oxid Med Cell Longev 2020:7370157

47. Krehl S, M Loewinger, S Florian, A Kipp, A Banning, L Wessjohann, M Brauer, R Iori, RS Esworthy, FF Chu, and R Brigelius-Flohé (2012) Glutathione peroxidase-2 and selenium decreased inflammation and tumors in a mouse model of inflammation-associated carcinogenesis whereas sulforaphane effects differed with selenium supply. Carcinogenesis 33:620–628

48. Koeberle SC, A Gollowitzer, J Laoukili, O Kranenburg, O Werz, A Koeberle, and AP Kipp (2020) Distinct and overlapping functions of glutathione peroxidases 1 and 2 in limiting NF-kappaB-driven inflammation through redox-active mechanisms. Redox Biol 28:101388

49. Brigelius-Flohé R, B Friedrichs, S Maurer, M Schultz, and R Streicher (1997) Interleukin-1-induced nuclear factor kappa B activation is inhibited by overexpression of phospholipid hydroperoxide glutathione peroxidase in a human endothelial cell line. Biochem J 328 (Pt 1):199–203

50. DiDonato JA, F Mercurio, and M Karin (2012) NF-kappaB and the link between inflammation and cancer. Immunol Rev 246:379–400

51. Wenk J, J Schüller, C Hinrichs, T Syrovets, N Azoitei, M Podda, M Wlaschek, P Brenneisen, LA Schneider, A Sabiwalsky, T Peters, S Sulyok, J Dissemond, M Schauen, T Krieg, T Wirth, T Simmet, and K Scharffetter-Kochanek (2004) Overexpression of phospholipid-hydroperoxide glutathione peroxidase in human dermal fibroblasts abrogates UVA irradiation-induced expression of interstitial collagenase/matrix metalloproteinase-1 by suppression of phosphatidylcholine hydroperoxide-mediated NFkappaB activation and interleukin-6 release. J Biol Chem 279:45634–45642

52. Lubos E, NJ Kelly, SR Oldebeken, JA Leopold, YY Zhang, J Loscalzo, and DE Handy (2011) Glutathione peroxidase-1 deficiency augments proinflammatory cytokine-induced redox signaling and human endothelial cell activation. J Biol Chem 286:35407–35417

53. Kretz-Remy C, P Mehlen, ME Mirault, and AP Arrigo (1996) Inhibition of I kappa B-alpha phosphorylation and degradation and subsequent NF-kappa B activation by glutathione peroxidase overexpression. J Cell Biol 133:1083–1093

54. Lee E, A Choi, Y Jun, N Kim, JI Yook, SY Kim, S Lee, and SW Kang (2020) Glutathione peroxidase-1 regulates adhesion and metastasis of triple-negative breast cancer cells via FAK signaling. Redox Biol 29:101391

55. An BC, YD Choi, IJ Oh, JH Kim, JI Park, and SW Lee (2018) GPx3-mediated redox signaling arrests the cell cycle and acts as a tumor suppressor in lung cancer cell lines. PLoS One 13:e0204170

56. Yu YP, G Yu, G Tseng, K Cieply, J Nelson, M Defrances, R Zarnegar, G Michalopoulos, and JH Luo (2007) Glutathione peroxidase 3, deleted or methylated in prostate cancer, suppresses prostate cancer growth and metastasis. Cancer Res 67:8043–8050

57. Chang SN, JM Lee, H Oh, and JH Park (2016) Glutathione peroxidase 3 inhibits prostate tumorigenesis in TRAMP mice. Prostate 76:1387–1398

58. Qi X, KT Ng, Y Shao, CX Li, W Geng, CC Ling, YY Ma, XB Liu, H Liu, J Liu, WH Yeung, CM Lo, and K Man (2016) The clinical significance and potential therapeutic role of GPx3 in tumor recurrence after liver transplantation. Theranostics 6:1934–1946

59. Zhao H, J Li, X Li, C Han, Y Zhang, L Zheng, and M Guo (2015) Silencing GPX3 expression promotes tumor metastasis in human thyroid cancer. Curr Protein Pept Sci 16:316–321

60. Barrett CW, W Ning, X Chen, JJ Smith, MK Washington, KE Hill, LA Coburn, RM Peek, R Chaturvedi, KT Wilson, RF Burk, and CS

Williams (2013) Tumor suppressor function of the plasma glutathione peroxidase gpx3 in colitis-associated carcinoma. Cancer Res 73:1245–1255

61. Heirman I, D Ginneberge, R Brigelius-Flohé, N Hendrickx, P Agostinis, P Brouckaert, P Rottiers, and J Grooten (2006) Blocking tumor cell eicosanoid synthesis by GPx 4 impedes tumor growth and malignancy. Free Radic Biol Med 40:285–294

62. Liu J, J Du, Y Zhang, W Sun, BJ Smith, LW Oberley, and JJ Cullen (2006) Suppression of the malignant phenotype in pancreatic cancer by overexpression of phospholipid hydroperoxide glutathione peroxidase. Hum Gene Ther 17:105–116

63. Dixon SJ, KM Lemberg, MR Lamprecht, R Skouta, EM Zaitsev, CE Gleason, DN Patel, AJ Bauer, AM Cantley, WS Yang, B Morrison, 3rd, and BR Stockwell (2012) Ferroptosis: an iron-dependent form of nonapoptotic cell death. Cell 149:1060–1072

64. Friedmann Angeli JP, M Schneider, B Proneth, YY Tyurina, VA Tyurin, VJ Hammond, N Herbach, M Aichler, A Walch, E Eggenhofer, D Basavarajappa, O Radmark, S Kobayashi, T Seibt, H Beck, F Neff, I Esposito, R Wanke, H Forster, O Yefremova, M Heinrichmeyer, GW Bornkamm, EK Geissler, SB Thomas, BR Stockwell, VB O'Donnell, VE Kagan, JA Schick, and M Conrad (2014) Inactivation of the ferroptosis regulator Gpx4 triggers acute renal failure in mice. Nat Cell Biol 16:1180–1191

65. Yang WS, R SriRamaratnam, ME Welsch, K Shimada, R Skouta, VS Viswanathan, JH Cheah, PA Clemons, AF Shamji, CB Clish, LM Brown, AW Girotti, VW Cornish, SL Schreiber, and BR Stockwell (2014) Regulation of ferroptotic cancer cell death by GPX4. Cell 156:317–331

66. Ding WQ and SE Lind (2007) Phospholipid hydroperoxide glutathione peroxidase plays a role in protecting cancer cells from docosahexaenoic acid-induced cytotoxicity. Mol Cancer Ther 6:1467–1474

67. Viswanathan VS, MJ Ryan, HD Dhruv, S Gill, OM Eichhoff, B Seashore-Ludlow, SD Kaffenberger, JK Eaton, K Shimada, AJ Aguirre, SR Viswanathan, S Chattopadhyay, P Tamayo, WS Yang, MG Rees, S Chen, ZV Boskovic, S Javaid, C Huang, X Wu, YY Tseng, EM Roider, D Gao, JM Cleary, BM Wolpin, JP Mesirov, DA Haber, JA Engelman, JS Boehm, JD Kotz, CS Hon, Y Chen, WC Hahn,

MP Levesque, JG Doench, ME Berens, AF Shamji, PA Clemons, BR Stockwell, and SL Schreiber (2017) Dependency of a therapy-resistant state of cancer cells on a lipid peroxidase pathway. Nature 547:453–457

68. Hangauer MJ, VS Viswanathan, MJ Ryan, D Bole, JK Eaton, A Matov, J Galeas, HD Dhruv, ME Berens, SL Schreiber, F McCormick, and MT McManus (2017) Drug-tolerant persister cancer cells are vulnerable to GPX4 inhibition. Nature 551:247–250

69. Lippmann J, K Petri, S Fulda, and J Liese (2020) Redox modulation and induction of ferroptosis as a new therapeutic strategy in hepatocellular carcinoma. Transl Oncol 13:100785

70. Jia M, D Qin, C Zhao, L Chai, Z Yu, W Wang, L Tong, L Lv, Y Wang, J Rehwinkel, J Yu, and W Zhao (2020) Redox homeostasis maintained by GPX4 facilitates STING activation. Nat Immunol 21:727–735

71. Friedmann Angeli JP, DV Krysko, and M Conrad (2019) Ferroptosis at the crossroads of cancer-acquired drug resistance and immune evasion. Nat Rev Cancer 19:405–414

72. Arnér ES (2009) Focus on mammalian thioredoxin reductases: important selenoproteins with versatile functions. Biochim Biophys Acta 1790:495–526

73. Gan L, XL Yang, Q Liu, and HB Xu (2005) Inhibitory effects of thioredoxin reductase antisense RNA on the growth of human hepatocellular carcinoma cells. J Cell Biochem 96:653–664

74. Yoo MH, XM Xu, BA Carlson, VN Gladyshev, and DL Hatfield (2006) Thioredoxin reductase 1 deficiency reverses tumor phenotype and tumorigenicity of lung carcinoma cells. J Biol Chem 281:13005–13008

75. Lee D, IM Xu, DK Chiu, J Leibold, AP Tse, MH Bao, VW Yuen, CY Chan, RK Lai, DW Chin, DF Chan, TT Cheung, SH Chok, CM Wong, SW Lowe, IO Ng, and CC Wong (2019) Induction of oxidative stress through inhibition of thioredoxin reductase 1 is an effective therapeutic approach for hepatocellular carcinoma. Hepatology 69:1768–1786

76. McLoughlin MR, DJ Orlicky, JR Prigge, P Krishna, EA Talago, IR Cavigli, S Eriksson, CG Miller, JA Kundert, VI Sayin, RA Sabol, J Heinemann, LO Brandenberger, SV Iverson, B Bothner, T Papagiannakopoulos, CT Shearn, ESJ Arner, and EE Schmidt (2019) TrxR1, Gsr, and oxidative stress determine hepatocellular

carcinoma malignancy. Proc Natl Acad Sci U S A 116:11408–11417

77. Carlson BA, MH Yoo, R Tobe, C Mueller, S Naranjo-Suarez, VJ Hoffmann, VN Gladyshev, and DL Hatfield (2012) Thioredoxin reductase 1 protects against chemically induced hepatocarcinogenesis via control of cellular redox homeostasis. Carcinogenesis 33:1806–1813

78. Stafford WC, X Peng, MH Olofsson, X Zhang, DK Luci, L Lu, Q Cheng, L Tresaugues, TS Dexheimer, NP Coussens, M Augsten, HM Ahlzen, O Orwar, A Ostman, S Stone-Elander, DJ Maloney, A Jadhav, A Simeonov, S Linder, and ESJ Arnér (2018) Irreversible inhibition of cytosolic thioredoxin reductase 1 as a mechanistic basis for anticancer therapy. Sci Transl Med 10

79. Bu L, W Li, Z Ming, J Shi, P Fang, and S Yang (2017) Inhibition of TrxR2 suppressed NSCLC cell proliferation, metabolism and induced cell apoptosis through decreasing antioxidant activity. Life Sci 178:35–41

80. Fink EE, S Mannava, A Bagati, A Bianchi-Smiraglia, JR Nair, K Moparthy, BC Lipchick, M Drokov, A Utley, J Ross, LP Mendeleeva, VG Savchenko, KP Lee, and MA Nikiforov (2016) Mitochondrial thioredoxin reductase regulates major cytotoxicity pathways of proteasome inhibitors in multiple myeloma cells. Leukemia 30:104–111

81. Liu X, Y Zhang, W Lu, Y Han, J Yang, W Jiang, X You, Y Luo, S Wen, Y Hu, and P Huang (2020) Mitochondrial TXNRD3 confers drug resistance via redox-mediated mechanism and is a potential therapeutic target in vivo. Redox Biol 36:101652

82. Shchedrina VA, Y Zhang, VM Labunskyy, DL Hatfield, and VN Gladyshev (2010) Structure-function relations, physiological roles, and evolution of mammalian ER-resident selenoproteins. Antioxid Redox Signal 12:839–849

83. Varlamova EG, MV Goltyaev, VI Novoselov, and EE Fesenko (2017) Cloning, intracellular localization, and expression of the mammalian selenocysteine-containing protein SELENOI (SelI) in tumor cell lines. Dokl Biochem Biophys 476:320–322

84. Lan X, J Xing, H Gao, S Li, L Quan, Y Jiang, S Ding, and Y Xue (2017) Decreased expression of selenoproteins as a poor prognosticator of gastric cancer in humans. Biol Trace Elem Res 178:22–28

85. Guerriero E, M Accardo, F Capone, G Colonna, G Castello, and S Costantini (2014) Assessment

of the Selenoprotein M (SELM) over-expression on human hepatocellular carcinoma tissues by immunohistochemistry. Eur J Histochem 58:2433

86. Jiang H, QQ Shi, LY Ge, QF Zhuang, D Xue, HY Xu, and XZ He (2019) Selenoprotein M stimulates the proliferative and metastatic capacities of renal cell carcinoma through activating the PI3K/AKT/mTOR pathway. Cancer Med 8:4836–4844

87. Reeves MA and PR Hoffmann (2009) The human selenoproteome: recent insights into functions and regulation. Cell Mol Life Sci 66:2457–2478

88. Varlamova EG, MV Goltyaev, and EE Fesenko (2019) Protein partners of selenoprotein SELM and the role of selenium compounds in regulation of its expression in human cancer cells. Dokl Biochem Biophys 488:300–303

89. Irons R, PA Tsuji, BA Carlson, P Ouyang, MH Yoo, XM Xu, DL Hatfield, VN Gladyshev, and CD Davis (2010) Deficiency in the 15-kDa selenoprotein inhibits tumorigenicity and metastasis of colon cancer cells. Cancer Prev Res (Phila) 3:630–639

90. Tsuji PA, S Naranjo-Suarez, BA Carlson, R Tobe, MH Yoo, and CD Davis (2011) Deficiency in the 15 kDa selenoprotein inhibits human colon cancer cell growth. Nutrients 3:805–817

91. Apostolou S, JO Klein, Y Mitsuuchi, JN Shetler, PI Poulikakos, SC Jhanwar, WD Kruger, and JR Testa (2004) Growth inhibition and induction of apoptosis in mesothelioma cells by selenium and dependence on selenoprotein SEP15 genotype. Oncogene 23:5032–5040

92. Tsuji PA, BA Carlson, S Naranjo-Suarez, MH Yoo, XM Xu, DE Fomenko, VN Gladyshev, DL Hatfield, and CD Davis (2012) Knockout of the 15 kDa selenoprotein protects against chemically-induced aberrant crypt formation in mice. PLoS One 7:e50574

93. Tsuji PA, BA Carlson, MH Yoo, S Naranjo-Suarez, XM Xu, Y He, E Asaki, HE Seifried, WC Reinhold, CD Davis, VN Gladyshev, and DL Hatfield (2015) The 15kDa selenoprotein and thioredoxin reductase 1 promote colon cancer by different pathways. PLoS One 10:e0124487

94. Bang J, M Jang, JH Huh, JW Na, M Shim, BA Carlson, R Tobe, PA Tsuji, VN Gladyshev, DL Hatfield, and BJ Lee (2015) Deficiency of the 15-kDa selenoprotein led to cytoskeleton remodeling and non-apoptotic membrane blebbing through a RhoA/ROCK pathway. Biochem Biophys Res Commun 456:884–890

95. Bang J, JH Huh, JW Na, Q Lu, BA Carlson, R Tobe, PA Tsuji, VN Gladyshev, DL Hatfield, and BJ Lee (2015) Cell proliferation and motility are inhibited by G1 phase arrest in 15-kDa seleno-protein-deficient Chang liver cells. Mol Cells 38:457–465

96. Koeberle SC and AP Kipp (2018) Selenium and inflammatory mediators. In *Selenium, Molecular and Integrative Toxicology*, B. Michalke, Editor. Springer

97. Li M, W Cheng, T Nie, H Lai, X Hu, J Luo, F Li, and H Li (2018) Selenoprotein K mediates the proliferation, migration, and invasion of human choriocarcinoma cells by negatively regulating human chorionic gonadotropin expression via ERK, p38 MAPK, and Akt signaling pathway. Biol Trace Elem Res 184:47–59

98. Ben SB, B Peng, GC Wang, C Li, HF Gu, H Jiang, XL Meng, BJ Lee, and CL Chen (2015) Overexpression of selenoprotein SelK in BGC-823 cells inhibits cell adhesion and migration. Biochemistry (Mosc) 80:1344–1353

99. Marciel MP, VS Khadka, Y Deng, P Kilicaslan, A Pham, P Bertino, K Lee, S Chen, N Glibetic, FW Hoffmann, ML Matter, and PR Hoffmann (2018) Selenoprotein K deficiency inhibits melanoma by reducing calcium flux required for tumor growth and metastasis. Oncotarget 9:13407–13422

100. Panee J, ZR Stoytcheva, W Liu, and MJ Berry (2007) Selenoprotein H is a redox-sensing high mobility group family DNA-binding protein that up-regulates genes involved in glutathione syn-thesis and phase II detoxification. J Biol Chem 282:23759–23765

101. Novoselov SV, GV Kryukov, XM Xu, BA Carlson, DL Hatfield, and VN Gladyshev (2007) Selenoprotein H is a nucleolar thioredoxin-like protein with a unique expression pattern. J Biol Chem 282:11960–11968

102. Bertz M, K Kühn, SC Koeberle, MF Müller, D Hoelzer, K Thies, S Deubel, R Thierbach, and AP Kipp (2018) Selenoprotein H controls cell cycle progression and proliferation of human colorec-tal cancer cells. Free Radic Biol Med 127:98–107

103. Cox AG, A Tsomides, AJ Kim, D Saunders, KL Hwang, KJ Evason, J Heidel, KK Brown, M Yuan, EC Lien, BC Lee, S Nissim, B Dickinson, S Chhangawala, CJ Chang, JM Asara, Y Houvras, VN Gladyshev, and W Goessling (2016) Selenoprotein H is an essential regulator of redox homeostasis that cooperates with p53 in

development and tumorigenesis. Proc Natl Acad Sci U S A 113:E5562-E5571

104. Hill KE, S Wu, AK Motley, TD Stevenson, VP Winfrey, MR Capecchi, JF Atkins, and RF Burk (2012) Production of selenoprotein P (Sepp1) by hepatocytes is central to selenium homeostasis. J Biol Chem 287:40414–40424

105. Takebe G, J Yarimizu, Y Saito, T Hayashi, H Nakamura, J Yodoi, S Nagasawa, and K Takahashi (2002) A comparative study on the hydroperoxide and thiol specificity of the gluta-thione peroxidase family and selenoprotein P. J Biol Chem 277:41254–41258

106. Tsutsumi R and Y Saito (2020) Selenoprotein P; P for plasma, prognosis, prophylaxis, and more. Biol Pharm Bull 43:366–374

107. Wang J, P Shen, S Liao, L Duan, D Zhu, J Chen, L Chen, X Sun, and Y Duan (2020) Selenoprotein P inhibits cell proliferation and ROX production in HCC cells. PLoS One 15:e0236491

108. Barrett CW, VK Reddy, SP Short, AK Motley, MK Lintel, AM Bradley, T Freeman, J Vallance, W Ning, B Parang, SV Poindexter, B Fingleton, X Chen, MK Washington, KT Wilson, NF Shroyer, KE Hill, RF Burk, and CS Williams (2015) Selenoprotein P influences colitis-induced tumorigenesis by mediating stemness and oxi-dative damage. J Clin Invest 125:2646–2660

109. Carlisle AE, N Lee, AN Matthew-Onabanjo, ME Spears, SJ Park, D Youkana, MB Doshi, A Peppers, R Li, AB Joseph, M Smith, K Simin, LJ Zhu, PL Greer, LM Shaw, and D Kim (2020) Selenium detoxification is required for cancer-cell survival. Nat Metab 2:603–611

110. Evans SO, PF Khairuddin, and MB Jameson (2017) Optimising Selenium for Modulation of Cancer Treatments. Anticancer Res 37:6497–6509

111. Cavalieri RR, KG Scott, and E Sairenji (1966) Selenite (75Se) as a tumor-localizing agent in man. J Nucl Med 7:197–208

112. Cavalieri RR and KG Scott (1968) Sodium sel-enite Se 75. A more specific agent for scanning tumors. JAMA 206:591–5

113. Walshe J, MM Serewko-Auret, N Teakle, S Cameron, K Minto, L Smith, PC Burcham, T Russell, G Strutton, A Griffin, FF Chu, S Esworthy, V Reeve, and NA Saunders (2007) Inactivation of glutathione peroxidase activity contributes to UV-induced squamous cell carci-noma formation. Cancer Res 67:4751–4758

CHAPTER EIGHTEEN

Hypoxia and Regulation of Cancer Stem Cells

Qun Lin and Zhong Yun

CONTENTS

18.1 INTRODUCTION

Molecular oxygen (O_2) is indispensable for all oxidative life forms on earth. In addition to generation of cellular energy in the form of adenosine triphosphate (ATP), O_2 actively participates in regulating various cellular functions including cell fate determination. In complex multicellular organisms, every tissue operates within a particular range of O_2 concentrations, known as physiological normoxia [1–3]. When O_2 concentrations decrease below the range of physiological normoxia in a given tissue, cells will experience insufficient oxygenation and then enter into a state of hypoxia. Under normal conditions of cell growth and tissue expansion, acute and moderate hypoxia may occur due to temporary limitation of O_2 diffusion to the newly generated cells distant from the nearest blood vessels. In contrast, severe and chronic hypoxia can occur under many pathological conditions that typically include cancer, cardiovascular disease, and ischemia [4–6].

Hypoxia is observed in all solid tumors, despite their different genetic backgrounds and tissue origins [2, 7–10]. Tumor hypoxia primarily results from defective blood vessel formation and compromised blood flow [2, 11]. Regions of hypoxia are often distributed randomly throughout a tumor mass. Both the intensity and duration of hypoxia can be highly variable from one area to another in the same tumor mass [11–13]. Importantly, tumor hypoxia is an independent prognostic factor for advanced malignancy, increased metastasis, and poor patient survival [4, 9, 14–16]. Fundamentally, hypoxia can induce a wide range of biological changes that are capable of promoting tumor aggression, therapy resistance, and selection of malignant tumor cell clones [17–19]. Recent advances have further found that hypoxia facilitates stem cell maintenance and regulates cell differentiation [17, 20–22]. From the perspective of cell fate plasticity, especially cancer stem cells (CSCs), the current findings have provided biologically important new insights into the role of hypoxia in the regulation of malignant progression. CSCs are defined as a distinct subpopulation of tumor cells with unlimited or significantly prolonged self-renewal and tumor-initiating potentials [23, 24]. Current studies have shown that CSCs are likely to

DOI: 10.4324/9781003204091-23

be the major cause of therapy resistance and tumor recurrence (23, 25). Because tumor initiation and continued growth require uninterrupted tumor cell replication, CSCs and the classically defined tumor-initiating cells (TICs) are essentially similar if not completely identical. For all intents and purposes, CSCs are considered herein as the sub-population of tumor cells with stem cell-like characteristics and tumor-initiating potentials.

Maintenance of CSCs and their cell fate plasticity are often regulated by both genetic and microenvironmental or niche factors (24, 26–28), similar to that of normal stem cells. Human tumor cells located in hypoxic regions appear poorly differentiated and express stem cell-associated genes (29–31). In primary pancreatic cancers, poorly differentiated pancreatic cancer cells show strong nuclear accumulation of the hypoxia-inducible factor 1α (HIF-1α) protein (32). Higher HIF-1α and/or HIF-2α protein levels are also found in the stem cell-like tumor cells in neuroblastomas (33, 34) and gliomas (35). When maintained under in vitro hypoxic conditions, tumor cells often exhibit enhanced clonogenicity (36–38). Using multiple cancer cell lines, Mathieu et al. have shown that hypoxia in vitro can activate an embryonic stem cell-like transcription program in a HIF-dependent manner (39). These data suggest that hypoxia has a direct and strong impact on the evolutionary fate of cancer cells, especially the fate

of CSCs. This chapter will briefly discuss several key mechanisms by which hypoxia exerts its impact on CSCs, with a focus on current advances of this rapidly advancing field (Figure 18.1).

18.2 HYPOXIA INDUCIBLE FACTOR (HIF)– DEPENDENT TRANSCRIPTION

18.2.1 Hypoxia-Induced HIF Activation

With decreasing O_2 concentrations in tissue microenvironment, human and most metazoan cells quickly respond by increasing transcription of a large cohort of genes primarily via the canonical HIF pathway (40–42). There have been many excellent reviews in the literature on the HIF pathway (5, 43–46). Only key background information germane to the topics in this chapter is briefly discussed.

The HIF transcription factor family consists of three heterodimers between HIF-1β and each of HIF-1α, HIF-2α, and HIF-3α, respectively. Both HIF-1β and HIF-1α are ubiquitously expressed, whereas HIF-2α has a more tissue-specific pattern of expression (47, 48). In contrast, HIF-3α, still poorly understood, appears to have a complex pattern of expression and is further regulated by alternative RNA splicing (49–51). Although all of these subunits are constitutively transcribed, HIF-1β protein maintains a constantly high steady-state

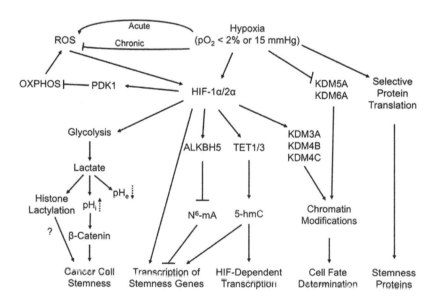

Figure 18.1 Hypoxia-dependent pathways and regulation of cancer cell stemness.

Representative pathways are selected from the current literature. See text for details. pHi, intracellular pH; pHe, extracellular pH.

level under physiological conditions, whereas the stability of HIF-α proteins, especially HIF-1α and HIF-2α, is tightly regulated by intracellular O_2 in a concentration-dependent manner.

Under physiologically normal O_2 conditions, HIF-α proteins are hydroxylated by HIF prolyl-hydroxylases (PHDs), a member of the dioxygenase family, at two conserved proline residues in the O_2-dependent degradation domain (ODD) of the HIF-α proteins (46). The hydroxylated HIF-α proteins interact with the von Hippel Lindau (vHL) protein, which leads to HIF-α ubiquitination and subsequent degradation by proteasomes. Under hypoxic conditions, the enzymatic activities of PHDs are suppressed due to reduced availability of O_2 molecules. The hydroxylation-free HIF-α proteins are stable and can then dimerize with HIF-1β in the nucleus to form functional transcription factors. Both HIF-1 and HIF-2 appear to bind the same core DNA sequence (RCGTG), termed hypoxia-responsible element (HRE), in promoters and/or enhancers found in a wide range of genes involved in many critical stress-response and survival pathways (40–42). However, HIF-1 and HIF-2 are activated at different pO_2 and display different time course of stabilization/accumulation (52, 53). Furthermore, HIF-1 appears to prefer binding in the promoter-proximal regions of the genome, whereas HIF-2 has a preference for the promoter-distant regions (54). These functional differences underscore the fact that HIF-1 and HIF-2 exhibit only partially overlapping patterns of transcription target genes (55, 56).

18.2.2 HIF-Regulated Stem Cell Genes

Several genes related to stem cell maintenance or cell fate determination are transcriptionally regulated by the HIF pathway. CD133, a penta-span transmembrane glycoprotein, is widely used as a CSC marker (23). In human glioma cells, hypoxia (1% O_2) increases CD133 expression, which leads to expansion of glioma CSCs (57–59). The hypoxia-induced CD133 expression appears to require both HIF-1 and HIF-2 because knocking down either HIF-1α (59) or HIF-2α (58) reduces the hypoxia-induced CD133 expression in glioma cells. Surprisingly, severe hypoxia (0.1% O_2) downregulates CD133 expression in gastric, colorectal, and lung cancer cell lines (60). These inconsistencies suggest that additional hypoxia-dependent mechanisms are also likely involved in the regulation of CD133 transcription at different levels of hypoxia.

Delta-like 1 homolog (*Drosophila*) or DLK1, a developmentally regulated gene encoding a type I transmembrane protein, is mainly expressed in embryonic tissues, and its expression is limited to immature cells during adulthood (61). However, DLK1 expression re-emerges in a wide range of tumors (62–67) and is found to enhance self-renewal and tumorigenicity of neuroblastoma cells (68, 69). Under hypoxic conditions, both HIF-1 and HIF-2 can bind to DLK1 promoter/enhancer to increase DLK1 transcription (69). Consistently, the DLK1-positve neuroblastoma cells are enriched in the hypoxic regions of xenograft tumors (68). These findings are consistent with the idea that cancer cells reactivate the embryonic program to survive and to disseminate (39, 70).

Expression of adenosine receptor 2B (A2BR) is upregulated by hypoxia in human breast cancer cells, which is dependent on HIF-1 (71). Genetic suppression of A2BR expression or pharmacological inhibition of its activity decreases hypoxia-induced breast cancer stem cell enrichment in vitro, which also strongly inhibits tumor initiation and lung metastasis in vivo (71). Similarly, adenosine receptor A3 (A3AR) promotes migration and invasion of glioblastoma stem-like cells under hypoxia (72). A recent study has found that A3AR can facilitate the hypoxia-induced transdifferentiation of glioblastoma stem-like cells to an endothelial cell-like phenotype (73). Because elevated levels of adenosine are frequently observed in hypoxic tumor microenvironment, the adenosine receptor pathway can exert profound impact on the maintenance and fate of cancer stem cells.

POU5F1 or OCT3/4 is an essential transcription factor that regulates pluripotency of stem cells but also facilitates tumorigenesis (74). Elevated levels of POU5F1 have been observed in several types of human cancers (75–78). In mouse embryonic stem cells, POU5F1 transcription is dependent on HIF-2α but not HIF-1α (79). However, the transcriptional regulation of POU5F1 remains to be determined in the human tumors.

The transcription co-activator TAZ, a component of the Hippo pathway, plays a critical role in body mass control and is overexpressed in human cancers (80). When exposed to hypoxia (1% O_2), human breast cancer cells upregulate TAZ expression, which requires HIF-1 but not HIF-2 (81). Because TAZ can promote the breast cancer stem cell phenotype, the hypoxia-dependent TAZ activation could constitute an important regulatory pathway in either converting non-CSCs to CSCs or maintaining the CSC state.

18.3 HYPOXIA-DEPENDENT CHROMATIN REMODELING

Dioxygenases, especially the 2-oxoglutarate-dependent dioxygenases (2-OGDD), are regarded as O_2 sensors because they directly bind O_2 molecules (82). In addition to the HIF-modifying PHDs, the 2-OGDD family members include Jumonji C (JmjC) domain histone lysine demethylases (KDM), the DNA/RNA demethylases, and dozens of other relatively poorly understood enzymes (82). The implied O_2-dependent modification of histones, DNA, and RNA strongly suggest that O_2 plays an essential role in reshaping the epigenetic landscape (83), which is expected to exert profound impact on cellular adaptation as well as cell fate in response to environmental hypoxia. Nonetheless, it should be noted that other hypoxia-regulated epigenetic pathways including the non-coding RNAs (84–88) can also affect CSC maintenance and phenotype plasticity of cancer cells. This chapter will focus on a few recently characterized demethylases only.

N(6)-methyladenosine (N^6-mA) is an important form of mRNA modification in mammalian cells and can regulate the stem cell fate of cancer cells (89–91). N^6-mA levels are concertedly regulated by methyltransferases (writers of methylation), demethylases (erasers of methylation), and methyl-binding proteins (readers of methylation). In breast cancer cells, hypoxia increases expression of ALKBH5 (eraser) and METTL14 (writer), while YTHDF3 (reader) level is significantly decreased compared to normoxia, leading to decreased N^6-mA and consequently increased expression of target transcripts in cancer cells (92). Similar results are reported in cervical cancer and liver cancer cell lines using transcriptome-wide N^6-mA analysis that also reveals extensive reprogramming of the N^6-mA epi-transcriptome under moderate hypoxia (93). Mechanistically, hypoxic breast cancer cells increase the expression of ALKB5 via HIF-dependent transcription (91). ALKBH5 then demethylates NANOG mRNA, leading to increased NANOG mRNA stabilization and synthesis of NANOG protein, a key pluripotency factor (91). The HIF-dependent ALKBH5 expression is mechanistically responsible for the enrichment of stem-like cell populations under hypoxia (91). Another member of the ALKBH family, ALKBH1

is involved in the dynamic regulation of N^6-mA levels in human glioblastoma cells, which is essential for maintaining glioma stemness (90). In addition to its impact on stemness genes, ALKBH1, strongly affects transcription of hypoxia-regulated genes in glioma stem cells (90). These observations suggest a potentially synergistic connection between hypoxia/HIF and N^6-mA modifications by ALKBHs.

The ten-eleven-translocation (TET) 5-methylcytosine dioxygenases convert 5-methylcytosine (5-mC) to 5-hydroxymethylcytosine (5-hmC) in the presence of O_2. In neuroblastoma cells, extended hypoxia (1% O_2 for 48 hours) induces TET1 transcription and global increase of 5-hmC levels (94). Interestingly, high densities of 5-hmC are found around the HIF-1 binding sites in the genome, which promotes HIF binding (94). TET1 is not only required for full transcriptional induction of hypoxia-inducible genes but also is transcriptionally regulated by HIF-1 (94). In addition to TET1, TET3 is also significantly upregulated by hypoxia in breast cancer cells in a HIF1-dependent manner (95). The hypoxia-induced TET1 and TET3 play a critical role in the regulation of breast cancer stemness. More importantly, elevated levels of TET1, TET3, and 5-hmC are correlated with breast tumor hypoxia, advanced malignancy, and poor patient survival (95). These studies underscore a synergistic mechanism between HIF-1 and TETs for the epigenetic regulation of hypoxia-induced transcriptome, as well as malignant progression.

Histone lysine demethylases (KDMs) constitute a major subgroup of the 2-OGDD family. Pollard et al. have shown that hypoxia (0.5% O_2) strongly induces the expression of KDM3A (JMJD1A), KDM4B (JMJD2B), and, to a lesser extent, KDM4C (JMJD2C) across a range of tumor cell lines in vitro (96). Both HIF-1α and HIF-2α can strongly bind to the HRE-containing DNA regions of the KDM3A and KDM4B genes, but with moderate binding of HIF-1α to KDM4C (96). Interestingly, KDM4C (JMJD2C) appears to preferentially interact with HIF-1α but not HIF-2α in breast cancer cells to enhance the transcription activities of HIF-1 (97). Paradoxically, hypoxia can enhance histone hypermethylation independent of HIF (98, 99). Remarkably, rapid induction of histone methylation can be seen upon acute exposure to hypoxia (1% O_2) in multiple human cell lines (100). These data strongly suggest that KDMs, as bona fide O_2

sensors, have the potential to modify chromatin in rapid and direct response to pO_2 fluctuations. Indeed, inactivation of KDM5A can, to a certain degree, recapitulate the hypoxia-induced histone methylation (100). Similar to the effects of hypoxia, loss of KDM6A results in sustained methylation of histone H3 lysine 27 (H3K27), which inhibits differentiation of C2C12 myoblasts (99). However, it should be noted that O_2 sensitivity or O_2 dependence varies significantly among individual KDMs (99), suggesting that each demethylase may function at different pO_2. Nonetheless, these observations demonstrate an essential role of the O_2-dependent epigenetic modifications not only in acute response to the hypoxia stress but also in long-term cell fate determination.

18.4 HYPOXIA-DEPENDENT TRANSLATIONAL REGULATION

In addition to transcriptional control of cell fate, selective or differential protein synthesis has recently emerged as a new mechanism of cell fate determination. Generally speaking, protein synthesis is reduced under hypoxic conditions, especially under severe hypoxia (101–104), much in line with suppressed cell proliferation. Nonetheless, a selective group of proteins are preferentially synthesized in response to hypoxia. Interestingly, levels of hypoxia-induced proteins appear to be more tightly controlled by translational efficiency than determined by mRNA levels (105). Several mechanisms have been proposed for hypoxia-selective mRNA translation (106) including selective activation of the upstream open-reading frame (uORF) in the 5′ untranslated regions (5′UTR), binding of the hypoxia-dependent translation initiation complex to the RNA hypoxia response elements (rHRE) in the target mRNA, and hypoxia-induced mRNA partitioning in the endoplasmic reticulum (107–110).

NANOG, SNAIL, and NODAL can regulate the stemness of human breast cancer cells, and their protein levels are strongly upregulated by hypoxia (91, 111, 112). Jewer et al. have recently found that multiple transcript isoforms of NANOG, SNAIL, and NODAL with different 5′UTRs are present in breast cancer cells (107). Hypoxia (1% O_2) appears to select specific mRNA isoforms for efficient translation. For both SNAIL and NANOG mRNAs, shorter 5′UTR isoforms without putative uORFs are preferred by hypoxia over their longer counterparts.

In contrast, NODAL mRNA isoforms with longer uORF-containing 5′UTRs are selected for translation under hypoxia (107). Functionally, such hypoxia-dependent differential translation of mRNA isoforms is accountable for cell fate plasticity and increased breast cancer stemness under hypoxia (107).

18.5 HYPOXIA-DEPENDENT METABOLIC REGULATION

Hypoxia can profoundly change cellular metabolism, with pronounced enhancement of glycolysis. Recent studies have suggested that hypoxia-induced metabolic reprogramming is not merely a simple response to the hypoxia stress by means of maintaining energy (ATP) homeostasis, but can rather serve as an essential adaptive mechanism to either reshape cell fate or maintain the stem cell state (113–116). Pluripotent stem cells (PSC) including embryonic stem cells (ESC), and induced pluripotent stem cells (iPSC) prefer glycolysis for ATP production, in a manner similar to cells under hypoxia or tumor cells exhibiting Warburg effects (115, 117, 118). The transition from mitochondrial respiration to glycolysis in PSCs is an active process of metabolic reprogramming that is necessary for maintaining self-renewal and pluripotency (115, 119–122).

Emerging evidence suggests that hypoxia-dependent metabolic regulation also plays a critical role in the maintenance of cancer stem cells. Posterior fossa A (PFA) ependymomas are epigenetically driven lethal malignancies of the hindbrain in infants and toddlers. Clinically, PFA tumors show a hypoxic gene signature that is associated with poor prognosis (123). Patient-derived primary cells maintained at 1% O_2 are able to continuously proliferate as primary cultures whereas cells cultured at 21% O_2 fail to sustain their growth (123). PFA ependymomas possess a unique metabolic profile and exhibit global histone hypomethylation. Interestingly, hypoxia is able to regulate the epigenome of PFA ependymomas by restricting the availability of specific metabolites required for histone methylation, thus leading to increased histone demethylation and acetylation at H3K27 (123). These findings demonstrate that the hypoxic tumor microenvironment facilitates propagation and malignant progression of PFA ependymoma via metabolite-dependent chromatin modifications.

Lactate production is a major consequence of hypoxia-induced glycolysis, which often leads to acidification of the hypoxic tumor microenvironment. An elegant study by Oginuma et al. (124) has recently shown that intracellular pH controls the WNT-β-catenin pathway downstream of glycolysis. The authors have found that there is a concomitant decrease of glycolytic and WNT activity with a progressive decrease of the intracellular pH (pH_i) in human pre-somitic mesoderm cells undergoing in vitro differentiation. Glycolysis increases extrusion of lactate by the monocarboxylate symporters, thus leading to elevated pH_i. The weakly basic pH_i promotes activation of the WNT pathway by facilitating non-enzymatic acetylation of β-catenin (124). In addition to its impact on pH, lactate itself has also been found to function directly as an epigenetic regulator. A rather surprising finding by Zhang et al. (125) has shown that lactate can modify chromatins via lactylation of histone lysine residues. Such unique epigenetic modification can directly stimulate gene transcription (125). Although their effects have not been examined in tumor cells, it is reasonable to predict that lactate and pH_i have the potential to make significant impact on the fate of cancer stem cells via their direct regulation of chromatins and specific stem cell pathways, especially under hypoxia.

18.6 HYPOXIA AND REACTIVE OXYGEN SPECIES (ROS)

ROS production is an integral part of energy metabolism. As a signal initiator and/or transducer, ROS strongly affects cellular response and adaptation to metabolic stresses. Notably, ROS exerts profound impact on the cell fate of both embryonal and adult stem cells (ESC and ASC). Elevated production of ROS facilitates ESC differentiation (126, 127) and is deleterious to self-renewal and repopulation potentials of hematopoietic stem cells (128, 129). In order to maintain their undifferentiated cell fate, stem cells deploy specific intracellular mechanisms to reduce ROS production and seek refuge in specialized niches from the harmful side effects of the diffusing O_2. Undifferentiated ESCs prefer glycolysis to mitochondrial respiration to maintain a low ROS status (130, 131). During iPSC formation, energy metabolism undergoes reprogramming from mitochondrial oxidative phosphorylation to glycolysis (132, 133). Such aversion to ROS drives stem cells to seek hypoxic niches for protecting their self-renewal potential. Human ESCs are more likely to maintain their full pluripotency and less likely to undergo spontaneous differentiation when cultured in vitro under hypoxia (134). Hematopoietic stem cells and other adult stem cells in vivo appear to favor hypoxic niches with reduced perfusion to maintain their undifferentiated states (22, 135). These studies strongly suggest that hypoxia, in general, has negative impact on ROS generation and is protective of undifferentiated stem cells.

However, the relationship between hypoxia and ROS has been a subject of intense debate (11, 136–139), with evidence either supporting ROS as a mediator of cellular response to hypoxia or arguing against an involvement of ROS in hypoxia response. This review will briefly discuss the direct connections between hypoxia and ROS. Readers are encouraged to read many expertly written reviews and treaties on this subject.

Early studies by Chandel, Schumacker, and colleagues found that ROS levels increased in response to moderate hypoxia of 1.5% O_2 (140) and mitochondrion-derived ROS was required for stabilization of the HIF-1α protein (141). Other groups also reported increased ROS production under moderate hypoxic conditions (142, 143). Genetic studies have further shown that loss of mitochondrial complex III (144), cytochrome C (145), or mitochondrial transcription factor TFAM (146) resulted in decreased ROS production and suppressed HIF activation under hypoxia, demonstrating an essential role of mitochondria-derived ROS in hypoxia response.

Contrary to these findings, decreased ROS production was observed under moderate hypoxic conditions (147, 148), and ROS was not involved in hypoxia-induced stabilization of the HIF-1α protein (148) or HIF functions (142). Chua et al. found that ROS production by mitochondrial complex III was not required for hypoxia (1% O_2)-induced HIF-1α stabilization (149). Chang et al. found that suppression of ROS could result in persistent stabilization of the HIF-1α protein in HUVECs exposed to 1% O_2 (150). Furthermore, stabilization of the HIF-1α protein appears to be intact in mitochondrion-deficient $ρ^0$ cells under near anoxic conditions (151). Although there are no convincing explanations for these discrepancies, different experimental conditions including different types of mitochondrion-deficient $ρ^0$ cells and different hypoxic conditions might partially contribute to these inconsistent observations.

It is worth noting that most of these experiments were performed under moderate hypoxic conditions with relatively short exposure time. It is highly probable that ROS is generated by mitochondria as an acute response to rapidly decreasing cellular O_2 concentrations, perhaps, in an attempt to maintain the previously established normoxic respiration rate. In addition to the impact of ROS on HIF-1, HIF-1 can reduce ROS production by upregulating glycolysis (116, 152) and by suppressing mitochondrial respiration via inducing expression of pyruvate dehydrogenase kinase 1 or PDK1 (153). PDK1 inactivates pyruvate dehydrogenase and thus reduces the supply of acetyl CoA to the TCA cycle, which results in decreased production of NADH for mitochondrial respiration. Furthermore, it would be interesting to examine both ROS production and its role in response or adaptation to long-term hypoxia (>24 hours at least), which is germane to both normal stem cell niches and tumor microenvironment.

18.7 CONCLUDING REMARKS

Stemness can be regarded as a universal survival mechanism against hypoxia stress. As discussed above, CSCs can utilize multiple hypoxia-regulated intracellular pathways to maintain their aggressive and therapy-resistant phenotype (Figure 18.1). It can be imagined that some of the pathways function in parallel, whereas others may function synergistically to either confer a new stemness feature or maintain an existing stem cell state. Despite such extensive impact of hypoxia on a multitude of cell fate-regulating pathways, these complex hypoxia-modulated signaling networks (Figure 18.1) can nonetheless offer many valuable therapeutic targets for reduction of the CSC population by means of induced differentiation or loss of stemness, or increased vulnerability of CSCs to anticancer therapies.

REFERENCES

1. Keeley TP, Mann GE. Defining physiological normoxia for improved translation of cell physiology to animal models and humans. Physiol Rev. 2019;99(1):161–234. Epub 2018/10/26. doi:10.1152/physrev.00041.2017. PubMed PMID: 30354965.

2. Vaupel P, Hockel M, Mayer A. Detection and characterization of tumor hypoxia using pO$_2$ histography. Antioxid Redox Signal. 2007;9(8):1221–1235. Epub 2007/06/01. doi:10.1089/ars.2007.1628. PubMed PMID: 17536958.

3. Carreau A, El Hafny-Rahbi B, Matejuk A, Grillon C, Kieda C. Why is the partial oxygen pressure of human tissues a crucial parameter? Small molecules and hypoxia. J Cell Mol Med. 2011;15(6):1239–1253. Epub 2011/01/22. doi:10.1111/j.1582-4934.2011.01258.x. PubMed PMID: 21251211; PMCID: PMC4373326.

4. Rankin EB, Giaccia AJ. Hypoxic control of metastasis. Science. 2016;352(6282):175–180. Epub 2016/04/29. doi:10.1126/science.aaf4405. PubMed PMID: 27124451; PMCID: PMC4898055.

5. Semenza GL. HIF-1: mediator of physiological and pathophysiological responses to hypoxia. J Appl Physiol (1985). 2000;88(4):1474–1480. Epub 2000/04/06. doi:10.1152/jappl.2000.88.4.1474. PubMed PMID: 10749844.

6. Bishop T, Ratcliffe PJ. HIF hydroxylase pathways in cardiovascular physiology and medicine. Circ Res. 2015;117(1):65–79. Epub 2015/06/20. doi:10.1161/CIRCRESAHA.117.305109. PubMed PMID: 26089364; PMCID: PMC4501273.

7. Quail DF, Joyce JA. Microenvironmental regulation of tumor progression and metastasis. Nat Med. 2013;19(11):1423–1437. Epub 2013/11/10. doi:10.1038/nm.3394. PubMed PMID: 24202395; PMCID: PMC3954707.

8. Vaupel P. Tumor microenvironmental physiology and its implications for radiation oncology. Semin Radiat Oncol. 2004;14(3):198–206. Epub 2004/07/16. doi:10.1016/j.semradonc.2004.04.008. PubMed PMID: 15254862.

9. Vaupel P, Mayer A. Hypoxia in cancer: significance and impact on clinical outcome. Cancer Metastasis Rev. 2007;26(2):225–239. Epub 2007/04/19. doi:10.1007/s10555-007-9055-1. PubMed PMID: 17440684.

10. Bhandari V, Hoey C, Liu LY, Lalonde E, Ray J, Livingstone J, Lesurf R, Shiah YJ, Vujcic T, Huang X, Espiritu SMG, Heisler LE, Yousif F, Huang V, Yamaguchi TN, Yao CQ, Sabelnykova VY, Fraser M, Chua MLK, van der Kwast T, Liu SK, Boutros PC, Bristow RG. Molecular landmarks of tumor hypoxia across cancer types. Nat Genet. 2019;51(2):308–318. Epub 2019/01/16. doi:10.1038/s41588-018-0318-2. PubMed PMID: 30643250.

11. Dewhirst MW, Cao Y, Moeller B. Cycling hypoxia and free radicals regulate angiogenesis and radiotherapy response. Nat Rev Cancer. 2008;8(6):425–437. Epub 2008/05/27. doi:10.1038/nrc2397. PubMed PMID: 18500244; PMCID: PMC3943205.

12. Brown JM. Evidence for acutely hypoxic cells in mouse tumours, and a possible mechanism of reoxygenation. Br J Radiol. 1979;52(620):650–656. Epub 1979/08/01. doi:10.1259/0007-1285-52-620-650. PubMed PMID: 486895.

13. Ljungkvist AS, Bussink J, Kaanders JH, van der Kogel AJ. Dynamics of tumor hypoxia measured with bioreductive hypoxic cell markers. Radiat Res. 2007;167(2):127–145. Epub 2007/03/30. PubMed PMID: 17390721.

14. Hockel M, Schlenger K, Aral B, Mitze M, Schaffer U, Vaupel P. Association between tumor hypoxia and malignant progression in advanced cancer of the uterine cervix. Cancer Res. 1996;56(19):4509–4515. Epub 1996/10/01. PubMed PMID: 8813149.

15. Aggerholm-Pedersen N, Sorensen BS, Overgaard J, Toustrup K, Baerentzen S, Nielsen OS, Maretty-Kongstad K, Nordsmark M, Alsner J, Safwat A. A prognostic profile of hypoxia-induced genes for localised high-grade soft tissue sarcoma. Br J Cancer. 2016;115(9):1096–1104. Epub 2016/10/26. doi:10.1038/bjc.2016.310. PubMed PMID: 27701385; PMCID: PMC5117798.

16. Nobre AR, Entenberg D, Wang Y, Condeelis J, Aguirre-Ghiso JA. The different routes to metastasis via hypoxia-regulated programs. Trends Cell Biol. 2018;28(11):941–956. Epub 2018/07/26. doi:10.1016/j.tcb.2018.06.008. PubMed PMID: 30041830; PMCID: PMC6214449.

17. Yun Z, Lin Q. Hypoxia and regulation of cancer cell stemness. Adv Exp Med Biol. 2014; 772:41–53.Epub2013/11/26.doi:10.1007/978-1-4614-5915-6_2. PubMed PMID: 24272353; PMCID: PMC4043215.

18. Lin Q, Yun Z. Impact of the hypoxic tumor microenvironment on the regulation of cancer stem cell characteristics. Cancer Biol Ther. 2010;9(12):949–956. Epub 2010/06/29. doi:10.4161/cbt.9.12.12347. PubMed PMID: 20581454; PMCID: PMC3637993.

19. Quail DF, Taylor MJ, Postovit LM. Microenvironmental regulation of cancer stem cell phenotypes. Curr Stem Cell Res Ther. 2012;7(3):197–216. Epub 2012/02/15. PubMed PMID: 22329582.

20. Lin Q, Lee YJ, Yun Z. Differentiation arrest by hypoxia. J Biol Chem. 2006;281(41):30678–30683. Epub 2006/08/24. doi:10.1074/jbc.C600120200. PubMed PMID: 16926163.

21. Simon MC, Keith B. The role of oxygen availability in embryonic development and stem cell function. Nat Rev Mol Cell Biol. 2008;9(4):285–296.

Epub 2008/02/21. doi:10.1038/nrm2354. PubMed PMID: 18285802; PMCID: PMC2876333.

22. Mohyeldin A, Garzón-Muvdi T, Quiñones-Hinojosa A. Oxygen in stem cell biology: a critical component of the stem cell niche. Cell Stem Cell. 2010;7(2):150–161. doi:10.1016/j.stem.2010.07.007. PubMed PMID: 20682444.

23. Visvader JE, Lindeman GJ. Cancer stem cells in solid tumours: accumulating evidence and unresolved questions. Nat Rev Cancer. 2008;8(10):755–768. Epub 2008/09/12. doi:10.1038/nrc2499. PubMed PMID: 18784658.

24. Kreso A, Dick JE. Evolution of the cancer stem cell model. Cell Stem Cell. 2014;14(3):275–291. Epub 2014/03/13. doi:10.1016/j.stem.2014.02.006. PubMed PMID: 24607403.

25. Liu C, Lin Q, Yun Z. Cellular and molecular mechanisms underlying oxygen-dependent radiosensitivity. Radiat Res. 2015;183(5):487–496. Epub 2015/05/06. doi:10.1667/RR13959.1. PubMed PMID: 25938770; PMCID: PMC4441855.

26. Ku SY, Rosario S, Wang Y, Mu P, Seshadri M, Goodrich ZW, Goodrich MM, Labbe DP, Gomez EC, Wang J, Long HW, Xu B, Brown M, Loda M, Sawyers CL, Ellis L, Goodrich DW. Rb1 and Trp53 cooperate to suppress prostate cancer lineage plasticity, metastasis, and antiandrogen resistance. Science. 2017;355(6320):78–83. Epub 2017/01/07. doi:10.1126/science.aah4199. PubMed PMID: 28059767; PMCID: PMC5367887.

27. Mu P, Zhang Z, Benelli M, Karthaus WR, Hoover E, Chen CC, Wongvipat J, Ku SY, Gao D, Cao Z, Shah N, Adams EJ, Abida W, Watson PA, Prandi D, Huang CH, de Stanchina E, Lowe SW, Ellis L, Beltran H, Rubin MA, Goodrich DW, Demichelis F, Sawyers CL. SOX2 promotes lineage plasticity and antiandrogen resistance in TP53- and RB1-deficient prostate cancer. Science. 2017;355(6320):84–88. Epub 2017/01/07. doi:10.1126/science.aah4307. PubMed PMID: 28059768; PMCID: PMC5247742.

28. Xin T, Greco V, Myung P. Hardwiring stem cell communication through tissue structure. Cell. 2016;164(6):1212–1225. Epub 2016/03/12. doi:10.1016/j.cell.2016.02.041. PubMed PMID: 26967287; PMCID: PMC4805424.

29. Das B, Tsuchida R, Malkin D, Koren G, Baruchel S, Yeger H. Hypoxia enhances tumor stemness by increasing the invasive and tumorigenic side population fraction. Stem Cells. 2008;26(7):1818–1830. Epub 2008/05/10. doi:10.1634/stemcells.2007-0724. PubMed PMID: 18467664.

30. Jogi A, Ora I, Nilsson H, Lindeheim A, Makino Y, Poellinger L, Axelson H, Pahlman S. Hypoxia alters gene expression in human neuroblastoma cells toward an immature and neural crest-like phenotype. Proc Natl Acad Sci U S A. 2002;99(10):7021–7026. Epub 2002/05/16. doi:10.1073/pnas.102660199. PubMed PMID: 12011461; PMCID: PMC124521.

31. Kim H, Lin Q, Glazer PM, Yun Z. The hypoxic tumor microenvironment in vivo selects the cancer stem cell fate of breast cancer cells. Breast Cancer Res. 2018;20(1):16. Epub 2018/03/08. doi:10.1186/s13058-018-0944-8. PubMed PMID: 29510720; PMCID: PMC5840770.

32. Couvelard A, O'Toole D, Turley H, Leek R, Sauvanet A, Degott C, Ruszniewski P, Belghiti J, Harris AL, Gatter K, Pezzella F. Microvascular density and hypoxia-inducible factor pathway in pancreatic endocrine tumours: negative correlation of microvascular density and VEGF expression with tumour progression. Br J Cancer. 2005;92(1):94–101. Epub 2004/11/24. doi:10.1038/sj.bjc.6602245. PubMed PMID: 15558070; PMCID: PMC2361752.

33. Pietras A, Gisselsson D, Ora I, Noguera R, Beckman S, Navarro S, Pahlman S. High levels of HIF-2alpha highlight an immature neural crest-like neuroblastoma cell cohort located in a perivascular niche. J Pathol. 2008;214(4):482–488. Epub 2008/01/15. doi:10.1002/path.2304. PubMed PMID: 18189331.

34. Pietras A, Hansford LM, Johnsson AS, Bridges E, Sjolund J, Gisselsson D, Rehn M, Beckman S, Noguera R, Navarro S, Cammenga J, Fredlund E, Kaplan DR, Pahlman S. HIF-2alpha maintains an undifferentiated state in neural crest-like human neuroblastoma tumor-initiating cells. Proc Natl Acad Sci U S A. 2009;106(39):16805–16810. Epub 2009/10/07. doi:10.1073/pnas.0904606106. PubMed PMID: 19805377; PMCID: PMC2745331.

35. Li Z, Bao S, Wu Q, Wang H, Eyler C, Sathornsumetee S, Shi Q, Cao Y, Lathia J, McLendon RE, Hjelmeland AB, Rich JN. Hypoxia-inducible factors regulate tumorigenic capacity of glioma stem cells. Cancer Cell. 2009;15(6):501–513. Epub 2009/05/30. doi:10.1016/j.ccr.2009.03.018. PubMed PMID: 19477429; PMCID: PMC2693960.

36. Desplat V, Faucher JL, Mahon FX, Dello Sbarba P, Praloran V, Ivanovic Z. Hypoxia modifies proliferation and differentiation of CD34(+) CML cells. Stem Cells. 2002;20(4):347–354. Epub 2002/07/12.

doi:10.1634/stemcells.20-4-347. PubMed PMID: 12110704.

37. Kim Y, Lin Q, Zelterman D, Yun Z. Hypoxia-regulated delta-like 1 homologue enhances cancer cell stemness and tumorigenicity. Cancer Res. 2009;69(24):9271–9280. Epub 2009/11/26. doi:10.1158/0008-5472.CAN-09-1605. PubMed PMID: 19934310; PMCID: PMC2828615.

38. Schmaltz C, Hardenbergh PH, Wells A, Fisher DE. Regulation of proliferation-survival decisions during tumor cell hypoxia. Mol Cell Biol. 1998;18(5):2845–2854. Epub 1998/05/05. PubMed PMID: 9566903; PMCID: PMC110663.

39. Mathieu J, Zhou W, Xing Y, Sperber H, Ferreccio A, Agoston Z, Kuppusamy KT, Moon RT, Ruohola-Baker H. Hypoxia-inducible factors have distinct and stage-specific roles during reprogramming of human cells to pluripotency. Cell Stem Cell. 2014;14(5):592–605. Epub 2014/03/25. doi:10.1016/j.stem.2014.02.012. PubMed PMID: 24656769; PMCID: PMC4028142.

40. Ivan M, Kondo K, Yang H, Kim W, Valiando J, Ohh M, Salic A, Asara JM, Lane WS, Kaelin WG, Jr. HIFα targeted for VHL-mediated destruction by proline hydroxylation: implications for O_2 sensing. Science. 2001;292(5516):464–468.

41. Jaakkola P, Mole DR, Tian YM, Wilson MI, Gielbert J, Gaskell SJ, von Kriegsheim A, Hebestreit HF, Mukherji M, Schofield CJ, Maxwell PH, Pugh CW, Ratcliffe PJ. Targeting of HIF-alpha to the von Hippel-Lindau ubiquitylation complex by O_2-regulated prolyl hydroxylation. Science. 2001;292(5516):468–472. Epub 2001/04/09. doi:10.1126/science.1059796. PubMed PMID: 11292861.

42. Semenza GL. Hypoxia-inducible factor 1 (HIF-1) pathway. Sci STKE. 2007;2007(407):cm8. Epub 2007/10/11. doi:10.1126/stke.4072007cm8. PubMed PMID: 17925579.

43. Majmundar AJ, Wong WJ, Simon MC. Hypoxia-inducible factors and the response to hypoxic stress. Mol Cell. 2010;40(2):294–309. Epub 2010/10/23. doi:10.1016/j.molcel.2010.09.022. PubMed PMID: 20965423; PMCID: PMC3143508.

44. Greer SN, Metcalf JL, Wang Y, Ohh M. The updated biology of hypoxia-inducible factor. EMBO J. 2012;31(11):2448–2460. Epub 2012/05/09. doi:10.1038/emboj.2012.125. PubMed PMID: 22562152; PMCID: PMC3365421.

45. Pugh CW, Ratcliffe PJ. Regulation of angiogenesis by hypoxia: role of the HIF system. Nat Med. 2003;9(6):677–684. Epub 2003/06/05.

doi:10.1038/nm0603-677. PubMed PMID: 12778166.

46. Schofield CJ, Ratcliffe PJ. Oxygen sensing by HIF hydroxylases. Nat Rev Mol Cell Biol. 2004;5(5):343–354. Epub 2004/05/04. doi:10.1038/nrm1366. PubMed PMID: 15122348.

47. Tian H, Hammer RE, Matsumoto AM, Russell DW, McKnight SL. The hypoxia-responsive transcription factor EPAS1 is essential for catecholamine homeostasis and protection against heart failure during embryonic development. Genes Dev. 1998;12(21):3320–3324. Epub 1998/11/10. PubMed PMID: 9808618; PMCID: PMC317225.

48. Wiesener MS, Jurgensen JS, Rosenberger C, Scholze CK, Horstrup JH, Warnecke C, Mandriota S, Bechmann I, Frei UA, Pugh CW, Ratcliffe PJ, Bachmann S, Maxwell PH, Eckardt KU. Widespread hypoxia-inducible expression of HIF-2alpha in distinct cell populations of different organs. FASEB J. 2003;17(2):271–273. Epub 2002/12/20. doi:10.1096/fj.02-0445fje. PubMed PMID: 12490539.

49. Duan C. Hypoxia-inducible factor 3 biology: complexities and emerging themes. Am J Physiol Cell Physiol. 2016;310(4):C260–269. Epub 2015/11/13. doi:10.1152/ajpcell.00315.2015. PubMed PMID: 26561641.

50. Makino Y, Cao R, Svensson K, Bertilsson G, Asman M, Tanaka H, Cao Y, Berkenstam A, Poellinger L. Inhibitory PAS domain protein is a negative regulator of hypoxia-inducible gene expression. Nature. 2001;414(6863):550–554. Epub 2001/12/06. doi:10.1038/35107085. PubMed PMID: 11734856.

51. Makino Y, Kanopka A, Wilson WJ, Tanaka H, Poellinger L. Inhibitory PAS domain protein (IPAS) is a hypoxia-inducible splicing variant of the hypoxia-inducible factor-3alpha locus. J Biol Chem. 2002;277(36):32405–32408. Epub 2002/07/18. doi:10.1074/jbc.C200328200. PubMed PMID: 12119283.

52. Holmquist-Mengelbier L, Fredlund E, Lofstedt T, Noguera R, Navarro S, Nilsson H, Pietras A, Vallon-Christersson J, Borg A, Gradin K, Poellinger L, Pahlman S. Recruitment of HIF-1α and HIF-2α to common target genes is differentially regulated in neuroblastoma. HIF-2α promotes an aggressive phenotype. Cancer Cell. 2006;10(5):413–423. PubMed PMID: 17097563.

53. Lin Q, Cong X, Yun Z. Differential hypoxic regulation of hypoxia-inducible factors 1alpha and 2alpha. Mol Cancer Res. 2011;9(6):757–765. Epub 2011/05/17. doi:10.1158/1541-7786.MCR-11-0053. PubMed PMID: 21571835; PMCID: PMC3117969.

54. Smythies JA, Sun M, Masson N, Salama R, Simpson PD, Murray E, Neumann V, Cockman ME, Choudhry H, Ratcliffe PJ, Mole DR. Inherent DNA-binding specificities of the HIF-1alpha and HIF-2alpha transcription factors in chromatin. EMBO reports. 2019;20(1). Epub 2018/11/16. doi:10.15252/embr.201846401. PubMed PMID: 30429208; PMCID: PMC6322389.

55. Hu CJ, Wang LY, Chodosh LA, Keith B, Simon MC. Differential roles of hypoxia-inducible factor 1alpha (HIF-1alpha) and HIF-2alpha in hypoxic gene regulation. Mol Cell Biol. 2003;23(24):9361–9374. Epub 2003/12/04. PubMed PMID: 14645546; PMCID: PMC309606.

56. Schodel J, Oikonomopoulos S, Ragoussis J, Pugh CW, Ratcliffe PJ, Mole DR. High-resolution genome-wide mapping of HIF-binding sites by ChIP-seq. Blood. 2011;117(23):e207–217. Epub 2011/03/31. doi:10.1182/blood-2010-10-314427. PubMed PMID: 21447827; PMCID: PMC3374576.

57. Griguer CE, Oliva CR, Gobin E, Marcorelles P, Benos DJ, Lancaster JR, Jr., Gillespie GY. CD133 is a marker of bioenergetic stress in human glioma. PLoS One. 2008;3(11):e3655. Epub 2008/11/06. doi:10.1371/journal.pone.0003655. PubMed PMID: 18985161; PMCID: PMC2577012.

58. Seidel S, Garvalov BK, Wirta V, von Stechow L, Schanzer A, Meletis K, Wolter M, Sommerlad D, Henze AT, Nister M, Reifenberger G, Lundeberg J, Frisen J, Acker T. A hypoxic niche regulates glioblastoma stem cells through hypoxia inducible factor 2α. Brain. 2010;133(Pt 4):983–995. Epub 2010/04/09. doi:awq042 [pii] 10.1093/brain/awq042. PubMed PMID: 20375133.

59. Soeda A, Park M, Lee D, Mintz A, Androutsellis-Theotokis A, McKay RD, Engh J, Iwama T, Kunisada T, Kassam AB, Pollack IF, Park DM. Hypoxia promotes expansion of the CD133-positive glioma stem cells through activation of HIF-1α. Oncogene. 2009. Epub 2009/09/01. doi:onc2009252 [pii] 10.1038/onc.2009.252. PubMed PMID: 19718046.

60. Matsumoto K, Arao T, Tanaka K, Kaneda H, Kudo K, Fujita Y, Tamura D, Aomatsu K, Tamura T, Yamada Y, Saijo N, Nishio K. mTOR signal and hypoxia-inducible factor-1α regulate CD133 expression in cancer cells. Cancer Res. 2009;69(18):7160–7164. Epub 2009/09/10.

REDOX REGULATION OF DIFFERENTIATION AND DE-DIFFERENTIATION

doi:0008-5472.CAN-09-1289 [pii] 10.1158/0008-5472.CAN-09-1289. PubMed PMID: 19738050.

61. Floridon C, Jensen CH, Thorsen P, Nielsen O, Sunde L, Westergaard JG, Thomsen SG, Teisner B. Does fetal antigen 1 (FA1) identify cells with regenerative, endocrine and neuroendocrine potentials? A study of FA1 in embryonic, fetal, and placental tissue and in maternal circulation. Differentiation. 2000;66(1):49–59. PubMed PMID: 10997592.

62. Jensen CH, Krogh TN, Hojrup P, Clausen PP, Skjodt K, Larsson LI, Enghild JJ, Teisner B. Protein structure of fetal antigen 1 (FA1). A novel circulating human epidermal-growth-factor-like protein expressed in neuroendocrine tumors and its relation to the gene products of dlk and pG2. Eur J Biochem. 1994;225(1):83–92. Epub 1994/10/01. PubMed PMID: 7925474.

63. Tornehave D, Jensen CH, Teisner B, Larsson LI. FA1 immunoreactivity in endocrine tumours and during development of the human fetal pancreas; negative correlation with glucagon expression. Histochem Cell Biol. 1996;106(6):535–542. Epub 1996/12/01. PubMed PMID: 8985741.

64. Yin D, Xie D, Sakajiri S, Miller CW, Zhu H, Popoviciu ML, Said JW, Black KL, Koeffler HP. DLK1: increased expression in gliomas and associated with oncogenic activities. Oncogene. 2006;25(13):1852–1861. Epub 2005/11/17. doi:10.1038/sj.onc.1209219. PubMed PMID: 16288219.

65. Sakajiri S, O'Kelly J, Yin D, Miller CW, Hofmann WK, Oshimi K, Shih LY, Kim KH, Sul HS, Jensen CH, Teisner B, Kawamata N, Koeffler HP. Dlk1 in normal and abnormal hematopoiesis. Leukemia. 2005;19(8):1404–1410. Epub 2005/06/17. doi:10.1038/sj.leu.2403832. PubMed PMID: 15959531.

66. Van Limpt VA, Chan AJ, Van Sluis PG, Caron HN, Van Noesel CJ, Versteeg R. High delta-like 1 expression in a subset of neuroblastoma cell lines corresponds to a differentiated chromaffin cell type. Int J Cancer. 2003;105(1):61–69. Epub 2003/04/03. doi:10.1002/ijc.11047. PubMed PMID: 12672031.

67. Li L, Forman SJ, Bhatia R. Expression of DLK1 in hematopoietic cells results in inhibition of differentiation and proliferation. Oncogene. 2005;24(27):4472–4476. PubMed PMID: 15806146.

68. Begum A, Kim Y, Lin Q, Yun Z. DLK1, delta-like 1 homolog (Drosophila), regulates tumor cell differentiation in vivo. Cancer Lett. 2012;318(1):26–33. Epub 2011/12/07. doi:S0304-3835(11)00733-6 [pii] 10.1016/j.canlet.2011.11.032. PubMed PMID: 22142700; PMCID: 3243111.

69. Kim Y, Lin Q, Zelterman D, Yun Z. Hypoxia-regulated delta-like 1 homologue enhances cancer cell stemness and tumorigenicity. Cancer Res. 2009;69(24):9271–9280. Epub 2009/11/26. doi:0008-5472.CAN-09-1605 [pii] 10.1158/0008-5472.CAN-09-1605. PubMed PMID: 19934310; PMCID: 2828615.

70. Ben-Porath I, Thomson MW, Carey VJ, Ge R, Bell GW, Regev A, Weinberg RA. An embryonic stem cell-like gene expression signature in poorly differentiated aggressive human tumors. Nat Genet. 2008;40(5):499–507. Epub 2008/04/30. doi:10.1038/ng.127. PubMed PMID: 18443585; PMCID: PMC2912221.

71. Lan J, Lu H, Samanta D, Salman S, Lu Y, Semenza GL. Hypoxia-inducible factor 1-dependent expression of adenosine receptor 2B promotes breast cancer stem cell enrichment. Proc Natl Acad Sci U S A. 2018;115(41):E9640–E9648. Epub 2018/09/23. doi:10.1073/pnas.1809695115. PubMed PMID: 30242135; PMCID: PMC6187157.

72. Torres A, Erices JI, Sanchez F, Ehrenfeld P, Turchi L, Virolle T, Uribe D, Niechi I, Spichiger C, Rocha JD, Ramirez M, Salazar-Onfray F, San Martin R, Quezada C. Extracellular adenosine promotes cell migration/invasion of Glioblastoma Stem-like Cells through A3 Adenosine Receptor activation under hypoxia. Cancer Lett. 2019;446:112–122. Epub 2019/01/21. doi:10.1016/j.canlet.2019.01.004. PubMed PMID: 30660649.

73. Rocha R, Torres A, Ojeda K, Uribe D, Rocha D, Erices J, Niechi I, Ehrenfeld P, San Martin R, Quezada C. The adenosine A(3) receptor regulates differentiation of glioblastoma stem-like cells to endothelial cells under hypoxia. Int J Mol Sci. 2018;19(4). Epub 2018/04/20. doi:10.3390/ijms19041228. PubMed PMID: 29670017; PMCID: PMC5979496.

74. Hochedlinger K, Yamada Y, Beard C, Jaenisch R. Ectopic expression of Oct-4 blocks progenitor-cell differentiation and causes dysplasia in epithelial tissues. Cell. 2005;121(3):465–477. Epub 2005/05/11. doi:10.1016/j.cell.2005.02.018. PubMed PMID: 15882627.

75. Cheng L. Establishing a germ cell origin for metastatic tumors using OCT4 immunohistochemistry. Cancer. 2004;101(9):2006–2010. Epub

2004/09/24. doi:10.1002/cncr.20566. PubMed PMID: 15386301.

76. Gidekel S, Pizov G, Bergman Y, Pikarsky E. Oct-3/4 is a dose-dependent oncogenic fate determinant. Cancer Cell. 2003;4(5):361–370. Epub 2003/12/12. PubMed PMID: 14667503.

77. Jones TD, Ulbright TM, Eble JN, Cheng L. OCT4: a sensitive and specific biomarker for intratubular germ cell neoplasia of the testis. Clin Cancer Res. 2004;10(24):8544–8547. Epub 2004/12/30. doi:10.1158/1078-0432.CCR-04-0688. PubMed PMID: 15623637.

78. Tai MH, Chang CC, Kiupel M, Webster JD, Olson LK, Trosko JE. Oct4 expression in adult human stem cells: evidence in support of the stem cell theory of carcinogenesis. Carcinogenesis. 2005;26(2):495–502. Epub 2004/10/30. doi:10.1093/carcin/bgh321. PubMed PMID: 15513931.

79. Covello KL, Kehler J, Yu H, Gordan JD, Arsham AM, Hu CJ, Labosky PA, Simon MC, Keith B. HIF-2alpha regulates Oct-4: effects of hypoxia on stem cell function, embryonic development, and tumor growth. Genes Dev. 2006;20(5):557–570. Epub 2006/03/03. doi:10.1101/gad.1399906. PubMed PMID: 16510872; PMCID: PMC1410808.

80. Pan D. The hippo signaling pathway in development and cancer. Dev Cell. 2010;19(4):491–505. Epub 2010/10/19. doi:10.1016/j.devcel.2010.09.011. PubMed PMID: 20951342; PMCID: PMC3124840.

81. Xiang L, Gilkes DM, Hu H, Takano N, Luo W, Lu H, Bullen JW, Samanta D, Liang H, Semenza GL. Hypoxia-inducible factor 1 mediates TAZ expression and nuclear localization to induce breast cancer stem cell phenotype. Oncotarget. 2014;5(24):12509–12527. Epub 2015/01/15. doi:10.18632/oncotarget.2997. PubMed PMID: 25587023; PMCID: PMC4350363.

82. Islam MS, Leissing TM, Chowdhury R, Hopkinson RJ, Schofield CJ. 2-Oxoglutarate-Dependent Oxygenases. Annu Rev Biochem. 2018;87:585–620. Epub 2018/03/02. doi:10.1146/annurev-biochem-061516-044724. PubMed PMID: 29494239.

83. Hancock RL, Dunne K, Walport LJ, Flashman E, Kawamura A. Epigenetic regulation by histone demethylases in hypoxia. Epigenomics. 2015;7(5):791–811. Epub 2015/04/03. doi:10.2217/epi.15.24. PubMed PMID: 25832587.

84. Camps C, Saini HK, Mole DR, Choudhry H, Reczko M, Guerra-Assuncao JA, Tian YM, Buffa FM, Harris AL, Hatzigeorgiou AG, Enright AJ,

Ragoussis J. Integrated analysis of microRNA and mRNA expression and association with HIF binding reveals the complexity of microRNA expression regulation under hypoxia. Mol Cancer. 2014;13:28. Epub 2014/02/13. doi:10.1186/1476-4598-13-28. PubMed PMID: 24517586; PMCID: PMC3928101.

85. Gee HE, Ivan C, Calin GA, Ivan M. HypoxamiRs and cancer: from biology to targeted therapy. Antioxid Redox Signal. 2014;21(8):1220–1238. Epub 2013/10/12. doi:10.1089/ars.2013.5639. PubMed PMID: 24111776; PMCID: PMC4142802.

86. Macharia LW, Wanjiru CM, Mureithi MW, Pereira CM, Ferrer VP, Moura-Neto V. MicroRNAs, hypoxia and the stem-like state as contributors to cancer aggressiveness. Front Genet. 2019;10:125. Epub 2019/03/08. doi:10.3389/fgene.2019.00125. PubMed PMID: 30842790; PMCID: PMC6391339.

87. Choudhry H, Harris AL, McIntyre A. The tumour hypoxia induced non-coding transcriptome. Mol Aspects Med. 2016;47–48:35–53. Epub 2016/01/26. doi:10.1016/j.mam.2016.01.003. PubMed PMID: 26806607.

88. Duguang L, Jin H, Xiaowei Q, Peng X, Xiaodong W, Zhennan L, Jianjun Q, Jie Y. The involvement of lncRNAs in the development and progression of pancreatic cancer. Cancer Biol Ther. 2017;18(12):927–936. Epub 2017/10/21. doi:10.1080/15384047.2017.1385682. PubMed PMID: 29053398; PMCID: PMC5718823.

89. Cui Q, Shi H, Ye P, Li L, Qu Q, Sun G, Sun G, Lu Z, Huang Y, Yang CG, Riggs AD, He C, Shi Y. m(6)A RNA methylation regulates the self-renewal and tumorigenesis of glioblastoma stem cells. Cell Rep. 2017;18(11):2622–2634. Epub 2017/03/16. doi:10.1016/j.celrep.2017.02.059. PubMed PMID: 28297667; PMCID: PMC5479356.

90. Xie Q, Wu TP, Gimple RC, Li Z, Prager BC, Wu Q, Yu Y, Wang P, Wang Y, Gorkin DU, Zhang C, Dowiak AV, Lin K, Zeng C, Sui Y, Kim LJY, Miller TE, Jiang L, Lee CH, Huang Z, Fang X, Zhai K, Mack SC, Sander M, Bao S, Kerstetter-Fogle AE, Sloan AE, Xiao AZ, Rich JN. N(6)-methyladenine DNA modification in glioblastoma. Cell. 2018;175(5):1228–1243 e20. Epub 2018/11/06. doi:10.1016/j.cell.2018.10.006. PubMed PMID: 30392959; PMCID: PMC6433469.

91. Zhang C, Samanta D, Lu H, Bullen JW, Zhang H, Chen I, He X, Semenza GL. Hypoxia induces the breast cancer stem cell phenotype by HIF-dependent and ALKBH5-mediated m(6)A-demethylation of NANOG mRNA. Proc Natl

Acad Sci U S A. 2016;113(14):E2047–2056. Epub 2016/03/24. doi:10.1073/pnas.1602883113. PubMed PMID: 27001847; PMCID: PMC4833258.

92. Panneerdoss S, Eedunuri VK, Yadav P, Timilsina S, Rajamanickam S, Viswanadhapalli S, Abdelfattah N, Onyeagucha BC, Cui X, Lai Z, Mohammad TA, Gupta YK, Huang TH, Huang Y, Chen Y, Rao MK. Cross-talk among writers, readers, and erasers of m(6)A regulates cancer growth and progression. Sci Adv. 2018;4(10):eaar8263. Epub 2018/10/12. doi:10.1126/sciadv.aar8263. PubMed PMID: 30306128; PMCID: PMC6170038.

93. Wang YJ, Yang B, Lai Q, Shi JF, Peng JY, Zhang Y, Hu KS, Li YQ, Peng JW, Yang ZZ, Li YT, Pan Y, Koeffler HP, Liao JY, Yin D. Reprogramming of m(6)A epitranscriptome is crucial for shaping of transcriptome and proteome in response to hypoxia. RNA Biol. 2020. Epub 2020/08/05. doi:10.1080/15476286.2020.1804697. PubMed PMID: 32746693.

94. Mariani CJ, Vasanthakumar A, Madzo J, Yesilkanal A, Bhagat T, Yu Y, Bhattacharyya S, Wenger RH, Cohn SL, Nanduri J, Verma A, Prabhakar NR, Godley LA. TET1-mediated hydroxymethylation facilitates hypoxic gene induction in neuro-blastoma. Cell Rep. 2014;7(5):1343–1352. Epub 2014/05/20. doi:10.1016/j.celrep.2014.04.040. PubMed PMID: 24835990; PMCID: PMC4516227.

95. Wu MZ, Chen SF, Nieh S, Benner C, Ger LP, Jan CI, Ma L, Chen CH, Hishida T, Chang HT, Lin YS, Montserrat N, Gascon P, Sancho-Martinez I, Izpisua Belmonte JC. Hypoxia drives breast tumor malignancy through a TET-TNFalpha-p38-MAPK signaling axis. Cancer Res. 2015;75(18):3912–3924. Epub 2015/08/22. doi:10.1158/0008-5472.CAN-14-3208. PubMed PMID: 26294212.

96. Pollard PJ, Loenarz C, Mole DR, McDonough MA, Gleadle JM, Schofield CJ, Ratcliffe PJ. Regulation of Jumonji-domain-containing his-tone demethylases by hypoxia-inducible factor (HIF)-1alpha. Biochem J. 2008;416(3):387–394. Epub 2008/08/21. doi:10.1042/BJ20081238. PubMed PMID: 18713068.

97. Luo W, Chang R, Zhong J, Pandey A, Semenza GL. Histone demethylase JMJD2C is a coactivator for hypoxia-inducible factor 1 that is required for breast cancer progression. Proc Natl Acad Sci U S A. 2012;109(49):E3367–3376. Epub 2012/11/07. doi:10.1073/pnas.1217394109. PubMed PMID: 23129632; PMCID: PMC3523832.

98. Shmakova A, Batie M, Druker J, Rocha S. Chromatin and oxygen sensing in the context of JmjC histone demethylases. Biochem J. 2014; 462(3):385–395. Epub 2014/08/26. doi:10.1042/BJ20140754. PubMed PMID: 25145438; PMCID: PMC4147966.

99. Chakraborty AA, Laukka T, Myllykoski M, Ringel AE, Booker MA, Tolstorukov MY, Meng YJ, Meier SR, Jennings RB, Creech AL, Herbert ZT, McBrayer SK, Olenchock BA, Jaffe JD, Haigis MC, Beroukhim R, Signoretti S, Koivunen P, Kaelin WG, Jr. Histone demethylase KDM6A directly senses oxygen to control chromatin and cell fate. Science. 2019;363(6432):1217–1222. Epub 2019/03/16. doi:10.1126/science.aaw1026. PubMed PMID: 30872525.

100. Batie M, Frost J, Frost M, Wilson JW, Schofield P, Rocha S. Hypoxia induces rapid changes to histone methylation and reprograms chromatin. Science. 2019;363(6432):1222–1226. Epub 2019/03/16. doi:10.1126/science.aau5870. PubMed PMID: 30872526.

101. Liu L, Simon MC. Regulation of transcription and translation by hypoxia. Cancer Biol Ther. 2004;3(6):492–497. Epub 2004/07/16. doi:10.4161/cbt.3.6.1010. PubMed PMID: 15254394.

102. Fahling M. Surviving hypoxia by modula-tion of mRNA translation rate. J Cell Mol Med. 2009;13(9A):2770–2779. Epub 2009/08/14. doi:10.1111/j.1582-4934.2009.00875.x. PubMed PMID: 19674191; PMCID: PMC4498934.

103. Koumenis C, Wouters BG. "Translating" tumor hypoxia: unfolded protein response (UPR)-dependent and UPR-independent pathways. Mol Cancer Res. 2006;4(7):423–436. Epub 2006/07/20. doi:10.1158/1541-7786.MCR-06-0150. PubMed PMID: 16849518.

104. Goetz AE, Wilkinson M. Stress and the nonsense-mediated RNA decay pathway. Cell Mol Life Sci. 2017;74(19):3509–3531. Epub 2017/05/16. doi:10.1007/s00018-017-2537-6. PubMed PMID: 28503708; PMCID: PMC5683946.

105. Ho JJD, Wang M, Audas TE, Kwon D, Carlsson SK, Timpano S, Evagelou SL, Brothers S, Gonzalgo ML, Krieger JR, Chen S, Uniacke J, Lee S. Systemic reprogramming of transla-tion efficiencies on oxygen stimulus. Cell Rep. 2016;14(6):1293–1300. Epub 2016/02/09. doi:10.1016/j.celrep.2016.01.036. PubMed PMID: 26854219; PMCID: PMC4758860.

106. Chee NT, Lohse I, Brothers SP. mRNA-to-protein translation in hypoxia. Mol Cancer. 2019;18(1):49. Epub 2019/03/31. doi:10.1186/s12943-019-0968-4. PubMed PMID: 30925920.

107. Jewer M, Lee L, Leibovitch M, Zhang G, Liu J, Findlay SD, Vincent KM, Tandoc K, Dieters-Castator D, Quail DF, Dutta I, Coatham M, Xu Z, Puri A, Guan BJ, Hatzoglou M, Brumwell A, Uniacke J, Patsis C, Koromilas A, Schueler J, Siegers GM, Topisirovic I, Postovit LM. Translational control of breast cancer plasticity. Nat Commun. 2020;11(1):2498. Epub 2020/05/20. doi:10.1038/s41467-020-16352-z. PubMed PMID: 32427827.

108. Uniacke J, Holterman CE, Lachance G, Franovic A, Jacob MD, Fabian MR, Payette J, Holcik M, Pause A, Lee S. An oxygen-regulated switch in the protein synthesis machinery. Nature. 2012;486(7401):126–129. Epub 2012/06/09. doi:10.1038/nature11055. PubMed PMID: 22678294; PMCID: PMC4974072.

109. Barbosa C, Romao L. Translation of the human erythropoietin transcript is regulated by an upstream open reading frame in response to hypoxia. RNA. 2014;20(5):594–608. Epub 2014/03/22. doi:10.1261/rna.040915.113. PubMed PMID: 24647661; PMCID: PMC3988562.

110. Staudacher JJ, Naarmann-de Vries IS, Ujvari SJ, Klinger B, Kasim M, Benko E, Ostareck-Lederer A, Ostareck DH, Bondke Persson A, Lorenzen S, Meier JC, Bluthgen N, Persson PB, Henrion-Caude A, Mrowka R, Fahling M. Hypoxia-induced gene expression results from selective mRNA partitioning to the endoplasmic reticulum. Nucleic Acids Res. 2015;43(6):3219–3236. Epub 2015/03/11. doi:10.1093/nar/gkv167. PubMed PMID: 25753659; PMCID: PMC4381074.

111. Lundgren K, Nordenskjold B, Landberg G. Hypoxia, Snail and incomplete epithelial-mesenchymal transition in breast cancer. Br J Cancer. 2009;101(10):1769–1781. Epub 2009/10/22. doi:10.1038/sj.bjc.6605369. PubMed PMID: 19844232; PMCID: PMC2778529.

112. Quail DF, Taylor MJ, Walsh LA, Dieters-Castator D, Das P, Jewer M, Zhang G, Postovit LM. Low oxygen levels induce the expression of the embryonic morphogen Nodal. Mol Biol Cell. 2011;22(24):4809–4821. Epub 2011/10/28. doi:10.1091/mbc.E11-03-0263. PubMed PMID: 22031289; PMCID: PMC3237624.

113. Rodriguez C, Puente-Moncada N, Reiter RJ, Sanchez-Sanchez AM, Herrera F, Rodriguez Blanco J, Duarte-Olivenza C, Turos-Cabal M, Antolin I, Martin V. Regulation of cancer cell glucose metabolism is determinant for cancer cell fate after melatonin administration. J Cell Physiol. 2020. Epub 2020/07/30. doi:10.1002/jcp.29886. PubMed PMID: 32725819.

114. Yadav UP, Singh T, Kumar P, Sharma P, Kaur H, Sharma S, Singh S, Kumar S, Mehta K. Metabolic adaptations in cancer stem cells. Front Oncol. 2020;10:1010. Epub 2020/07/17. doi:10.3389/fonc.2020.01010. PubMed PMID: 32670883; PMCID: PMC7330710.

115. Nishimura K, Fukuda A, Hisatake K. Mechanisms of the Metabolic Shift during Somatic Cell Reprogramming. Int J Mol Sci. 2019;20(9). Epub 2019/05/10. doi:10.3390/ijms20092254. PubMed PMID: 31067778; PMCID: PMC6539623.

116. Semenza GL. Hypoxia-inducible factors: coupling glucose metabolism and redox regulation with induction of the breast cancer stem cell phenotype. EMBO J. 2017;36(3):252–259. Epub 2016/12/23. doi:10.15252/embj.201695204. PubMed PMID: 28007895; PMCID: PMC5286373.

117. Zhou W, Choi M, Margineantu D, Margaretha L, Hesson J, Cavanaugh C, Blau CA, Horwitz MS, Hockenbery D, Ware C, Ruohola-Baker H. HIF1alpha induced switch from bivalent to exclusively glycolytic metabolism during ESC-to-EpiSC/hESC transition. EMBO J. 2012;31(9):2103–2116. Epub 2012/03/27. doi:10.1038/emboj.2012.71. PubMed PMID: 22446391; PMCID: PMC3343469.

118. Vander Heiden MG, Cantley LC, Thompson CB. Understanding the Warburg effect: the metabolic requirements of cell proliferation. Science. 2009;324(5930):1029–1033. Epub 2009/05/23. doi:10.1126/science.1160809. PubMed PMID: 19460998; PMCID: PMC2849637.

119. Folmes CD, Terzic A. Energy metabolism in the acquisition and maintenance of stemness. Semin Cell Dev Biol. 2016;52:68–75. Epub 2016/02/13. doi:10.1016/j.semcdb.2016.02.010. PubMed PMID: 26868758; PMCID: PMC4905551.

120. Shyh-Chang N, Daley GQ. Metabolic switches linked to pluripotency and embryonic stem cell differentiation. Cell Metab. 2015;21(3):349–350. Epub 2015/03/05. doi:10.1016/j.cmet.2015.02.011. PubMed PMID: 25738450.

121. Somasundaram L, Levy S, Hussein AM, Ehnes DD, Mathieu J, Ruohola-Baker H. Epigenetic metabolites license stem cell states. Curr Top Dev Biol. 2020;138:209–240. Epub 2020/03/30. doi:10.1016/bs.ctdb.2020.02.003. PubMed PMID. 32220298.

122. Nakamura-Ishizu A, Ito K, Suda T. Hematopoietic stem cell metabolism during development

and aging. Dev Cell. 2020;54(2):239–255. Epub 2020/07/22. doi:10.1016/j.devcel.2020.06.029. PubMed PMID: 32693057.

123. Michealraj KA, Kumar SA, Kim LJY, Cavalli FMG, Przelicki D, Wojcik JB, Delaidelli A, Bajic A, Saulnier O, MacLeod G, Vellanki RN, Vladoiu MC, Guilhamon P, Ong W, Lee JJY, Jiang Y, Holgado BL, Rasnitsyn A, Malik AA, Tsai R, Richman CM, Juraschka K, Haapasalo J, Wang EY, De Antonellis P, Suzuki H, Farooq H, Balin P, Kharas K, Van Ommeren R, Sirbu O, Rastan A, Krumholtz SL, Ly M, Ahmadi M, Deblois G, Srikanthan D, Luu B, Loukides J, Wu X, Garzia L, Ramaswamy V, Kanshin E, Sanchez-Osuna M, El-Hamamy I, Coutinho FJ, Prinos P, Singh S, Donovan LK, Daniels C, Schramek D, Tyers M, Weiss S, Stein LD, Lupien M, Wouters BG, Garcia BA, Arrowsmith CH, Sorensen PH, Angers S, Jabado N, Dirks PB, Mack SC, Agnihotri S, Rich JN, Taylor MD. Metabolic regulation of the epigenome drives lethal infantile ependymoma. Cell. 2020;181(6):1329–1345 e24. Epub 2020/05/24. doi:10.1016/j.cell.2020.04.047. PubMed PMID: 32445698.

124. Oginuma M, Harima Y, Tarazona OA, Diaz-Cuadros M, Michaut A, Ishitani T, Xiong F, Pourquie O. Intracellular pH controls WNT downstream of glycolysis in amniote embryos. Nature. 2020;584(7819):98–101. Epub 2020/06/26. doi:10.1038/s41586-020-2428-0. PubMed PMID: 32581357.

125. Zhang D, Tang Z, Huang H, Zhou G, Cui C, Weng Y, Liu W, Kim S, Lee S, Perez-Neut M, Ding J, Czyz D, Hu R, Ye Z, He M, Zheng YG, Shuman HA, Dai L, Ren B, Roeder RG, Becker L, Zhao Y. Metabolic regulation of gene expression by histone lactylation. Nature. 2019;574(7779):575–580. Epub 2019/10/28. doi:10.1038/s41586-019-1678-1. PubMed PMID: 31645732; PMCID: PMC6818755.

126. Ji AR, Ku SY, Cho MS, Kim YY, Kim YJ, Oh SK, Kim SH, Moon SY, Choi YM. Reactive oxygen species enhance differentiation of human embryonic stem cells into mesendodermal lineage. Exp Mol Med. 2010;42(3):175–186. Epub 2010/02/19. doi:10.3858/emm.2010.42.3.018. PubMed PMID: 20164681; PMCID: PMC2845002.

127. Sauer H, Wartenberg M. Reactive oxygen species as signaling molecules in cardiovascular differentiation of embryonic stem cells and tumor-induced angiogenesis. Antioxid Redox Signal. 2005;7(11–12):1423–1434. Epub 2005/12/17. doi:10.1089/ars.2005.7.1423. PubMed PMID: 16356105.

128. Jang YY, Sharkis SJ. A low level of reactive oxygen species selects for primitive hematopoietic stem cells that may reside in the low-oxygenic niche. Blood. 2007;110(8):3056–3063. Epub 2007/06/28. doi:10.1182/blood-2007-05-087759. PubMed PMID: 17595331; PMCID: PMC2018677.

129. Tothova Z, Gilliland DG. FoxO transcription factors and stem cell homeostasis: insights from the hematopoietic system. Cell Stem Cell. 2007;1(2):140–152. Epub 2008/03/29. doi:10.1016/j.stem.2007.07.017. PubMed PMID: 18371346.

130. Rafalski VA, Mancini E, Brunet A. Energy metabolism and energy-sensing pathways in mammalian embryonic and adult stem cell fate. J Cell Sci. 2012;125(Pt 23):5597–5608. Epub 2013/02/20. doi:10.1242/jcs.114827. PubMed PMID: 23420198; PMCID: PMC3575699.

131. Shyh-Chang N, Daley GQ, Cantley LC. Stem cell metabolism in tissue development and aging. Development. 2013;140(12):2535–2547. Epub 2013/05/30. doi:10.1242/dev.091777. PubMed PMID: 23715547; PMCID: PMC3666381.

132. Folmes CD, Nelson TJ, Martinez-Fernandez A, Arrell DK, Lindor JZ, Dzeja PP, Ikeda Y, Perez-Terzic C, Terzic A. Somatic oxidative bioenergetics transitions into pluripotency-dependent glycolysis to facilitate nuclear reprogramming. Cell Metab. 2011;14(2):264–271. Epub 2011/08/02. doi:10.1016/j.cmet.2011.06.011. PubMed PMID: 21803296; PMCID: PMC3156138.

133. Shyh-Chang N, Locasale JW, Lyssiotis CA, Zheng Y, Teo RY, Ratanasirintrawoot S, Zhang J, Onder T, Unternaehrer JJ, Zhu H, Asara JM, Daley GQ, Cantley LC. Influence of threonine metabolism on S-adenosylmethionine and histone methylation. Science. 2013;339(6116):222–226. Epub 2012/11/03. doi:10.1126/science.1226603. PubMed PMID: 23118012; PMCID: PMC3652341.

134. Ezashi T, Das P, Roberts RM. Low O2 tensions and the prevention of differentiation of hES cells. Proc Natl Acad Sci U S A. 2005;102(13):4783–4788. Epub 2005/03/18. doi:10.1073/pnas.0501283102. PubMed PMID: 15772165; PMCID: PMC554750.

135. Winkler IG, Barbier V, Wadley R, Zannettino AC, Williams S, Levesque JP. Positioning of bone marrow hematopoietic and stromal cells relative to blood flow in vivo: serially reconstituting hematopoietic stem cells reside in distinct nonperfused niches. Blood. 2010;116(3):375–385. Epub 2010/04/16. doi:10.1182/blood-2009-07-233437. PubMed PMID: 20393133.

136. Galanis A, Pappa A, Giannakakis A, Lanitis E, Dangaj D, Sandaltzopoulos R. Reactive oxygen species and HIF-1 signalling in cancer. Cancer Lett. 2008;266(1):12–20. Epub 2008/04/02. doi:10.1016/j.canlet.2008.02.028. PubMed PMID: 18378391.

137. Movafagh S, Crook S, Vo K. Regulation of hypoxia-inducible factor-1a by reactive oxygen specics: new developments in an old debate. J Cell Biochem. 2015;116(5):696–703. Epub 2014/12/30. doi:10.1002/jcb.25074. PubMed PMID: 25546605.

138. Pouyssegur J, Mechta-Grigoriou F. Redox regulation of the hypoxia-inducible factor. Biol Chem. 2006;387(10–11):1337–1346. Epub 2006/11/04. doi:10.1515/BC.2006.167. PubMed PMID: 17081104.

139. Tormos KV, Chandel NS. Inter-connection between mitochondria and HIFs. J Cell Mol Med. 2010;14(4):795–804. Epub 2010/02/18. doi:10.1111/j.1582-4934.2010.01031.x. PubMed PMID: 20158574; PMCID: PMC2987233.

140. Chandel NS, Maltepe E, Goldwasser E, Mathieu CE, Simon MC, Schumacker PT. Mitochondrial reactive oxygen species trigger hypoxia-induced transcription. Proc Natl Acad Sci U S A. 1998;95(20):11715–11720. Epub 1998/09/30. doi:10.1073/pnas.95.20.11715. PubMed PMID: 9751731; PMCID: PMC21706.

141. Chandel NS, McClintock DS, Feliciano CE, Wood TM, Melendez JA, Rodriguez AM, Schumacker PT. Reactive oxygen species generated at mitochondrial complex III stabilize hypoxia-inducible factor-1alpha during hypoxia: a mechanism of O2 sensing. J Biol Chem. 2000;275(33):25130–25138. Epub 2000/06/02. doi:10.1074/jbc.M001914200. PubMed PMID: 10833514.

142. Enomoto N, Koshikawa N, Gassmann M, Hayashi J, Takenaga K. Hypoxic induction of hypoxia-inducible factor-1alpha and oxygen-regulated gene expression in mitochondrial DNA-depleted HeLa cells. Biochem Biophys Res Commun. 2002;297(2):346–352. Epub 2002/09/19. doi: 10.1016/s0006-291x(02)02186-1. PubMed PMID: 12237125.

143. Killilea DW, Hester R, Balczon R, Babal P, Gillespie MN. Free radical production in hypoxic pulmonary artery smooth muscle cells. Am J Physiol Lung Cell Mol Physiol. 2000;279(2):L408–412. Epub 2000/08/05. doi: 10.1152/ajplung.2000.279.2.L408. PubMed PMID: 10926565.

144. Guzy RD, Hoyos B, Robin E, Chen H, Liu L, Mansfield KD, Simon MC, Hammerling U, Schumacker PT. Mitochondrial complex III is required for hypoxia-induced ROS production and cellular oxygen sensing. Cell Metab. 2005;1(6):401–408. Epub 2005/08/02. doi:10.1016/j.cmet.2005.05.001. PubMed PMID: 16054089.

145. Mansfield KD, Guzy RD, Pan Y, Young RM, Cash TP, Schumacker PT, Simon MC. Mitochondrial dysfunction resulting from loss of cytochrome c impairs cellular oxygen sensing and hypoxic HIF-alpha activation. Cell Metab. 2005;1(6):393–399. Epub 2005/08/02. doi:10.1016/j.cmet.2005.05.003. PubMed PMID: 16054088; PMCID: PMC3141219.

146. Hamanaka RB, Weinberg SE, Reczek CR, Chandel NS. The mitochondrial respiratory chain is required for organismal adaptation to hypoxia. Cell Rep. 2016;15(3):451–459. Epub 2016/04/14. doi:10.1016/j.celrep.2016.03.044. PubMed PMID: 27068470; PMCID: PMC4838509.

147. Fandrey J, Frede S, Jelkmann W. Role of hydrogen peroxide in hypoxia-induced erythropoietin production. Biochem J. 1994;303 (Pt 2):507–510. Epub 1994/10/15. doi:10.1042/bj3030507. PubMed PMID: 7980410; PMCID: PMC1137356.

148. Vaux EC, Metzen E, Yeates KM, Ratcliffe PJ. Regulation of hypoxia-inducible factor is preserved in the absence of a functioning mitochondrial respiratory chain. Blood. 2001;98(2):296–302. Epub 2001/07/04. doi:10.1182/blood.v98.2.296. PubMed PMID: 11435296.

149. Chua YL, Dufour E, Dassa EP, Rustin P, Jacobs HT, Taylor CT, Hagen T. Stabilization of hypoxia-inducible factor-1alpha protein in hypoxia occurs independently of mitochondrial reactive oxygen species production. J Biol Chem. 2010;285(41):31277–31284. Epub 2010/08/03. doi:10.1074/jbc.M110.158485. PubMed PMID: 20675386; PMCID: PMC2951202.

150. Chang TC, Huang CJ, Tam K, Chen SF, Tan KT, Tsai MS, Lin TN, Shyue SK. Stabilization of hypoxia-inducible factor-1{alpha} by prostacyclin under prolonged hypoxia via reducing reactive oxygen species level in endothelial cells. J Biol Chem. 2005;280(44):36567–36574. Epub 2005/08/24. doi:10.1074/jbc.M504280200. PubMed PMID: 16115891.

151. Schroedl C, McClintock DS, Budinger GR, Chandel NS. Hypoxic but not anoxic stabilization of HIF-1alpha requires mitochondrial

reactive oxygen species. Am J Physiol Lung Cell Mol Physiol. 2002;283(5):L922–931. Epub 2002/10/12. doi:10.1152/ajplung.00014.2002. PubMed PMID: 12376345.

152. Denko NC. Hypoxia, HIF1 and glucose metabolism in the solid tumour. Nat Rev Cancer. 2008;8(9):705–713. Epub 2009/01/15. doi:10.1038/nrc2468. PubMed PMID: 19143055.

153. Kim JW, Tchernyshyov I, Semenza GL, Dang CV. HIF-1-mediated expression of pyruvate dehydrogenase kinase: a metabolic switch required for cellular adaptation to hypoxia. Cell Metab. 2006;3(3):177–185. Epub 2006/03/07. doi:10.1016/j.cmet.2006.02.002. PubMed PMID: 16517405.

Redox Medicine

CHAPTER NINETEEN

Redox Homeostasis and Diseases of Cellular Differentiation

Leilei Zhang and Kenneth D. Tew

CONTENTS

19.1 INTRODUCTION

Over billions of years, life has adapted to the presence of atmospheric oxygen, to the point where it is now an indispensable component of aerobic existence. Cellular homeostasis requires redox equilibrium. Oxygen consumption by mitochondria leads to the formation of a variety of ROS/RNS, which under fluctuating physiological conditions can influence cell survival, proliferation, differentiation, and migration [1]. In some instances, imbalances in those pathways that utilize oxygen can lead to accumulation of ROS and/or RNS, contributing to the development of a broad range of human pathologies associated with aging, exemplified by diabetes, neurodegenerative diseases, cardiovascular disease, and cancer [2]. Within cells, there is a narrow concentration threshold that determines whether ROS/RNS

activate discrete signaling pathways or disrupt redox homeostasis. Ironically, evolution has subverted oxygen and accompanying ROS metabolites, using them in physiological pathways that control signaling events critical to growth and differentiation. The interplay between oxygen and sulfur serves to balance electrophilic and nucleophilic "stresses" and, within the context of cell metabolism, deviations in levels of ROS can influence the functioning of a variety of transcription factors and create redox switches that eventuate critical signaling events that impact downstream cellular programs [3]. Strategies to manipulate these events have led to disciplines in drug discovery and development that target pathways controlling redox homeostasis with limited clinical success thus far. The present chapter summarizes how the apparently

DOI: 10.4324/9781003204091-25

opposite and contrary properties of oxygen and sulfur can contribute complementarity and interconnectivity in the study of a variety of human pathologies.

19.2 MAINTENANCE OF CELLULAR REDOX HOMEOSTASIS

19.2.1 Sources of ROS and RNS

Under physiological conditions, normal cells maintain ROS/RNS levels through controlling a layered range of antioxidant systems. High levels of ROS/RNS destroy biological macromolecules and cause damage that can lead to mutagenesis, carcinogenesis, or various conditions that lead to cell damage and/or death. As is discussed elsewhere in this volume, the family of ROS is represented by a series of oxygen-containing molecules, some of which are by-products of cell metabolism. These include superoxide anion ($O_2^{\cdot-}$), hydroxyl radical ($OH^{\cdot}$), and hydrogen peroxide (H_2O_2). H_2O_2 is co-opted as a second messenger in regulating pro-survival and pro-proliferation signaling events exemplified by PI3K/AKT, HIF, and MAPK/ERK pathways [4] as well as apoptotic pathways controlled by JNK or p38/MAPK [5]. Major forms of RNS are nitric oxide ($NO^{\cdot}$) and the peroxynitrite anion ($ONOO^-$) produced by reaction of O_2 and NO. Endogenous ROS production in cells occurs primarily from mitochondria or reduced nicotinamide adenine dinucleotide phosphate (NADPH) oxidase (NOX) reactions, where O_2 reacts with NADPH to form O_2^-, which through electron leakage is released by mitochondrial respiratory chain complex I–III. In the mitochondrial matrix it is converted to H_2O_2 by superoxide dismutase 2 (SOD2) [6]. $O_2^{\cdot-}$ can be released from mitochondria through voltage-dependent anion channels (VDAC), which can operate as a component of a switch system for global control of mitochondrial metabolism. Within this family, VDAC 1 and 2 are crucial mediators of oxidative stress response and are targeted by some anticancer drugs, inducing apoptosis [7]. Other cell components that can produce ROS include cytochrome P450 (CYP), monoamine oxidase, xanthine oxidase, cyclo oxygenase (COX), glycolate oxidase, hydroxy-acid oxidase, aldehyde oxidase, and amino acid oxidase [3] (Figure 19.1A). As a vasodilator and neurotransmitter, $NO^{\cdot}$ is biosynthesized from arginine by nitric oxide synthase (NOS) and can react with superoxide to form $ONOO^-$ [8]. In the context of human behavior, drugs, environmental chemicals, and radiation can produce exogenous ROS/RNS. The impact of these species on biological macromolecules can be significant and these are causally related to cell damage and disease progression.

19.2.2 Antioxidant Defenses

To counter excessive ROS/RNS accumulation, cells can activate a battery of antioxidant pathways. Activation of the transcription factor nuclear factor erythroid 2-related factor 2 (Nrf2) is regarded as one of the primary mechanisms (Figure 19.1B). During redox homeostasis, Nrf2 binds to Kelch-like ECH-associated protein 1 (Keap1) and is continuously ubiquitinated and degraded. While generally considered a protective transcription factor, cancer cells can subvert its activities by stabilizing Nrf2 protein, creating an anomalous capacity to thrive during irregular conditions of redox flux [9].

Superoxide dismutases (SODs) are catalytic antioxidants present in the cytosol (Cu-Zn SOD or SOD1), mitochondria (Mn SOD, SOD2), or extracellularly (Ec SOD or SOD3), each of which mediates cellular signaling by regulating $O_2^{\cdot-}$ levels and producing H_2O_2. Both SOD1- and SOD2 are increased in many cancers, and mice deficient in either have enhanced propensities to develop tumors as a result of increased levels of oxidative damage [10]. The SOD inhibitor, 2-methoxyestradiol, increased $O_2^{\cdot-}$ production and caused cell death in leukemia cells [11]. Catalases, also as catalytic antioxidants, convert H_2O_2 to water and can also be upregulated in cancer cells and operate in conjunction with SODs to provide tumor cells with additional growth advantages by mediating resistance against increased ROS-mediated apoptosis [12].

Glutathione peroxidases (GPx) use GSH to detoxify peroxides. In tumors, increased levels of one of this family, GPx-4, promote cell survival and resistance to a non-apoptotic form of cell death, ferroptosis (characterized by lipid peroxide accumulation and iron dependency) [13].

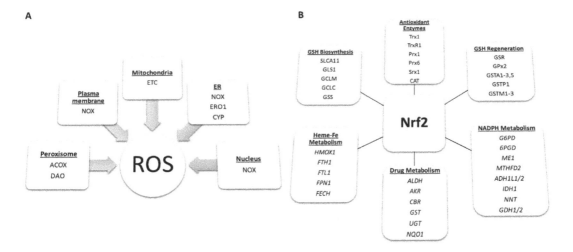

Figure 19.1 Reactive oxygen species and antioxidant systems.

(A) Subcellular sites of ROS generation. Key redox systems generating O_2^- and H_2O_2 at various subcellular sites. (B) Nrf2 is a transcription factor governing antioxidant response and regulates expression of several antioxidant genes, such as GSH biosynthesis, GSH regeneration, heme-Fe metabolism, NADPH metabolism, and drug metabolism.

Drug-induced ferroptosis entails VDAC-mediated mitochondrial dysfunction and oxidative stress induced as a consequence of iron accumulation, presumably through enhanced capacity for Fenton reactions. Heme oxygenase 1 (HO-1) regulates catabolism of heme and cellular iron levels [14], and our own recent studies established that the cytotoxic anticancer isoflavone ME-344 binds to both HO-1 and VDAC, induces mitochondrial dysfunction and in this regard shares characteristics with another drug, erastin [7, 15]. Erastin has been reported to promote pathogenesis of idiopathic pulmonary fibrosis by increasing lipid peroxidation and inhibiting the expression of GPx-4 [16]. GPx-4, as one of the four selenium GPxs (GPx-1–4), is much more efficient at detoxifying peroxides than non-selenium members of the family (GPx-5–8) [17].

Peroxiredoxins (Prx) catalyze H_2O_2 reduction and a range of organic hydroperoxides [18], and also directly, and indirectly, reduce RNS such as NO˙ and ONOO⁻ [19]. Prxs are maintained in their reduced state by thioredoxin1 (Trx1), a member of a family of proteins that reduce cysteine residues in transcription factors and numerous metabolic enzymes. Oxidized Trxs are reduced by NADPH-dependent Trx reductases (TrxR1). The expression of both Trx1 and TrxR1 are regulated by Nrf2.

19.3 REDOX REGULATORY SWITCHES

19.3.1 Cysteine as a Redox Switch

Although the pKa of the thiol group on free cysteine is between 8 and 9, in proteins the impact of proximal basic residues can modify this value to ~4–5 [20]. Such reactive cysteines become thiolate anions and can be sequentially oxidized to sulfenic, sulfinic, and sulfonic acids by molecules like hydrogen peroxide, peroxynitrite, and various hydroperoxides (RSOH) [21, 22]. However, cysteine is susceptible to a wide range of post-translational modifications including disulfide (RSSR), S-glutathionylation (RSSG) and nitrosylation (RSNO), each of which can be a critical regulator of intracellular signaling pathways [23, 24]. In peroxiredoxins, specific reduction of oxidized cysteine can be achieved by sulfiredoxin (Srx) [25], although Srx can also carry out deglutathionylation reactions in other proteins [26].

The most abundant intracellular form of cysteine is GSH, synthesized de novo in two ATP-consuming reactions that primarily occur in the cytosol, with subsequent transport into the mitochondria, nucleus, and endoplasmic reticulum (ER), where there are independent pools (Figure 19.2).

Whereas ratios of GSH:GSSG are generally in the 10:1 range, this value is shifted towards a more oxidized state in the ER [27]. As a redox buffer, GSH

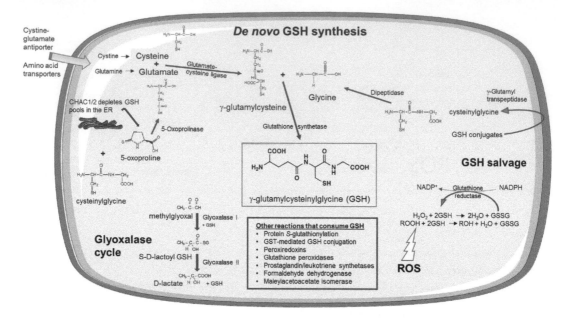

Figure 19.2 Pathways that contribute to glutathione homeostasis.

De novo synthesis follows import of the constituent amino acids. Salvage occurs through recycling of components from the breakdown of GSH, GSSG, or various endogenous or exogenous GS-conjugates. The glyoxalase cycle achieves a GSH net neutral status when GSH is both consumed and created equally.

concentrations are generally approximately two logs higher than the next most prevalent Trx [28]. The multiple functionalities of the GSH system have been discussed elsewhere in this volume, however here we discuss in more detail the importance of the post-translational modification S-glutathionylation. This modification can protect proteins from further deleterious oxidation events and may frequently change both their structure and function, increasing molecular mass and introducing a net negative charge from the tripeptide. Cysteines on the surfaces of globular proteins are exposed to GSH and GSSG and are prone to spontaneous S-glutathionylation [29]. While the total numbers of proteins susceptible to this modification are not yet known, it is likely to be restricted to a small percentage of the proteome. Indeed, the importance of a cluster of functional proteins that regulate cellular events through S-glutathionylation has been reviewed elsewhere [30, 31]. The forward reaction can occur spontaneously or be catalyzed by GSTP [32]. GSTs can effectively lower the apparent pKa of the glutathione cysteine, causing the formation of GS− at the active site [33], and can bind and activate GSH in a variety of reactions resulting in proton donation [34]. Of the GST isozymes, GSTP is the most effective at enhancing the rate and magnitude of S-glutathionylation. GSTP mutant cells that lack catalytic activity have decreased S-glutathionylation levels in response to ROS or RNS [32]. GSTP has also been shown to regulate S-glutathionylation of 1-Cys peroxiredoxin (1-cysPrx), where oxidation of the cysteine inactivates peroxidase activity. Reactivation of oxidized 1-cysPrx is accomplished by heterodimerization with GSTP, S-glutathionylating the oxidized cysteine to regenerate peroxidase activity [35]. A schematic of the S-glutathionylation cycle is shown in Figure 19.3.

Human polymorphisms in enzymes such as GSTP can contribute to the efficacy of S-glutathionylation and may predispose individuals to sensitivities to oxidative stress and disease [36], a factor that may be critical to ethnic differences in response to various chemical exposures [37].

19.3.2 Transcription Factors Sensitive to Redox

In eukaryotes, a number of transcription factors are subject to regulation by cellular exposure to ROS/RNS. Redox-sensitive transcription factors can participate as master regulators in various

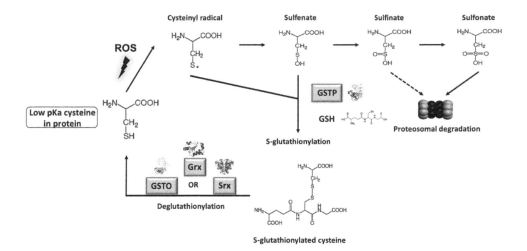

Figure 19.3 S-glutathionylation cycle.

Under conditions of mild oxidative stress, certain low pK cysteine residues in basic regions of a protein become deprotonated, forming a thiolate anion. Additional oxidation sequentially produces sulfenic, sulfinic, and sulfonic acids, the latter of which results in proteolytic degradation. Disulfide formation of the thiolate anion with GSH results in S-glutathionylation, catalyzed by GSTP. Deglutathionylation by Grx, Srx, or GSTO occurs when oxidative stress diminishes. Sulfenyl amide intermediates have been suggested to occur for certain phosphatases.

aspects of cell homeostasis, controlling broad ranges of biological functions. Some of the more important are now discussed in the contexts of their physiological importance (Figure 19.4). A more in-depth analysis can be found in a recent review [38].

Nuclear factor erythroid 2-related factor 2 (NRF2) is a master transcription factor that forms heterodimers with small musculoaponeurotic fibrosarcoma (MAF) protein and recognizes the antioxidant response elements (AREs) in promoter regions to transactivate a wide spectrum of genes implicated in redox homeostasis [39]. Interestingly, it does not regulate SOD1 or SOD2, and so its activation does not necessarily contribute to direct quenching of $O_2^{\cdot-}$-based redox signaling. While the molecular events underlying the interaction between NRF2 and pro-oxidant suppression are not entirely defined, it does downregulate NOX4, thereby suppressing the expression of interleukin-1β (IL-1β) and IL-6 [40]. NRF2 also regulates a number of genes pertinent to intermediary metabolism, and some (e.g., serine biosynthesis through activating transcription factor 4 (ATF4)) may have indirect effects on GSH and nucleic acid biosynthesis [41].

BTB and CNC homology 1 (BACH1) binds to DNA as a heterodimer with small MAF proteins [42] and inhibits the transcription of many oxidative stress-response genes. It increases the expression of CXC-chemokine receptor 4 (CXCR4) and matrix metalloproteinases (MMPs) 1, 9, 13 in cancer metastasis [43]. Moreover, it targets mitochondrial metabolism by suppressing transcription of tricarboxylic acid (TCA) cycle and electron transport chain (ETC) intermediates, both of which are potential targets for cancer therapy [44].

Activator protein 1 (AP-1) refers to dimeric transcription factors that contain members of the JUN, FOS, ATF, and MAF protein families. As a redox-sensitive transcription factor, stimulation of AP-1 can activate antioxidant enzymes, inhibit H_2O_2, and induce GSH generation [45, 46]. There are layers of complexity in its function in that it may trigger apoptosis as an oncogenic complex or induce survival as an anti-oncogenic complex, contingent upon the genetic and/or cellular context [47].

Forkhead box O (FOXO) is one member of the forkhead transcription factor family consisting of FOXO1, FOXO3, FOXO4, and FOXO6, each with a role in general cell functioning and homeostasis [48]. They activate the transcription of genes

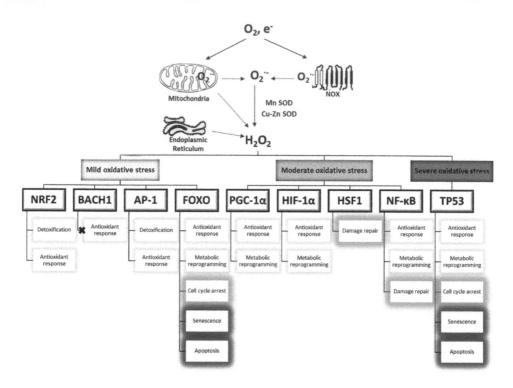

Figure 19.4 Hierarchical responses of antioxidant transcription factors to ROS.

The ETC and NOX take up O_2 and generate O_2^{-} dismutated mainly in the mitochondria and ER by SOD1 and/or SOD2 to generate H_2O_2, levels which contribute to mild, moderate, or severe oxidative stress. The antioxidant transcription factors are activated at different ROS thresholds, thereby causing appropriate hierarchical responses that enact specific detoxification, antioxidant, metabolic reprogramming, damage repair, or cell cycle arrest; senescence and apoptosis-stimulating genes may be induced at the appropriate severity of stress. For greater discussion of this topic, see [38].

encoding SOD2, Prx3, and Prx5 in mitochondria and catalase in peroxisomes, and also regulate metal ion chelation by increasing the expression of metallothioneins (MT) [49]. They also influence metabolic balance and energy metabolism via glycogenolysis and gluconeogenesis [50, 51]. Moreover, FOXOs regulate the induction of GADD45A, p27^{KIPI}, and Bim to promote cell cycle arrest and apoptosis, respectively [52].

Peroxisome proliferator-activated receptor γ coactivator 1α (PGC-1α) as a regulated coactivator plays a crucial role in the transcriptional control of mitochondrial biogenesis and respiratory functions. It binds to, and interacts with, peroxisome proliferator-activated receptor γ (PPAR γ) to increase oxidative phosphorylation and decrease mitochondrial production of ROS by stimulating mitochondrial biogenesis [53, 54].

Hypoxia-inducible factor 1α (HIF-1α) acts as a master regulator of adaptive responses to hypoxia, stimulating GSH synthetic pathways under hypoxic conditions [55] and regulating genes involved in glycolysis and lactate metabolism to support cell survival and equilibrate energy metabolism [56, 57]. Stabilization and activation of HIF-1α promotes angiogenesis by inducing vascular endothelial growth factor (VEGF), associated with cancer spread and metastases [58].

Heat shock factor 1 (HSF1) transactivates genes that encode heat shock protein (HSP) chaperones. However, it also induces antioxidant genes and responds to ROS [59]. The capacity of cancer cells to harness HSF1 for metastatic progression highlights the plasticity of this factor in contributing to the rewiring of an oncogenic phenotype [60].

Nuclear factor κB (NF-κB) is part of a family of transcription factors that comprises protein homo- or heterodimers from the Rel homology family (p50, p52, p65/Rel A, Rel B, c-Rel), having a central role in regulating immune and inflammation responses [61]. Activation induces expression of immunoreceptors, cytokines, chemokines, and growth factors by increasing ROS to mediate cellular stress [62]. In some cases, ROS levels are influenced by increased expression of antioxidant proteins [63], but in cancer cells it promotes their survival, proliferation, and metastasis by increasing expression of cyclins, MMPs, cell adhesion, and pro-angiogenic and anti-apoptotic genes [64]. It also facilitates some aspects of metabolic conversion to glycolysis and controls the microenvironment by directing the tumor-promotion role of immune cells [65].

Tumor protein p53 (TP53) has been associated with a wide variety of roles in cancer. It is redox-sensitive and can activate transcription of genes that, for instance, scavenge ROS, boost GSH synthesis and NADPH production, detoxify xenobiotics, and inhibit pro-oxidant enzymes [66, 67]. NADPH availability is regulated through TP53-induced glycolysis and apoptosis regulator (TIGAR), which promotes NADPH production via the PPP and through TP53-mediated inhibition of glucose-6-phosphate dehydrogenase (G6PD) that controls the rate-limiting step of PPP [68]. However, its targets with pro-antioxidant functions, such as protein TP53 inducible protein 3 (PIG3; a NADPH-quinone oxidoreductase) can generate harmful oxidant species, allowing it to function as a DNA damage response sensor [69].

19.4 HUMAN DISEASES LINKED WITH REDOX IMBALANCE

Imbalances in redox homeostasis have been causally linked with the pathogenesis of many human diseases, encompassing most of the major organs of the human body. These include progressive aging, cancer, cardiovascular disease, diabetes, and neurodegenerative disorders. In principle, each of these conditions is more prevalent in older individuals, and the generally accepted viewpoint is that accumulation of macromolecular damage resulting from chronic ROS exposure, together with decreasing capacities to mount an antioxidant response, can shift the redox environment to a more oxidized state. Because cell proliferation and differentiation are impacted by changes in this homeostasis, the following section discusses some information pertinent to specific diseases.

19.4.1 Type 2 Diabetes Mellitus

Type 2 diabetes mellitus (T2DM) is caused by a combination of insulin resistance and dysfunction of pancreatic β-cells. Pancreatic β-cells originate from embryonic stem cells through cellular differentiation; however, under certain conditions, mature β-cells can lose their differentiated features and regress to a less differentiated state, termed de-differentiation, which contributes to the loss of functional β-cell mass in T2DM [70]. While physiological glucose stimulation is a major contributor to the β-cell differentiated phenotype, exposure to elevated glucose concentrations can cause β-cell de-differentiation. Hyperglycemia is a major contributor leading to the progressive decline of functional β-cell mass. Hyperglycemia has been considered as one of the primary factors responsible for induction of ROS in β-cells. Because of high energy consumption and comparatively low levels of antioxidant enzymes in β-cells, they are susceptible to oxidative stress and subsequent damage [71]. Oxidative stress is also a major inducer of the pathogenesis of Friedreich's ataxia, where affected patients frequently advance to a diabetic condition. Frataxin protein is in the mitochondrial matrix and participates in iron-sulfur cluster assembly, and its levels are decreased in Friedreich's ataxia [72]. Loss of this protein in β-cells in mice impaired glucose tolerance, impacted oxidative energy flux, and led eventually to overt T2DM [73]. Moreover, targets for oxidative stress in β-cells are likely to include the key transcription factors pancreatic and duodenal homeobox 1 (PDX-1) and V-maf musculoaponeurotic fibrosarcoma oncogene homolog A (MAFA), which play key roles in pancreas differentiation and maintaining normal β-cell function [74, 75]. Exposure of rat islets to H_2O_2 reduced the nuclear expression and DNA binding activities of PDX-1 and MAFA, consequently decreasing insulin expression. Treatment of the islets of diabetic rats with the GPx mimetic ebselen restored the expression of PDX-1 and MAFA caused by oxidative stress [76]. Oxidative stress-mediated alterations of PDX-1 and MAFA involve the activation of c-Jun N-terminal kinase (JNK) [77] and p38 MAPK [78] pathways, respectively. Furthermore, NO production and S-nitrosylation also regulate glucokinase (GK) localization and activity of glucose-stimulated insulin secretion in β cells [79]. Due to the high demand to synthesize insulin in response to increases in circulating glucose, β-cells are also vulnerable to ER stress. Indeed, genetic disruption of some UPR genes, including perk, eIF2α, ire1, and xbp1 causes β-cell death and the development of insulin-dependent diabetes [80]. In response, a cycle between oxidative stress and ER stress can be created, leading to disruption of ER function. Collectively, these observations suggest that understanding involvement of redox in the maintenance and well-being of β-cells could be a viable route for possible treatment for T2DM. For example, studies of the classical therapeutic agents of diabetes suggest that an ideal drug for diabetes would not only have anti-hyperglycemia activity but also enhance the antioxidant defense system [81]. Edaravone, as one of the new therapeutic agents for diabetes, is an effective scavenger of free radicals and carbonyls and an inhibitor of lipid peroxidation. The drug has been tested in a variety of mouse models, including streptozotocin-induced disease. Treatment of edaravone inhibited hyperglycemia and scavenged free radicals [82, 83].

19.4.2 Alzheimer's Disease

Alzheimer's disease (AD) results from the aggregation of neurotoxic β-amyloid peptide (Aβ) [84]. Oxidative stress has been shown to contribute to the pathology of AD, and there is a link between Aβ and redox active metal dysregulation, also implicated through high concentrations of copper, iron, and zinc in the human brain [85]. This imbalance of metal homeostasis occurring in concert with accumulation of Aβ implies the involvement of Fenton chemistry in generation of ROS [86]. Moreover, Aβ can integrate into the membranes of astrocytes (higher cholesterol content compared to neurons) and form pores that stimulate Ca^{2+} influx [87]. Enhanced Ca^{2+} concentrations then activate NADPH oxidase in astrocytes, thereby inhibiting GSH synthesis and mitochondrial depolarization and causing cell death [88]. It has been suggested that even picomolar concentrations of Aβ oligomers can increase Ca^{2+} influx in astrocytes, stimulating ROS production and then caspase-3 activation not just in astrocytes but also adjacent neurons [89]. Aβ-induced ROS generation, in combination with nitric oxide production, can stimulate peroxynitrite production—also a trigger for neurotoxicity [90]. Melatonin can act as an antioxidant in these circumstances, but this hormone decreases with age, thereby reducing it capacity to protect neurons and astrocytes against Aβ toxicity [91]. Lipid peroxides are also found at high levels in the brain of AD patients and these can form peroxyl radicals, subsequently converted to hydroxynonenal, which can form covalent adducts with cysteine, lysine, or histidine residues.

Hydroxynonenal can also induce Aβ formation [92]. Collectively, these data suggest that targeting Aβ-induced ROS/RNS production in astrocytes may have a positive therapeutic impact. At present, while therapeutic interventions in this disease are not optimal, acetylcholinesterase inhibitors are an example of one of the few approved drugs causing increased acetylcholine levels [93]. There are antioxidant properties in 7-methoxytacrine [94], and other derivatives such as tacrine-coumarin [95], tacrine-trolox [96], and tacrine-melatonin [97] reduce oxidative stress through their antioxidant and metal-chelation properties.

19.4.3 Importance of Redox Regulation in Normal and Aberrant Hematopoiesis

19.4.3.1 Redox, Hematopoietic Stem/Progenitor Cells (HSPCs) and Bone Marrow (BM)

The bone marrow compartment is responsible for producing all differentiated hematopoietic cells in peripheral blood. In the marrow environment, HSPCs and mature plasma cells each have their own environmental niches related to factors that maintain and affect terminal differentiation and total numbers of hematopoietic stem cells (HSCs) [98]. HSCs are self-renewing and capable of producing each type of mature blood cell. At any given time, 75% of HSCs are in phase G_0 of cell cycle [99]. They are either located at the bone–bone marrow junction (osteoblastic niche), where the microenvironment facilitates quiescence, or in proximity to blood vessels (vascular niche), where the microenvironment facilitates proliferation and differentiation [100]. Each population is identified by the expression of adhesive molecules, plus cytokines and chemokine signaling molecules, all of which contribute to "stability" in a particular niche. For example, the chemokine CXCL12 regulates migration of HSCs to the vascular niche, and absence of CXCR4 leads to decrease of HSCs in the vascular niche, indicating the importance of each in niche maintenance [101]. In addition, calcium gradients in the endosteum regulate HSC migration [102]. As a tissue, bone marrow maintains a relatively hypoxic environment (1%–2% O_2), and between the osteoblastic and the vascular niche there exists an oxygen gradient [103]. The hypoxic osteoblastic environment supports the quiescence of HSCs, and the migration of HSCs to the more oxygenated vascular niche promotes differentiation, supplying myeloid and lymphoid cells to the peripheral blood. Aged mice show accumulation of HSCs in the endosteum as well as increased levels of endogenous DNA damage in their HSCs, perhaps due to the accumulation of oxidative stress and accompanying damage associated with the aging process [104]. In 1953, cysteines and thiols were reported to stimulate bone marrow cell proliferation [105]. More recently, we have shown that GSH, like oxygen, exists in gradient niches within the marrow compartment, and these can influence the localization of HPCs and their pathways to maturity/release into the peripheral circulation [106]. In this regard, GSTP1 plays a role in regulating bone marrow proliferation and differentiation pathways in bone marrow–derived dendritic cells (BMDDCs). In these cells, GSTP1 induces S-glutathionylation of critical cysteines within estrogen receptor α (ERα) and influences both development and maturation of BMDDCs. Depletion of GSTP1 decreases S-glutathionylation of ERα and serves to emphasize the fact that redox differences can regulate gene expression in bone marrow [107].

Telintra is a peptidomimetic inhibitor of GSTP and has demonstrated clinical activity in patients with myelodysplastic syndrome [108]. Redox regulation by MGST1, another member of the GST family that shares functional characteristics with GPx4, also promotes HSC differentiation to more mature and dedicated hematopoietic cells. Knock down of MGST1 in mice confirmed a negative control of HIF1α signaling, influencing energy balance that impacted glycolysis and subsequently promoted dendritic cell differentiation [109]. The involvement of MGST1 in myeloproliferation pathways appears to be evolutionarily conserved, because similar activities have been ascribed to it a teleost zebrafish [109].

19.4.3.2 ROS-Mediated Signaling in Hematological Disease

Long-term, self-renewing HSCs have lower intracellular ROS levels. However, the occurrence of high ROS during chemotherapy may lead to senescence, apoptosis, and failure of self-renewal. The transcription factor, FOXOs upregulates antioxidant genes including MnSOD, catalase, and GADD45 to protect quiescent HSCs from ROS, and FOXOs are important for the transition of HSCs to myeloid progenitor cells. Conditional FOXO knockout precedes an increase in ROS and causes defects in cell cycle and HSC regeneration. NAC treatment restores these defects and FOXO transcription [110], implying that redox-sensitive transcriptional programs have exclusive roles in each sub-population of HSC. Thus, differences in ROS levels between myeloid progenitor cells and quiescent HSCs may drive differentiation of HSC. More detailed information on the importance of ROS in HSC function can be found elsewhere [111].

Lack of cellular differentiation is a characteristic of tumorigenesis and is pertinent to the leukemic phenotype. Acute myeloid leukemia (AML) is characterized by a loss of differentiation and an accumulation of blast cells [112]. Myelodysplastic syndromes (MDS) are a group of premalignant hematopoietic disorders characterized by increased risk of transformation to AML [113]. Clinically, patients with AML who are treated with conventional chemotherapy and stem cell transplantation still have a substantial risk of relapse and progression of their disease to a more malignant condition. A number of studies have shown that intracellular and extracellular ROS are significantly higher in AML cells than normal leukocytes. Overexpression of the mda-7/IL-24 gene increases ROS production and causes differentiation of leukemic cells [114]. Inhibition of the oncoprotein mucin 1-C also promotes a differentiated myeloid phenotype in AML cells through a ROS-dependent mechanism, and increased extracellular superoxide is found in 65% of primary AML cells [115]. Constitutively activated serine-threonine kinase Akt and increased ROS levels have been assessed in primary AML cells [116]. Other studies have identified a role for mutant receptor kinases in promoting ROS production in AML. Ras mutations occur in 10%–15% of AML and are accompanied by excessive ROS, and Ras activation has increased ROS in myeloid progenitor cells transduced with H-Ras or N-Ras [117]. Fms-like tyrosine kinase 3 (FLT3) receptor mutations are found in 30%–35% of AML [118], also associated with increased ROS production. Taken together, these data suggest that mutant receptor kinases can promote intracellular ROS levels in AML and may be cause/effect related to disease formation and progression.

ROS impair the function of T cells and natural killer (NK) cells but can be reversed by interfering with, or inhibiting NOX activities [119]. In a pre-clinical trial, NOX2-derived superoxide drove transfer of mitochondria from bone marrow stromal cells to AML blasts [120]. Consolidation therapy with interleukin 2 and histamine dihydrochloride has been proposed to reduce ROS and improve the functions of T and NK cells in the tumor microenvironment, present in a post-intensive induction chemotherapy setting [121]. One proposed hypothesis is that increased ROS levels push thresholds of sensitivity. Arsenic trioxide has shown activity in the therapy of the disease, and it increases expression of NOX2 and ROS in acute promyelocytic leukemia (APL). Combination with phorbol myristate acetate (PMA), a known NOX2 activator, showed synergistic cytotoxicity in APL cells [122]. In addition, most anticancer drugs used for AML either directly, or indirectly, increase ROS and induce apoptosis. Given that NOX family members have been identified as the drivers of ROS in AML, clinical trials with NOX inhibitors used in AML may provide a meaningful clinical benefit.

A number of extracellular and intracellular oxidative stress markers have shown potential as redox biomarkers and targets for redox therapeutics. For example, a growing body of evidence indicates that serine protease inhibitor (serpin) family members are important in myeloproliferation and HPC mobilization [123]. Oxidation of sensitive cysteine residues can result in inhibition of their activities. Exposure to ROS results in S-glutathionylation of either Cys256 of serpin A1 or Cys263 of serpin A3 and, while temporarily interfering with enzyme activities, ultimately prevents their over-oxidation and proteolysis [124]. Downregulation of serpins A1 and A3 in bone marrow occurs during progenitor cell mobilization and influences the marrow microenvironment and migratory behavior of HPCs. Our earlier work in mice showed that S-glutathionylation of serpin A1 and A3 with drugs such as NOV-002 or PABA/NO altered the structure of each protein and impacted their functional roles in myeloproliferation [125]. In addition, these posttranslationally modified serpins can act as pharmacodynamic biomarkers in buccal cells following exposure to H_2O_2-containing mouthwashes [126]. More recently, a clinical trial in prostate cancer patients identified correlations between elevated levels of S-glutathionylated serpin A1 and A3 and increased duration and dose exposure to radiotherapy treatments [127]. Thus, from a redox standpoint, there are indications that S-glutathionylated serum proteins can be useful as candidate biomarkers in clinical situations where patients are exposed to various types of electrophilic stress, particularly with drugs that produce ROS.

19.5 CONCLUSIONS

Cellular redox homeostasis is a pervasive event in determining life and death events in cells. In this chapter, we have summarized how the balance of ROS and GSH pathways influence normal cell functions and how aberrations in these pathways can translate into human pathologies. Bone marrow provides a useful benchmark as a tissue that is quite dependent upon redox conditions, either for normal myeloproliferation and differentiation or in premalignant and malignant diseases. GSTP-mediated S-glutathionylation regulates bone marrow progenitor cell functions and S-glutathionylation of plasma protein can be used as biomarkers that may eventuate the optimized design of therapeutic trials for cancer patients. The cyclical nature of S-glutathionylation cycle is enabling of regulation of signaling pathways resulting from variations in ROS/RNS. These events can be referred to as redox switches. Either genetic or pharmaceutical manipulation of GSTP supports the notion that GSTP and S-glutathionylation are targetable by small molecules. Ongoing efforts to manage pathologies determined by redox and cell differentiation should allow us to enhance our understanding of how to manipulate redox homeostasis and perhaps develop clinical effective ROS-based treatment strategies.

19.6 ABBREVIATIONS

1-Cys peroxiredoxin (1-cysPrx); 6-phosphogluconate dehydrogenase (6PGD); β-amyloid peptide (Aβ); acyl-CoA oxidase (ACOX); Alzheimer's disease (AD); aldo-keto reductase (AKR); acute myeloid leukemia (AML); aldehyde dehydrogenase 1 family, and member L1 and L2 (ADH1L1/2); activator protein 1 (AP-1); acute promyelocytic leukemia (APL); antioxidant response element (ARE); activating transcription factor 4 (ATF4); BTB and CNC homology 1 (BACH1); bone marrow (BM); bone marrow–derived dendritic cell (BMDDC); catalase (CAT); carbonyl reductase (CBR); cyclooxygenase (COX); CXC-chemokine receptor 4 (CXCR4); cytochrome P450–dependent monooxygenase (CYP); D-amino acid oxidase (DAO); endoplasmic reticulum (ER); estrogen receptor α (ERα); endoplasmic reticulum oxidoreductin 1 (ERO1); electron transport chain (ETC); ferrochelatase (FECH); forkhead box O (FOXO); FMS-like tyrosine kinase 3 (FLT3); ferritin light chain 1 (FTL1); ferritin heavy chain 1 pseudogene 12, 17 (FTHL12,17); glucose-6-phosphate dehydrogenase (G6PD); glutamate-cysteine ligase catalytic subunit

(GCLC); glutamate-cysteine ligase modifier subunit (GCLM); glutamate dehydrogenase 1/2 (GDH1/2); glucokinase (GK); glutaminase1 (GLS1); glutathione peroxidase 2 (GPx2); glutathione (GSH); glutathione-disulfide reductase (GSR); glutathione synthetase (GSS); glutathione S-transferase alpha (GSTA1-3,5); glutathione S-transferase mu (GSTM1-3); glutathione S-transferase pi (GSTP1); hypoxia-inducible factor 1α (HIF-1α); heme oxygenase 1 (HO-1/HMOX1); hydrogen peroxide (H_2O_2); hematopoietic stem cells (HSCs); heat shock factor 1 (HSF1); heat shock protein (HSP); hematopoietic stem/progenitor cells (HSPCs); isocitrate dehydrogenase (IDH1); interleukin-1β (IL-1β); c-Jun N-terminal kinase (JNK); Kelch-like ECH-associated protein 1 (Keap1); musculoaponeurotic fibrosarcoma (MAF); musculoaponeurotic fibrosarcoma oncogene homolog A (MAFA); myelodysplastic syndromes (MDS); malic enzyme (ME1); matrix metalloproteinases (MMPs); metallothioneins (MT); methylenetetrahydrofolate dehydrogenase (MTHFD2); reduced nicotinamide adenine dinucleotide phosphate (NADPH); nuclear factor κB (NF-κB); natural killer (NK); nicotinamide nucleotide transhydrogenase (NNT); nitric oxide (NO‾); nitric oxide synthase (NOS); NADPH oxidases (NOX); NAD(P)H quinone dehydrogenase 1 (NQO1); nuclear factor erythroid 2-related factor 2 (Nrf2); superoxide anion (O_2^{-}); hydroxyl radical (OH‾); peroxynitrite anion ($ONOO^{-}$); pancreatic and duodenal homeobox 1 (PDX-1); peroxisome proliferator-activated receptor γ coactivator 1α (PGC-1α); protein TP53 inducible protein 3 (PIG3); phorbol myristate acetate (PMA); peroxisome proliferator-activated receptor γ (PPAR γ); peroxiredoxins (Prx); reactive nitrogen species (RNS); reactive oxygen species (ROS); nitrosylation (RSNO); hydroperoxides (RSOH); S-glutathionylation (RSSG); disulfide (RSSR); serine protease inhibitor (serpin); solute carrier family 7 member 11 (SLC7A11); superoxide dismutase (SOD); sulfiredoxin (Srx); type 2 diabetes mellitus (T2DM); tricarboxylic acid (TCA); TP53-induced glycolysis and apoptosis regulator (TIGAR); tumor protein p53 (TP53); thioredoxin 1 (Trx1); thioredoxin reductase 1 (TrxR1); UDP-glucuronosyl transferase (UGT); voltage-dependent anion channels (VDAC); vascular endothelial growth factor (VEGF).

REFERENCES

1. Holmstrom, K.M. and T. Finkel, *Cellular mechanisms and physiological consequences of redox-dependent signalling*. Nat Rev Mol Cell Biol, 2014. **15**(6): p. 411–421.

2. Sies, H., *Oxidative stress: a concept in redox biology and medicine*. Redox Biol, 2015. **4**: p. 180–183.

3. Sies, H. and D.P. Jones, *Reactive oxygen species (ROS) as pleiotropic physiological signalling agents*. Nat Rev Mol Cell Biol, 2020. **21**(7): p. 363–383.

4. Weinberg, F. and N.S. Chandel, *Reactive oxygen species-dependent signaling regulates cancer*. Cell Mol Life Sci, 2009. **66**(23): p. 3663–3673.

5. Glasauer, A., et al., *Targeting SOD1 reduces experimental non-small-cell lung cancer*. J Clin Invest, 2014. **124**(1): p. 117–128.

6. Balaban, R.S., S. Nemoto, and T. Finkel, *Mitochondria, oxidants, and aging*. Cell, 2005. **120**(4): p. 483–495.

7. Zhang, L., et al., *Voltage-dependent anion channels influence cytotoxicity of ME-344, a therapeutic isoflavone*. J Pharmacol Exp Ther, 2020. **374**(2): p. 308–318.

8. Luiking, Y.C., M.P. Engelen, and N.E. Deutz, *Regulation of nitric oxide production in health and disease*. Curr Opin Clin Nutr Metab Care, 2010. **13**(1): p. 97–104.

9. Kansanen, E., et al., *The Keap1-Nrf2 pathway: mechanisms of activation and dysregulation in cancer*. Redox Biol, 2013. **1**: p. 45–49.

10. Gill, J.G., E. Piskounova, and S.J. Morrison, *Cancer, oxidative stress, and metastasis*. Cold Spring Harb Symp Quant Biol, 2016. **81**: p. 163–175.

11. Huang, P., et al., *Superoxide dismutase as a target for the selective killing of cancer cells*. Nature, 2000. **407**(6802): p. 390–395.

12. Bauer, G., *Tumor cell-protective catalase as a novel target for rational therapeutic approaches based on specific intercellular ROS signaling*. Anticancer Res, 2012. **32**(7): p. 2599–2624.

13. Hangauer, M.J., et al., *Drug-tolerant persister cancer cells are vulnerable to GPX4 inhibition*. Nature, 2017. **551**(7679): p. 247–250.

14. Suttner, D.M. and P.A. Dennery, *Reversal of HO-1 related cytoprotection with increased expression is due to reactive iron*. FASEB J, 1999. **13**(13): p. 1800–1809.

15. Zhang, L., et al., *Isoflavone ME-344 disrupts redox homeostasis and mitochondrial function by targeting heme oxygenase 1*. Cancer Res, 2019. **79**(16): p. 4072–4085.

16. Gong, Y., et al., *Lipid peroxidation and GPX4 inhibition are common causes for myofibroblast differentiation and ferroptosis*. DNA Cell Biol, 2019. **38**(7): p. 725–733.

17. Ye, Z.W., et al., *Oxidative stress, redox regulation and diseases of cellular differentiation*. Biochim Biophys Acta, 2015. **1850**(8): p. 1607–1621.

18. Elko, E.A., et al., *Peroxiredoxins and beyond; Redox systems regulating lung physiology and disease.* Antioxid Redox Signal, 2019. **31**(14): p. 1070–1091.

19. Valko, M., et al., *Free radicals and antioxidants in normal physiological functions and human disease.* Int J Biochem Cell Biol, 2007. **39**(1): p. 44–84.

20. Finkel, T., *Signal transduction by reactive oxygen species.* J Cell Biol, 2011. **194**(1): p. 7–15.

21. Manta, B., et al., *The peroxidase and peroxynitrite reductase activity of human erythrocyte peroxiredoxin 2.* Arch Biochem Biophys, 2009. **484**(2): p. 146–154.

22. Winterbourn, C.C., *Reconciling the chemistry and biology of reactive oxygen species.* Nat Chem Biol, 2008. **4**(5): p. 278–286.

23. Marnett, L.J., J.N. Riggins, and J.D. West, *Endogenous generation of reactive oxidants and electrophiles and their reactions with DNA and protein.* J Clin Invest, 2003. **111**(5): p. 583–593.

24. Smith, B.C. and M.A. Marletta, *Mechanisms of S-nitrosothiol formation and selectivity in nitric oxide signaling.* Curr Opin Chem Biol, 2012. **16**(5–6): p. 498–506.

25. Jonsson, T.J., et al., *Structural basis for the retroreduction of inactivated peroxiredoxins by human sulfiredoxin.* Biochemistry, 2005. **44**(24): p. 8634–8642.

26. Findlay, V.J., et al., *A novel role for human sulfiredoxin in the reversal of glutathionylation.* Cancer Res, 2006. **66**(13): p. 6800–6806.

27. Kumar, C., et al., *Glutathione revisited: a vital function in iron metabolism and ancillary role in thiol-redox control.* EMBO J, 2011. **30**(10): p. 2044–2056.

28. Kakkar, P. and B.K. Singh, *Mitochondria: a hub of redox activities and cellular distress control.* Mol Cell Biochem, 2007. **305**(1–2): p. 235–253.

29. Ghezzi, P., *Regulation of protein function by glutathionylation.* Free Radic Res, 2005. **39**(6): p. 573–580.

30. Zhang, J., et al., *Pleiotropic functions of glutathione S-transferase P.* Adv Cancer Res, 2014. **122**: p. 143–175.

31. Grek, C.L., et al., *Causes and consequences of cysteine S-glutathionylation.* J Biol Chem, 2013. **288**(37): p. 26497–26504.

32. Townsend, D.M., et al., *Novel role for glutathione S-transferase pi. Regulator of protein S-Glutathionylation following oxidative and nitrosative stress.* J Biol Chem, 2009. **284**(1): p. 436–445.

33. Graminski, G.F., Y. Kubo, and R.N. Armstrong, *Spectroscopic and kinetic evidence for the thiolate anion of glutathione at the active site of glutathione S-transferase.* Biochemistry, 1989. **28**(8): p. 3562–3568.

34. Atkins, W.M., et al., *The catalytic mechanism of glutathione S-transferase (GST). Spectroscopic determination of the pKa of Tyr-9 in rat alpha 1-1 GST.* J Biol Chem, 1993. **268**(26): p. 19188–19191.

35. Townsend, D.M., *S-glutathionylation: indicator of cell stress and regulator of the unfolded protein response.* Mol Interv, 2007. **7**(6): p. 313–324.

36. Manevich, Y., et al., *Allelic variants of glutathione S-transferase P1-1 differentially mediate the peroxidase function of peroxiredoxin VI and alter membrane lipid peroxidation.* Free Radic Biol Med, 2013. **54**: p. 62–70.

37. Zhang, J., et al., *Racial disparities, cancer and response to oxidative stress.* Adv Cancer Res, 2019. **144**: p. 343–383.

38. Hayes, J.D., A.T. Dinkova-Kostova, and K.D. Tew, *Oxidative stress in cancer.* Cancer Cell, 2020.

39. Tonelli, C., I.I.C. Chio, and D.A. Tuveson, *Transcriptional regulation by Nrf2.* Antioxid Redox Signal, 2018. **29**(17): p. 1727–1745.

40. Ma, Q., *Role of nrf2 in oxidative stress and toxicity.* Annu Rev Pharmacol Toxicol, 2013. **53**: p. 401–426.

41. DeNicola, G.M., et al., *NRF2 regulates serine biosynthesis in non-small cell lung cancer.* Nat Genet, 2015. **47**(12): p. 1475–1481.

42. Oyake, T., et al., *Bach proteins belong to a novel family of BTB-basic leucine zipper transcription factors that interact with MafK and regulate transcription through the NF-E2 site.* Mol Cell Biol, 1996. **16**(11): p. 6083–6095.

43. Liang, Y., et al., *Transcriptional network analysis identifies BACH1 as a master regulator of breast cancer bone metastasis.* J Biol Chem, 2012. **287**(40): p. 33533–33544.

44. Weinberg, S.E. and N.S. Chandel, *Targeting mitochondria metabolism for cancer therapy.* Nat Chem Biol, 2015. **11**(1): p. 9–15.

45. Soriano, F.X., et al., *Transcriptional regulation of the AP-1 and Nrf2 target gene sulfiredoxin.* Mol Cells, 2009. **27**(3): p. 279–282.

46. Glorieux, C., et al., *Chromatin remodeling regulates catalase expression during cancer cells adaptation to chronic oxidative stress.* Free Radic Biol Med, 2016. **99**: p. 436–450.

47. Eferl, R. and E.F. Wagner, *AP-1: a double-edged sword in tumorigenesis.* Nat Rev Cancer, 2003. **3**(11): p. 859–868.

48. Farhan, M., et al., *FOXO signaling pathways as therapeutic targets in cancer.* Int J Biol Sci, 2017. **13**(7): p. 815–827.

49. Klotz, L.O., et al., *Redox regulation of FoxO transcription factors.* Redox Biol, 2015. **6**: p. 51–72.

50. Brown, A.K. and A.E. Webb, *Regulation of FOXO factors in mammalian cells.* Curr Top Dev Biol, 2018. **127**: p. 165–192.

51. Webb, A.E. and A. Brunet, *FOXO transcription factors: key regulators of cellular quality control.* Trends Biochem Sci, 2014. **39**(4): p. 159–169.

52. Weidinger, C., et al., *FOXO3 is inhibited by oncogenic PI3K/Akt signaling but can be reactivated by the NSAID sulindac sulfide.* J Clin Endocrinol Metab, 2011. **96**(9): p. E1361–1371.

53. St-Pierre, J., et al., *Suppression of reactive oxygen species and neurodegeneration by the PGC-1 transcriptional coactivators.* Cell, 2006. **127**(2): p. 397–408.

54. Valle, I., et al., *PGC-1alpha regulates the mitochondrial antioxidant defense system in vascular endothelial cells.* Cardiovasc Res, 2005. **66**(3): p. 562–573.

55. Stegen, S., et al., *HIF-1alpha promotes glutamine-mediated redox homeostasis and glycogen-dependent bioenergetics to support postimplantation bone cell survival.* Cell Metab, 2016. **23**(2): p. 265–279.

56. Semenza, G.L., *Oxygen-dependent regulation of mitochondrial respiration by hypoxia-inducible factor 1.* Biochem J, 2007. **405**(1): p. 1–9.

57. Kim, J.W., et al., *HIF-1-mediated expression of pyruvate dehydrogenase kinase: a metabolic switch required for cellular adaptation to hypoxia.* Cell Metab, 2006. **3**(3): p. 177–185.

58. Goodwin, M.L., et al., *Lactate and cancer: revisiting the Warburg effect in an era of lactate shuttling.* Front Nutr, 2014. **1**: p. 27.

59. Kovacs, D., et al., *HSF1Base: a comprehensive database of HSF1 (Heat Shock Factor 1) target genes.* Int J Mol Sci, 2019. **20**(22).

60. Vihervaara, A. and L. Sistonen, *HSF1 at a glance.* J Cell Sci, 2014. **127**(Pt 2): p. 261–266.

61. Hoffmann, A., T.H. Leung, and D. Baltimore, *Genetic analysis of NF-kappaB/Rel transcription factors defines functional specificities.* EMBO J, 2003. **22**(20): p. 5530–5539.

62. Mercurio, F. and A.M. Manning, *NF-kappaB as a primary regulator of the stress response.* Oncogene, 1999. **18**(45): p. 6163–6171.

63. Morgan, M.J. and Z.G. Liu, *Crosstalk of reactive oxygen species and NF-kappaB signaling.* Cell Res, 2011. **21**(1): p. 103–115.

64. Perkins, N.D., *The diverse and complex roles of NF-kappaB subunits in cancer.* Nat Rev Cancer, 2012. **12**(2): p. 121–132.

65. Taniguchi, K. and M. Karin, *NF-kappaB, inflammation, immunity and cancer: coming of age.* Nat Rev Immunol, 2018. **18**(5): p. 309–324.

66. Maillet, A. and S. Pervaiz, *Redox regulation of p53, redox effectors regulated by p53: a subtle balance.* Antioxid Redox Signal, 2012. **16**(11): p. 1285–1294.

67. Nguyen, T.T., et al., *Revealing a human p53 universe.* Nucleic Acids Res, 2018. **46**(16): p. 8153–8167.

68. Eriksson, S.E., et al., *p53 as a hub in cellular redox regulation and therapeutic target in cancer.* J Mol Cell Biol, 2019. **11**(4): p. 330–341.

69. Sablina, A.A., et al., *The antioxidant function of the p53 tumor suppressor.* Nat Med, 2005. **11**(12): p. 1306–1313.

70. Rahier, J., et al., *Pancreatic beta-cell mass in European subjects with type 2 diabetes.* Diabetes Obes Metab, 2008. **10**(Suppl 4): p. 32–42.

71. Martin-Gronert, M.S. and S.E. Ozanne, *Metabolic programming of insulin action and secretion.* Diabetes Obes Metab, 2012. **14**(Suppl 3): p. 29–39.

72. Marmolino, D., *Friedreich's ataxia: past, present and future.* Brain Res Rev, 2011. **67**(1–2): p. 311–330.

73. Ristow, M., et al., *Frataxin deficiency in pancreatic islets causes diabetes due to loss of beta cell mass.* J Clin Invest, 2003. **112**(4): p. 527–534.

74. Ohlsson, H., K. Karlsson, and T. Edlund, *IPF1, a homeodomain-containing transactivator of the insulin gene.* EMBO J, 1993. **12**(11): p. 4251–4259.

75. Matsuoka, T.A., et al., *Regulation of MafA expression in pancreatic beta-cells in db/db mice with diabetes.* Diabetes, 2010. **59**(7): p. 1709–1720.

76. Mahadevan, J., et al., *Ebselen treatment prevents islet apoptosis, maintains intranuclear Pdx-1 and MafA levels, and preserves beta-cell mass and function in ZDF rats.* Diabetes, 2013. **62**(10): p. 3582–3588.

77. Kawamori, D., et al., *Oxidative stress induces nucleocytoplasmic translocation of pancreatic transcription factor PDX-1 through activation of c-Jun NH(2)-terminal kinase.* Diabetes, 2003. **52**(12): p. 2896–2904.

78. El Khattabi, I. and A. Sharma, *Preventing p38 MAPK-mediated MafA degradation ameliorates beta-cell dysfunction under oxidative stress.* Mol Endocrinol, 2013. **27**(7): p. 1078–1090.

79. Rizzo, M.A. and D.W. Piston, *Regulation of beta cell glucokinase by S-nitrosylation and association with nitric oxide synthase.* J Cell Biol, 2003. **161**(2): p. 243–248.

80. Han, J., et al., *Antioxidants complement the requirement for protein chaperone function to maintain beta-cell function and glucose homeostasis.* Diabetes, 2015. **64**(8): p. 2892–2904.

81. Erejuwa, O.O., et al., *Comparison of antioxidant effects of honey, glibenclamide, metformin, and their combinations in the kidneys of streptozotocin-induced diabetic rats.* Int J Mol Sci, 2011. **12**(1): p. 829–843.

82. Kawai, H., et al., *Effects of a novel free radical scavenger, MCI-186, on ischemic brain damage in the rat distal middle cerebral artery occlusion model.* J Pharmacol Exp Ther, 1997. **281**(2): p. 921–927.

83. Satoh, K., et al., *Edarabone scavenges nitric oxide*. Redox Rep, 2002. **7**(4): p. 219–222.

84. Hardy, J. and D.J. Selkoe, *The amyloid hypothesis of Alzheimer's disease: progress and problems on the road to therapeutics*. Science, 2002. **297**(5580): p. 353–356.

85. Liu, Y., et al., *Metal ions in Alzheimer's disease: a key role or not?* Acc Chem Res, 2019. **52**(7): p. 2026–2035.

86. La Penna, G., et al., *Identifying, by first-principles simulations, Cu[amyloid-beta] species making Fenton-type reactions in Alzheimer's disease*. J Phys Chem B, 2013. **117**(51): p. 16455–16467.

87. Abramov, A.Y., et al., *Membrane cholesterol content plays a key role in the neurotoxicity of beta-amyloid: implications for Alzheimer's disease*. Aging Cell, 2011. **10**(4): p. 595–603.

88. Abramov, A.Y., L. Canevari, and M.R. Duchen, *Beta-amyloid peptides induce mitochondrial dysfunction and oxidative stress in astrocytes and death of neurons through activation of NADPH oxidase*. J Neurosci, 2004. **24**(2): p. 565–575.

89. Narayan, P., et al., *Rare individual amyloid-beta oligomers act on astrocytes to initiate neuronal damage*. Biochemistry, 2014. **53**(15): p. 2442–2453.

90. Giacovazzi, R., et al., *Copper-amyloid-beta complex may catalyze peroxynitrite production in brain: evidence from molecular modeling*. Phys Chem Chem Phys, 2014. **16**(21): p. 10169–10174.

91. Ionov, M., et al., *Mechanism of neuroprotection of melatonin against beta-amyloid neurotoxicity*. Neuroscience, 2011. **180**: p. 229–237.

92. Gwon, A.R., et al., *Oxidative lipid modification of nicastrin enhances amyloidogenic gamma-secretase activity in Alzheimer's disease*. Aging Cell, 2012. **11**(4): p. 559–568.

93. Galimberti, D. and E. Scarpini, *Old and new acetylcholinesterase inhibitors for Alzheimer's disease*. Expert Opin Investig Drugs, 2016. **25**(10): p. 1181–1187.

94. Minarini, A., et al., *Multifunctional tacrine derivatives in Alzheimer's disease*. Curr Top Med Chem, 2013. **13**(15): p. 1771–1786.

95. Sun, Q., et al., *Syntheses of coumarin-tacrine hybrids as dual-site acetylcholinesterase inhibitors and their activity against butylcholinesterase, Abeta aggregation, and beta-secretase*. Bioorg Med Chem, 2014. **22**(17): p. 4784–4791.

96. Nepovimova, E., et al., *Tacrine-trolox hybrids; a novel class of centrally active, nonhepatotoxic multi-target-directed ligands exerting anticholinesterase and antioxidant activities with low in vivo toxicity*. J Med Chem, 2015. **58**(22): p. 8985–9003.

97. Rodriguez-Franco, M.I., et al., *Novel tacrine-melatonin hybrids as dual-acting drugs for Alzheimer disease, with improved acetylcholinesterase inhibitory and antioxidant properties*. J Med Chem, 2006. **49**(2): p. 459–462.

98. Lo Celso, C., J.W. Wu, and C.P. Lin, *In vivo imaging of hematopoietic stem cells and their microenvironment*. J Biophotonics, 2009. **2**(11): p. 619–631.

99. Cheshier, S.H., et al., *In vivo proliferation and cell cycle kinetics of long-term self-renewing hematopoietic stem cells*. Proc Natl Acad Sci U S A, 1999. **96**(6): p. 3120–3125.

100. Nilsson, S.K., H.M. Johnston, and J.A. Coverdale, *Spatial localization of transplanted hemopoietic stem cells: inferences for the localization of stem cell niches*. Blood, 2001. **97**(8): p. 2293–2299.

101. Sugiyama, T., et al., *Maintenance of the hematopoietic stem cell pool by CXCL12-CXCR4 chemokine signaling in bone marrow stromal cell niches*. Immunity, 2006. **25**(6): p. 977–988.

102. Adams, G.B., et al., *Stem cell engraftment at the endosteal niche is specified by the calcium-sensing receptor*. Nature, 2006. **439**(7076): p. 599–603.

103. Cipolleschi, M.G., P. Dello Sbarba, and M. Olivotto, *The role of hypoxia in the maintenance of hematopoietic stem cells*. Blood, 1993. **82**(7): p. 2031–2037.

104. Kohler, A., et al., *Altered cellular dynamics and endosteal location of aged early hematopoietic progenitor cells revealed by time-lapse intravital imaging in long bones*. Blood, 2009. **114**(2): p. 290–298.

105. Baldini, M. and C. Sacchetti, *Effect of cystine and cysteine on human bone marrow cultured in medium deficient in amino acids*. Rev Hematol, 1953. **8**(1): p. 3–19.

106. Grek, C.L., D.M. Townsend, and K.D. Tew, *The impact of redox and thiol status on the bone marrow: pharmacological intervention strategies*. Pharmacol Ther, 2011. **129**(2): p. 172–184.

107. Zhang, J., et al., *S-Glutathionylation of estrogen receptor alpha affects dendritic cell function*. J Biol Chem, 2018. **293**(12): p. 4366–4380.

108. Raza, A., et al., *Phase 1-2a multicenter dose-escalation study of ezatiostat hydrochloride liposomes for injection (Telintra, TLK199), a novel glutathione analog prodrug in patients with myelodysplastic syndrome*. J Hematol Oncol, 2009. **2**: p. 20.

109. Brautigam, L., et al., *MGST1, a GSH transferase/peroxidase essential for development and hematopoietic stem cell differentiation*. Redox Biol, 2018. **17**: p. 171–179.

110. Tothova, Z., et al., *FoxOs are critical mediators of hematopoietic stem cell resistance to physiologic oxidative stress*. Cell, 2007. **128**(2): p. 325–339.

111. Naka, K., et al., *Regulation of reactive oxygen species and genomic stability in hematopoietic stem cells.* Antioxid Redox Signal, 2008. **10**(11): p. 1883–1894.

112. Zhou, F., Q. Shen, and F.X. Claret, *Novel roles of reactive oxygen species in the pathogenesis of acute myeloid leukemia.* J Leukoc Biol, 2013. **94**(3): p. 423–429.

113. Montalban-Bravo, G. and G. Garcia-Manero, *Myelodysplastic syndromes: 2018 update on diagnosis, risk-stratification and management.* Am J Hematol, 2018. **93**(1): p. 129–147.

114. Yang, B.X., et al., *Novel functions for mda-7/IL-24 and IL-24 delE5: regulation of differentiation of acute myeloid leukemic cells.* Mol Cancer Ther, 2011. **10**(4): p. 615–625.

115. Hole, P.S., et al., *Overproduction of NOX-derived ROS in AML promotes proliferation and is associated with defective oxidative stress signaling.* Blood, 2013. **122**(19): p. 3322–3330.

116. Xu, Q., et al., *Survival of acute myeloid leukemia cells requires PI3 kinase activation.* Blood, 2003. **102**(3): p. 972–980.

117. Aydin, E., et al., *NOX2 inhibition reduces oxidative stress and prolongs survival in murine KRAS-induced myeloproliferative disease.* Oncogene, 2019. **38**(9): p. 1534–1543.

118. Staudt, D., et al., *Targeting oncogenic signaling in mutant FLT3 acute myeloid leukemia: the path to least resistance.* Int J Mol Sci, 2018. **19**(10).

119. Hellstrand, K., *Histamine in cancer immunotherapy: a preclinical background.* Semin Oncol, 2002. **29**(3 Suppl 7): p. 35–40.

120. Marlein, C.R., et al., *NADPH oxidase-2 derived superoxide drives mitochondrial transfer from bone marrow stromal cells to leukemic blasts.* Blood, 2017. **130**(14): p. 1649–1660.

121. Romero, A.I., et al., *Post-consolidation immunotherapy with histamine dihydrochloride and interleukin-2 in AML.* Scand J Immunol, 2009. **70**(3): p. 194–205.

122. Chou, W.C., et al., *Role of NADPH oxidase in arsenic-induced reactive oxygen species formation and cytotoxicity in myeloid leukemia cells.* Proc Natl Acad Sci U S A, 2004. **101**(13): p. 4578–4583.

123. Winkler, I.G., et al., *Serine protease inhibitors serpina1 and serpina3 are down-regulated in bone marrow during hematopoietic progenitor mobilization.* J Exp Med, 2005. **201**(7): p. 1077–1088.

124. Tyagi, S.C., *Reversible inhibition of neutrophil elastase by thiol-modified alpha-1 protease inhibitor.* J Biol Chem, 1991. **266**(8): p. 5279–5285.

125. Grek, C.L., et al., *S-glutathionylated serine proteinase inhibitors as plasma biomarkers in assessing response to redox-modulating drugs.* Cancer Res, 2012. **72**(9): p. 2383–2393.

126. Grek, C.L., et al., *S-glutathionylation of buccal cell proteins as biomarkers of exposure to hydrogen peroxide.* BBA Clin, 2014. **2**: p. 31–39.

127. Zhang, L., et al., *S-glutathionylated serine proteinase inhibitors as biomarkers for radiation exposure in prostate cancer patients.* Sci Rep, 2019. **9**(1): p. 13792.

CHAPTER TWENTY

Gas Plasma

INNOVATIVE CANCER THERAPY AND CELLULAR DIFFERENTIATION IN IMMUNO-ONCOLOGY

Sander Bekeschus

CONTENTS

20.1 GAS PLASMA—AN INTRODUCTION

Gas plasma is the so-called fourth state of matter, after solid, liquid, and gas. The plasma, being a partially ionized gas, is generated when adding energy to a gas so that a (partial) dissociation of electrons from ions takes place while remaining overall electrically neutral [1]. Gas plasmas also go by other terms such as cold physical plasma, cold atmospheric (pressure) plasma (CAP), non-thermal plasma (NTP), tissue-tolerable plasma (TTP), and several others. Gas plasma is a generic term covering all types of plasmas, but in this chapter, it refers to those types and devices operated at body temperature and at atmospheric pressure to allow biomedical applications without thermal harm to cells and tissues. This is an important segregation because, for many years now, there has been a plasma device already marketed in the field of

surgery: the argon plasma coagulator (APC) [2]. The APC is essentially an electrical discharge that cauterizes the tissue (i.e., makes it necrotic) [3]. By contrast, medical gas plasma technology operated at body temperature does not induce thermal tissue damage and relies on its effects being dominated by the release of various reactive oxygen and nitrogen species (ROS/RNS) [4]. Its applications in medicine, therefore, aim at modulating physiological processes in the target cells and tissues. Those responses are a consequence of either oxidative eustress or oxidative distress [5]. This places the field of plasma medicine and the use of medical gas plasma technology devices for biomedical applications at the heart of the field of applied redox medicine, in which the use of ROS/RNS is proposed for the therapy of human and veterinary diseases [6]. From a historical perspective, it should

DOI: 10.4324/9781003204091-26

be mentioned that gas plasma devices were used experimentally by many medical doctors in the early 20th century [7], but it took another century to re-spark the utilization of gas plasma devices in medicine. Today, several gas plasma devices have been accredited as a medical product class IIa, based on clinical evidence, electrical characterization, and toxicity and safety studies [8]. Clinical success is achieved primarily in dermatology because gas plasma treatment can be repeated several times until the clinical benefit has manifested as desired [9]. The reader is referred to recent reviews on the different gas plasma device types, geometries, and principles from the physics point of view [10–12].

20.1.1 Gas Plasma Components

Gas plasma is a multi-component system. It consists of (mild) thermal radiation, electric fields, charged species, UV radiation, and ROS/RNS generation. Via the deposition of ROS/RNS, a decrease of pH is also a consequence of gas plasma treatment of non-buffered liquids [13] and 3D matrices [14]. This is due to acidic species stemming from the precursor species such as nitric oxide generating nitric and nitrous acids. The drop in pH is likely also relevant during gas plasma treatment in the clinical setting, as observed using the Franz-Kammer model system [13]. Nevertheless, absolute changes in pH in the gas plasma-treated target tissue likely depend on the plasma device used as well as the type of target tissue, as a mostly dry and lipid-rich target such as the skin has a different overall composition than, for example, an ulcerating, infected, wet tumor surface. Sparging (de-gassing) effects are also observed in gas plasma–treated liquids, which, however, might be less pronounced in vivo. The impact of the electric fields generally is considered to be modest. For the kINPen, its electric field is not strong enough to induce the transient permeabilization of the cell membrane in vitro [14]. In vivo, enhanced penetration of (lipophilic) compounds into the stratum corneum has been observed in gas plasma-treated human skin [15–17] as well as in mice [18] using a gas plasma jet accredited for medical purposes. However, the latter study's action mechanisms were not related to electropermeabilization but rather rapid changes in the junctional network within the upper skin layers. This might also explain similar findings made with other types of plasma sources [19–21]. Of note, it should be mentioned that the topic of

gas plasma–enhanced drug penetration is gaining momentum in the field of dermatology and cancer treatment [22], regardless of the mechanism being dominated by electric fields of physiological changes in the skin. UV radiation is usually low. For the accredited atmospheric pressure argon plasma jet kINPen, the UV radiation emitted during the clinically recommend 1-minute treatment time per square centimeter is only 5% of the maximum allowed as defined by the International Commission on Non-Ionizing Radiation Protection (ICNIRP) [23]. It has also been found that multiple gas plasma–treatment cycles of wounds on regular skin in mice are not sufficient to induce DNA damage [24] or tumorogenesis, as shown in a 1-year follow-up study [25]. This is important for identifying not only the mechanistic basis but also the safety of the approach. A 1-year follow up study in volunteers of plasma-treated wounds [26] reached a similar conclusion [27]. Nevertheless, the reactive plasma moieties (e.g., argon metastables) generated in the UV range contribute to the ROS generation observed due to hydrolysis and subsequent hydroxyl radical and superoxide generation [28].

20.1.2 Gas Plasma Reactive Species Generation

In biomedical gas plasma application, the current dogma is that ROS/RNS are the primary mediators of the effects observed. This follows from direct evidence and also from indirect evidence that the other physical gas plasma components do not have significant effects as outlined above. The ROS/RNS generation can be illustrated well using the kINPen plasma jet, with hundreds of publications on the most studied gas plasma device. Many of these studies and the detailed principles on the physics and chemistry of this plasma jet have been outlined previously [29]. In principle, a noble gas is driven into the jet, and a high-frequency electric field partially ionizes this gas. The feed gas used is mostly argon, but helium, neon, and even synthetic air have also been used before. Partial ionization means that some argon molecules become highly reactive while others are expelled in their native, non-reactive state. The reactive molecules subsequently leave the argon jet into the ambient air, mix with oxygen and nitrogen, and generate reactive oxygen and nitrogen species, respectively (Figure 20.1). The composition of the mixture of ROS/RNS can be modulated in multiple ways; for example, the admixture of other molecular gases,

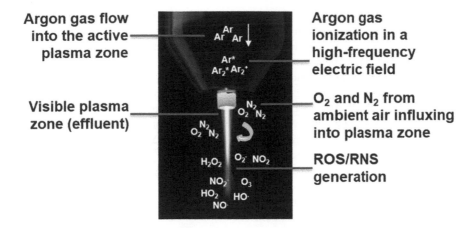

Figure 20.1 Illustration of the principle of gas plasma generation of the kINPen argon plasma device.

The type, as well as the location of ROS/RNS generation, is only schematic. The entire visible effluent contains several dozen ROS/RNS with distinct spatiotemporal profiles already.

(Figure adapted and reproduced from [37].)

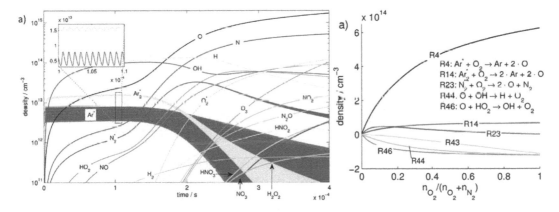

Figure 20.2 Temporal resolution of the density of selected kINPen gas plasma-derived ROS/RNS as determined using computational modeling (left). Computational modeling example of multiple reaction kinetics leading to the formation of atomic oxygen in relation to the oxygen-nitrogen-ratio in the immediate vicinity to the active plasma zone, where positive and negative values are attributed to the generation and destruction of atomic oxygen, respectively (right).

(Figures adapted and reproduced from [39].)

including water (humidity) directly into the argon, or shielding the plasma effluent from the regular ambient air [30,31]. This way, gas plasmas can be generated that are highly enriched for, among others, nitrogen metastables [32], NOx species [33], hydroxyl radicals [34,35], and atomic and singlet delta oxygen [35,36].

Several techniques allow studying the ROS/RNS mixture of gas plasmas; for instance, the plasma gas phase, optical emission spectroscopy, Fourier-transformed infrared spectroscopy, (time-absorbed) laser-induced fluorescence, and molecular beam mass spectrometry to analyze several types of ROS/RNS or reaction kinetics and partners of individual species, such as nitric oxide [38]. However, a complete picture of all reactive species being generated is not possible, even when combining several methods.

To circumvent this limitation, computational modeling is frequently used to include hundreds of reactions (Figure 20.2, left). This allows predicting not only the generation of distinct types of ROS/RNS but also their temporal resolution along the axis of the gas plasma jet effluent [39]. As multiple reaction pathways can lead to the formation of an individual species, their contribution to the generation and destruction of that individual product can also be modeled in dependence on the oxygen-nitrogen-ratio in the immediate vicinity of the plasma effluent (Figure 20.2, right). Subsequent experimental measurement of that species can then determine the degree of agreement with the modeling data. This approach helps maximize the enrichment of individual species in the gas plasma effluent, which is an academic endeavor and has practical consequences in downstream biomedical applications. As shown in the example, for atomic oxygen, maximizing its production via adjustment of the feed gas composition led to improved anti-melanoma effects in vivo [40]. For the best antimicrobial efficacy of gas plasma treatment, nitrogen in the ambient air is essential, as its absence leads to full abrogation of microbial inactivation, presumably because of defective peroxynitrite generation pathways [41]. Moreover, it has been shown that room air humidity is a negligible parameter in gas plasma mediated toxicity in vitro. In contrast, humidification of the argon gas itself (i.e., after leaving the gas bottle but before entering the plasma jet and its active plasma zone) is utterly decisive in terms of in vitro cytotoxicity [34,35,42]. Notwithstanding the potential of shaping the ROS/RNS mixture in the plasma gas phase, the multi-ROS/RNS composition remains a big challenge when attempting unambiguously attributing any biochemical or biological effect single type of ROS/RNS from the plasma gas phase. This is illustrated by the high complexity and dynamics as well as multiple reaction and destruction pathways of short-lived reactive species, ultimately affecting the species density at a given position of the plasma jet. Nevertheless, the kINPen argon plasma jet harbors the most defined plasma chemistry of all gas plasma devices. However, the complexity is even enhanced once the plasma jet connects to the biological target tissue. The latter not only disturbs the known flow velocities and hence species distributions but also acts as a reaction partner itself, scavenging some types of ROS/RNS over others, likely also depending on the types of proteins and lipids and their ratios. Moreover, technical tools are lacking in redox biology to unambiguously identify the consequences of individual species, partly also because dissimilar types of ROS/RNS sometimes have similar biochemical effects in terms of oxidation, nitration, and so forth. This is the current knowledge gap in applied plasma medicine. The plasma jet gas phase chemistry is well resolved and many effects in cells and tissues have been document over the past decade. The interphase, however, where the gas plasma hits the target and dozens of ROS/RNS either decompose or even give rise to other sets of secondary species, is largely a black box that only recently has begun to be revealed using oxidative post-translational modification (oxPTM) mapping to disentangle putative reaction pathways [18].

20.1.3 ROS/RNS in Gas Plasma–Treated Liquids

As another degree of freedom, liquids come into play as another reaction partner because they are ubiquitously present in vitro as well as in target tissues surrounding the cells. Several reviews have covered the reaction pathways and types of ROS/RNS that can be detected in plasma-treated liquids void of biomolecules [43–46]. Essentially, all molecules dissolved in those liquids potentially affect how the plasma treatment eventually affects the cells. This mainly includes dissolved oxygen and nitrogen, amino acids, salts and minerals, and proteins. All components potentially scavenge the ROS/RNS according to reported rate constants as well as putatively contribute to the formation of secondary ROS/RNS. In addition to this, all types of ROS/RNS have their individual reported lifetimes and diffusion distances in liquids. Using mass spectrometry and relatively simple biomolecules such as cysteine and glutathione, much information has been obtained recently to elucidate the chemistry in gas plasma–treated liquids. Mapping oxPTMs has allowed, for example, control for nitrosylation versus oxidation [47], fingerprint cysteine and glutathione in relation to the gas phase ROS/RNS mixture and chemical reaction pathways in liquids [48,49], and identifying the generation of reactive sulfur species [50]. It has also been found that gas plasma-treated glutathione exerts biological effects in cells in vitro [51]. For in vitro experiments, antioxidants and enzymes can be added to confirm the decisive action of ROS/RNS on the effects observed. Especially long-lived products such as hydrogen peroxide are frequently found to have dominating effects [52,53]. While it is tempting to attribute most findings to these molecules, multiple evidence has shown that hydrogen peroxide alone either cannot fully explain or replicate the findings made with gas plasma treatment,

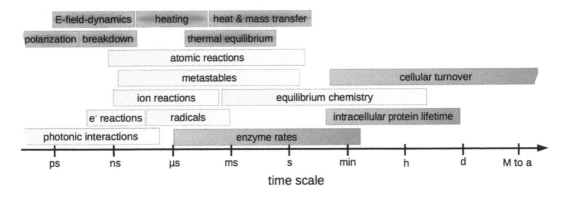

Figure 20.3 Time scales in gas plasma generation and bio-applications.

(Reproduced from [29].)

as shown, for instance, in neutrophil extracellular trap formation [54], tumor spheroid elimination [55], tumor cell inactivation [56], and physiological changes in cancer cells [57]. This is likely much more pronounced in the situation of a gas plasma–treated target tissue, as here the biomolecule-to-liquid ratio is much higher than in in vitro setups, effectively reducing the diffusion distances of short-lived ROS/RNS to proteins and lipids and therefore discouraging ROS/RNS deterioration and formation of long-lived reaction products. For example, this is well illustrated by the fact that almost all hydrogen peroxide found in liquids is a secondary product of other, primary plasma gas phase–derived ROS such as hydroxyl radicals and superoxide. The same is true for another long-lived oxidant, hypochlorous acid, a secondary product formed by plasma gas phase–derived atomic and singlet delta oxygen in chloride-containing liquids [58,59]. In the absence of sufficient amounts of liquids, as in relatively dry tumor tissue in vivo, these species might directly attack the cancer cells.

20.1.4 Gas Plasma—Summary

Gas plasmas are partially ionized gases that expel a multitude of ROS/RNS into the ambient air. The characterization of these ROS/RNS in the plasma gas phase is well established and can be modified to enrich some types of ROS/RNS over others. Depending on the type of application, these species subsequently interact with liquids and their abiotic or biotic content to either deteriorate, form secondary species, or form modifications such as oxPTMs on proteins and lipids. This process only

occurs as long as the gas plasma device is switched on and treating the target. The biochemical and biological consequences subsequently follow the timelines known in redox biology (Figure 20.3).

20.2 ANTICANCER ACTION OF GAS PLASMA THERAPY

Gas plasma technology emits several types of ROS/RNS simultaneously. The concepts of oxidative eustress/distress and hormesis predict that those species might induce effects that differ, depending on the concentration applied [5,60]. This concept is well illustrated in a study of gas plasma–treated human skin ex vivo, in which short treatment times using the kINPen induce a slight increase in pro-liferating keratinocytes, while extended exposure abolishes this effect [61]. Like non-malignant cells, most malignant cells succumb to gas plasma–derived ROS/RNS-induced cell death [4]. This includes different cancer cell types culturable in vitro, for instance, melanoma [62–64], glioblastoma [65–67], leukemia [68–70], colorectal [71–73], liver [74], pancreatic [75–77], prostate [78–80], neuro-blastoma [81–83], ovarian [84–86], gastric [87–89], kidney [90–92], breast [93–95], and squamous cell carcinoma subtypes [96–98]. In vitro, where ROS/RNS and more long-lived oxidants dominate, can-cer cell death is usually regulated. Accordingly, sev-eral types of regulated cell death [99] have been observed. This includes intrinsic apoptosis [100–102], autophagy [103–105], ferroptosis [106–108], pyroptosis [109], and extrinsic apoptosis pathways [110–112]. Necrosis rather appears if the feed gas flux dries out the cells or if gas plasma treatment times or energy deposition extends what would be realistically achievable in the clinical situation [113,114]. So far, cancer cell death has been observed in several types of model systems, such as 2D cultures [115–117], 3D cultures [118–120], in ovo tumor cultures [121–123], in human tumor tissue gas plasma-treated ex vivo [124–126], in vivo (predominantly in mice) [127–129], and in a limited number of head and neck cancer patients in the palliative stage in Greifswald [130–132]. Overall, apoptosis is the most frequently observed type of gas plasma–induced type of regulated cell death in tumor tissue, although it should be mentioned that investigating cell death modalities in vivo is not as clean-cut as in vitro, as several types of regulated cell death can occur simultaneously [133–135]. The reason for the plethora of observations, which might seem to contradict in some circumstances, is the usage of different cell culture model systems, time points of investigations, and types of gas plasma sources that have different physicochemical properties and ROS mixtures that they generate. It is undisputed, however, that gas plasma technology is capable of inducing tumor cell death. The selectivity of this approach has been suggested in many studies [136–138]. However, it should be mentioned that selectivity is mainly an issue for systemic therapies to reduce side effects in non-target tissues. In local(ized) therapies, selectivity is less needed if the goal is to either remove or ablate or to change the function and immunogenic properties of tumor cells. Examples include cryo- and thermal ablation [139], photodynamic therapy [140], electrochemotherapy [141], radiotherapy [142], surgery, and hyperthermia [143]. Gas plasma cancer treatment aims not at ablating large bulks of tumor mass but rather to supplement existing therapeutic regimens by another mode of action. While ROS/RNS are a suitable weapon to induce tumor cell death, their penetration depth of long-lived candidates exogenously applied to tissues is several dozen micrometers [144,145]. Hence, many treatment cycles would be needed to achieve significant tumor reduction. While this principle is possible for topical tumors, such as head and neck squamous cell carcinoma, with a significant survival benefit for several patients as recently shown [130], non-topical tumors may not benefit from this option. This includes all tumors that do not multiply or metastasize to or grow through the skin, as observed regularly with skin and breast cancer entities, for example. Vice versa, gas plasma treatment in a localized manner of cancer tissues certainly does affect the cancer cells and hence will affect their cellular phenotype and/or the phenotype of adjacent cell types in the tumor microenvironment.

20.3 GAS PLASMA TREATMENT AND DIFFERENTIATION OF CELLS RELEVANT IN THE TUMOR MICROENVIRONMENT

The tumor microenvironment (TME) is composed of many cell types [146]. Besides cancer cells, this includes stromal cells such as fibroblasts and, in some types of tumors, also keratinocytes. Endothelial cells are involved as well, being a target for anti-angiogenic therapies [147]. Immune cells are major contributors, too. These include, for instance, neutrophils, macrophages, dendritic cells, NK cells, and CD4+ and CD8+ T cells [148]. Three crucial aspects need to be noted when discussing the TME. First, the TME shows a certain degree of plasticity. It is known that many anticancer agents change the composition and phenotype of several types of cells within the TME [149]. Hence, it can be assumed that tumor cell death-induced gas plasma therapy does so as well. Second, oxygen and hypoxia, and hence also differential regulation of ROS/RNS production, are critical factors in the TME and outcome of therapies [150–152]. In superficial and repeatedly exposed tumors, the gas plasma–derived ROS/RNS might interfere with these processes. Third, tumor cells are known to promote the differentiation of several cell types in the TME by hijacking their growth-supporting functions in favor of cancer growth. This is known in the case of, for example, tumor-associated (M2) macrophages (TAM) [153], cancer-associated fibroblasts (CAFs) [154], and tumor-associated neutrophils (TANs) [155]. Other immune cell types contributing to the phenome of immunosenescence in the TME are regulatory T-cells (T_{regs}) [156] and myeloid-derived suppressor cells (MDSCs) [157]. It is one of the main aims of oncology to therapeutically target those detrimental cellular phenotypes either by removal or reprogramming, which involves differentiation and de-differentiation processes. Many reports and reviews have appreciated the role of ROS/RNS as critical primary or secondary agents to achieve this aim [158–163]. Although gas plasma is a relatively new technology, some reports have pointed at its modulatory role in the differentiation of cell types relevant to the TME.

Similar to other cell types, cancer cells can succumb to irreversible cell cycle arrest, a signal transduction program called cellular senescence. Hallmarks of senescence include the lack of proliferation and disability of mitogens for stimulation, although cells continue to be metabolically active [164]. This is caused either by replicative senescence (shortening of telomers) or exogenous acute or chronic stress signals, but many types of tumor cells often bypass these processes in the course of oncogenesis [165]. Nevertheless, cellular senescence can be re-initiated in cancer cells and is often preceded by autophagy, known as autophagy-senescence-transition (AST) [166], which involves p53 activation [167].

Several studies have linked gas plasma treatment of cancer cells to autophagy and senescence. In human Mel Im and Mel Juso melanoma cells, cell cycle arrest and p21 upregulation was demonstrated following gas plasma exposure [168]. The same group later demonstrated the involvement of calcium signaling in cellular senescence induction [169]. The group of Adachi has demonstrated in human A549 cells that the gas plasma–derived ROS/RNS were capable of inducing cellular senescence via activation of the ATM/p53 pathway and in a zinc-mediated fashion [170]. Regarding the induction of autophagy, the reports are controversial. While some studies failed to document autophagy [171], others did show full or partial effects of autophagy or autophagy inhibitors in cancer cells following gas plasma treatment [104,105,172,173]. For instance, in human melanoma and osteosarcoma cells, gas plasma–derived ROS was equally potent to TRAIL, an important autophagy inducer [174], in terms of promoting autophagy [175]. In murine melanomas targeted with gas plasma, a decrease of tumor growth was accompanied by an upregulation of gene expression linked to autophagic processes [128].

Another well-known feature of cancer cells is their ability to undergo epithelial-mesenchymal-transition (EMT) [176]. This process promotes cancer cells' motility and allows them to leave the primary tumor site to enter the vasculature and disseminate in the body as a metastatic lesion. Several markers are critically involved in EMT and plasticity of tumor cells: ZEB1 [177], LEF1 [178], and many others [179]. Because the involvement of inflammation and ROS/RNS has been proposed in EMT [180], it was natural to investigate the effect of gas plasma treatment in the onset of EMT.

Treatment of murine 6606 pancreatic ductal adenocarcinoma cells with gas plasma-derived ROS in vivo did not increase peritoneal metastasis [76]. Similar findings were made for CT26 colorectal cancer cells, and the treatment decreased cellular motility while leaving several EMT markers such as ZEB1, LEF1, αSMA, vimentin, and β-catenin unchanged [55]. Using four human pancreatic cancer cell lines, these findings were confirmed for a number of adhesion-related markers such as integrins and cadherins [77]. Moreover, this study addressed a more physical parameter of gas plasma treatment of tumor mass: the question of whether the electric discharge together with a gas flux of liters per minute leads to expulsion of viable tumor cells away from the primary tumor to the surrounding area. Besides, and using advanced high-content imaging techniques, the gas plasma-treated tumor spheroids were embedded in Matrigel and allowed to grow. Subsequently, the distances between tumor cells outside the tumor spheroid region to the core tumor spheroid were calculated using algorithm-driven quantitative image analysis. An increase of distance that would argue for increased motility and possibly EMT was not observed. A similar finding was made in gas plasma-treated human melanoma cells, also in terms of unchanged expression of EMT markers in 3D tumor spheroids following gas plasma exposure [120]. To extend these findings to the crosstalk with TME-relevant cells, human pancreatic cancer cells were co-cultured together with pancreatic stellate cells in 3D tumor spheroids. After gas plasma treatment, the spheroids were embedded in Matrigel, and the outgrowth characteristics of fluorescently labeled pancreatic cancer and stellate cells, respectively, were investigated. A stellate cell's growth-supporting role to the tumor cells was not observed, and the number of stellate cells was stable while those the tumor cells declined, suggesting plasma-derived ROS/RNS to be potentially useful in targeting pancreatic cancer.

Another rather recent feature of differentiation-like changes of tumor cells is their immunogenicity. This is a timely matter because cancer immunotherapy is the most promising new anticancer treatment modality in recent years [181–183], as evident by the Nobel Prize in Medicine or Physiology awarded in 2018 to immune checkpoint antibodies. This topic has several dimensions (Figure 20.4), which are only briefly discussed here. First, tumor cells can succumb to

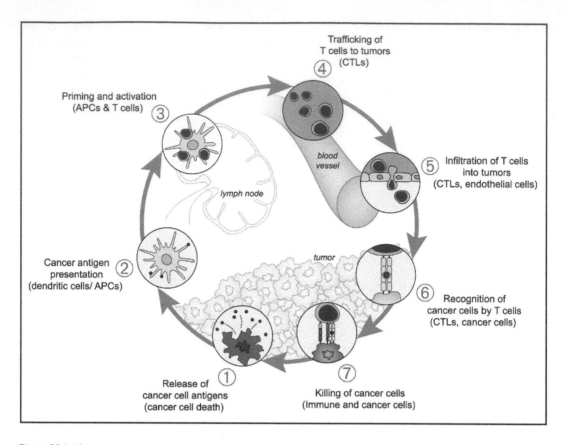

Figure 20.4 The cancer-immunity cycle.

After a cytotoxic event (1), for example, gas plasma treatment, cancer cell antigens are being released together with inflammation-promoting molecules in the context of immunogenic cancer cell death (ICD). Next, professional antigen-presenting cells (APCs, such as dendritic cells) invade the TME and take up to tumor antigen (2). Subsequently, APCs migrate to the draining lymph node where they (cross-)present tumor antigens to CD4+ and CD8+ T cells, respectively (3). If T cells recognize their cognate antigen, they activate and proliferate before entering the circulation (4). At the tumor site, the tumor-specific T cells extravasate (5) and recognize tumor cells (6) prior to eliciting their cytotoxic action upon them (7). This fuels the cancer-immunity cycle by freeing more antigen and spurring inflammation, auto-amplifying the initial toxic stimulus. It is hypothesized that the anticancer effect of gas plasma treatment is promoted in this way, as increased numbers of intratumoral T cells in gas plasma–treated tumors suggest [40,55,202–204].

(Figure reproduced from [208].)

treatment-induced immunogenic cell death (ICD) [184]. Consequently, myeloid cells such as dendritic cells are activated, differentiate, and thus become more efficient in (cross-)presenting antigens to T cells. The latter subsequently differentiate and proliferate to attack cancer cells at remote sites of metastatic tumor lesions [185]. ICD indirectly affects cellular differentiation, as it comes with general alterations of the inflammatory micro-milieu via the release of damage-associated molecular patterns (DAMPs) [186–188]. Second, these DAMPs contribute to changes in the differentiation of TME-resident myeloid cells that critically promote (e.g., M2 macrophages) or abolish (e.g., M1 macrophages) tumor growth [189–191]. Third, this comes with alterations in hypoxia and hence tumor cell phenotype, which can consequently be rendered more immunogenic [192]. This is evident by, for example, the unfolded protein and upregulation of the machinery of antigen processing [193] and stress of the endoplasmic reticulum [194], ultimately allowing more tumor-associated antigens to be presented. This was underlined by findings using ROS-promoting selenium-containing drugs that came with an upregulation of major histocompatibility class (MHC) class I molecules [195], a key target of cytotoxic CD8$^+$ T cells. These findings illustrate that ICD is not a differentiation event per se, but its inflammatory consequences affect the TME and its hypoxia and cellular metabolism, which are known to be intimately linked [196].

Several studies have provided evidence of ICD-induction in tumor cells following exposure to gas plasma–derived ROS. For the kINPen plasma jet, the treatment was shown to promote the externalization of calreticulin [35] and an increase of MHC class I surface expression [197] as well as of the DAMP heat-shock protein 70 [40] in B16F10 melanoma cells. kINPen plasma treatment of melanoma cells also promoted a release of the pro-chemotactic DAMP ATP [198,199]. Vaccination with gas plasma–killed melanoma cells leads to enhanced immunoprotection from tumor growth in syngeneic C57BL/6 mice [40]. Similar findings were also made on another type of gas plasma source (DBD) that is not operated with a noble gas. Here, ATP release and subsequent ER stress responses were noted [200]. Mechanistically, CRT translocation and ATP release were dependent on exogenous ROS/RNS deposition. At the same time, the former but not the latter was also related to auto-amplification of ROS in gas plasma–treated A549 lung cancer cells, as experiments using the NOX-inhibitor DPI suggested [201]. A series of additional experiments in this study also provided evidence that ROS were the major drivers for the effects observed, whereas RNS, UV radiation, and electric fields were of minor importance. These findings were reiterated in melanoma cells in vitro, and vaccination with gas plasma–killed melanoma cells provided immunoprotection from tumor growth in vivo [57]. Studies using another type of gas plasma source but investigating the same tumor model confirmed the increased immunogenicity of gas plasma–treated B16 melanoma cells in vivo [202,203]. Using colorectal cancer cells, ICD and immunoprotection of gas plasma–treated CT26 colorectal cancer cells were shown for CRT translocation and ATP release in vitro. Treatment of the tumors in vivo led to an increase of CRT and HMGB1 in a syngeneic animal model [204]. In vitro, exposure with gas plasma–treated liquids has provided evidence of increased CRT translocation and ATP release using a DBD plasma source [171]. For gas plasma–treated saline, similar findings and nuclear translocation of HMGB1 were found in colorectal cancer cells in vitro [55] and increased CRT staining in pancreatic cancer tissues in vivo [205]. These findings were supported by another recent study using four malignant and three non-malignant cell types, suggesting a tumor-selective effect [206] despite a potential contribution of relatively long-lived ROS/RNS in those liquids [207].

20.3.2 Immune Cells

As outlined above, myeloid cells are decisive in modulating the inflammatory profile of the TME and by promoting either tumor growth or anti-tumor immune responses [209]. Neutrophils are gaining attention as critical effectors in the TME [210]. Gas plasma treatment of neutrophils elicits extracellular trap (NET) formation and IL-8 release [54] as well as extracellular vesicles (EVs), enriched DAMPs (e.g., Annexin A1, Cathepsins, CRT, and HSPs), and ROS and redox-related proteins and enzymes (e.g., peroxiredoxins, superoxide dismutase 2, and myeloperoxidase) [211]. Increased presence of neutrophils and NETs was found in gas plasma–treated liquids targeting pancreatic cancer in vivo [205], but their differentiated phenotype (e.g., TANs) remains to be established.

Monocytes are a critical population evading from the vasculature into the TME, where they differentiate into macrophages and MDSCs, for example [212]. While there are no published studies on MDSCs yet in the field of gas plasma cancer treatment, we have performed several studies using monocytes and the kINPen plasma jet. Co-culture of splenocytes with gas plasma–treated tumor cells showed a rather modest modulation in differentiation markers in splenic monocytes [198]. Gas plasma treatment of monocytes promotes a differentiation-like phenotype of both M1 and M2 cells when investigating murine bone marrow–derived cells [213] and primary PBMC-derived as well as THP-1 monocytes [214]. The latter was accompanied by a change in intracellular ROS generation, morphological changes, difference in the differentiation marker gene expression and cell surface profile, cytokine release, and tumor toxic activity. Proteomic studies revealed an increase in intracellular thiol content, IL-8 release, and expression of IL-8 and the transcription factor EPAS1 [215], along with other transcriptional and protein translation responses [59,69]. When co-cultured with gas plasma–treated tumor cells or their supernatants, a differentiation-like surface marker response was observed when compared to non-treated controls [86]. Similar findings were made for HL-60 cells, and both THP-1 monocytes and HL-60 cells show enhanced tumor-toxic action when cultured with gas plasma–treated compared to untreated tumor cell supernatants [80]. A study using another type of gas plasma source has found that exposure of monocytes generates a mixed M1-M2 phenotype and promotes intracellular ROS formation [216], underlining our results. The same study found gas plasma treatment to have a subtle but significant stimulating effect in human macrophages to show an M1 phenotype. Additive killing effects of macrophages were found to depend on enhanced TNFα production when co-cultured with gas plasma–treated cancer cells [217]. Interestingly, the potency of a mitogenic stimulus (such as PMA) can be enhanced when the cells are exposed to gas plasma [218]. The authors found an augmented NF-κB activation, cytokine release, and resistance to toxic concentrations of the anticancer drugs doxorubicin and 5-fluorouracil, when using cells pre-treated with gas plasma prior to differentiation. Interestingly, exposure of RAW.264 macrophages to gas plasma–treated liquids rich in RNS promoted the expression of M1 over M2 genes, surface expression markers, and cytokines. Intratumoral application of this liquid in a syngeneic model of melanoma in vivo led to a significant decline in tumor mass together with an increase of intratumoral iNOS [219], a marker known to be increased in M1 macrophages [220]. Along similar lines but using a different type of gas plasma source, direct exposure to its ROS/RNS not only enhanced monocyte-to-macrophage differentiation as evident by transcriptomic expression and protein data but also promoted the tumor-toxic macrophage phenotype both in vitro and in vivo [221]. Finally, investigating the macrophages in the TME exposed to gas plasma–treated liquids [55,213] or direct gas plasma exposure [40] revealed a pronounced increase in macrophage influx. A detailed phenotypic analysis, however, remains to be established, but it should be noted that their increase was accommodated by a decrease in tumor mass in all cases.

Dendritic cells are the most critical cells at the interphase between tumor cells and adaptive immune responses [222]. For this reason, this cell type is envisioned for cell therapy vaccination studies in oncoimmunology [223]. Studies investigating dendritic cells in gas plasma cancer treatment are scarce, however. An increased number of intratumoral DCs was previously noted in directly gas plasma–treated subcutaneous syngeneic tumor models [40,204]. In vitro, co-culture of gas plasma–treated cancer cells with human monocyte-derived DCs increased the expression of differentiation marker CD86 as well as the release of the antitumor cytokines IFNγ and TNFα [206]. Several studies are underway in the field to further explore and harness the power of dendritic cells in promoting the immunogenicity of gas plasma–killed tumor cells.

REFERENCES

1. Weltmann, K.D.; Polak, M.; Masur, K.; von Woedtke, T.; Winter, J.; Reuter, S. Plasma processes and plasma sources in medicine. *Contrib. Plasma Phys.* **2012**, 52, 644–654, 10.1002/ctpp. 201210061.

2. Cipolletta, L.; Bianco, M.A.; Rotondano, G.; Piscopo, R.; Prisco, A.; Garofano, M.L. Prospective comparison of argon plasma coagulator and heater probe in the endoscopic treatment of major peptic ulcer bleeding. *Gastrointest. Endosc.* **1998**, 48, 191–195, 10.1016/s0016-5107(98)70163-4.

3. Bekeschus, S.; Brüggemeier, J.; Hackbarth, C.; von Woedtke, T.; Partecke, L.-I.; van der Linde, J. Platelets are key in cold physical plasma-facilitated blood coagulation in mice. *Clin. Plas. Med.* **2017**, 7–8, 58–65, 10.1016/j.cpme.2017.10.001.

4. Privat-Maldonado, A.; Schmidt, A.; Lin, A.; Weltmann, K.D.; Wende, K.; Bogaerts, A.; Bekeschus, S. Ros from physical plasmas: Redox chemistry for biomedical therapy. *Oxid. Med. Cell. Longev.* **2019**, 2019, 9062098, 10.1155/2019/9062098.

5. Sies, H.; Berndt, C.; Jones, D.P. Oxidative stress. *Annu. Rev. Biochem.* **2017**, 86, 715–748, 10.1146/annurev-biochem-061516-045037.

6. von Woedtke, T.; Schmidt, A.; Bekeschus, S.; Wende, K.; Weltmann, K.D. Plasma medicine: A field of applied redox biology. *In Vivo* **2019**, 33, 1011–1026, 10.21873/invivo.11570.

7. Napp, J.; Daeschlein, G.; Napp, M.; von Podewils, S.; Gumbel, D.; Spitzmueller, R.; Fornaciari, P.; Hinz, P.; Junger, M. On the history of plasma treatment and comparison of microbiostatic efficacy of a historical high-frequency plasma device with two modern devices. *GMS Hyg. Infect. Control.* **2015**, 10, Doc08, 10.3205/dgkh000251.

8. Bekeschus, S.; Schmidt, A.; Weltmann, K.-D.; von Woedtke, T. The plasma jet kinpen: A powerful

tool for wound healing. *Clin. Plas. Med.* **2016**, 4, 19–28, 10.1016/j.cpme.2016.01.001.

9. Bernhardt, T.; Semmler, M.L.; Schafer, M.; Bekeschus, S.; Emmert, S.; Boeckmann, L. Plasma medicine: Applications of cold atmospheric pressure plasma in dermatology. *Oxid. Med. Cell. Longev.* **2019**, 2019, 3873928, 10.1155/2019/3873928.

10. Gentile, R.D.; McCoy, J.D. Pulsed and fractionated techniques for helium plasma energy skin resurfacing. *Facial Plast. Surg. Clin. North Am.* **2020**, 28, 75–85, 10.1016/j.fsc.2019.09.007.

11. Brandenburg, R. Dielectric barrier discharges: Progress on plasma sources and on the understanding of regimes and single filaments. *Plasma Sources Sci. T.* **2017**, 26, 053001, 10.1088/1361-6595/aa6426.

12. Lu, X.; Naidis, G.V.; Laroussi, M.; Reuter, S.; Graves, D.B.; Ostrikov, K. Reactive species in non-equilibrium atmospheric-pressure plasmas: Generation, transport, and biological effects. *Phys. Rep.-Rev. Sect. Phys. Lett.* **2016**, 630, 1–84, 10.1016/j.physrep.2016.03.003.

13. Heuer, K.; Hoffmanns, M.A.; Demir, E.; Baldus, S.; Volkmar, C.M.; Rohle, M.; Fuchs, P.C.; Awakowicz, P.; Suschek, C.V.; Oplander, C. The topical use of non-thermal dielectric barrier discharge (dbd): Nitric oxide related effects on human skin. *Nitric Oxide* **2015**, 44, 52–60, 10.1016/j.niox.2014.11.015.

14. Wolff, C.M.; Kolb, J.F.; Weltmann, K.D.; von Woedtke, T.; Bekeschus, S. Combination treatment with cold physical plasma and pulsed electric fields augments ros production and cytotoxicity in lymphoma. *Cancers (Basel)* **2020**, 12, 845, 10.3390/cancers12040845.

15. Lademann, J.; Patzelt, A.; Richter, H.; Lademann, O.; Baier, G.; Breucker, L.; Landfester, K. Nanocapsules for drug delivery through the skin barrier by tissue-tolerable plasma. *Laser Phys. Lett.* **2013**, 10, 083001, 10.1088/1612-2011/10/8/083001.

16. Lademann, O.; Richter, H.; Kramer, A.; Patzelt, A.; Meinke, M.C.; Graf, C.; Gao, Q.; Korotianskiy, E.; Rühl, E.; Weltmann, K.D.; Lademann, J.; Koch, S. Stimulation of the penetration of particles into the skin by plasma tissue interaction. *Laser Phys. Lett.* **2011**, 8, 758–764, 10.1002/lapl.201110055.

17. Lademann, O.; Richter, H.; Meinke, M.C.; Patzelt, A.; Kramer, A.; Hinz, P.; Weltmann, K.D.; Hartmann, B.; Koch, S. Drug delivery through the skin barrier enhanced by treatment with tissue-tolerable plasma. *Exp. Dermatol.* **2011**, 20, 488–490, 10.1111/j.1600-0625.2010.01245.x.

18. Schmidt, A.; Liebelt, G.; Striesow, J.; Freund, E.; von Woedtke, T.; Wende, K.; Bekeschus, S. The molecular and physiological consequences of cold plasma treatment in murine skin and its barrier function. *Free Radic. Biol. Med.* **2020**, 161, 32–49, 10.1016/j.freeradbiomed.2020.09.026.

19. Gelker, M.; Müller-Goymann, C.C.; Viöl, W. Permeabilization of human stratum corneum and full-thickness skin samples by a direct dielectric barrier discharge. *Clin. Plas. Med.* **2018**, 9, 34–40, 10.1016/j.cpme.2018.02.001.

20. Gelker, M.; Müller-Goymann, Christel, C.; Viöl, W. Plasma permeabilization of human excised full-thickness skin by μs- and ns-pulsed dbd. *Skin Pharmacol. Physiol.* **2020**, 33, 69–76, 10.1159/000505195.

21. Shimizu, K.; Hayashida, K.; Blajan, M. Novel method to improve transdermal drug delivery by atmospheric microplasma irradiation. *Biointerphases* **2015**, 10, 10.1116/1.4919708.

22. Gouarderes, S.; Mingotaud, A.F.; Vicendo, P.; Gibot, L. Vascular and extracellular matrix remodeling by physical approaches to improve drug delivery at the tumor site. *Expert Opin Drug Deliv* **2020**, 1–24, 10.1080/17425247.2020.1814735.

23. Bussiahn, R.; Lembke, N.; Gesche, R.; von Woedtke, T.; Weltmann, K.-D. Plasma sources for biomedical applications. *Hyg. Med.* **2013**, 38, 212–216,

24. Pasqual-Melo, G.; Nascimento, T.; Sanches, L.J.; Blegniski, F.P.; Bianchi, J.K.; Sagwal, S.K.; Berner, J.; Schmidt, A.; Emmert, S.; Weltmann, K.D.; von Woedtke, T.; Gandhirajan, R.K.; Cecchini, A.L.; Bekeschus, S. Plasma treatment limits cutaneous squamous cell carcinoma development in vitro and in vivo. *Cancers (Basel)* **2020**, 12, 1993, 10.3390/cancers12071993.

25. Schmidt, A.; Woedtke, T.V.; Stenzel, J.; Lindner, T.; Polei, S.; Vollmar, B.; Bekeschus, S. One year follow-up risk assessment in skh-1 mice and wounds treated with an argon plasma jet. *Int. J. Mol. Sci.* **2017**, 18, 10.3390/ijms18040868.

26. Metelmann, H.-R.; von Woedtke, T.; Bussiahn, R.; Weltmann, K.-D.; Rieck, M.; Khalili, R.; Podmelle, F.; Waite, P.D. Experimental recovery of co2-laser skin lesions by plasma stimulation. *Am. J. Cosmetic Surg.* **2012**, 29, 52–56, 10.5992/ajcs-d-11-00042.1.

27. Metelmann, H.-R.; Vu, T.T.; Do, H.T.; Le, T.N.B.; Hoang, T.H.A.; Phi, T.T.T.; Luong, T.M.L.; Doan,

V.T.; Nguyen, T.T.H.; Nguyen, T.H.M., *et al.* Scar formation of laser skin lesions after cold atmospheric pressure plasma (cap) treatment: A clinical long term observation. *Clin. Plas. Med.* **2013**, 1, 30–35, 10.1016/j.cpme.2012.12.001.

28. Jablonowski, H.; Bussiahn, R.; Hammer, M.U.; Weltmann, K.D.; von Woedtke, T.; Reuter, S. Impact of plasma jet vacuum ultraviolet radiation on reactive oxygen species generation in bio-relevant liquids. *Phys. Plasmas* **2015**, 22, 122008, 10.1063/1.4934989.

29. Reuter, S.; von Woedtke, T.; Weltmann, K.D. The kinpen—a review on physics and chemistry of the atmospheric pressure plasma jet and its applications. *J. Phys. D: Appl. Phys.* **2018**, 51, 10.1088/1361-6463/aab3ad.

30. Reuter, S.; Tresp, H.; Wende, K.; Hammer, M.U.; Winter, J.; Masur, K.; Schmidt-Bleker, A.; Weltmann, K.D. From rons to ros: Tailoring plasma jet treatment of skin cells. *IEEE Trans. Plasma Sci.* **2012**, 40, 2986–2993, 10.1109/Tps.2012.2207130.

31. Reuter, S.; Winter, J.; Schmidt-Bleker, A.; Tresp, H.; Hammer, M.U.; Weltmann, K.D. Controlling the ambient air affected reactive species composition in the effluent of an argon plasma jet. *IEEE Trans. Plasma Sci.* **2012**, 40, 2788–2794, 10.1109/Tps.2012.2204280.

32. Iseni, S.; Bruggeman, P.J.; Weltmann, K.-D.; Reuter, S. Nitrogen metastable (n2(a3 σu+)) in a cold argon atmospheric pressure plasma jet: Shielding and gas composition. *Appl. Phys. Lett.* **2016**, 108, 10.1063/1.4948535.

33. Schmidt-Bleker, A.; Bansemer, R.; Reuter, S.; Weltmann, K.-D. How to produce an nox- instead of ox-based chemistry with a cold atmospheric plasma jet. *Plasma Process. Polym.* **2016**, 13, 1120–1127, 10.1002/ppap.201600062.

34. Winter, J.; Wende, K.; Masur, K.; Iseni, S.; Dunnbier, M.; Hammer, M.U.; Tresp, H.; Weltmann, K.D.; Reuter, S. Feed gas humidity: A vital parameter affecting a cold atmospheric-pressure plasma jet and plasma-treated human skin cells. *J. Phys. D: Appl. Phys.* **2013**, 46, 295401, 10.1088/0022-3727/46/29/295401.

35. Bekeschus, S.; Schmidt, A.; Niessner, F.; Gerling, T.; Weltmann, K.D.; Wende, K. Basic research in plasma medicine—a throughput approach from liquids to cells. *J. Vis. Exp.* **2017**, e56331, 10.3791/56331.

36. Wende, K.; Williams, P.; Dalluge, J.; Gaens, W.V.; Aboubakr, H.; Bischof, J.; von Woedtke,

T.; Goyal, S.M.; Weltmann, K.D.; Bogaerts, A.; Masur, K.; Bruggeman, P.J. Identification of the biologically active liquid chemistry induced by a nonthermal atmospheric pressure plasma jet. *Biointerphases* **2015**, 10, 029518, 10.1116/1.4919710.

37. Gumbel, D.; Bekeschus, S.; Gelbrich, N.; Napp, M.; Ekkernkamp, A.; Kramer, A.; Stope, M.B. Cold atmospheric plasma in the treatment of osteosarcoma. *Int. J. Mol. Sci.* **2017**, 18, 2004, 10.3390/ijms18092004.

38. Iseni, S.; Zhang, S.; van Gessel, A.F.H.; Hofmann, S.; van Ham, B.T.J.; Reuter, S.; Weltmann, K.D.; Bruggeman, P.J. Nitric oxide density distributions in the effluent of an rf argon appj: Effect of gas flow rate and substrate. *New J. Phys.* **2014**, 16, 123011, 10.1088/1367-2630/16/12/123011.

39. Schmidt-Bleker, A.; Winter, J.; Bosel, A.; Reuter, S.; Weltmann, K.D. On the plasma chemistry of a cold atmospheric argon plasma jet with shielding gas device. *Plasma Sources Sci. T.* **2016**, 25, 015005, 10.1088/0963-0252/25/1/015005.

40. Bekeschus, S.; Clemen, R.; Niessner, F.; Sagwal, S.K.; Freund, E.; Schmidt, A. Medical gas plasma jet technology targets murine melanoma in an immunogenic fashion. *Adv. Sci. (Weinh)* **2020**, 7, 1903438, 10.1002/advs.201903438.

41. Jablonowski, H.; Hansch, M.A.; Dunnbier, M.; Wende, K.; Hammer, M.U.; Weltmann, K.D.; Reuter, S.; Woedtke, T. Plasma jet's shielding gas impact on bacterial inactivation. *Biointerphases* **2015**, 10, 029506, 10.1116/1.4916533.

42. Winter, J.; Tresp, H.; Hammer, M.U.; Iseni, S.; Kupsch, S.; Schmidt-Bleker, A.; Wende, K.; Dunnbier, M.; Masur, K.; Weltmannan, K.D.; Reuter, S. Tracking plasma generated H_2O_2 from gas into liquid phase and revealing its dominant impact on human skin cells. *J. Phys. D: Appl. Phys.* **2014**, 47, 285401, 10.1088/0022-3727/47/28/285401.

43. Jablonowski, H.; von Woedtke, T. Research on plasma medicine-relevant plasma–liquid interaction: What happened in the past five years? *Clin. Plas. Med.* **2015**, 3, 42–52, 10.1016/j.cpme.2015.11.003.

44. Bruggeman, P.J.; Kushner, M.J.; Locke, B.R.; Gardeniers, J.G.E.; Graham, W.G.; Graves, D.B.; Hofman-Caris, R.C.H.M.; Maric, D.; Reid, J.P.; Ceriani, E., *et al.* Plasma-liquid interactions: A review and roadmap. *Plasma Sources Sci. T.* **2016**, 25, 053002, 10.1088/0963-0252/25/5/053002.

45. Wende, K.; von Woedtke, T.; Weltmann, K.D.; Bekeschus, S. Chemistry and biochemistry of

cold physical plasma derived reactive species in liquids. *Biol. Chem.* **2018**, 400, 19–38, 10.1515/hsz-2018-0242.

46. Gorbanev, Y.; Privat-Maldonado, A.; Bogaerts, A. Analysis of short-lived reactive species in plasma-air-water systems: The dos and the do nots. *Anal. Chem.* **2018**, 10.1021/acs.analchem.8b03336.

47. Lackmann, J.W.; Bruno, G.; Jablonowski, H.; Kogelheide, F.; Offerhaus, B.; Held, J.; Schulz-von der Gathen, V.; Stapelmann, K.; von Woedtke, T.; Wende, K. Nitrosylation vs. Oxidation—how to modulate cold physical plasmas for biological applications. *PLoS One* **2019**, 14, e0216606, 10.1371/journal.pone.0216606.

48. Lackmann, J.W.; Wende, K.; Verlackt, C.; Golda, J.; Volzke, J.; Kogelheide, F.; Held, J.; Bekeschus, S.; Bogaerts, A.; Schulz-von der Gathen, V.; Stapelmann, K. Chemical fingerprints of cold physical plasmas—an experimental and computational study using cysteine as tracer compound. *Sci. Rep.* **2018**, 8, 7736, 10.1038/s41598-018-25937-0.

49. Klinkhammer, C.; Verlackt, C.; Smilowicz, D.; Kogelheide, F.; Bogaerts, A.; Metzler-Nolte, N.; Stapelmann, K.; Havenith, M.; Lackmann, J.W. Elucidation of plasma-induced chemical modifications on glutathione and glutathione disulphide. *Sci. Rep.* **2017**, 7, 13828, 10.1038/s41598-017-13041-8.

50. Bruno, G.; Heusler, T.; Lackmann, J.-W.; von Woedtke, T.; Weltmann, K.-D.; Wende, K. Cold physical plasma-induced oxidation of cysteine yields reactive sulfur species (rss). *Clin. Plas. Med.* **2019**, 14, 10.1016/j.cpme.2019.100083.

51. Heusler, T.; Bruno, G.; Bekeschus, S.; Lackmann, J.-W.; von Woedtke, T.; Wende, K. Can the effect of cold physical plasma-derived oxidants be transported via thiol group oxidation? *Clin. Plas. Med.* **2019**, 14, 10.1016/j.cpme.2019.100086.

52. Bekeschus, S.; Kolata, J.; Winterbourn, C.; Kramer, A.; Turner, R.; Weltmann, K.D.; Broker, B.; Masur, K. Hydrogen peroxide: A central player in physical plasma-induced oxidative stress in human blood cells. *Free Radic. Res.* **2014**, 48, 542–549, 10.3109/10715762.2014.892937.

53. Conway, G.E.; Casey, A.; Milosavljevic, V.; Liu, Y.; Howe, O.; Cullen, P.J.; Curtin, J.F. Non-thermal atmospheric plasma induces ros-independent cell death in u373mg glioma cells and augments the cytotoxicity of temozolomide. *Br. J. Cancer* **2016**, 114, 435–443, 10.1038/bjc.2016.12.

54. Bekeschus, S.; Winterbourn, C.C.; Kolata, J.; Masur, K.; Hasse, S.; Broker, B.M.; Parker, H.A. Neutrophil extracellular trap formation is elicited in response to cold physical plasma. *J. Leukoc. Biol.* **2016**, 100, 791–799, 10.1189/jlb.3A0415-165RR.

55. Freund, E.; Liedtke, K.R.; van der Linde, J.; Metelmann, H.R.; Heidecke, C.D.; Partecke, L.I.; Bekeschus, S. Physical plasma-treated saline promotes an immunogenic phenotype in ct26 colon cancer cells in vitro and in vivo. *Sci. Rep.* **2019**, 9, 634, 10.1038/s41598-018-37169-3.

56. Bauer, G. Intercellular singlet oxygen-mediated bystander signaling triggered by long-lived species of cold atmospheric plasma and plasma-activated medium. *Redox Biol.* **2019**, 26, 101301, 10.1016/j.redox.2019.101301.

57. Lin, A.; Gorbanev, Y.; De Backer, J.; Van Loenhout, J.; Van Boxem, W.; Lemiere, F.; Cos, P.; Dewilde, S.; Smits, E.; Bogaerts, A. Non-thermal plasma as a unique delivery system of short-lived reactive oxygen and nitrogen species for immunogenic cell death in melanoma cells. *Adv. Sci. (Weinh)* **2019**, 6, 1802062, 10.1002/advs.201802062.

58. Gorbanev, Y.; Van der Paal, J.; Van Boxem, W.; Dewilde, S.; Bogaerts, A. Reaction of chloride anion with atomic oxygen in aqueous solutions: Can cold plasma help in chemistry research? *Phys. Chem. Chem. Phys.* **2019**, 21, 4117–4121, 10.1039/c8cp07550f.

59. Bekeschus, S.; Wende, K.; Hefny, M.M.; Rodder, K.; Jablonowski, H.; Schmidt, A.; Woedtke, T.V.; Weltmann, K.D.; Benedikt, J. Oxygen atoms are critical in rendering thp-1 leukaemia cells susceptible to cold physical plasma-induced apoptosis. *Sci. Rep.* **2017**, 7, 2791, 10.1038/s41598-017-03131-y.

60. Schieber, M.; Chandel, N.S. Ros function in redox signaling and oxidative stress. *Curr. Biol.* **2014**, 24, R453–462, 10.1016/j.cub.2014.03.034.

61. Hasse, S.; Hahn, O.; Kindler, S.; Woedtke, T.v.; Metelmann, H.-R.; Masur, K. Atmospheric pressure plasma jet application on human oral mucosa modulates tissue regeneration. *Plasma Med.* **2014**, 4, 117–129, 10.1615/PlasmaMed.2014011978.

62. Lee, H.J.; Shon, C.H.; Kim, Y.S.; Kim, S.; Kim, G.C.; Kong, M.G. Degradation of adhesion molecules of g361 melanoma cells by a non-thermal atmospheric pressure microplasma. *New J. Phys.* **2009**, 11, 115026, 10.1088/1367-2630/11/11/115026.

63. Schmidt, A.; Bekeschus, S.; von Woedtke, T.; Hasse, S. Cell migration and adhesion of a human melanoma cell line is decreased by cold

plasma treatment. *Clin. Plas. Med.* **2015**, 3, 24–31, 10.1016/j.cpme.2015.05.003.

64. Zirnheld, J.L.; Zucker, S.N.; DiSanto, T.M.; Berezney, R.; Etemadi, K. Nonthermal plasma needle: Development and targeting of melanoma cells. *IEEE Trans. Plasma Sci.* **2010**, 38, 948–952, 10.1109/Tps.2010.2041470.

65. Tanaka, H.; Mizuno, M.; Ishikawa, K.; Nakamura, K.; Kajiyama, H.; Kano, H.; Kikkawa, F.; Hori, M. Plasma-activated medium selectively kills glioblastoma brain tumor cells by down-regulating a survival signaling molecule, AKT kinase. *Plasma Med.* **2011**, 1, 265–277, 10.1615/PlasmaMed.2012006275.

66. Koritzer, J.; Boxhammer, V.; Schafer, A.; Shimizu, T.; Klampfl, T.G.; Li, Y.F.; Welz, C.; Schwenk-Zieger, S.; Morfill, G.E.; Zimmermann, J.L.; Schlegel, J. Restoration of sensitivity in chemo-resistant glioma cells by cold atmospheric plasma. *PLoS One* **2013**, 8, e64498, 10.1371/journal.pone.0064498.

67. Cheng, X.Q.; Sherman, J.; Murphy, W.; Ratovitski, E.; Canady, J.; Keidar, M. The effect of tuning cold plasma composition on glioblastoma cell viability. *PLoS One* **2014**, 9, 10.1371/journal.pone.0098652.

68. Wang, C.; Zhang, H.X.; Xue, Z.X.; Yin, H.J.; Niu, Q.; Chen, H.L. The relation between doses or post-plasma time points and apoptosis of leukemia cells induced by dielectric barrier discharge plasma. *AIP Adv.* **2015**, 5, 127220, 10.1063/1.4938546.

69. Schmidt, A.; Rodder, K.; Hasse, S.; Masur, K.; Toups, L.; Lillig, C.H.; von Woedtke, T.; Wende, K.; Bekeschus, S. Redox-regulation of activator protein 1 family members in blood cancer cell lines exposed to cold physical plasma-treated medium. *Plasma Process. Polym.* **2016**, 13, 1179–1188, 10.1002/ppap.201600090.

70. Bundscherer, L.; Wende, K.; Ottmuller, K.; Barton, A.; Schmidt, A.; Bekeschus, S.; Hasse, S.; Weltmann, K.D.; Masur, K.; Lindequist, U. Impact of non-thermal plasma treatment on MAPK signaling pathways of human immune cell lines. *Immunobiology* **2013**, 218, 1248–1255, 10.1016/j.imbio.2013.04.015.

71. Plewa, J.M.; Yousfi, M.; Frongia, C.; Eichwald, O.; Ducommun, B.; Merbahi, N.; Lobjois, V. Low-temperature plasma-induced antiproliferative effects on multi-cellular tumor spheroids. *New J. Phys.* **2014**, 16, 043027, 10.1088/1367-2630/16/4/043027.

72. Irani, S.; Shahmirani, Z.; Atyabi, S.M.; Mirpoor, S. Induction of growth arrest in colorectal cancer cells by cold plasma and gold nanoparticles. *Arch. Med. Sci.* **2015**, 11, 1286–1295, 10.5114/aoms.2015.48221.

73. Han, D.; Cho, J.H.; Lee, R.H.; Bang, W.; Park, K.; Kim, M.S.; Shim, J.H.; Chae, J.I.; Moon, S.Y. Antitumorigenic effect of atmospheric-pressure dielectric barrier discharge on human colorectal cancer cells via regulation of sp1 transcription factor. *Sci. Rep.* **2017**, 7, 43081, 10.1038/srep43081.

74. Duan, J.; Lu, X.; He, G. The selective effect of plasma activated medium in an in vitro co-culture of liver cancer and normal cells. *J. Appl. Phys.* **2017**, 121, 013302, 10.1063/1.4973484.

75. Brulle, L.; Vandamme, M.; Ries, D.; Martel, E.; Robert, E.; Lerondel, S.; Trichet, V.; Richard, S.; Pouvesle, J.M.; Le Pape, A. Effects of a non thermal plasma treatment alone or in combination with gemcitabine in a mia paca2-luc orthotopic pancreatic carcinoma model. *PLoS One* **2012**, 7, 10.1371/journal.pone.0052653.

76. Liedtke, K.R.; Bekeschus, S.; Kaeding, A.; Hackbarth, C.; Kuehn, J.P.; Heidecke, C.D.; von Bernstorff, W.; von Woedtke, T.; Partecke, L.I. Non-thermal plasma-treated solution demonstrates antitumor activity against pancreatic cancer cells in vitro and in vivo. *Sci. Rep.* **2017**, 7, 8319, 10.1038/s41598-017-08560-3.

77. Bekeschus, S.; Freund, E.; Spadola, C.; Privat-Maldonado, A.; Hackbarth, C.; Bogaerts, A.; Schmidt, A.; Wende, K.; Weltmann, K.D.; von Woedtke, T.; Heidecke, C.D.; Partecke, L.I.; Kading, A. Risk assessment of kinpen plasma treatment of four human pancreatic cancer cell lines with respect to metastasis. *Cancers (Basel)* **2019**, 11, 1237, 10.3390/cancers11091237.

78. Weiss, M.; Gumbel, D.; Hanschmann, E.M.; Mandelkow, R.; Gelbrich, N.; Zimmermann, U.; Walther, R.; Ekkernkamp, A.; Sckell, A.; Kramer, A.; Burchardt, M.; Lillig, C.H.; Stope, M.B. Cold atmospheric plasma treatment induces antiproliferative effects in prostate cancer cells by redox and apoptotic signaling pathways. *PLoS One* **2015**, 10, e0130350, 10.1371/journal.pone.0130350.

79. Ishaq, M.; Bazaka, K.; Ostrikov, K. Intracellular effects of atmospheric-pressure plasmas on melanoma cancer cells. *Phys. Plasmas* **2015**, 22, 122003, 10.1063/1.4933366.

80. Bekeschus, S.; Ressel, V.; Freund, E.; Gelbrich, N.; Mustea, A.; Stope, M.B. Gas plasma-treated

prostate cancer cells augment myeloid cell activity and cytotoxicity. *Antioxidants (Basel)* **2020**, 9, 323, 10.3390/antiox9040323.

81. Pai, K.K.; Singarapu, K.; Jacob, J.D.; Madihally, S.V. Dose dependent selectivity and response of different types of mammalian cells to surface dielectric barrier discharge (sdbd) plasma. *Plasma Process. Polym.* **2015**, 12, 666–677, 10.1002/ppap.201400134.

82. Hara, H.; Taniguchi, M.; Kobayashi, M.; Kamiya, T.; Adachi, T. Plasma-activated medium-induced intracellular zinc liberation causes death of sh-sy5y cells. *Arch. Biochem. Biophys.* **2015**, 584, 51–60, 10.1016/j.abb.2015.08.014.

83. Yan, X.; Meng, Z.; Ouyang, J.; Qiao, Y.; Yuan, F. New application of an atmospheric pressure plasma jet as a neuro-protective agent against glucose deprivation-induced injury of sh-sy5y cells. *J. Vis. Exp.* **2017**, 10.3791/56323.

84. Utsumi, F.; Kajiyama, H.; Nakamura, K.; Tanaka, H.; Mizuno, M.; Ishikawa, K.; Kondo, H.; Kano, H.; Hori, M.; Kikkawa, F. Effect of indirect nonequilibrium atmospheric pressure plasma on anti-proliferative activity against chronic chemo-resistant ovarian cancer cells in vitro and in vivo. *PLoS One* **2013**, 8, e81576, 10.1371/journal.pone.0081576.

85. Kajiyama, H.; Utsumi, F.; Nakamura, K.; Tanaka, H.; Mizuno, M.; Toyokuni, S.; Hori, M.; Kikkawa, F. Possible therapeutic option of aqueous plasma for refractory ovarian cancer. *Clin. Plas. Med.* **2016**, 4, 14–18, 10.1016/j.cpme.2015.12.002.

86. Bekeschus, S.; Wulf, C.; Freund, E.; Koensgen, D.; Mustea, A.; Weltmann, K.-D.; Stope, M. Plasma treatment of ovarian cancer cells mitigates their immuno-modulatory products active on thp-1 monocytes. *Plasma* **2018**, 1, 201–217, 10.3390/plasma1010018.

87. Torii, K.; Yamada, S.; Nakamura, K.; Tanaka, H.; Kajiyama, H.; Tanahashi, K.; Iwata, N.; Kanda, M.; Kobayashi, D.; Tanaka, C., *et al.* Effectiveness of plasma treatment on gastric cancer cells. *Gastric Cancer* **2015**, 18, 635–643, 10.1007/s10120-014-0395-6.

88. Chen, Z.; Lin, L.; Cheng, X.; Gjika, E.; Keidar, M. Treatment of gastric cancer cells with non-thermal atmospheric plasma generated in water. *Biointerphases* **2016**, 11, 031010, 10.1116/1.4962130.

89. Takeda, S.; Yamada, S.; Hattori, N.; Nakamura, K.; Tanaka, H.; Kajiyama, H.; Kanda, M.; Kobayashi, D.; Tanaka, C.; Fujii, T.; Fujiwara, M.; Mizuno, M.; Hori, M.; Kodera, Y. Intraperitoneal administration of plasma-activated medium: Proposal of a novel treatment option for peritoneal metastasis from gastric cancer. *Ann. Surg. Oncol.* **2017**, 24, 1188–1194, 10.1245/s10434-016-5759-1.

90. Kim, S.J.; Joh, H.M.; Chung, T.H. Production of intracellular reactive oxygen species and change of cell viability induced by atmospheric pressure plasma in normal and cancer cells. *Appl. Phys. Lett.* **2013**, 103, 10.1063/1.4824986.

91. Mohades, S.; Laroussi, M.; Maruthamuthu, V. Moderate plasma activated media suppresses proliferation and migration of mdck epithelial cells. *J. Phys. D: Appl. Phys.* **2017**, 50, 10.1088/1361-6463/aa678a.

92. Iuchi, K.; Morisada, Y.; Yoshino, Y.; Himuro, T.; Saito, Y.; Murakami, T.; Hisatomi, H. Cold atmospheric-pressure nitrogen plasma induces the production of reactive nitrogen species and cell death by increasing intracellular calcium in hek293t cells. *Arch. Biochem. Biophys.* **2018**, 654, 136–145, 10.1016/j.abb.2018.07.015.

93. Adil, B.H.; Al-Shammari, A.M.; Murbat, H.H. Breast cancer treatment using cold atmospheric plasma generated by the fe-dbd scheme. *Clin. Plasma Med.* **2020**, 19–20, 10.1016/j.cpme.2020.100103.

94. Lafontaine, J.; Boisvert, J.S.; Glory, A.; Coulombe, S.; Wong, P. Synergy between non-thermal plasma with radiation therapy and olaparib in a panel of breast cancer cell lines. *Cancers (Basel)* **2020**, 12, 10.3390/cancers12020348.

95. Liu, Y.; Tan, S.; Zhang, H.; Kong, X.; Ding, L.; Shen, J.; Lan, Y.; Cheng, C.; Zhu, T.; Xia, W. Selective effects of non-thermal atmospheric plasma on triple-negative breast normal and carcinoma cells through different cell signaling pathways. *Sci. Rep.* **2017**, 7, 7980, 10.1038/s41598-017-08792-3.

96. Guerrero-Preston, R.; Ogawa, T.; Uemura, M.; Shumulinsky, G.; Valle, B.L.; Pirini, F.; Ravi, R.; Sidransky, D.; Keidar, M.; Trink, B. Cold atmospheric plasma treatment selectively targets head and neck squamous cell carcinoma cells. *Int. J. Mol. Med.* **2014**, 34, 941–946, 10.3892/ijmm.2014.1849.

97. Welz, C.; Emmert, S.; Canis, M.; Becker, S.; Baumeister, P.; Shimizu, T.; Morfill, G.E.; Harreus, U.; Zimmermann, J.L. Cold atmospheric plasma: A promising complementary therapy for squamous head and neck cancer. *PLoS One* **2015**, 10, e0141827, 10.1371/journal.pone.0141827.

98. Witzke, K.; Seebauer, C.; Jesse, K.; Kwiatek, E.; Berner, J.; Semmler, M.L.; Boeckmann, L.; Emmert, S.; Weltmann, K.D.; Metelmann, H.R.; Bekeschus, S. Plasma medical oncology: Immunological interpretation of head and neck squamous cell carcinoma. *Plasma Process. Polym.* **2020**, 17, e1900258, 10.1002/ppap.201900258.

99. Galluzzi, L.; Vitale, I.; Aaronson, S.A.; Abrams, J.M.; Adam, D.; Agostinis, P.; Alnemri, E.S.; Altucci, L.; Amelio, I.; Andrews, D.W., *et al.* Molecular mechanisms of cell death: Recommendations of the nomenclature committee on cell death 2018. *Cell Death Differ.* **2018**, 25, 486–541, 10.1038/s41418-017-0012-4.

100. Kaushik, N.; Lee, S.J.; Choi, T.G.; Baik, K.Y.; Uhm, H.S.; Kim, C.H.; Kaushik, N.K.; Choi, E.H. Non-thermal plasma with 2-deoxy-d-glucose synergistically induces cell death by targeting glycolysis in blood cancer cells. *Sci. Rep.* **2015**, 5, 8726, 10.1038/srep08726.

101. Adachi, T.; Tanaka, H.; Nonomura, S.; Hara, H.; Kondo, S.; Hori, M. Plasma-activated medium induces a549 cell injury via a spiral apoptotic cascade involving the mitochondrial-nuclear network. *Free Radic. Biol. Med.* **2015**, 79, 28–44, 10.1016/j.freeradbiomed.2014.11.014.

102. Ishaq, M.; Bazaka, K.; Ostrikov, K. Pro-apoptotic noxa is implicated in atmospheric-pressure plasma-induced melanoma cell death. *J. Phys. D: Appl. Phys.* **2015**, 48, 464002, 10.1088/0022-3727/48/46/464002.

103. Shi, L.; Wang, Y.; Ito, F.; Okazaki, Y.; Tanaka, H.; Mizuno, M.; Hori, M.; Richardson, D.R.; Toyokuni, S. Biphasic effects of l-ascorbate on the tumoricidal activity of non-thermal plasma against malignant mesothelioma cells. *Arch. Biochem. Biophys.* **2016**, 605, 109–116, 10.1016/j.abb.2016.05.016.

104. Shi, L.; Ito, F.; Wang, Y.; Okazaki, Y.; Tanaka, H.; Mizuno, M.; Hori, M.; Hirayama, T.; Nagasawa, H.; Richardson, D.R.; Toyokuni, S. Non-thermal plasma induces a stress response in mesothelioma cells resulting in increased endocytosis, lysosome biogenesis and autophagy. *Free Radic. Biol. Med.* **2017**, 108, 904–917, 10.1016/j.freeradbiomed.2017.04.368.

105. Yoshikawa, N.; Liu, W.; Nakamura, K.; Yoshida, K.; Ikeda, Y.; Tanaka, H.; Mizuno, M.; Toyokuni, S.; Hori, M.; Kikkawa, F.; Kajiyama, H. Plasma-activated medium promotes autophagic cell death along with alteration of the mtor pathway. *Sci. Rep.* **2020**, 10, 1614, 10.1038/s41598-020-58667-3.

106. Furuta, T.; Shi, L.; Toyokuni, S. Non-thermal plasma as a simple ferroptosis inducer in cancer cells: A possible role of ferritin. *Pathol. Int.* **2018**, 68, 442–443, doi:10.1111/pin.12665.

107. Sato, K.; Shi, L.; Ito, F.; Ohara, Y.; Motooka, Y.; Tanaka, H.; Mizuno, M.; Hori, M.; Hirayama, T.; Hibi, H.; Toyokuni, S. Non-thermal plasma specifically kills oral squamous cell carcinoma cells in a catalytic fe(ii)-dependent manner. *J. Clin. Biochem. Nutr.* **2019**, 65, 8–15, 10.3164/jcbn.18–91.

108. Okazaki, Y.; Toyokuni, S. Induction of cancer cell-specific ferroptosis by non-thermal plasma exposure. *Jpn. J. Appl. Phys.* **2020**, 59, 10.35848/1347-4065/abbc56.

109. Yang, X.; Chen, G.; Yu, K.N.; Yang, M.; Peng, S.; Ma, J.; Qin, F.; Cao, W.; Cui, S.; Nie, L.; Han, W. Cold atmospheric plasma induces gsdme-dependent pyroptotic signaling pathway via ros generation in tumor cells. *Cell Death Dis.* **2020**, 11, 295, 10.1038/s41419-020-2459-3.

110. Hwang, S.Y.; Nguyen, N.H.; Kim, T.J.; Lee, Y.; Kang, M.A.; Lee, J.S. Non-thermal plasma couples oxidative stress to trail sensitization through dr5 upregulation. *Int. J. Mol. Sci.* **2020**, 21, 10.3390/ijms21155302.

111. Xia, J.; Zeng, W.; Xia, Y.; Wang, B.; Xu, D.; Liu, D.; Kong, M.G.; Dong, Y. Cold atmospheric plasma induces apoptosis of melanoma cells via sestrin2-mediated inos signaling. *J. Biophotonics* **2018**, e201800046, 10.1002/jbio.201800046.

112. Turrini, E.; Laurita, R.; Stancampiano, A.; Catanzaro, E.; Calcabrini, C.; Maffei, F.; Gherardi, M.; Colombo, V.; Fimognari, C. Cold atmospheric plasma induces apoptosis and oxidative stress pathway regulation in t-lymphoblastoid leukemia cells. *Oxid. Med. Cell. Longev.* **2017**, 2017, 4271065, 10.1155/2017/4271065.

113. Hirst, A.M.; Simms, M.S.; Mann, V.M.; Maitland, N.J.; O'Connell, D.; Frame, F.M. Low-temperature plasma treatment induces DNA damage leading to necrotic cell death in primary prostate epithelial cells. *Br. J. Cancer* **2015**, 112, 1536–1545, 10.1038/bjc.2015.113.

114. Virard, F.; Cousty, S.; Cambus, J.P.; Valentin, A.; Kemoun, P.; Clement, F. Cold atmospheric plasma induces a predominantly necrotic cell death via the microenvironment. *PLoS One* **2015**, 10, e0133120, 10.1371/journal.pone.0133120.

115. Zucker, S.N.; Zirnheld, J.; Bagati, A.; DiSanto, T.M.; Des Soye, B.; Wawrzyniak, J.A.; Etemadi, K.; Nikiforov, M.; Berezney, R. Preferential induction of apoptotic cell death in melanoma

cells as compared with normal keratinocytes using a non-thermal plasma torch. *Cancer Biol. Ther.* **2012**, 13, 1299–1306, 10.4161/cbt.21787.

116. Naciri, M.; Dowling, D.; Al-Rubeai, M. Differential sensitivity of mammalian cell lines to non-thermal atmospheric plasma. *Plasma Process. Polym.* **2014**, 11, 391–400, 10.1002/ppap.201300118.

117. Privat-Maldonado, A.; Gorbanev, Y.; Dewilde, S.; Smits, E.; Bogaerts, A. Reduction of human glioblastoma spheroids using cold atmospheric plasma: The combined effect of short- and long-lived reactive species. *Cancers (Basel)* **2018**, 10, 10.3390/cancers10110394.

118. Chauvin, J.; Judee, F.; Merbahi, N.; Vicendo, P. Effects of plasma activated medium on head and neck fadu cancerous cells: Comparison of 3d and 2d response. *Anticancer Agents Med. Chem.* **2018**, 18, 776–783, 10.2174/1871520617666170801111055.

119. Griseti, E.; Kolosnjaj-Tabi, J.; Gibot, L.; Fourquaux, I.; Rols, M.P.; Yousfi, M.; Merbahi, N.; Golzio, M. Pulsed electric field treatment enhances the cytotoxicity of plasma-activated liquids in a three-dimensional human colorectal cancer cell model. *Sci. Rep.* **2019**, 9, 7583, 10.1038/s41598-019-44087-5.

120. Hasse, S.; Meder, T.; Freund, E.; von Woedtke, T.; Bekeschus, S. Plasma treatment limits human melanoma spheroid growth and metastasis independent of the ambient gas composition. *Cancers (Basel)* **2020**, 12, 2570, 10.3390/cancers12092570.

121. Liedtke, K.R.; Diedrich, S.; Pati, O.; Freund, E.; Flieger, R.; Heidecke, C.D.; Partecke, L.I.; Bekeschus, S. Cold physical plasma selectively elicits apoptosis in murine pancreatic cancer cells in vitro and in ovo. *Anticancer Res.* **2018**, 38, 5655–5663, 10.21873/anticanres.12901.

122. Shaw, P.; Kumar, N.; Hammerschmid, D.; Privat-Maldonado, A.; Dewilde, S.; Bogaerts, A. Synergistic effects of melittin and plasma treatment: A promising approach for cancer therapy. *Cancers (Basel)* **2019**, 11, 10.3390/cancers11081109.

123. Freund, E.; Spadola, C.; Schmidt, A.; Privat-Maldonado, A.; Bogaerts, A.; von Woedtke, T.; Weltmann, K.-D.; Heidecke, C.-D.; Partecke, L.-I.; Käding, A.; Bekeschus, S. Risk evaluation of emt and inflammation in metastatic pancreatic cancer cells following plasma treatment. *Front. Phys.* **2020**, 8, 10.3389/fphy.2020.569618.

124. Bekeschus, S.; Eisenmann, S.; Sagwal, S.K.; Bodnar, Y.; Moritz, J.; Poschkamp, B.; Stoffels, I.; Emmert, S.; Madesh, M.; Weltmann, K.D.;

von Woedtke, T.; Gandhirajan, R.K. Xct (slc7a11) expression confers intrinsic resistance to physical plasma treatment in tumor cells. *Redox Biol* **2020**, 30, 101423, 10.1016/j.redox.2019.101423.

125. Bekeschus, S.; Moritz, J.; Helfrich, I.; Boeckmann, L.; Weltmann, K.D.; Emmert, S.; Metelmann, H.R.; Stoffels, I.; von Woedtke, T. Ex vivo exposure of human melanoma tissue to cold physical plasma elicits apoptosis and modulates inflammation. *Appl. Sci.-Basel* **2020**, 10, 10.3390/app10061971.

126. Seebauer, C.; Freund, E.; Hasse, S.; Miller, V.; Segebarth, M.; Lucas, C.; Kindler, S.; Dieke, T.; Metelmann, H.R.; Daeschlein, G.; Jesse, K.; Weltmann, K.D.; Bekeschus, S. Effects of cold physical plasma on oral lichen planus: An in-vitro study. *Oral Dis.* **2020**, 10.1111/odi.13697.

127. Saadati, F.; Mahdikia, H.; Abbaszadeh, H.A.; Abdollahifar, M.A.; Khoramgah, M.S.; Shokri, B. Comparison of direct and indirect cold atmospheric-pressure plasma methods in the b16f10 melanoma cancer cells treatment. *Sci. Rep.* **2018**, 8, 7689, 10.1038/s41598-018-25990-9.

128. Alimohammadi, M.; Golpur, M.; Sohbatzadeh, F.; Hadavi, S.; Bekeschus, S.; Niaki, H.A.; Valadan, R.; Rafiei, A. Cold atmospheric plasma is a potent tool to improve chemotherapy in melanoma in vitro and in vivo. *Biomolecules* **2020**, 10, 10.3390/biom10071011.

129. Binenbaum, Y.; Ben-David, G.; Gil, Z.; Slutsker, Y.Z.; Ryzhkov, M.A.; Felsteiner, J.; Krasik, Y.E.; Cohen, J.T. Cold atmospheric plasma, created at the tip of an elongated flexible capillary using low electric current, can slow the progression of melanoma. *PLoS One* **2017**, 12, e0169457, 10.1371/journal.pone.0169457.

130. Metelmann, H.-R.; Seebauer, C.; Miller, V.; Fridman, A.; Bauer, G.; Graves, D.B.; Pouvesle, J.-M.; Rutkowski, R.; Schuster, M.; Bekeschus, S., et al. Clinical experience with cold plasma in the treatment of locally advanced head and neck cancer. *Clin. Plas. Med.* **2018**, 9, 6–13, 10.1016/j.cpme.2017.09.001.

131. Schuster, M.; Seebauer, C.; Rutkowski, R.; Hauschild, A.; Podmelle, F.; Metelmann, C.; Metelmann, B.; von Woedtke, T.; Hasse, S.; Weltmann, K.D.; Metelmann, H.R. Visible tumor surface response to physical plasma and apoptotic cell kill in head and neck cancer. *J. Craniomaxillofac. Surg.* **2016**, 44, 1445–1452, 10.1016/j.jcms.2016.07.001.

132. Metelmann, H.R.; Seebauer, C.; Rutkowski, R.; Schuster, M.; Bekeschus, S.; Metelmann, P. Treating cancer with cold physical plasma: On the way to evidence-based medicine. *Contrib. Plasma Phys.* **2018**, 58, 415–419, 10.1002/ctpp.201700085.

133. Galluzzi, L.; Bravo-San Pedro, J.M.; Vitale, I.; Aaronson, S.A.; Abrams, J.M.; Adam, D.; Alnemri, E.S.; Altucci, L.; Andrews, D.; Annicchiarico-Petruzzelli, M., *et al.* Essential versus accessory aspects of cell death: Recommendations of the nccd 2015. *Cell Death Differ* **2015**, 22, 58–73, 10.1038/cdd.2014.137.

134. Bock, F.J.; Tait, S.W.G. Mitochondria as multifaceted regulators of cell death. *Nat. Rev. Mol. Cell Biol.* **2020**, 21, 85–100, 10.1038/s41580-019-0173-8.

135. Lei, P.; Bai, T.; Sun, Y. Mechanisms of ferroptosis and relations with regulated cell death: A review. *Front. Physiol.* **2019**, 10, 139, 10.3389/fphys.2019.00139.

136. Wang, M.; Holmes, B.; Cheng, X.; Zhu, W.; Keidar, M.; Zhang, L.G. Cold atmospheric plasma for selectively ablating metastatic breast cancer cells. *PLoS One* **2013**, 8, e73741, 10.1371/journal.pone.0073741.

137. Keidar, M.; Walk, R.; Shashurin, A.; Srinivasan, P.; Sandler, A.; Dasgupta, S.; Ravi, R.; Guerrero-Preston, R.; Trink, B. Cold plasma selectivity and the possibility of a paradigm shift in cancer therapy. *Br. J. Cancer* **2011**, 105, 1295–1301, 10.1038/bjc.2011.386.

138. Xiang, L.; Xu, X.; Zhang, S.; Cai, D.; Dai, X. Cold atmospheric plasma conveys selectivity on triple negative breast cancer cells both in vitro and in vivo. *Free Radic. Biol. Med.* **2018**, 124, 205–213, 10.1016/j.freeradbiomed.2018.06.001.

139. Chu, K.F.; Dupuy, D.E. Thermal ablation of tumours: Biological mechanisms and advances in therapy. *Nat. Rev. Cancer* **2014**, 14, 199–208, 10.1038/nrc3672.

140. Agostinis, P.; Berg, K.; Cengel, K.A.; Foster, T.H.; Girotti, A.W.; Gollnick, S.O.; Hahn, S.M.; Hamblin, M.R.; Juzeniene, A.; Kessel, D., *et al.* Photodynamic therapy of cancer: An update. *CA Cancer J. Clin.* **2011**, 61, 250–281, 10.3322/caac.20114.

141. Cabula, C.; Campana, L.G.; Grilz, G.; Galuppo, S.; Bussone, R.; De Meo, L.; Bonadies, A.; Curatolo, P.; De Laurentiis, M.; Renne, M., *et al.* Electrochemotherapy in the treatment of cutaneous metastases from breast cancer: A multicenter cohort analysis. *Ann. Surg. Oncol.* **2015**, 22 Suppl 3, S442–450, 10.1245/s10434-015-4779-6.

142. Apetoh, L.; Ghiringhelli, F.; Tesniere, A.; Criollo, A.; Ortiz, C.; Lidereau, R.; Mariette, C.; Chaput, N.; Mira, J.P.; Delaloge, S.; Andre, F.; Tursz, T.; Kroemer, G.; Zitvogel, L. The interaction between hmgb1 and tlr4 dictates the outcome of anticancer chemotherapy and radiotherapy. *Immunol. Rev.* **2007**, 220, 47–59, 10.1111/j.1600-065X.2007.00573.x.

143. Frey, B.; Weiss, E.M.; Rubner, Y.; Wunderlich, R.; Ott, O.J.; Sauer, R.; Fietkau, R.; Gaipl, U.S. Old and new facts about hyperthermia-induced modulations of the immune system. *Int. J. Hyperthermia* **2012**, 28, 528–542, 10.3109/02656736.2012.677933.

144. Partecke, L.I.; Evert, K.; Haugk, J.; Doering, F.; Normann, L.; Diedrich, S.; Weiss, F.U.; Evert, M.; Huebner, N.O.; Guenther, C., *et al.* Tissue tolerable plasma (ttp) induces apoptosis in pancreatic cancer cells in vitro and in vivo. *BMC Cancer* **2012**, 12, 10.1186/1471-2407-12-473.

145. Fluhr, J.W.; Sassning, S.; Lademann, O.; Darvin, M.E.; Schanzer, S.; Kramer, A.; Richter, H.; Sterry, W.; Lademann, J. In vivo skin treatment with tissue-tolerable plasma influences skin physiology and antioxidant profile in human stratum corneum. *Exp. Dermatol.* **2012**, 21, 130–134, 10.1111/j.1600-0625.2011.01411.x.

146. Hirata, E.; Sahai, E. Tumor microenvironment and differential responses to therapy. *Cold Spring Harb. Perspect. Med.* **2017**, 7, a026781, 10.1101/cshperspect.a026781.

147. Al-Abd, A.M.; Alamoudi, A.J.; Abdel-Naim, A.B.; Neamatallah, T.A.; Ashour, O.M. Anti-angiogenic agents for the treatment of solid tumors: Potential pathways, therapy and current strategies—a review. *J. Adv. Res.* **2017**, 8, 591–605, 10.1016/j.jare.2017.06.006.

148. Schreiber, R.D.; Old, L.J.; Smyth, M.J. Cancer immunoediting: Integrating immunity's roles in cancer suppression and promotion. *Science* **2011**, 331, 1565–1570, 10.1126/science.1203486.

149. Holzel, M.; Bovier, A.; Tuting, T. Plasticity of tumour and immune cells: A source of heterogeneity and a cause for therapy resistance? *Nat. Rev. Cancer* **2013**, 13, 365–376, 10.1038/nrc3498.

150. Balamurugan, K. Hif-1 at the crossroads of hypoxia, inflammation, and cancer. *Int. J. Cancer* **2016**, 138, 1058–1066, 10.1002/ijc.29519.

151. Policastro, L.L.; Ibanez, I.L.; Notcovich, C.; Duran, H.A.; Podhajcer, O.L. The tumor

microenvironment: Characterization, redox considerations, and novel approaches for reactive oxygen species-targeted gene therapy. *Antioxid. Redox Signal.* **2013**, 19, 854–895, 10.1089/ars.2011.4367.

152. Rohwer, N.; Cramer, T. Hypoxia-mediated drug resistance: Novel insights on the functional interaction of hifs and cell death pathways. *Drug Resist. Updat.* **2011**, 14, 191–201, 10.1016/j.drup.2011.03.001.

153. Ruffell, B.; Affara, N.I.; Coussens, L.M. Differential macrophage programming in the tumor microenvironment. *Trends Immunol.* **2012**, 33, 119–126, 10.1016/j.it.2011.12.001.

154. Bello, I.O.; Vered, M.; Dayan, D.; Dobriyan, A.; Yahalom, R.; Alanen, K.; Nieminen, P.; Kantola, S.; Laara, E.; Salo, T. Cancer-associated fibroblasts, a parameter of the tumor microenvironment, overcomes carcinoma-associated parameters in the prognosis of patients with mobile tongue cancer. *Oral Oncol.* **2011**, 47, 33–38, 10.1016/j.oraloncology.2010.10.013.

155. Galdiero, M.R.; Varricchi, G.; Loffredo, S.; Mantovani, A.; Marone, G. Roles of neutrophils in cancer growth and progression. *J. Leukoc. Biol.* **2018**, 103, 457–464, 10.1002/JLB.3MR0717-292R.

156. Wang, H.; Franco, F.; Ho, P.C. Metabolic regulation of tregs in cancer: Opportunities for immunotherapy. *Trends Cancer* **2017**, 3, 583–592, 10.1016/j.trecan.2017.06.005.

157. Nakamura, K.; Smyth, M.J. Myeloid immunosuppression and immune checkpoints in the tumor microenvironment. *Cell. Mol. Immunol.* **2020**, 17, 1–12, 10.1038/s41423-019-0306-1.

158. Bailly, C. Regulation of pd-l1 expression on cancer cells with ros-modulating drugs. *Life Sci.* **2020**, 246, 117403, 10.1016/j.lfs.2020.117403.

159. Chen, X.F.; Song, M.J.; Zhang, B.; Zhang, Y. Reactive oxygen species regulate t cell immune response in the tumor microenvironment. *Oxid. Med. Cell. Longev.* **2016**, 2016, 10.1155/2016/1580967.

160. Gu, H.; Huang, T.; Shen, Y.; Liu, Y.; Zhou, F.; Jin, Y.; Sattar, H.; Wei, Y. Reactive oxygen species-mediated tumor microenvironment transformation: The mechanism of radioresistant gastric cancer. *Oxid. Med. Cell. Longev.* **2018**, 2018, 5801209, 10.1155/2018/5801209.

161. Kotsafti, A.; Scarpa, M.; Castagliuolo, I.; Scarpa, M. Reactive oxygen species and antitumor immunity-from surveillance to evasion. *Cancers (Basel)* **2020**, 12, 10.3390/cancers12071748.

162. Narayanan, D.; Ma, S.; Ozcelik, D. Targeting the redox landscape in cancer therapy. *Cancers (Basel)* **2020**, 12, 10.3390/cancers12071706.

163. Peng, X.; Gandhi, V. Ros-activated anticancer prodrugs: A new strategy for tumor-specific damage. *Ther. Deliv.* **2012**, 3, 823–833, 10.4155/tde.12.61.

164. Roninson, I.B. Tumor cell senescence in cancer treatment. *Cancer Res.* **2003**, 63, 2705–2715,

165. Shay, J.W.; Roninson, I.B. Hallmarks of senescence in carcinogenesis and cancer therapy. *Oncogene* **2004**, 23, 2919–2933, 10.1038/sj.onc.1207518.

166. Capparelli, C.; Guido, C.; Whitaker-Menezes, D.; Bonuccelli, G.; Balliet, R.; Pestell, T.G.; Goldberg, A.F.; Pestell, R.G.; Howell, A.; Sneddon, S., et al. Autophagy and senescence in cancer-associated fibroblasts metabolically supports tumor growth and metastasis via glycolysis and ketone production. *Cell Cycle* **2012**, 11, 2285–2302, 10.4161/cc.20718.

167. Sui, X.; Han, W.; Pan, H. P53-induced autophagy and senescence. *Oncotarget* **2015**, 6, 11723–11724, 10.18632/oncotarget.4170.

168. Arndt, S.; Wacker, E.; Li, Y.F.; Shimizu, T.; Thomas, H.M.; Morfill, G.E.; Karrer, S.; Zimmermann, J.L.; Bosserhoff, A.K. Cold atmospheric plasma, a new strategy to induce senescence in melanoma cells. *Exp. Dermatol.* **2013**, 22, 284–289, 10.1111/exd.12127.

169. Schneider, C.; Gebhardt, L.; Arndt, S.; Karrer, S.; Zimmermann, J.L.; Fischer, M.J.M.; Bosserhoff, A.K. Cold atmospheric plasma causes a calcium influx in melanoma cells triggering cap-induced senescence. *Sci. Rep.* **2018**, 8, 10048, 10.1038/s41598-018-28443-5.

170. Hara, H.; Kobayashi, M.; Shiiba, M.; Kamiya, T.; Adachi, T. Sublethal treatment with plasma-activated medium induces senescence-like growth arrest of a549 cells: Involvement of intracellular mobile zinc. *J. Clin. Biochem. Nutr.* **2019**, 65, 16–22, 10.3164/jcbn.19–17.

171. Azzariti, A.; Iacobazzi, R.M.; Di Fonte, R.; Porcelli, L.; Gristina, R.; Favia, P.; Fracassi, F.; Trizio, I.; Silvestris, N.; Guida, G.; Tommasi, S.; Sardella, E. Plasma-activated medium triggers cell death and the presentation of immune activating danger signals in melanoma and pancreatic cancer cells. *Sci. Rep.* **2019**, 9, 4099, 10.1038/s41598-019-40637-z.

172. Zhen, X.; Sun, H.N.; Liu, R.; Choi, H.S.; Lee, D.S. Non-thermal plasma-activated medium

induces apoptosis of aspc1 cells through the ros-dependent autophagy pathway. *In Vivo* **2020**, 34, 143–153, 10.21873/invivo.11755.

173. Adhikari, M.; Adhikari, B.; Ghimire, B.; Baboota, S.; Choi, E.H. Cold atmospheric plasma and silymarin nanoemulsion activate autophagy in human melanoma cells. *Int. J. Mol. Sci.* **2020**, 21, 10.3390/ijms21061939.

174. Mills, K.R.; Reginato, M.; Debnath, J.; Queenan, B.; Brugge, J.S. Tumor necrosis factor-related apoptosis-inducing ligand (trail) is required for induction of autophagy during lumen formation in vitro. *Proc. Natl. Acad. Sci. U. S. A.* **2004**, 101, 3438–3443, 10.1073/pnas.0400443101.

175. Ito, T.; Ando, T.; Suzuki-Karasaki, M.; Tokunaga, T.; Yoshida, Y.; Ochiai, T.; Tokuhashi, Y.; Suzuki-Karasaki, Y. Cold psm, but not trail, triggers autophagic cell death: A therapeutic advantage of psm over trail. *Int. J. Oncol.* **2018**, 53, 503–514, 10.3892/ijo.2018.4413.

176. Kalluri, R.; Weinberg, R.A. The basics of epithelial-mesenchymal transition. *J. Clin. Invest.* **2009**, 119, 1420–1428, 10.1172/JCI39104.

177. Krebs, A.M.; Mitschke, J.; Lasierra Losada, M.; Schmalhofer, O.; Boerries, M.; Busch, H.; Boettcher, M.; Mougiakakos, D.; Reichardt, W.; Bronsert, P., *et al.* The emt-activator zeb1 is a key factor for cell plasticity and promotes metastasis in pancreatic cancer. *Nat. Cell Biol.* **2017**, 19, 518–529, 10.1038/ncb3513.

178. Blazquez, R.; Rietkötter, E.; Wenske, B.; Wlochowitz, D.; Sparrer, D.; Vollmer, E.; Müller, G.; Seegerer, J.; Sun, X.; Dettmer, K., *et al.* Lef1 supports metastatic brain colonization by regulating glutathione metabolism and increasing ros resistance in breast cancer. *Int. J. Cancer* **2020**, 146, 3170–3183, 10.1002/ijc.32742.

179. Tiwari, N.; Gheldof, A.; Tatari, M.; Christofori, G. Emt as the ultimate survival mechanism of cancer cells. *Semin. Cancer Biol.* **2012**, 22, 194–207, 10.1016/j.semcancer.2012.02.013.

180. Jiang, J.; Wang, K.; Chen, Y.; Chen, H.; Nice, E.C.; Huang, C. Redox regulation in tumor cell epithelial-mesenchymal transition: Molecular basis and therapeutic strategy. *Signal. Transduct. Target Ther.* **2017**, 2, 17036, 10.1038/sigtrans.2017.36.

181. Azoury, S.C.; Straughan, D.M.; Shukla, V. Immune checkpoint inhibitors for cancer therapy: Clinical efficacy and safety. *Curr. Cancer Drug Targets* **2015**, 15, 452–462, 10.2174/15680096150 6150805145120.

182. Desrichard, A.; Snyder, A.; Chan, T.A. Cancer neoantigens and applications for immuno-therapy. *Clin. Cancer Res.* **2016**, 22, 807–812, 10.1158/1078-0432.CCR-14-3175.

183. Galluzzi, L.; Zitvogel, L.; Kroemer, G. Immuno-logical mechanisms underneath the efficacy of cancer therapy. *Cancer Immunol. Res.* **2016**, 4, 895–902, 10.1158/2326-6066.CIR-16-0197.

184. Galluzzi, L.; Vitale, I.; Warren, S.; Adjemian, S.; Agostinis, P.; Martinez, A.B.; Chan, T.A.; Coukos, G.; Demaria, S.; Deutsch, E., *et al.* Consensus guidelines for the definition, detection and interpretation of immunogenic cell death. *J. Immunother. Cancer* **2020**, 8, 10.1136/jitc-2019-000337.

185. Kroemer, G.; Galluzzi, L.; Kepp, O.; Zitvogel, L. Immunogenic cell death in cancer therapy. *Annu. Rev. Immunol.* **2013**, 31, 51–72, 10.1146/annurev-immunol-032712-100008.

186. Garg, A.D.; Agostinis, P. Cell death and immunity in cancer: From danger signals to mimicry of pathogen defense responses. *Immunol. Rev.* **2017**, 280, 126–148, 10.1111/imr.12574.

187. Garg, A.D.; Dudek, A.M.; Agostinis, P. Cancer immunogenicity, danger signals, and damps: What, when, and how? *BioFactors* **2013**, 39, 355–367, 10.1002/biof.1125.

188. Hou, W.; Zhang, Q.; Yan, Z.; Chen, R.; Zeh Iii, H.J.; Kang, R.; Lotze, M.T.; Tang, D. Strange attractors: Damps and autophagy link tumor cell death and immunity. *Cell Death Dis.* **2013**, 4, e966, 10.1038/cddis.2013.493.

189. Schcolnik-Cabrera, A.; Oldak, B.; Juarez, M.; Cruz-Rivera, M.; Flisser, A.; Mendlovic, F. Calreticulin in phagocytosis and cancer: Opposite roles in immune response outcomes. *Apoptosis* **2019**, 24, 245–255, 10.1007/s10495-019-01532-0.

190. Zhou, M.; Wang, X.; Lin, S.; Liu, Y.; Lin, J.; Jiang, B.; Zhao, X.; Wei, H. Combining photothermal therapy-induced immunogenic cell death and hypoxia relief-benefited m1-phenotype macrophage polarization for cancer immunotherapy. *Adv. Ther.* **2020**, 10.1002/adtp.202000191.

191. Sethuraman, S.N.; Singh, M.P.; Patil, G.; Li, S.; Fiering, S.; Hoopes, P.J.; Guha, C.; Malayer, J.; Ranjan, A. Novel calreticulin-nanoparticle in combination with focused ultrasound induces immunogenic cell death in melanoma to enhance antitumor immunity. *Theranostics* **2020**, 10, 3397–3412, 10.7150/thno.42243.

192. Yan, W.; Lang, T.; Qi, X.; Li, Y. Engineering immunogenic cell death with nanosized drug

delivery systems improving cancer immuno-therapy. *Curr. Opin. Biotechnol.* **2020**, *66*, 36–43, 10.1016/j.copbio.2020.06.007.

193. Rufo, N.; Garg, A.D.; Agostinis, P. The unfolded protein response in immunogenic cell death and cancer immunotherapy. *Trends Cancer* **2017**, *3*, 643–658, 10.1016/j.trecan.2017.07.002.

194. Garg, A.D.; Martin, S.; Golab, J.; Agostinis, P. Danger signalling during cancer cell death: Origins, plasticity and regulation. *Cell Death Differ.* **2014**, *21*, 26–38, 10.1038/cdd.2013.48.

195. Lennicke, C.; Rahn, J.; Bukur, J.; Hochgrafe, F.; Wessjohann, L.A.; Lichtenfels, R.; Seliger, B. Modulation of mhc class i sur-face expression in b16f10 melanoma cells by methylseleninic acid. *Oncoimmunology* **2017**, *6*, e1259049, 10.1080/2162402X.2016.1259049.

196. Riera-Domingo, C.; Audige, A.; Granja, S.; Cheng, W.C.; Ho, P.C.; Baltazar, F.; Stockmann, C.; Mazzone, M. Immunity, hypoxia, and metab-olism-the menage a trois of cancer: Implications for immunotherapy. *Physiol. Rev.* **2020**, *100*, 1–102, 10.1152/physrev.00018.2019.

197. Bekeschus, S.; Rodder, K.; Fregin, B.; Otto, O.; Lippert, M.; Weltmann, K.D.; Wende, K.; Schmidt, A.; Gandhirajan, R.K. Toxicity and immunogenicity in murine melanoma fol-lowing exposure to physical plasma-derived oxidants. *Oxid. Med. Cell. Longev.* **2017**, *2017*, 4396467, 10.1155/2017/4396467.

198. Rödder, K.; Moritz, J.; Miller, V.; Weltmann, K.-D.; Metelmann, H.-R.; Gandhirajan, R.; Bekeschus, S. Activation of murine immune cells upon co-cul-ture with plasma-treated b16f10 melanoma cells. *Appl. Sci.* **2019**, *9*, 660, 10.3390/app9040660.

199. Sagwal, S.K.; Pasqual-Melo, G.; Bodnar, Y.; Gandhirajan, R.K.; Bekeschus, S. Combination of chemotherapy and physical plasma elic-its melanoma cell death via upregulation of slc22a16. *Cell Death Dis.* **2018**, *9*, 1179, 10.1038/s41419-018-1221-6.

200. Lin, A.; Truong, B.; Pappas, A.; Kirifides, L.; Oubarri, A.; Chen, S.Y.; Lin, S.J.; Dobrynin, D.; Fridman, G.; Fridman, A.; Sang, N.; Miller, V. Uniform nanosecond pulsed dielectric barrier discharge plasma enhances anti-tumor effects by induction of immunogenic cell death in tumors and stimulation of macrophages. *Plasma Process. Polym.* **2015**, *12*, 1392–1399, 10.1002/ppap.201500139.

201. Lin, A.; Truong, B.; Patel, S.; Kaushik, N.; Choi, E.H.; Fridman, G.; Fridman, A.; Miller, V. Nanosecond-pulsed dbd plasma-generated reac-tive oxygen species trigger immunogenic cell death in a549 lung carcinoma cells through intracellular oxidative stress. *Int. J. Mol. Sci.* **2017**, *18*, 966, 10.3390/ijms18050966.

202. Mizuno, K.; Shirakawa, Y.; Sakamoto, T.; Ishizaki, H.; Nishijima, Y.; Ono, R. Plasma-induced sup-pression of recurrent and reinoculated melanoma tumors in mice. *IEEE TRPMS* **2018**, *2*, 353–359, 10.1109/trpms.2018.2809673.

203. Mizuno, K.; Yonetamari, K.; Shirakawa, Y.; Akiyama, T.; Ono, R. Anti-tumor immune response induced by nanosecond pulsed streamer discharge in mice. *J. Phys. D: Appl. Phys.* **2017**, *50*, 12LT01, 10.1088/1361-6463/aa5dbb.

204. Lin, A.G.; Xiang, B.; Merlino, D.J.; Baybutt, T.R.; Sahu, J.; Fridman, A.; Snook, A.E.; Miller, V. Non-thermal plasma induces immunogenic cell death in vivo in murine ct26 colorectal tumors. *Oncoimmunology* **2018**, *7*, e1484978, 10.1080/2162402X.2018.1484978.

205. Liedtke, K.R.; Freund, E.; Hackbarth, C.; Heidecke, C.-D.; Partecke, L.-I.; Bekeschus, S. A myeloid and lymphoid infiltrate in murine pancreatic tumors exposed to plasma-treated medium. *Clin. Plas. Med.* **2018**, *11*, 10–17, 10.1016/j.cpme.2018.07.001.

206. Van Loenhout, J.; Flieswasser, T.; Freire Boullosa, L.; De Waele, J.; Van Audenaerde, J.; Marcq, E.; Jacobs, J.; Lin, A.; Lion, E.; Dewitte, H., *et al.* Cold atmospheric plasma-treated pbs eliminates immunosuppressive pancreatic stellate cells and induces immunogenic cell death of pancreatic cancer cells. *Cancers (Basel)* **2019**, *11*, 10.3390/cancers11101597.

207. Bauer, G. The synergistic effect between hydrogen peroxide and nitrite, two long-lived molecular species from cold atmospheric plasma, triggers tumor cells to induce their own cell death. *Redox Biol.* **2019**, *26*, 101291, 10.1016/j.redox.2019.101291.

208. Chen, D.S.; Mellman, I. Oncology meets immu-nology: The cancer-immunity cycle. *Immunity* **2013**, *39*, 1–10, 10.1016/j.immuni.2013.07.012.

209. Bekeschus, S.; Clemen, R.; Metelmann, H.-R. Potentiating anti-tumor immunity with physical plasma. *Clin. Plas. Med.* **2018**, *12*, 17–22, 10.1016/j.cpme.2018.10.001.

210. Berger-Achituv, S.; Brinkmann, V.; Abed, U.A.; Kuhn, L.I.; Ben-Ezra, J.; Elhasid, R.; Zychlinsky, A. A proposed role for neutrophil extracellular traps in cancer immunoediting. *Front. Immunol.* **2013**, *4*, 48, 10.3389/fimmu.2013.00048.

211. Bekeschus, S.; Moritz, J.; Schmidt, A.; Wende, K. Redox regulation of leukocyte-derived microparticle release and protein content in response to cold physical plasma-derived oxidants. Clin. Plas. Med. **2017**, 7–8, 24–35, 10.1016/j.cpme.2017.07.001.

212. Awad, R.M.; De Vlaeminck, Y.; Maebe, J.; Goyvaerts, C.; Breckpot, K. Turn back the time: Targeting tumor infiltrating myeloid cells to revert cancer progression. Front. Immunol. **2018**, 9, 1977, 10.3389/fimmu.2018.01977.

213. Bekeschus, S.; Scherwietes, L.; Freund, E.; Liedtke, K.R.; Hackbarth, C.; von Woedtke, T.; Partecke, L.-I. Plasma-treated medium tunes the inflammatory profile in murine bone marrow-derived macrophages. Clin. Plas. Med. **2018**, 11, 1–9, 10.1016/j.cpme.2018.06.001.

214. Freund, E.; Moritz, J.; Stope, M.; Seebauer, C.; Schmidt, A.; Bekeschus, S. Plasma-derived reactive species shape a differentiation profile in human monocytes. Appl. Sci.-Basel **2019**, 9, 10.3390/app9122530.

215. Bekeschus, S.; Schmidt, A.; Bethge, L.; Masur, K.; von Woedtke, T.; Hasse, S.; Wende, K. Redox stimulation of human thp-1 monocytes in response to cold physical plasma. Oxid. Med. Cell. Longev. **2016**, 2016, 5910695, 10.1155/2016/5910695.

216. Crestale, L.; Laurita, R.; Liguori, A.; Stancampiano, A.; Talmon, M.; Bisag, A.; Gherardi, M.; Amoruso, A.; Colombo, V.; Fresu, L. Cold atmospheric pressure plasma treatment modulates human monocytes/macrophages responsiveness. Plasma **2018**, 1, 261–276, 10.3390/plasma1020023.

217. Kaushik, N.K.; Kaushik, N.; Min, B.; Choi, K.H.; Hong, Y.J.; Miller, V.; Fridman, A.; Choi, E.H. Cytotoxic macrophage-released tumour necrosis factor-alpha (tnf-alpha) as a killing mechanism for cancer cell death after cold plasma activation. J. Phys. D: Appl. Phys. **2016**, 49, 084001, 10.1088/0022-3727/49/8/084001.

218. Maikho, T.; Patwardhan, R.S.; Das, T.N.; Sharma, D.; Sandur, S.K. Cold atmospheric plasma-modulated phorbol 12-myristate 13-acetate-induced differentiation of u937 cells to macrophage-like cells. Free Radic. Res. **2018**, 52, 212–222, 10.1080/10715762.2017.1423069.

219. Lee, C.B.; Seo, I.H.; Chae, M.W.; Park, J.W.; Choi, E.H.; Uhm, H.S.; Baik, K.Y. Anticancer activity of liquid treated with microwave plasma-generated gas through macrophage activation. Oxid. Med. Cell. Longev. **2020**, 2020, 2946820, 10.1155/2020/2946820.

220. Benner, B.; Scarberry, L.; Suarez-Kelly, L.P.; Duggan, M.C.; Campbell, A.R.; Smith, E.; Lapurga, G.; Jiang, K.; Butchar, J.P.; Tridandapani, S.; Howard, J.H.; Baiocchi, R.A.; Mace, T.A.; Carson, W.E. Generation of monocyte-derived tumor-associated macrophages using tumor-conditioned media provides a novel method to study tumor-associated macrophages in vitro. J. Immunother. Cancer **2019**, 7, 10.1186/s40425-019-0622-0.

221. Kaushik, N.K.; Kaushik, N.; Adhikari, M.; Ghimire, B.; Linh, N.N.; Mishra, Y.K.; Lee, S.J.; Choi, E.H. Preventing the solid cancer progression via release of anticancer-cytokines in co-culture with cold plasma-stimulated macrophages. Cancers (Basel) **2019**, 11, 10.3390/cancers11060842.

222. Apetoh, L.; Locher, C.; Ghiringhelli, F.; Kroemer, G.; Zitvogel, L. Harnessing dendritic cells in cancer. Semin. Immunol. **2011**, 23, 42–49, 10.1016/j.smim.2011.01.003.

223. Bol, K.F.; Schreibelt, G.; Gerritsen, W.R.; de Vries, I.J.; Figdor, C.G. Dendritic cell-based immunotherapy: State of the art and beyond. Clin. Cancer Res. **2016**, 22, 1897–1906, 10.1158/1078-0432. CCR-15-1399.

CHAPTER TWENTY-ONE

Nutrition-Based Redox Regulation in Fish

IMPLICATIONS FOR GROWTH, DEVELOPMENT, HEALTH, AND FLESH QUALITY

Kristin Hamre, Sofie Remø, and Rune Waagbø

CONTENTS

21.1 INTRODUCTION

Modern fish farming is a product of systematic research and development during several decades in disciplines of farming techniques, relevant fish species, environmental coping, diseases, feed, and feeding. The development of suitable and sustainable feeds and feeding regimes has been accompanied by solving nutrition-based redox challenges that were at the base of many of the early production-related diseases (Hardy, 2012; Oliva-Teles, 2012; Waagbø, 2006; Waagbø, 2008). Management of feed oxidation and supplementation of correct levels and combinations of antioxidants are still important topics in research to assure good fish health and welfare. Recently, we have also adopted the ideas of redox signaling (Jones and Sies, 2015) in our research on fish nutrition.

21.1.1 Classical Nutritional Disorders— What Went Wrong

In the early history of aquaculture, feed-associated problems were experienced both from the chosen or available feed ingredients, unbalanced feed composition and single-nutrient deficiencies. The major feed ingredients at the time were marine trash fish and fish meals and oils, with highly unsaturated fatty acids that needed protection from auto-oxidation both in the feed and in the

DOI: 10.4324/9781003204091-27

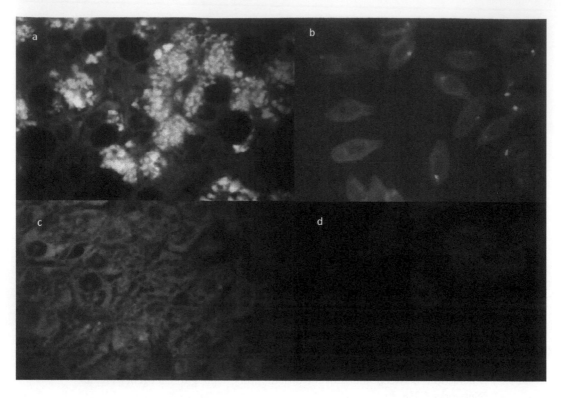

Figure 21.1 Auto-fluorescent oxidized material in the liver (a) and in red blood cells (b) of fish with vitamin E deficiency (c) and (d) are controls.

(Liver and red blood cell images are from Hamre et al. (1994) with permission from *Fish Physiology and Biochemistry*.)

fish. A major problem associated with rancid diets and lack of antioxidants is the development of lipoid liver degeneration (Figure 21.1), manifesting oxidative damage to exposed organs like the liver and erythrocytes (Hamre et al., 1994; Hardy, 2012; Oliva-Teles, 2012; Waagbø, 2006; Waagbø, 2008). The most probable nutrient deficiencies included the unstable antioxidants, vitamins C (ascorbic acid, AA) and E (tocopherol, TOH). These deficiencies may have occurred due to ignorance, omitted supplementation, or through heavy losses in feed production and storage. By feeding rancid diets, both feed levels and tissue levels of tocopherol and ascorbic acid were reduced (Baker and Davies, 1997; Hardy, 2012). Thus, lipid oxidation could easily be prevented by supplementation of synthetic antioxidants and the antioxidant vitamins.

The development towards intensive farming included more efficient energy-dense feeds, mainly by adding (coating) more lipids to extruded pellets. The feed production with extrusion techniques and higher lipid levels with polyunsaturated fatty acids (PUFA) caused the need for antioxidant protection to increase (Hardy, 2012; Oliva-Teles, 2012; Waagbø, 2006).

The early oxidative diseases and problems put forward scientific approaches to determine the dietary requirements of the most sensitive nutrients (Halver and Hardy, 2002; NRC, 2011). Because the natural inherent and the supplemented vitamins in the feed ingredients and feeds were easily lost during production and storage in the early days, both vitamin additives were soon replaced by the chemically protected forms, such as α-tocopheryl acetate and ascorbyl phosphate. The protected and stable vitamins have no function as antioxidants in the feed, because the chemical side groups need to be hydrolysed in the intestine before uptake of the active vitamins. The stable forms made more accurate vitamin requirement studies possible.

21.2 NUTRIENTS CENTRAL TO REDOX REGULATION IN FISH— CURRENT KNOWLEDGE

21.2.1 Requirements

Most teleost fishes need vitamin C (ascorbic acid, AA) in the diet due to the lack of enzymes for endogenous AA synthesis (Maeland and Waagbø, 1998). The requirement varies between species, farming conditions, and interacting nutrients (Waagbø et al., 2000). Classical deficiency signs of AA in fish, like anaemia, bone deformities (lordosis and scoliosis; Figure 21.2), and mortalities can easily be induced by omitting AA supplementation, because pre-treatment of most feed ingredients destroys the natural AA content. The minimum requirement for AA ranges between 10 and 100 mg/kg dry diet, using the stable AA-phosphate forms and a panel of somatic and biochemical biomarkers such as weight gain and feed efficiency, absence of deficiency signs and mortality, OH-Pro concentration, and maximum tissue AA stores (Dabrowski, 2001; Halver and Hardy, 2002; NRC, 2011).

Similarly, vitamin E deficiency causes oxidative damage to organs (muscular dystrophy, oedema, anaemia, depigmentation, and ceroid accumulation in the liver; Figure 21.1) and lowered growth and higher mortalities (Hamre and Lie, 1995; Hardy, 2012; NRC, 2011). The requirement and display of deficiency symptoms depend on interactions with other nutrients, such as vitamin C, selenium, and polyunsaturated fatty acids (Hamre, 2011; Hamre et al., 1997). Rancid diets increase the requirement (Baker and Davies, 1997; Hamre, 2011; Hung

et al., 1981), and the environment may also have an effect (Hamre et al., submitted; Hamre et al., 2016). According to NRC (2011), the requirement for vitamin E in 16 species of fish ranges between 25 and 200 mg/kg using the stable α-tocopheryl acetate form, mostly based upon weight gain, absence of deficiency signs, maximum liver tocopherol storage, and peroxidation measures. By exerting antioxidant properties, excess feed levels of both tocopherols and ascorbic acid have been extensively examined for their immunomodulatory properties in farmed fish species (Hamre, 2011; Kiron, 2012; Pohlenz and Gatlin, 2014; Trichet, 2010; Waagbø, 2006), as discussed in more detail below.

Astaxanthin has a natural part in the coloration of the growing salmonid muscle, skin, and egg (NRC, 2011). The pigment originates from the use of natural crustacean meals and oils as feed ingredients, additives based on microorganisms or synthetically produced products (da Costa and Miranda, 2020). Despite no defined essential role as a vitamin, astaxanthin is a powerful antioxidant and a provitamin A source (NRC, 2011). The antioxidative mechanisms include singlet oxygen quenching, where cis isomers exhibit higher antioxidant activity than the all-trans isomer (Liu and Osawa, 2007). Salmon feed normally contains approximately 50 mg/kg astaxanthin (Sanden et al., 2017), while the current safe upper limit is 100 mg/kg (EC, No 1415/2015). There is an annual variation in muscle astaxanthin in Atlantic salmon, where lower values in the spring with rising water temperatures have been related to increasing growth rates, described in more detail below.

Figure 21.2 Vertebral deformities in fish with vitamin C deficiency, lordosis, and scoliosis (left) and control fish with sufficient vitamin C (right). The deformities are due to insufficient connective tissue due to lack of hydroxyproline.

Several other nutrients take part in the multicomponent antioxidant defence system in fish, such as minerals (Oliva-Teles, 2012) and free amino acids and metabolites (Andersen et al., 2016). For the minerals, feed supplementation above the requirement may impact fish physiology, immunity, and welfare (Baeverfjord et al., 2019; Berntssen et al., 2000; Chanda et al., 2015; Kiron, 2012; Prabhu et al., 2020; Waagbø and Remø, 2020). Based on the best knowledge at the time, the feed law has upper safe limits for both essential and heavy metals. Because the law should consider both human and fish safety and environmental concerns, there may be a conflict between fulfilling the requirements for growth and welfare and, at the same time, minimizing the output to the environment. The solution lays in increasing dietary mineral availability (Prabhu et al., 2019; Silva, 2019).

Synthetic antioxidants are supplemented in aquafeeds as a technical additive to prevent oxidation in the production, transportation, and storage of feed ingredients and complete feeds. By doing so, they spare antioxidant nutrients. However, trace amounts of the compounds or metabolites may be found in the edible products (Bohne et al., 2008; Lundebye et al., 2010; Yamashita et al., 2009), and due to fish health and safety concerns, there is continuous re-evaluation of their use and upper safe levels in fish feed and seafood (EFSA/FEEDAP, 2015).

The developing scene for fish feed during the last decennials include a shift from 70% fishmeal and fish oil to 70%–80% of protein and lipid ingredients from plant seeds and grains (Aas et al., 2019).

Further, novel feed ingredients used in aquafeeds include animal by-products, new protein- and fat-rich alternatives among plant ingredients, and ingredients based on intensively grown bacteria, yeast, and insects. While animal by-products and insect meals and fats need to be carefully protected against oxidation in the ingredients and feeds, plant meals and oils may contain relatively high levels of natural antioxidants, mostly α-TOH and γ-TOH.

21.2.2 Interactions between Antioxidant Nutrients and Coupling to Endogenous Redox Networks

The hypothesis that oxidized vitamin E is recycled by ascorbic acid was first proposed by Tappel (1962) and was intensively researched in the ensuing years (Chan et al., 1991; Kagan et al., 1992; Niki, 1987; Wefers and Sies, 1988). However, it was difficult to prove that the same mechanism was active in vivo (Chen, 1989; Burton et al., 1990; Draper, 1993). Later it became clear that the antioxidant nutrients are part of the redox network, described for example by Jones and Sies (2015). Märtensson and Meister (1991) and Meister (1994) showed that dehydro-ascorbic acid was recycled to AA by GSH. It is well-known that GSSG is recycled to GSH by NADPH (Jones and Sies, 2015). In Atlantic salmon, indications that vitamin E was recycled by vitamin C were presented by Hamre et al. (1997), because both vitamin E status and survival of vitamin E deficient fish increased with increasing vitamin C in the diet (Figure 21.3). Furthermore, vitamin C

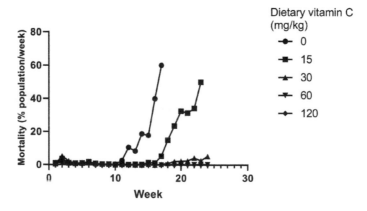

Figure 21.3 The mortality in vitamin E–deficient Atlantic salmon juveniles is dependent on dietary vitamin C.

(Modified from Hamre et al. (1997) with permission from *Free Radical Biology and Medicine*.)

REDOX REGULATION OF DIFFERENTIATION AND DE-DIFFERENTIATION

deficiency appeared earlier in fish fed high levels of vitamin E. These interactions were confirmed in yellow perch by Lee and Dabrowski (2003, 2004) and Yildirim-Aksoy et al. (2008) in channel catfish. In this way, the functions of vitamins C and E in Atlantic salmon are coupled to the redox network and energy metabolism (Hamre, 2011).

There is also a strong interaction between vitamin E and selenium in fish nutrition, where tocopherols are membrane-bound antioxidants and selenium exerts its effect through the selenium-dependent GSH-PXs. In early fish nutrition history, it was shown that both nutrients were needed to prevent muscular dystrophy in Atlantic salmon (Poston et al., 1976). Because this interaction was not seen in rainbow trout and channel catfish, it was suggested that it could depend on fish species and age as well as on dietary lipid level and water temperature.

Several other vitamin-vitamin and vitamin-mineral interactions have major effects on the redox system in farmed fish, such as vitamin E–PUFA (Watanabe et al., 1981), vitamin C–iron (Maage et al., 1990), and astaxanthin–antioxidant vitamins (Christiansen et al., 1995).

21.2.3 Oxidized Dietary Lipids: Feed Intake, Absorption, and Antioxidant Status

Atlantic salmon avoid eating rancid feed. When oxidized lipids are eaten, aldehydes, but not lipid peroxides, seem to be absorbed (Hamre et al., 2001). Vertebrates appear to be protected from uptake of lipid hydroperoxides by gastrointestinal glutathione peroxidase (GI-GPX), which reduces fatty acid hydroperoxides to hydroxy fatty acids at the expense of reduced glutathione, thereby preventing the decomposition to fatty acid alkoxy radicals (Aw et al., 1992; Esworthy et al., 1998). Oxidized lipids may also lead to lowered levels of antioxidants in the fish tissues and pathologies, which can be overcome by feeding excess antioxidants to the fish (Baker and Davies, 1997; Cowey et al., 1984; Hung et al., 1981).

21.2.4 Redox Regulation in Development of Fish Embryo and Larvae

The concentrations of nutrients in the egg at fertilization, including nutrients involved in redox regulation, depend on broodstock nutrition (Fernandez-Palacios et al., 2011). In a study on zebrafish, selenium in the yolk increased linearly with increasing levels in the broodstock diet (Penglase et al., 2014). Similarly, ascorbic acid and α-tocopherol in eggs from Atlantic salmon depended on the dietary concentrations (Lie et al., 1994; Waagbo et al., 1989). Broodstock diets for fish tend to contain ample amounts of vitamin C and E, which is reflected in the concentrations in the yolk. Accordingly, Rønnestad et al. (1999) followed the transfer of vitamins C and E from the yolk to the body of Atlantic halibut larvae. They found that at the onset of first feeding, when the yolk had been absorbed, there was a sharp increase of antioxidant vitamins in the body of the larvae that were then rapidly excreted, indicating a surplus of antioxidant vitamins in the broodstock diet. There is still a need for further research on the nutrient requirements in fish broodstock.

Embryonic development is an orchestrated process with well-defined events that are used for staging (Kimmel et al., 1995). It is accompanied by regulations of the concentrations of reduced and oxidized glutathione (GSH/GSSG) and the resulting glutathione-based redox potential (Eh_{GSH}). In zebrafish, Eh_{GSH} was relatively reduced at fertilization (−230mV) and reached a maximum (became more oxidized) after the blastula and cleavage stages. Then it stayed at −118 to −120 mV during organ differentiation and decreased to approximately −230 mV at hatching (Timme-Laragy et al., 2013). This is in line with cell studies that show that proliferation happens at relatively reduced redox potentials and differentiation in more oxidized conditions (Hoffman et al., 2008; Lillig et al., 2008). A similar trajectory was found for cod embryos (Skjaerven et al., 2013), in zebrafish expressing roGFP (Zhang et al., 2019), and in mice (Timme-Laragy et al., 2013). The studies also measured the complex development of GSH/GSSG concentrations and genes involved in synthesis and utilization of GSH/GSSG and other redox couples. The small size of embryos hampers detailed studies of redox regulation at the tissue level of embryonic development. However, development of transgene zebrafish expressing redox sensitive probes for measurements of redox potentials and H_2O_2 production in selected organs or in the whole body will give great opportunities for further work (Gutscher et al., 2008; Morgan et al., 2011).

Larvae of marine fish are not fully developed at hatching and first feeding; they are very small (3–12 mm standard length); and most often must be given live feed for 1–8 weeks before they can survive on a standard commercial diet. The feed organisms used in commercial hatcheries are rotifers and *Artemia* enriched with specific nutrients, while in the wild, fish larvae mostly feed on copepods, which are very nutritious compared to the hatchery feeds (Hamre et al., 2013; Karlsen et al., 2015). Feeding copepods to cod larvae in the hatchery gave a large increase in larval growth (Busch et al., 2010; Karlsen et al., 2015; Koedijk et al., 2010), which was accompanied by increased concentration of GSSG, a more oxidized Eh_{GSH}, and altered expression of many redox-related genes (Penglase et al., 2015). In an earlier study of the ontogeny of redox regulation in cod larvae fed rotifers and *Artemia* (Hamre et al., 2014), the GSH concentration of the larvae during metamorphosis was approximately double that of the larvae fed *Artemia* in the study of Penglase et al. (2015), the GSSG concentration was low and stable, and the Eh_{GSH} was very reduced. Later it was discovered that standard rotifer cultures at the time had selenium levels far below the requirement in fish, which could have triggered a compensatory synthesis of glutathione in the larvae. In the study of Penglase et al. (2015), rotifers were enriched with selenium to levels above the requirement.

Vitamin C and E have been in focus in research on marine fish larval culture. Natural levels in *Artemia* are very high at 500–1000 mg/kg DM; still both rotifers and *Artemia* are enriched with ample amounts so that they may contain up to 3 g/kg DM (Kolkovski et al., 2000; Merchie, 1997). Similar high levels may be found in compound diets for marine fish larvae (Atalah et al., 2008). However, copepods as the natural diet for fish larvae contain only 500 and 110 mg/kg DM of vitamin C and E, respectively (Hamre et al., 2013; Karlsen et al., 2015). It is not known how the current supplementation of antioxidant nutrients in live and particulate feed affect the redox regulation in fish larvae.

21.2.5 Effects of Nutrient Variation and Environmental Conditions on GSH Metabolism

Several studies indicate that the GSH/GSSG concentrations and the redox potential are strictly regulated in fish. In a two-way regression study on Atlantic salmon juveniles (Hamre et al., 1997), dietary vitamin C was gradually increased from 0 to 60 mg/kg at three dietary levels of vitamin E

(0, 150, and 300 mg/kg). Fish fed the diets with no vitamin supplementation became deficient and there was a clear decline in the whole-body concentrations of both vitamins in response to deficiency. The concentrations of GSH, cysteine, and cysteine-glycine changed with time and stage of the fish, but there was very little effect of the dietary variation. Similar results were found in Atlantic salmon smolts in the sea, which were fed diets with two levels of seven nutrients involved in redox metabolism (vitamins C and E, astaxanthin, marine lipid, iron, copper, and manganese), according to a reduced factorial design (Hamre et al., 2010). The low levels were just above the minimum requirement and the high levels were below the assumed toxicity levels. There was very little effect of the dietary variation on GSH/GSSG concentrations and the Eh_{GSH}. Furthermore, the GSH/GSSG concentrations and Eh_{GSH} were stable in Atlantic salmon smolt in seawater fed a nutrient package with all micronutrients varying from no addition to a supplementation up to four times the recommended supplementation (Hamre et al., 2016; NRC, 2011). There was some variation in GSH status in the freshwater phase, with more reduced fish when micronutrient supplementation was below the requirement. This may indicate that the fish compensated for lowered micronutrient supplementation by producing more GSH and/or reducing GSSG.

The above indicates that the Eh_{GSH} in healthy Atlantic salmon is strictly regulated and that dietary manipulation must be severe, including for example nutrient deficient diets, to have an effect on redox status. However, the GSH/GSSG concentrations and Eh_{GSH} seem to vary with time and stage of the fish. For example, the Eh_{GSH} in Atlantic salmon liver did not respond to dietary variation in amino acids, fatty acids, and B vitamins in the fresh water phase but became more reduced after transfer of the fish to seawater (Sissener et al., 2021).

Environmental conditions affect the redox state in Atlantic salmon. An example is the increase in daylight and temperature in spring, which causes a burst in fish growth, apparently accompanied by increased oxidative stress and increased utilisation of antioxidants, such as vitamin C, E, and astaxanthin (Figure 21.4; Hamre et al., submitted; Nordgarden et al., 2003). The GSH/GSSG concentrations and Eh_{GSH} in liver, muscle, and lens varied over time. The redox potential of the lens was very oxidized in the beginning of spring and then

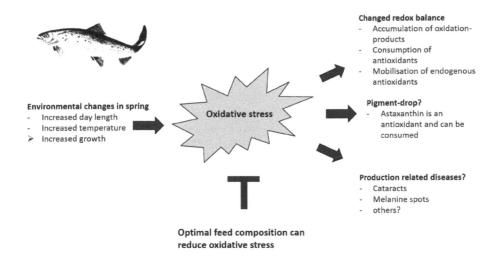

Figure 21.4 Causes, consequences, and prevention of oxidative stress in Atlantic salmon in spring. See text for further explanations.

became more reduced after a burst in production of GSH and lowered GSSG. This change may represent a compensatory mechanism to protect the lens. The period of oxidized lens Eh_{GSH} corresponded with an increase in development of cataract, which is a well-known effect of oxidative stress in mammals (Lou, 2003; Spector, 1995; Williams, 2006).

21.3 DISEASES AND REDOX REGULATION IN FISH

Essential nutrients take part in redox reactions as substrates, oxidants, or antioxidants. In addition to the antioxidant vitamins described in detail above, several nutrients have antioxidant functions or influence the antioxidant system, and both nutrient deficiencies and toxicities resulting in increased oxidative stress have been associated with the occurrence of production-related disorders in fish farming (Waagbø, 2006; Waagbø and Remø, 2020). In addition, the environmental impact on fish from sub-optimal water temperatures or hypoxic or hyperoxic conditions can result in oxidative stress (Waagbø et al., 2020). The combined effects of sub-optimal nutrition, challenging environmental conditions, and genetic predisposition can increase the risk for production-related disorders in fish (Waagbø et al., 2020). Of the disorders relevant today, anaemia, bone deformities, and cataracts may be caused by alterations in redox regulation and oxidative stress, resulting in specific and visible symptoms.

21.3.1 Nutritional Anaemias

Anaemia is defined as a reduction in haemoglobin to an extent that impacts oxygen transport, where the fish develop pale gills and low blood haemoglobin concentration. Fish may experience anaemias related to bleeding, infectious diseases, nutrient deficiencies (iron, selenium, vitamin C, vitamin E, vitamin D, vitamin K, B vitamins, and essential fatty acids) and toxicologies (metals, oxidized fish oil, contaminants; Roberts and Rodger, 2012; Waagbø, 2006).

The most frequently observed form is the haemolytic anaemia. It appears during infectious diseases, caused by bacterial haemolysins (e.g., from *Vibrio anguillarum*) or intraerythrocytic viruses (the iridovirus causing viral erythrocytic necrosis; Roberts and Rodger, 2012). Deficiencies of the antioxidant vitamins C and E caused haemolytic anaemia by oxidative destabilization of cell membranes (Hamre and Lie, 1995; Sandnes et al., 1990). Erythrocytes are specifically sensitive to membrane damage due to the presence of haem iron and the role as oxygen transporter. On the other hand, feeding antioxidant vitamins in excess may protect the fish cell membranes from oxidative damage during infectious diseases (Waagbø, 1994). For any reasons of anaemia, it is important that the fish have the possibility to recover through increased haematopoiesis. This means that essential nutrients like folate and vitamin B12, iron, and vitamin C, needed for cell division and haemoglobin synthesis, should be supplied in sufficient amounts. Iron is transported and incorporated into the erythrocytes in the reduced state. Because ascorbic acid is needed for this reduction, AA deficiency will cause an iron deficiency–like hypochromic (but normocytic) anaemia (Hamre et al., 1994).

21.3.2 Nutrition-Related Bone Deformities

Bone deformities often occur in farmed fish and at any stages of the life cycle. The disorders are of multifactorial origin, however they are often caused by nutritional imbalance (Waagbø, 2006; Waagbø et al., 2020). Nutrients in deficiency and excess may impact bone development and maintenance (remodelling) through impairing bone cell differentiation and function, matrix composition, and bone mineralization (Waagbø, 2006). Of the classical bone deformities, scoliosis and lordosis appear in vitamin C–deficient fish (Dabrowski, 2001) and are related to missing post-translatory OH-proline formation in collagen peptides.

Excess of vitamin A has been demonstrated to be bone toxic to the developing fish. The mechanism seems to be through the role of vitamin A metabolites (retinoic acid) in bone cell development and mineralization (Dedi, 1997; Haga et al., 2002; Hernandez and Hardy, 2020). By being part of the factors influencing bone formation and remodelling, selected nutrients can be used to cure bone disorders.

21.3.3 Cataracts

Cataract is characterized as opacities of the lens that result in reduced vision, caused by both changes in the epithelial tissues surrounding the lens fibres and the composition and structure of the lens fibres (Bjerkås et al., 2006). Oxidative stress is one of the major risk factors for cataract development in both animals and humans (Spector, 1995; Williams, 2006), and the lens is dependent on a balanced redox state to maintain transparency (Lou, 2003). Sub-optimal levels of several nutrients have been associated with cataract development in fish, including methionine, tryptophan, riboflavin, zinc, and manganese (Bjerkås et al., 2006), as well as balanced levels of pro- and antioxidants (Waagbø et al., 2003). In the latter study, supplementation of vitamin C and astaxanthin was beneficial for minimizing cataract development, while vitamin E had no effect. In the last decades, cataract development in Atlantic salmon has been linked

specifically to sub-optimal dietary levels of the histidine, where a higher level than the requirement for growth is needed to minimize cataract development after seawater transfer and the second year in the sea (Breck et al., 2005; Remo et al., 2014; Waagbø et al., 2010). Both histidine and histidine derivatives such as N-acetylhistidine (NAH) have well-known antioxidant functions (Wade and Tucker, 1998). Histidine also influences the gene expression of glutaredoxin in lenses *ex vivo* (Remo et al., 2011) and antioxidant enzymes in the liver (Remo et al., 2014).

Cataract development has also been related to rapid growth rates and elevated temperatures, possibly in relation to underlying mechanisms resulting in increased oxidative pressure and oxidative stress (Waagbø et al., 2020; Waagbø and Remø, 2020). In a comparative study with rainbow trout (*Oncorhynchus mykiss*) and Atlantic salmon, high temperature was shown to influence the GSH metabolism in lenses. Higher levels of both GSH and GSSG were seen in the rainbow trout lenses, indicating a higher synthesis, while salmon had a large variation in lens GSH level as well as changes in precursor and intermediates of GSH synthesis. This may suggest a lower ability to synthetize or regenerate GSH in salmon than in rainbow trout. Both species also had increased levels of ophthalmate, a suggested marker for GSH depletion and oxidative stress, in lenses at high temperature (Remo et al., 2017).

21.3.4 Infectious Diseases and Immune Response

Tissue injuries after infectious diseases and inflammation may be caused by oxidative stress and a rate of formation of free radicals that exceeds the capacity of the antioxidant defence to remove them. The formation of ROS in the respiratory burst is an essential part of the response to an infection, actively eliminating invading pathogens and acting as signal molecules in inflammation and migration of leukocytes (Biller and Takahashi, 2018). The respiratory burst can cause damage to surrounding cells and tissues and the nutritional status at the time of infection, and the ability to maintain redox homeostasis, is crucial in determining the efficiency of the immune response and the resulting severity of pathology.

Alterations in antioxidant status in tissues, transcriptomic expression of antioxidant enzymes and oxidative injuries have been seen in a wide range of infectious diseases in fish. For instance, lower GSH concentrations were seen in liver of Atlantic salmon suffering from infectious salmon anaemia (ISAV; Hjeltnes et al., 1992) and in ASK cells exposed to ISAV in vitro, along with transcriptional indications of oxidative stress (Schiotz et al., 2008). In Atlantic salmon diagnosed with pancreas disease (SAV), effects were also seen on the oxidative stress response in the heart, as well as in lowered muscle astaxanthin levels (Larsson et al., 2012). Alterations in the redox status indicated by changes in the transcriptional levels of glutathione S-transferase, glutathione reductase and glutathione peroxidase, for instance, were seen in Atlantic salmon macrophages after *Piscirickettsia salmonis* infection (Rise et al., 2004). Furthermore, SOD and CAT activities were altered in skin and mucus of Atlantic salmon challenged with *Aeromonas salmonicida* (Du et al., 2015). Oxidative stress was also observed as a response to amoebic gill disease, either as a cause or a consequence of the development of lesions in the gills (Marcos-Lopez et al., 2018).

Dietary supplementation of nutrients with antioxidant properties, or nutrients that can modulate the antioxidant system, can also modulate disease resistance. Effects are detected at different biological levels; for example, strengthening barrier tissues, humoral and cellular components of the immune system, non-specific and acquired immune responses, and mortality (Kiron, 2012; Pohlenz and Gatlin, 2014; Trichet, 2010; Waagbø, 2006; Waagbø, 2008; Waagbø and Remø, 2020). On one hand, antioxidants may take part in radical formation as protective measures towards pathogens, while on the other hand they take part in cell and tissue self-protection by scavenging oxygen-rich free radicals. Several studies have shown that dietary supplementation of nutrients with antioxidant functions such as vitamins C and E and carotenoids can enhance the phagocytic respiratory burst activity (Pohlenz and Gatlin, 2014; Trichet, 2010). Dietary vitamin E and C supplementation have also been shown to increase lysozyme activity (Saurabh and Sahoo, 2008; Waagbø et al., 1993a) and sufficient levels are necessary to support alternative complement pathway activity (Boshra et al., 2006; Waagbø et al., 1993a). In mrigal (*Cirrhinus mrigala*) it was shown that vitamin C supplementation modulated the inflammatory response after injection with *A. hydrophila*, induced a quicker infiltration of phagocytic cells, and reduced lesions at the injection site in skeletal muscle (Sobhana et al.,

2002). Dietary vitamin C has also been shown to improve the wound-healing process in rainbow trout (Wahli et al., 2003).

In our experiment with the aim to examine the overall effects of several important dietary pro- and antioxidants (vitamin E, vitamin C, astaxanthin, total lipid, iron, copper, and manganese) and their interactions on protective mechanisms and health in Atlantic salmon post-smolts, the results on immunity were reported by Lygren et al. (1999). As for the glutathione system described above, the immune system was relatively robust to the wide ranges of the selected nutrients. The head kidney macrophage respiratory burst activities were mostly influenced by the dietary vitamin E, while the tissue endogenous antioxidants and antioxidant enzymes (catalase, superoxide dismutase, GSH, and total mercaptans) were influenced by dietary vitamin E, astaxanthin, and lipid levels.

Similar to responses during an infection, vaccination has been shown to cause oxidative stress in fish, as indicated by, for instance, increased liver MDA in rainbow trout (Tkachenko et al., 2016) and reduced liver α-TOH and AA concentrations in Atlantic salmon (Lygren et al., 2001). Dietary supplementation of antioxidant vitamins has been shown to boost the response to vaccines, with increased antibody response after vaccination in Atlantic salmon supplemented with vitamin C (Waagbø et al., 1993a). In the study by Tkachenko et al. (2016), the vaccinated fish also had increased antioxidant enzyme activities in the liver, possibly related to the oxidative stimuli due to the vaccine. Thus, being able to maintain the redox homeostasis after an oxidative insult such as vaccination may improve the overall response.

In summary, research shows that the antioxidant status may be a determinant for the initiation, progression, and recovery from infectious diseases in fish. Thus, antioxidant remedies are widely used as health promoting and immunomodulating supplements in health feeds for aquaculture (Waagbø, 2006; Waagbø and Remø, 2020).

21.4 ANTIOXIDANTS AND FLESH QUALITY

Fish fed diets with relatively high concentrations of n-3 polyunsaturated fatty acids accumulate these in the fillet, which becomes susceptible to lipid peroxidation post-harvest (Boggio et al., 1985; Frigg et al., 1990; Gatlin et al., 1992; Sigurgisladottir et al., 1994). In cooled storage, bacterial degradation is more important than lipid oxidation in reducing flesh quality. However, fish in frozen storage accumulate lipid peroxidation products and lose antioxidants over time. Increased dietary supplementation with vitamins C and E will inhibit this development (Hamre et al., 1998; Ruff et al., 2003; Secci and Parisi, 2016). It is therefore a long tradition to feed extra antioxidant vitamins to increase the shelf life of cultured fish (Waagbo et al., 1993b). In salmonids, which are fed astaxanthin to achieve the red flesh colour, extra antioxidant vitamins may protect against colour deterioration both in vivo and post-harvest. Other nutrients in the diet may also affect the absorption and retention of astaxanthin (Chimsung et al., 2014).

21.5 SUMMARY AND CONCLUSIONS

From the discussion above, farmed fish experience direct and indirect nutrition-based impacts on the integrated redox system, especially affecting development, growth, and metabolism in sensitive stages. Historically, the impact of severe redox imbalances has been related to reduced immunity and welfare, development of production related disorders and mortalities.

The positive aspect here is that many of the challenges can be alleviated with knowledge-based nutritional measures through both balanced supplementation of antioxidants and feeding regimes. Feed antioxidants and feeding regimes should also be adjusted according to scientific knowledge on life cycle, health status, and environmental conditions. Excess of supplemented antioxidants, either as a mix of pure chemical compounds or as complex feed additives, are among popular ingredients in health feeds for aquaculture (Waagbø & Remø, 2020). It is, however, important to scientifically validate the biological effects of the remedies, in concentration and in time.

REFERENCES

Aas, T.S., Ytrestøyl, T., Åsgård, T., 2019. Utilization of feed resources in the production of Atlantic salmon (Salmo salar) in Norway: An update for 2016. Aquaculture Reports 15.

Andersen, S.M., Waagbø, R., Espe, M., 2016. Functional amino acids in fish nutrition, health and welfare. Front Biosci (Elite Ed) 8, 143–169.

Atalah, E., Hernandez-Cruz, C.M., Montero, D., Ganuza, E., Benitez-Santana, T., Ganga, R., Roo, J.,

Fernandez-Palacios, H., Izquierdo, M.S., 2008. Enhancement of Gilthead Seabream and sea bass larval growth by dietary vitamin E in relation to different levels of essential fatty acids, XIII International Symposium on Fish Nutrition and Feeding, June 1–5, Florianopolis, Brazil.

Aw, T.Y., Williams, M.W., Gray, L., 1992. Absorption and lymphatic transport of peroxidized lipids by rat small-intestine in vivo—role of mucosal GSH. American Journal of Physiology 262, G99–G106.

Baeverfjord, G., Prabhu, P.A.J., Fjelldal, P.G., Albrektsen, S., Hatlen, B., Denstadli, V., Ytteborg, E., Takle, H., Lock, E.J., Berntssen, M.H.G., Lundebye, A.K., Asgard, T., Waagbø, R., 2019. Mineral nutrition and bone health in salmonids. Reviews in Aquaculture 11, 740–765.

Baker, R.T.M., Davies, S.J., 1997. Modulation of tissue alpha-tocopherol in African catfish, Clarias gariepinus (Burchell), fed oxidized oils, and the compensatory effect of supplemental dietary vitamin E. Aquaculture Nutrition 3, 91–97.

Bell, J.G., Cowey, C.B., Adron, J.W., Shanks, A.M., 1985. Some effects of vitamin E and selenium deprivation on tissue enzyme levels and indices of tissue peroxidation in rainbow trout (Salmo gairdneri). British Journal of Nutrition 53, 149–157.

Berntssen, M.H.G., Lundebye, A.K., Hamre, K., 2000. Tissue lipid peroxidative responses in Atlantic salmon (Salmo salar L.) parr fed high levels of dietary copper and cadmium. Fish Physiology and Biochemistry 23, 35–48.

Biller, J.D., Takahashi, L.S., 2018. Oxidative stress and fish immune system: Phagocytosis and leukocyte respiratory burst activity. Anais da Academia Brasileira de Ciências 90, 3403–3414.

Bjerkås, E., Breck, O., Waagbø, R., 2006. The role of nutrition in cataract formation in farmed fish. CAB Reviews 1(33), 1–16.

Boggio, S.M., Hardy, R.W., Babbitt, J.K., Brannon, E.L., 1985. The influence of dietary lipid source and alpha-tocopheryl acetate level on product quality of rainbow trout (Salmo gairdneri). Aquaculture 51, 13–24.

Bohne, V.J.B., Lundebye, A.K., Hamre, K., 2008. Accumulation and depuration of the synthetic antioxidant ethoxyquin in the muscle of Atlantic salmon (Salmo salar L.). Food and Chemical Toxicology 46, 1834–1843.

Boshra, H., Li, J., Sunyer, J.O., 2006. Recent advances on the complement system of teleost fish. Fish & Shellfish Immunology 20, 239–262.

Breck, O., Bjerkas, E., Campbell, P., Rhodes, J.D., Sanderson, J., Waagbø, R., 2005. Histidine nutrition

REDOX REGULATION OF DIFFERENTIATION AND DE-DIFFERENTIATION

and genotype affect cataract development in Atlantic salmon, Salmo salar L. Journal of Fish Diseases 28, 357–371.

Burton, G.W., Wronska, U., Stone, L., Foster, D.O., Ingold, K.U., 1990. Biokinetics of dietary RRR-α-tocopherol in the male guinea pig at three dietary levels of vitamin C and two levels of vitamin E. Evidence that vitamin C does not 'spare' vitamin E in vivo. Lipids 25, 199–210.

Busch, K.E.T., Falk-Petersen, I.B., Peruzzi, S., Rist, N.A., Hamre, K., 2010. Natural zooplankton as larval feed in intensive rearing systems for juvenile production of Atlantic cod (Gadus morhua L.). Aquaculture Research 41, 1727–1740.

Chan, A.C., Tran, K., Raynor, T., Ganz, P.R., Chow, C.K., 1991. Regeneration of vitamin-E in human platelets. Journal of Biological Chemistry 266, 17290–17295.

Chanda, S., Paul, B.N., Ghosh, K., Giri, S.S., 2015. Dietary essentiality of trace minerals in aquaculture—a review. Agricultural Review 36, 100–112.

Chen, L.H., 1989. Interaction of vitamin E and ascorbic acid. In Vivo 3: 199–209.

Chimsung, N., Tantikitti, C., Milley, J.E., Verlhac-Trichet, V., Lall, S.P., 2014. Effects of various dietary factors on astaxanthin absorption in Atlantic salmon (Salmo salar). Aquaculture Research 45, 1611–1620.

Christiansen, R., Glette, J., Lie, Ø., Torrissen, O.J., Waagbø, R., 1995. Antioxidant status and immunity in Atlantic salmon, Salmo salar L, fed semi-purified diets with and without astaxanthin supplementation. Journal of Fish Diseases 18, 317–328.

Cowey, C.B., Degener, E., Tacon, A.G.J., Youngson, A., Bell, J.G., 1984. The effect of vitamin E and oxidized fish oil on the nutrition of rainbow trout (Salmo gairdneri) grown on natural varying water temperatures. British Journal of Nutrition 51.

Dabrowski, K., 2001. History, present, and future of ascorbic acid research in aquatic organisms, in: K. Dabrowski (Ed.), Ascorbic Acid in Aquatic Organisms—Status and Perspectives. CRC Press, Boca Raton, 255–277.

da Costa, D.P., Miranda, K.C., 2020. The use of carotenoid pigments as food additives for aquatic organisms and their functional roles. Reviews in Aquaculture 12, 1567–1578.

Dedi, J., Takeuchi, T., Seikai, T., Watanabe, T., Hosoya, K., 1997. Hypervitaminosis a during vertebral morphogenesis in larval Japanese flounder. Fisheries Science 63, 466–473.

Draper, H.H., 1993. Interrelationships of Vitamin E with Other Nutrients. Marcel Dekker, Inc, New York, Basel, Hong Kong.

Du, Y.S., Yi, M.M., Xiao, P., Meng, L.J., Li, X., Sun, G.X., Liu, Y., 2015. The impact of Aeromonas salmonicida infection on innate immune parameters of Atlantic salmon (Salmo salar L). Fish & Shellfish Immunology 44, 307–315.

EC, No 1415/2015. COMMISSION REGULATION (EU) 1415/ 2015 of 20 August 2015 concerning the authorization of astaxanthin as a feed additive for fish, crustaceans and ornamental fish, in: EU (Ed.), Brussels, 1–4.

EFSA/FEEDAP, 2015. Scientific opinion on the safety and efficacy of ethoxyquin (6-ethoxy-1,2-dihydro-2,2,4-trimethylquinoline) for all animal species. EFSA Journal 13, 4272.

Esworthy, R.S., Swiderek, K.M., Ho, Y.S., Chu, F.F., 1998. Selenium-dependent glutathione peroxidase-GI is a major glutathione peroxidase activity in the mucosal epithelium of rodent intestine. BBA-General Subjects 1381, 213–226.

Fernandez-Palacios, H., Norberg, B., Izquierdo, M., Hamre, K., 2011. Effects of broodstock diets on eggs and larvae, in: G.J. Holt (Ed.), Fish Larval Nutrition. Wiley-Blackwell, Chichester, UK, 153–181.

Frigg, M., Prabucki, A.L., Ruhdel, E.U., 1990. Effect of dietary vitamin E levels on oxidative stability of trout fillets. Aquaculture 84, 145–158.

Gatlin, D.M., Bai, S.C., Erickson, M.C., 1992. Effects of dietary vitamin-E and synthetic antioxidants on composition and storage quality of channel catfish, Ictalurus punctatus. Aquaculture 106, 323–332.

Gutscher, M., Pauleau, A.L., Marty, L., Brach, T., Wabnitz, G.H., Samstag, Y., Meyer, A.J., Dick, T.P., 2008. Real-time imaging of the intracellular glutathione redox potential. Nature Methods 5, 553–559.

Haga, Y., Takeuchi, T., Seikai, T., 2002. Influence of all-trans retinoic acid on pigmentation and skeletal formation in larval Japanese flounder. Fisheries Science 68, 560–570.

Halver, J.E., Hardy, R.W., 2002. Fish Nutrition, 3rd ed. Academic Press, San Diego.

Hamre, K., 2011. Metabolism, interactions, requirements and functions of vitamin E in fish. Aquaculture Nutrition 17, 98–115.

Hamre, K., Berge, R.K., Lie, O., 1998. Oxidative stability of Atlantic salmon (Salmo salar L.) fillet enriched in alpha-, gamma-, and delta-tocopherol through dietary supplementation. Food Chemistry 62, 173–178.

Hamre, K., Hjeltnes, B., Kryvi, H., Sandberg, S., Lorentzen, M., Lie, Ø., 1994. Decreased concentration of hemoglobin, accumulation of lipid oxidation products and unchanged skeletal muscle in Atlantic salmon (Salmo salar) fed low dietary vitamin E. Fish Physiology and Biochemistry 12, 421–429.

Hamre, K., Kolås, K., Sandnes, K., Julshamn, K., Kiessling, A., 2001. Feed intake and absorption of lipid oxidation products in Atlantic salmon (*Salmo salar*) fed diets coated with oxidised fish oil. *Fish Physiology and Biochemistry* 25, 209–219.

Hamre, K., Lie, Ø., 1995. Minimum requirement of vitamin E for Atlantic salmon, *Salmo salar* L., at first feeding. *Aquaculture Res* 26, 175–184.

Hamre, K., Micallef, G., Hillestad, M., Johansen, J., Remø, S.C., Zhang, W., Ødegård, E., Araujo, P., Prabhu, A.J., Ørnsrud, R., Waagbø, R., submitted. Conditions during spring and early summer cause increased growth, consumption of antioxidants, increased oxidation of astaxanthin and onset of cataracts in Atlantic salmon (*Salmo salar*) reared in sea cages.

Hamre, K., Penglase, S.J., Rasinger, J.D., Skjaerven, K.H., Olsvik, P.A., 2014. Ontogeny of redox regulation in Atlantic cod (*Gadus morhua*) larvae. *Free Radical Biology and Medicine* 73, 337–348.

Hamre, K., Sissener, N.H., Lock, E.J., Olsvik, P.A., Espe, M., Torstensen, B.E., Silva, J., Johansen, J., Waagbø, R., Hemre, G.I., 2016. Antioxidant nutrition in Atlantic salmon (*Salmo salar*) parr and post-smolt, fed diets with high inclusion of plant ingredients and graded levels of micronutrients and selected amino acids. *PeerJ* 4.

Hamre, K., Torstensen, B.E., Maage, A., Waagbø, R., Berge, R.K., Albrektsen, S., 2010. Effects of dietary lipid, vitamins and minerals on total amounts and redox status of glutathione and ubiquinone in tissues of Atlantic salmon (*Salmo salar*): a multivariate approach. *British Journal of Nutrition* 104, 980–988.

Hamre, K., Waagbo, R., Berge, R.K., Lie, Ø., 1997. Vitamins C and E interact in juvenile Atlantic salmon (*Salmo salar* L). *Free Radical Biology and Medicine* 22, 137–149.

Hamre, K., Yufera, M., Rønnestad, I., Boglione, C., Conceicao, L.E.C., Izquierdo, M., 2013. Fish larval nutrition and feed formulation: Knowledge gaps and bottlenecks for advances in larval rearing. *Reviews in Aquaculture* 5, S26–S58.

Hardy, R.W., 2012. The nutritional pathology of teleosts, in: R.J. Roberts (Ed.), In Fish Pathology. Wiley-Blackwell, Oxford, 402–424.

Hernandez, L.H., Hardy, R.W., 2020. Vitamin A functions and requirements in fish. *Aquaculture Research* 51, 3061–3071.

Hjeltnes, B., Samuelsen, O.B., Svardal, A.M., 1992. Changes in plasma and liver glutathione levels in Atlantic salmon *Salmo salar* suffering from infectious salmon anemia (ISA). *Diseases of Aquatic Organisms* 14, 31–33.

Hoffman, A., Spetner, L.M., Burke, M., 2008. Ramifications of a redox switch within a normal cell: Its absence in a cancer cell. *Free Radical Biology and Medicine* 45, 265–268.

Hung, S.S.O., Cho, C.Y., Slinger, S.J., 1981. Effect of oxidized fish oil, dl-a-tocopheryl-acetate and etoxyquin supplementation on the vitamin E nutrition of Rainbow trout (*Salmo gairdneri*) fed practical diets. *Journal of Nutrition* 111, 648–657.

Jones, D.P., Sies, H., 2015. The redox code. *Antioxid Redox Sign* 23, 734–746.

Kagan, V.E., Serbinova, E.A., Forte, T., Scita, G., Packer, L., 1992. Recycling of vitamin E in human low density lipoproteins. *Journal of Lipid Research* 33, 385–397.

Karlsen, Ø., van der Meeren, T., Rønnestad, I., Mangor-Jensen, A., Galloway, T.F., Kjørsvik, E., Hamre, K., 2015. Copepods enhance nutritional status, growth and development in Atlantic cod (*Gadus morhua* L.) larvae—can we identify the underlying factors? *PeerJ* 3, e902. https://doi.org/10.7717/peerj.902

Kimmel, C.B., Ballard, W.W., Kimmel, S.R., Ullmann, B., Schilling, T.F., 1995. Stages of embryonic-development of the zebrafish. *Developmental Dynamics* 203, 253–310.

Kiron, V., 2012. Fish immune system and its nutritional modulation for preventive health care. *Animal Feed Science and Technology* 173, 111–133.

Koedijk, R.M., Folkvord, A., Foss, A., Pittman, K., Stefansson, S.O., Handeland, S., Imsland, A.K., 2010. The influence of first-feeding diet on the Atlantic cod *Gadus morhua* phenotype: Survival, development and long-term consequences for growth. *Journal of Fish Biology* 77, 1–19.

Kolkovski, S., Czesny, S., Yackey, C., Moreau, R., Cihla, F., Mahan, D., Dabrowski, K., 2000. The effect of vitamins C and E in (n-3) highly unsaturated fatty acids-enriched *Artemia nauplii* on growth, survival and stress resistance of fresh water walley *Stizostedion vitreum* larvae. *Aquaculture Nutrition* 6, 199–206.

Larsson, T., Krasnov, A., Lerfall, J., Taksdal, T., Pedersen, M., Mørkøre, T., 2012. Fillet quality and gene transcriptome profiling of heart tissue of Atlantic salmon with pancreas disease (PD). *Aquaculture* 330, 82–91.

Lee, K.J., Dabrowski, K., 2003. Interaction between vitamins C and E affects their tissue concentrations, growth, lipid oxidation, and deficiency symptoms in yellow perch (*Perca flavescens*). *British Journal of Nutrition* 89, 589–596.

Lee, K.J., Dabrowski, K., 2004. Long-term effects and interactions of dietary vitamins C and E on growth and reproduction of yellow perch, *Perca flavescens*. *Aquaculture* 230, 377–389.

Lie, Sandvin, A., Waagbø, R., 1994. Transport of alpha-tocopherol in Atlantic salmon (*Salmo salar*) during vitellogenesis. *Fish Physiol and Biochem* 13, 241–247.

Lillig, C.H., Berndt, C., Holmgren, A., 2008. Glutaredoxinsystems. *Biochim Biophys Acta* 780, 1304–1317.

Liu, X.B., Osawa, T., 2007. Cis astaxanthin and especially 9-cis astaxanthin exhibits a higher antioxidant activity in vitro compared to the all-trans isomer. *Biochemical and Biophysical Research Communications* 357, 187–193.

Lou, M.F., 2003. Redox regulation in the lens. *Progress in Retinal and Eye Research* 22, 657–682.

Lundebye, A.K., Hove, H., Maage, A., Bohne, V.J., Hamre, K., 2010. Levels of synthetic antioxidants (ethoxyquin, butylated hydroxytoluene and butylated hydroxyanisole) in fish feed and commercially farmed fish. *Food Addit Contam Part A Chem Anal Control Expo Risk Assess* 27, 1652–1657.

Lygren, B., Hamre, K., Waagbø, R., 1999. Effects of dietary pro- and antioxidants on some protective mechanisms and health parameters in Atlantic Salmon. *Journal of Aquatic Animal Health* 11, 211–221.

Lygren, B., Hjeltnes, B., Waagbø, R., 2001. Immune response and disease resistance in Atlantic salmon (*Salmo salar* L.) fed three levels of dietary vitamin E and the effect of vaccination on the liver status of antioxidant vitamins. *Aquaculture International* 9, 401–411.

Maage, A., Waagbø, R., Olsson, P.E., Julshamn, K., Sandnes, K., 1990. Ascorbate-2-sulfate as a dietary vitamin C source for Atlantic salmon (*Salmo salar*): 2. Effects of dietary levels and immunisation on the metabolism of trace elements. *Fish Physiology and Biochemistry* 8, 429–436.

Mæland, A., Waagbø, R., 1998. Examination of the qualitative ability of some cold water marine teleosts to synthesise ascorbic acid. *Comparative Biochemistry and Physiology Part A* 121, 249–255.

Marcos-Lopez, M., Espinosa, C.R., Rodger, H.D., O'Connor, I., MacCarthy, E., Esteban, M.A., 2018. Oxidative stress is associated with late-stage amoebic gill disease in farmed Atlantic salmon (*Salmo salar* L.). *Journal of Fish Diseases* 41, 383–387.

Märtensson, J., Meister, A., 1991. Glutation deficiency decreases tissue ascorbate levels in newborn rats: Ascorbate spares glutathion and protects. *Proceedings of the National Academy of Sciences of the United States of America* 88, 4656–4660.

Meister, A., 1994. Glutathione ascorbic acid antioxidant system in animals. *Journal of Biological Chemistry* 269, 9397–9400.

Merchie, G., Lavens, P., Sorgeloos, P., 1997. Optimization of dietary vitamin C in fish and crustacean larvae: A review. *Aquaculture* 155, 165–181.

Morgan, B., Sobotta, M.C., Dick, T.P., 2011. Measuring E(GSH) and H_2O_2 with roGFP2-based redox probes. *Free Radical Biology and Medicine* 51, 1943–1951.

Niki, E., 1987. Antioxidants in relation to lipid peroxidation. *Chemistry and Physics of Lipids* 44.

Nordgarden, U., Ørnsrud, R., Hansen, T., Hemre, G.I., 2003. Seasonal changes in selected muscle quality parameters in Atlantic salmon (*Salmo salar* L.) reared under natural and continuous light. *Aquaculture Nutrition* 9, 161–168.

NRC, 2011. *Nutrient Requirements of Fish and Shrimp*. The Natioanal Academic Press, Washington, DC.

Oliva-Teles, A., 2012. Nutrition and health of aquaculture fish. *Journal of Fish Diseases* 35, 83–108.

Penglase, S., Edvardsen, R.B., Furmanek, T., Rønnestad, I., Karlsen, Ø., van der Meeren, T., Hamre, K., 2015. Diet affects the redox system in developing Atlantic cod (*Gadus morhua*) larvae. *Redox Biology* 5, 308–318.

Penglase, S., Hamre, K., Ellingsen, S., 2014. Selenium and Mercury have a synergistic negative effect on fish reproduction. *Aquatic Toxicology* 149, 16–24.

Pohlenz, C., Gatlin, D.M., 2014. Interrelationships between fish nutrition and health. *Aquaculture* 431, 111–117.

Poston, H.A., Combs Jr., G.F., Leibovitz, L., 1976. Vitamin E and Selenium interrelations in the diet of Atlantic salmon (*Salmo salar*): Gross, histological and biochemical deficiency signs. *Journal of Nutrition* 106.

Prabhu, P.A.J., Holen, E., Espe, M., Silva, M.S., Holme, M.H., Hamre, K., Lock, E.J., Waagbø, R., 2020. Dietary selenium required to achieve body homeostasis and attenuate pro-inflammatory responses in Atlantic salmon post-smolt exceeds the present EU legal limit. *Aquaculture* 526.

Prabhu, P.A.J., Lock, E.J., Hemre, G.I., Hamre, K., Espe, M., Olsvik, P.A., Silva, J., Hansen, A.C., Johansen, J., Sissener, N.H., Waagbø, R., 2019. Recommendations for dietary level of micro-minerals and vitamin D-3 to Atlantic salmon (*Salmo salar*) parr and post-smolt when fed low fish meal diets. *PeerJ* 7.

Remø, S.C., Hevrøy, E.M., Breck, O., Olsvik, P.A., Waagbø, R., 2017. Lens metabolomic profiling as a tool to understand cataractogenesis in Atlantic salmon and rainbow trout reared at optimum and high temperature. *PLoS ONE* 12.

Remø, S.C., Hevrøy, E.M., Olsvik, P.A., Fontanillas, R., Breck, O., Waagbø, R., 2014. Dietary histidine requirement to reduce the risk and severity of cataracts is higher than the requirement for growth in

Atlantic salmon smolts, independently of the dietary lipid source. *British Journal of Nutrition* 111, 1759–1772.

Remø, S.C., Olsvik, P.A., Torstensen, B.E., Amlund, H., Breck, O., Waagbø, R., 2011. Susceptibility of Atlantic salmon lenses to hydrogen peroxide oxidation ex vivo after being fed diets with vegetable oil and methylmercury. *Experimental Eye Research* 92, 414–424.

Rise, M.L., Jones, S.R.M., Brown, G.D., von Schalburg, K.R., Davidson, W.S., Koop, B.F., 2004. Microarray analyses identify molecular biomarkers of Atlantic salmon macrophage and hematopoietic kidney response to *Piscirickettsia salmonis* infection. *Physiological Genomics* 20, 21–35.

Roberts, R.J., Rodger, H.D., 2012. The pathophysiology and systematic pathology of teleosts, in: R.J. Roberts (Ed.), *Fish Pathology*, 4th ed. Wiley Blackwell Ltd, West Sussex, UK, 62–143.

Rønnestad, I., Hamre, K., Lie, Ø., Waagbø, R., 1999. Ascorbic acid and alpha-tocopherol levels in larvae of Atlantic halibut before and after exogenous feeding. *Journal of Fish Biology* 55, 720–731.

Ruff, N., Fitzgerald, R.D., Cross, T.F., Hamre, K., Kerry, J.P., 2003. The effect of dietary vitamin E and C level on market-size turbot (*Scophthalmus maximus*) fillet quality. *Aquaculture Nutrition* 9, 91–103.

Sanden, M., Hemre, G.-I., Maage, A., Lunestad, B.-T., Espe, M., Lie, K.K., Lundebye, A.-K., Amlund, H., Waagbø, R., Ørnsrud, R., 2017. *Program for overvåkning av fiskefôr. Årsrapport for prøver innsamlet i 2016, Mattilsynets overvåkningsprogram.* NIFES, ISBN 978-82-91065-46-5, Bergen.

Sandnes, K., Hansen, T., Killie, J.E.A., Waagbø, R., 1990. Ascorbate-2-sulfate as a dietary vitamin-C source for Atlantic salmon (*Salmo salar*).1. Growth, bioactivity, hematology and humoral immune-response. *Fish Physiology and Biochemistry* 8, 419–427.

Saurabh, S., Sahoo, P.K., 2008. Lysozyme: An important defence molecule of fish innate immune system. *Aquaculture Research* 39, 223–239.

Schiotz, B.L., Jørgensen, S.M., Rexroad, C., Gjøen, T., Krasnov, A., 2008. Transcriptomic analysis of responses to infectious salmon anemia virus infection in macrophage-like cells. *Virus Research* 136, 65–74.

Secci, G., Parisi, G., 2016. From farm to fork: Lipid oxidation in fish products. A review. *Italian Journal of Animal Science* 15, 124–136.

Sigurgisladottir, S., Parrish, C.C., Lall, S.P., Ackman, R.G., 1994. Effects of feeding natural tocopherols and astaxanthin on Atlantic salmon (*Salmo salar*) fillet quality. *Food Research International* 27, 23–32.

Silva, M.S., 2019. Development of novel methods to evaluate availability of zinc, selenium and manganese in Atlantic salmon (*Salmo salar*). PhD Thesis, University of Bergen, Bergen, 133.

Sissener, N.H., Hamre, K., Fjelldal, P.G., Phillip, A.J., Espe, M., Miao, L., Høglund, E., Sørensen, C., Skjærven, K.H., Holen, E., Subramanian, S., Vikeså, V., Nordberg, B., Remø, S.C., 2021. Can improved nutrition for Atlantic salmon in freshwater increase fish robustness, survival and growth after seawater transfer? *Aquaculture*, 542. https://doi.org/10.1016/j.aquaculture.2021.736852

Skjærven, K.H., Penglase, S., Olsvik, P.A., Hamre, K., 2013. Redox regulation in Atlantic cod (*Gadus morhua*) embryos developing under normal and heat-stressed conditions. *Free Radical Biology and Medicine* 57, 29–38.

Sobhana, K.S., Mohan, C.V., Shankar, K.M., 2002. Effect of dietary vitamin C on the disease susceptibility and inflammatory response of mrigal, *Cirrhinus mrigala* (Hamilton) to experimental infection of *Aeromonas hydrophila*. *Aquaculture* 207, 225–238.

Spector, A., 1995. Oxidative stress-induced cataract—mechanism of action. *The FASEB Journal* 9, 1173–1182.

Tappel, A.L., 1962. Vitamin E as the biological lipid antioxidant. *Vitamins and Hormones* 20, 493–510.

Timme-Laragy, A., Goldstone, J., Imhoff, B., Stegeman, J., Hahn, M., Hansen, J., 2013. Glutathione redox dynamics and expression of glutathione-related genes in the developing embryo. *Free Radical Biology & Medicine* 65, 89–101.

Tkachenko, H., Grudniewska, J., Pekala, A., Terech-Majewska, E., 2016. Oxidative stress and antioxidant defence markers in muscle tissue of rainbow trout (*Oncorhynchus mykiss*) after vaccination against *Yersinia ruckeri*. *Journal of Veterinary Research* 60, 25–33.

Trichet, V.V., 2010. Nutrition and immunity: An update. *Aquaculture Research* 41, 356–372.

Waagbø, R., 1994. The impact of nutritional factors on the immune system in Atlantic salmon, *Salmo salar* L.: A review. *Aquaculture and Fisheries Management* 25, 175–197.

Waagbø, R., 2006. Feeding and disease resistance in fish, in: R. Mosenthin, J. Zentek, T. Zebrowska (Eds.), *Biology of the Growing Animal*. Elsevier, London, 387–415.

Waagbø, R., 2008. Reducing production related diseases in farmed fish, in: Ø. Lie (Ed.), *Improving Farmed Fish Quality and Safety*. VS Woodhead Publishing, Cambridge, UK, 363–398.

Waagbo, R., Glette, J., Raanilsen, E., Sandnes, K., 1993a. Dietary vitamin-C, immunity and disease resistance in Atlantic salmon (*Salmo salar*). *Fish Physiology and Biochemistry* 12, 61–73.

Waagbo, R., Hamre, K., Bjerkås, E., Berge, R., Wathne, E., Lie, Ø., Torstensen, B., 2003. Cataract formation in Atlantic salmon, *Salmo salar* L., smolt relative to dietary pro- and antioxidants and lipid level. *Journal of Fish Diseases* 26, 213–229.

Waagbø, R., Hamre, K., Maage, A., 2000. The impact of micronutrients on the requirement of ascorbic acid in crustaceans and fish, in: K. Dabrowski (Ed.), *Ascorbic Acid in Aquatic Organisms—Status and Perspectives*. CRC Press, Boca Raton, 105–131.

Waagbø, R., Olsvik, P.A., Remø, S.C., 2020. Nutritional and metabolic disorders, in: P.T.K. Woo, G.K. Iwama (Eds.), *Climate Change and Non-infectious Fish Disorders*. Cabi, Egham.

Waagbø, R., Remø, S.C., 2020. Functional diets in fish health management, in: F. Kibenge, M. Powell (Eds.), *Aquaculture Health Management*. Elsevier, Amsterdam, 187–234.

Waagbø, R., Sandnes, K., Torrissen, O.J., Sandvin, A., Lie, O., 1993b. Chemical and sensory evaluation of fillets from Atlantic salmon (*Salmo salar*) Fed 3 levels of N-3 polyunsaturated fatty acids at 2 levels of vitamin-E. *Food Chemistry* 46, 361–366.

Waagbø, R., Thorsen, T., Sandnes, K., 1989. Role of dietary ascorbic-acid in vitellogenesis in rainbow-trout (*Salmo gairdneri*). *Aquaculture* 80, 301–314.

Waagbø, R., Trøsse, C., Koppe, W., Fontanillas, R., Breck, O., 2010. Dietary histidine supplementation prevents cataract development in adult Atlantic salmon, *Salmo salar* L., in seawater. *British Journal of Nutrition* 104, 1460–1470.

Wade, A.M., Tucker, H.N., 1998. Antioxidant characteristics of L-histidine. *Journal of Nutritional Biochemistry* 9, 308–315.

Wahli, T., Verlhac, V., Girling, P., Gabaudan, J., Aebischer, C., 2003. Influence of dietary vitamin C on the wound healing process in rainbow trout (*Oncorhynchus mykiss*). *Aquaculture* 225, 371–386.

Watanabe, T., Takeuchi, T., Wada, M., Uehara, R., 1981. The relationship between dietary lipid levels and a-Tocopherol requirement of rainbow trout. *Bull. Jap. Soc. Scie. Fish.* 47, 1463–1471.

Wefers, H., Sies, H., 1988. The protection of ascorbate and glutathion against microsomal lipid peroxidation is dependent on vitamin E. *European Journal of Biochemistry* 174, 353–357.

Williams, D.L., 2006. Oxidation, antioxidants and cataract formation: A literature review. *Vet Ophthalmol* 9, 292–298.

Yamashita, Y., Katagiri, T., Pirarat, N., Futami, K., Endo, M., Maita, M., 2009. The synthetic antioxidant, ethoxyquin, adversely affects immunity in tilapia (*Oreochromis niloticus*). *Aquaculture Nutrition* 15, 144–151.

Yildirim-Aksoy, M., Lim, C., Li, M.H., Klesius, P.H., 2008. Interaction between dietary levels of vitamins C and E on growth and immune responses in channel catfish, *Ictalurus punctatus* (Rafinesque). *Aquaculture Research* 39, 1198–1209.

Zhang, W., Berndt, C., Sæle, Ø., Mykkeltvedt, E., Hamre, K., 2019. Ontogeny of the reox potential in cytosol and mitochondria in zebrafish embryos as measured by fluorescent roGFP, *Paris Redox 2019—21st International Conference on Antioxidants*, Université Pierre et Marie Curie, Paris.

INDEX

A

AA supplementation, 357
abscisic acid (ABA), 24
actin
    actin-related protein (ARP), 212
    de-polymerizing factor (ADF), 214
    dynamics, 213
    filaments, 212
    oxidized methionyl residues, 213
    redox modifications of, 211
    S-glutathionylation of, 149
actinomycin D, 151
activating transcription factor 4 (ATF4), 321
acute promyelocytic leukemia (APL), 326
adenosine receptor 2B (A2BR), 299
adenosine triphosphate (ATP), 297
    conformational changes, 253
    production, 70
adipocytes, 195
ADP-glucose pyrophosphorylase, 23
ADP-transferases, 233
Aedes aegypti, 179
aerobic energy production system, 272
Aeromonas hydrophila, 363
Aeromonas salmonicida, 363
African sleeping sickness, 37
age-related deterioration, in multicellular organisms, 250
age-related protein, 256
aging, yeast replicative aging model of
    deregulated nutrient sensing, 257–260
    epigenetic alterations, 253
    facile genetics, 249
    genomic instability, 252–253
    hallmarks of, 252
    H_2O_2 toxicity *versus* signaling, 252
    integrative theory of, 260–261
    mitochondrial dysfunction, 256–257
    proteostasis, loss of, 253–256
    reactive oxygen species, in shaping yeast replicative
        aging, 251–252

    replicative aging, 250
    yeast replicative aging, 250–251
airway basal stem cells (ABSCs), 179
alcohol dehydrogenase (ADH), 154
aldehyde oxidase, 318
algorithm-driven quantitative image analysis, 339
ALKBH5, hypoxia increases expression of, 300
all-*trans* retinoic acid (atRA), 139
alternative oxidase (AOX), 43
Alzheimer's disease (AD), 216, 324
amino acid starvation-unrelated activation, 256
AMP-activated protein kinase (AMPK), 43, 277
amyotrophic lateral sclerosis (ALS), 217
Angiosperm species, 23
anti-apoptotic genes, 97
antibodies, 38
anticancer treatment, 217
anti-hyperglycemia, 324
anti-inflammatory functions, 276
antioxidant enzymes, 283
antioxidant response elements (AREs), 172, 321
antioxidant vitamins, 361
    dietary supplementation of, 363
    lipid oxidation, 356
    vitamins C and E, 361

α-phenyl-N-t-butylnitrone(PBN), 153, 154
apical ectodermal ridge (AER), 153
apical meristems, redox regulation of, 21
apolipoprotein E receptor 2 (APOER2), 289
AP-1 transcription factor, 148
Arabidopsis thaliana, 5
argon plasma coagulator (APC), 333
ascorbate peroxidase (APX), 16
ascorbic acid (AsA), 21, 22
ATM/p53 pathway, 339
autism spectrum disorder (ASD), 96, 100
auto-fluorescent proteins (AFPs), 82
auto-oxidation, 355
autophagy-senescence-transition (AST), 339
auxin/PLETHORA (PLT) signaling pathways, 22

O

OH-Pro concentration, 357
oligodendroglial progenitor cells (OPCs), 216
organogenesis-stage rat embryos, 138
osteoblast, 111, 112
osteoblasts build osteoid, 108
osteoclastogenesis, 112
osteoclasts, 107, 110
osteocytes, 108, 109
osteoprotegerin (OPG), 107
OxICAT approach, 125
oxidant exposure, 146
oxidative diseases, 356
oxidative eustress/redox signaling, 60
oxidative phosphorylation (OXPHOS), 134
oxidative post-translational modifications (oxPTM), 123
 cysteine, profiling of, 126
 cysteine sulfenic/sulfinic acids, 123
 cysteine sulfenic/sulfonic acid, 127
 direct detection, 123
 mapping oxPTMs, 336
oxidative post-translational protein modifications (ox-PTM), 121
oxidative stress, 171
 definition of, 3
 GPx4-deficient mice, 140
 oxidized GSH-conjugates, 132
oxidized cysteines, 275
2-oxoglutarate-dependent dioxygenases (?-OGDD), 300
8-oxoguanine DNA glycosylase (OGG1), 230
oxygen-containing molecules, 318
oxygenic photosynthesis, 5
oxygen sensing, 27
OxyR family, 82

P

pancreas disease, 363
parathyroid hormone (PTH), 108
Parkinson's disease (PD), 216
 hippocampal neurogenesis, 95
 patients, cell-based therapy of, 97
peroxidatic cysteine (Cp), LiPrx1m lacking, 45
peroxiredoxin (Prx), 6, 61, 213, 275, 319
 T. brucei, the cytosolic (TbcPrx), 46
 zebrafish development, 63
peroxisome proliferator-activated receptor-g (PPAR-g), 195
peroxisome proliferator-activated receptor g coactivator 1a (PGC-1a), 95, 322
persulfides, 272
phenobarbital (PB), 288
phorbol myristate acetate (PMA), 326
phosphoinositide 3-kinase (PI3K), antimycin-A-induced, 108
phospholipid biosynthesis, 135
photosynthetic bacteria, 274
Piscirickettsia salmonis infection, 363
plant development, redox regulation of
 apical meristems, 21
 cell division, 17–22
 differentiation/meristems, 17–22
 dormancy, acquisition of, 23
 epigenetic/redox regulation, 27–28
 germination, 23–24
 reactive oxygen species (ROS), 15, 16

redox mutants, developmental defect, 18–20
 reproductive development, 22–23
 root development, 24–26
 stress, developmental adaptation, 26–27
plasma membrane transporters, 132
plastidial NTRC controls, 25
platelet-derived growth factor (PDGF), 109
pLink-SS approach, 124
pluripotent stem cells (PSC), 301
posterior fossa A (PFA), 301
post-translational modifications (PTM), 15, 213
 cysteine oxidative, 123, 124
 regulatory control, 144
p47phox knockout mice, bone density of, 111
precursor cells, IL4 stimulation of, 112
progeria-related Sgs1 RecQ helicase homologue, 252
programmed cell death (PCD), 26
progress zone (PZ) mesenchyme, 153
prolyl-hydroxylases (PHDs), 212, 299
prostate cancer, 286
protein cysteinyl, redox modifications of, 208
protein disulfide isomerase (PDI), 217
protein disulphides (Prot-S2), 40
protein kinase A (PKA), 146
 cAMP-dependent, 258
 deficiency, 146
 pathway, 259
 signaling, 259
protein kinase C, 74
protein redox switches, glutathione regulation of, 145
proteins, S-nitrosylation of, 91, 92–95
protein sulfenic acid (PrSOH), 145
protein synthesis, 253
protein thiols (PrSH), 5, 40, 145
protein translation responses, 341
protein-tyrosine phosphatase 1B (PTP1B), 275
protein tyrosine phosphatases (PTPs), 149

Q

quiescent center (QC), 21

R

Rac1, TRAF6-dependent recruitment of, 111
RANK ligand (RANKL), 107
 OPG determines osteoclastogenesis, 108
 osteoclast differentiation, 112
 stimulation, 112
Ras GTP-exchange factor (GEF), 259
Ras-related C3 botulinum toxin substrate, 212
rat embryos, buthionine-(S,R)-sulfoximine, 152
RBBP7, S-Nitrosylation of, 93
RBOH-derived ROS, 25
reactive nitrogen species (RNS), 91, 189, 333
reactive oxygen species (ROS), 15, 69, 85, 132, 171, 189, 271, 276, 283, 302, 333
 accumulation, 276
 cell-to-cell transmission of, 26
 dependent cellular metabolism, 16
 hydrogen peroxide (H$_2$O$_2$), 81
 nervous system, 70
 NOX2-mediated, 73
 production, 70

V-maf musculoaponeurotic fibrosarcoma oncogene homolog
  A (MAFA), 323
von Hippel Lindau (vHL) protein, 299

W

Wiskott-Aldrich syndrome proteins (WASP), 212
Wnt/β-catenin signaling, 87, 108
Wnt signaling, 179, 284
worm (*Caenorhabditis elegans*), 95

X

xenograft formation, 288
Xenopus laevis, 27, 81, 82

Y

yeast (*Saccharomyces cerevisiae*), 95

metacaspase Mca1, 255
replicative aging, mechanisms of, 250, 258

Z

zebrafish
  adult heart regeneration, 86
  development, redox regulation, 57–63
  disease models, 59
  embryos, 57–58, 60, 62, 83
  generation cycle of, 59
  larva, 86
  midbrain-hindbrain boundary (MHB), 84
  model, 60
  redox systems, 61
Zebrafish International Resource Center (ZIRC), 58
zinc-finger protein, 42

9 781032 068428